# 25th IUPAP Conference on Computational Physics

# (CCP 2013)

AA002771

Journal of Physics: Conference Series Volume 510

## Moscow, Russia
## 20 – 24 August 2013

ISBN: 978-1-63439-290-7
ISSN: 1742-6588

**Printed from e-media with permission by:**

Curran Associates, Inc.
57 Morehouse Lane
Red Hook, NY 12571

**Some format issues inherent in the e-media version may also appear in this print version.**



Printed by Curran Associates, Inc. (2014)

For permission requests, please contact the Institute of Physics
at the address below.

Institute of Physics
Dirac House, Temple Back
Bristol BS1 6BE UK

Phone: 44 1 17 929 7481
Fax: 44 1 17 920 0979

techtracking@iop.org

**Additional copies of this publication are available from:**

Curran Associates, Inc.
57 Morehouse Lane
Red Hook, NY 12571 USA
Phone: 845-758-0400
Fax: 845-758-2634
Email: curran@proceedings.com
Web: www.proceedings.com

# TABLE OF CONTENTS

## ASTROPHYSICS AND COSMOLOGY



## MONTE CARLO METHODS



## DENSITY FUNCTIONAL THEORY

## QUANTUM PHYSICS AND LOW TEMPERATURE PHYSICS



## MOLECULAR DYNAMICS



## PLASMA PHYSICS

## HYDRODYNAMICS

Journal of Physics: Conference Series **510** (2014) 011001 doi:10.1088/1742-6596/510/1/011001

# Preface

## Computational Physics at CCP2013

**Lev N. Shchur[1,2] and Lev Yu. Barash[1]**

[1] Landau Institute for Theoretical Physics, 142432 Chernogolovka, Russia

[2] National Research University Higher School of Economics, 101000 Moscow, Russia

Participants of the XXV IUPAP Conference on Computational physics came to Moscow at the end of the August during a hot period. It was not a hot period because of the summer; in fact, the weather was quite comfortable. It was a hot period for the atmosphere amidst scientific society in Russia, especially for scientists working for the Russian Academy of Sciences.

Four years ago, the C20 IUPAP Commission on Computational Physics and Computational Physics Group of the European Physical Society chose Moscow for several reasons. The first reason was connected to the high level and deep traditions of computational physics in Russia. It is known from experience at the former CCP conferences that native participants contribute about half of the presentations which form the solid scientific background of the conference, and the good level of domestic science makes the conference interesting and successful. The second reason was due to the fact that for the last twenty years there were not many IUPAP conferences in Russia, and it was a time to open more places for information exchange and intensify scientific collaboration. Thirdly, it was common opinion four years ago that the situation in Russia had become stable enough after the transition to a modern society, which took almost a quarter of a century.

In fact, the first and second reasons were completely proven. We hope that discussions during presentations and after the conference triggered new collaborations, new ideas, and new research, and that therefore the main purpose of CCP2013 was achieved. Unfortunately, the third reason – regarding the stable situation – changed unexpectedly. One of the changes is that all institutions of the Russian Academy of Sciences were moved out of the Academy to the new state body, the Federal Agency of Scientific Organizations. The conference was organized by two bodies of the Russian Academy of Sciences: Landau Institute for Theoretical Physics and Department of Physical Sciences. The reform of the Russian Academy of Sciences is still in progress, and it is too early to discuss its actual results.

We had 142 presentations by computational physicists from 30 countries. The conference was organized around invited plenary talks, and began with the story on how computational facilities and computational techniques were used in the hunt for the Higgs boson. Professor Dezso Horvath (Wigner Research Centre for Physics, Budapest, Hungary; Atomki, Debrecen, Hungary; and CMS Collaboration, CERN) presented in his lecture the story of the CMS and ATLAS projects, both of which discovered five events of the suspected boson. The last invited plenary talk was given by professor Stefan Gottlöber (Leibniz Institute for Astrophysics, Potsdam, Germany) on the formation of structure in the Universe and new approaches to cosmological simulations. These two lectures reflect the wide spectrum of scales, from the microscale of the Higgs boson at about 125 GeV, to the astronomic scale of galaxies.

Professor Ian T. Foster (Argonne National Laboratory, Argonne, USA) who is an internationally recognized guru in parallel computing and who invented GRID architecture, which is the base infrastructure in data storing and analysis for the CMS and ATLAS experiments, presented in his invited plenary talk his view on the further development of computing for science and for all in his lecture entitled "Preparing for the Computer Revolution."

Professor Isaak M. Khalatnikov (Landau Institute for Theoretical Physics, Chernogolovka, Russia) told stories in his invited plenary talk on the early days of using computers in physics research. Most of the participants did not know that the marching algorithm for the numerical solution of partial differential equations, which is an essential part of university courses on numerical simulations and is widely used in scientific research and engineering, was developed in the early fifties of the 20th century by Isaak Khalatnikov and Lev Landau within the framework of the Soviet atomic project. Professor Khalatnikov said he was happy to present the topic at the conference, it was presented by him for the first time in sixty years.

**Figure 1.** Two pioneers in physics on computers, Berni J. Alder (left) and Isaak M. Khalatnikov (right) after the invited plenary talk by I.M. Khalatnikov

One of the invited plenary talks shed light on what happened in Chelyabinsk in 2013. Doctor Natalia Artemieva (Planetary Science Institute, USA) presented simulations of a meteorite entering the suburban atmosphere at a slight angle and high speed. It is possible that the Tunguska event that happened a century ago was of the same type, although with a larger meteorite and without webcams filming the amazing impact.

Two invited plenary talks were connected with classical computational physics. Professor Kurt Binder (University of Mainz, Germany) presented on the aspects of simulation and physics of interfacial phenomena in soft condensed matter. Professor Wolfgang Paul (University of Halle-Wittenberg, Germany) gave a lecture on the simulation of semi-flexible polymers.

Journal of Physics: Conference Series **510** (2014) 011001 doi:10.1088/1742-6596/510/1/011001

Two invited plenary talks were connected with quantum computational physics. Professor Ali Alavi (Trinity College, England) presented a lecture on the ground state eigenvalue problem of many-electron systems and Professor Carlo Pierleoni (Universitá degli Studi dell'Aquila, Italy) on the simulation of high-pressure hydrogen from the first principles.

Professor Morten Hjorth-Jensen (University of Oslo, Norway) gave a lecture on computational and algorithmic challenges in nuclear physics. Professor Tomo-Hiko Watanabe (National Institute for Fusion Science, Japan) gave a lecture on the simulation of turbulence in magnetic fusion plasmas.

Each session was headlined by the invited talk. There were 11 invited talks representing new achievements in algorithms and methods of computational physics, as well as new scientific results. The list of invited plenary talks and invited talks can be found in the informational article below. Interested readers may find abstracts of all contributions at the conference website: <`http://ccp2013.ac.ru/files/book5x.pdf`>

In the present volume of the Journal of Physics: Conference Series, we have collected articles of CCP2013 presentations. We hope that the reader will find many interesting scientific results in this volume, in which fifty articles were collected after several months of a careful and professional review process, carried out thanks to the activity of reviewers and members of the international advisory board.

Papers collected in the present volume reflect the interests of Russian participants, which contributed about 60 percent of the articles. These primarily comprise Monte Carlo methods, quantum physics, plasma physics and hydrodynamics, density functional theory, and information technology and high-performance computing in physics and education.

We hope this volume will be helpful and scientifically interesting to readers.

Journal of Physics: Conference Series **510** (2014) 011002 doi:10.1088/1742-6596/510/1/011002

# XXV IUPAP Conference on Computational Physics (CCP2013): Preface

XXV IUPAP Conference on Computational Physics (CCP2013) was held from 20–24 August 2013 at the Russian Academy of Sciences in Moscow, Russia. The annual Conferences on Computational Physics (CCP) present an overview of the most recent developments and opportunities in computational physics across a broad range of topical areas. The CCP series aims to draw computational scientists from around the world and to stimulate interdisciplinary discussion and collaboration by putting together researchers interested in various fields of computational science. It is organized under the auspices of the International Union of Pure and Applied Physics and has been in existence since 1989. The CCP series alternates between Europe, America and Asia-Pacific. The conferences are traditionally supported by European Physical Society and American Physical Society. This year the Conference host was Landau Institute for Theoretical Physics. The Conference contained 142 presentations, and, in particular, 11 plenary talks with comprehensive reviews from airbursts to many-electron systems.

We would like to take this opportunity to thank our sponsors: International Union of Pure and Applied Physics (IUPAP), European Physical Society (EPS), Division of Computational Physics of American Physical Society (DCOMP/APS), Russian Foundation for Basic Research, Department of Physical Sciences of Russian Academy of Sciences, RSC Group company.

## CCP2013 Committees

### Conference Chair:

Prof. Lev Shchur (Landau Institute for Theoretical Physics)

### Conference Vice-Chairs:

Prof. Ravil Nazirov (Institute for Space Research),

Prof. Mikhail Romanovsky (Prokhorov General Physics Institute)

### Members of Local Organizing Committee:

| | | |
|---|---|---|
| Dr Lev Barash | Dr Alexander Russkov | Svetlana Shikota |
| Dr Marat Biktimirov | Alina Kovaleva | Tamara Kozlova |
| Prof Akai Murtazaev | Dr Sergei Krashakov | Dr Anton Menshutin |

Journal of Physics: Conference Series **510** (2014) 011002 doi:10.1088/1742-6596/510/1/011002

## International Scientific Advisory Committee



## History of the Conference on Computational Physics

The Conference on Computational Physics was organized for the first time in 1989 in Boston, and since 2003 it takes place every year, rotating between Europe-Africa, North-South America, and Asia-Oceania.

Past and future editions of the Conference on Computational Physics:

2014 - Boston, USA
2013 - Moscow, Russia
2012 - Kobe, Japan
2011 - Gatlinburg, USA
2010 - Trondheim, Norway
2009 - Kaohsiung, Taiwan
2008 - Ouro Preto, Brazil
2007 - Brussels, Belgium
2006 - Gyeongju, Republic of Korea
2005 - Los Angeles, California, USA
2004 - Genoa, Italy
2003 - Beijing, China
2002 - San Diego, California, USA
2001 - Aachen, Germany
2000 - Brisbane, Australia
1999 - Atlanta, Georgia, USA
1998 - Granada, Spain
1997 - Santa Cruz, USA
1996 - Cracow, Poland
1995 - Pittsburgh, USA
1994 - Lugano, Switzerland
1993 - Albuquerque, USA
1992 - Prague, Czechoslovakia
1991 - San Jose, USA
1990 - Amsterdam, the Netherlands
1989 - Boston, USA

Journal of Physics: Conference Series **510** (2014) 011002 doi:10.1088/1742-6596/510/1/011002

## List of Plenary Talks

1. Dezso Horvath, *Search for the Higgs Boson: a Numerical Adventure of Exclusion and Discovery*

2. Kurt Binder, *Simulations of Interfacial Phenomena in Soft Condensed Matter and Nanoscience*

3. Ali Alavi, *Quantum Monte Carlo approach to the ground state eigenvalue problem of many-electron systems*

4. Ian T. Foster, *Preparing for the Computer Revolution*

5. Wolfgang Paul, *Monte Carlo Simulations of Semi-flexible Polymers: From Single Chains to Nematic Melts*

6. Isaak M. Khalatnikov, *Numerical Methods for Partial Differential Equations and Early Days of Computational Physics*

7. Tomo-Hiko Watanabe, *Exploring phase space turbulence in magnetic fusion plasmas*

8. Carlo Pierleoni, *First-principle calculations of high pressure hydrogen*

9. Morten Hjorth-Jensen, *Living at the edge of stability, understanding the limits of the nuclear landscape: Computational and algorithmic challenges*

10. Natalia Artemieva, *Airbursts - from Tunguska to Chelyabinsk*

11. Stefan Gottloeber, *Formation of structure in the Universe*

## List of Invited Talks

1. Norbert Attig,*Supercomputing Infrastructures in Europe*

2. Luca Baiotti, *Fully general-relativistic simulations of binary neutron-star mergers*

3. Marcia Barbosa, *Enhancement Flow in Nanoconfined Water*

4. Sara Bonella, *Quantum time correlation functions visa noisy Monte Carlo and classical trajectories*

5. Chin-Kun Hu, *Computational approach to synchronization of nonlinear coupled systems*

6. Nail Inogamov, *Ultrafast lasers, highly excited solids, and DFT-EAM-MD simulations*

7. Wolfhard Janke, *Generalized Ensemble Simulations of Polymer Adsorption in an attractive Spherical Cage*

8. David Landau, *A New Paradigm for Petascale Monte Carlo: Replica Exchange Wang-Landau Sampling*

9. Anthony Maggs, *Constrained statistical mechanics for charges and spins* **(EPS Invited Talk)**

10. Vladimir Voevodin, *Supercomputing Center of Moscow State University: Computational Factory and Education*

11. Martin Weigel, *Simulating spin models on GPU: a tour*

# European Physical Society Awards

The European Physical Society encourages recognition of the scientific merit of early career researchers. Best poster prize should be awarded to researchers in the PhD phase of his/her career in recognition of an excellent poster at the conference. EPS Best poster prize was awarded at CCP2013 to Ivanov Momchil Nikolaev, Leipzig University, Germany, who presented his work *"Polymer Adsorption onto a Stripe-Patterned Surface"*.

Presentation of doctor Anthony C. Maggs (Laboratory de Physicochimie théorique, Gulliver, CNRS-ESPCI, Paris, France) *"Constrained statistical mechanics for charges and spins"* was chosen as EPS Invited Talk.

**Figure 1.** Ivanov Momchil Nikolaev

Also, the European Physical Society makes grants available to organisers of Europhysics conferences to allow for the participation of young scientists who otherwise would not be able to attend. EPS Europhysics Conference Grants were awarded at CCP2013 to the following young researchers: Eren Metin Elci, Coventry University, England; Guillaume Jeanmairet, École Normale Supérieure, France; Najafi Amin, Islamic Azad University, Iran.

**Figure 2.** Dr Anthony Maggs

Journal of Physics: Conference Series **510** (2014) 011002 doi:10.1088/1742-6596/510/1/011002

## Young Investigators from United States of America

We would like to thank Professor Mark Novotny for his support in the selection process, which allowed four young investigators from United States of America to attend the Conference with funding from US National Science Foundation. The selected Young Investigators are Lee Ricketson, University of California at Los Angeles; Zack Terranova, University of Notre Dame; Christina Torres, University of Texas at Brownsville; Ehsan Khatami, University of California, Davis.

**Figure 3.** Young researchers from United States of America (from left to right) Lee Ricketson, Christina Torres, and Ehsan Khatami with conference chair Lev Shchur (second from right).

## Berni J Adler CECAM Prize

The prize recognizes exceptional contributions to the field of microscopic simulation of matter. The prize is meant to honor an individual scientist; exceptionally it can be awarded to at most three scientists having equally contributed to the specific topic for which the prize is granted. This is the most prestigious European prize for computer simulations in condensed matter physics/chemistry, statistical physics and physical chemistry. The 5000 Euros prize, awarded every three years by CECAM, was originally created in 1999. The selection committee consists of scientists appointed by the CECAM Council, the Computational Physics Group of the EPS, and the C20 Commission of IUPAP.

The joint winners of 2013 award are Professors Herman Berendsen (Groningen University, the Netherlands) and Jean-Pierre Hansen (Universite Pierre et Marie Curie, Paris, France and Department of Chemistry, Cambridge, UK).

Both these scientists have made outstanding contributions to the developments in molecular dynamics and related simulation methods for understanding the properties of materials ranging from simple and complex liquids and solids to proteins, ionic liquids and plasmas.

Professor Berendsen, through the development of the GROMOS package and the associated force fields, has used simulation to study fundamental processes in biochemistry and molecular biology. His important methodological contributions include the weak coupling strategy for constant pressure and temperature dynamics and the SHAKE algorithm for constraint dynamics. His force fields for water are extensively used in bimolecular simulations.

Professor Hansen was among the first to demonstrate the power of computer simulation as a reliable test-bed for simple theories that can be used to understand complicated phenomena. He has used simulation to probe the freezing of liquids and the structure and dynamics of charged fluids and plasmas. He has validated integral equation theories of strongly coupled fluids by appeal to simulation. He has also made important contributions to the field of charged and neutral colloids through this approach.

**Figure 4.** Professors Jean-Pierre Hansen, Berni J. Adler, and Herman J.C. Berendsen.

# Peer review statement

All papers published in this volume of *Journal of Physics: Conference Series* have been peer reviewed through processes administered by the proceedings Editors. Reviews were conducted by expert referees to the professional and scientific standards expected of a proceedings journal published by IOP Publishing.

Journal of Physics: Conference Series **510** (2014) 012001 doi:10.1088/1742-6596/510/1/012001

# Search for the Higgs boson: a statistical adventure of exclusion and discovery

**Dezső Horváth (on behalf of the CMS Collaboration)**[1]

Wigner Research Centre for Physics, Budapest, Hungary
and Atomki, Debrecen, Hungary

E-mail: horvath.dezso@wigner.mta.hu

**Abstract.** The 40 years old Standard Model, the theory of particle physics, seems to describe all experimental data very well. All of its elementary particles were identified and studied apart from the Higgs boson until 2012. For decades many experiments were built and operated searching for it, and finally, the two main experiments of the Large Hadron Collider at CERN, CMS and ATLAS, in 2012 observed a new particle with properties close to those predicted for the Higgs boson. In this talk we describe the search process: the exclusion of the Higgs boson at LEP, the Large Electron Positron collider, and the observation at LHC of a new boson with properties close to those predicted for the Higgs boson of the Standard Model. We try to pay special attention on the statistical methods used.

## 1. Introduction: the Standard Model

The Standard Model, the general theory of particle physics was established more than 40 years ago. It describes our world as consisting of two kinds of elementary particles, fermions and bosons, differing by their spin, intrinsic angular momentum: fermions have half-integer, bosons have integer spins measured in units of $\hbar$, the reduced Planck constant. The elementary fermions have three families, each consisting of one pair of quarks and one pair of leptons. All fermions have antiparticles of opposite charges. The leptons can propagate freely, but the quarks are confined in hadrons: they can only exist in bound states of three quarks, baryons (like the proton and neutron) or those of a quark and an antiquark, mesons (like the pion). Three antiquarks make antibaryons like the antiproton.

In the Standard Model the three basic particle interactions, the strong interaction holding the quarks in the nucleons and the nucleons in the atomic nucleus, the weak interaction, responsible for the decay of nuclei and of the neutron, and the well known electromagnetic interaction, are all derived from local gauge symmetries. A gauge symmetry is a freedom to define the coordinate system measuring the strength of an interaction, the best known example of which is the freedom to choose the potential zero of an electric field. A local symmetry is its modified form when the gauge is changing in space-time according to a known function. The three basic interactions are mediated by elementary bosons: the strong nuclear force by 8 gluons, the weak interactions by the three heavy weak bosons and the electromagnetism by the photon. In order to cancel

[1] Invited plenary lecture at CCP-2013, the XXV IUPAP Conference on Computational Physics, Landau Institute for Theoretical Physics, Moscow, 21 August 2013.

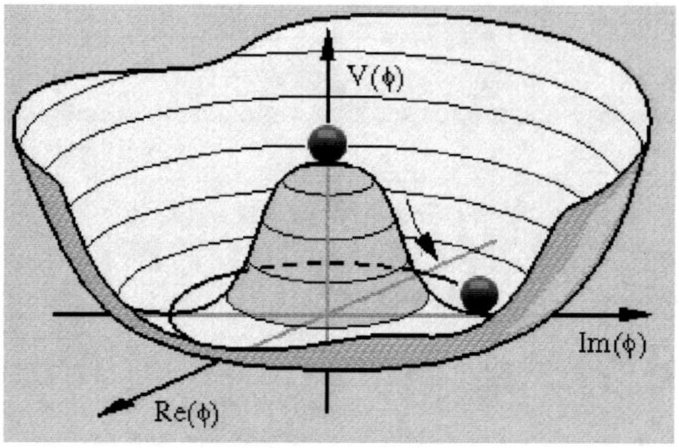


**Figure 1.** Spontaneous symmetry breaking: the Higgs potential. The axial symmetry of the potential is not violated by putting a ball on the top at zero, but it will be spontaneously broken when the ball rolls down in the valley. However, the coordinate system can always be chosen so that the ball were at point $Im(\Phi) = 0$.

uncomfortable terms from equations the theory also needs the existence of an additional scalar boson, a particle with all its quantum numbers like charges and spin zero.

Local gauge symmetries give correct answers to important questions except the mass of elementary particles: one has to violate them in order to introduce their masses. This spontaneous symmetry breaking (SSB) was introduced in several steps to particle physics and it is now an integral part of the Standard Model. It is called, somewhat unjustified, also the Higgs mechanism, although it is the product of several people, so it could also be called *Higgs–Englert–Brout–Guralnik–Hagen–Kibble* mechanism [1].

The spontaneous symmetry breaking mechanism consists of adding to vacuum a potential which breaks its perfect symmetry. This is well illustrated by a Mexican hat (Fig. 1). Its axial symmetry is not violated by putting a ball on its top, however, the ball will eventually go down and break the original symmetry. SSB makes it possible to introduce masses in the theoretical equations: masses of the heavy weak bosons, W+, W- and Z0 mediating the weak interaction and also masses for the basic fermions, the quarks and leptons. The masses of the weak bosons are predicted by the Standard Model, whereas the fermion masses are not, those are free, adjustable parameters. Note that the masses of our macroscopic world are mostly due to the energy content of the proton and the neutron and not due to the SSB mechanism.

Since almost 40 years, more and more precise new data were acquired at the particle accelerators and all seem to agree very well with the predictions of the Standard Model. Hundreds of experiments are summarized in Fig. 2 according to the LEP Electroweak Working Group [2]. It shows the 2012 situation of the analysis of electroweak data: all experimental data and theoretical estimates agree within the statistical boundaries. The only parameter which deviates at more than 2 uncertainties is the forward-backward asymmetry of the decay of the Z boson to two b quarks.

Figure 3 shows the various formation processes of the SM Higgs boson in p-p collisions at LHC. The dominant reaction is gluon fusion and vector boson fusion is also significant.

The Higgs boson of the Standard Model is the only scalar particle: all of its quantum numbers are zero, its only property is mass. Fitting experimental data predicts that the Higgs mass should be around 100 GeV (between 80 and 160 GeV within 95% confidence). All constituents of the Standard Model were identified and studied experimentally before the launch of the LHC, apart from the Higgs boson, that is how it became the most wanted particle. As Peter Higgs himself

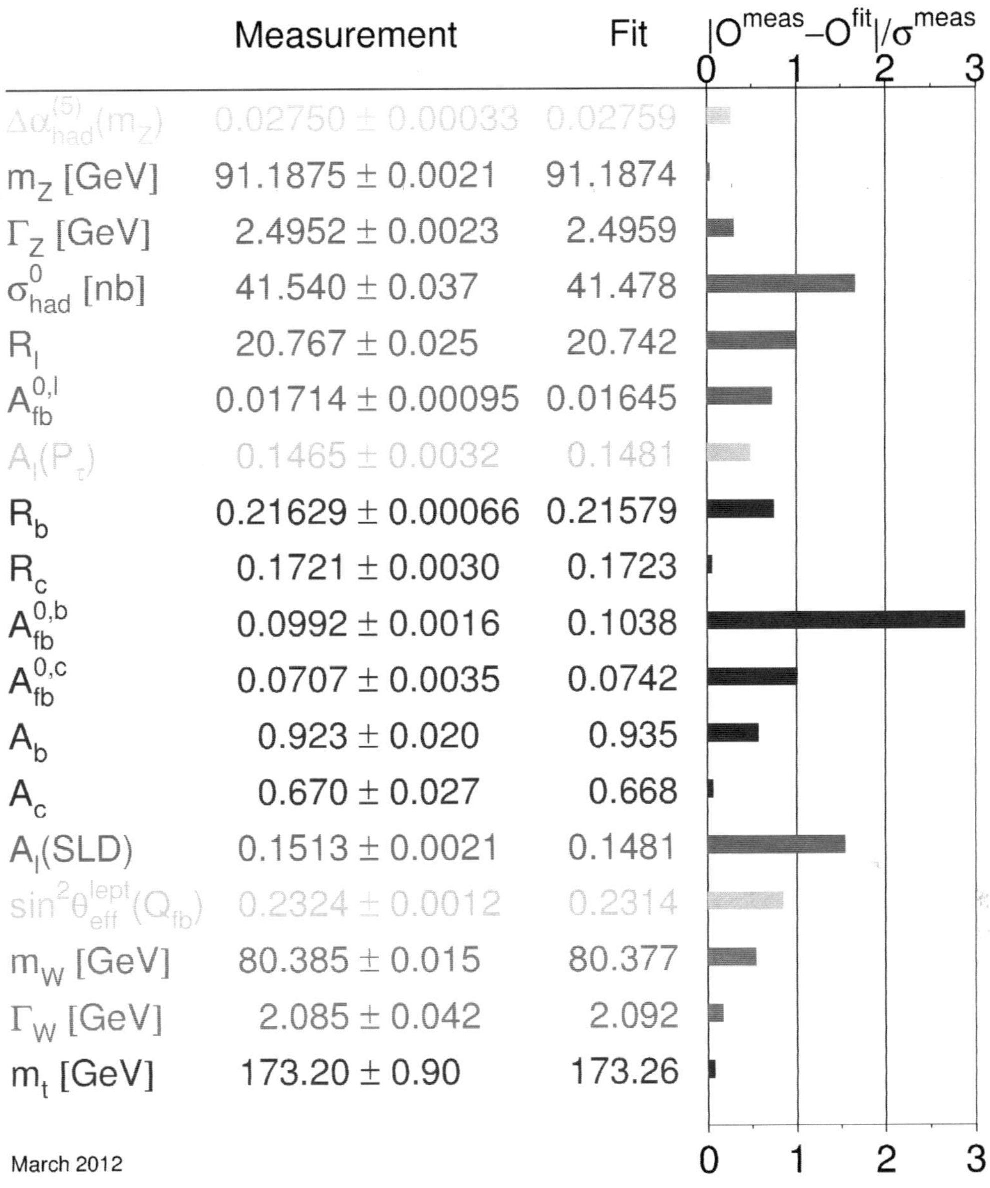


**Figure 2.** Parameters of the Standard Model [2] as determined by the experiment (2nd column) with uncertainties (3rd column), its prediction or fit by the Standard Model (4th column) and a bar plot showing the difference between theory and experiment divided by the experimental uncertainty. The agreement if purely statistical as the difference is in only one case more than 2 uncertainties.

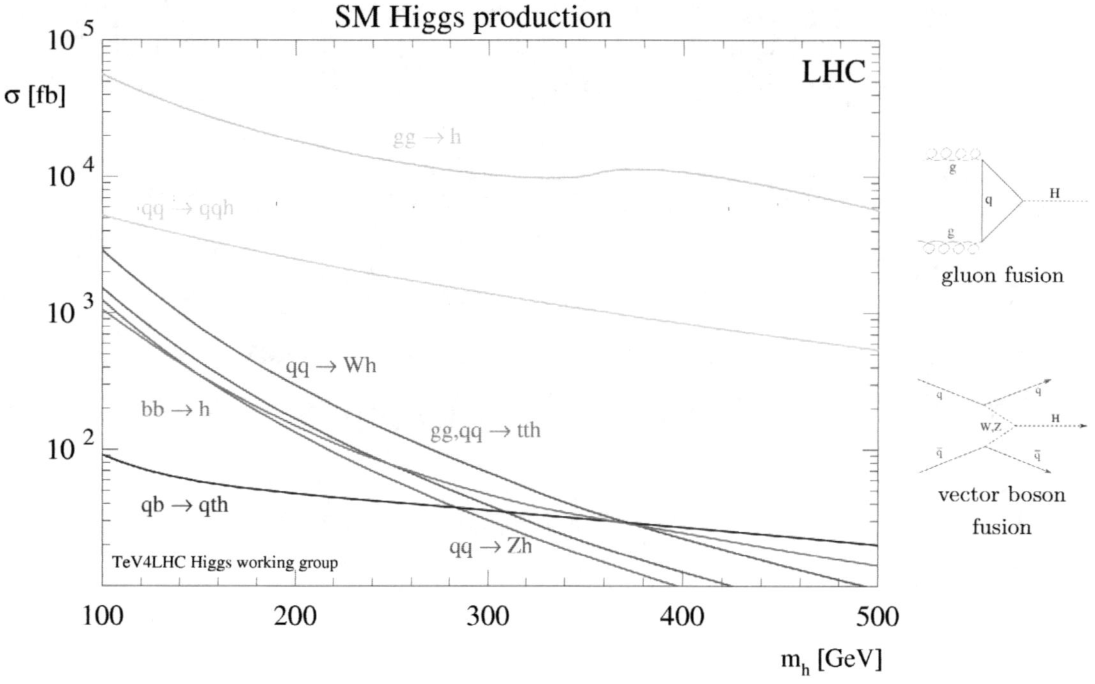


**Figure 3.** Formation of the SM Higgs boson in p-p collisions at LHC.

told [3] "It was in 1972 ... that my life as a boson really began".

## 2. Statistical Concepts of Particle Physicists

Those are as different from the *official* mathematical statistics as mechanical engineering from the Lagrangian or Hamiltonian formulation of theoretical mechanics. At the same time statistics is extremely important for data analysis in particle physics: every few years international workshops are organized by particle physicists working at the Large Hadron Collider to exchange ideas on statistical methods, the last one having been in 2011 [4]. In the Appendix of that volume *Eilam Gross* defines the aim of his paper *LHC Statistics for Pedestrians*: *"A pedestrian's guide ... to help the confused physicist to understand the jargon and methods used by HEP phystatisticians. ... A phystatistician is a physicist who knows his way in statistics and knows how Kendall's advanced theory of statistics book looks like.*

Every high-energy collaboration has phystatistician experts and they all have quite different ideas how to analyze data. In order to avoid confusion, the large LHC collaborations have Statistics Committees which publish home pages of recommendations how to do things. The Statistics Committees of both CMS and ATLAS have several members who published text books on statistics for physicists and ATLAS and CMS have a joint such committee as well.

As in high energy physics the experimental data are basically event counts, the basic concepts are Poisson-like. The data follow the Poisson distribution ($n_i$ events in bin $i$): $\mathcal{P}(n_i|\mu_i) = \frac{\mu_i^{n_i} e^{-\mu_i}}{n_i!}$ and the result is usually expressed in terms of the Poisson likelihood: $\mathcal{L} = \Pi_i \mathcal{P}(n_i|\mu_i)$, where the expected number of events is $\tilde{\mu}_i = \sum_j L\sigma_j\epsilon_{ji}$, $L$ is the integral luminosity collected, $\sigma_j$ is the cross section of source $j$ and $\epsilon_{ji}$ is the efficiency (determined by Monte Carlo simulation) of source $j$ in bin $i$.

Luminosity is the rate of collecting data for colliders, similar to the flux of fixed-target experiments. It is defined as $L = fn\frac{N_1 N_2}{A}$ where $f$ is the circulation frequency of the colliding beams; $n$ the number of particle bunches in the ring; $N_1, N_2$ are the numbers of particles in the two kinds of bunches; $A$ is the spatial overlap of the colliding bunches. The total number of collisions is characterized by the integrated luminosity: $\int_{t_1}^{t_2} Ldt$ which is usually measured in units of inverse cross-section, at LHC in $[pb^{-1},\ fb^{-1}]$. The expected rate of a reaction with cross section $\sigma$ at $\epsilon$ detection efficiency is $R = \epsilon\sigma L$.

According to the general convention in accelerator experiments a given new phenomenon is excluded if we can show it not appearing at a $\geq 95\%$ confidence level and observed if it exceeds $> 5\sigma$ above background where now, for a change, $\sigma$ is the experimental uncertainty according to the best honest guess of the experimentalist.

That $\sigma$ uncertainty has a *statistical component* from the number of observed events and *systematic* ones from various sources, like Monte Carlo statistics and inputs, experimental calibration factors, detection efficiencies, etc, with the common name *nuisance parameters.* To get a rough estimation of the total error the systematic uncertainties could be added quadratically to the statistical one: $\sigma = \sqrt{\sigma_{stat}^2 + \sigma_{syst}^2}$, but in practice we derive the final uncertainty via marginalizing (integrating out) [5] the nuisance parameters $\Theta$ in likelihood $\mathcal{L}$ using the related probability distributions $\mathcal{W}$: $\mathcal{L}(P;x) = \mathcal{W}(x|P) = \int \mathcal{W}(x|P,\Theta)\mathcal{W}(\Theta|P)d\Theta$.

Another important feature of high-energy data analysis is the *blind analysis* [6]: "A blind analysis is a measurement which is performed without looking at the answer. Blind analyses are the optimal way to reduce or eliminate experimenter's bias, the unintended biasing of a result in a particular direction." It came from medical research and the idea is to optimize, prove and publish your analysis technique using simulations and earlier data only before touching new data in the critical region. For instance, in Spring and early Summer, the 2012 CMS data were blinded in the $110 < M_H < 140$ GeV (where $M_H$ is the simulated Higgs mass) because of the $3\sigma$ excess observed in 2011. The same procedure was used again in Autumn 2012. The methods had to be fixed and approved by the collaboration before simultaneous *unblinding* for all analysis channels.

## 3. Search for the Higgs boson

What we usually try to observe is a resonance. For a particle with lifetime $\tau = \Gamma^{-1}$ and decay rate $\Gamma$ the event rate against the invariant mass of the decay products is $|\chi(E)|^2 = \frac{1}{(E-M)^2+\Gamma^2/4}$, i.e. a Lorentz curve (Breit-Wigner resonance). It shows a peak at the $M$ invariant mass of the decaying system with a full width at half maximum $\Gamma$. We claim the discovery of a new particle if we see a resonance at the invariant mass of the particle in all expected decay channels, by all related experiments.

The search involves several consecutive steps.

- Compose a complete *Standard Model background* using Monte Carlo simulation taking into account all types of possible events normalized to their cross-sections.
- Compose *Higgs signals*, simulations of all possible production and decay processes with all possible Higgs-boson masses.
- Put all these through the *detector simulation* to get events analogous to the expected measured ones.
- *Optimize the event selection* via reducing the $B$ background and enhancing the $S$ signal via maximizing e.g. $N_S/\sqrt{N_B}$ or $2 \cdot (\sqrt{N_S + N_B} - \sqrt{N_B})$ [7]
- *Check the background*, i.e. the description of data by the simulation for the given luminosity: the simulation should reproduce the observed background distributions in all details. For instance, you can check the background of the decay of a neutral particle to charged leptons by selecting lepton pairs of identical charges.

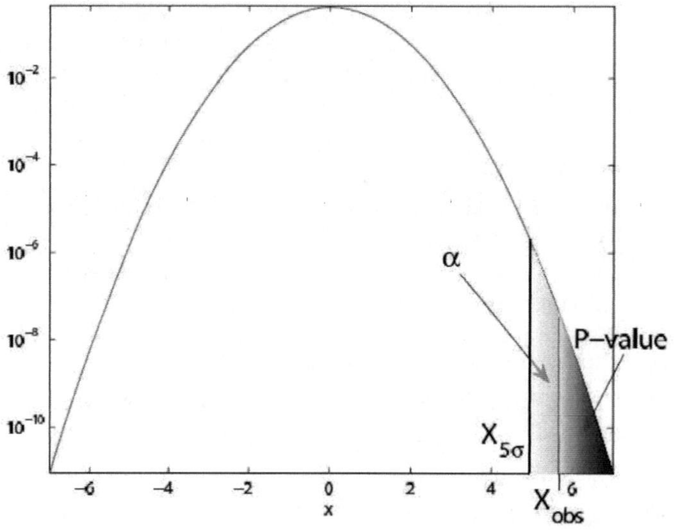


**Figure 4.** The definition of the $p$-value: how much is the observed excess above background as compared to the $\sigma$ uncertainty. We claim a discovery if that excess is above $5\sigma$.

- *Check the signal:* does it agree with the expectation by the theoretical model?

Once you are happy with the simulations and the event selection, you must chose a test statistic. That could be any kind of probability variable characteristic of the given phenomenon: probabilities for having background only, signal or combinations. One of the favorite is the $Q$ likelihood ratio of signal + background over background: $Q = \mathcal{L}_{s+b}/\mathcal{L}_b$. As you see, although our basic approach is definitely frequentist there is a certain Bayesian influence as well. What most frequently plotted is

$$-2\ln Q(m_H) = 2\sum_{k=1}^{N_{ch}}\left[s_k(m_H) - \sum_{j=1}^{n_k}\ln\left(1 + \frac{s_k(m_H)S_k(x_{jk};m_H)}{b_kB_k(x_{jk})}\right)\right]$$

Here the variables are the following:

- $n_k$: events observed in channel $k$, $k = 1 \dots N_{ch}$.
- $s_k(m_H)$ and $b_k$: signal and background events in channel $k$ for Higgs mass $m_H$.
- $S_k(x_{jk};m_H)$ and $B_k(x_{jk})$: probability distributions for events for Higgs mass $m_H$ at test point $x_{jk}$.
- $x_{jk}$: position of event $j$ of channel $k$ on the plane of its reconstructed Higgs mass and cumulative testing variable constructed of various special features of the event like b-tagging, signal likelihood, neural network output, etc.

Several other testing variables can be constructed on the same basis, the most frequently used ones are probabilities of NOT having the expected signal on the basis of the expected background and the collected data:

- $CL_b$, the signal confidence level assuming background only, i.e. the complete absence of the signal, or
- The so-called $p-value$: the probability of obtaining a test statistic at least as extreme as the one that was actually observed, assuming that the null hypothesis is true (Fig. 4). Translated to our language that means the probability that random fluctuation of the measured background could give the observed excess.

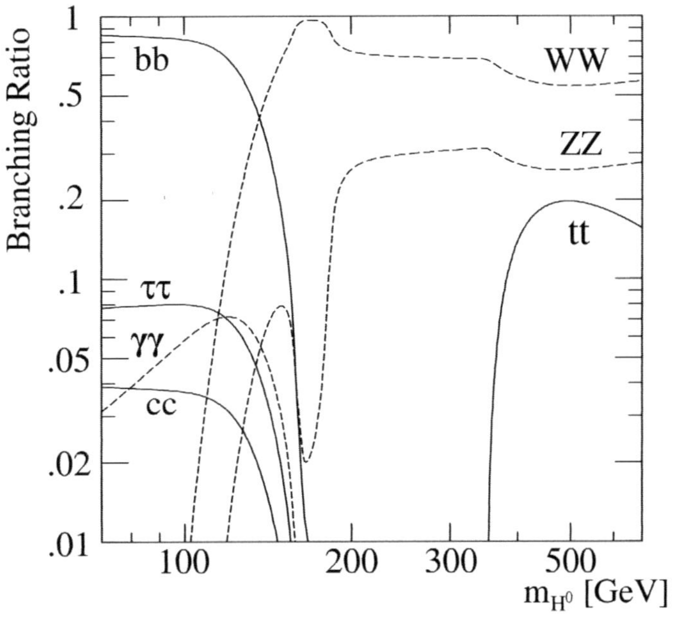


**Figure 5.** The various decay channels of the Higgs boson according to the Standard Model. Below 120 GeV the H→ $b\bar{b}$ decay channel dominates.

## 4. Exclusion at LEP

Although the four large experiments at the Large Electron Positron (LEP) collider saw no new physics, no deviation from the Standard Model, LEP provided an incredible amount of very precise measurements, some of which are presented in Fig. 2. In its last two years of working, LEP was mostly devoted to the search for the Higgs boson, collecting more luminosity at higher energies than in the previous 10 years together.

At LEP the dominant formation process is *Higgs-strahlung* $e^-e^+ \to ZH$ (the name comes from the funny *English* word Bremsstrahlung) and the dominant Higgs decay is to 2 b-quarks. The various channels are different only due to the various decay processes of the accompanying Z boson.

LEP had 4 large experiments in the 4 interaction points of the electron-positron collider, ALEPH, DELPHI, L3 and OPAL (the present author was in OPAL). The structure of the large high-energy detectors are very similar, consisting of onion-like layers. A sensitive pixel detector right around the beam pipe, a tracking system of multiwire chambers or semiconductor detectors of minimal weight material following the tracks of the particles in the magnetic field of the detector, then an electromagnetic calorimeter, something heavy absorbing all electrons and photons, outside of that an even heavier hadron calorimeter, absorbing the pions, protons, neutrons, etc., and finally, muon chambers, identifying the path of muons leaving the system. All detectors have huge magnets encompassing as much as possible of the detector parts.

Statistics played a rough joke at LEP: one of the experiments, ALEPH, saw in one of the possible Higgs decay channels a very significant signal corresponding to a Higgs boson of a mass of 115 GeV/$c^2$, while the rest of LEP have not seen anything (Fig. 6, [8]). ALEPH saw the excess in the 4-jet events only, in those events where the Higgs boson decays to a pair of b quarks and the accompanying Z boson also decays to a quark pair. The b quark is identified by its long lifetime leading to a secondary decay vertex in the event. Another strange thing was that the

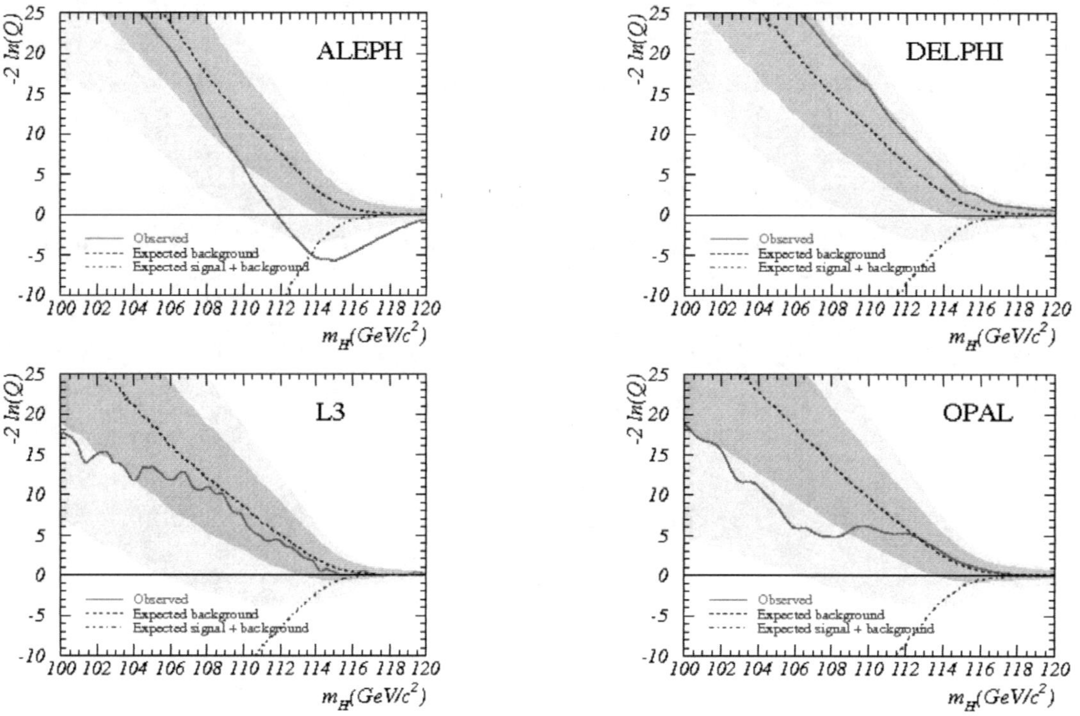


**Figure 6.** Exclusion of the Higgs boson at LEP. The test statistic, $-2\ln Q$ shows a significant signal for ALEPH and nothing for the other 3 LEP experiments at equivalent statistical and experimental circumstances. The observed signal of ALEPH by far exceeds the expectations of the Standard Model.

Higgs signal seen by ALEPH by far exceeded the expectations of the Standard Model. Also, the observed Higgs mass was critical as it coincided with the average kinematic limit of LEP: in 2000 the average collision energy of LEP was about 206 GeV and the observed resonance was found at 115 GeV/$c^2$, the difference is very close to the mass of the Z boson, 91 GeV/$c^2$.

A quite interesting feature of data analysis was the plotting of *spaghetti diagrams.* Those are signal weight distributions of each selected event as a function of the assumed Higgs mass. Fig. 7 shows the weight distributions of 17 selected Higgs-like candidate events observed by the 4 LEP experiments [8]. The ALEPH events crowd around 115 GeV/$c^2$ whereas for the other 3 experiments there are less of them with a rather random mass distribution. This caused quite an excitement at LEP: many physicists signed the petition to the Director General of CERN to extend the life of LEP by another year, but that was refused: the simulated projections were not very promising for a discovery of the SM Higgs boson (the effect seen by ALEPH only was far too large, much higher than the prediction of the Standard Model), and the contractors for building LHC were already prepared to start.

## 5. Observation at LHC

Just like LEP had, the Large Hadron Collider has also 4 interaction points with a major experiment (and possibly smaller ones) in each. The two largest ones, ATLAS [9] and CMS [10] were designed with the main aim of discovering the Higgs boson, ALICE [11] is specialized on heavy ion collisions and LHCb [12] on studying rare processes involving b quarks. The

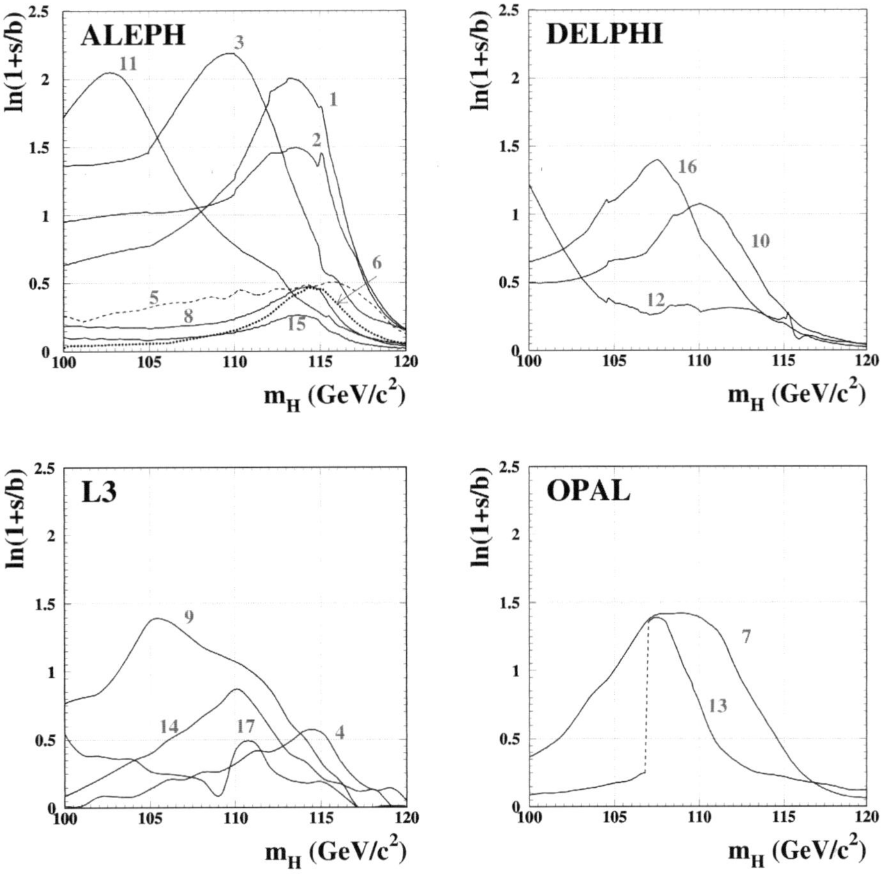


**Figure 7.** *Spaghetti diagrams* of 17 selected events for the 4 LEP experiments: signal weights against the simulated Higgs mass [8]. The ALEPH events crowd around 115 GeV/$c^2$ whereas for the other 3 experiments there are less of them with a rather random mass distribution.

author belongs to CMS, so most of the results we mention are due to CMS, but all will be compared to those of ATLAS pointing out the similarities and the (very few and not significant) differences. These collaborations are huge. According to the official statistics in 2012 CMS had 3275 physicists (incl. 1535 students) and 790 engineers and technicians from 179 institutions of 41 countries. The largest participant country was the USA, then Italy, Germany and Russia.

The design of LHC and its experiments started well before the actual start of LEP, which means that the construction of the LHC detectors took two decades of hard work before the actual data acquisition started. Its first two years LHC devoted to development rather than data taking, that really started in 2011 only.

Even before LHC started the parameter fitting of the Standard Model pointed toward a light Higgs boson, with a mass around 100 GeV/$c^2$. As LEP excluded the Higgs boson below 114 GeV/$c^2$ the LHC experiments had to be prepared for detecting the Higgs boson in the most complicated mass region, around 120 GeV/$c^2$, with several competing decay channels. It was shown very early that the best channel to observe a light Higgs boson at LHC should be the decay to 2 photons because of the very high hadron background. Thus both large experiments, CMS and ATLAS designed their electromagnetic calorimeters with this in mind. The CMS one consists of 75,848 $PbWO_4$ single crystal scintillators, whereas the electromagnetic calorimeter ATLAS is a sampling one based on liquid argon shower detectors.

By the beginning of 2012, when all 2011 data were analyzed, the possible mass of the SM Higgs boson was already confined to the region of $114 < M_H < 127$ GeV/$c^2$ by CMS [13] (with very similar results from ATLAS). In that region $2 \ldots 3\sigma$ excesses were found at ~125 GeV/$c^2$ in the two main decay channels, H $\to \gamma\gamma$ and H $\to$ ZZ. After reanalyzing their data the Tevatron experiments, CDF and D0 also found an excess at this mass (after the LHC started the Tevatron accelerator of Fermilab was stopped). It seemed more and more probable that the Higgs boson will be observed at LHC in 2012, it was even decided by the CERN administration to extend the data taking scheduled for 2012 before the long shutdown for accelerator development if necessary for the discovery.

July 4th, the beginning of the large annual high-energy physics congress in Melbourne, the spokespersons of ATLAS and CMS gave talks from CERN (in internet connection to the whole word, including, of course, the main auditorium of the Australian conference) on Higgs search. They announced that at LHC collision energies 7 and 8 TeV, in two decay channels H $\to \gamma\gamma$ and H $\to$ ZZ $\to \ell^+\ell^-\ell^+\ell^-$, at an invariant mass of $m \approx 126$ GeV a new boson is seen at a convincing statistical significance of $5\sigma$ confidence level each with properties corresponding to those of the Standard Model Higgs boson. The fact that the new particle could decay to two photons or Z bosons, confined its spin to an even integer, i.e. a boson of $S = 0$ or $S = 2$. Of course, as the data analysis was optimized to find the SM Higgs, it was very unlikely to find something very different. Nevertheless, the two experiments emphasized that it has to be studied, whether or not its spin is really zero with a + parity (the pseudo-scalar mesons have spin 0 with negative parity), and that its decay probabilities to various final states follow the predictions of the Standard Model.

## 6. Reactions of the Media

The saying that *three people can keep something secret only if two of them are dead* is attributed to Benjamin Franklin. As any result of a collaboration has to be approved by all members before it is made public, the more than 6000 participants of two large experiments knew well in advance the developing result. Thus two days before the 4th July announcement, *Nature Online* already reported the result [14]. Of course, the fact that CERN invited all leading scientists of the field including those who developed spontaneous symmetry breaking for the Standard Model also helped people to guess that something dramatic will be announced.

CERN produced some figures concerning the media echo of the day: 55 media organizations were represented at the talks of 4 July, the talks were broadcasted via close to half a million internet connections (many of them conference rooms in partner institutions, e.g. three in Hungary with quite an audience in each), 1034 TV stations devoted 5016 news broadcasts to the event for more than a billion ($10^9$) people. Many-many news articles and even more blogs and talks discussed the conditions and importance of the discovery.

## 7. The observations

On 31 July the two experiments submitted papers of the discovery to Physics Letters B, they were published 14 August [15, 16]. Both papers are 15 pages long followed by 16 pages of close to 3000 authors and both are dedicated to the memory of those participants who could not live to see the result of the more than two decades of construction work. Fig. 8 shows the di-photon spectra obtained by CMS in July 2012, after analyzing about a quarter of the data to be collected in 2012. The 4-lepton spectra are quite similar with less background and signal, for CMS it is shown in Fig. 9.

What was really convincing of the observation was the distribution of the p-values of the events selected in the various analyzed decay channels of the hypothetical Higgs boson. For

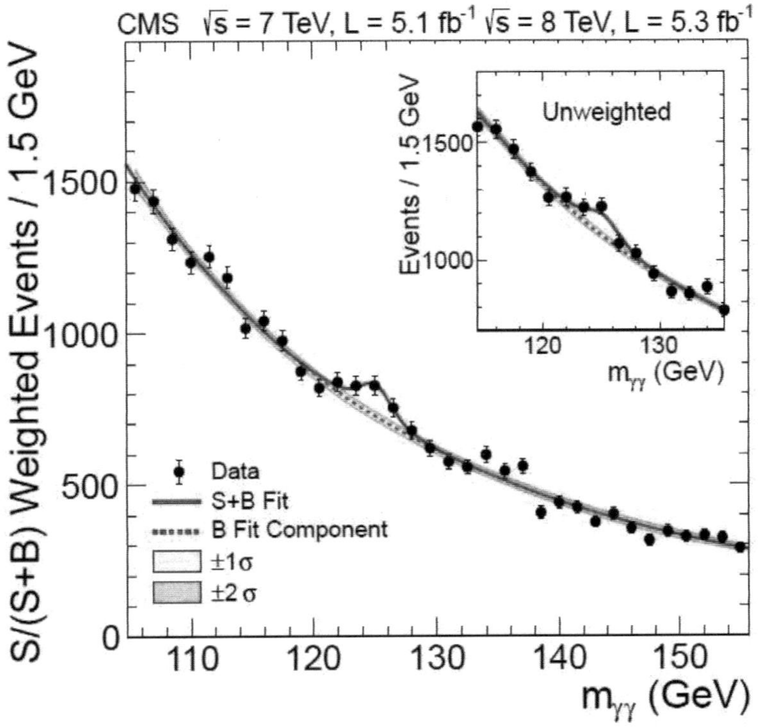


**Figure 8.** Observation of the Higgs-like boson by CMS [16] in the $\gamma\gamma$ invariant mass distribution at 125 GeV/$c^2$. The amplitude of the observed signal is close to the expectations of the Standard Model.

CMS data it is shown in Fig. 10 for all available data: the significance is already as high as $6.9\sigma$. It was a joke of statistics that in July 2012 adding together two decay channels, $\mathrm{H} \to \gamma\gamma$ and $\mathrm{H} \to 4\ell$ gave the same $5\sigma$ significance for both ATLAS and CMS whereas adding to it the WW channel increased the significance to $6\sigma$ for ATLAS and reduced it to $4.9\sigma$ for CMS.

## 8. Is it the Higgs boson?

Analyzing most of the data collected in 2012 led to the conclusion that all observed properties of the newly discovered particle are within statistics close to those predicted for the Higgs boson of the Standard Model. The fact that it decays to two photons points to its having spin 0 or 2. The charged lepton spectra bears the features of its having $S = 0^+$ [17]. Its mass as determined by CMS [18] by the average of all decay channels is $< M_X >= 125.7 \pm 0.3(stat) \pm 0.3(syst)$. The ATLAS result is almost exactly the same: $125.5 \pm 0.2(stat) \left\{ \begin{array}{c} +0.5 \\ -0.6 \end{array} \right\} (syst)$. The difference in the uncertainties are due to the facts that (i) ATLAS had more signal-like data, but (ii) got more different masses in the two main channels. The signal strengths of the new particle is also compatible with that expected for the Standard Model Higgs boson: for CMS it is $0.80 \pm 0.14$ and for ATLAS $1.43 \pm 0.16$ (stat) $\pm 0.14$ (syst). As a theoretician remarked whenever ATLAS has an excess CMS comes up for everybody's annoyance with a deficit, bringing the average close to the SM prediction.

The LHC experiments studied the cross sections of the processes connected to the new particle. Fig. 11 shows the signal strengths of production and decay in various possible channels

Journal of Physics: Conference Series **510** (2014) 012001 doi:10.1088/1742-6596/510/1/012001

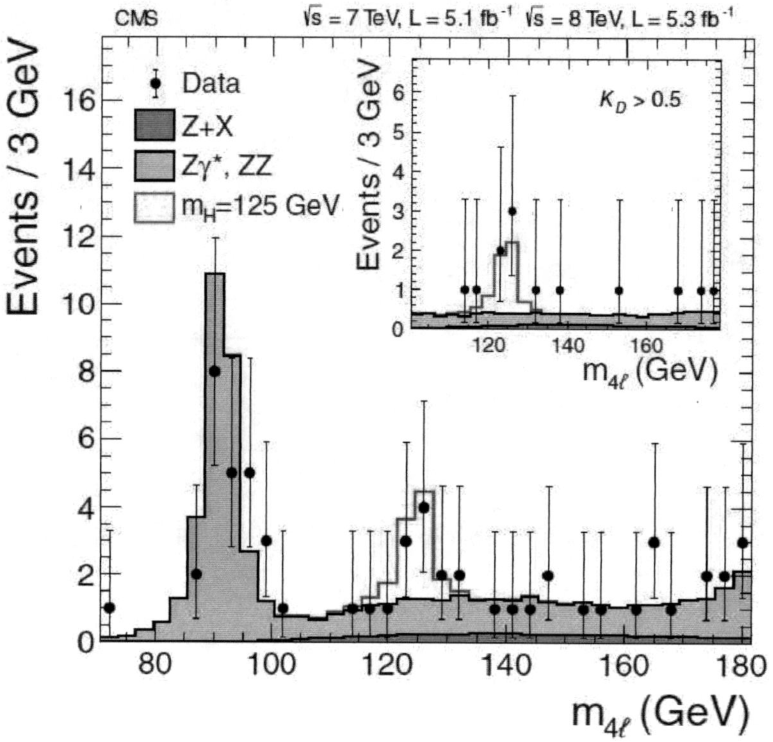


**Figure 9.** Observation of the Higgs-like boson by CMS [16] in the $\ell^+\ell^-\ell^+\ell^-$ invariant mass distribution at 125 GeV/$c^2$. The amplitude of the observed signal is close to the expectations of the Standard Model.

of the Higgs-like boson measured by CMS [18] as compared to those predicted by the Standard Model for the Higgs boson with a mass of 125 GeV/$c^2$. The amplitudes of all observed signals are in agreement with the expectations of the Standard Model.

Thus what we found is very likely the Standard Model Higgs boson. On one hand this is a great success of particle physics. On the other hand this is somewhat of a disappointment as the SM has theoretical shortcomings which need new physics to resolve. Just to list a few of them: it cannot unite the interactions at large energies, cannot account for the dark matter of the Universe and cannot explain neutrino oscillations. There are many extensions of the theory which should result in deviations from the Standard Model. All those problems can be resolved e.g. by *supersymmetry*, but none of its predicted phenomena could be found yet experimentally. The observables of the Higgs boson should be sensitive to some of the features of new physics and these studies will be the main job of ATLAS and CMS in the future, from 2015 when the LHC will restart with twice the energy and luminosity of 2012.

It is very interesting that the 126 GeV mass of the Higgs boson seems to be exciting for theoreticians, there was even a special workshop [19] organized to discuss this mass in 2013. The reason is that $M_H = 126$ GeV is at the border line of the stability of electroweak vacuum on the plane of top mass against Higgs mass, see e.g. [20]. At the Madrid workshop the apparent fine tuning of the Standard Model compelled some physicists to recall the anthropic principle.

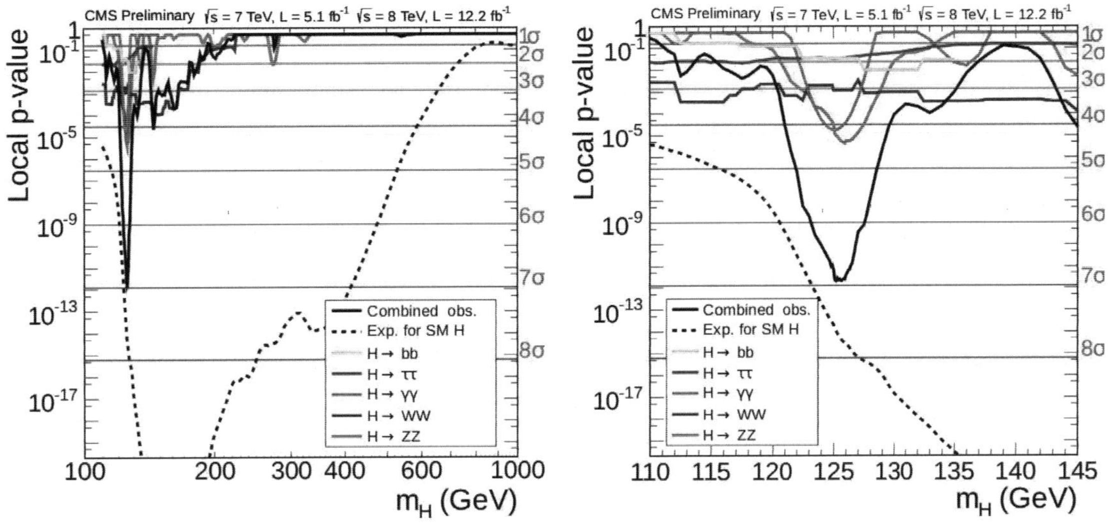


**Figure 10.** Observation of the Higgs-like boson by CMS [18] in the invariant mass distribution of p-values in a wide range and closer to 125 GeV/$c^2$ as based on the data collected at LHC in 2011 and 2012. The amplitude of the observed signal is close to the expectations of the Standard Model.

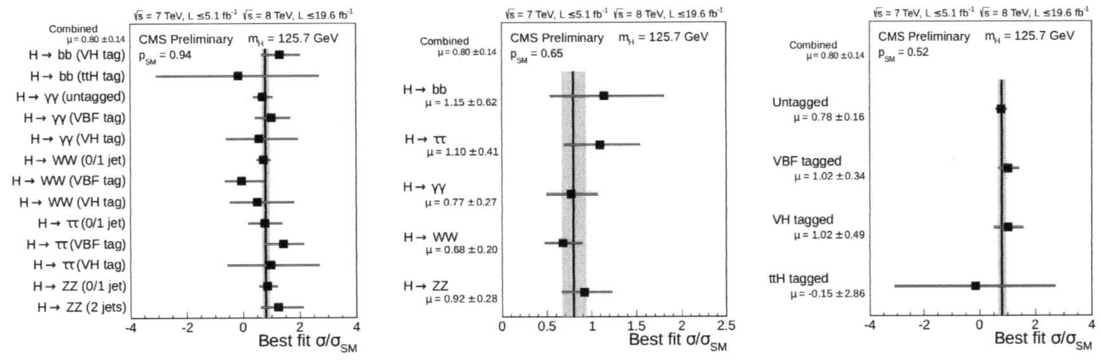


**Figure 11.** Signal strengths of production and decay in various possible channels of the Higgs-like boson measured by CMS [18] as compared to those predicted by the Standard Model for the Higgs boson with a mass of 125 GeV/$c^2$.

## 9. Acknowledgments

The author is indebted to the organizers of the CCP-2013 conference for the invitation with a substantial financial support. The results reported here are due to the Higgs groups of LEP and of the CMS Collaboration. Tommaso Dorigo, the chairperson of the CMS Statistics Committee has read the first draft of the slides of this talk and improved it by very useful suggestions. This work was supported by the Hungarian National Science Foundation OTKA via contracts NK-81447 and K-103917.

## References

[1] Higgs P W 1964 *Phys.Lett.* 12 132-133 and *Phys.Rev.Lett.* 13 508-509;
Englert F and Brout R 1964 *Phys.Rev.Lett.* 13 321-323;
Guralnik G S, Hagen C S and Kibble T W B 1964 *Phys.Rev.Lett.* 13 585-587.
[2] The LEP Electroweak Working Group, *http://lepewwg.web.cern.ch*
[3] Higgs P W 2002 *Int. J. Mod. Phys. A* **17** Suppl. 86-88.
[4] Prosper H B and Lyons L, editors 2011 *Proceedings of the PHYSTAT 2011 Workshop on Statistical Issues Related to Discovery Claims in Search Experiments and Unfolding*, CERN, Geneva, Switzerland, 1720 January 2011; CERN-2011-006.
[5] Cousins R D and Highland V L 1992 *Nucl. Instrum. Meth. A* **320** 331.
[6] Klein J R and Roodman A 2005 *Ann. Rev. Nucl. Part. Sci.* **55** 141.
[7] Bityukov S I and Krasnikov N V 2000 *Nucl. Instrum. Meth. A* **452** 518.
[8] The LEP Experiments, ALEPH, DELPHI, L3 and OPAL 2003, *Phys. Lett. B* **565** 61.
[9] ATLAS Collaboration 2013, A Toroidal Lhc ApparatuS, *http://atlas.web.cern.ch/Atlas/Collaboration/*
[10] CMS Collaboration 2013, Compact Muon Solenoid, *http://cms.web.cern.ch/*
[11] ALICE Collaboration 2013, A Large Ion Collider Experiment, *http://aliweb.cern.ch/*
[12] LHCb Collaboration 2013, *http://lhcb.web.cern.ch/lhcb/*
[13] The CMS Collaboration 2012A, *Physics Letters B* **710** 26-48.
[14] Chalmers M 2012, *http://www.nature.com/news/physicists-find-new-particle-but-is-it-the-higgs-1.10932*, 2 July 2012,
[15] The ATLAS Collaboration 2012, *Physics Letters B* **716** 1–29.
[16] The CMS Collaboration 2012B, *Physics Letters B* **716** 30–61.
[17] The CMS Collaboration 2013A, *Phys. Rev. Letters* **110** 081803.
[18] The CMS Collaboration 2013, *CMS PAS HIG-13-005*
[19] International Workshop, 2013, *Why* $M_H = 126$ *GeV?*, Madrid, Spain, 25-27 Sep. 2013, `http://workshops.ift.uam-csic.es/WMH126/index.html`
[20] Alekhin S, Djouadi A, Moch S, 2012, *Physics Letters B* **716** 214.

# Simulations of Interfacial Phenomena in Soft Condensed Matter and Nanoscience

**Kurt Binder**

Institut für Physik, Johannes Gutenberg Universität Mainz, Staudinger Weg 7, D-55099 Mainz, Germany

E-mail: kurt.binder@uni-mainz.de

**Abstract.** Computation of interfacial free energies between coexisting phases (e.g. saturated vapor coexisting with liquid) is a fundamental problem of classical statistical mechanics: the standard approach (dating back to van der Waals, Ginzburg-Landau, Cahn-Hillard $\cdots$) is based on the continuation of the free energy of homogeneous states throughout the two phase coexistence region. Beyond mean field this continuation does not exist, nor does an "intrinsic profile" of the interface exist! These problems can be overcome by computer simulation: one popular method is based on sampling the order parameter distribution function in the two-phase coexistence region, which yields information on the surface tension of planar interfaces (from "slab configurations") and of curved interfaces (from states containing "droplets"), elucidating the problem of the "Tolman length". Another method (suitable also for solid-liquid interfaces) analyzes the capillary wave broadening or the capillary wave spectrum; all these methods require a careful assessment of finite size effects. Related problems occur for excess free energies due to walls, needed to describe wetting phenomena, capillary condensation, heterogenous nucleation, etc. As an example, a thermodynamic integration method (based on "mixing" systems with and without walls) will be described, and an application to understand phase separation in nanoconfinement will be mentioned.

## 1. Introduction

Interfacial phenomena are familiar from every day life: water droplets may spread on a glass surface, or they may not spread on a lotus leaf [1]. As we shall see, droplet spreading on surfaces reflect the wetting properties [2, 3, 4] of such substrates, and these wetting properties are ruled by the competition of three surface tensions [5]: the wall-vapor surface excess free energy, the wall-liquid surface excess free energy, and the liquid-vapor interfacial excess free energy (=interfacial tension). Very common and known to everybody also are interfacial phenomena in partially miscible more-component fluids (e.g. oil droplets in water, etc.). A less known system, which is nevertheless very popular for scientists because it is so well suited for experimental investigation, are colloid-polymer mixtures [6] (dispersed in a common solvent). Such systems are useful for the study of wetting phenomena, one can see individual colloidal particles in a microscope, and thus study the structure of wetting layers coating the walls of a container on the particle level [7, 8].

While some aspects of interfacial phenomena are understood since decades [5, 9] other aspects are not understood at all and have remained controversial until today [4, 10, 11, 12, 13, 14]. Only on a macroscopic level of description, the situation is clear-cut: the standard problem that

Journal of Physics: Conference Series **510** (2014) 012002 doi:10.1088/1742-6596/510/1/012002

is considered is a saturated gas exposed to a flat planar wall. In the case of "partial wetting" (also called "incomplete wetting") [1, 2, 3, 4, 5] the macroscopic droplet attached to the wall will not spread out but attain an equilibrium shape, namely (in the absence of gravity) a sphere cap. The liquid-gas interface that delineates the surface of this droplet then meets the wall along the "contact line" under the contact angle $\theta$, which is given by the Young equation [15].

$$\gamma_{wv}(T) - \gamma_{w\ell}(T) = \gamma_{\ell v}(T)\cos\theta \quad . \tag{1}$$

Of course, $|\cos\theta| \leq 1$, and hence partial wetting requires $|\gamma_{wv}(T) - \gamma_{w\ell}(T)| < \gamma_{\ell v}(T)$. When $\gamma_{wv}(T) - \gamma_{w\ell}(T)$ tends towards $\gamma_{\ell v}(T)$, $\cos\theta \to 1$, and we encounter a wetting transition. For $\gamma_{wv}(T) - \gamma_{w\ell}(T) > \gamma_{\ell v}(T)$, the wall is coated by a (macroscopically thick) wetting layer, and therefore the vapor avoids extending up to the wall. For this situaiton of "complete wetting" (where a droplet is unstable and spreads out into a wetting layer) one simply has [1, 2, 3, 4, 5]

$$\gamma_{wv} = \gamma_{w\ell} + \gamma_{\ell v} \quad . \tag{2}$$

The fact that raising the temperature may cause a surface phase transition (i.e., a singularity of the surface excess free energy of the fluid) [16] from a partially wet to a completely wet state of the surface is only known since 1977 [17]. Consequently, many aspects of wetting transitions are still under exploration today [4, 10, 11, 12, 13, 14].

Another important problem is the fact that viewing interfaces as purely geometrical objects, infinitely thin dividing surfaces between volume regions containing different phases, is a drastic simplification [9, 10, 18, 19, 20, 21, 22, 23]. Such a coarse-grained picture breaks down at the nanoscale, where interfaces may look very rough or diffuse-depending on the resolution that is used [21, 24]. The local structure in which particles are arranged across an interface may be very irregular, and it may be rather difficult to identify where the local position of the interface actually is, and to distinguish genuinely interfacial fluctuations from bulk fluctuation in the phases close by the interface. Identifying a criterion that corresponds to the dividing surface locally, such that one phase is below this interface and the other phase above it, still is a matter of debate [24, 25, 26, 27, 28]. There occur large-scale fluctuations of the local interfacial height, the so-called capillary waves [29, 30]; for short wavelengths, however, it is again difficult to ascertain what is a capillary wave and what is a bulk fluctuation [9, 18, 19, 20, 21, 22, 23, 24, 25, 26, 27, 28].

Before these problems are discussed in depth, a reminder of the standard approach in statistical thermodynamics to deal with interfaces [5, 9] is useful. The first step always is the formulation of a mean field theory, and then one discusses how to improve it. For the vapor-liquid transition of a simple fluid this starting point is the van der Waals theory [31]. Near the critical temperature $T_c$, the van der Waals theory can be reduced to the Landau theory [32], using the concept of an order parameter density $\psi$, the reduced density difference from the critical density $\rho_c$,

$$\psi = (\rho - \rho_c)(/(2\rho_c) \quad , \tag{3}$$

so that the free energy density $\Delta f_{\rm Landau}(\rho, T)$ can be written as

$$\Delta f_{\rm Landau}(\rho, T) = \frac{3}{2} k_B T_c \rho_c \Big[\frac{1}{2}\Big(\frac{T}{T_c} - 1\Big)\psi^2 + \frac{1}{4}\psi^4\Big] \quad . \tag{4}$$

Remember that this function describes the free energy density of a homogeneous system, although one passes the two-phase coexistence region, in between the saturated vapor density $\rho_v = \rho_c(1 - 2\psi_{\rm coex})$ and the liquid density $\rho_\ell = \rho_c(1 + 2\psi_{\rm coex})$, so $\rho_v, \rho_\ell$ correspond to $\pm\psi_{\rm coex}$ with $\psi_{\rm coex} = \sqrt{1 - T/T_c}$. Of course, the true free energy density in the regime $\rho_v < \rho < \rho_\ell$ is not given by Eq. (4), but rather described by the double-tangent construction: i.e., $\Delta f_{\rm Landau}(\rho, T)$ is constant $\Big( = -\frac{3}{8} k_B T_c \rho_c \Big)$ throughout the two-phase coexistence regime, and the hump described by Eq. (4) in between the two minima is a mean field artefact [33, 34]. Similarly, the van der Waals loop for the reduced chemical potential

$$\frac{\Delta\mu}{k_BT} = \frac{\partial\Delta f}{\partial\rho} \propto \phi - \phi^3, \quad \phi = \psi/\psi_{\rm coex} \tag{5}$$

is a mean-field artefact as well, and interpreting the extrema ("spinodal points") of this loop as limits of metastability for the supersaturated vapor (or undersaturated liquid, respectively) is a "folklore" interpretation [33, 34, 35] which lacks any justification from statistical mechanics (for systems with short-range forces, as considered here) [36]. The true behavior is given by the Maxwell construction, which represents a description of a state of two-phase equilibrium [32]. The amounts of the two phases contained in these coexisting regions are given by the lever rule, and these regions are separated by an interface (e.g., a droplet surrounded by vapor) [32, 33, 34].

It is the description of this interface on which we now focus. When we continue to rely on Landau theory, we need to account for the cost of order parameter inhomogeneities across the interface. The traditional approach is to amend Eq. (4) by a $(\nabla\psi)^2$ term, to obtain the Landau free energy functional [37] $\Delta F\{\psi(\vec{x})\}$, namely

$$\Delta F\{\psi(\vec{x})\} = \int d^d\vec{x}[\Delta f_{\rm Landau}(\psi) + (R^2/2d)(\nabla\psi)^2] \quad , \tag{6}$$

where $d$ is the dimensionality of space, and $R$ the range of interaction between the particles. The task is to find solutions $\psi(\vec{x})$ that minimize $\Delta F\{\psi(\vec{x})\}$. Of course, the trivial solution $\nabla\psi \equiv 0$ yields the homogeneous phases, $\psi = \pm\psi_{\rm coex}$. To describe an interface, one looks for a solution $\psi(z)$ that satisfies $\psi(z \to \pm\infty) = \pm\psi_{\rm coex}$ with $\psi(z = 0) = 0$, such that $z = 0$ describes the center of the interface. The condition that

$$\delta\Delta F\{\psi(z)\}/\delta\psi(z) = 0 \tag{7}$$

yields the Ginzburg-Landau equation {recall $\phi = \psi/\psi_{\rm coex}$ }

$$\phi - \phi^3 = 2\frac{d^2\phi}{dZ^2} \quad , \quad Z = z/\xi, \tag{8}$$

where $\xi$ is the correlation length for order parameter fluctuations, given as

$$\xi^2 = R^2/[d(1 - T/T_c)] \quad . \tag{9}$$

The solution of Eq. (8) is the well-known tanh-profile for the order parameter variation across the interface,

$$\phi(z) = \tanh(Z/2), \quad \psi(z) = \psi_{\rm coex}\tanh(z/2\xi) \quad . \tag{10}$$

So the (intrinsic) width of the interface, according to this theory, is nothing but twice the correlation length. Using then this solution in $\Delta F\{\psi(z)\}$, one obtains the interfacial tension from the normalized free energy excess ($A$ is the interfacial area)

$$F_{\rm int}/(k_BTA) = \frac{1}{3}\psi^4_{\rm coex}\xi \quad (\propto (1 - T/T_c)^{3/2} \text{ as } T \to T_c) \quad . \tag{11}$$

Of course, the critical exponent $\mu = 3/2$ is a mean field result, different from the correct value predicted by Widom's scaling law, $F_{\rm int}/k_BTA \propto (1 - T/T_c)^\mu$ with $\mu = (d-1)\nu$, $\nu$ being the critical exponent of the correlation length {$\xi \propto (1 - T/T_c)^{-\nu}$, with $\nu \approx 0.63$ in $d = 3$ dimensions}.

However, the failure of the Ginzburg-Landau theory of interfaces is more fundamental than just the fact that it predicts a wrong critical exponent: the problem is that the $\tanh(Z)$ profile is a direct mathematical consequence of the double well free energy function, Eq. (4), or the equivalent van der Waals-type loops Eq. (5): neither of these functions are well defined!

It is illuminating to compute both the free energy density $f(\rho)$ of the fluid (with short-range interactions among the particles) or its derivative $\mu(\rho) = (\partial f(\rho)/\partial\rho)_T$ by more reliable methods of (classical) statistical mechanics. For a fluid phase where intermolecular interactions are simply modelled by the Lennard-Jones potential, Monte Carlo methods yield numerically exact results (apart from statistical errors, which can be made as small as desired, if enough computer time ressources are invested). However, Monte Carlo simulations (as well as Molecular Dynamics simulations) always deal with finite systems, while analytical methods of statistical thermodynamics address the thermodynamic limit, the particle number $N$ is taken to infinity. So the typical geometry of a simulated fluid is a $L \times L \times L$ box (in $d = 3$ spatial dimensions), with periodic boundary conditions.

When one obtains the true free energy density $f_L(\rho, T)$ and its derivative $\mu_L(\rho, T) = (\partial f_L(\rho, T)/\partial\rho)_T$ from such a simulation, one does find a double-well potential but the shape of the hump in between the two minima at $\rho_v$ and $\rho_\ell$ is not described by Eq. (4): in particular the height of the maximum scales like $1/L$, and as $L \to \infty$ the whole part of $f_L(\rho, T)$ for $\rho_v < \rho < \rho_\ell$ smoothly converges toward the horizontal straight line provided by the double-tangent construction. Conversely, the parts of $f_L(\rho, T)$ for $\rho \leq \rho_v$ and $\rho \geq \rho_\ell$ rapidly converge towards size-independent results, as soon as $L \gg \xi$ [33, 34].

The loop of $\mu_L(\rho, T)$ as function of $\rho$ tells the same story: while it resembles the van der Waals loop (Eq. (5)) for $L \leq \xi$, its shape is very different from Eq. (5) for $L \gg \xi$: both extrema positions shift with increasing size towards the coexistence curve location ($\rho_v$, $\rho_\ell$), the height (depth) of the extrema (as well as the whole loop) systematically decrease as $L$ increases: for $L \to \infty$, the straight line of the Maxwell construction results automatically [33, 34].

In fact, both $f_L(\rho, T)$ and $\mu_L(\rho, T)$ reflect the true equilibrium behavior of the system, one must not interpret the part of the loop where $\mu_L(\rho, T)$ decreases with increasing $\rho$ as being due to "unstable states", as is commonly done in the context of the van der Waals equation: rather this peculiar behavior of $f_L(\rho, T)$ and $\mu_L(\rho, T)$ is simply due to interfacial phenomena. As we shall see in the next section, such data do yield direct information on the interfacial tension between the coexisting phases, and there is no need to locate the interface on the particle level for this purpose.

## 2. Extracting Interfacial Free Energies from the Order Parameter Distribution

We now discuss in more detail the behavior of the loop $\mu_L(\rho)$ versus $\rho$ that is observed in a simulation when one crosses the vapor-liquid two-phase coexistence region of a fluid (see figures two, three and four of Ref. [34] for illustrative explicit examples). In the thermodynamic limit ($L \to \infty$), two-phase coexistence starts immediately when $\rho$ exceeds $\rho_v$; in the finite simulation box, however, the system stays homogeneous up to some density $\rho_1 > \rho_v$, where $\mu_L(\rho)$ has a maximum (which becomes singular in the limit L $\to \infty$, while $\rho_1(L \to \infty) \to \rho_v$). The physical significance of this maximum is the "droplet evaporation/condensation transition" [38, 39, 40, 41, 42, 43, 44]. For $\rho > \rho_1$ the state of the system now is inhomogeneous, for $\rho_1 < \rho < \rho_2$ a droplet (with on average spherical shape) coexists with surrounding vapor (which is supersaturated in comparison to the saturated vapor that occurs for $L \to \infty$ at $\rho = \rho_v$). At $\rho = \rho_2$ the shape of the droplet changes from spherical to cylindrical (stabilized by the periodic boundary conditions), and this shows up via a (rounded) kink in the $\mu_L(\rho)$ vs. $\rho$ curve. Most important in the present context, however, is the region in between $\rho_3$ and $\rho_4$, in which $\mu_L(\rho)$ is a constant, independent of $L$, and this is the chemical potential characteristic for two-phase coexistence in the thermodynamic limit. The rounded kinks in the $\mu_L(\rho)$ vs. $\rho$ curve that occur near $\rho_3$ and $\rho_4$ are due to the transition from a cylindrical droplet (or bubble, for $\rho > \rho_4$) to a "slab configuration", where the periodic boundary conditions stabilize a liquid-slab, separated from saturated vapor by two planar interfaces. Note that the area of these interfaces is just $L^2$, and so the total free energy cost associated with these interfaces is (in $d$ dimensions)

$$\Delta F/k_BT = 2L^{(d-1)}\gamma_{\ell v}(L) \quad . \tag{12}$$

Here we have taken into account that the periodic boundary conditions on the length scale $L$ constrain the interfacial fluctuations, and so the interfacial tension associated with these interfaces may deviate from the associated bulk value by some finite size correction. This finite size dependence of the interfacial tension $\gamma_{\ell v}(L)$ in practice is rather weak, and carrying out simulations over a range of values of the linear dimension $L$ and an extrapolation of $\gamma_{\ell v}(L)$ verus $1/L$ allows to estimate $\gamma_{\ell v} \equiv \gamma_{\ell v}(\infty)$.

At this point it is useful to recall how $f_L(\rho, T)$ and $\mu_L(\rho, T)$ are obtained from the simulations in practice. Typically it is most convenient to work in the grand- canonical $\mu VT$ ensemble ($V = L^d$ is the volume of the $d$-dimensional simulation box). Using successive umbrella sampling [45] and histogram reweighting [46] methods the distribution function $P_{\mu VT}(N)$ of the particle number $N$ can be sampled very accurately [45]. Then the free energy $Vf_L(\rho, T)$ discussed above is given by

$$Vf_L(\rho, T) = F(N, V, T) \equiv -k_BT \ln P_{\mu VT}(N) + \mu N + F_0 \quad , \tag{13}$$

where $F_0$ is an unimportant constant. The chemical potential function $\mu_L(\rho, T)$ simply is the derivative

$$\mu_L(\rho, T) = \Big(\frac{\partial F(N, V, T)}{\partial N}\Big)_{T,V} \tag{14}$$

as usual. Phase coexistence (yielding the chemical potential $\mu_{\text{coex}}(\rho, T)$ to which $\rho_v$ and $\rho_\ell$ belong) simply is obtained from the equal weight rule for $P_{\mu VT}(N)$ [47, 48]. Even simpler is the problem for symmetric binary (A,B) Lennard-Jones mixtures, where one works in the semi-grandcanonical ensemble, carrying out identity switches between the particles ($A \to B$ or $B \to A$, respectively) [49, 50, 51], in addition to standard displacement moves of particle coordinates ($\vec{r}_i \to \vec{r}_i\,' = \vec{r}_i + \vec{\delta}$, $\vec{\delta}$ being a random small displacement), as usually done in NVT simulations [52]. Then phase coexistence between segregated A-rich and B-rich phases of such a mixture occurs automatically at chemical potential difference $\Delta\mu = 0$ between the particles, due to the symmetry of the model.

In all such systems one observes a flat horizontal region of $f_L$ in the center of the two-phase coexistence region, and $\mu_L$ is constant, equal to the bulk value of the chemical potential at two-phase coexistence. The fact that the plateau of $f_L(T, \rho)$ must be strictly horizontal is due to the fact that a variation of the order parameter in this region just means that the amount of liquid phase in the liquid slab changes, but as long as the two planar interfaces are far enough apart from each other, so they have only a negligible interaction, the free energy density does not change. Since $f_L(\rho, T) = F(N, V, T)/L^d$, one obtains from Eq. (12)

$$\gamma_{\ell v}(L) = Lf_\ell(\rho, T)/2 \quad . \tag{15}$$

This method was first invented for the two-dimensional Ising (lattice gas) model [53] and now is established as a first principles approach that has proven useful for large variety of models of fluids [10, 20, 24, 34, 51, 54, 55]. This method is particularly useful near the critical point, where $\gamma_{\ell,v}$ vanishes, and traditional methods advocated in the textbooks [5, 52], such as using the anisotropy of the pressure tensor $p_{\alpha\beta}(z)$ across an interface, are difficult to apply, because the anisotropy then is small, and fluctuations are huge. Defining normal ($p_n(z)$) and tangential components ($p_T(z)$) as

$$p_n(z) = p_{zz}(z) \quad , \quad p_T(z) = \frac{1}{2}\Big(p_{xx}(z) + p_{yy}(z)\Big), \tag{16}$$

and for a system with periodic boundary conditions, containing a liquid slab with two liquid-vapor interfaces oriented along the $z$-axis, one has

$$\gamma_{\ell v} = \frac{1}{2}\int_0^L dz\Big[p_n(z) - p_T(z)\Big] \quad . \tag{17}$$

Eq. (17) is only preferable to Eq. (15) at low temperatures (e.g. close to the triple point), where the grand-canonical Monte Carlo algorithms become less efficient, or when one uses Molecular Dynamic methods [52, 56].

Eq. (15) was tested for the Ising (lattice gas) model since there the interfacial tension is known exactly from transfer matrix methods in $d = 2$ dimensions [57]. The rigorous approach to interfacial free energies for such lattice models considers the ratio of partition functions of two systems, one containing an interface, the other ontaining no interfaces. In the Ising model, we can enforce the presence of an interface by flipping the sign of the exchange constant between the spins in two adjacent rows (in $d = 2$) or planes (in $d = 3$), respectively, still using periodic boundary conditions; denoting the partition sum of such a system as $Z_-$, while the partition sum of the system with only ferromagnetic bonds and no interfaces is $Z_+$, we obtain (A is the interfacial area, and the limit $A \to \infty$ is considered)

$$\gamma_{\ell v} = \frac{1}{k_B T A}\ln\Big(Z_-/Z_+\Big) \quad . \tag{18}$$

Alternatively, one can also use systems with fixed spin boundary conditions, where in one case ($Z_{++}$) all boundary spins are fixed as $S_i = +1$, while in the other case ($Z_{+-}$) only the boundary spins along the left half of the system are $S_i = +1$ while the boundary spins along the right half are all $S_i = -1$. Then

$$\gamma_{\ell v} = \frac{1}{k_B T A}\ln\Big(Z_{+-}/Z_{++}\Big) \tag{19}$$

Clearly, the merit of Eq. (15), for $L \to \infty$, is that it is a generalization of such first principle approaches that can be carried over to phase coexistence in any systems, off-lattice models, and is also not restricted to liquid-vapor type systems, but can also be used for phase coexistence in liquid crystalline systems (interfaces between isotropic and nematic phases [58]), solid-liquid interfaces [59], etc.

## 3. Interfacial Profiles and Capillary Waves

In the previous section, we have emphasized that one should not rely on the van der Waals-Ginzburg Landau-Cahn Hillard mean field theory to compute interfacial tensions between coexisting phases reliably, avoiding uncontrolled errors. In this section we focus on another aspect of interfacial phenomena, namely interfacial profiles.

For this problem, the failure of mean field theory is much worse: it is very doubtful, whether an "intrinsic" interfacial profile can be observed at all; and if such an intrinsic profile can be defined, it does not have the simple tanh-form of Eq. (10). The problem is to understand precisely the broadening of the profile due to capillary waves [9, 10, 18, 19, 20, 21, 22, 23, 24, 25, 26, 27, 28, 29, 30]. This broadening has first been seen in simulations of various Ising-type lattice models [19, 60] and models for symmetrical binary polymers [20], and particular extensive studies were carried through for colloid-polymer mixtures, both experimentally [7, 8, 61] and by model simulations [24]. These effects can also be seen in model studies of nematic-isotropic interfaces [62] of liquid-crystalline systems, and even for interfaces between the fluid and the crystal at the melting transition [63, 64, 65, 66, 67, 68, 69]. Here we shall focus on the work on symmetrical polymer blends [20], since there a very complete study was performed, including also an estimation of the interfacial tension with the method based on the order parameter probability distribution, as explained in Sec. 2. Simulations were performed [20] for the bond fluctuation model of polymers [70, 71], using chain lengths of $N_A = N_B = N = 32$ monomers, and interactions of square-well type, of strength $\varepsilon_{AA} = \varepsilon_{BB} = -\varepsilon_{AB} = \varepsilon$ and range $\sqrt{6}$ lattice spacings. In separate work it had already been found that the model exhibits phase separation into A-rich and B-rich phases when $\varepsilon$ exceeds a critical value $\varepsilon_c \approx 0.14$ (a factor $1/k_BT$ is absorbed in the definition of $\varepsilon$ here). Most calculations were done in the strongly segregated case $\varepsilon = 0.3 (T/T_c \approx 0.48)$. Choosing a system geometry $L \times L \times D$ with $D = 128$ lattice spacings the lateral linear dimension $L$ was varied from $L = 16$ to $L = 256$. The system simulated contained 50% A chains and 50% B chains, with a single interface (in $z$-direction a variant of an "antiperiodic" boundary condition was used, i.e. A monomers leaving the box at one side reenter as B-monomers at the opposite side, or vice versa). Recording the interfacial profile, a very pronounced broadening of the profile with increasing lateral linear dimension $L$ is observed, and it is clear that one suspects that this size-dependent broadening needs to be attributed to the capillary waves. In order to study them systematically, the interfacial configurations of the model system in the course of the simulation need to be carefully analysed. A view of snapshot pictures of the system, where the positions of all monomers of one kind are drawn, reveals the fact already mentioned above, that the interface on the single particle level is very rugged, and that it is difficult to distinguish what is a fluctuation of the local height of the interface, and what is a bulk composition fluctuation [20]. So it normally is decided [20, 24, 25] to use a block analysis method to coarse-grain the interface. I.e., if the interface normally is the $z$-axis, one divides the $L \times L$ planes normal to the $z$-axis into $n_B^2$ squares of linear dimension $B \times B$ each, with $B = L/n_B$. One then considers rectangular columns $B \times B \times \Delta z$, with $\Delta z = z_+ - z_-$ and the values $z_+$, $z_-$ are chosen such that the interface position $z_G$ is in between $z_-$ and $z_+$ for all blocks. Here $z_G$ is then defined as a Gibbs dividing surface in each block. In this way a local height variable $h(x, y) = z_G$ is obtained, but it turns out that the snapshots of those coarse-grained interface configurations depend very much on the resolution of the "microscope" with which the interface is viewed: choosing a small block size, $B = 2$ lattice spacings, so that one is able to see small local details of the interface configuration, the picture is extremely rugged; choosing a larger block size, e.g. $B = 8$ (a length comparable to the gyration radius $R_g \approx 7$ of the polymer chains considered in this study), one obtains smoother variations, the picture looks like a hilly landscape with mountain ridges and valleys. Still it is clear, that strong fluctuations of the local height $h(x, y)$ are present, with many wavelengths $\lambda$ (with $\lambda \geq B$, of course; smaller-scale structures by construction cannot be resolved).

The standard assumption now is that such fluctuations of the interfacial height $h(x, y)$ can be described by the capillary wave Hamiltonian,

$$\mathcal{H}_{CW} = \frac{1}{2} k_B T \tilde{\gamma} \int_0^L dx \int_0^L dy |\nabla h(x, y)|^2 \quad , \tag{20}$$

where the "interfacial stiffness" $\tilde{\gamma}$ for a liquid-vapor interface simply can be identified with the interface tension $\gamma_{v\ell}$, and for a fluid binary (AB) mixtures $\tilde{\gamma} = \gamma_{AB}$, the interface tension between coexisting A-rich and B-rich phases.

Due to the periodic boundary conditions in the $x, y$-direction parallel to the interface, introduction of Fourier transforms

$$h(x,y) = \sum_{\vec{q}} h(\vec{q}) \exp[i(q^x x + q_y y)] \tag{21}$$

diagonalizes the capillary wave Hamiltonian, reducing it to a problem analogous to independent harmonic oscillators,

$$\mathcal{H}_{CW} = \frac{1}{2} k_B T \tilde{\gamma} \sum_{\vec{q}} q^2 |h(\vec{q})|^2 \quad . \tag{22}$$

For such problems we can immediately invoke the equipartition theorem [32]

$$\frac{k_B T \tilde{\gamma}}{2} q^2 \langle |h(\vec{q})|^2 \rangle_T = \frac{1}{2} k_B T \tag{23}$$

and hence we conclude that the capillary wave spectrum is described by

$$\langle |h(\vec{q}|^2 \rangle = 1/(\tilde{\gamma} q^2) \quad . \tag{24}$$

Numerical tests of this relation [20, 24] have shown very good agreement with this relation, over all accessible wavelengths $\lambda = 2\pi/q$ from $\lambda_{\text{min}} = B$ to $\lambda_{\text{max}} = L$. Eq. (24) can also be used to predict the mean-square width of this coarse-grained interface, using the convention that $\langle |h(x,y)| \rangle = 0$,

$$s_L^2 = \langle |h(x,y)|^2 \rangle = \sum_{\vec{q}} \langle |h(q)|^2 \rangle = \frac{1}{\tilde{\gamma}} \sum_{\vec{q}} q^{-2} \quad . \tag{25}$$

Converting the sum over discrete values of $\vec{q}$ to an integral yields

$$s_L^2 = (2\pi\tilde{\gamma})^{-1} \int_{2\pi/L}^{2\pi/B} dq/q = \ln(L/B)/(2\pi\tilde{\gamma}) \quad . \tag{26}$$

When we wish to compare this result to actual observations of averaged interfacial profiles and their widths, we need to take into account that the above calculation includes only the contribution of the fluctuations in height of the dividing surface, but not a possible "intrinsic width" $w_0$ (which is attributed to fluctuations on length scales smaller than the cutoff length B). This intrinsic width $w_0$ is included in terms of a convolution approximation of the intrinsic profile with the Gaussian distribution, $P_L(h) = (2\pi s_L^2)^{-1/2} \exp(-h^2/2s_L^2)$, to find [18]

$$w^2 = w_0^2 - (4\tilde{\gamma})^{-1} \ln B + (4\tilde{\gamma})^{-1} \ln L \quad . \tag{27}$$

This logarithmic variation of $w^2$ with the lateral linear dimension $L$ has been verified experimentally and for various models [19, 20, 21, 22, 23, 24, 60]. The prefactor $(4\tilde{\gamma})^{-1}$ of the $\ln L$ term has also been nicely confirmed [20, 24], for a broad range of temperatures. But we emphasize that there is no possibility whatsoever to distinguish in the term $w_0^2 - (4\tilde{\gamma})^{-1} \ln B$ what is $w_0$ and what is the cutoff contribution $-(4\tilde{\gamma})^{-1} \ln B$: since the choice of the cutoff $B$ to some extent is arbitrary, more or less of the short wavelength fluctuation spectrum of the interface is attributed to the intrinsic width. Of course, when the cutoff $B$ is of the order of a few particle diameters, there is no reason to expect that Eq. (20) is still accurate; it has been discussed [24, 73, 74] that Eq. (24) may need a generalization in terms of a wavevector-dependent interfacial stiffness $\tilde{\gamma}(q)$; but this generalization is still controversial, and it seems to depend on the way in which the fluctuating interface as a "dividing surface" is defined [26, 28]. So while the long wavelength fluctuations of interfaces can be understood quantitatively in terms of capillary wave excitations very well, their short wavelength fluctuations are still much less well understood.

Particularly useful is this capillary wave method for solid-liquid interfaces, since there the method based on the free energy hump to the slab configuration is difficult to apply. First successful analyses of the capillary wave spectrum were done for liquid-solid interfaces of metallic systems [63, 64, 65, 66, 67, 69] and later work addressed colloidal crystals [67, 68]. Also evidence for the size-dependence of the mean square interfacial width $w^2$ could be obtained [67, 68], and it could be shown that meaningful estimates of the interfacial stiffness $\tilde{\gamma}$ can be extracted [67, 68].

However, for solid-liquid interfaces there clearly enters another problem, namely the interface tension depends on the orientation of the interface normal relative to the axes of the crystal lattice. As a consequence, there is a systematic distinction to be drawn between interfacial tension and interfacial stiffness (see e.g. Privman [75] for a pedagogic account of this problem). This problem is already well-known for lattice models such as the Ising/lattice gas models, where (in $d = 3$ dimensions) one can show that $\tilde{\gamma}$ diverges to infinity when the temperature $T$ approaches the roughening transition temperature $T_R$ [76], while $\gamma$ at $T_R$ is only weakly singular (it exhibits the so-called "essential singularity", i.e. at $T = T_R$ the radius of a Taylor expansion of $\gamma$ in powers of $T - T_R$ is zero). For $T < T_R$, there are no longer long wavelength capillary waves present, and the interfacial width $w$ no longer diverges with increasing $L$. However, these phenomena are outside the scope of the present paper, and we only mention these problems to warn the reader that the extension of concepts for interfaces between fluid phases to fluid-solid interfaces need to be considered with caution. Also the identification of the local height $h(x, y)$ of a fluid-solid interface in terms of local order parameters for the solid is difficult [63, 64, 65, 66, 67, 68, 69].

## 4. Curved Interfaces

Let us return in this section to the vapor to liquid transition of fluids, and the estimation of the associated interfacial tension. So far (Sec. 2) only the slab configuration of two-phase coexistence in cubic simulation boxes with periodic boundary conditions was exploited; but the descending part of the loop of the $\mu_L(\rho, T)$ versus $\rho$ curve contains information on the interfacial free energy of droplets or bubbles. The droplets in the region $\rho_1 < \rho < \rho_2$ have on average a spherical shape, while the rounded kink in this function at $\rho_2$ is due to a transition to a cylindrical shape of the droplet (facilitated by the periodic boundary condition in the direction of the cylinder axis). This cylindrical morphology in fact is stable in the regime $\rho_2 < \rho < \rho_3$, while near $\rho_3$ a rounded step of $\mu_L(\rho, T)$ towards a constant indicates the onset of the slab configuration, see Sec. 2. Similar observations can be stated with respect to bubbles for the liquid-rich part of the two-phase coexistence region. As described in details elsewhere [10, 34, 43, 77, 78, 79] one can use the knowledge on $\mu_L(\rho, T)$ and $f_L(\rho, T)$ to extract information on the curvature-dependence of the surface tension of spherical droplets and bubbles with radius $R$. The standard description of this problem is due to Tolman [80]

$$\gamma_{\ell v}(T, R) = \gamma_{\ell v}(T, \infty)/\{1 + 2\delta(T)/R + 2[\ell(T)/R]^2\}, \quad R \to \infty \tag{28}$$

where $\gamma_{\ell v}(T, \infty) = \gamma_{\ell v}(T)$ is the standard interface tension of a planar interface, and $\delta(T)$, $\ell(T)$ are two (phenomenologically introduced) characteristic lengths. Tolman (1949) proposed three hypotheses: (i) the length $\delta(T)$, now commonly referred to as the "Tolman length", should be positive for droplets (and hence negative for bubbles: note that in a two-phase droplet and vapor configuration an interchange of the phases means a change in sign of the radius of curvature $R$ of the interface). (ii) $|\delta(T)|$ should be of the same order as $\sigma$, the molecular diameter. (iii) The length $\ell(T)$ appearing in the second-order term should be of the same order, $|\ell(T)| \approx |\delta(T)|$. These hypotheses have remained controversial until recently; however, the analysis of $\mu_L(\rho, T)$ and $f_L(\rho, T)$ due to Block et al. [77, 78, 79] based on the finite size effects modifying the "lever rule" of two-phase coexistence yielded compelling evidence that all three hypotheses in fact are wrong. Thus, the finding that $\gamma_{v\ell}(T,R)$ for bubbles is always smaller than $\gamma_{v\ell}(T,R)$ for droplets already shows that $\delta(T) > 0$ for bubbles and hence $\delta(T) < 0$ for droplets. Secondly, $\delta(T)$ is an order of magnitude smaller than $\ell(T)$. While $\ell(T)$ far below the critical point is of order of molecular lengths, it has been shown [78] that near $T_c$ this length is proportional to the correlation length, $\ell(T) \propto \xi(T)$, while $|\delta(T)|$ increases much slower when $T \to T_c$. This is due to the fact that $\delta(T) \equiv 0$ in symmetric systems (such as the Ising-lattice gas model), and the asymmetry of the liquid-vapor phase diagram near criticality is only related to a correction to scaling, and not the leading scaling behavior [81]. So the term $\delta(T)/R$ in Eq. (28) does not enter the universal scaling function for $\gamma_{\ell v}(T,R)$ in the critical region, and was estimated in Ref. [78]. A further interesting consequence of these findings [77, 78, 79] concerns the theory of homogeneous nucleation [33, 35, 82]: according to the standard "capillarity approximation" of the conventional theory of nucleation, the curvature dependence of $\gamma_{\ell v}(T,R)$ is neglected altogether when one predicts the free energy barrier $\Delta F^*/k_BT$ that needs to be overcome in a nucleation event. For typical supersaturations $S(S = \rho/\rho_\ell < 1$ for bubble nucleation while $S = \rho/\rho_v > 1$ for droplet nucleation), for which the barrier $\Delta F^*/k_BT$ is in the range from 20 to 50, one finds an error of about 20 for bubbles and 10 for droplets, i.e. the error is of the same order as the barrier itself [77, 78, 79]. Of course, the simulation data are consistent with the expectation that for $R \to \infty$ the capillarity approximation becomes valid; but it becomes numerically accurate only for barriers $\Delta F^*/k_BT = 100$ or larger, i.e. a regime that is not practically relevant.

## 5. Wall Tensions

As a last topic of this paper, we discuss the estimation of surface excess free energies due to walls. This is necessary to allow us to deal with problems such as wetting phenomena (mentioned already in the introduction) and heterogeneous nucleation [83]. There exists an enormous range of methods that have been developed to deal with such problems (see e.g. [10] for references). Here we shall focus on a single approach that was recently developed by Winkler et al. [84, 85]. It is a variant of a thermodynamic integration method (see also [86]) and has been termed "ensemble switch method".

One considers a fluid in the canonical NVT ensemble, assuming a volume $V = L \times L \times D_w$, with two walls a distance $D_w$ apart, and at the same time one considers a system with the same particle number $N$ (at the same temperature) and the same linear dimensions but now without any walls, taking periodic boundary conditions in all three directions. The free energies of these two systems will just differ by twice the wall excess free energy, their bulk free energies being the same.

Now the idea is to construct the free energy difference between these systems by finding a path in configuration space $\{\vec{X}\}$ that leads from the one system to the other. Such a path can be constructed in terms of a mixed Hamiltonian: using a parameter $\kappa$, we take $\kappa$ times the Hamilitonian $\mathcal{H}_{\text{bulk+walls}}(\vec{X})$ of the system with walls and adding $(1-\kappa)$ times the Hamiltonian $\mathcal{H}_{\text{bulk}}(\vec{X})$ of the system without walls,

$$\mathcal{H}(\vec{X}) = (1-\kappa)\mathcal{H}_{\text{bulk}}(\vec{X}) + \mathcal{H}_{\text{bulk+walls}}(\vec{X}) \quad . \tag{29}$$

Since the microstates (phase space points) $\vec{X}$ for both systems are the same, Eq. (29) is a well-defined Hamiltonian, suitable for Monte Carlo-simulation, even though such a system never can be created by an experimentalist in the laboratory. However, such possibilities of simulation to go beyond experiment are a particular strength of Monte Carlo simulation.

Discretizing then the interval $[0, 1]$ in which $\kappa$ is defined into 100 discrete steps $\kappa_i$, it makes sense to treat $\kappa_i$ as a dynamical variable which is included in the Monte Carlo sampling. Apart from the usual Monte Carlo moves in the canonical ensemble, where one attempts to displace particles from their old position in the simulation box to a new one, one also considers "switches" where $\kappa_i$ changes to a neighboring value ($\kappa_{i-1}$ or $\kappa_{i+1}$). The corresponding free energy differences between these states then can be gotten by Wang-Landau sampling [87].

In the following we shall briefly review an application example [84, 85], namely the Asakura-Oosawa model [88] of colloid-polymer mixtures [6]. In this model, colloids are described as hard spheres of radius $R_c$, which must neither overlap themselves nor the polymers which are treated as soft spheres of radius $R_p$. Originally the polymers were just treated as ideal gas-like (since random-walk like polymer coils can interpenetrate each other with little energy cost), but it is also possible to use a slightly different model with a soft polymer-polymer repulsion [89]. Using a ratio $q = R_p/R_c = 0.8$, the model exhibits a phase separation into a vapor-like polymer-rich phase and a liquid-like colloid-rich phase, if the normalized polymer fugacity (denoted as "polymer reservoir packing fraction" $\eta_p^r$ [6, 89, 90]) exceeds a critical value $\eta_{p,\text{crit}}^r = 0.767$ [90]. Note that $\eta_p^r$ is an intensive thermodynamic variable, analogous to inverse temperature for an ordinary binary mixture.

To avoid any problems due to large correlation lengths associated with the critical point, the methodology was tested deeply in the strongly segregated regime, e.g. for $\eta_p^r = 0.94$, choosing colloid packing fractions $\eta_c$ outside the coexistence curve. Free energy differences between neighboring states $\kappa_i$, $\kappa_{i+1}$ are obtained from observations of the probabilities $P(i)$ to find the state $i$, $\Delta F(\kappa_{i+1}) - \Delta F(\kappa_i) = k_B T[\ln P(i) - \ln P(i+1)]$, and for typical choice of parameters (e.g. $L = 6.735$ colloid diameters at $\eta_c = 0.319$) one finds $\Delta F(1) - \Delta F(0) \approx 125 k_B T$, but there is a distinct dependence on the distance $D_w$ between the walls, which repel the particles with a Weeks-Chandler-Andersen potential [91]; the range ($\sigma_w$) of this potential is used as a control parameter. However, it was shown that this size-dependence is due to a $1/D_w$ correction [84, 85] and carrying out an extrapolation towards $D_w \to \infty$ very precise results for the wall tensions $\gamma_{wv}$ and $\gamma_{w\ell}$ of both liquid-like and vapor-like phase could be obtained [84, 85]. For this model, both the colloid packing fractions $\eta_{c,v}, \eta_{c,\ell}$ at the "vapor-liquid" coexistence curve and the interface tension $\gamma_{\ell v}$ are accurately known [90]. Thus, using Young's equation [15], Eq. (1), the contact angle could be reliably predicted. It turns out that varying $\sigma_w$ one can change the wetting behavior of the system all the way from complete wetting (for $\sigma_w < 0.59$ colloid diameters) to complete drying (for $\sigma_w > 0.79$).

## 6. Application to Nanoconfinement

The knowledge on the phase behavior of the Asakura-Oosawa model in the bulk [90] together with the knowledge of its walls tensions [84] can be used to clarify the properties of this model system under conditions of confinement in a spherical cavity [85]. Phase behavior of various systems under conditions of confinement on the nanoscale has become a topic of widespread interest, and hence the present system allows to carry out a corresponding model study.

When one chooses a cavity radius $R = 10$ colloid diameters, and complete wetting conditions ($\sigma_w = 0.5$), under very strong segregation conditons ($\eta_p^r = 1.4$) one finds that nevertheless there is no wetting layer at the surface of the vapor-like polymer rich phase [85]. This surprising result is due to the fact that under nanoconfinement it can happen that the relevant particle number simply is too small: for $R = 10$ in the vapor-like phase at coexistence only $N_c = 177$ colloids are present in the whole volume of the nanocavity (while in the liquid-like phase $N_c = 2909$). There are simply too few colloids in the vapor-like phase to allow the formation of a wetting layer in such a system. When one studies the free energy $F(N_c)$ as a function of the number of colloids $N_c$ in the system, one finds a double-well structure. However, again it does not mean that we have a Landau-type free energy: still the peak is due to interfacial effects, the system forms a core-shell structure (the polymer-rich phase forms a core, which on average is spherical and in the center, the colloids form the shell). Conversely, if we consider neutral walls, the peak of $F(N_c)$ is due to a "Janus particle" structure, the system forms an essentially planar interface, with one hemisphere formed by the colloid-rich phase and the other hemisphere by the polymer-rich phase. In both cases the chemical potential $\mu(\eta_c)$ of the colloids exhibits loops, again reflecting two-phase coexistence, and hence must not be interpreted as van-der Waals-type loops. It is remarkable that in this situation of nanoconfinement the loops of $\mu(\eta_c)$ do not develop any sharp peaks and rounded kinks, as is the case with the periodic boundary conditions: these features were due to discontinuous transitions in the geometry of the coexisting phases (from the droplet to the slab configuration, etc.) These structures were stabilized by the periodic boundary conditions, and do not occur for experimentally more relevant boundary effects considered in the present section.

It is very interesting to analyze how these finite size effects on phase coexistence change when one varies the radius $R$ of the spheres. One then finds that the loops in the $\mu(\eta_c)$ curves get gradually flatter, so the system also converges to an isotherm, with a horizontal part from the pure vapor-like to the pure liquid-like phase as $R \to \infty$. A very interesting aspect is that for finite $R$ the different ensembles of statistical mechanics yield different results, although they become equivalent for $R \to \infty$. A loop in the $\mu(\eta_c)$ curve is only seen when $\eta_c$ is the control parameter and $\mu_c$ is obtained from the Monte Carlo sampling. However, carrying out a simulation in the grand-canonical ensemble, $\mu$ is the control parameter and $\eta_c$ is sampled: then one finds the isotherm $\eta_c(\mu_c)$ is monotonous; while in the thermodynamic limit this isotherm has a perpendicular part, for finite $R$ the slope of the isotherm is finite but large, the maximum slope scales like $R^3$. In contrast, the shift of the inflection point of the isotherm (where this maximum slope occurs) relative to the location of the bulk scales like $1/R$. Thus the rounding (proportional to the inverse of the maximum slope, i.e $R^{-3}$) scales differently from the shift.Thus for phase coexistence the finite size effects behave very differently from what one knows about finite size effects on critical phenomena (where shift and rounding scale with the same exponent, $R^{-1/\nu}$ where $\nu$ is the critical exponent of the correlation length [92]).

A consequence of the fact that isotherms in the grandcanonical ensemble are monotonic, while isotherms (plotting the same variable) in the canonical ensemble are nonmonotic (due to their loops) also means that they are not simply related by a Legendre transformation, as in the thermodynamic limit.

The fact that the shift of the transition scales like $1/R$ can easily be understood from the surface excess contributions to the free energy. For spherical geometry one obtains for the location of the transition [84, 85]

$$\mu_c(R) - \mu_c(\infty) = \frac{3}{R}\frac{\gamma_{w\ell} - \gamma_{wv}}{\rho_{c,\ell} - \rho_{c,v}} \quad , R \to \infty \tag{30}$$

where $\rho_{c,\ell}$ and $\rho_{c,v}$ are the colloid densities at the coexisting liquid-like and vapor-like phases in the bulk colloid-polymer mixture. Analyzing the shift observed in the simulation hence is a second method for estimation of the wall free energy difference $\gamma_{w\ell} - \gamma_{wv}$, that also enters the Young equation for the contact angle. It was found [85] that for $0 < \theta < 180^o$ the agreement with the thermodynamic integration method (the "ensemble switch method") of the previous section is very good. Interestingly, for the regime of complete wetting one simply finds the result

$$\gamma_{wv} = \gamma_{w\ell} + \gamma_{\ell v} \tag{31}$$

due to the liquid wetting layer intruding in between the wall and the vapor region in the thermodynamic limit. Thus, the analysis confirms the location of the wetting transition, at least roughly. Thus, we see that for simple systems, like the Asakura-Oosawa model, one can understand all aspects of phase coexistence in nanosystems in terms of an interplay of surface vs. bulk behavior and finite size effects.

These studies have been carried over to liquid-solid coexistence in the confined hard sphere system and the Asakura-Oosawa model (at high colloid density) [93, 94]. The "ensemble switch method" was found to be useful also in this case, and together with estimations of the interfacial free energy between coexisting fluid and solid phases (obtained from interfacial broadening due to capillary waves, see Sec. 3) predictions for the contact angle could be made [94]. This prediction was found to be in reasonable agreement with direct observations of the contact angle of wall-attached crystalline "droplets" at planar structureless walls. Although these results clearly are encouraging, they can be considered as a promising first step only, and a much more complete study is required.

## 7. Concluding remarks

In this paper, it was emphasized that interfaces between coexisting phases in condensed matter are important for many phenomena: nucleation, wetting, phase separation kinetics, phase coexistence in nano-materials; etc. It was also emphasized that analytical theories are quite misleading: concepts such as van der Waals loops, Landau double well free energies, intrinsic interfacial profiles of tanh form, etc., all are mean field artefacts.

Interfaces have structure from the atomistic-scale to the mesoscopic scale: but well-equilibrated large-scale interfaces are a challenge for simulation. The interfacial profiles depend on the system geometry, and it is difficult to disentangle interfacial from bulk fluctuations, and to give the "intrinsic profile" a precise meaning. This issue has been controversial over decades and still is today. Particularly many open problems concern solid-liquid interfaces (faceting transitions; equilibrium shape of nanocrystals in different environments, etc.) and also the "line tension" associated with the contact line where three interfaces meet. Thus, the subject will remain active and exciting for many years to come.

Acknowledgement: The author is grateful to many coworkers with whom he has collaborated on the subject outlined in this paper, for a pleasant and fruitful interaction. In particular, he thanks B.J. Block, S.K. Das, D. Deb, J. Horbach, M. Müller, F. Schmid, A. Statt, A. Tröster, R.L.C. Vink. P. Virnau, A. Werner, A. Winkler, and T. Zykova-Timan. This research was supported by the Deutsche Forschungsgemeinschaft and the John von Neumann Institute Jülich.

[1] DeGennes PG, Brochard-Wyart F, Quere D 2003 *Capillarity and Wetting Phenomena: Drops, Bubbles, Pearls, Waves* (New York: Springer)
[2] Dietrich S 1988 *Wetting Phenomena in Phase Transitions and Critical Phenomena, Vol 12* ed Domb and JL Lebowitz (London: Academic) pp. 1-218.
[3] Clarke DR, Rühle M, Tomsia AP 2008 eds. *Low and High-Temperature Wetting: State of the Art* (Ann. Rev. Mater. Res. Vol 38) (Palo Alto: Ann. Revs.)
[4] Bonn D, Eggers J, Indekeu J, Meunier J, Ratley E 2009 *Rev. Mod. Phys.* **81**, 739.
[5] Rowlinson JS, Widom B 1982 *Molecular Theory of Capillarity* (Oxford: Clarendon)
[6] Poon W 2002 *J. Phys.: Condens. Matter* **14**, R598.
[7] Aarts DGAL, Schmidt M, and Lekkerkerker HNW 2004 *Science* **304** 847.
[8] Hennequin Y, Aarts DGAL, Indekeu JO, Lekkerkerker HNW, and Bonn D 2008 *Phys. Rev. Lett.* **100** 178305.
[9] Widom B 1972 in *Phase Transitions and Critical Phenomena, Vol 2*, ed. C. Domb and MS Green (London: Academic) p. 79.
[10] Binder K, Block B, Das SK, Virnau P and Winter D 2011 *J. Stat. Phys.* **144**, 690.
[11] Pang L, Landau DP, and Binder K 2011 *Phys. Rev. Lett.* **106** 236102.
[12] Troester A and Binder K 2011 *Phys. Rev. Lett.* **107** 265701.
[13] Albano EV and Binder K 2012 *Phys. Rev. Lett.* **1090** 036101.
[14] Bryk P and Binder K 2013 *Phys. Rev. E* **88**, 030401 (R).
[15] Young T 1805 *Phil. Trans. Roy. Soc. London* **95** 65.
[16] Binder K 1983 *Critical Behavior at Surfaces* in *Phase Transitions and Critical Phenomena, Vol 8* ed. C. Domb and JL Lebowitz (London: Academic) pp. 1-145.
[17] Cahn JW 1977 *J. Chem. Phys.* **66** 3667.
[18] Jasnow D 1984 *Rep. Progr. Phys.* **47** 1059.
[19] Schmid F and Binder K 1992 *Phys. Rev. B* **46** 13553.
[20] Werner A, Schmid F, Müller M and Binder K 1999 *Phys. Rev. E* **59** 728.
[21] Binder K and Müller M 2000 *Int. f. Mod. Phys. C* **11** 1093.
[22] Binder K, Müller M, Schmid F and Werner A 2001 *Advanc. Coll. Interfaces Sci.* **94** 237.
[23] Binder K 2001 *Simulations of Interfaces Between Coexisting Phases: Multiscale Aspects in Multiscale Computational Methods in Chemistry and Physics* A Brandt, J Bernholc and K Binder (Amsterdam: IOS Press)
[24] Vink RLC, Horbach J and Binder K 2005 *J. Chem. Phys.* **122** 134905.
[25] Weeks JD 1977: *J. Chem. Phys.* **67** 3106.
[26] Chacón E and Trazona P 2003 *Phys. Rev. Lett.* **91** 166103.
[27] Tarazona P and Chacón E 2004 *Phys. Rev. B* **70** 235407.
[28] Fernandez EM, Chacón E, Tarazona R, Parry AO and Rascón C 2013 *Phys. Rev. Lett.*, in press
[29] Buff FP, Lovett RA and Stillinger 1965 *Phys. Rev. Lett* **15** 621.
[30] Weeks JD, van Searloss W, Bedeaux D and Blokhuis E 1989 *J. Chem. Phys.* **91** 6494.
[31] Van der Waals JD 1893 *Verhandl. Koningh. Akad. van Wetenschappen* (Amsterdam), 1979. eng. translation in *J. Stat. Phys.* **20**, 197.
[32] Landau LD and Lifshitz EM 1980 *Statistical Physics, Part I, 3rd ed.* (Oxford: Pergamon)
[33] Binder K 1987 *Rep. Progr. Phys.* **50** 783.
[34] Binder K, Block BJ, Virnau P and Troester A 2012 *Am. J. Phys.* **80** 1099.
[35] Binder K and Staufffer D 1976 *Advanc. Phys.* **25** 343.
[36] Lebowitz JL and Penrose O 1966 *J. Math. Phys.* **7** 98.
[37] Cahn JW and Hillard JE 1958 *J. Chem. Phys.* **28** 258.
[38] Binder K and Kalos MH 1980 *J. Stat. Phys.* **22** 363.
[39] Furukawa H and Binder K 1982 *Phys. Rev. A* **26** 556.
[40] Biskup M, Chayes L, and Kotecky R 2002 *Europhys. Lett* **60** 21.
[41] Binder K 2003 *Physica A* **319** 99.
[42] MacDowell LG, Virnau P, Müller M, and Binder K 2004 *J. Chem. Phys.* **120** 5293.
[43] MacDowell LG, Shen VK, and Errington JR 2006 *J. Chem. Phys.* **125** 034705.
[44] Schrader M, Virnau P and Binder K 2009 *Phys. Rev. E* **79** 061104.
[45] Virnau P and Müller M 2004 *J. Chem. Phys.* **120** 10925.
[46] Ferrenberg AM and Swendsen RH 1988 *Phys. Rev. Lett.* **61** 2635.
[47] Binder K and Landau DB 1984 *Phys. Rev. B* **30** 1477.
[48] Borgs C and Kotecky R 1990 *J. Stat. Phys.* **61** 79.
[49] Binder K, Lebowitz JL, Phani MK, and Kalos MH 1981 *Acta Metall.* **29** 1655.
[50] Das SK, Horbach J, Binder K, Fisher, ME, and Sengers JV 2006 *J. Chem. Phys.* **125** 024506.
[51] Das SK and Binder K 2010 *EPL* **92** 26006.

[52] Frenkel D and Smit B 2002 *Understanding Molecular Simulations: From Algorithms to Applications, 2nd ed.* (San Diego: Academic Press)
[53] Binder K 1982 *Phys. Rev. A* **25** 1699.
[54] Potoff I and Panagiotopoulos A 2000 *J. Chem. Phys.* **112** 6411.
[55] Mognetti BM, Virnau P, Yelash L, Paul W, Binder K, Müller M, and MacDowell LG 2009 *J. Chem. Phys.* **130** 044101.
[56] Allen MP and Tildesley DJ 1989 *Computer Simuation of Liquids* (Oxford: Clarendon Press)
[57] Onsager L 1944 *Phys. Rev.* **65** 117.
[58] Vink RLC and Schilling T 2005 *Phys. Rev. E* **71** 051716.
[59] Limmer D and Chandler D 2011 *J. Chem Phys.* **135** 134503.
[60] Mon KK, Landau DP and Stauffer D 1990 *Phys. Rev. B* **42** 545.
[61] Jamie EAG, Davies GJ, Howe MD, Dullens RPA and Aarts DG AL 2008 *J. Phys.: Condens. Matter* **29** 393231.
[62] Wolfsheimer S, Tanase C, Shundvak K, van Roij R, and Schilling T 2006 *Phys. Rev. E* 061703.
[63] Hoyt JJ. Asta M, and Karma A 2001 *Phys. Rev. Lett.* **86** 5530.
[64] Asta M, Hoyt JJ and Karma A 2002 *Phys. Rev. B* **66** 100101 (R).
[65] Morris JR 2002 *Phys. Rev. B* **66** 144104.
[66] Sun DY, Asta M and Hoyt JJ 2004 *Phsy. Rev. B* **69** 174103.
[67] Zykova-Timan T, Rozas RE, Horbach J and Binder K 2009 *J. Phys.: Condens Matter* **21** 064102.
[68] Zykova-Timan T, Horbach J, and Binder K 2010 *J. Chem. Phys.* **133** 014705.
[69] Rozas RE and Horbach J 2011 *EPL*
[70] Carmesin I and Kremer K 1988 *Macromolecules* **21** 2819.
[71] Paul W, Binder K, Heermann DW, and Kremer K 1991 *J. Phys. II (France)* **1** 37.
[72] Deutsch H-P and Binder K 1992 *Macromolecules* **25** 6214.
[73] Mecke K and Dietrich S 1999 *Phys. Rev. E***59** 6766.
[74] Milchev A and Binder K *Europhys. Lett.*
[75] Privman VP 1992 *Int. J. Mod. Phys. C* **3** 857.
[76] Weeks JD 1980 in *Ordering in Strongly Fluctuating Condensed Matter Systems*, ed. Riste T (Plenum Press: New York) p. 293.
[77] Block BJ, Das SK, Oettel M, Virnau P and Binder K 2010 *J. Chem. Phys.* **135** 154702.
[78] Das SK and Binder K 2011 *Phys. Rev. Lett.* **107** 235702.
[79] Troester A, Oettel M, Block BJ, Virnau P and Binder K 2012 *J. Chem. Phys.* **136** 064709.
[80] Tolman RC 1949 *J. Chem. Phys.* **17 17** 333.
[81] Anisimov MP 2007 *Phys. Rev. Lett.* **98** 035702.
[82] Kashchiev D 2000 *Nucleation: Basic Theory and Applications* (Butterworth-Heinemann: Oxford)
[83] Winter D, Virnau P and Binder K 2009 *Phys. Rev. Lett.* **103** 225703.
[84] Statt A, Winkler A, Virnau P, and Binder K 2012 *J. Phys.: Condens. Matter* **24** 464122.
[85] Winker A, Statt A, Virnau P and Binder K 2013 *Phys. Rev. E* **87** 032307.
[86] Heni M and Loewen H 1999 *Phys. Rev. E* **60** 7057.
[87] Wang F and Landau DP 2001 *Phys. Rev. Lett.* **86** 2050.
[88] Asakura S and Oosawa F 1958 *J. Polym. Sci.* **33** 183.
[89] Zausch J, Virnau P, Binder K, Horbach J, and Vink RLC 2009 *J. Chem. Phys.* **130** 064906.
[90] Vink RLC and Horbach J 2004 *J. Phys.: Condens. Matter* **16** 3807.
[91] Weeks JD, Chandler D and Andersen HC 1971 *J. Chem. Phys.* **54** 5237.
[92] Binder K 1992 in *Computational Methods in Field Theory* eds. H. Gausterer and C.B. Lang (Springer: Berlin), p.59.
[93] Deb, D, Wilms D, Winkler, A, Virnau P and Binder K 2012 *Int. J. Mod. Phys. C* **23**, 1240011.
[94] Deb, D, Winkler A, Virnau P and Binder K 2012 *J. Chem. Phys.* **136** 134710.

# How do I simulate problem X?

**Joan Adler, Yaron Artzi, Liron ben Bashat, Tzipora Yael Izraeli, Meital Kreif, Ido Lavi, Alexander Leibenzon, Adam Levi, Itai Schlesinger, Elad Toledano, Uria Peretz, Yonatan Weisler and Alon Yagil**

Technion - IIT, Haifa, Israel, 32000

E-mail: phr76ja@tx.technion.ac.il

**Abstract.** We present reports on projects from the Technion Physics Department's annual Computational Physics class that are in, near or far away from its lecturer's personal expertise.

## 1. Introduction

Learning numerical techniques of relevance to their theoretical or experimental research project is a valid reason for a graduate student to take a Computational Physics class. Most universities are not able to provide separate computational classes geared towards each sub-discipline and so lecturers can find themselves supervising projects falling outside their personal research expertise.

At the Technion last fall some 20 graduate and final year undergradate students took the Computational Physics class. It is relatively easy to find simulation projects loosely connected to the lecturer's research area, in that case atomistic simulations of condensed matter or statistical physics. However it is important that each student carries out a project that is relevant and interesting for them, and can be completed with the appropriate effort. When the class includes students from areas such as relativity, optics, plasma physics and types of quantum physics which fall outside the lecturer's expertise in either topic or algorithm (or both) this must be accounted for. In some cases their supervisors hope said projects will jumpstart a computational component in their research groups. We present projects from the fall 2012 class and describe our pitfalls and successes.

In the next section articles based on presentations of projects from the class in recent years will be described. Details of the course structure were presented there and do not need to be repeated. Then some educational projects, designed for class use in undergraduate courses at the Technion will be introduced. Three groups of research projects - in condensed matter/statistical physics, close to condensed matter/statistical physics, and those beyond the lecturer's personal research experience will then be presented. Not all projects will be described in detail here, but a full list can be found on the class website [1].

## 2. Previous presentations and generalities

A presentation about the class and its syllabus [2] explained the process whereby each student submits a research proposal halfway through the course in order to decide on topic, langauge, operating system, visualization tools and algorithms. Although the majority of projects described here are from the class (70% of the class grade), a few are final undergraduate projects of a slightly larger scope, and in some cases project students are asked to attend the class lectures to widen their algorithmic knowledge. In [3] two successful class projects from earlier years are described, and in [4], [5] and [6] three projects closely related to the Computational Physics group's research interests are introduced. The last is from the recent class, but is an integral part of the Computational Physics group's research and thus was published seperately.

In all projects submission as a website is a requisite step. Code has to be ready for download with help files, default parameters must be proposed and the project must run on a LINUX computer (generally either Red Hat or Ubuntu at the moment). When appropriate, windows versions can also be presented. Visualization is heavily emphasised and the website on the classroom server (known as PHELAFEL) is avaliable for student websites. Since the Technion Computational Physics Group has developed a Mesa/Qt based C++ Atomistic Visualization code, known as AViz [7, 8, 9], this is recommended when appropriate, but for two-dimensional systems PGPLOT is often used and sometimes OpenGL is appropriate. When the visualization does not involve discrete objects, MATLAB, MATHEMATICA or gnuplot are used. The PGPLOT codes usually create the graphics while the code is being run, but most of the AViz visualizations involve preparation of data files which are then visualized post preparation. Several of the projects provide instructions on how to output results from different programs into the .xyz AViz input files. Each year some new codes/systems enter the mix; this year it was python, php and C# (C-sharp) as well as additional supplements to OpenGL such as GLEW and GLUT.

## 3. Educational projects

Three dimensional visualizations of hydrogen atom wavefunctions are very helpful for teaching Modern Physics or Quantum Mechanics classes. The concept of electronic density is hard to grasp. Earlier [10] 3d Hydrogen Schroedinger wavefunctions with binned color and rotating images on a website were helpful and used in Modern Physics class. This approach uses an "off-label" application of AViz, wherein the dot representation, originally created to enable quick selection of viewing angle etc for atomistic samples, is used to create a translucent effect whereby the sample's interiors are visible. Subsequent to the analglyphic stereo enhancement to AViz described in [4, 11], these were redone. The new website also has improved explanations of the generation process. While two color glasses give even more insight, just the images on a white background are quite useful. This project used MATHEMATICA to generate the data and AViz to view it. Two examples are given in Figures 1 and 2, the former of the $n = 3$, $l = 1$, $m = 1$ orbital and the latter of the $n = 4$, $l = 3$, $m = 2$ orbital of the H atom. On the website all images can be rotated to further aid in depth perception.

Another concept of Quantum Mechanics that is difficult for students to grasp is the notion of energy levels of infinite wells and how bosons and fermions take their different states therein. Since they are going to sit in class with smartphones and tablets above or below the desk, we may as well encourage them to use these tools for something beyond Facebook. The energy level app enables them to enter some number of different particles (selected both from different preassigned choices and added "on the fly") into a well and see how many select each energy state. Following on the talk at CCP2013 a realtime test was made to introduce the Higgs particle described earlier in the day; it took less than 1 minute on an iPhone. This project is written in php and accessible from every smartphone and tablet that it was tried on. A typical output image is given in Figure 3.

To complete the trio of elementary Quantum Mechanics education projects, a website with downloadable executable code to simulate the Stern-Gerlach experiment was prepared. The code is OpenGL with GLUT and comes with executable for both Windows 7 and Ubuntu/RedHat. The former requires another file (provided on the website) to be downloaded to run, the latter

Journal of Physics: Conference Series **510** (2014) 012003 doi:10.1088/1742-6596/510/1/012003

some installation, again with instructions on the website. There are also sample outputs and movies on the website. The code shows atoms released on the left of a variable magnetic field that they pass through on their way to a screen on the right. Figure 4 shows how atoms with classical spins simply cover some region of the screen more or less uniformly. Figure 5 shows how spins quantized with $\pm$ 1/2 spins form two lines, exactly as on the famous postcard [12] Gerlach sent to Niels Bohr in 1922.

The final project in this section is for a somewhat more advanced course, and concerns quantum scattering preconfigured for the Yukawa potential. It has a good interactive interface, and is python based, but is a little less straightforward to run for non-experts than the above three projects. Images can be seen at [1] and the link thereon.

## 4. Condensed matter/statistical physics projects

The simplest project in this group is one concerned with percolation on fractals using C++ to create the datafiles and AViz to visualize the results. Percolation, fractals and the Mersenne twister random number generator are introduced and the code is well documented. One result for the Sierpinski square pyramid in shown in Figure 6.

Two projects have been carried out using the LAMMPS [13] molecular dynamics code. Both include general instructions as well as Technion-specific files, which are probably also useful for other parallel machines that use the pbs queueing system. One concerns visualization of defects in graphene, and presents an additional C++ code to indentify defects as well as instructions on extracting the .xyz AViz input file from LAMMPS. The second was an exercise in creating results with LAMMPS that verify a research groups' earlier calculations with the IMD code. The system studied here was crack propogation in silicon. Again here examples of script files and input to that groups preferred visualization tool, atomeye, are given. Script files for the Technion Parallel computer, TAMNUN are also presented, and images are shown on [1].

The final project here concerns the temperature dependence of conductivity in graphene using MATHEMATICA and MATLAB. This project is a detailed numerical realization of a study by Hwang and Das Sarma [14], with downloadable, well documented, MATHEMATICA and MATLAB codes.

## 5. Biophysics projects

The projects in this section all apply techniques of statistical physics to biophysics models. They all also happen to invovlve a slightly more advanced level of programming than some of the others and provide optimized code with excellent instructions for installation and compilation of the codes.

The first is an significant extension of the concept from the animated simulated annealing PGPLOT code of Adler and Silverman [15] to a model for confined water molecules. In both cases the idea is to view the sample while carrying out annealing in order to monitor the progress to the required state. This project was made within cygwin [16] and used Fortran90, pushing the envelope of previous class experience. One snapshot is given in Figure 7.

Another interesting project is the simulation of a dilute polymer solution with the disspative particles dynamics method. Here MATLAB with the parallel toolbox and GPUs was used and a visualization in MATLAB has been prepared for youtube. Both this and the previous project simulate and visualize concurrently.

The last project with an image featured herein describes intermittent reaction paths of tfs (transcription factors). This project uses C# for the code; although an executable is also given on the website and AViz for the visualization, and gives instructions for both on Ubuntu, with monodevelop recommended for the former.

## 6. Optics and relativity

The three projects in this section are closely connected to specific research groups in the Physics Department, and more details may be published by those groups in the future. Two use MATLAB, all are at a high level. The first concerns solution of the Schroedinger equation and the second light propogation on glial cells.

The third and final project to be described is in relativity, and concerns the numerical solution of the CGHS equations [17]. Although the equations date from 1991, and are considered to be a "toy model", a non-pertubative numerical solution has been evasive. Unlike the "toy model" of statistical physics, viz. the Ising model, which has been studied ad nauseum, little intensive numerical effort has been made here. However, since the equations can be turned into difference equations and solved with a second order predictor corrector, it is amenable to a full numerical solution, if the intial conditions are chosen with care and sufficient computer power can be found. This is a classic example of applying very basic numerical methods to a problem outside the range of previous applications in the class programme, and achieving success. It was fun to present at CCP2013 where all teachers of basic numerical analysis enjoyed hearing from I. M. Khalatnikov about the early days of numerical methods for differential equations.

The C# equation solving code was originally designed for windows, but can be ported to the monodevelop in Ubuntu that was installed for the DNA project described above. The data plotting is done in MATLAB. Both parts have a nice interactive interfaces.

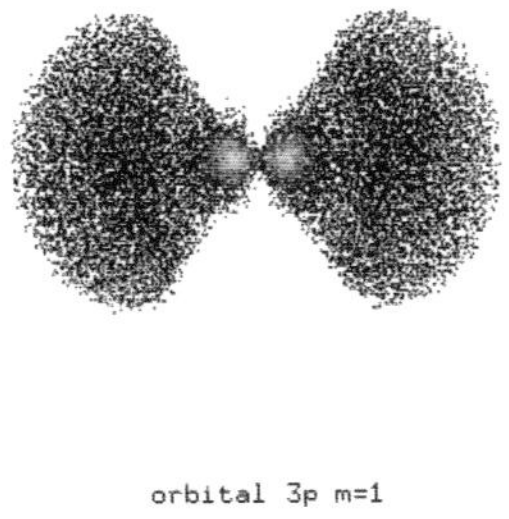


**Figure 1.** Image of the $n = 3$, $l = 1$, $m = 1$ orbital.

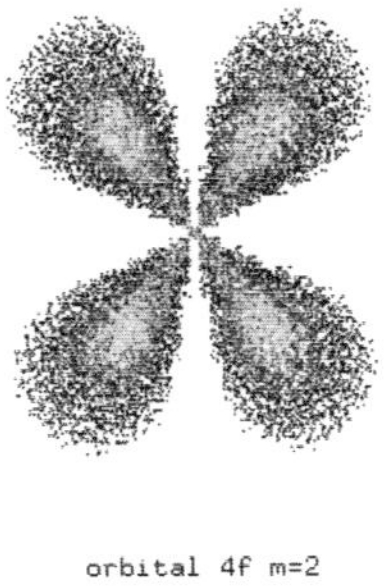


**Figure 2.** Image of the the $n = 4$, $l = 3$, $m = 2$ orbital.

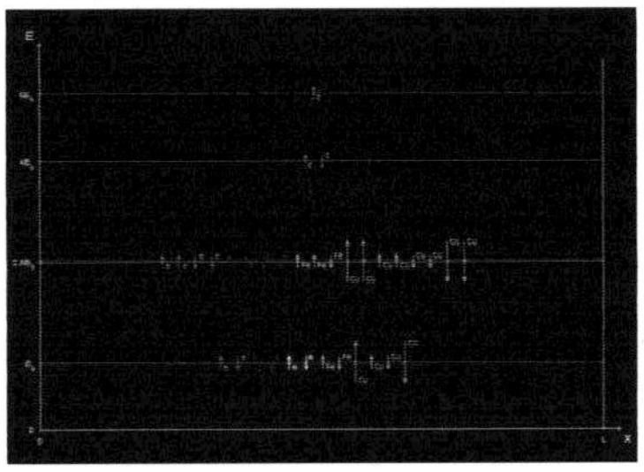

**Figure 3.** Energy levels in an infinite well.

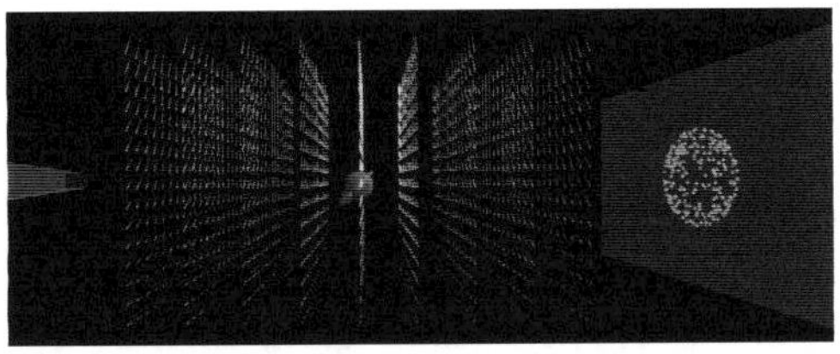

**Figure 4.** Stern-Gerlach experiment with classical spins.

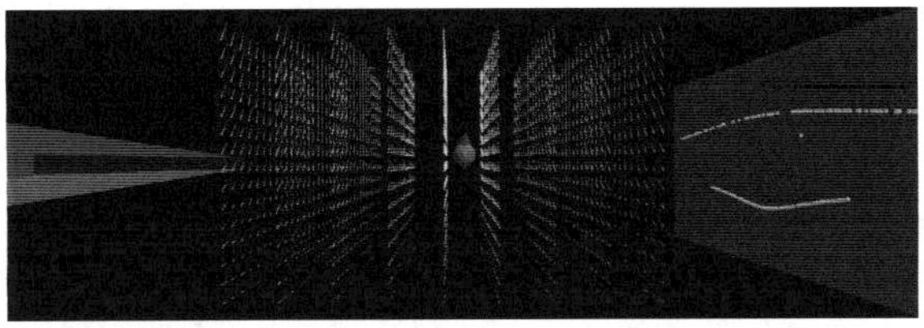

**Figure 5.** Stern-Gerlach experiment with $s = \pm(1/2)$ spins.

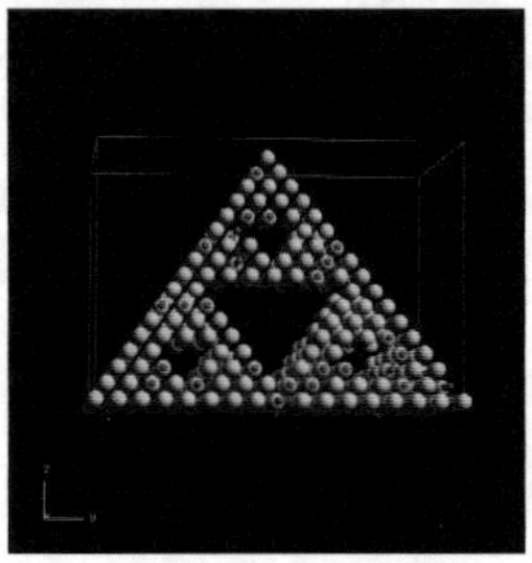

**Figure 6.** Sierpinski square pyramid (green balls) with 67 sites removed and indicated as small red balls.

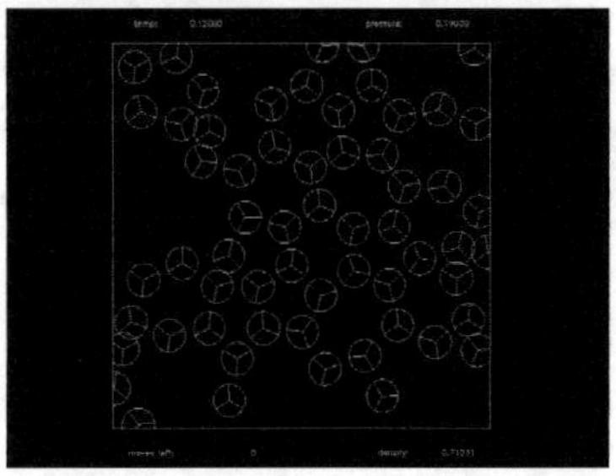

**Figure 7.** A "good enough" two dimensional ice model system with temperature of 0.12 and a pressure of 0.19.

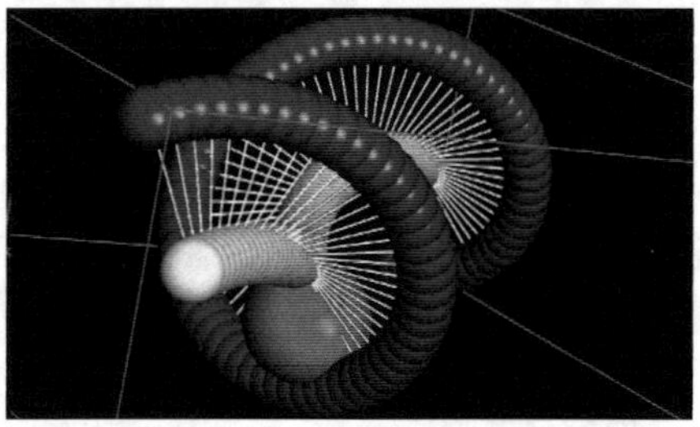

**Figure 8.** Double stranded DNA with diffusing protein (red ball).

## 7. Conclusions

The codes described above cover many areas of Computational Physics and we expect they will be useful both in Physics Education for future students, in Computational Physics Education and in Computational aspects of Physics research. Future extensions will no doubt be made to some of these, and already several new undergraduate physics projects are in progress, some of which extend these directions. Relevant updates will be placed on [1] as appropriate.

As well as usable codes and websites which should enable others to run and extend the codes, many participants learnt new tools, and some of those who already came with good computer skills learnt something about documentation and transfer to other operating systems. There was a lot of cooperation within the class. A word on pitfalls: while the end result in all cases was excellent, some flexibility in topics and goals was needed in a few cases, and everyone worked very hard throughout and after the semester.

## Acknowledgments

We thank those class students whose projects are on the webpage but not mentioned explicitly above (due to separate publication or confidentiality) for also being part of the experience, and also acknowledge other departmental graduate students for helping class members in many ways. We thank J. Halupovich and D. Pomerancblum for help with OpenGL, and R. Kraizelman and Y. Katz for support and installation on classroom computers.

## References

[1] http://phycomp.technion.ac.il/~comphy/projects2012.html
[2] J. Adler, "Educating the next generation of Computational Physicists" Physics Procedia, to appear (Athens, 2013).
[3] D. Mazvovsky, G. Halioua and J. Adler, "The role of projects in (Computational) Physics Education", Physics Procedia, **34**, p 1-5, (2012).
[4] D. Peled, A. Silverman and J. Adler "3D visualization of atomistic simulations on every desktop', IOP Conference Series, **454**, p. 012076, (2013), CCP2012.
[5] J. Adler and O. Cohen, "Solid Hydrogen - New twists on an old problem" in "Recent Developments in Computer Simulation Studies in Condensed Matter Physics, X", edited by D. P. Landau, S. P. Lewis and B. Schuttler, Physics Procedia **6**, p 2-6, (2010).
[6] J. Adler, N. Ierushalmi, A. Silverman and R. Kalish, IOP Conference Series, to appear (Joao Pessoa, 2013).
[7] http://phycomp.technion.ac.il/~aviz
[8] J. Adler, A. Hashibon, N. Schreiber, A. Sorkin, S. Sorkin, and G. Wagner, "Visualization of MD and MC simulations for atomistic modeling", Computer Physics Communications, **147** pp 665-9, (2002).

[9] J. Adler, "Visualization in Atomistic and Spin Simulations", Computers in Science and Engineering, **5** pp 61-65, (2003).
[10] J. Adler, J. Fox, R. Kalish, T. Mutat, A. Sorkin and E. Warszawski, "The essential role of visualization for modeling nanotubes and nanodiamond", Computer Physics Communications **177** pp 19-20, (2007).
[11] J. Adler, Y. Koenka and A. Silverman, "Adventures in carbon visualization with AViz", Physics Procedia, **15** pp 7-16, (2011).
[12] A. J. G. Hey and P. Walters, "The quantum universe", Cambridge University Press, (1987).
[13] http://lammps.sandia.gov
[14] E. H. Hwang and S. Das Sarma, "Screening induced temperature-dependent transpot in two-dimensional graphene", Phys Rev. B **79** p.1654044, (2009).
[15] A. Silverman and J. Adler, "Animated simulated annealing", Computers in Physics, **6**, pp. 277-281, (1992).
[16] http://www.cygwin.com
[17] C. Callan, S. Giddings, J. Harvey, A. Strominger, "Evanescent black holes", Phys. Rev. D **45** pp. 1005-9, (1992).

Journal of Physics: Conference Series **510** (2014) 012004 doi:10.1088/1742-6596/510/1/012004

# Experimental research control software system

**I A Cohn, A G Kovalenko, A N Vystavkin**

Kotel'nikov Institute of Radio Engineering and Electronics of RAS, 7-11 Mokhovaya str., Moscow, 125009, Russia

**Abstract.** A software system, intended for automation of a small scale research, has been developed. The software allows one to control equipment, acquire and process data by means of simple scripts. The main purpose of that development is to increase experiment automation easiness, thus significantly reducing experimental setup automation efforts. In particular, minimal programming skills are required and supervisors have no reviewing troubles. Interactions between scripts and equipment are managed automatically, thus allowing to run multiple scripts simultaneously. Unlike well-known data acquisition commercial software systems, the control is performed by an imperative scripting language. This approach eases complex control and data acquisition algorithms implementation. A modular interface library performs interaction with external interfaces. While most widely used interfaces are already implemented, a simple framework is developed for fast implementations of new software and hardware interfaces. While the software is in continuous development with new features being implemented, it is already used in our laboratory for automation of a helium-3 cryostat control and data acquisition. The software is open source and distributed under Gnu Public License.

## 1. Introduction

Currently two different approaches are widely used to automate experimental equipment in small (single PC) experimental setups. Either the whole system is programmed using commercial software like LabVIEW, Stateflow, etc., or there is a custom solution that is designated to a single experiment. Both ways have drawbacks, which complicate the automation task.

Commercial software costs money that has better ways to be spent. The major applications use the combination of a dataflow paradigm with visual programming. Visual programming significantly complicates supervision and support while the dataflow paradigm is not studied in computer science class.

Custom solutions are commonly built using general purpose programming languages. This means that the programmer has to deal with all the interfaces, including hardware, user interaction, output, - all on his own. Thus, good programming skills are required as well as a significant amount of labor. Moreover, this approach lacks flexibility. Slight change in the experiment's algorithm may cause substantial program changes. Hardware access collisions and stability issues must be solved each time from scratch.

## 2. Goals

Here we present a new software solution that lacks the above drawbacks. To meet the requirements of an experiment automation system that are imposed in our laboratory, a universal experiment control software system is proposed. Here are the goals to be achieved:

- First of all, there should not be a requirement of strong programming skills to automate an experimental setup. Unfortunately experimental physicists are not the best programmers, especially students.
- The system should be portable, to allow easy transition between different platforms.
- There should be an easy way to extend the software to operate new hardware and software interfaces.
- High flexibility in terms of experimental setup automation is essential. New hardware installation, experiment algorithm change etc. should be easily handled, not affecting experiments in action.
- Automation scripts should be easily reviewed and corrected by a supervisor. That would allow handing automation tasks to students without a worry.

## 3. Architecture

To accomplish those goals we have come up with the following program architecture: the experiment is controlled by separate scripts that run simultaneously in separate threads. The scripts interact with the hardware using a semaphore synchronized layer. Each script can access a private logging interface. In such an arrangement, operating the hardware and logging is quite straightforward, with no worry for access collision and data mess-up.

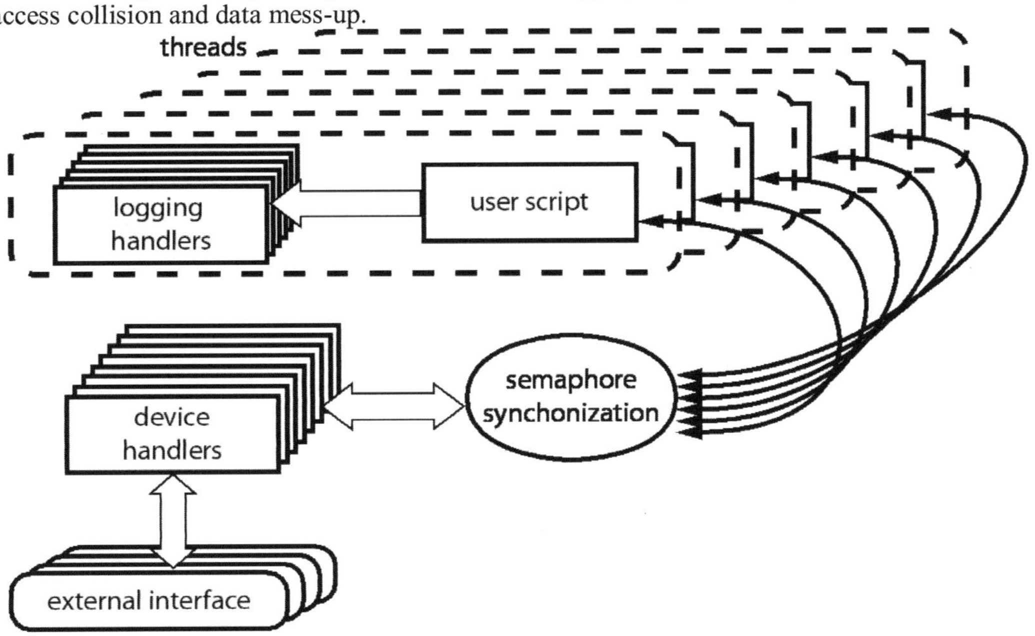


Application architecture

## 4. Tools

As the main programming language we have chosen C++. The reasons for that are good OS multithreading and IPC capabilities utilization as well as hardware access library support. To meet portability requirements we have chosen the wxWidgets C++ framework [1]. It provides all of the cross platform abstraction needed to implement our software architecture. Unlike QT it does not use C++ extensions and promises better portability through the wxUniversal implementation.

To compile our code for Windows operating system we have chosen OpenWatcom compiler [2]. Though it hasn't been changed much in the last couple of years, it is a free open source C/C++ compiler that produces a nice code and is available to everyone. Unlike the current version of Microsoft Visual C Express it runs under windows XP. The alternative is gcc with MinGW [3] environment. It is open source and free as well, but there is a glitch with symbols been generated in binary files. The compiler adds an underscore before each symbol, so it complicates run-time dynamic linking. For POSIX platforms gcc is an obvious choice.

As the scripting language we have chosen REXX [4]. It was developed by IBM as a glue language to automate mainframes, now it serves us to automate experiments. It is an imperative language, very easy to learn, very forgiving. It has natural means of software integration [5]. Thus there should not be a problem for a student with a beginner programming skill to write a simple script.

## 5. Features

While the application is in continuous development, today it can already operate with NI hardware interfaces, especially GPIB, via the NI VISA library [6]. There is a separate handler for the serial port as well. Data logging uses CSV format for easier parsing. There also is a simple graphical output facility to display the measurement results in real-time. User scripts can communicate with each other by a messaging system. Finally, a script can start other scripts.

## 6. The handlers mechanism

In order to provide a convenient way to add new functionality, such as new hardware support or new logging capabilities, there is a software interface based on the concept of separate handlers. A handler is an object that serves a specific task. If it handles a hardware interface that should be shared among user scripts, it would be called “a device”. If it handles a logging mechanism it is called “a logger”.

All handlers are dynamically loaded from shared libraries and are identified by short names. The handler for the serial port is identified with “SERIAL”. The handler for the graphical logging mechanism is identified with “PLOT”. The GPIB bus as well as VXI, etc. are accessed through the “VISA” handler.

One can build his own handler to replace existing functionality as well as add a new one. The shared library should provide C++ objects inherited from an interface base class. Thus, not more than a basic understanding of virtual function is required to build such a library. The software interface uses pure C++, and the wxWidgets framework is not needed to build a handler library.

## 7. Operation example

As an example of how our software works, let us take a look at the experimental setup operating in our laboratory. We study superconductor samples at low temperatures, about 0.3K. To reach such temperatures we use a He-3 absorption cryostat. The cryostat cool down process is carried out in several stages. We use a chain of scripts to operate the cool down process. Each script is designated to a separate cool down stage. When the conditions to proceed to the next stage are met, the script starts another script, which operates the next stage. A script that records all of the cryostat sensors data runs in parallel with the cool down control.

We have installed a blackbody radiation source [7] into the cryostat. A separate script runs a PID regulation cycle to set the blackbody temperature at the desired point. When the cryostat reaches the working temperature point, we launch another script that, for instance, measures the IV curves of our samples. As a set of IV curves with different blackbody radiation temperatures is required, the experiment script communicates the blackbody script to set different blackbody temperatures via the messaging system. The cryostat control devices are connected to the computer by the GPIB bus. The IV measurement is done by two devices that use the GPIB bus as well. The blackbody source is controlled via a USB ADC/DAC board. All that communication with different commands to different devices is handled by the application so no hardware access collisions occur.

## 8. A comparative look

Let us have a closer look at the differences between our software and Labview. A typical Labview programs represents a set of interconnected instruments drivers and program control elements, passing data to each other.

To compose such a program, one should acquire the necessary instrument drivers for his equipment, figure out how to represent the experiment's algorithm in the dataflow paradigm and connect all the elements in the appropriate way.

In the case of our software, the program is a set of scripts. Instead of creating or downloading an instruments driver, one just has to look for the corresponding command in the device's manual. In case of Labview the effort is just the same as searching for the appropriate instruments driver in the Internet. The scripts are imperative that makes algorithms implementation straightforward.

One can states that visual programming offers supreme usability. However, studies show [8] that text-style programming is in fact more effective.

## 9. Further developments

Concerning further developments, we consider the following development objectives to be the most interesting:

- The graphical user interface should be separated from the program core. It is not clear yet, whether it is better to have a separate build with a console-style control, a program interface that loads UIs from a shared library, or some kind of a network protocol that can attach different UIs. Anyway that would give new opportunities to the user, such as using an ARM board for automating the experiment.
- As the dataflow paradigm is quite popular with experimental scientists, we think that a sort of descriptive JSON [9] based dataflow scripting could be implemented.

## 10. Summary

To summarize, an open source software system has been developed to aid researchers in experiment automation. The program has been designed with an intention to make the experiment automation tasks easy for inexperienced users to accomplish, and clear for supervisors to review. Common hardware interfaces have been implemented and an API is provided that allows the implementation of a new hardware support, as well as introduction of additional output methods.

The software is available on Source Forge http://tirelessworker.sf.net

## Reference

[1] http://wxwidgets.org/
[2] http://openwatcom.org/index.php/Main_Pagehttp://www.beercapital.ru
[3] http://www.mingw.org/
[4] http://www-01.ibm.com/software/awdtools/rexx/
[5] http://publibz.boulder.ibm.com/cgi-bin/bookmgr_OS390/BOOKS/E04A2A01/CCONTENTS
[6] http://www.ni.com/visa/
[7] Andrey V. Uvarov, Ilya A. Cohn, Sergey V. Shitov, Alexander N. Vystavkin, Oleg V. Koryukin, Alla G. Kovalenko, Oleg S. Bol'shakov, Vyacheslav F. Vdovin, Cryogenic submillimeter wave blackbody source for tes-bolometer calibration, 2-nd International Conference "Terahertz and Microwave radiation: Generation, Detection and Applications", June 20 - 22, 2012, Abstract book, p.50.
[8] Green, Thomas R. G., and Marian Petre. "Usability analysis of visual programming environments: a 'cognitive dimensions' framework." 1996 Journal of Visual Languages & Computing **7** (2) 131-174.
[9] http://json.org/

# Automated Code Generation for Lattice Quantum Chromodynamics and beyond

**Denis Barthou**
University of Bordeaux
LaBRI / INRIA Bordeaux Sud-Ouest
E-mail: `firstname.lastname@inria.fr`

**Olivier Brand-Foissac,Olivier Pène**
University of Paris Sud
Laboratory of Theoretical Physics
E-mail: `firstname.lastname@th.u-psud.fr`

**Gilbert Grosdidier**
University of Paris Sud
Laboratoire de l'Accelerateur Lineaire
E-mail: `firstname.lastname@cern.ch`

**Romain Dolbeau**
CAPS Entreprise
Rennes, France
E-mail: `firstname.lastname@caps-entreprise.com`

**Christina Eisenbeis, Michael Kruse, Konstantin Petrov (speaker)**
INRIA
Saclay, France
E-mail: `firstname.lastname@inria.fr`

**Claude Tadonki**
Mines ParisTech
Fontainebleau, France
E-mail: `firstname.name@mines.paristech.fr`

**Abstract.** We present here our ongoing work on a Domain Specific Language which aims to simplify Monte-Carlo simulations and measurements in the domain of Lattice Quantum Chromodynamics. The tool-chain, called Qiral, is used to produce high-performance OpenMP C code from LaTeX sources. We discuss conceptual issues and details of implementation and optimization. The comparison of the performance of the generated code to the well-established simulation software is also made.

## 1. Introduction

Quantum Chromodynamics (QCD) is a fundamental theory within Standard Model used to describe strong intercations inside a nuclei. As the coupling constant for this force is large, the perturbation theory cannot be used in most cases. Instead, a number of non-perturbative methods have been developed, with the most prominent of them being Lattice QCD (LQCD). It is formulated in discrete space-time, with the matter fields residing on sites and the interaction fields (gluons) live on links. This setup allows us to simulate it on a computer by means of Monte-Carlo simulations. Such analysis already provided many insights into the nature of strong interactions and delivered a number of precise results to confront experimental data. With the arrival of teraflop installations, LQCD is now regarded as the most reliable method of solving QCD, as the simulations can be systematically improved. It has also been extended to help solve theories other than QCD.

The most time-consuming part of these simulations is the frequent inversion of an immense matrix (Hopping Matrix), which encodes interactions within the simulated system. To do this efficiently, thousands of people over the globe are working for years, inventing new methods for the inversion. At the same time, new Lagrangians appear, which, while preserving physical properties, have different mathematical ones which are supposed to make Monte-Carlo simulations faster and more reliable.

The Hopping Matrix is actually a representation of a tensor called Wilson-Dirac operator. The matrix is always sparse and structured, so the iterative methods are definitely considered. Therefore, the procedure of the application of this operator, resulting in a vector-matrix product, appears as a critical computation kernel that should be optimized as much as possible. Due to the size of the matrix, it has to be recalculated on every iteration. Therefore, in the simplest case, evaluating the Wilson-Dirac operator involves each node and 8 neighbors. Such configuration is really hindering in terms of computation, as memory access is very far from sequential and standard methods fail miserably. For current and future generation of supercomputers the hierarchical memory structure makes it next to impossible for a physicist to write an efficient code.

But even for computer scientists, the rapid change in parallel architectures makes the design of an optimized LQCD simulation a real challenge. This requires to design, select and combine iterative methods and preconditioners adapted to the problem and the target architecture, to optimize data layout and organize parallelism between nodes, cores, accelerators and SIMD units. In order to harness all resources of the hardware, orchestrating the work on many cores and accelerators, using different levels of parallelism, complex memory hierarchies and interconnect networks takes a large part of the tuning time, often at the expense of the exploration of new algorithms/preconditioners. Indeed, testing new methods can only be achieved with large enough data sets, requiring efficient parallel codes. Several codes and libraries have been designed for Lattice QCD and many works have been published on code optimization for Lattice QCD, among them the studies on Blue Gene/Q [4], Intel Xeon Phi [8] or clusters of GPUs [7]. They offer some degree of flexibility, but usually only focus at a small subset of existing architectures. However, designing new iterative methods, combining existing ones, changing data layout within these frameworks and tools is difficult and requires a significant code rewriting effort. This clearly hinders the adaptation of code to the new parallel machines, limiting performance and the expected scientific results.

This paper proposes a domain-specific language (DSL), QIRAL, for the description of Lattice QCD simulations, and its compiler to generate parallel code. A DSL can help to separate the high level aspects of the simulation from machine-dependent issues. The contribution of QIRAL is to address this twofold challenge:

- Propose to physicists a domain-specific language expressive enough to enable the description of different models and algorithms, and more importantly, expressive enough to enable algorithmic exploration by composing different algorithms and preconditioners as well as the design of new algorithms.

- Generate from this description efficient codes for parallel machines. Explicit parallelism and data layout are automatically generated and can be guided by the user. The code generated by QIRAL targets shared memory parallel machines, corresponding to one node of larger Lattice QCD simulations. This code uses OpenMP and a library for efficient SIMD operations.

With a higher level description of the Lattice QCD formulation we achieve multiple goals. It becomes easier to try new algorithmic ideas, the high level code is easier to maintain and develop. This makes numerical simulation accessible to a large number of users, not necessarily high performance computing experts. We show on several architectures, from Nehalem-EX with 128 cores to the Xeon Phi accelerator that the code generated with QIRAL competes in terms of parallel efficiency and performance with tmLQCD, while QIRAL provides an easier framework for the writing of algorithms and the adaptation to new architectures.

This paper is organized as follows: first we describe the DSL in Section 2, describe the high-level compiler in Section 3. Then the optimizations for locality, parallelism and SIMDization are presented in Section 4. Benchmarks, comparing with tmLQCD and describing strong scalability are shown in Section 5.

The whole project, under the name PetaQCD [1], was partly funded by a grant from ANR, through the program COSINUS-2008, from 2009 up to 2011.

## 2. The QIRAL Domain-Specific Language

As one of the purposes of the QIRAL DSL is to give scientists a familiar tool to describe the problem in scope, it makes sense to take an existing system of symbolic notation as the basic language. There are two such systems in most disciplines, LaTeX and Mathematica. While we are not attached to a particular one, we chose to use LaTeX-like syntax where certain additional macros have been defined. Therefore the QIRAL description can be processed either using the QIRAL compiler to produce program source code or alternatively, by the LaTeX typesetter to produce its documentation. This means we revive the principle of *literate programming* coined by Donald Knuth [9]. For instance, the algorithm in Figure 3 is a QIRAL program included into this document as processed by LaTeX. The description of the language given in the following complements a description previously presented by the authors [2].

QIRAL is a language for describing linear algebra objects and operations, with a specialization in manipulation of sparse matrices defined through tensor products and direct sums of dense matrices as they appear in Lattice QCD. It is well-adapted to the particular features of Lattice QCD, or, rather, any realistic field theory. All interactions in such theories are local, therefor it is well-suited for the so-called stensil computation, involving a point and its nearest neighbours only. It assumes that the underlying manifold is a 4D space-time, cut into a cartesian mesh (the lattice). The Dirac operator, used for this inversion, is a sparse but regularly structured matrix that can be seen as a diagonal of dense matrices. This operator describes interaction between matter fields and guides the evolution of the system. The matter fields are represented by spinors, which are complex matrices having four spin and three color. Hence for a $24^3 \times 48$ lattice, the Dirac operator is a matrix of $(24^3 \times 48 \times 12)^2$ complex values. Due to locality of the interaction, the resulting matrix is sparce, and QIRAL takes advantge of this structure to reduce the unnecessary computation.

Elements of the language are declarations, equations, algorithms and the goal. Declarations declare symbols and functions with their type. Basic types are boolean, integers, real ($\mathbb{R}$), complex ($\mathbb{C}$) vectors ($\mathtt{V}$), matrices ($\mathtt{M}$), indices and index sets. Vectors and matrices are defined over index sets either defined through the notation $\mathtt{V1}[is]$, where $is$ is the possibly multi-dimensional index set for vector `V1`, or deduced through type inference. A particular element of a vector is accessed by the use of an index: `V1[I1]`. Figure 1 shows the declaration of the constants used for Lattice QCD, and the definition of Dirac operator as a matrix. The two other matrices, $P_e$ and $P_o$ are projections, keeping only black or white elements of the lattice, like a 4D checkerboard.

Equations are used to define variables or functions. Figure 2 describes nearly all properties and definitions on the constant and functions used for the simulation. For instance, the function "invertible" is defined only for some expressions.

**Constant:**

$$Dirac, P_e, P_o, \gamma_5 \in M,$$
$$L, S, C, even \in Indexset,$$
$$\gamma \in Index- > M,$$
$$U \in Index- > M,$$
$$\kappa, \mu, \in \mathbb{R},$$
$$D \in Indexset$$

**Variable:** $s \in Index, d \in Index$

$$\begin{aligned} Dirac = & I_{L\otimes C\otimes S} \\ & + 2 * i * \kappa * \mu * I_{L\otimes C} \otimes \gamma_5 \\ & + -\kappa * \sum_{d\in D} ((J_L^{-d} \otimes I_C) * \bigoplus_{s\in L} U[s \otimes d]) \otimes (I_S - \gamma[d]) \\ & + -\kappa * \sum_{d\in D} ((J_L^{d} \otimes I_C) * \bigoplus_{s\in L} U[s \otimes -d]) \otimes (I_S + \gamma[d]) \end{aligned}$$

$$P_e = P_{even,L} \otimes I_{C\otimes S}$$
$$P_o = P_{!even,L} \otimes I_{C\otimes S}$$

Figure 1: Definitions of the Dirac matrix on a Lattice $L$ in QIRAL, and the two projections for even and odd elements ($P_e$ and $P_o$ respectively) of this lattice.

**Constant:**

$$dx, dy, dz, dt \in Index$$

$$\begin{aligned} D &= \{dx, dy, dz, dt\} \\ \text{isPeriodic}(L) &= true \\ U[s \otimes d]^\dagger &= U[(s+d) \otimes -d] \\ U[s \otimes -d]^\dagger &= U[(s + -d) \otimes d] \\ \text{Preconditioner1}(Dirac) &= Pe \\ \text{Preconditioner2}(Dirac) &= Po \\ \gamma[d]^\dagger &= \gamma[d] \\ \text{diagonal}(\gamma_5) &= true \\ \gamma_5 * \gamma_5 &= I_S \\ \gamma_5 * \gamma[d] &= -\gamma[d] * \gamma_5 \\ \text{invertible}(I_S + c * \gamma_5) &= true \\ \text{invertible}(I_S - c * \gamma_5) &= true \\ \text{invertible}(-(c * I_S) + i * \gamma_5) &= true \\ \gamma_5^\dagger &= \gamma_5 \\ \text{type}(\gamma[d]) &= S \times S \\ \text{type}(U[s \otimes d]) &= C \times C \\ \text{type}(\gamma_5) &= S \times S \\ vol(S) &= 4 \\ vol(C) &= 3 \end{aligned}$$

Figure 2: Identities of constants used for Lattice QCD.

Algorithms are given as possible definitions for statements or expressions. For instance, the conjugate gradient algorithm in Figure 3 provides the code that computes expressions of the form $x = A^{-1} * b$, when $A$ and $b$ are given. It outputs the value of $x$, i.e. solves the linear system $Ax = b$.

The initial statement, in the **Match** clause, is then defined (and replaced) by the pseudo-code. The **Var** keyword declares the type of local variables. This algorithm is written using the "algorithm2e" package in LaTeX, and is not specific to Lattice QCD. The user has the possibility to write new algorithms for Lattice QCD or any other algorithm found in common literature. The QIRAL compiler finds automatically how to compute for instance $A * p$ when $A$ is instantiated with the Dirac operator.

Most often the validity of an algorithm depends on prerequisites, special properties the inputs

**Input** : $A \in M, b \in V$
**Output** : $x \in V$
**Constant**: $\epsilon \in \mathbb{R}$
**Match** : $x = A^{-1} * b$
**Var** : $r, p, Ap, z \in V, \alpha, \beta, n_r, n_z, n_{z1} \in \mathbb{R}$
$r = b$ ;
$z = A^{\dagger} * r$;
$p = z$ ;
$x = 0$ ;
$n_z = (z \mid z)$ ;
$n_r = (r \mid r)$ ;
**while** $(n_r > \epsilon)$ **do**
    $Ap = A * p$;
    $\alpha = n_z/(Ap \mid Ap)$ ;
    $x = x + \alpha * p$ ;
    $r = r - \alpha * Ap$ ;
    $z = A^{\dagger} * r$ ;
    $n_{z1} = (z \mid z)$ ;
    $\beta = n_{z1}/(n_z)$ ;
    $p = z + \beta * p$ ;
    $n_z = n_{z1}$ ;
    $n_r = (r \mid r)$ ;

Figure 3: Conjugate Gradient, normal residual method (CGNR).

**Input** : $A, P_e, P_o \in M, b \in V$
**Output** : $x \in V$
**Match** : $x = A^{-1} * b$
**Constant**: $D_{11}, D_{12}, D_{21}, D_{22} \in M$
**Var** : $v_1, v_2, x_1, x_2 \in V$
**Require** : invertible$(P_e * A * P_e^t)$
$D_{21} = P_o * A * P_e^t$ ;
$D_{11} = P_e * A * P_e^t$ ;
$D_{22} = P_o * A * P_o^t$ ;
$D_{12} = P_e * A * P_o^t$ ;
$v_1 = P_e * b$ ;
$v_2 = P_o * b$ ;
$x_2 = (D_{22} - D_{21} * D_{11}^{-1} * D_{12})^{-1} * (v_2 - D_{21} * D_{11}^{-1} * v_1)$ ;
$v_1 = P_e * (2 * \kappa * b)$ ;
$x_1 = D_{11}^{-1} * (v_1 - D_{12} * x_2)$ ;
$x = P_e^t * x_1 + P_o^t * x_2$ ;

Figure 4: Definition of Schur complement method.

must have. These prerequisites are declared in a clause **Require** and is proved by the QIRAL compiler. The following example illustrates this prerequisite mechanism. Figure 4 describes the Schur complement method that is used as a preconditioner for the conjugate gradient and the conjugate residual. The condition invertible($P_e * A * P_e^t$) is proved automatically by the compiler when $A$ matches the matrix *Dirac*. To prove this, the property defined previously for the function "invertible" is used. If the compiler is not able to prove the requirements attached to an algorithm, the algorithm is not applied and an error is generated. Notice that on this preconditioning, the statements involve computation of inverse matrices. For the expression $D_{11}^{-1}$, the QIRAL compiler can prove automatically that $D_{11}$ is diagonal (when $A$ is the Dirac operator), and knows how to invert this matrix. For the computation of the expression $(D_{22} - D_{21} * D_{11}^{-1} * D_{12})^{-1}$, an iterative method has to be applied.

The goal defines the initial code and the list of algorithms to apply. The algorithms are composed from right to left.

**Input** : $bb \in V$
**Output** : $xx \in V$
**Templates**: CGNR schur
$xx[L \otimes C \otimes S] = Dirac^{-1} * bb[L \otimes C \otimes S]$ ;

For this goal here the preconditioner `schur` is applied on the initial statement, and then the CGNR algorithm. It is possible to chain multiple algorithms, used to apply preconditions before the solvers. The index set $L \otimes C \otimes S$ represents the Cartesian product of these sets and the domain for the vector $bb$. At this level, there is no implicit data layout for vectors and matrices. The vector $bb$ could be either a 4D array of structures, one dimension for each dimension of L and the structure representing elements indexed by $C$ and $S$, or a 1D array of structures, or just a large 1D array of complex values. This is orthogonal to the expression of the algorithm.

The output of QIRAL compiler is a function in C and OpenMP pragmas representing the computation described in the goal, and taking as parameters $bb$ and $xx$. All other constant values (in particular constant matrices) are assumed to be global.

## 3. Implementation Details

The QIRAL compiler is based on a rewriting system, Maude [3]. The different steps of this transformation are explained in this section.

### *3.1. Algorithms composition and expression simplification*

Algorithms are translated into rules of the rewriting system, while equations define the equational theory for the rewriting system. The first step consists in transforming LaTeX input into a Maude program. Additional modules, defining usual algebraic simplifications are added to this code.

The first step parses the LaTeX input and captures only what is described in predefined environments, for algorithms, definitions and the goal. Syntactic verification as well as type checking is performed. The output generated is a Maude module, with equations corresponding to definitions, rules corresponding to algorithms, and a unique Maude statement, corresponding to the goal.

The algorithms declared in the goal are applied, in turn, to the statements provided. These statements are terms for Maude. The **Match** clause is the left-hand side (lhs) of the rule, while the pseudo-code corresponds to a term that is the right-hand side (rhs) of the rule. Any statement matching the lhs will then be rewritten in the rhs. If a **Require** clause exists, it constitutes the condition for the rewriting. The first statement is provided by the goal, then algorithms are applied successively.

Definitions and properties are considered by Maude as defining the equational theory for the rewriting system. Actually, these equations are handled as automatic rewriting rules: Maude automatically applies all possible equations, rewriting their lhs into rhs, until the term is normalized. For instance, the property $x * (y + z) = x * y + x * z$, stating the distributivity of $+$ over $*$, will only be used to distribute the operators, not to factorize terms.

The main objective of this formal rewriting is to eliminate all terms that are equal to zero. In Lattice QCD, the Dirac matrix used in the problem is sparse, but built from dense matrices with a regular structure. To obtain such simplifications, an additional module defines properties for the linear algebra operators, on complex, vectors and matrices: Addition is commutative and associative, multiplication is associative and distributes over the addition, binary subtraction is converted to unary minus, transposition distributes over the addition and multiplication, etc. Moreover, some properties are also defined for permutation and projection matrices, in particular to handle Schur preconditioning.

## 4. Code Optimizations

### *4.1. Improving Locality*

Loops fusion is a transformation to reduce reuse distances, hence improving locality. To check if fusion is valid, a simple dependence analysis, based on dependence distance, is computed. The fusion method is applied on consecutive independent loops that share the same iterators, and is applied on all code until no more fusion is possible. This simple strategy is sufficient for Lattice QCD generated codes.

Following this fusion, the regions of arrays that are written/read by all loops, and the regions that are inputs/outputs of loops are computed. All arrays that are used only in one loop are scalar promoted. The resulting values are allocated on the stack, and aligned for further vectorization. This reduces memory consumption.

Both transformations are applied automatically.

### *4.2. Versioning Matrix Multiplication*

The computation involve many multiplications of vectors by constant matrices, accounting for transformations on spin and color ($S$ and $C$ index sets respectively). These matrices are small, of size $3 \times 3$ and $4 \times 4$ respectively, and the latest have only 2 non-null elements per line, these elements being among $\{1, -1, i, -i\}$. Therefore specialization of these products is necessary in order to obtained better performance. These matrix-vector multiplications appear in expressions of the form $(M_1 \otimes M_2) \cdot V$ with $M_1$ and $M_2$ the two matrices multiplied by a tensor product, and $V$ is a 12 element vector. In this case the QIRAL compiler uses the identity

$$(M_1 \otimes M_2) * V = M_1 * V * M_2^t$$

where on the lhs, $V$ is considered a matrix of size $3 \times 4$ and $*$ stands for the matrix product. Therefore, instead of using general matrix multiplication, QIRAL compiler finds these occurrences and calls versioned matrix multiplications. Specializing such multiplications for these particular sizes, in particular performing SIMDization, is essential for performance. These functions correspond to the hot-spot of the codes generated by QIRAL. The codes of these functions are hand-written in `libqiral` library as presented in Figure **??**.

Other expressions can be replaced by library calls, and QIRAL changes expressions on vectors and matrices into BLAS calls (or specialized BLAS). The fact that the QIRAL compiler automatically identifies these functions in the code generated from the different algorithms facilitates the optimization process and is an asset of QIRAL. The optimization of these functions in `libqiral`, specializations of BLAS, can indeed be achieved by an expert in high-performance computing, independent of any Lattice QCD context.

Journal of Physics: Conference Series **510** (2014) 012005 doi:10.1088/1742-6596/510/1/012005

## 5. Performance Results

Several iterative methods are written with QIRAL. Figure 5 presents some of these methods, for two architectures: CGNR, CRNE, MCR1 and MCR2 with some preconditioners: Schur and preMCR. We observe that while MCR2 exhibits the best time per iteration, the method takes

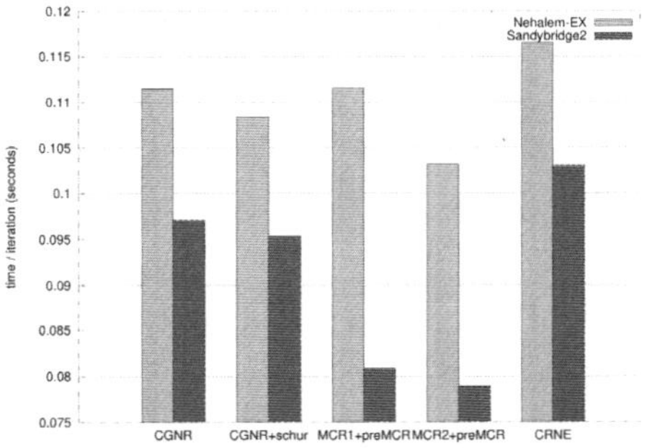


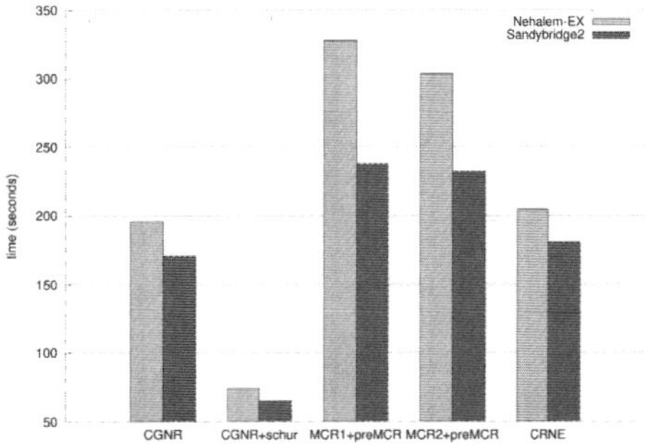


Figure 5: Comparison between different iterative methods, on Nehalem-EX and Sandybridge 2 architectures. Left figure: Time in seconds per iteration. Right figure: Total execution time.

more time to converge than CGNR and Schur. This shows that the best method cannot be determined only by benchmarking a single iteration, but it is necessary to run all iterations. Besides, the second plot of Figure 5 shows that the relative difference may vary according to the architecture. While the absolute best method is still the same (here CGNR combined with Schur), this stresses the fact that the algorithmic solution may be chosen depending on the target architecture.

In order to compare tmLQCD with the code generated by QIRAL, the same algorithm is used for both (CGNR and Schur preconditioning). Performance is displayed for all architectures as the total execution time multiplied by the number of cores. Due to the fact that tmLQCD is using MPI, there is no version for Xeon Phi. Besides, the tmLQCD code uses in-line assembly code with SSE3 instructions. Adapting this code for newer SIMD extensions is more difficult than adapting intrinsics as used by QIRAL. Indeed for intrinsics, part of the optimization work still relies on the compiler: register allocation, generation of FMAs, scheduling. The code generated by QIRAL has been quickly ported to these architectures, and then code tuning has focused on the library used by QIRAL (with versioned BLAS), using intrinsics and aggressive in-lining.

Figure 6 presents timing results on different architectures, comparing tmLQCD code with QIRAL generated code. For QIRAL, the "hand-optimized library" corresponds to the best version obtained, using intrinsics (AVX, AVX2, Xeon Phi) for Sandybridge, Haswell and Xeon Phi architectures. The Nehalem EX version does not use SSE SIMD intrinsics. This explains why QIRAL/Nehalem EX version is more than two times slower than tmLQCD. For Xeon Phi, the performance displayed corresponds to the use of all the 60 cores, and a linear speed-up can be observed by using an increasing number of cores. The ISPC compiler has been used to generate SIMD version of matrix multiplication of size $3 \times 4$ on complexes. The compiler is still under heavy development and does not fully work for Xeon Phi. Figure 6 shows that the level of performance reached with ISPC is not competing with the level for hand-tuned intrinsics.

The strong scalability of the code generated by QIRAL is evaluated on Xeon Phi and Nehalem-EX architectures. Figure 6, right, shows efficiency results for different number of cores. Note that the size of the lattice is different for both architectures, reflecting the need for different granularity. The efficiency for the Xeon Phi is compared to the run on 4 cores, with 4 threads each. This explains why for some number of threads, the efficiency goes beyond 1. The code

Journal of Physics: Conference Series **510** (2014) 012005 doi:10.1088/1742-6596/510/1/012005

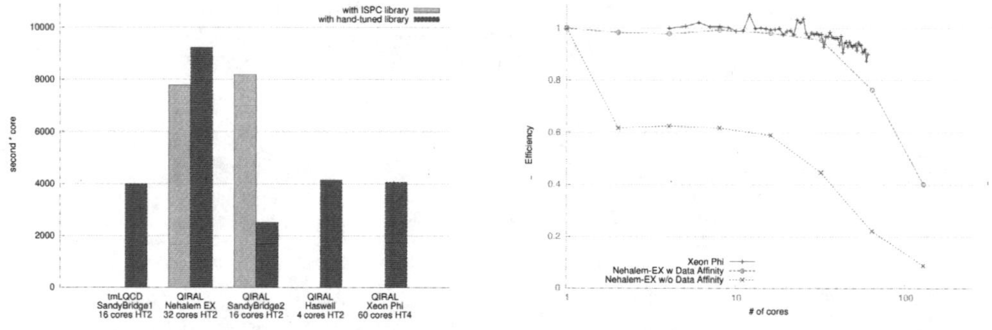


Figure 6: Left: Normalized performance for the inversion on different architectures, with QIRAL and tmLQCD codes. Performance is shown in sec*core (ie "seconds times number of cores"), lower is better. The execution time is obtained by dividing this performance by the number of cores. The same method, a conjugate gradient with Schur preconditioning is used in all cases, with a lattice of size $24^3 * 48$ and an error of $10^{-14}$. Right: Efficiency of the code generated by QIRAL on Xeon Phi and Nehalem EX, according to the number of cores used. For both architectures, the method used is the conjugate gradient with Schur preconditioning. The lattice size for the Xeon Phi is $24^3 * 48$ and for the Nehalem-EX, $64^3 * 128$. On the Nehalem-EX the efficiency is measured with and without NUMA-aware memory allocation.

scales well up to the 60 cores (240 threads). For the Nehalem-EX machine, the efficiency is higher than 95% up to 32 cores, and then drops quickly. The reason is that a 128-core node is structured with 4 groups of 4 octo-cores, connected through a switch. Going through the switch has a high penalty in terms of performance.

## 6. Conclusion

The contribution of this paper is a new domain-specific language, QIRAL, for the automatic code generation of OpenMP codes for Lattice QCD simulations. QIRAL language offers to physicists the possibility to implement iterative methods and preconditioners, literally "from the book" using LaTeX, or design new ones, and test them on large parallel shared memory machines or on accelerators such as the Xeon Phi. The language enables the composition of preconditioners and iterative methods, and the compiler checks automatically the validity of application for each method. This makes possible a more systematic exploration of the algorithmic space: indeed, it removes from the physicists the burden of long and stressful validations of their new code since it will be automatically generated, then safer, and the time-to-market for a viable product will be much shorter. The QIRAL compiler generates OpenMP parallel code using BLAS or specialized versions of BLAS functions. Further hand-tuning is possible on the code generated by QIRAL, and we have shown that the performance on various multi-core architectures and on Xeon Phi accelerator it compares or outperforms the performance of a hand-tuned Lattice QCD application, tmLQCD.

Among the perspectives of this work, the automatic generation of a communication code for multi-node computation would enable to run Lattice QCD simulations on a larger scale. Besides, the fine tuning of the library functions used by QIRAL on different architectures, in particular their SIMDization, could be improved.

Journal of Physics: Conference Series **510** (2014) 012005 doi:10.1088/1742-6596/510/1/012005

## 7. The bibliography


[1] J. Angles D'Auriac, D. Barthou, D. Becirevic, R. Bilhaut, F. Bodin, P. Boucaud, O. Brand-Foissac, J. Carbonell, C. Eisenbeis, P. Gallard, G. Grosdidier, P. Guichon, P. Honore, G. Le Meur, P. Pene, L. Rilling, P. Roudeau, A. Seznec, A. Stocchi, and F. Touze. Towards the petaflop for Lattice QCD simulations the PetaQCD project. In J. G. . M. Lokajicek, editor, *J. of Physics Conf. Series*, volume 219, page 052021, Prague, Tchèque, République, 2010. IOP Publishing. LAL 09-58, LPSC09233.

[2] D. Barthou, G. Grosdidier, M. Kruse, O. Pène, and C. Tadonki. QIRAL: A High Level Language for Lattice QCD Code Generation. *CoRR*, abs/1208.4035, 2012.

[3] M. Clavel, F. Durán, S. Eker, P. Lincoln, N. Martí-Oliet, J. Meseguer, and C. Talcott. The Maude 2.0 System. In R. Nieuwenhuis, editor, *Rewriting Techniques and Applications (RTA 2003)*, number 2706 in Lecture Notes in Computer Science, pages 76–87. Springer-Verlag, June 2003.

[4] J. Doi. Peta-scale lattice quantum chromodynamics on a Blue Gene/Q supercomputer. In *Proceedings of the International Conference on High Performance Computing, Networking, Storage and Analysis*, SC '12, pages 45:1–45:10, Los Alamitos, CA, USA, 2012. IEEE Computer Society Press.

[5] B. Joó and M. A. Clark. Lattice QCD on GPU clusters, using the QUDA library and the Chroma software system. *Int. J. High Perform. Comput. Appl.*, 26(4):386–398, Nov. 2012.

[6] B. Jo, D. Kalamkar, K. Vaidyanathan, M. Smelyanskiy, K. Pamnany, V. Lee, P. Dubey, and I. Watson, William. Lattice QCD on Intel Xeon Phi Coprocessors. In J. Kunkel, T. Ludwig, and H. Meuer, editors, *Supercomputing*, volume 7905 of *Lecture Notes in Computer Science*, pages 40–54. Springer Berlin Heidelberg, 2013.

[7] D. E. Knuth. Literate programming. *The Computer Journal*, 27:97–111, 1984.

# Multi-agent tasks scheduling system in software defined networks

P.O. Skobelev[1], O.N. Granichin[2], D.S. Budaev[1], V.B. Laryukhin[1], I.V. Mayorov[1]
[1]SEC "Smart Solutions", Ltd, Samara, Russia
[2]Saint Petersburg State University, Saint Petersburg, Russia

Key words: multi-agent system, real-time scheduling, adaptability, software defined networks.

Abstract: In this paper a multi-agent tasks scheduling system in software defined networks is considered. This system is designed for distribution simulation and tasks implementation on computational resources including network dynamic characteristics and topology.

## 1 Introduction

Until now, computer networks were generally developing in an extensive way: more and more devices were joined into local and global networks, network equipment was being improved, data communication channels capacity was increasing, and response time of servers and network components was increasing.

However, conceptually new solutions were not widely integrated in practice and gradually, the existing networks could not cope with the increasing number of devices and data flows. The structure of the network schedule changed because each inquiry creates a big amount of references to application servers, databases, other storage and information processing systems. Complexity of the modern network infrastructure and the scheduled growth lead to the increasing number of the rejections and in the course of the current tendencies can cause a deterioration of the speed characteristics of data transfer.

At the same time the existing network architecture is still oriented towards the group of old enough protocols to provide the interaction of complex network devices, which makes the scaling difficult. Adding new devices takes high effort, so does the resetting of commutators, routers and firewalls. The network structure remains static and therefore significant means for its support are needed.

Such innovations, as cloud computing and virtualizing, demand new approaches to the network organization.

In this regard in the last few years there has grown an interest to the new architecture of Software Defined Networks (SDN) that should enable to overcome the outlined crisis [1].

SDN concept defines the paradigm shift in the network architecture when the network management layer is programmed directly and is separated from the direct packet routing. This management transfer provides the abstraction of the basic network for the top layers applications, which allows them to see the network as a logical or virtual entity. The developed protocol OpenFlow (OF) that is now applied in many network devices, allows transferring the functions of network routers management into the central module that is called controller or network operational system [2].

Essentially, controller is an intermediate member that connects real network devices and some applications, realizing equipment operation logic. OpenFlow protocol defines the possibility and the scope of the interaction between the controller and the network equipment. Packet routing functions stay inside the routers that are called forwarders or commutators for the reason that they do not make decisions about the route selection, they simply follow the routing rules set by the controller. The controller (together with the applications) can be regarded as system intellectual center, because the route decisions are made on this level for the packet protocols instead of the packets individually and then the network devices configure via the OpenFlow protocol. The protocol allows providing the complete network presence on the controller, monitoring all the resources and networking visualization that allows creating complex solutions for the network management [3].

In that regard the problem remains up to date and valuable for the simulation and research of different variants of network organization.

Journal of Physics: Conference Series **510** (2014) 012006 doi:10.1088/1742-6596/510/1/012006

## 2 Problem statement

One of the rapidly developing concepts nowadays is a distributed computing model. It is based on the network access provision to the common computing network resources, which considerably reduce the users expenses on the creation of the corresponding infrastructure and has flexible reaction to the changes of computational requirements. Distributed computing controlled by SDN can be more efficient from the side of the operation speed, as the network can be optimized by multiple special algorithms, considering topology and network dynamic characteristics. One of the computing optimization tasks in the distributed system of the network devices is a load balancing task that is being actively investigated today. In the paper [4] there is considered a games theory application to the task to provide the required level of service in the system of the applications that compete for the resources. A service quality optimization problem in distributed dynamic computing systems [5] is investigated with the help of the theory of probability. It is suggested to calculate the local times of the tasks completion, queue lengths, waiting times and other metrics. There was carried out a resources availability simulation depending on the scheduling algorithms and computing system characteristics. The approach using randomized algorithms is considered in [6]. Load balancing task of the computing network nodes in the stochastic statement is reviewed in [7].

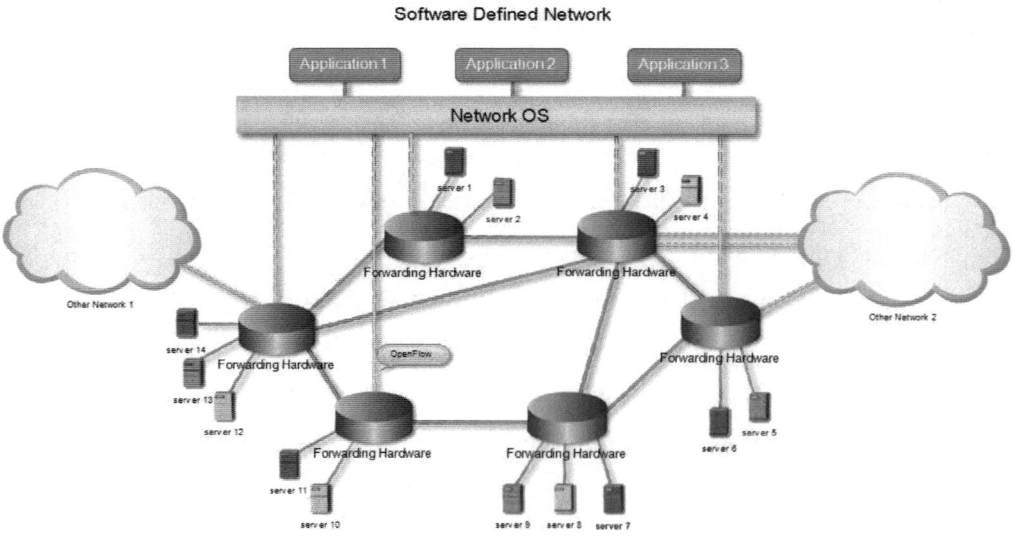


Figure 1 – Software defined network ontology

Journal of Physics: Conference Series **510** (2014) 012006 doi:10.1088/1742-6596/510/1/012006

In the current paper a multi-agent approach to the computing problems scheduling in the distributed system based on the developed model of tasks and resources agents network is considered [8, 9].

**3 Solution method and multi-agent scheduling system description**

There is a system, consisting of several computational servers that are connected via the OpenFlow commutators to the network, managed by the SDN controller. The servers are designed to complete the tasks of different types with different speed, depending on the task type and resource (server) workload. A task flow enters the system and each one of the tasks can be executed on the bounded subset of servers (Fig. 1).

The compliance of the servers and tasks types is defined in the description of each task. Tasks arrive at any time and redistribute in the system. As a result a tasks queue for the execution is formed on every server. The order of tasks redistribution is defined on the logic layer of the controller. In that case dynamic characteristics of data transfer channels, queue statuses on each network device (commutators, servers) and type compatibility of tasks and servers are considered. The SND controller dynamically defines the most convenient servers for the tasks execution, including the consideration of tasks routes in the system. A dynamic rescheduling can be performed if necessary (for example, in case of unexpected network nodes turn-off). The global goal of modeling is to provide the tasks execution for the shortest time with the balanced load of all the network nodes.

It is suggested to fulfill the dynamic distribution of tasks over the network servers with the multi-agent approach. Basic system entities are associated with the program agent, task agent, resource (server) agent and commutator agent. Agents are described by attributes that include the most essential fields, properties and interaction rules. Each agent has its own goal and aspires to reach it, by sending and receiving messages from other agents. To solve the optimization problems of agents states "virtual costs" are introduced, which allow applying market approach in multi-agent system. At the same time there are changes in the agents attribute values, therefore at each moment of time there exists some state of multi-agent system that is responsible for the local optimum decision.

## 4 System logical architecture

System consists of the knowledge base – ontology, which includes descriptions of the concepts types: task agents, resource agents and commutator agents. Attributes, properties and relations are set between them. Virtual world scene serves for the agents' instances interaction through the message exchange. Generator of external events creates flows of the tasks entering the system with different static characteristics, for example events of productivity and servers types change, their inaccessibility periods, connection and disconnection of the highlighted network fragments.

User interface subsystem allows starting the events simulation, visualizing network graph, tasks cost dynamics, tasks distribution and queues on servers, transfer channels loading and others.

### Task agent

Task agent represents a computing task that needs to be scheduled and executed with the smallest expenses on one of the servers in the SDN network.

At the moment of each task arrival into the system, task agent is created and in this domain area its attributes will be the following fields:

- task arrival time,
- task data volume,
- task type (correlates with the server type),
- task computational complexity, if it is known,
- task priority (high, medium, low), if it is known,
- desirable task starting time, if it is known,
- desirable critical time of task completion, if it is known,
- actual task implementation starting time on server,
- actual task implementation ending time on server,
- current virtual account state.

Every task agent connects with the piece linear function of bonuses/fines, which changes the current virtual account of task agent depending on the correlation of the desirable and actual time, priority and computational cost of the task. Computational cost and correspondingly the critical ending time represent estimated upper limits that are not certain until the task implementation start on server. Actual values are defined at the start of real execution based on some statistics.

**Resource (server) agent**

Resource agent provides a server to the system, on which the tasks can be executed. Server agent goal is to schedule as much as possible tasks and get the maximum virtual profit from the tasks execution.

Resource agent attributes in this domain area will be the following fields:

- set of task types that can be executed on this server,
- resource productivity that defines the estimated time of tasks execution,
- current task list that are scheduled on the resource,
- allowable task queue volume,
- resource time unit cost table,
- current virtual resource account state.

Each resource agent associates with a penalty function that depends on the amount of the scheduled tasks on the specified scheduling horizon.

**Commutator agent**

Commutator agents provide actual information about network state to the scene and they are designed to provide the efficient tasks allocation to the nodes of the further movement.

Commutator agent is described with the following fields:

- allowable task queue volume,
- each connection capacity with the closest commutator agents,
- current task list for the transfer via any connection to commutator agents,
- current virtual commutator account state.

**5 System operation**

Initially the system has certain topology of connections, for example, ring. Tasks arrive in a random way on one or several commutators from the external networks. When a new task arrives, system automatically performs the scheduling in three stages:

1) Conflict-free scheduling stage

Task agent searches for the resource descriptions in the scene of a certain type, which are the closest from the side of the delivery time through the commutators on its way. The logical entity of the scene of multi-agent system corresponds to the physical entity of SDN controller. As a result of the scene operation, task agent creates a list of the most appropriate resources. Then the agent forms a message with a request to define the conditions of the initial task allocation on the resource taking into account its parameters. This message is sent to all the agents of the chosen resources. Resource agents analyze their schedule and calculate the possibility and cost of task allocation and form counter-offers with the shifted terms concerning the requested ones. As a result for all the sent messages resource agent gets a set of possible tasks allocation variants on the resources including allocation cost, which later on should be written-off from the task virtual account and charged to the chosen resource account. From the existing table of variants task agent chooses the best option by the initially set criteria with their weights.

If the task is successfully allocated to the resource it is fixed by a "contract" (relation) in the scene and task budget updates, becoming equal to the operation cost excluding resource cost. Resource agent also updates its account, increasing the budget on the specified cost and adds a job into the schedule. Times of start, end, lead and delay are being defined.

2) Task proactivity stage

Tasks proactivity consists in the activation of those tasks that are already scheduled, but are not executed. Their position in the resource schedule can be unprofitable from the point of view of the task agent. If task agent has enough amount of virtual money it can dynamically try to reschedule on the other appropriate resource, compensating possible deterioration of other task agents' states in the course of conflicts solving about tasks positions in the schedule.

If there are penalty functions, task agents' proactivity is a process controlled by the difference of the cost in the resource cost table, penalty functions representation that are involved in the task agents' conflict and dependent on a current virtual operations budget. Proactivity stage is designed to form the updated balanced state of the system with the quantitative improvement of tasks condition.

3) Resources proactivity stage and routes definition

Resource proactivity consists in the tasks allocation schedule change that begins by the resource initiative. When the schedule is not concentrated enough or vice versa the queues are too long, when there are events of tasks arrival, the resource can take off the schedule part of low-gain for him tasks. The routes of tasks movement in the network are defined by the choice of the least loaded channels. If it is necessary to transfer the profitable task to the resource and there is a sufficient virtual budget, then new channels are created. Dynamically created channels, that provide local routing optimum, are defined by the operation of the logical virtual layer of software defined network.

## 6 Advantages and restrictions

The advantages of the suggested multi-agent approach are the following:

- ontological description of the world of computing resources, tasks and network configuration scene allows to introduce new entities of SND and set their characteristics without significant expenses on reprogramming;
- dynamic management of the state of all SDN world scene objects with the help of the scenarios (scripts);
- changes in network topology can be made in real-time mode;
- dynamic tasks scheduling on servers in multi-agent system that is provided by the constant tasks and resources proactivity, on basis of self-organization based on agents aspire to maximize the virtual profit in the interaction acts through the message exchange;
- local optimality and consistency of the current schedule;
- local optimality of packets routing ways and tasks queues;
- the account of changes in the network surrounding and in statistic task flow characteristics in real-time (adaptivity) does not lead to the full schedule redesign, which reduces computational costs of simulation system;
- dynamic traffic redistribution with the account of channels loading;
- simplicity of agents interaction logic based on virtual profit maximization;

The developed system is a platform for simulation and experimentation on the choice of equipment operation logic and program system of network management.

Journal of Physics: Conference Series **510** (2014) 012006 doi:10.1088/1742-6596/510/1/012006

## 7 Implementation

The system is developed as an OS Windows application using the .NET Framework platform on the programming language C#. Prototype kernel is implemented as a library (build) and the existing simulation logic can be extended with the use of it. User interface development is performed with the WPF technology. To display the network graph a Graph# library was used.

The current system implementation allows visualizing the studied network in the form of graph, executing initial network configuration and defining the parameters of individual devices: volume for task queues for individual commutators and servers, communication channels capacity, processing servers' productivity, devices turned on and off (Fig. 3).

In the course of the modeling process there is possible a modeling time scaling concerning the real, which allows to speed-up or slow down the simulation process at system user's will. Also in the top part of the user interface a system change log is displayed. Each event (new task arrival, task execution on the resource) is recorded in the log and is available for the analysis at any moment of simulation.

The system allows saving the previously formed network configuration to file. Moreover, the system can save the current state of the simulation process itself at any moment of time. At the same time simulation devices save all their current characteristics and in particular lists of all the executed and transferred tasks for each network device. Therefore, user has an opportunity to save the partial results of simulation on different stages and next time continue the simulation from this stage by loading the saved file.

Besides the network configuration, saving and loading options, there is a possibility to form simulation script that defines parameters of the flow of the arriving to the network tasks (tasks distribution rule by time on each device) and individual parameters of the arriving tasks (volume, computational cost, type, priority and others). After the modeling scripts are formed, they can be saved and loaded in future. Therefore, network configurations are separated from the modeling scripts, which allows to carry out a series of various simulation experiments for one or different network topologies. At the moment with the help of the developed system prototype the research of advantages and restrictions of multi-agent approach continues.

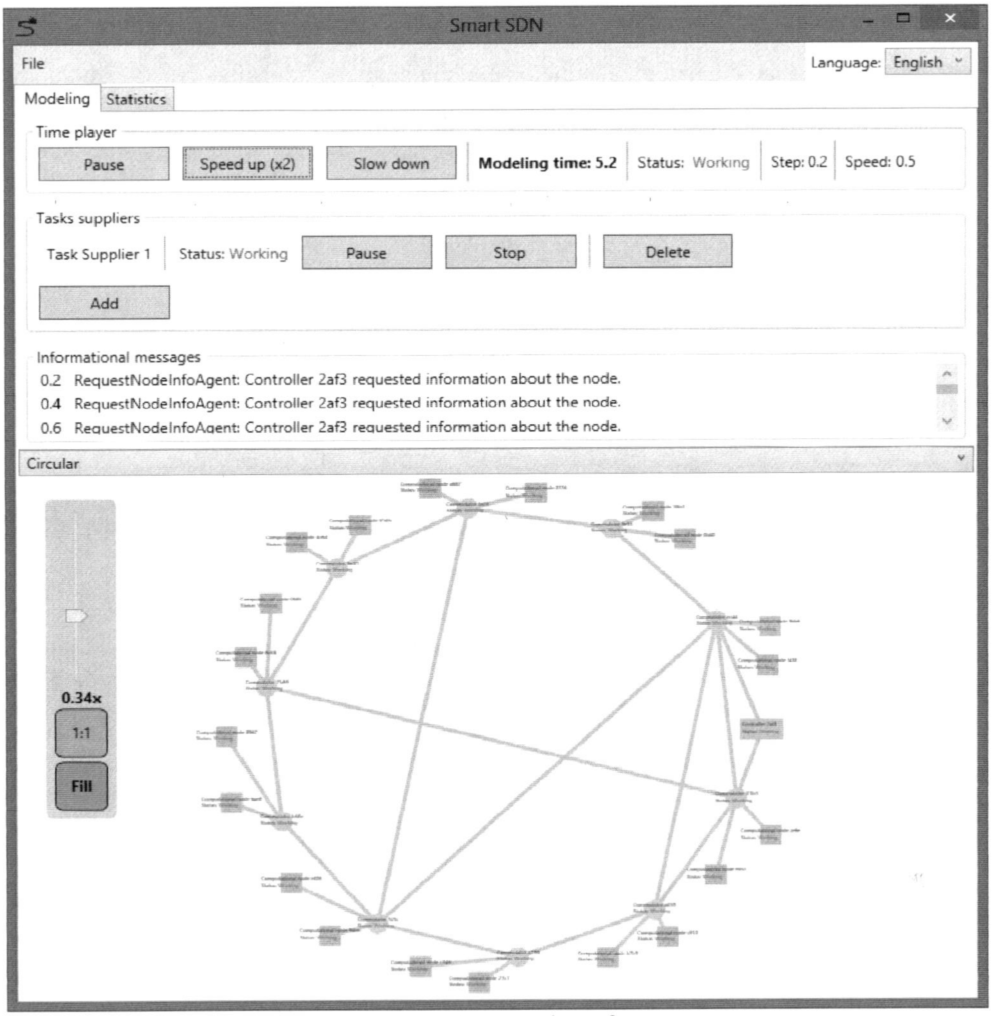


Fig. 3 – System interface

## 8 Conclusion

The developed multi-agent tasks scheduling system in software defined networks is based on the dynamic improvement of network state by the local optimization of servers and commutators load on the basis of the "market" approach by expressing all of the characteristics for concessions solving by artificial currency. Task, resource and channel agents tend to concordantly raise their virtual profit that should lead to the system characteristics improvement.

In future it is planned to carry out the research of task queues dynamics in SDN, servers load characteristics, channels capacity and the effect of topology change on the given parameters.

This work is performed under the state contract № 14.514.11.4086.

Journal of Physics: Conference Series **510** (2014) 012006 doi:10.1088/1742-6596/510/1/012006

## 9 References

1. N. McKeown, T. Anderson, H. Balakrishnan, G. Parulkar, L. Peterson, J. Rexford, S. Shenker, and J. Turner. Openflow: Enabling innovation in campus networks // SIGCOMM CCR, vol. 38, no. 2, pp. 69–74, 2008.

2. OpenFlow http://www.openflow.org/documents/openflow-spec-v1.1.0.pdf, last accessed date 19.08.2013

3. N. Gude, T. Koponen, J. Pettit, B. Pfaff, M. Casado, N. McKeown, and S. Shenker. NOX: Towards an operating system for networks // SIGCOMM CCR, vol. 38, no. 3, 2008.

4. Martina Maggio, Enrico Bini, Georgios C. Chasparis, Karl-Erik Årzén, A Game-Theoretic Resource Manager for RT Applications // Proceedings of the 25th Euromicro Conference on Real-Time Systems, Paris, France, July 2013.

5. Zhoujun Hu, Zhigang Hu, and Zhenhua Liu , Resource Availability Evaluation in Service Grid Environment // Proceedings of the Asia-Pacific Services Computing Conference (APSCC,). – IEEE. – 2007. – P. 232-238.

6. Volkovic Ya.V., Granichin O.N., Adaptive server optimization that processes tasks queue // Stochastic optimization in informatics, Vol. 1, p.p. 17-28, 2005.

7. Amelina N.O., Fradkov A.L., Approximate consensus in the dynamic stochastic network with incomplete information and measurement delays // Avtomat. i Telemekh. 2012. № 11. p.p. 6-29.

8. Vittikh V.A., Skobelev P.O. Method of contiguous interactions for real time resource allocation management // Autometry, 2009. Vol. 45, № 2, p. 78

9. Skobelev P.O. Multi-Agent Technology for Industrial Applications: Towards 20 years Anniversary of Samara Scientific School of Multi-Agent Systems // Mechatronics, automation, management. – 2010, №12.

Journal of Physics: Conference Series **510** (2014) 012007 doi:10.1088/1742-6596/510/1/012007

# Automatic post processing algorithm for passive seismic monitoring data

**K Nepeina**

PhD student, Institute of Geosphere Dynamics Russian Academy of Scince, Moscow, Russian Federation

E-mail: nepeina.k@mail.ru

**Abstract**. The problem of monitoring of different types of seismic events - geoacoustic precursors of earthquakes, industrial and field explosions, places fragments fall of separating parts of rockets-carriers, etc. is one of the key in the modern ecology of the environment. The peculiarity of this kind of monitoring is that it is mobile seismic groups, which should be based in the proposed area of occurrence of events. One of the most important steps for solving the problems connected with the detection and identification of recorded data from passive sensors in mobile seismic array (MSA). The task of determining the nature of the source and its' coordinates lies in the basis of direction, referred to as the geoacoustic location. Using a new approach (not by location but by neural classification of waveform "portraits") usability of algorithm which based on quantitative parameters of signal will be demonstrated.

## 1. Introduction

The increase on the demand for Passive Seismic especially in the domains of exploration and production has most recently become a challenging field of opportunity. The method based on uncontrolled sources, requires an understanding and analysis of what we call passive events in a careful manner, so as not to be misinterpreted leading to a possible damage of its credibility [1]. The problem of earthquake occurrence in platform territories, in spite of numerous examples of such seismic events, remains still unsolved in many aspects. This problem is particularly important for densely populated southern European Russia [2].

**The first** problem of monitoring of different types of seismic events - geoacoustic precursors of earthquakes, industrial and field explosions, places fragments fall of separating parts of rockets-carriers, etc. is one of the key in the modern ecology of the environment. The peculiarity of this kind of monitoring is that it is mobile seismic groups, which should be based in the proposed area of occurrence of events. One of the most important steps for solving the problems connected with the detection and measurement of parameters of sources of seismic events on the basis of registration of seismic signals with the help of spatially distributed sensors mobile seismic group. The task of determining the geographical coordinates of the source lies in the basis of direction, referred to as the geoacoustic location [3].

**The second** fundamental research focuses on the main differences between processes leading to earthquakes and the seismic rock mass response to mining. In earthquake seismology the driving forces are fairly constant and relatively slow facilitating the processes of self-organization. An important agent in the development of spatial and temporal correlations is seismic activity itself. Mining is not a spontaneous process. It induces stresses at a particular place, at a particular time and at a particular rate, which are all highly variable compared to the tectonic regime This process is controlled more by the commodity prices than by the plate tectonics. Consequently, unlike earthquakes, the seismic rock mass response to mining can be controlled to a certain degree [3].

**The third** objective of seismic monitoring is to detect and locate underground nuclear explosions. Nuclear explosion monitoring is as important today as it was at the dawn of the atomic age. Data resulting from seismic monitoring are used to distinguish between an underground nuclear explosion and the numerous natural and man-made seismic events that occur every day, such as earthquakes and mining explosions [3].

The passive seismic monitoring is a very important method for obtain geodata which is useful for academic science and practical applications. The main targets of passive seismic monitoring are:

- précising the movements of Earth's crust by understanding frequency of the seismic events
- detecting the character of seismic events and its localization
- zoning the endogenic epicentres and focusing on the territories: understanding nature
- seismic stability for constructions nuclear power stations
- forecasting and statistics of seismic events
- Comprehensive Nuclear-Test-Ban Treaty (CTBT) verification regime: location underground nuclear explosions

The Government has ratified the Federal Program - "Seismic safety of territory of Russia" (2002-2010). The purpose of this Program is the maximal increase of seismic safety of the population, reduction of social, economic, ecological risk in seismically dangerous areas of the Russian Federation, decrease of damages from destructive earthquakes by certification, strengthening and reconstruction of existing buildings and constructions, and also preparation of cities and other settlements, transport, power constructions, pipelines for strong earthquakes. The actual frequency of large earthquakes in Northern Eurasia is three and more times higher than previously assumed. The use of straight plots in past years resulted in significant overestimation of the return time of large earthquakes, hence, in underestimation of seismic hazard practically in all the regions of the former USSR [2].

## 2. Prerequisites

Array stations allow for a better estimation of the azimuth of incoming signals, i.e. they identify the direction from which the signal arrived. The spatial distribution of the sensors also permits an estimation of the seismic waves' speed. Information on both the direction and speed of the incoming seismic waves is crucial when identifying the source of a particular event. Three-component stations employ only one seismic sensor that measures the three spatial components of seismic waves– up-down, East-West and North-South. This set-up provides information on the depth and strength of an event. In order to eliminate the influence of seismic noise, seismic stations are usually built in remote areas, preferably on the outcrops of geological hard rock and as far away as possible from human activity [4].

These were difficult to search when seeking past events for comparison with new events and expensive to maintain. Catalogs devised by many individual observatories around the planet were inconsistent in format and difficult to collate and reconcile manually. Automatic catalog reconciliation is now a reality for scores of formats. Databases are used to integrate and index waveform observations from thousands of stations in many formats into structured collections of uniform data amenable to sophisticated data mining approaches. Increasingly the construction of databases will be automated with autonomous agents scouring the Web for new catalog results and waveform data collected across the Internet in continuous streams. The convergence of communication technology, storage technology, sophisticated software tool development and database technology will put terabytes of organized information in the hands of researchers through largely automated processes [5].

*2.1. Research object*

Modern seismological monitoring of the EEP is carried out by the Geophysical Survey of Russian Academy of Sciences (GS RAS) in close contact with different regional organizations, including academic and university research units. Local networks and processing centers are set up near the Russian cities of Voronezh, Perm, Saint-Petersburg, and Arkhangelsk. Some of these stations use an automatic signal detecting and identification application module. But identification criteria are different in various territories. Thus we should consider territorial features (depth of the sediment layers, body wave velocities etc.). The small aperture seismic array (MSA) Mikhnevo is situated in Moscow area on East-European platform. An advantage of this array is the noise reduction by coupling the receivers. Using SM3-KV instruments we collected data from 2004 year (Table 1). The interpreters made catalogs of seismic events and produced database of seismic events: among others there were earthquakes with magnitudes about 2.5-3.0.

| SM3-KV | Dimension | Value |
|---|---|---|
| Measurement frequency domain | Hz | 0.5 – 100 |
| Conversion coefficient | V × sec / m | 135 ± 20 |
| Threshold of sensivity | m / sec | 3 × 10-9 |
| Self-noise | m / sec | 3 × 10-10 |
| Natural period | sec | 2.0 ± 0.1 |
| T | °C | -20 … +40 |

Table 1. Technical characteristics of seismic receivers SM3-KV.

The objective of this presentation is to look at one of the methodologies that benefit from passive sources, and justify its use with case studies from MSA Mikhnevo. The targets of MSA are the low-magnitude events (M > 1.5-4.5). The data of the Mikhnevo 3-*C* station from the geofon network are added to these observations for some earthquakes. The average value of microseismic noise spectral density for displacements is equal to 2 $nm^2/Hz$ at the frequency of 1 Hz, which is comparable to array conditions at other sites [Harjes, 1990; Nevskiy et al., 2003]. The array configuration is similar to the NORES type arrays. The low-frequency acoustic/seismic background noise (between 0.5-10.0 Hz), which is actively emitted by the earth, lies near the central frequency of wanted signal (body waves: 4-8 Hz – *P*-waves, 2-3 Hz – *S*-waves, 0.5-0.8 Hz – Surface waves). In 2007, MHVAR independently recorded 623 local events with magnitudes ($M_L$ = 0.79–3.24) which are identified as quarry blasts. The seismic monitoring of the territory of the East European Platform (EEP) is a part of the complex study of dynamics of tectonic processes, manmade influence and geo-ecological effects. The main feature of MHVAR is that the constantly operating small aperture array is set up in the region with the presence of sedimentary cover which is 1.5–2 km thick [6].

Reviewing datasets we can observe some different seismic events. They could have nature from: natural hazard (landslides, earthquakes etc.) or anthropogenic activity (mining exploration, munitions disposal etc.). We need to produce an automatic algorithm which can indicate a seismic event nature. After this procedure operator will use results for next processing (such seismic event location). This is of particular importance in passive seismic systems where *STA/LTA* algorithms require an input signal. Typical casing failure events have large relative amplitude, a body waves *P/S* ratio close to unity.

Then we estimate P/S ratios, Hilbert transform, signal lasting and Fourier spectra – these are waveform "portraits".

## 3. Methods

Trends in instrumentation are providing us with progressively more and better data, and at lower cost. Mathematics was translated to assembler code in the early days of the computer and through the evolution of the computer the research and business world has seen FORTRAN, BASIC, PASCAL, C++ and many other languages. Automatic catalog reconciliation is now a reality for scores of formats. Databases are used to integrate and index waveform observations from thousands of stations in many formats into structured collections of uniform data amenable to sophisticated data mining approaches [5].

The object of this work is to construct a post-processing algorithm for waveform identification. At first raw data is a rather large massive. For programming and quick automatic identification we need to describe waveform by limited set of parameters. We provide an application CASE (Classification Algorithm for Seismic Events) employing Matlab 6.1. and internal «Fast Artificial Neural Network Library» (Figure 1). CASE can make solutions like an expert system providing an interpretator choice. Neural algorithms are the most powerful for classification and identification. The complex database includes waveform "portraits" (2004-2013 years) is used for network training. This algorithm contains three computational blocks. The first one is raw data processing and detection of main parameters. Then we could observe data limits and make decisions about difference between close events. After this we construct a simple neural network with weight matrice for machine teaching. The third block is made for identification a text file with some set of parameters and it is presented as a neural network too.

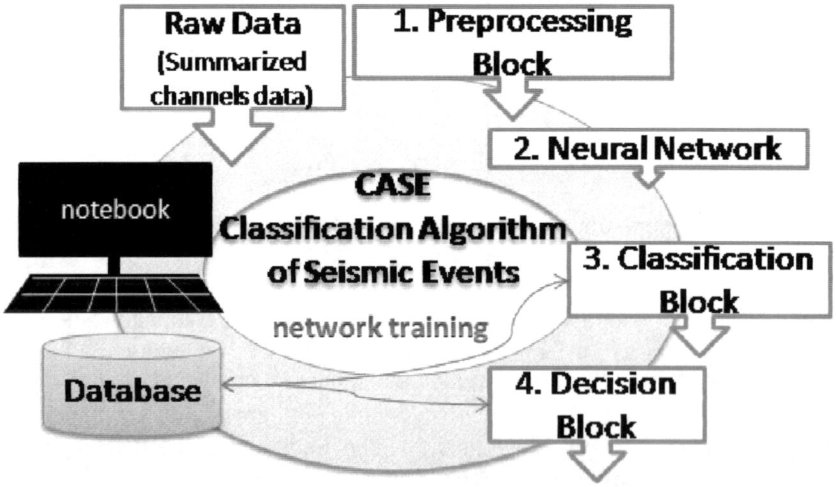


Fig. 1. A structure of CASE algorithm: the main steps and processing block relationships

Modeling this network is rather difficult task. At the end of this work we expect to have one of the answers: noise, known event and unknown event (Figure 2).

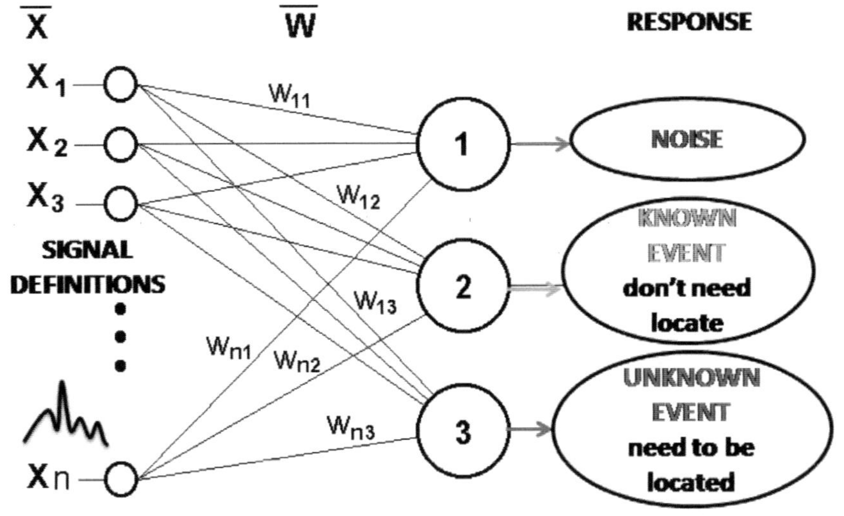


Fig. 2. Possible three simpled results after neural network response. Here vectors $\overline{X}$ and $\overline{W}$ means set of parameters and weight function correspondingly

3.1. *A signal detecting.*

The algorithm *STA/LTA* is processing already filtered signal with the help of two moving time window («moving average»): a Short Time Average window (*STA*) and a Long Time Average (*LTA*).

$$STA_i = \frac{1}{NS}\sum_{j=i-NS}^{i}|X_j|, \qquad LTA_i = \frac{1}{NL}\sum_{j=i-NL}^{i}|X_j| \qquad (1)$$

$$r_i = STA_i / LTA_i, \qquad (2)$$

where *NS* и *NL* – number of digitizing points in a short and in a long time "windows" ? consequently (*NS* << *NL*), $r_i$ – ratio *STA/LTA*. In this work after some investigations the ratio was equal 1,4. The time of start is the time of primary wave (P). This fact fits well with the article Plenkers, Ritter & Schindler [2012].

3.2. *Waveform parameters for neural network system.*

At the beginning we should install rules of signal selection and rejection. This question is close to clasterization problem, because it is important to classify variety of patterns. The most intense parameter for cluster determination is a time delay between *S*- and *P*-wave: close or far event. Then there is a possibility to define a surface wave's existence. If yes – it should be an explosion, otherwise it is an earthquake. For more accurate definition a body wave's spectrum amplitude ratio and envelope function need to be employed. For distinct separation scientist should have a data statistics, which is laid in neural network operation. Hence the waveform database from 2004 is so required. Thus four main parameters and three ongoing parameters were selected for description image (Table 2).

In this investigation an algorithm with machine teaching was applied. Feed-back neural networks - the input information defines the initial activity state of a system. The first output of the system is taken as the new input, which produces a new output and so on. The first input pattern is presented to an initially randomized network, and the weights and thresholds adjusted in all the links. Other patterns are then presented in succession, and the weights and thresholds adjusted from the previously determined values. This process continues until all patterns in the training set are exhausted. It is generally accepted that this procedure is independent of the order in which the example patterns are presented. However, a final check can be performed by looking at the pattern error which is defined as the square of difference between desired output and neural network output for each pattern and the system error which is defined as the average of all pattern errors, to determine whether the final network solution satisfies all of the patterns presented to it within a certain error. The set of weights and thresholds in the network are now specifically tailored to "remember" each input and output pattern, and can consequently be used to recognize or generate new patterns given an unknown input [7].

| Parameters | Nature etalons | Ranges |
|---|---|---|
| Existence P-wave | ✓ 50 career blasts | yes/no |
| Time P-wave | ✓ noise: | hh:mm:ss |
| Time S-wave | • Landslides | hh:mm:ss |
| Delta Time S-P-waves | • Thunderstorms | 00:00:ss |
| Signal Length | • Avalanches | 00:mm:ss |
| Existence L-wave | • Falling meteorites | yes/no |
| Polynomial coefficients for Hilbert transform | • Anthropogenic noise | $A_1 \ldots A_{10}$ |
| Spectral amplitude ratio P/S waves | from transport, building | 0-1,5 |
| Extremes of spectral amplitude ratio P/S waves function | etc.<br>✓ endogenic processes | $B_1 \ldots B_3$ |

Table 2. Used parameters for neural network teaching and work

During comparison author decided use Euclidian measure (Figure 3) – the simplest way for definition and is computed as an equation (3).

$$\rho(x_i, x_j) = \sqrt{(x_i^1 - x_j^1)^2 + (x_i^2 - x_j^2)^2 + \ldots + (x_i^P - x_j^P)^2} \quad (3).$$

Referred to a monograph [7] for the group of "per sample" it is more suitable to apply Euclidian measure for computing and k-means method or LVQ (Linear Vector Quantization). The first one and the simplest is the k-means algorithm. This algorithm clusters observations into k groups, where k is provided as an input parameter. It then assigns each observation to clusters based upon the observation's proximity to the mean of the cluster. The cluster's mean is then recomputed and the process begins again. Here's how the algorithm works:

1. The algorithm arbitrarily selects k points as the initial cluster centers ("means").
2. Each point in the dataset is assigned to the closed cluster, based upon the Euclidean distance between each point and each cluster center.
3. Each cluster center is recomputed as the average of the points in that cluster.
4. Steps 2 and 3 repeat until the clusters converge. Convergence may be defined differently depending upon the implementation, but it normally means that either no observations change clusters when steps 2 and 3 are repeated or that the changes do not make a material difference in the definition of the clusters.

LVQ directly define class boundaries based on prototypes, a nearest-neighbour rule and a winner-takes-it-all paradigm. The main idea is to cover the input space of samples with 'codebook vectors' (CVs), each representing a region labelled with a class. A CV can be seen as a prototype of a class member, localized in the centre of a class or decision region ('Voronoï cell') in the input space. Note that every set of parameters (waveform portraits) should be normalized by etalon characteristics (certain average object). Whereupon minimize error function is fitted.

The most efficient for solving the main problem is k-means algorithm, because we can predict a number of clusters. Moreover during the processing we obtain different values, such is available for means estimation. So Euclidian measure is a distances $L_{1\text{-}0}$ and $L_{2\text{-}0}$ (Figure 3) either between cluster cores, or average waveforms portraits and between, also or new unknown events and etalon (a standard vector line).

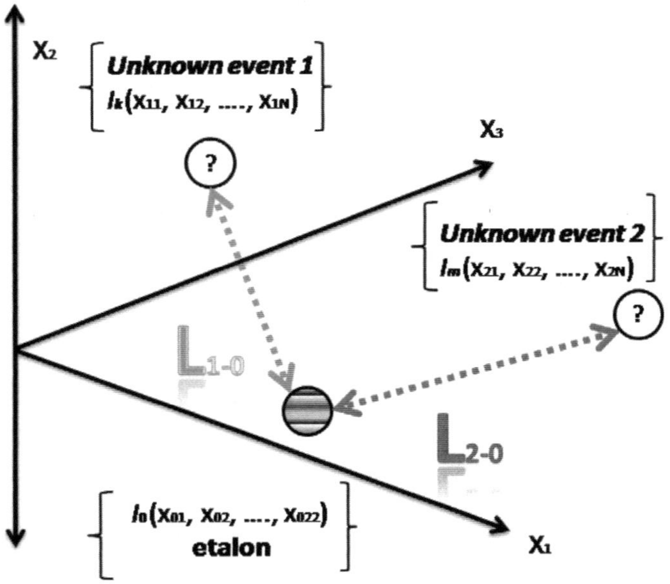


Fig. 3. Euclidian measure for waveform "portraits" comparison in N-space. There is a simple example where one (grey colour) point is an etalon used as a basis for comparison; points with question-mark are unknown events. $L_{1-0}$ and $L_{2-0}$ are the distances between observed set of parameters $I_k$ ($x_{11}$, $x_{12}$, ..., $x_{1N}$) and $I_m$($x_{21}$, $x_{22}$, ..., $x_{2N}$)

## 4. Results

For numerical calculations was used Matlab 6.1 and its Neural Networks Toolbox. The multi layer perceptron is the most employed, because it holds a good computational power and capability to learning. It is referred to **feed-forward back propagation** algorithm. For the learning rule was applied **trainlm** function. For adaptation that **learngdm**. In accordance with chosen network the activation function was **tansig** and performance criterion – **mse** (mean-square error). All description you may find in Matlab manuals. It was decided provide two layers for this construction. One of the main decision rule was delta time S-P-waves: in case it is less than 0,20 sec, 0,20-0,30 and more than 0,30 sec.

After some manipulation with input and output vectors errors were estimated. And we obtained three big clusters. As a result of this conception it was possible to classify different rock response from the region seismic activity. It is very crucial to correctly find a set of parameters for each etalon. However it remains a problem of retraining. It is worth to specify deviations from the standard. After all procedures classes were derived among which were allocated by the characteristics as career explosions but earthquakes were simply discarded. The most recognizable were carrier blasts from Novogurovskyi, Kovrov, Shurovskoy ore mining department. They differ in general view (Figure 4) and consequently in portrait parameters (Table 3). Thus various portraits "points" got into different clusters. This information is the newest one for near Moscow region. More than 10 events were referred to unknown. They strongly attracted to the cluster with Kovrov explosions.

| Name | $T_{signal}$, sec | Ts-Tp, sec | $maxA_P/maxA_S$ |
|---|---|---|---|
| Novogurovskyi | 28-35 | 6-7 | 1,02-1,88 |
| Shurovskoy | 40-60 | 8-9 | 0,8-1,12 |
| Kovrov | 14-20 | 29-30 | 0,9-1,20 |

Table 3. Some resulted main parameters for carrier explosions

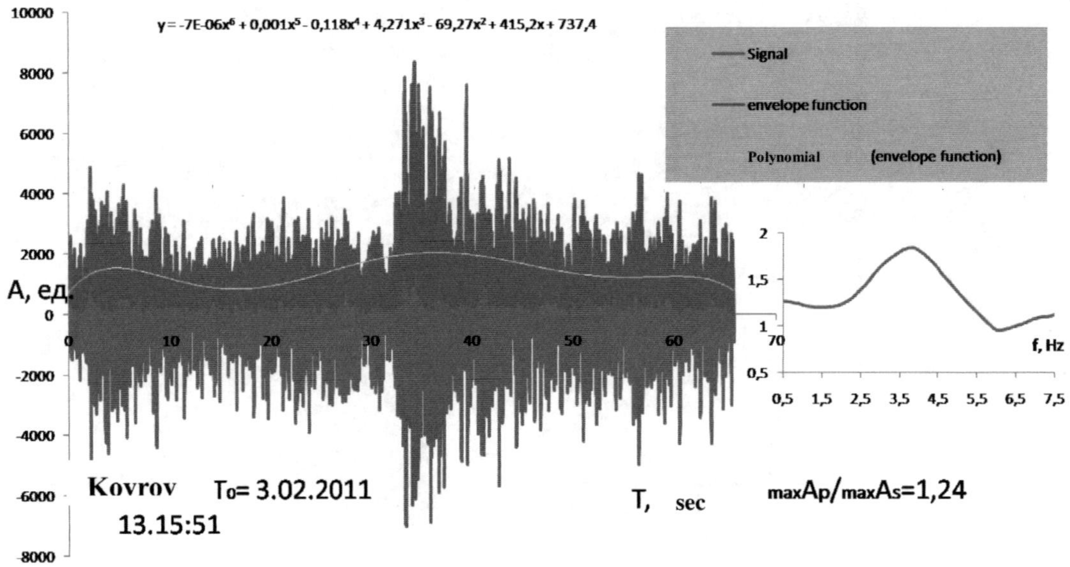


Fig. 4. A waveform portrait of Kovrov carrier explosion 03.02.2011. Lines: blue – natural signal, red – Hilbert transform (envelope function), yellow – polynomial estimation for envelope function. Mini graph: the maximum spectral amplitudes ratio

## 5. Conclusions

A new approach without procedure of epicenter location is innovative method and is capable to be programmed. This algorithm optimizes human work and help to observe a large data more quickly. Automatic detection facilitates processing. Other researchers could use this software in future for bulletins and catalogs of events. Basing on these results scientists could make forecast using probability and statistics of events timing.

It is widely known that neural networks are ideally suited to situations where standard algorithms cannot be used as the mathematical relationship is uncertain, provided a supply of many examples of what is required can be generated. The performance depends on the training data and its ability to generate solutions cannot lie too far outside its experience-in our case. Although, in general, extra training sets should help with this problem, for this application it is too time-consuming to train the network to recognize more complicated portraits, and pre-processing for the source function appears necessary. Consequently, application of a network can often be used as a tool to highlight relationships which exist between certain quantities, and help gain a greater insight into the physical process [8].

## Acknowledgements

I would like to thank IDG RAS for collected data and my academic tutor Sanina I.A. and Institute of Geosphere Dynamics for seismic data.

## References

[1] Kapotas S, Martakis N, Tselentis A 2006 The Place of Passive Seismic in E&P *Passive Seismic: Exploration and Monitoring Applications* **A21** (Dubai)

[2] Ulomov V I 2007 Seismogeodynamics and seismic hazard prediction National report to the International Association of Seismology and Physics of the Earth's Interior of the International Union of Geodesy and Geophysics 2003 – 2006 (Moscow) pp 45-49

[3] Tkhak K 2013 Past, Present and Future Passive Seismic Tasks for Sustainable Development Sustainable Earth Sciences (DOI: 10.3997/2214-4609.20131639)

[4] http://www.ctbto.org/verification-regime/monitoring-technologies-how-they-work/seismic-monitoring/
[5] Anderson D N *et al* 26th Seismic Research review – trends in nuclear explosion monitoring (www.ees.lanl.gov/pdfs/01-01.pdf)
[6] Sanina I A *et al* 2011 The Mikhnevo small aperture array enhances the resolution property of seismological observations on the East European Platform *J. Seismol.* **15** (DOI 10.1007/s10950-010-9211-x) pp 545–556
[7] Ovchinnikov P E 2012 An artificial neural networks application for signals processing Nizhnii Novgorod p 32 (http://window.edu.ru/resource/846/77846/files/Ovch_ANN.pdf)
[8] Dai H, MacBeth C 1994 Split shear-wave analysis using an artificial neural network? *First break* Vol 12 **12** pp 605-613

Journal of Physics: Conference Series **510** (2014) 012008 doi:10.1088/1742-6596/510/1/012008

# Fourier Monte Carlo Simulation of Hexatic Membranes

**Andreas Tröster**

Vienna University of Technology, Wiedner Hauptstrasse 8-10/136, A-1040 Wien, Austria

E-mail: `andreas.troester@tuwien.ac.at`

**Abstract.** Hexatic membranes are difficult to study both in theory as well as in simulations. We present new results for the flat phase of hexatic membranes using a unique simulation approach based on an recently developed optimization [1] of our Fourier Monte Carlo algorithm [2, 3] which enables us to drastically reduce critical slowing down and observe critical behavior with excellent statistical accuracy. In detail, we calculate the correlation function $\langle|\tilde{f}(\boldsymbol{q})|\rangle^2 = G(\boldsymbol{q})$ and the related mean squared displacement $\langle(\Delta f)^2\rangle$ of the membrane's out-of-plane deformations in the Monge parametrization $f(\boldsymbol{x})$ and give a finite size scaling analysis of these data. In our present treatment, the case of hexatic membranes is formally found to closely resemble that of solid membranes for which our algorithm has already proven to be quite successful [1].

## 1. Introduction

The properties of membranes constitute an active research area of modern science. Membranes are ubiquitous in a biological context, they play an important role in advanced chemical synthesis and technology, and the research on graphene has recently been rewarded by a Nobel prize in physics for its exciting new perspectives on condensed matter physics and electronics. Motivated by the continuing interest in graphene, we have recently published high precision simulation data on the universal elastic fluctuation behavior of solid membranes.

Due to its strong covalent bondings, at low temperatures the properties of defect-free graphene come quite close to those of a "tethered membrane", i.e. a membrane model system whose bonds are by definition unbreakable. The caution implicit in this statement is based on the theoretical result that in reality there is nothing like a 2d "tethered" or "crystalline" membrane, since the Wagner-Mermin-Hohenberg theorem [4] assures us that in two dimensions true long-range translational order for systems with short-ranged interactions cannot exist. While the energy required to create an isolated dislocations in a strict 2d system grows logarithmically with system size, the energy of bound dislocations pairs in 2d turns out to be finite. Thus, at any low but positive $T$ there is a nonzero density of such pairs. This implies that even at the lowest temperatures there is only quasi-long-range order in the density-density correlations, i.e. an algebraic decay with distance $r$ exhibiting a $T$-dependent exponent. According to the Kosterlitz-Thouless-Halperin-Nelson (KTHNY) theory of 2d melting [5, 6, 7, 8, 9], above a certain critical temperature $T_c$ the dislocation pairs break up, resulting in a gas of free dislocations and complete loss of translational order, but quasi-long-range *orientational* order of bonds persist in this *hexatic* phase. Free dislocations can, however, be in turn regarded as bound pairs of orientational defects, so-called *disclinations* [10]. Below $T_c$ the energy required to create an isolated disclination grows quadratically with system size, but once in the hexatic phase, the presence of free dislocations mediates a screening of this elastic energy that reduces this growth to a logarithmic one and thus facilitates the breaking-up of dislocation pairs at a still higher temperature $T_m$, resulting in complete destruction of the remaining orientational order. According to this KTNHY scenario, it is only above this second phase transition that the system assumes a true liquid state.

Membranes should, however, not be confused with true 2d systems in the above sense. Rather, they can be regarded as two-dimensional manifolds immersed in a 3d space, and whithin this embedding space they possess the additional freedom to bend and buckle, which allows them to drastically reduce the elastic energy required to create defect configurations. Even though taking into account the wave-vector dependent renormalization of the membrane's elastic constants considerably complicate the calculations, one can show [11] that for a membrane of size larger than a certain buckling radius $R_b$ the energy of an isolated disclination is reduced from $\sim R_b$ to $\sim \ln R_b$, while that of a isolated dislocation decreases from $\sim \ln R_b$ to a finite constant. Thus, at arbitrary low but nonzero temperature there is always a finite concentration of free dislocations, and one concludes that any large enough would-be crystalline membrane is inevitably in a hexatic state, provided its size exceeds $R_b$.

Despite of these theoretical findings that guarantee the existence of hexatic membranes, it remains a challenge to study their properties, be it in experiment, theory or by computer simulation. In a real solid membrane like graphene, $R_b$ may be prohibitively large, such that the hexatic nature of a specimen remains undetected in experiment and beyond reach in simulations based on microscopic or atomistic model Hamiltonians. In particular, this is true when it comes to precisely determining the universal long wavelength elastic properties of membranes, since for such simulations also large system sizes are required, and critical slowing down becomes a major obstacle. Thus, it comes as no surprise that the literature does not abound with computational studies of hexatic membranes [12]. In a recent paper [1] we have shown how to calculate the exponent $\eta$ governing the asymptotic elastic behavior of solid membranes with great precision. The reason why this became possible was that we did not conduct simulations based on a "microscopic" Hamiltionian formulated on a lattice in direct space, but instead carried out Fourier Monte Carlo (FMC) simulations based on an effective Hamiltonian that focuses only on the long wavelength collective degrees of freedom. In addition, this algorithm also allows to efficiently suppress critical slowing down by performing an initial tuning step. It is the purpose of the present paper to extend this powerful approach to hexatic membranes in the low temperature flat phase.

## 2. Effective Hamiltonian

On the level of effective Hamiltonians, it turns out that even if their derivation proceeds quite differently in the solid and hexatic cases, the resulting expressions are extremely similar. In the so-called Monge parametrization of the low temperature flat phase, the "height" of a square membrane of projectted area $U = L \times L$ with coordinates $\boldsymbol{x} = (x_1, x_2) \in U$ is measured by a scalar function $f(\boldsymbol{x})$. Similar to the case of solid and liquid membranes, a hexatic membrane is subject to a harmonic bending energy

$$\mathcal{H}^{(h)}_{\text{hex}}[f] = \frac{\kappa_0}{2} \int_U d^2x (\Delta f)^2(\boldsymbol{x}) \tag{1}$$

In addition, the hexatic membrane Hamiltonian has an anharmonic fourth order contribution of a structure that is formally quite similar to that for a solid membrane, although it originates from a rather different mechanism. Since the full derivation makes heavy use of differential geometry here we content ourselves with a quick sketch and refer the reader to the literature on

Riemannian manifolds (of which accessible accounts can be found e.g. in Refs. [13] or [14]) the original derivations [15, 16, 17] and the very nice reviews [18, 19] for further details.

The hexatic state is characterized by a unit vector field $\boldsymbol{n}(\boldsymbol{x})$ encoding the directional order of bonds, which, as we are targeting an underlying triangular lattice order, is defined only up to local 60° rotations. In writing down an effective continuum Hamiltonian, consider first such a field residing on a flat membrane. Then, the lowest order contribution would actually resemble a XY type of model, for which deviations from perfect parallelism of neighboring hexatic "directors" are subject to an energy penalty proportional to the squared gradient of $\boldsymbol{n}$. Introduction of (Gaussian) curvature leads, however, to a frustration of the field $\boldsymbol{n}(\boldsymbol{x})$, since the energetically most favorable parallel state cannot be maintained globally. Expressed in the Monge parametrization, the resulting coupling between the field $\boldsymbol{n}(\boldsymbol{x})$ and the Gaussian curvature of the membrane can be shown to be encoded in the anharmonic hexatic energy contribution

$$\mathcal{H}^{(a)}_{\text{hex}}[f,\theta] = \frac{K_A}{2}\int_U d^2x\left(\partial_\mu\theta - \frac{1}{2}\epsilon_{\mu\sigma}\partial_\sigma\left[P^T_{\alpha\beta}(\partial_\alpha f)(\partial_\beta f)\right]\right)^2 \tag{2}$$

where $P^T_{\alpha\beta} = \delta_{\alpha\beta} - \Delta^{-1}\partial_\alpha\partial_\beta$ is the so-called transverse projector ($\Delta^{-1}$ is, of course, the inverse Laplacian in 2d), $\theta = \theta(\boldsymbol{x})$ denotes the local bond angle (i.e. $\boldsymbol{n}(\boldsymbol{x}) = (\cos\theta(\boldsymbol{x}), \sin\theta(\boldsymbol{x}))$ with respect to a local orthonormal frame of the membrane's tangent bundle) and $K_A$ is termed the hexatic stiffness.

In our present treatment, we assume that the field $\boldsymbol{n}(\boldsymbol{x})$ is single-valued and thus defect-free, which means that we neglect configurations with disclination defects. Thus, we expect to correctly describe the flat hexatic phase, but admittedly we will miss the disclination unbinding transition. The advantage of this assumption is, however, that it allows to completely eliminate the field $\boldsymbol{n}(\boldsymbol{x})$ from the partition function by virtue of a Gaussian integration. Up to an unimportant constant, one then obtains

$$\mathcal{H}^{(a)}_{\text{hex}}[f] = \frac{K_A}{2}\int_U d^2x\left(\frac{1}{2}\partial_\sigma P^T_{\alpha\beta}(\partial_\alpha f)(\partial_\beta f)\right)^2 \tag{3}$$

Comparing this expression to the corresponding anharmonic contribution for solid membranes (see Eqn. (2.14) of Ref. [20]), we note a close formal correspondence. In fact, the harmonic contributions are identical in structure, while the anharmonic contributions only differ because of the extra derivatives $\partial_\sigma$ appearing in (3). Now recall [1] that in the solid case, we had been able to recast the anharmonic part of the membrane's effective Hamiltonian in a form

$$\mathcal{H}^{(a)}_{\text{sol}}[f] = \frac{K_0}{8}\int\frac{d^2Q}{(2\pi)^2}|\tilde{\mathcal{F}}_{\text{sol}}(\boldsymbol{Q})|^2 \tag{4}$$

that is diagonal in Fourier space, albeit not with respect to the amplitudes $\tilde{f}(\boldsymbol{q})$, but with respect to the generalized convolution amplitude

$$\tilde{\mathcal{F}}_{\text{sol}}(\boldsymbol{Q}) = \int\frac{d^2q}{(2\pi)^2}(\hat{\boldsymbol{Q}}\times\boldsymbol{q})^2\tilde{f}(\boldsymbol{q})\tilde{f}(\boldsymbol{Q}-\boldsymbol{q}) \tag{5}$$

($K_0$ denoted the solid membrane's 2d Young's modulus). Upon one partial integration in (3), the extra derivatives $\partial_\sigma$ appearing in the present hexatic case translate into an extra squared wave vector $\boldsymbol{Q}^2 = |\boldsymbol{Q}| \cdot |\boldsymbol{Q}|$ that can be equally distributed between the two $\mathcal{F}$-factors. Thus we finally end up with an effective Hamiltonian

$$\mathcal{H}_\Lambda[f] = \frac{\kappa_\Lambda}{2} \int \frac{d^2q}{(2\pi)^2} q^4 |\tilde{f}(\boldsymbol{q})|^2 + \frac{K_\Lambda}{8} \int \frac{d^2Q}{(2\pi)^2} |\tilde{\mathcal{F}}(\boldsymbol{Q})|^2 \tag{6}$$

for the hexatic system that looks exactly like that for a solid one, except that the amplitude

$$\tilde{\mathcal{F}}(\boldsymbol{Q}) := \int \frac{d^2q}{(2\pi)^2} |\boldsymbol{Q}| (\hat{\boldsymbol{Q}} \times \boldsymbol{q})^2 \tilde{f}(\boldsymbol{q}) \tilde{f}(\boldsymbol{Q} - \boldsymbol{q}) \tag{7}$$

inherits an extra factor $|\boldsymbol{Q}|$ in comparison to version (5) appropriate for solids. Note that in these last two formulas (6) and (7) we have dropped the explicit reference to the hexatic case, but have explicitly indicated a dependence of the coupling parameters on the presence of a wavevector cutoff $\Lambda$. Similar to the case of crystalline membranes, the anharmonic contribution to $\mathcal{H}_\Lambda[f]$ is again of a non-local nature, encoding an effective long-range anharmonic effective self-interaction of the field $f(\boldsymbol{x})$ mediated by the hexatic director field $\boldsymbol{n}(\boldsymbol{x})$ that effectively stiffens up the membrane at long wavelengths and thus stabilizes a low temperature flat phase.

## 3. Fourier Monte Carlo Approach

Criticality in systems with long-ranged interactions is very difficult to assess in a conventional Monte Carlo simulation based on a spins model defined on the direct lattice. In addition, the fact that the above model is formulated in terms of Fourier amplitudes rather suggests to employ our FMC algorithm [2, 3] which is tailor-made for such a situation. In a FMC simulation the continuous Fourier transform is replaced by a discrete Fourier transformation, for which we shall use the convention

$$f(\boldsymbol{x}) = \frac{1}{N} \sum_{\boldsymbol{q}} \tilde{f}(\boldsymbol{q}) e^{i\boldsymbol{q}\boldsymbol{x}}, \quad \tilde{f}(\boldsymbol{q}) = \sum_{\boldsymbol{x}} f(\boldsymbol{x}) e^{-i\boldsymbol{q}\boldsymbol{x}} \tag{8}$$

For periodic boundary conditions the admissible wave vectors $\boldsymbol{q}$ have components $|q_i| < \Lambda \leq \pi$ parametrized by integers $m_i \in \mathbb{Z}$ as $q_i = 2\pi m_i/L$, where we have put the lattice constant $a$ of the underlying simple cubic lattice of $L^d = N$ sites to $a = 1$. A FMC move consists of picking a wave vector $\boldsymbol{q}_0$ and a complex number $\epsilon \in \mathbb{C}$ with $|\epsilon| \leq r$ at random, where $r > 0$ is some prescribed radius of a circle in the complex plane around zero. One then shifts

$$\tilde{f}(\boldsymbol{q}_0) \to \tilde{f}(\boldsymbol{q}_0) + \epsilon, \quad \tilde{f}(-\boldsymbol{q}_0) \to \tilde{f}(-\boldsymbol{q}_0) + \epsilon^*,$$

For the harmonic part $1/(2N) \sum_{|\boldsymbol{q}| \leq \Lambda} \kappa(\boldsymbol{q}) \tilde{f}(\boldsymbol{q}) \tilde{f}(-\boldsymbol{q})$ of a spin model Hamiltonian given in terms of the amplitudes $\tilde{f}(\boldsymbol{q})$, the accompanying energy change is easy to compute. The hard part is to calculate the change in the anharmonic part of the Hamiltonian in an efficient way. Since we have already reviewed the nuts and bolts of FMC in several publications [21, 22, 23], here we content ourselves with the following general statements.

FMC is certainly far from being a universal weapon for performing general simulation tasks in statistical mechanics. Rather than that, it has more of a specialized surgical tool that is capable of solving certain types of problems that had remained difficult to assess with other methods (see e.g. [2, 3]). In particular, FMC is a powerful strategy for studying critical behavior in systems with effective or microscopic long-range, since the Fourier transform diagonalizes all interactions irrespectively of their range as long as they are translationally invariant [24], and the move scheme (8) is nonlocal and collective. Of course, the algorithm becomes only efficient to use if a rather small cutoff $\Lambda$ enclosing the critical wave vector, which is usually $\boldsymbol{q}_c = \boldsymbol{0}$, is chosen, and all "fast" models outside the cutoff are regarded as having been integrated out from the partition function, such their effect is absorbed in the remaining effective "Landau-Ginzburg" Hamiltonian governing the behavior of the "slow modes" $\tilde{f}(\boldsymbol{q})$ with $|\boldsymbol{q}| < \Lambda$. In an application to crystalline membranes, it was recently even demonstrated [1] that by slightly modifying the above scheme one may obtain an algorithm that is practically free of critical slowing down in the sense that at criticality the integrated autocorrelation times of the amplitudes $\tilde{f}(\boldsymbol{q})$ show only negligible growth when approaching $\boldsymbol{q} \to \boldsymbol{0}$. This major improvement that catapults FMC into the top league of algorithms for studying criticality in interactions with long range is achieved by simply allowing for different values $r(\boldsymbol{q})$ for the radius $r$ governing the maximum change in amplitude in the MC move (8), and optimizing these radii $r(\boldsymbol{q})$ separately for, say, $30 - 40\%$ acceptance rates during the start-up of the simulation. A further advantage of FMC is that it is parallelizeable using standard MPI and OpenMP approaches.

## 4. Results and Discussion

Similar to the approach used in Ref. [1], we have calculated the out-of-plane displacement correlation function

$$\tilde{G}(\boldsymbol{q}) = \langle|\tilde{f}(\boldsymbol{q})|^2\rangle \tag{9}$$

and the related out-of-plane mean squared displacement

$$\langle(\Delta f)^2\rangle = \int \frac{d^2q}{(2\pi)^2} G(\boldsymbol{q}) \tag{10}$$

for the effective Hamiltonian (6). The largest system accessible to our available computer resources at a wave vector cutoff $\Lambda = \pi/8$ had a linear size $L = 1096$ (thus effectively corresponding to a real space system of linear size $8 \times 1096 = 8552$), and for finite size scaling we used the sequence $L = 296, 329, 360, \ldots, 1096$ of system sizes at the same cutoff. Our choice of parameters $\kappa_\Lambda = 1.0$ and $K_\Lambda = 10.0$ was motivated by the desire to have an approximate balance between harmonic and anharmonic energy contributions in the simulation. In Fig. 1 we present the results for $G(\boldsymbol{q})$ in a double logarithmic plot. We recall that for a solid membrane one expects $G^{-1}(\boldsymbol{q}) \sim \kappa(\boldsymbol{q})q^4$, with a renormalized $\boldsymbol{q}$-dependent bending stiffness $\kappa(\boldsymbol{q}) \sim q^{-\eta}$ that shows an infra-red power law divergence with a critical exponent $\eta$ that characterizes the nontrivial critical elastic behavior of the membrane [1]. For the hexatic case at hand, however, we report that the best fit to the data was obtained for a scaling ansatz

$$\tilde{G}(\boldsymbol{q}) \sim \frac{k_B T}{\kappa_R(q)q^4}, \qquad \kappa_R(q) = \kappa_0 + \gamma \ln \frac{\delta}{q} \tag{11}$$

of logarithmic type rather than a "true" power law ($\kappa_0$, $\gamma$ and $\delta$ denote fit parameters). In other words, our present findings would thus suggest a value $\eta = 0_{\log}$. However, our precision is hampered by the finite size of the system and the lack of a concrete crossover ansatz capturing the crossover from mean field to critical behavior for $\boldsymbol{q} \to 0$. Thus, a small nonzero value of the exponent $\eta$ cannot be ruled out. As to $(\Delta f)^2$, such a logarithmic behavior is compatible with a finite size scaling ansatz of type

$$(\Delta f)^2 \sim \omega + \alpha^2 L^2(1 + \beta \log L + \rho/L)$$

with parameters $\omega, \alpha, \beta, \rho$. Indeed, a corresponding fit of our data for $(\Delta f)^2$ shown in Fig. 2 seems to support this reasoning. Unfortunately it is not straightforward to compare our results to analytical predictions for the exponent $\eta$. In Ref. [16], a field-theoretic renormalization group

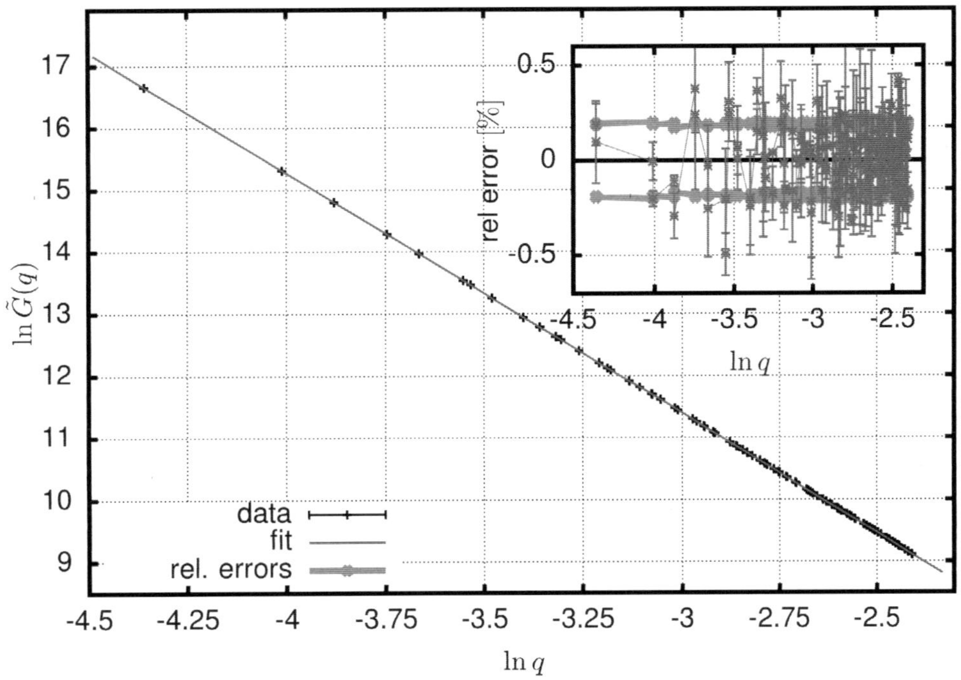


**Figure 1.** Top: fit of simulation results for $G(\boldsymbol{q})$ at linear system size $L = 1096$ and cutoff $\Lambda = \pi/8$. Error bars are smaller than symbol size. Bottom: Relative deviations of simulation data from the fit. Red lines are a guide to the eye. Gray lines indicate the relative statistical errors of the simulation data in percent.

(RG) analysis based on expansions in inverse powers of $K_\Lambda$ was worked out to first order in $1/K_\Lambda$, and a line of fixed points with a value $\eta$ continuously depending on the parameter $K_\Lambda$ is found. For the corresponding exponent $\eta$ that describes this so-called *crinkled phase*, the result

$$\eta = O(1/K_\Lambda^2) \tag{12}$$

was obtained, which obviously is compatible with our above findings. However, as recently shown by Codello and Zanusso [25] using the machinery of the exact RG, $\eta$ is not only nonzero to second order in $1/K_\Lambda$, but moreover also receives contributions from the fixed point value $\mu_\Lambda^*$ of the (bare) *surface tension* coefficient $\mu_\Lambda$ to this order. To the accuracy of the Monge parametrization, the contribution of out-of-plane fluctuations $\tilde{f}(\boldsymbol{q})$ on the surface tension is captured by the term

$$\mathcal{H}_\Lambda^{(s)} = \mu_\Lambda \int \frac{d^2q}{(2\pi)^2} q^2 |\tilde{f}(\boldsymbol{q})|^2 \tag{13}$$

and near the crinkled phase fixed point $(\kappa_\Lambda^*, \mu_\Lambda^*)$, the parameter $\mu_\Lambda$ represents a *relevant* coupling. Thus, to obtain an accuracy beyond $O(1/K_\Lambda)$ for the exponent $\eta$, it is necessary to consider the energy (13) in addition to the membrane Hamiltonian (6). Furthermore, in order to reach the crinkled phase fixed point, the bare surface tension parameter $\mu_\Lambda$ must be tuned such as to be located on the critical manifold $\mathcal{C}$ of the crinkled phase fixed point, which represents a line in the projected renormalization flow plane with coordinates $(\kappa_\Lambda, \mu_\Lambda)$. Thus, even though

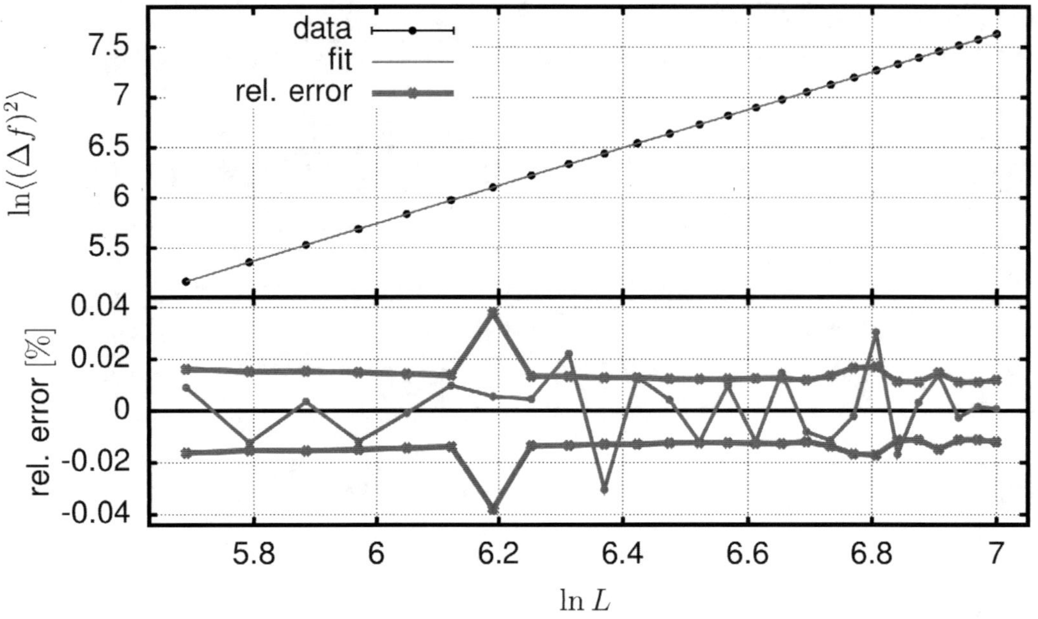


**Figure 2.** Top: fit of FSS ansatz (12) to simulation results for $\langle(\Delta f)^2\rangle$ obtained for $L = 296, 329, 360, \ldots, 1096$ and $\Lambda = \pi/8$. Error bars are smaller than symbol size. Bottom: Relative deviations of simulation data from the fit. Red lines are a guide to the eye. Gray lines indicate the relative statistical errors of the simulation data in percent.

the slope of this line in the $(\kappa_\Lambda, \mu_\Lambda)$-plane may not be steep, it may well be that our above numerical efforts, which are aimed at describing the critical behavior of the crinkled phase, have actually been contaminated from an (hopefully small) accumulated nonzero value for $\mu_\Lambda$ picked up during integrating out the implicit integration of modes in the FMC simulation.

Since the Hamiltonian (13) is also harmonic, it is principle straightforward to incorporate it into our simulations. The trouble is, however, that it is difficult to actually make sure that the system at hand was indeed critical. That is, for constant $K_\Lambda$ and a given a fixed value $\kappa_\Lambda$, one must find an accurate calculation scheme for determining the specific value $\mu_\Lambda^{(c)} = \mu_\Lambda^{(c)}(\kappa_\Lambda, K_\Lambda)$ of the bare surface tension $\mu_\Lambda$ such that the point $(\kappa_\Lambda, \mu_\Lambda^{(c)})$ is located on the critical manifold $\mathcal{C}$. For the moment, ignoring this difficulty by putting $\mu_\Lambda \equiv 0$ was all we could do. This problem is, however, not specific to our present approach. Rather than that, any simulation approach that attempts to study the crinkled phase is challenged by this difficulty. Yet, in contrast to other simulation strategies, FMC also offers a way out of this difficulty, as FMC in principle also allows to implement a momentum shell RG scheme from which the RG flow of the parameters $\kappa_\Lambda, \mu_t$ can actually be calculated [26]. Knowledge of this flow then makes it possible to assess the desired crinkled fixed point and determine its critical manifold $\mathcal{C}$. As shown in Refs. [27, 28], such a RG scheme does not merely represent a qualitative description of what is going on, but can be further optimized to even produce decent quantitative numerical results. While beyond the scope of the present paper, work in this direction is currently in progress.

## Acknowledgments
We would like to thank A. Codello, N. Hasselmann, U. Pedersen, A. Travesset and O. Zanusso for discussions and acknowledge support by the Austrian Science Fund (FWF) Project P22087-N16. Major parts of our computations were performed on the Vienna Scientific Cluster (VSC2).

## References
[1] Tröster A 2013 *Phys. Rev. B* **87**(10) 104112 URL http://link.aps.org/doi/10.1103/PhysRevB.87.104112
[2] Tröster A 2007 *Phys. Rev. B* **76** 012402
[3] Tröster A 2008 *Phys. Rev. Lett.* **100** 140602
[4] Mermin N and Wagner H 1966 *Phys. Rev. Lett.* **17** 1133
[5] Kosterlitz J M and Thouless D J 1973 *Journal of Physics C: Solid State Physics* **6** 1181
[6] Coleman S 1973 *Communications in Mathematical Physics* **31** 259–264 ISSN 0010-3616 URL http://dx.doi.org/10.1007/BF01646487
[7] Young A P 1979 *Phys. Rev. B* **19**(4) 1855–1866 URL http://link.aps.org/doi/10.1103/PhysRevB.19.1855
[8] Halperin B and Nelson D R 1978 *Phys. Rev. Lett.* **41** 121
[9] Nelson D R and Halperin B I 1979 *Phys. Rev. B* **19**(5) 2457–2484 URL http://link.aps.org/doi/10.1103/PhysRevB.19.2457
[10] Nelson D R 2002 *Defects and Geometry in Condensed Matter Physics* (Cambridge, UK: Cambridge University Press)
[11] Carraro C and Nelson D R 1993 *Phys. Rev. E* **48**(4) 3082–3090 URL http://link.aps.org/doi/10.1103/PhysRevE.48.3082
[12] Gompper G and Kroll D M 1997 *Phys. Rev. Lett.* **78**(14) 2859–2862 URL http://link.aps.org/doi/10.1103/PhysRevLett.78.2859
[13] Carroll S M 2004 *Spacetime and Geometry: An Introduction to General Relativity* (San Francisco: Addison Wesley)
[14] Frankel T 2012 *The Geometry of Physics* 3rd ed (New York: Cambridge University Press)
[15] Nelson D and Peliti L 1987 *J. Phys. (Paris)* **48** 1085
[16] David F, Guitter E and Peliti L 1987 *J. Physique* **48** 2059
[17] Nelson D, Piran T and Weinberg S (eds) 2004 *Statistical Mechanics of Membranes and Surfaces, Second Edition* (Singapore: World Scientific)
[18] Bowick M J and Travesset A 2001 *Physics Reports* **344** 255 – 308 ISSN 0370-1573 URL http://www.sciencedirect.com/science/article/pii/S0370157300001289
[19] Bowick M J and Giomi L 2009 *Advances in Physics* **58** 449–563 (*Preprint* http://www.tandfonline.com/doi/pdf/10.1080/00018730903043166) URL http://www.tandfonline.com/doi/abs/10.1080/00018730903043166
[20] Nelson D 2004 *Statistical Mechanics of Membranes and Surfaces, Second Edition* (Singapore: World Scientific) chap 6. Theory of the Crumpling Transition, p 131
[21] Tröster A and Dellago C 2007 *Ferroelectrics* **354** 225
[22] Tröster A and Dellago C 2010 *Physics Procedia* **6** 106 – 116 ISSN 1875-3892 URL http://www.sciencedirect.com/science/article/pii/S1875389210006504
[23] Tröster A 2008 *Comput. Phys. Commun.* **179** 30–33
[24] Tröster A 2010 *Phys. Rev. B* **81** 012406
[25] Codello A and Zanusso O 2013 *Phys. Rev. E* **88**(2) 022135 URL http://link.aps.org/doi/10.1103/PhysRevE.88.022135
[26] Tröster A 2009 *Phys. Rev. E* **79** 036707
[27] Tröster A 2010 *Phys. Rev. B* **81** 125135
[28] Tröster A 2011 *Comput. Phys. Commun.* **182** 1837 – 1841 ISSN 0010-4655 URL http://www.sciencedirect.com/science/article/pii/S0010465510004510

# Sync or anti-sync – dynamical pattern selection in coupled self-sustained oscillator systems

**Larissa Davidova, Szeréna Újvári and Zoltán Néda**

Babeş-Bolyai University, Department of Physics, RO-400084, Cluj-Napoca, Romania

**Abstract.** The dynamics of similar, self-sustained oscillators coupled by a common platform exhibits fascinating collective behavior. Experiments performed with pendulum clocks and metronomes reported both the absence of synchronization, in-phase synchronization, anti-phase synchronization, beat-death phenomenon, or even chaotic dynamics. Here we present a numerical study on two identical self-sustained oscillators placed on a common movable platform. As order parameter for synchronization we use the Pearson correlation coefficient between the oscillators coordinates. As a function of the relevant physical parameters of this system we reproduce all the experimentally reported dynamics. We provide conditions for obtaining stable and emergent in-phase or anti-phase synchronization.

## 1. Introduction

Spontaneous synchronization of similar units is a very common form of collective behaviour. No matter if the system is biological, electrical or mechanical, synchronization of oscillatory dynamics will emerge under certain favoring circumstances.[1]. When similar units exhibiting self-sustained periodic oscillations interact, the strength of the interaction will decide whether the system exhibits or not an emerging synchronization. The synchronized state is likely to appear if the strength of interaction exceeds a particular value. This threshold value depends on how different the oscillators are: the more the oscillators are different higher threshold values are necessary to achieve synchronization.

The most simple system that can exhibit spontaneous synchronization is formed by two coupled self-sustained oscillators. The simplest way to imagine such system is by connecting somehow two pendulum clocks as the dutch scientist Christian Huygens did in his legendary experiment. Huygens observed that two pendulum clocks suspended on a common support synchronized in anti-phase, meaning that during their dynamics they had a stable $180^0$ phase-difference. He called this phenomena "odd kind of sympathy", and carried out a series of studies to investigate the conditions which lead to the emergence of spontaneous synchronization.[2] This simple phase-locked state of the pendulum clocks, was found to be stable through time until an external force perturbed the system. However, after the perturbation ceased, in around thirty minutes, the system was able to return to this stable, anti-synchronized state. In his letters, he predicted that in-phase motion should also be a stable state of the pendulum system, but later he never mentioned to have observed such state. Strangely though, but even today we do not have a simple criteria determining when two coupled and similar self-sustained oscillators will end up in an in-phase or anti-phase synchronized state.

Recently, the problem of emerging synchronization in coupled mechanical oscillator systems entered in the focus of the scientific community again [3]. Huygens' experiment was reconsidered by several groups using modern experimental techniques. The first who re-examined Huygens two pendulum clocks experiment was Benett and his group [4], reporting somehow similar results. They found that the clocks synchronized in anti-phase when the system parameters were close to the value indicated by Huygens. Stable anti-phase motion was obtained however only for very accurately matching frequencies. When the frequency difference was slightly increased, anti-phase synchronization disappeared, and the clock's swing seemingly in an uncorrelated manner. Huygens had some luck thus with his experiments by choosing two clocks with closely matching natural frequencies. Benett and his group found also a "beat death" phenomenon, when due to the interaction one of the clocks stopped working. Panteleone used metronomes instead of pendulum clocks [5]. His experimental setup consisted of a light platform, placed on two empty soda cans. In this system he observed that for small intrinsic frequency differences, the metronomes were likely to synchronize in-phase with a small phase difference. He explained the appearance of the in-phase synchronization with the presence of large oscillation amplitudes, which disturbs the equilibrium state, for which the anti-phase synchronization would emerge. By adding large damping to the motion of the platform, seemingly the possibility of reaching anti-phase synchronization increases considerably [5]. In response to the experiments carried out by Panteleone, Ulrichs and his group, studied the system with computer simulations [6]. For modeling the dynamics of the coupled system they used the equations of motions derived by Pantaleone. They found that the critical coupling strength for large number of metronomes appears to be almost the same as for only two metronomes. In agreement with the observation of the real metronome system, they reported only in-phase synchronization, and confirmed the absence of the anti-phase synchronization. Czolczynski et. al [7] modeled an array of pendulum clocks hanging from an elastically fixed horizontal beam, and reported many types of collective behavior: in-phase synchronization of the pendulums, synchronization in clusters and anti-phase synchronization in pairs. By performing an energy balance analysis they derived conditions for different synchronization modes for the case of two [9] and many pendulum [8] system. Very recently Boda et. al. [10] performed well-controlled experiments with metronomes placed on a rotating platform, and considered also a realistic mechanical model for this system. Both the experiments and computer simulations for realistically chosen system parameters yield only in-phase synchronization.

The present work adhere to the path drawn by the preceding studies and aims to investigate theoretically the simplest coupled self-sustained oscillator system. The main question we intend to clarify concerns the conditions under which such a system will end up in in-phase or anti-phase synchronized states. The Pearson type correlation between the motion of the two oscillators will be used to quantify the nature of the achieved synchronized state, and the parameter state of the system will be thoroughly mapped. For the sake of simplicity we consider the case of two perfectly identical oscillators coupled through a common freely movable platform. Such system could be approximated either by considering two pendulum clocks hanging on a horizontally movable beam (like in the setup in [7]) or by two metronomes placed on a movable platform (like in the setup from [5, 10]).

## 2. Ideal oscillators coupled through a common platform

We consider a platform of mass $M$ with two identical mass $m$ attached to it through ideal springs with spring constant $k$. This system represents two ideal harmonic oscillators with mass

$m$ coupled by the common platform with mass $M$ (Figure 1). We denote by $x_1$ and $x_2$ the spring deformation values and $x_3$ is the absolute coordinate (relative to the chosen inertial Reference frame) of the platform. As a first study, let us assume that friction and driving are absent.

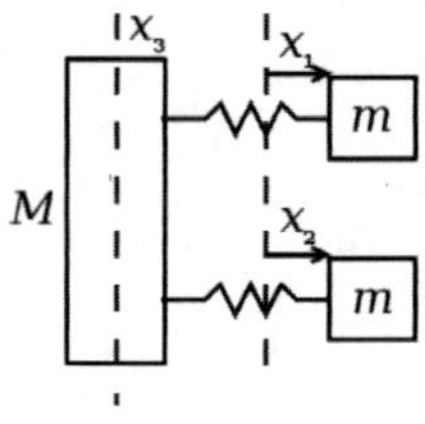


**Figure 1.** *The considered coupled oscillator system.*

The Lagrange function for this system is

$$L = \frac{1}{2}M\dot{x}_3^2 + \frac{1}{2}m(\dot{x}_1 + \dot{x}_3)^2 + \frac{1}{2}m(\dot{x}_2 + \dot{x}_3)^2 - \frac{1}{2}kx_1^2 - \frac{1}{2}kx_2^2, \tag{1}$$

where the first term is the kinetic energy of the platform, the second and third terms stands for the kinetic energy of the oscillators relative to the chosen inertial reference frame, and the last two terms are the potential energies of the oscillators. The Euler-Lagrange equations of motion are:

$$\begin{aligned} M\ddot{x}_3 &= k(x_1 + x_2) \\ m\ddot{x}_1 &= -kx_1 - m\ddot{x}_3 \\ m\ddot{x}_2 &= -kx_2 - m\ddot{x}_3 \end{aligned} \tag{2}$$

Eliminating the $\ddot{x}_3$ terms, we can derive a system of coupled differential equations yielding the dynamical evolution of the two masses $m$:

$$\begin{aligned} kx_1 + m(\tfrac{m+M}{2m+M})\ddot{x}_1 - \tfrac{m^2}{2m+M}\ddot{x}_2 = 0 \\ kx_2 + m(\tfrac{m+M}{2m+M})\ddot{x}_2 - \tfrac{m^2}{2m+M}\ddot{x}_1 = 0 \end{aligned} \tag{3}$$

This system allows for an exact analytical solution. Assuming the initial positions of the oscillators $x_1(0) = 1, x_2(0) = a$, and that they are in rest relative to the platform ($\dot{x}_1(0) = 0$, $\dot{x}_2(0) = 0$) the exact solutions for $x_1(t)$ and $x_2(t)$ are:

$$\begin{aligned} x_1(t) &= \frac{1}{2}\left((1-a)\cos\frac{\sqrt{k}t}{\sqrt{m}} + (1+a)\cos\frac{k(2m+M)t}{\sqrt{kmM(2m+M)}}\right) \\ x_2(t) &= \frac{1}{2}\left((a-1)\cos\frac{\sqrt{k}t}{\sqrt{m}} + (1+a)\cos\frac{k(2m+M)t}{\sqrt{kmM(2m+M)}}\right) \end{aligned} \tag{4}$$

The Pearson correlation coefficient will be used as a measure of the synchronization level for the two oscillators. It is important to note, that this coefficient will not distinguish between strong (phase-locked) and weak forms of synchronization. This order parameter will be denoted

Journal of Physics: Conference Series **510** (2014) 012009 doi:10.1088/1742-6596/510/1/012009

by $r$ and take values between [-1,1]. For a completely in-phase synchronized state $r = 1$ and for a completely anti-phase synchronized state $r = -1$. Mathematically it is defined as:

$$r = \frac{\langle x_1 x_2 \rangle_t - \langle x_1 \rangle_t \langle x_2 \rangle_t}{\sqrt{\langle x_1^2 \rangle_t - \langle x_1 \rangle_t^2} \sqrt{\langle x_2^2 \rangle_t - \langle x_2 \rangle_t^2}} \quad (5)$$

Here we denoted by $\langle x \rangle_t$ the time-average of quantity $x$. Taking into account that $\langle \cos(\alpha t) \rangle_t = 0$ and $\langle \cos(\alpha t) \cos(\beta t) \rangle_t = 0$ for all $\alpha \neq 0$, $\beta \neq 0$, by simple algebra one gets:

$$r = \frac{2a}{a^2 + 1} \quad (6)$$

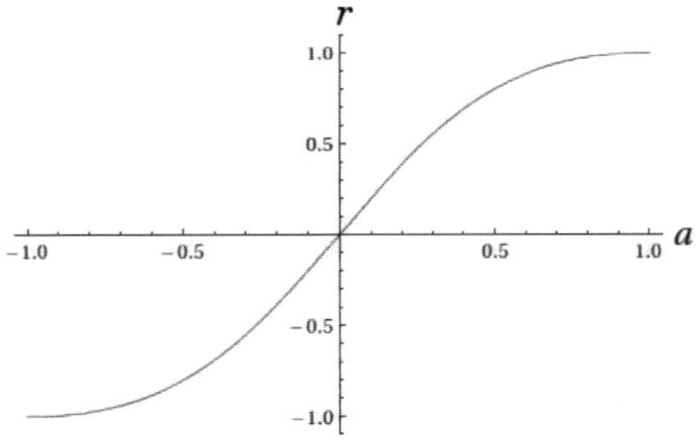


**Figure 2.** *Pearson correlation of the two oscillators coordinate as a function of the initial position of one of the oscillators ($x_1(0) = 1$, $\dot{x}_1(0) = 0$, $\dot{x}_2(0) = 0$ and $x_2(0) = a$, no friction and no driving).*

This means that in this simple system, the synchronization order parameter $R$ (measured through the Pearson correlation) of the oscillators depends only on their initial relative phases and does not depend on any other physical parameters of this system. This universal $r(a)$ curve is plotted in Figure 2. The phase portraits corresponding to different $a$ values are Lissajous curves, and in Figure 3 we illustrate them for some specific $a$ values. The obtained results suggests that complete in-phase synchronization is possible only if the oscillators start with the same initial phases. Similarly, complete anti-phase synchronization will be obtained only if the oscillators are initially in anti-phase. In all other situations only a weak form of synchronization, without phase-locking is possible.

## 3. Coupled oscillators with damping and driving

Adding damping and driving to the system makes the collective behavior more interesting. In such cases strong synchronization (through phase-locking) can be observed under more general conditions. The equations of motion can be derived from the ones given in (2). Adding friction and driving terms, the equations of motion becomes:

$$\begin{aligned} M\ddot{x}_3 &= k(x_1 + x_2) + C_1(\dot{x}_1 + \dot{x}_2) - C_0\dot{x}_3 - F_1 - F_2 \\ m\ddot{x}_1 &= -kx_1 - m\ddot{x}_3 - C_1\dot{x}_1 + F_1 \\ m\ddot{x}_2 &= -kx_2 - m\ddot{x}_3 - C_1\dot{x}_2 + F_2 \end{aligned} \quad (7)$$

$C_0$ and $C_1$ are friction coefficients for the motion of the platform and of the oscillators, respectively, and $F_1$ and $F_2$ denote driving forces applied to each of the two masses $m$. A pulse-like driving acting at the moment when $x_i = 0$ ($i \in \{1, 2\}$) in the direction of the oscillators

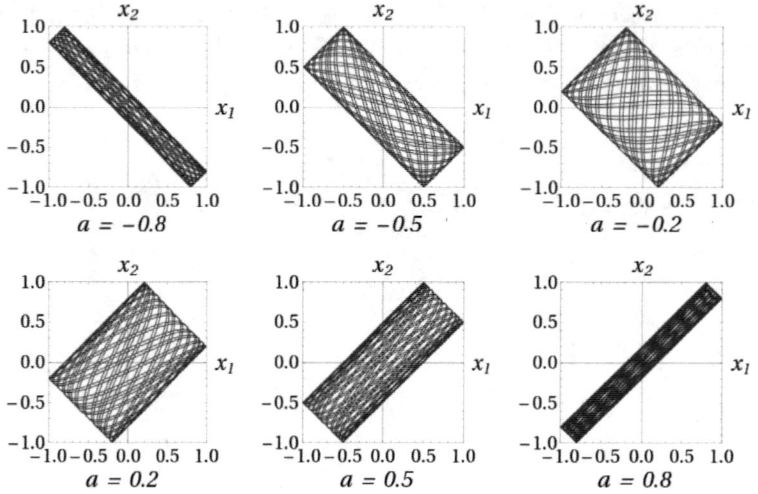


**Figure 3.** *Phase portraits $x_1(t), x_2(t)$ of two identical oscillators for various $x_2(0) = a$ values. ($x_1(0) = 1$, $\dot{x}_1(0) = 0$, $\dot{x}_2(0) = 0$, no friction and no driving).*

motion is used. This is similar with the driving experienced in pendulum clocks or metronomes. We assume thus:

$$F_i = P\dot{x}_i\delta(x_i) \tag{8}$$

$P$ is a coefficient, which characterizes the strength of the given pulses and $\delta(x)$ is the Dirac function. The term $\dot{x}_i$ is needed in order to ensure a constant momentum input, independently of the metronomes' amplitude. It also insures that the excitation is given in the good direction (direction of the motion). It is easy to see that the total momentum transferred, $M_{trans}$, to the metronomes in a half period ($T/2$) is always $P$:

$$M_{trans} = \int_t^{t+T/2} P\delta(x_i)\dot{x}_i dt = \int_{-x_{max}}^{x_{max}} P\delta(x_i)dx_i = P \tag{9}$$

The equations of motion with damping and driving cannot be solved analytically, we consider thus a numerical integration by using the Mathematica software. Since the Dirac delta function is not suitable for use with numerical methods, we approximated it with a steeply decaying exponential of the form: $F_i = P\dot{x}_i \exp(-10x_i^2)$.

For setting the units for the mass we fix $m = 1$. We also consider $P = 10$, all over the simulations. This later condition together with the fixed initial condition $x_1(0) = 1$ fixes the units for forces and distances. The spring-constant value and the friction coefficients are given through this unit definition. The dynamics of the system is initialized always as $x_1(0) = 1$, $x_2(0) = a$, $\dot{x}_1(0) = 0$ and $\dot{x}_2(0) = 0$.

The numerical integration for the equations of motions (7) exhibits a high dependence on its parameters rather than the initial position of one of the oscillators. For simulation time $t \gg T$ (where T is the natural period of the oscillators alone) the system shows either no synchronization or a stable in-phase or anti-phase synchronization, depending on parameters $M, k, C_0, C_1$.

Let us fix as a first example $C_0 = C_1 = 1$ and study the behavior of the system as a function of $M$ and $k$ alone. For $k = 1$ we illustrate some phase-portraits in the $x_1, x_2$ plane. In the case of $M = 2$ for example, the time-evolution of the system for various initial conditions (quantified by the value of $a$) is sketched in Figure 4. One will observe that regardless of the initial starting condition, the system ends up in a complete anti-phase synchronized state. As another example, in the $M = 1$ case (Figure 5) the system never synchronize. The fact that the behavior of the self-sustained oscillators shows little dependence on the initial position $a$, allows us to average the correlation coefficient across all values of $a$ ($R = \langle r \rangle_a$) and to get the general behavior of the system in the $M, k$ parameter plane. We have mapped the $R$ averaged correlation for the $M \in [0, 100]$ and $k \in [0, 100]$ interval, and the results are plotted in Figure 6.

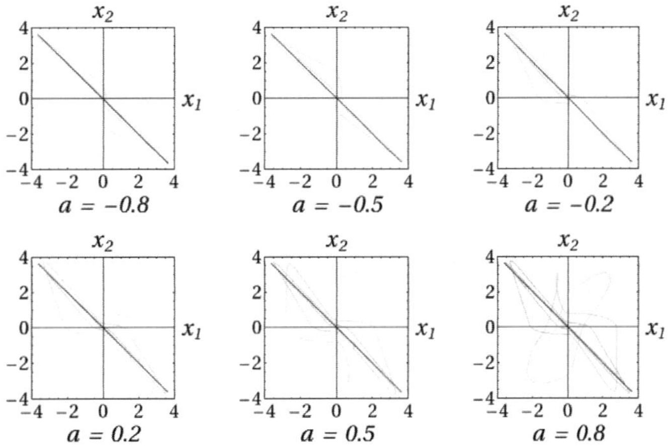


**Figure 4.** *Phase diagrams for $x_1(t)$, $x_2(t)$ with $M = 2, k = 1$ show convergence to anti-phase synchrony regardless of the initial position $a$. Darker color denotes greater simulation time. ($C_0 = C_1 = 1$)*

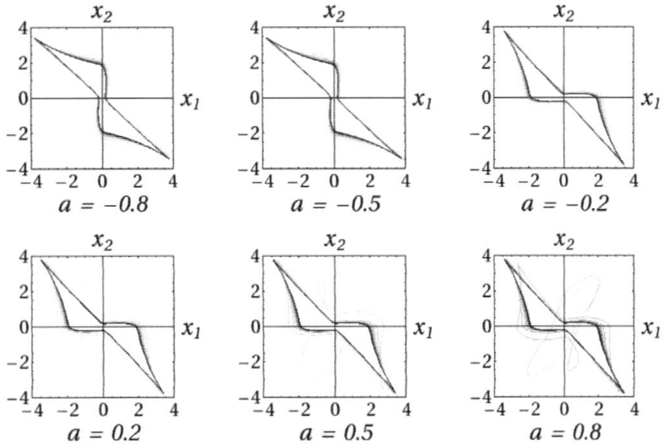


**Figure 5.** *Phase diagrams for $x_1(t)$, $x_2(t)$ for $M = 1, k = 1$ shows non synchronized state of the system regardless of the initial position $a$. Darker color denotes greater simulation time. ($C_0 = C_1 = 1$).*

For the chosen friction coefficients $C_0 = C_1 = 1$, one can observe that only the anti-phase synchronized state ($R = -1$) is stable. As a function of $M$ and $k$, either the phase-locked and complete anti-phase-synchronized state is obtained, or no synchronization is detected ($R = 0$). In the intermediate states $R \in (0, 1)$ the nature of the final stable dynamics depends on the chosen initial condition. As a function of $a$, either in-phase synchronized states, anti-phase synchronized states or no synchronization is achieved. The $R > 0$ values suggests however that anti-phase synchronization is more often obtained than in-phase synchronization.

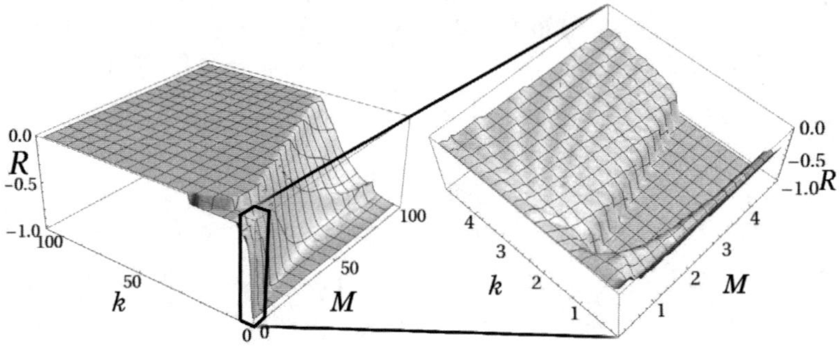


**Figure 6.** *Averaged correlation, R, as a function of M and k. ($C_0 = C_1 = 1$).*

We turn now our attention to the $C_0$ and $C_1$ friction coefficients. The value of the friction coefficient of the platform $C_0$ does not affect qualitatively the motion of the oscillators very differently than the mass $M$. By increasing either of them, the oscillation amplitude of the platform will decrease. Changing however the $C_1$ friction coefficient of the oscillators the picture becomes largely different, and one can achieve also a state of total in-phase synchrony in the system. Increasing the value of the $C_1$ damping without changing the driving force leads to much lower amplitudes of the oscillations and large numerical errors accumulates in our calculations. Meaningful results for $k = 1$ were obtained thus for friction coefficient values $C_1 < 15$. Numerical results obtained for $R$ as a function of $C_1$ and $M$ are plotted in Figure 7. This plot suggests that complete in-phase synchrony can be achieved with high values of the $C_1$ friction values. Seemingly thus the selection of the phase locked complete in-phase or anti-phase synchronized states are mainly governed by the choice of the $C_1$ parameter. For experimentally reasonable situations with $M > m$, we found that low values of $C_1$ are favoring the in-phased synchronized states and high values of $C_1$ will lead to anti-phase synchronized states. This simple picture can be nicely illustrated by considering now a multi-dimensionally averaged correlation across all $M \in [1, 50]$, $k \in [1, 50]$ and $a \in [-1, 1]$ ($R' = \langle r \rangle_{a,M,k}$) and plot it as a function of $C_1$. The results plotted on Figure 8 shows clearly that there are region of $C_1$ where the nature of the achieved synchrony (in-phase or anti-phase) is independent of any other parameters than solely the value of $C_1$.

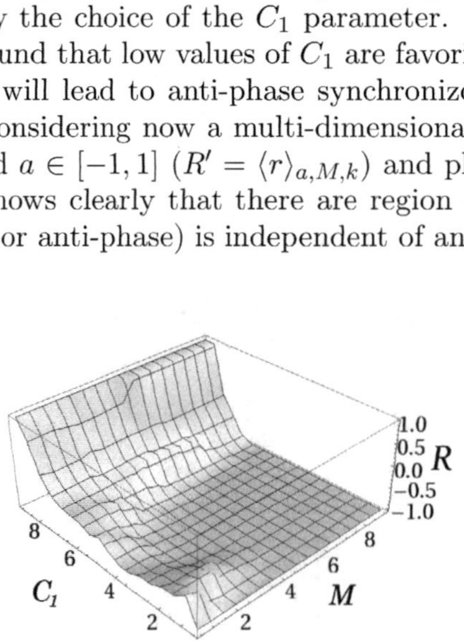


**Figure 7.** *Averaged correlation, R, as a function of M and $C_1$. ($C_0 = 1, k = 1$).*

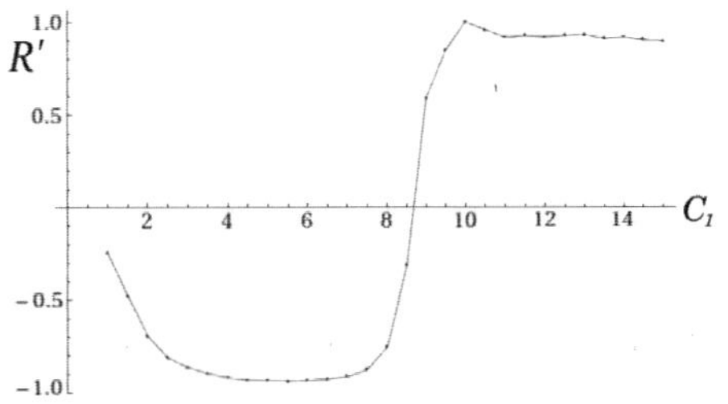


**Figure 8.** *Correlation averaged across all* $M \in [1, 50]$, $k \in [1, 50]$ *and* $a \in [-1, 1]$ *(*$R' = \langle r \rangle_{a,M,k}$*) as a function of* $C_1$*. (*$C_0 = 1$*).*

## 4. Conclusions

Spontaneous synchronization in a system composed of two identical self-sustained oscillators coupled through a common platform was investigated. Such systems can be realized experimentally using pendulum clocks suspended on a common beam or metronomes placed on a movable platform. The equations of motions were numerically integrated, and a Pearson type correlation coefficient was used for quantitatively characterizing the emerging collective behavior. Particularly, we were interested in determining the conditions under which stable phase-locked in-phase and anti-phase synchronization emerges. Simulations suggested that depending on the system's parameters, the dynamics of the two oscillators will show either a stable in-phase or anti-phase synchrony, or the system will fail to converge to any type of synchrony at all. We found that spontaneous synchronization is present in a relatively large part of the parameter space. The main parameter that will decide whether the in-phase or the anti-phase synchrony is selected is the ratio of the oscillators friction coefficient and the platforms friction coefficient ($C_1/C_0$). For $3 < C_1/C_0 < 7$ mainly the anti-phase synchrony is the stable dynamics, while for $C_1/C_0 > 10$ the in-phase synchronized collective behavior is dominant. This selection can be also understood by assuming the convergence to an energy balanced state with zero total momentum, the nature of which is determined by the friction coefficients value, as it was studied in [9]. A simple and elegant explanation for the observed dynamical pattern selection is still lacking however.

## Acknowledgments

Work supported by the Romanian Ideas research grant: PN-II-ID-PCE-2011-3-0348

## References

[1] Strogatz S H 2003 Sync: the emerging science of spontaneous order *Hyperion*
[2] Huygens C 1665 Letter to de sluse, in: Oeuveres completes de christian huygens *Societe Hollandaise Des Sciences*
[3] Kapitaniak M, Czolczynski K, Perlikowski P, Stefanski A, and Kapitaniak T 2012 *Physics Reports* **517**
[4] Bennett M, Schatz M F, Rockwood H and Wiesenfeld K 2002 *Proc. R. Soc. London* **458** 563-579
[5] Pantaleone J 2002 *Am. J. Phys.* **70** 921000
[6] Ulrichs H, Mann A and Parlitz U 2009 *Chaos* 19:043120
[7] Czolczynski K, Perlikowski P, Stefanski A, Kapitaniak T 2009 *Physica A* **388** 5013
[8] Czolczynski K, Perlikowski P, Stefanski A, Kapitaniak T 2013 *Commun Nonlinear Sci Numer Simulat* **18** 386-400
[9] Czolczynski K, Perlikowski P, Stefanski A, Kapitaniak T 2011 *Chaos* **21(2)** 023129-023129
[10] Boda Sz, Neda Z, Tyukodi B and Tunyagi A 2013 *Eur. Phys. J. B* **86** 213

Journal of Physics: Conference Series **510** (2014) 012010 doi:10.1088/1742-6596/510/1/012010

# Radiation damage thin coating of silicon carbide

**Anna Bondareva[1], Galina Zmievskaya[1] and Tatjana Levchenko[2]**
[1]*M.V. Keldysh Institute of applied mathematics, RAS, Moscow, Russia*

[2]*VNI Geosystem Russian Federal Center, Moscow, Russia*

E-mail: bal310775@yandex.ru

**Abstract.** This paper presents computer experiments studies interaction of radiation (ions Xe++ gas with keV energy) with crystal lattice of "SiC/metal" substrate. Properties of substrate are altered due to porosity and stresses formation into thin layers covering silicon carbide as a result of inert gas flux penetration. First, the stochastic model of non-equilibrium phase transitions (or nucleation of defects) in nanomaterials is presented. Behavior of gaseous bubbles in thin layers are tested and corresponding details computer modeling are discussed. Secondly, the analytical models of Gibbs free energy implanted microdefects as well as their Brownian motion in long-range potentials of indirect interaction between gaseous bubbles into lattice through acoustic phonons are concretized. These models predict stress from gas penetration as a result of damaging in form of gaseous bubbles (named here "blisters"). New emphasis is on numerical modeling includes stochastic model clustering of defects as the diffusion of Markov processes. Finally, the mechanism of porosity formation in materials such as silicon carbide can be accounted as the ways to improve solidity and fracture toughness and hardening of materials.

## 1. Introduction

Transformations in solids at ion irradiation are accompanied strongly nonequilibrium processes. Large amplitude of atoms oscillations, acoustic and nonlinear waves appear due to radiation defects formation. It significantly limits the possibilities to apply the classical theories of phase transitions. Known to help in understanding the nature of those or other transformations, as well as in their forecasting can give a solution of the kinetic equations in the partial derivatives describing phase transitions in solids. Study of radiation stimulated damaging can be carried out in computer simulations of models of bilayers, for example, Si/Mo which for metal mirrors of future X-Ray lithography [1-3]. Damaging of silicon carbide to action of radiation fluxes can be considered for prediction of materials properties for plasma engine and the thermonuclear reactor divertor as well as in case of interaction radiation fluxes with new kind semiconductors in the junction with metal nanowire (such as SiC/Mo). Modeling blistering (stretched gaseous implanted structures arising in solids) was carried out by stochastic kinetic code [1-8]. Peculiarities clustering non-point wise defects are follows: alteration of micro structure of porosity into thin layers of devices under irradiation as well as formation tension within anti-diffusion "barrier" layer into multilayer structure of mirrors. Instability phase transition in "open physical systems" (under constant flux of gas) during impulse time into bilayer SiC/Mo is discussed. New kind of self-organization defects in bilayer appears from indirect elastic interaction between defects through perturbations acoustic phonons which can be complemented by the accounting of interaction of defects with the internal interfaces. We are able to put attention on the kinetic description of defects formation (voids or blisters) structures under

Journal of Physics: Conference Series **510** (2014) 012010 doi:10.1088/1742-6596/510/1/012010

condition of many interfaces between metal layers and dielectric layers but here we study only bilayer. The processes of blistering are characterized by the kinetic distribution function /DF/ of defects which depends on Cartesian coordinates and velocities as well defects sizes. The blistering in substrate SiC/Mo is considered under $Xe^{++}$ fluxes similar to parameters of plasma engine. We put main attention on the kinetic description of defects formation (voids or blisters) structures as the heterogeneous first-order phase transition at non-equilibrium stage. It can be represented by model of vacancy-gaseous clustering. Many problems of probability nature described by statistical mechanics, kinetic theory and stochastic dynamics are also in good agreement with models governed by Fokker-Planck-Kolmogorov and Smolukhivskyi-Einstein equations. Non-steady state stochastic model of gas nucleation in lattice bilayer takes into consideration influence on growing (or degradation) sizes of defects the elastic forces in model of Gibbs free energy for cluster formation. This process of phase transition can be complemented by their Brownian motion into lattice. The interaction between the surface of bilayer and flow $Xe^{++}$ can be described by the equations mathematical physics [1-8], and by their stochastic analogues equations Ito-Stratonovich [9] in the presenting numerical models. We show the relation of the both problems in one-, two- and three-dimensional cases the inference of coefficients stochastic differential equations /SDEs/. The solution of SDEs gives evaluation of mathematical expectation of average values of cluster's sizes and defects coordinates as well their dispersions. We additionally are able calculate density of transient probabilities of stochastic processes /DTPSP/ as well find macroscopic characteristics of damaging SiC in $Xe^{++}$ fluxes, such as porosity SiC and stress of their lattice. DTPSP reflect (in form of DF) the evolution of clustering and Brownian motion. Modification of the surface layers properties is happening at the atomic level as a result of formation of spatial structures, for example, in the material thin film coating. Change of stress in thin surface layers in result of the development of porosity under ion implantation can be calculated. The nonlinearity of DF of blisters from their sizes is presented in this paper. This phenomenon consists in appearance distribution defects on sizes which is different from lognormal during initial non-equilibrium stage of phase transition (during several microseconds).

## 2. Kinetic and stochastic equations

General physical and physic-chemical problems of ion implantation and radiation physics of solid body related with physical problems of inert gases ions implantation in crystal lattice are considered in computer experiments. This models are focused on research of radiation-thermal modification of properties of materials. Kinetic equations of Kolmogorov–Feller and Einstein–Smoluchowskiy had been formulated in general. The radiation damaging are able to be described in terms of probability density of distribution defects into lattice and this approach is applied to description of initial stage of first-order phase transition. The equations in partial derivative are solved using Stochastic Simulation Method /SSM/ which is similar to molecular dynamics but different from the approach of imitation real particles of lattice and Newtonian dynamics of motion. We have used set of stochastic differential equations Ito-Stratonovich which equivalent to kinetic equations in partial derivatives. Gaseous defects (pores) appear as a result of penetration into lattice of inert gas ions of high energy (about several keV). This model is used for prediction of radiation damage (in case of cover silicon carbide). Also the same model is considered for creation of new materials using radiation stimulated structures of porosity [3]. Method of stochastic analogue [9] is based on theory according to which the kinetic equations of parabolic type are uniquely linked with stochastic differential equations /SDE/ Ito and with the density of transition probabilities of Markov random process, the solution of SDE's Ito can be interpreted as the distribution function /DF/ of the corresponding kinetic equation [1-9]. Coefficients of the kinetic equations depend on probability density defects distribution namely the thermodynamic potential of nucleation (or Gibbs free energy) and long-range potentials of indirect elastic interaction of lattice defects each with other (occurs through the perturbation of acoustic phonons lattice defects and Friedel oscillations of electrons density). Diffusion in the phase space of defects sizes {G} and diffusion in phase space {R} of crystalline lattices are accounted also as coefficients depended on DF. So, Brownian motion of radiation defects in thin layers of silicon carbide occurs under the influence

of long-range forces. Stable solution of linear SDE's Ito-Stratonovich [10] is modified on a case quasilinear equations of model as stable numerical method the second order of accuracy with infinite area of sustainability (according to the new definitions and theorems of [10]) on a regular grid of time (without the limitations on the time step).
The scheme of numerical experiment is presented on fig.1.

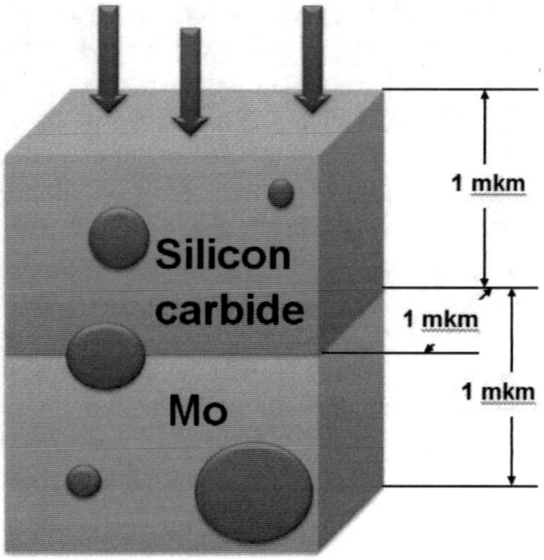


Fig.1 The scheme of numerical experiment is presented.

The kinetic Fokker-Plank-Kolmogorov equation is

$$\begin{cases} \dfrac{\partial f(X,t)}{\partial t} = -\sum\limits_{i=1}^{N} \dfrac{\partial}{\partial X_i} a_i(X,t) f(X,t) + \dfrac{1}{2}\sum\limits_{i=1}^{N}\sum\limits_{j=1}^{N} \dfrac{\partial^2}{\partial X_i \partial X_j} b_{ij}(X,t) f(X,t) \\ f(X,t)\big|_{t=0} = f_0(X) \\ \oint \left[ a_i(X,t) f(X,t) + \dfrac{1}{2}\sum\limits_{j=1}^{N} \dfrac{\partial}{\partial X_j} b_{ij} f(X,t) \right] d\Gamma = 0, \qquad i = 1,..m \end{cases}$$

Where $f(X,t)$ is probability density of a Markovian diffusion process $X(t)$. The coefficients $a(X, t)$ and $b(X, t)$ are drift and diffusion coefficients or local characteristics of the process.

The trajectories of $n_y$-dimensional diffusion process with the one-dimensional probability density satisfying the Fokker-Planck-Kolmogorov equation, can be found as solution of the initial value problem for SDEs in the Ito sense

$$X(t) = X(t_0) + \int_{t_0}^{t} \underbrace{H(\tau, X(\tau))d\tau}_{drift} + \int_{t_0}^{t} \underbrace{\sigma(\tau, X(\tau))dW(\tau)}_{stochastic\ diffusion},$$

$$X\big|_{t=0} = X_0 \quad t \in [t_0, T_{end}]$$

$dW$ is increment of stochastic Wiener process.
The association of kinetic and stochastic equation is following for one-dimensional case:

$$a_i = \underset{\Delta t \to 0}{l.i.m.} \frac{1}{\Delta t} \mathrm{E}\left[X(t+\Delta t) - X(t) \middle| X(t) = x\right]$$

$$b_{ij} = \underset{\Delta t \to 0}{l.i.m.} \frac{1}{\Delta t} \mathrm{E}\left[\left[X(t+\Delta t) - X(t)\right]^2 \middle| X(t) = x\right]$$

and

$$H(X,t) = a(X,t) - \frac{1}{2}\frac{\partial\sigma(X,t)}{\partial X}\sigma(X,t),$$

$$\left(\frac{\partial\sigma(X,t)}{\partial X}\sigma(X,t)\right)_i = \sum_{k=1}^{n}\sum_{j=1}^{n}\frac{\partial\sigma_{ij}(X,t)}{\partial y_k}\sigma_{kj}(X,t)$$

$$\sigma(E(X),t) = \sqrt{\frac{1}{q}b(E(X),t)}$$

σ in two-dimensional case is $\sigma = \begin{pmatrix} \sqrt{b_{11}} & 0 \\ \frac{b_{21}}{\sqrt{b_{11}}} & \sqrt{b_{22} - \frac{b_{21}^2}{b_{11}}} \end{pmatrix}$

in three- dimentional case

$$\sigma = \begin{pmatrix} \sqrt{b_{11}} & 0 & 0 \\ \frac{b_{21}}{\sqrt{b_{11}}} & \sqrt{b_{22} - \frac{b_{21}^2}{b_{11}}} & 0 \\ \frac{b_{31}}{\sqrt{b_{11}}} & \frac{(b_{32} - \frac{b_{31}b_{21}}{b_{11}})}{\sqrt{b_{22} - \frac{b_{21}^2}{b_{11}}}} & \sqrt{b_{33} - \frac{b_{31}^2}{b_{11}} - \frac{(b_{32} - \frac{b_{31}b_{21}}{b_{11}})^2}{b_{22} - \frac{b_{21}^2}{b_{11}}}} \end{pmatrix}$$

So kinetic equations for examined problem are presented below.

The Kolmogorov–Feller equation for blister size evolution at the point with the coordinate r is as follows:

$$\frac{\partial f_r(g,t)}{\partial t} = \frac{\partial\left[D_g(g,t)\frac{\partial f_r(g,t)}{\partial g}\right]}{\partial g} + \frac{1}{kT}\frac{\partial\left[D_g(g,t)f_r(g,t)\frac{\partial\{\Delta\Phi(g,\vec{r},t)\}}{\partial g}\right]}{\partial g} + S_\alpha, \tag{1}$$

$$f_r(g,0) = f_{0g}, \quad \left.\frac{df_r(g,t)}{dg}\right|_{g=2} = 0, \quad f_r(g,t)|_{g<2} = 0.$$

Here $S_\alpha(f_\alpha)$ is the source of particles forming a nucleus; $S(f_\alpha)$ is the source of "monomers" (vacancy or gas particles) with $f_\alpha$ - Maxwellian DF, $f_r(g, t)$ is the size of nuclei`s DF at the point r of the lattice volume; $T$ is the sample`s temperature; $D_g$ is the diffusion coefficient in the phase space defined according to nucleus sizes; $\Delta\Phi(g, r, t)$ is the thermodynamic potential of the nucleus formation, $\Delta\Phi$ is measured in $kT$, $k$ is Boltzmann constant. Total potential energy of cluster formation $\Delta\Phi$ may consist of a number of additive terms:

$$\Delta\Phi = \Delta\Phi_{\mu} + \Delta\Phi_{\sigma} + \Delta\Phi_{el} + \Delta\Phi_{b} + \Delta\Phi_{p} + \Delta\Phi_{t} \qquad (2)$$

$\Delta\Phi_{\mu}$ corresponding to difference of chemical potentials of the phases (monomers of vacancy/gas in lattice and blisters); $\Delta\Phi_{\sigma}$ taking into account the surface tension at the bubble–lattice interface and spherical form of blister; $\Delta\Phi_{el}$ which takes into consideration elastic site bonds in the lattice; the rupture of relations of the atoms in the lattice around a blister is accounted by $\Delta\Phi_{b}$, $\Delta\Phi_{r}$ is a part taking into account the difference between the nodes and internodes in the array; $\Delta\Phi_{p}$ which takes into consideration place of blister in tension fields occurring because of disparity of parameters of lattices *Mo* and *SiC*. $D_g$ is functional-coefficient of diffusion in space of cluster sizes, which depends from the frequency of oscillations of the lattice atom, lattice parameter of material, radius of injected gas atom, modulus of rigidity of lattice material and expectation of blister size in lattice point {x,y,z}.

The common form of Gibbs energy (kJoul/mole) of blister located in *SiC* on distance from *SiC/Mo* surface 2.5 lattice parameters is presented in Fig.2.

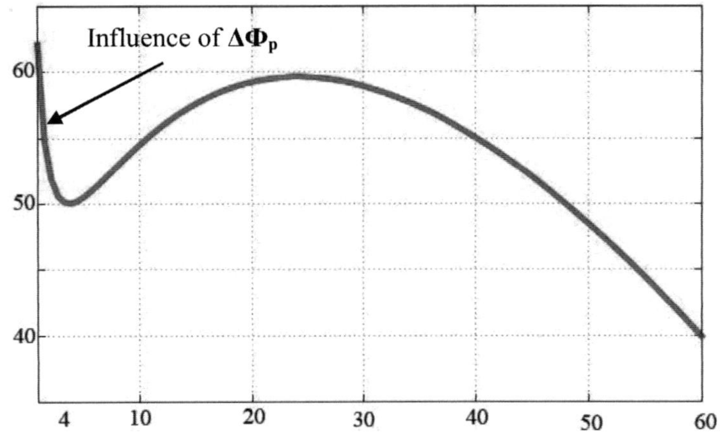


Fig. 2 The dependence of Gibbs energy (kJoul/mole) of cluster of nuclei from blister size is shown. Blister size is measured in unit volums (*Xe* atom volum). Blisters locate in 3C-SiC. The distance from *SiC/Mo* surface equals 2.5 lattice parameters of *SiC*.

The Einstein–Smoluchowski equation for BPs with the mass $M_g$ obtained from Eq. (1)

$$\frac{\partial f_g(\vec{r},t)}{\partial t} = \frac{\partial \left[ D_r(\vec{r},t) \frac{\partial f_g(\vec{r},t)}{\partial r} \right]}{\partial \vec{r}} - \frac{\partial \left[ \frac{\vec{F}(\vec{r},t)}{M_g \gamma} f_g(\vec{r},t) \right]}{\partial \vec{r}} - Q,$$

$$f_g(\vec{r},t)|_{t=0} = f_{0r},$$

$$f_g(\vec{r},t)|_{x=x_{left}} = f_g(\vec{r},t)|_{x=x_{right}}, \quad f_g(\vec{r},t)|_{y=y_{left}} = f_g(\vec{r},t)|_{y=y_{right}}, \qquad (3)$$

here $f_g(r, t)$ is the distribution function of a BP with mass $M_g$ over the substrate lattice coordinates. Blister mass $M_g$ is measured in injected atom mass. Drains of the monomers and blisters are designated in (3) as Q; their location depends on model, here it had been chosen boundary between layers and boundaries of bilayer; $\gamma$ is coefficient of friction. Projection on the z-axis of the total force acting on BP from other particles and boundaries is located in the right part of the equation (3) and calculated as $F_z = -\partial U(x, y, z)/\partial z$, where $U(x,y,z)$ is the superposition of long-range and sign-variable potentials of the indirect elastic interaction of BPs (through the perturbation of the lattice acoustic phonons and Friedel oscillations of the lattice electron density) among themselves, with layer boundaries, and so on. The superposition of potentials of the indirect elastic interaction of BPs only through the perturbation of the lattice acoustic phonons is presented on fig. 3.

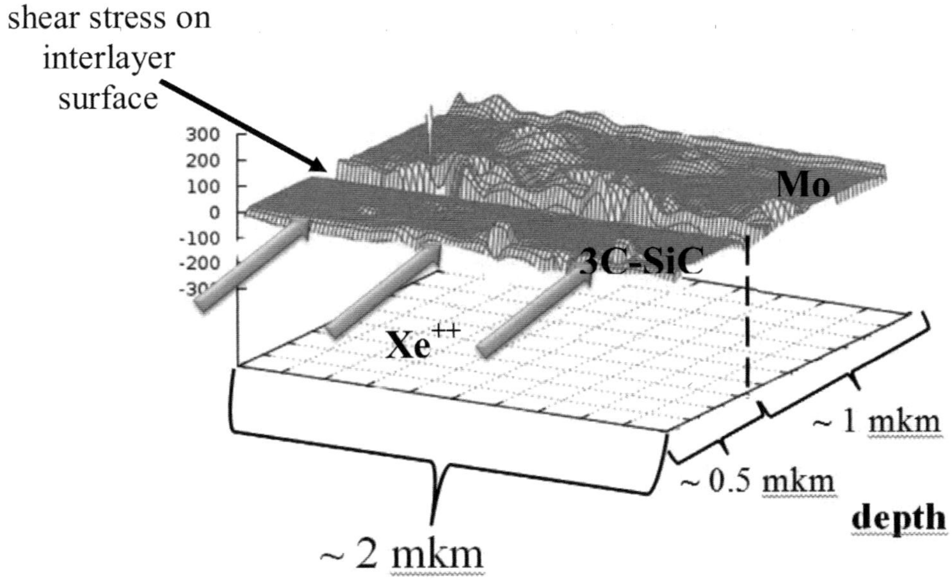


Fig. 3 The superposition of potentials of the indirect elastic interaction of BPs only through the perturbation of the lattice acoustic phonons is presented. Potential is normed on $\left(\frac{N_e V_F(2k_F)}{2\pi\varepsilon^2} + W_{elastic}\right)\frac{\Omega}{kT}$ where $\Omega$ - volume of elementary cell, $N_e$ - electron density, $\varepsilon$ - permittivity, $W_{elastic}$ ~ elastic moduli of the material, $k_F$ is Fermi momentum of electron, $V_F(2k_F)$ is the Fourier component of the potential interaction of the electron with a point defect, $k$ is Boltzman constant, $T$ is temperature.

$D_r$ is functional-coefficient of diffusion in space of crystal lattice which depending from physical parameters of problem and potential energy of cluster formation $\Delta\Phi(g,\vec{r},t)$, cluster mass and expectation of blister size in lattice point $\vec{r}$.

Numerical method of non-linear SDE solution is based on Taylor series expansion of exact solution of Ito SDE [11-13]. The generalized type Rozenbrok methods are used for the numerical solution of the SDEs in the Stratonovich form. In this work SDEs in Ito-Stratonovich form is solved using a modified method having a second-order (or higher) accuracy [10]. Every step of time functional-coefficients (diffusion in space of blister size, diffusion in lattices space, Gibbs free energy, interaction potential) of problem are iterated taking account data about DF on previous step of time.

Example of numerical scheme for calculation of blister size and its depth for $i$-trajectory of Markovian random process is following:

$$\begin{pmatrix} g \\ z \end{pmatrix}_{n+1}^{i} = \begin{pmatrix} g_{n+0.9} \\ z_n \end{pmatrix}^{i} + \left[ \hat{I} - \begin{pmatrix} \frac{h_g}{2} \frac{\partial H^i_{n+0.9}}{\partial g} \\ \frac{h_z}{2} \cdot \frac{\partial H^i_{zn}}{\partial z_i} \end{pmatrix} \right]^{-1} \begin{pmatrix} K_{(n+0.9)g} \\ K_{nz} \end{pmatrix}^{i}$$

$$K^i_{(n+0.9)g} = h_g H^i_{(n+0.9)g} + \sqrt{h}\sigma^i_{(n+0.9)g}\xi_{(n+0.9)g} + \frac{h_g}{2}\frac{\partial \sigma^i_{(n+0.9)g}}{\partial g}\sigma^i_{(n+0.9)g}\xi^2_{(n+0.9)g}$$

$$K^i_{nz} = hH^i_{zn} + \sqrt{h_z}\sigma^i_{zn}\xi_{zn} + \frac{h_z}{2}\frac{\partial \sigma^i_{nz}}{\partial z}\sigma^i_{nz}\xi^2_{nz}$$

The time step for bubble size calculation is $h_g = 0.1 h_r$, here $h_r = h_z = h_x = h_y$ is time step for calculation position in bilayer, $I$ – the identity matrix.

$$H^i_{(n+0,9)g} = -\frac{1}{kT} D^i_g\left(g^i_{n+0.9}, t_n\right)\frac{\partial \Delta\Phi^i\left(g^i_{n+0.9}, x^i_n, y^i_n, z^i_n, t_n\right)}{\partial g^i_{n+0.9}} - \frac{1}{2}\frac{\partial D^i\left(g^i_{n+0.9}, t_n\right)}{\partial g^i_{n+0.9}},$$

$$\sigma^i_{(n+0.9)g} = \frac{1}{q}\sqrt{2D^i_{n+0.9}\left(g^i_{n+0.9}, t_n\right)}.$$

$$H^i_{zn} = \frac{1}{M(g^i_{n+1})\gamma}\cdot\frac{\partial U^i(x^i_n, y^i_n, z^i_n)}{\partial z} - \frac{1}{2}\frac{\partial D^i_{zn}}{\partial z}\xi^2_{zn}, \quad \sigma^i_{zn} = \frac{1}{q}\sqrt{2D^i_{zn}}$$

$q$- noise intensity, $\alpha_1$ and $\alpha_1$ are random numbers uniformly distributed in the interval (0;1].

## 3. Results of numerical simulation

The color visualization of distribution function of blisters from its radiuses and time are presented on fig 4. The initial bubble radius in silicon carbide is 10Å approximately, temperature is 1600 K, irradiation dose is $10^{15}$ $cm^{-2}$, $Xe^{++}$ energy is 5 keV. As can see from this picture, DF is not lognormal distribution during fluctuation stage of first order phase transition. The blisters with radiuses from 3 A to 28 Å develop. The expectation of blister size in lattice point is $\langle g \rangle = \int_{g_{min}=2} g f(g,\vec{r},t)dg$. The thermodynamic critical blister size ($g_{cr}$) is calculated from $\partial\Delta\Phi/\partial g\big|_{g=g_{cr}} = 0$. Despite the fact that initial blister size is bigger that critical blister size, two branch of DF are develop (see fig. 4). The initial bubbles radiuses locate in the region of instability in sizes of first order phase transition. The domain of instability in sizes is determined by the equation $\Delta\Phi\big|_{g_{min}} = \Delta\Phi\big|_{g_{max}} = \Delta\Phi\big|_{g_{cr}} - kT$ Blisters with radius from 3 Å to 6 Å form one branch of DF. Big blisters form second branch of DF. Number of trajectories is $10^6$. The time step for calculation is $5\cdot 10^{-7}$ c, number of time step is $10^3$, finish time is 0.5 mc. In future, so different typical sizes of nucleus will lead to the formation of large and small blisters which are observed in microscope.

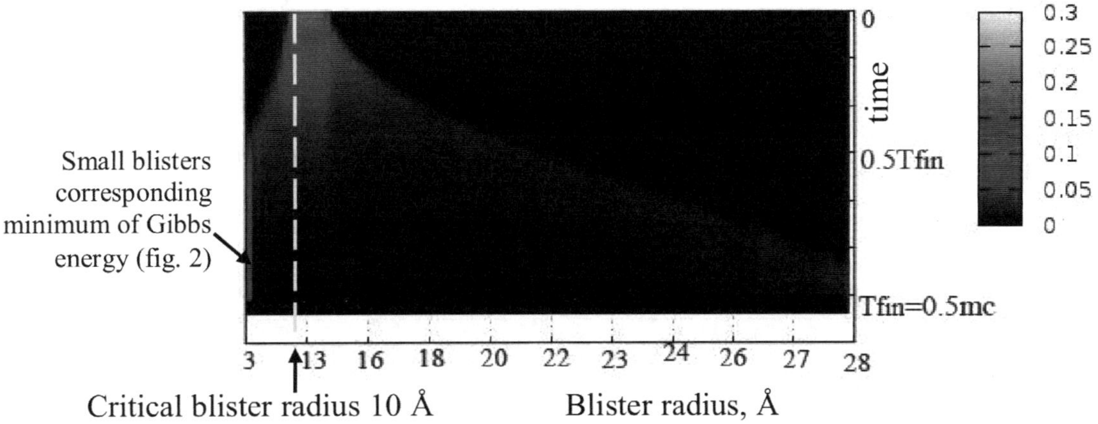


Fig. 4 The color visualization of distribution function of blisters from its radiuses and time are presented

The important characteristic is the stress resulting from the ions implantation of poorly soluble gases into bilayer. The stress resulting from blistering can influence on efficiency of devices under plasma irradiation (for example, surfaces of the plasma engines and nano-optical devices). Stress corresponding vacancy-gas pores formation and mismatch lattices layers is presented on fig. 5. Temperature is 1530 K, dose is $10^{15}$ $cm^{-2}$.

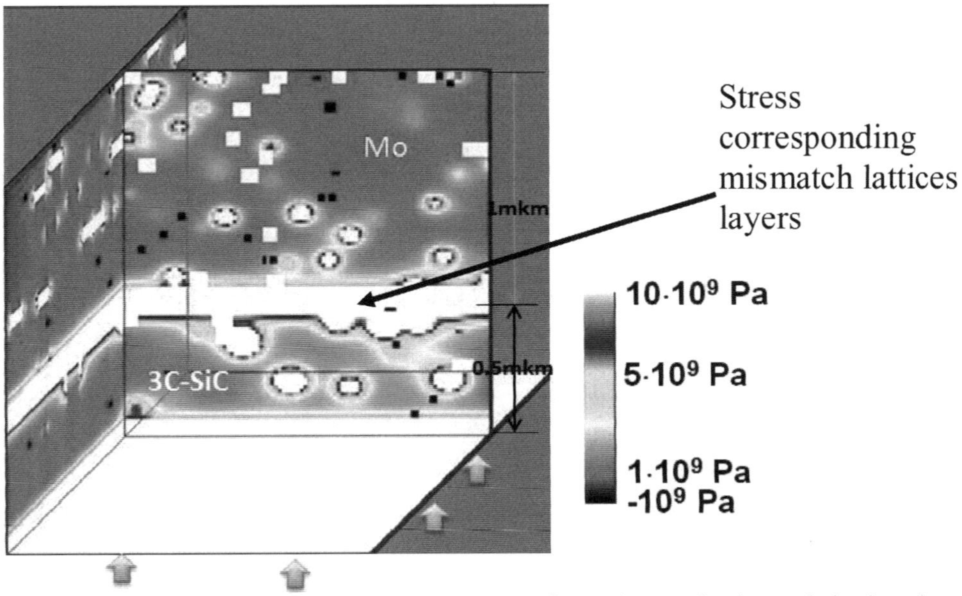


Fig. 5 Stress corresponding vacancy-gas pores formation and mismatch lattices layers is presented. Scale corresponding color and value is presented on right part of picture. The sign «-» corresponds to tensile stresses lattice which due to the vacancy clusters defects, « + » sign by squeeze (stress of mismatch layers and stress due to the growth of gas defects).

Porous silicon carbide is recognized promising materials for nanooptics now. Therefore, the formation and development of porosity in a layer of silicon carbide are examined in this article. The porosity of bilayer changes during fluctuation stage. First, single blisters and vacancies form and develop/ degrade. Next, blisters and vacancies clusters form long-sized structures of defects (nuclei cracks), after cracks develop. These crake can have output on the irradiated surface and can be fill in others nano-materials for obtaining the required optical properties. The development of long-sized structures of defects for several time moments is presented on fig. 6.

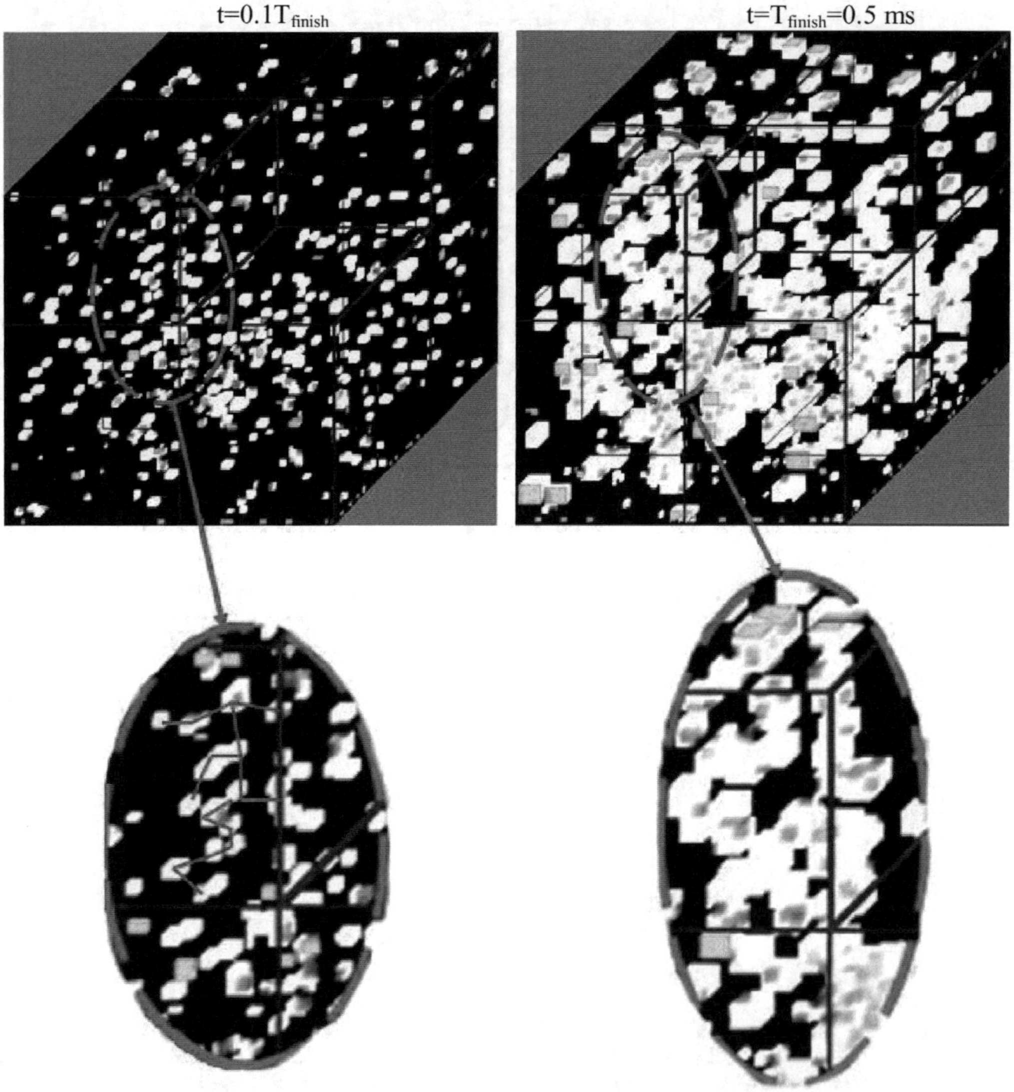


Fig.6 The development of long-sized structures of defects is presented. Red line on left callout corresponds possible directions of development of long-sized structure. The realization of this script of crack formation is presented on right callout.

The correspondence between porosity due to the development of vacancy-gas defects and stresses in the sample is shown from the comparison of fig. 5 and 6. Stress corresponding blisters formation is comparable to stress corresponding mismatch lattices layers during fluctuation stage of blistering. So, this stress must be taken into account for creation of disordered porous medium as a basis for the one-dimensional photonic crystal [14]. It should be note that porosity of layer of silicon carbide increases at 19 times during fluctuation stage, while the average stress in this layer changes only 4 times.

## 4. Conclusions

Calculations have shown, that the porosity formation in a layer of silicon carbide depends on its thickness, doses, temperature and degree of discrepancy of parameters lattices layer of the coating and the substrate. Elastic stress from defects in the layers of mkm- thickness can reach value of stress corresponding discrepancy between lattice parameters of layers «coating-substrate» during of initial stage of nucleation (ms). Study of the mechanisms of phase transition which are non-equilibrium at short time (about 1 mc) carry out by SSM. Calculations are important in fusion reactor materials science, electrical propulsion engines of the spacecraft, and also in the creation of porous semiconducting and dielectric materials [14].

The work is partially supported by the program of RAS 3.5 and grant RFBR № 14-01-00787, № 12-01-00490, 12-01-00708). Authors are thankful to V.D. Levchenko, A.V. Ivanov and S.A. Khilkov for fruitful discussions.

## 5. References

[1] Bondareva A.L., Zmievskaya G.I. 2010 Computer simulation of blistering in multylayer mirrors for EUV lithography *Journal of Surface Investigation. X-ray, Synchrotron and Neutron Techniques* vol. 4 No 3 pp 480-487

[2] Bondareva A.L., Levchenko T.V., Zmievskaya G.I. 2010 Computer simulation model for first-order phase transition fluctuation stage *Defect and Diffusion Forum* vol. 297–301: *Diffusion in Solids and Liquids V* (Switzerland: Trans Tech Publications) pp 502-507.

[3] Zmievskaya G.I., Bondareva A.L 2012 Numerical simulation of porosity development kinetics into multy-layer media *Selected scientific works of the M.V. Keldysh Institute of applied mathematics of Russian Academy of Sciences "Physical and mathematical models of plasma and plasma-like media"* Edited by Giuseppe Maino and Galina I. Zmievskaya pp 16-30 (rus) (ISBN 978-5-98354-009-5 )

[4] Zmievskaya G.I., Bondareva A.L., Levchenko V.D., Levchenko T.V. 2007 A kinetic stochastic model of blistering and nanofilm islands deposition: self organization problem *Journ. of Phys.* ***D**: Appl.Phys.* vol. 40 pp 4842–4849.

[5] Bondareva A.L., Zmievskaya G.I., Levchenko V.D. 2006 Stochastic model applied to plasma-surface interaction's simulation *Eur.Phys.J.D* vol. 38, pp 143-149

[6] Bondareva A.L., Zmievskaya G.I. 2005 Stochastic simulation of fluctuation stage of thin films formation *Doklady Akademii Nauk.* vol. 401. № 4. (Moscow: "Nauka") p. 471-475

[7] Zmievskaya G.I., Bondareva A.L. 2011 Crystalline islands of semiconductors films *Plasma Physics Reports* vol. 37, No. 1, pp. 87–95.

[8] Zmievskaya G.I., Bondareva A.L 2010 Thin film semiconductors islands and numerical experiment *Journal of Surface Investigation. X-ray, Synchrotron and Neutron Techniques* vol.4 No 5 pp 836-844

[9] Zmievskaya G.I. 1997 *Plasma Physics Reports* vol. 23 No. 4 p 45.

[10] Artem'ev S.S., Averina T.A. 1997 Numerical analysis of systems of ordinary and stochastic differential equations (The Netherlands: Utrecht) pp. 176.

[11] Kloeden P.E., Platen E. 1992 Numerical Solution of Stochastic Differential Equations (New York: Springer) p. 36.

[12] Kloeden P.E., Platen E., Schurz, H. 1994 Numerical Solution of SDE Through Computer Experiments (Berlin: Springer) p. 292.

[13] Pugachev V., Sinitsin I. 1987 Stochastic Differential Systems: Analysis and Filtering (New York: Wiley)

[14] Diederik S. Wiersma 2013 Disordered photonics *Nature Photonics* vol. 7. pp 188–196

Journal of Physics: Conference Series **510** (2014) 012011 doi:10.1088/1742-6596/510/1/012011

# Numerical code for multi-component galaxies: from $N$-body to chemistry and magnetic fields

**S.A. Khoperskov [1,2], E.O. Vasiliev[3], A.V. Khoperskov[4], V.N. Lubimov[1]**

Institute of Astronomy of the Russian Academy of Sciences, Pyatnitskaya st., 48, 119017, Moscow, Russia[1]
Sternberg Astronomical Institute, Moscow M.V. Lomonosov State University, Universitetskij pr., 13, Moscow, 119992, Russia[2]
Institute of Physics, Department of Physics, Southern Federal University, Stachki Ave. 194, 344090 Rostov-on-Don, Russia[3]
Volgograd State University, Universitetsky pr., 100, Volgograd 400062, Russia[4]

E-mail: `khoperskov@inasan.ru`

**Abstract.** We present a numerical code for multi-component simulation of the galactic evolution. Our code includes the following parts: $N$-body is used to evolve dark matter, stellar dynamics and dust grains, gas dynamics is based on TVD-MUSCL scheme with the extra modules for thermal processes, star formation, magnetic fields, chemical kinetics and multi-species advection. We describe our code in brief, but we give more details for the magneto-gas dynamics. We present several tests for our code and show that our code have passed the tests with a reasonable accuracy. Our code is parallelized using the MPI library. We apply our code to study the large scale dynamics of galactic discs.

## 1. Introduction

An explosive growth of observational data on dynamics and chemical composition of various components (subsystems) of galaxies is seen in last decade. To interpret the huge data cubes we need reasonable models included more physical processes like magnetic fields, star formation, feedback, chemical kinetics, dust grain dynamics and so on. Spatial scales of the processes in a galaxy vary from sub-parsecs for turbulent structures in the warm interstellar medium to several kiloparsec size for galactic disc and winds. Formation of molecules and distribution of dust grains strongly affect on the efficiency of star formation, which, in turn, changes the thermal state of a gas through stellar winds and supernova explosions [1]. Then, to study the interaction between various components numerically someone needs a code with a lot of physical modules.

In recent years several open source codes were developed for astrophysical applications, such as ZEUS[1], RAMSES[2], GADGET[3], Athena[4], FLASH[5] and many others. These codes are general-purpose ones, so that each of them has own advantages and disadvantages in the application

[1] http://www.astro.princeton.edu/ jstone/zeus.html
[2] http://irfu.cea.fr/Phocea/Vie_des_labos/Ast/ast_sstechnique.php?id_ast=904
[3] http://www.mpa-garching.mpg.de/gadget/
[4] https://trac.princeton.edu/Athena/
[5] http://flash.uchicago.edu/site/

to the galactic evolution [2, 3, 4, 5]. However galaxies are multi-component systems, and its simulation requires a robust model included special physical processes. So the main goal of this project is to develop the numerical code for simulation of the evolution of disc galaxies including generation of spiral structure, physics of interstellar medium, formation of clouds and stars [6, 7].

This paper is organized as follows. In Section 2 the main numerical methods are described. Section 3 presents the results of tests. Section 4 describes the application of our code to the evolution of stellar-gaseous galactic disc. In Section 5 we summarize our results.

## 2. Description of the code

### *2.1. Gas dynamics*

The gas dynamics equations for an ideal magnetized gas can be written in the conserved variables **U** in Cartesian coordinates

$$\mathbf{U} = [\rho,\ \rho v_x,\ \rho v_y,\ \rho v_z,\ E,\ B_x,\ B_y,\ B_z]^T . \tag{1}$$

Then, the conservation laws can be written in a compact form

$$\frac{\partial \mathbf{U}}{\partial t} + \frac{\partial \mathbf{F}}{\partial x} + \frac{\partial \mathbf{G}}{\partial y} + \frac{\partial \mathbf{H}}{\partial z} = \mathbf{S}, \tag{2}$$

where **F**, **G**, and **H** are vectors of fluxes in $x$, $y$, and $z$-directions, respectively and **S** is the source vector. The components of these vectors are

$$\mathbf{F} = \begin{bmatrix} \rho v_x \\ \rho v_x^2 + P^* - B_x^2 \\ \rho v_x v_y - B_x B_y \\ \rho v_x v_z - B_x B_z \\ (E+P^*)v_x - (\mathbf{B}\cdot\mathbf{v})B_x \\ 0 \\ B_y v_x - B_x v_y \\ B_z v_x - B_x v_z \end{bmatrix} \mathbf{G} = \begin{bmatrix} \rho v_y \\ \rho v_y v_x - B_y B_x \\ \rho v_y^2 + P^* - B_y^2 \\ \rho v_y v_z - B_y B_z \\ (E+P^*)v_y - (\mathbf{B}\cdot\mathbf{v})B_y \\ B_x v_y - B_y v_x \\ 0 \\ B_z v_y - B_y v_z \end{bmatrix} \mathbf{H} = \begin{bmatrix} \rho v_z \\ \rho v_z v_x - B_z B_x \\ \rho v_z v_y - B_z B_y \\ \rho v_z^2 + P^* - B_z^2 \\ (E+P^*)v_z - (\mathbf{B}\cdot\mathbf{v})B_z \\ B_x v_z - B_z v_x \\ B_y v_z - B_z v_y \\ 0 \end{bmatrix} \tag{3}$$

where $P^* = P + \dfrac{B^2}{2}$ and total gas energy $E = \dfrac{P}{\gamma - 1} + \dfrac{1}{2}\rho v^2 + \dfrac{B^2}{2}$. For pure gas dynamics we set $\mathbf{B} \equiv 0$.

In multi-dimensional magneto-gas dynamics to obtain zero-value for the magnetic field divergence we apply the Constrained Transport technique for magnetic field transport through computational domain [8, 9]. In this approach magnetic field strength is defined at faces of a cell, while other gas dynamical variables are defined at the center of a cell.

We solve the equations (2) in Cartesian coordinates $(x, y, z)$. The center of each cell is located at the position $(x_i,\ y_j,\ z_k)$. For each cell the faces being normal to the $x$ direction have coordinates $x_{i\pm1/2}$ and sizes $\delta x$, $\delta y$ and $\delta z$. The gas dynamic variables (density, momentum, energy) are volume-averaged and defined at the center of a cell. For example, the density value can be written as follow

$$\rho_{i,j,k} = \frac{1}{\delta x \delta y \delta z} \int_{x_{i-1/2}}^{x_{i+1/2}} \int_{y_{j-1/2}}^{y_{j+1/2}} \int_{z_{k-1/2}}^{z_{k+1/2}} \rho(x', y', z') dy' dz' . \tag{4}$$

While the magnetic field components are surface-averaged and found at faces of a cell:

$$B_{x,i-1/2,j,k} = \frac{1}{\delta y \delta z} \int_{y_{j-1/2}}^{y_{j+1/2}} \int_{z_{k-1/2}}^{z_{k+1/2}} B_x(x_{i-1/2}, y', z') dy' dz' . \tag{5}$$

To calculate the momentum and energy fluxes we apply the second-order cell-centered averages for magnetic field, e.g. for $B_x$:

$$B_{x,i,j,k} = \frac{1}{2}(B_{x,i+1/2,j,k} + B_{x,i-1/2,j,k}) \,. \tag{6}$$

We use a finite-volume discretization for gas dynamics equations and finite-area discretization for magnetic field transport equations. In this case gas-dynamical variables at time $t + \delta t$ can be found as follow

$$\mathbf{U}^{n+1}_{i,j,k} = \mathbf{U}^{n}_{i,j,k} - \frac{\delta t}{\delta x}\left[\mathbf{F}^{n+1/2}_{i+1/2,j,k} - \mathbf{F}^{n+1/2}_{i-1/2,j,k}\right] - \frac{\delta t}{\delta y}\left[\mathbf{G}^{n+1/2}_{i,j+1/2,k} - \mathbf{G}^{n+1/2}_{i,j-1/2,k}\right] - \frac{\delta t}{\delta z}\left[\mathbf{H}^{n+1/2}_{i,j,k+1/2} - \mathbf{H}^{n+1/2}_{i,j,k-1/2}\right], \tag{7}$$

where the vectors $\mathbf{F}^{n+1/2}_{i+1/2,j,k}$, $\mathbf{G}^{n+1/2}_{i,j+1/2,k}$, $\mathbf{H}^{n+1/2}_{i,j,k+1/2}$ are the time- and area-averaged fluxes through the $x$, $y$ and $z$-faces of a cell, correspondingly.

Using the Stokes' law for the magnetic field transport equation we obtain time-evolution of magnetic field equation:

$$B^{n+1}_{x,i-1/2,j,k} = B^{n}_{x,i-1/2,j,k} - \frac{\delta t}{\delta y}\left[\mathcal{E}^{n+1/2}_{z,i-1/2,j+1/2,k} - \mathcal{E}^{n+1/2}_{z,i-1/2,j-1/2,k}\right] + \frac{\delta t}{\delta z}\left[\mathcal{E}^{n+1/2}_{z,i-1/2,j,k+1/2} - \mathcal{E}^{n+1/2}_{z,i-1/2,j,k-1/2}\right], \tag{8}$$

where $\mathcal{E}_x$, $\mathcal{E}_y$ and $\mathcal{E}_z$ are the components of electric field (or electro-magnetic force):

$$\mathcal{E} = -\mathbf{v} \times \mathbf{B} \,. \tag{9}$$

The approximation of the electro-magnetic force can be written as follow:

$$\mathcal{E}_{x,j-1/2,k-1/2} = \frac{1}{4}\left[\mathcal{E}_{x,j-1/2,k} + \mathcal{E}_{x,j-1/2,k+1} + \mathcal{E}_{x,j,k-1/2} + \mathcal{E}_{x,j+1,k-1/2}\right] +$$
$$+\frac{\delta y}{8}\left[\left(\frac{\partial \mathcal{E}_x}{\partial y}\right)_{j-1/4,k-1/2} - \left(\frac{\partial \mathcal{E}_x}{\partial y}\right)_{j-3/4,k-1/2}\right] + \frac{\delta z}{8}\left[\left(\frac{\partial \mathcal{E}_x}{\partial z}\right)_{j-1/2,k-1/4} - \left(\frac{\partial \mathcal{E}_x}{\partial z}\right)_{j-1/2,k-3/4}\right], \tag{10}$$

where the derivative of $\mathcal{E}_x$ for each grid cell face is computed by selecting the upwind direction according to the contact mode, namely,

$$\left(\frac{\partial \mathcal{E}_x}{\partial y}\right)_{k-1/2} = \begin{cases} (\partial \mathcal{E}_x/\partial y)_{k-1} & \text{if } v_{z,k-1/2} > 0 \\ (\partial \mathcal{E}_x/\partial y)_{k} & \text{if } v_{z,k-1/2} < 0 \\ \frac{1}{2}\left[\left(\frac{\partial \mathcal{E}_x}{\partial y}\right)_{k-1} + \left(\frac{\partial \mathcal{E}_x}{\partial y}\right)_{k}\right] & \text{otherwise.} \end{cases} \tag{11}$$

Note that in 3D the expressions similar to the above are required to convert the $x$- and $y$-components of the electric field to the appropriate cell corners. These expressions might be obtained directly cyclic permutation of the $(x, y, z)$ and $(i, j, k)$.

Godunov-type methods are widely used for solving of hyperbolic equations. An idea is to use the Riemann solution for the decay of an arbitrary discontinuity. In this case it is assumed that the solution might be non-smooth and discontinuities might be in every computational cell. That is why, this kind of numerical schemes is a universal tool for simulation of gasdynamical problems of evolution shocks and contact discontinuities. The main problem of Godynov-type methods is a necessity of solving system of nonlinear equations. Usually this leads to iterative procedure, which requires additional computational resources. To resolve this issue there are a lot of approximate solvers of Riemann problem. Our code includes two types of solvers for pure gas dynamics: Harten-Lax-van Leer (HLL) or Harten-Lax-van Leer-Contact (HLLC), and one for magneto-gas dynamics, namely, Harten-Lax-van Leer-Discontinuities (HLLD). These methods are described by [10] in detail.

An exact or approximate Riemann solver requires values for the conserved variables at left and right interfaces of each cell: $U_L = U_i - 1/2$, $U_R = U_i - 1/2$. These values can be reconstructed from cell centered values with second-order or third-order accuracy, on a choice. The interpolation of the values is carried out for primitive variables $w \equiv \{\rho, P, \mathbf{v}, \mathbf{B}\}$ as follow

$$dw_i = w_{i+1} - w_i\,, \tag{12}$$

$$dW_i = K_2 \text{minmod}(dw_{i-1}, dw_i) + K_1 \text{minmod}(dw_i, b dw_{i-1})\,, \tag{13}$$

$$dW_{i+1} = K_2 \text{minmod}(dw_{i+1}, dw_i) + K_1 \text{minmod}(dw_i, b dw_{i+1})\,, \tag{14}$$

and then $w_L = w_i + dW_i$ $w_R = w_{i+1} - dW_{i+1}$, where $K_1 = \dfrac{1}{4}(1-\kappa)$, $K_2 = \dfrac{1}{4}(1+\kappa)$, $b = (3-\kappa)/(1-\kappa)$ and minmod is a limiter function [11]. The parameter $\kappa$ can be equal to $-1$, 0, or 1/3, which represent second-order fully upwind, second-order upwind biased, and third-order upwind biased cases, respectively. Below we use $\kappa = 1/3$.

### 2.2. *N-body dynamics*

We use the "particle-in-cell" method for the dynamics of both dark matter halo ("live" halo) and stellar discs. To obtain the second-order accuracy on time we apply the "flip-flop" integrator. The main advantage of this approach is that it requires only one calculation of forces per time step for each particle. More details about our $N$-body solver for galaxy dynamics can be found in [12].

### 2.3. *Poisson solver*

To take into account self-gravity of a gas we solve the Poisson equation with the gas dynamics conservation laws (2):

$$\Delta\Psi(x, y, z) = 4\pi G\rho(x, y, z)\,. \tag{15}$$

In our code the Poisson equation can be solved using three different methods: FFT-based, TreeCode [13] and over-relaxation Gauss-Seidel method. To calculate self-gravity in $N$-body/gasdynamical problems we use regular Cartesian grid. For interpolation of particle density to a mesh we use the second order finite-volume interpolation, which conserves the total mass of the system with a good accuracy.

## 3. Test problems

Below we present several tests for our gasdynamical code in 1D, 2D and 3D.

### 3.1. *The Sod shock tube test*

This is the most simple test demonstrated the formation of shock wave, contact discontinuity and rarefaction wave. The initial distribution of gasdynamical values is taken as usual: $\rho, v_x, P = \{1, 0, 1\}$ at $x < 0$; $\rho, v_x, P = \{0.125, 0, 0.1\}$ at $x > 0$. Initially the velocity is equal to zero. Figure 1 presents the result of numerical simulation for different resolution and the exact solution. One can see a good coincidence of the numerical solution with the exact one even for low resolution.

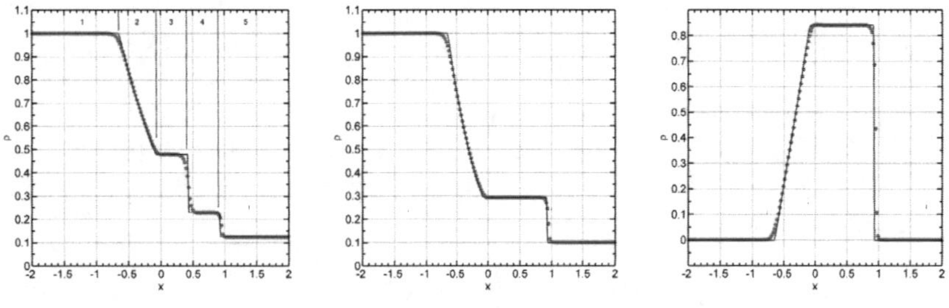

**Figure 1.** The Sod shock tube. The exact solution is shown by black line, the other lines represent the distribution at $t = 0.5$ with grid resolution $N = 200$ (green), $N = 500$ (red), $N = 1500$ (blue). One can find five regions in the solution: the regions 1 and 5 correspond to the initial unperturbed state of a gas, the rarefaction wave is found in the region 2, the regions 3 and 4 are separated by contact discontinuity, and the shock wave is located between regions 4 and 5.

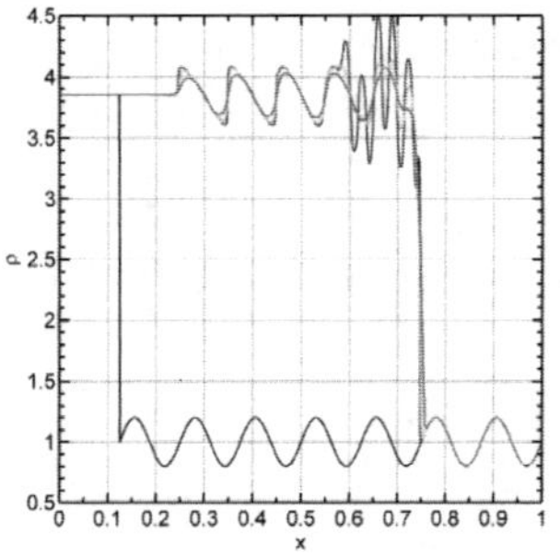

**Figure 2.** The Shu-Osher test. The initial density distribution is shown by black line, the other lines represent the distribution at $t = 0.18$ with grid resolution $N = 200$ (green), $N = 500$ (red), $N = 1500$ (blue).

### *3.2. The Shu-Osher problem*

The Shu-Osher problem tests a shock-capturing scheme's ability to resolve small-scale flow features. It shows the numerical viscosity of the method. The initial distribution is: $\rho = 3.857143$, $p = 10.33333$, $u = 2.629369$ at $x < 0.125$ and $\rho = 1 + 0.2\sin(16\pi x)$, $p = 1$, $u = 0$ otherwise, $\gamma$ equals 1.4.

Figure 2 shows the initial density distribution (black line). We study the dependence on spatial resolution: the case for $N = 200$ is exactly smooth, whereas oscillations appear for $N = 500$, for further increase of resolution ($N = 1500$) our result becomes closer to the exact solution [14].

### *3.3. The bow shock simulation*

A bow shock usually forms due to supersonic motion of star (or planet) through interstellar medium [15] or a galaxy through the intercluster medium [16], so that this is one of the most important astrophysical problem. To test our code we simulate a supersonic gas flow through potential well. The initial distribution of the parameters are homogeneous: $\rho = 1.0$, $P = 1.0$, $v_x = 1.0$, and $\gamma$ equals 1.4. The external potential well is set in the form $\Psi(x) = -\Psi_0 \exp(-(x/x_0)^2)$, where $\Psi_0 = 5$ and $x_0 = 0.1$. The boundary conditions are free at the right and the steady inflow at the left.

At $t = 0.051$ the shock wave is formed on the far edge of the well (towards to the flow) due to supersonic falling of the gas to the potential well (red line). Note that this configuration is unstable. So the shock wave moves upstream through the potential well and after crossing the minimum of the potential it becomes steady-state at front edge of the potential well (blue line).

### *3.4. The Brio-Wu problem*

The Brio-Wu test is one dimensional magneto-gas dynamics problem. The solution of this test consists of the fast rarefaction wave, that moves to the left, the intermediate shock wave, the slow rarefaction wave, the contact discontinuity, slow shock wave and another fast rarefaction wave, that moves to the right (see figure 4). The intermediate shock and slow rarefaction waves

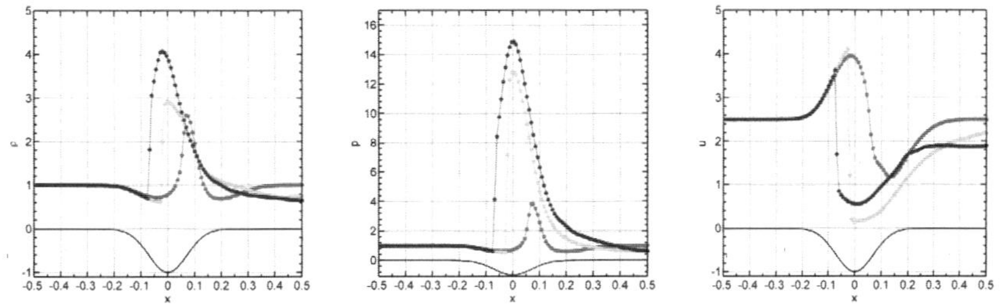

**Figure 3.** The bow shock test. The density distribution is at $t = 0.051$ (red), 0.226 (green) and 0.375 (blue). The potential well $\Psi(x)/\Psi_0$ is shown by black solid line.

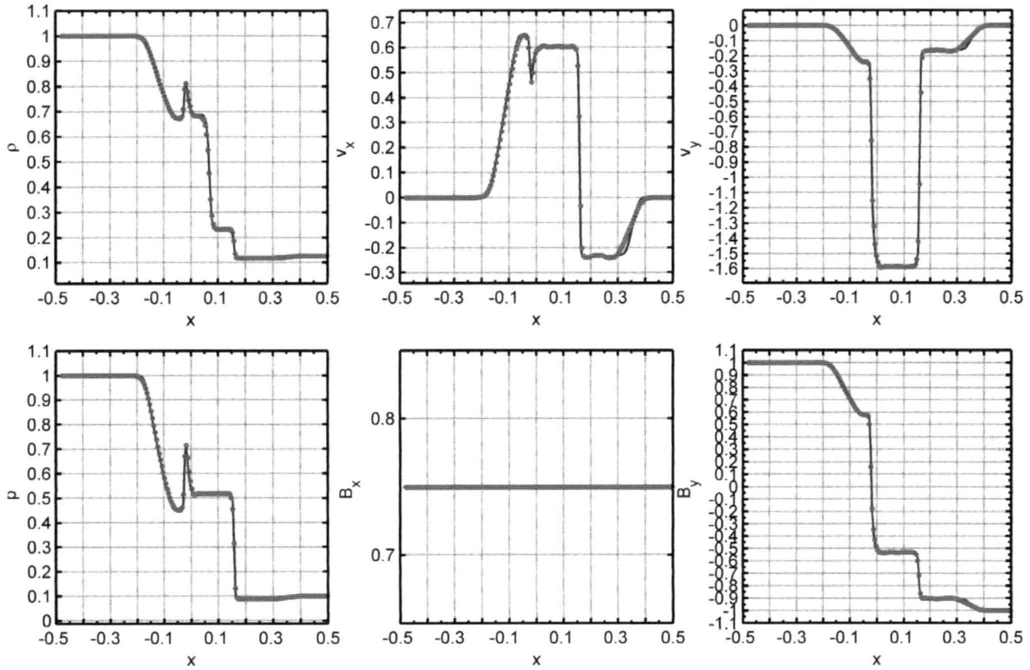

**Figure 4.** The Brio-Wu problem. Our result is depicted by red dots, the result obtained in the Athena code is shown by blue line.

form a structure called the compound wave [17]. The initial distribution for this test is: $\rho = 1$, $P = 1$, $B_y = 1$ at $x < 0$, $\rho = 0.125$, $P = 0.1$, $B_y = -1$ for $x > 0$. The longitudinal component of magnetic field $B_x = 0.75$ is constant over the grid.

This test (and the next one, the Ryu-Jones problem) is compared with the results obtained by the Athena code [18]. For both simulations we set the same spatial resolution $N = 200$ (figure 4). One can see a good agreement between the results, but the artificial oscillations can be found for the longitudinal velocity component for both solutions.

### *3.5. The Ryu-Jones problem*

The Ryu-Jones problem tests the rotation of the magnetic field components. The solution consists of two fast shock waves with velocities 1.22 and 1.28 Mach numbers, two slow shock

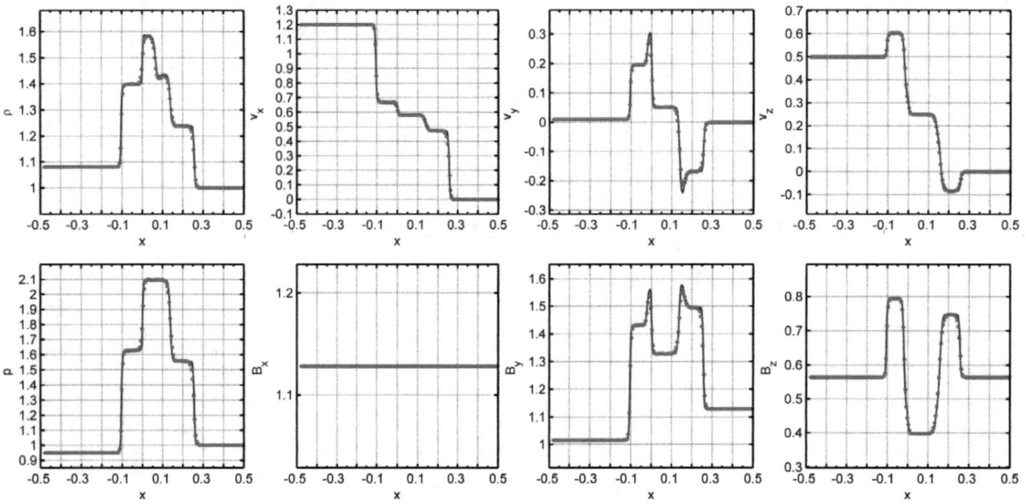

**Figure 5.** The Ryu-Jones problem. Our result is depicted by red dots, the result obtained in the Athena code is shown by blue line.

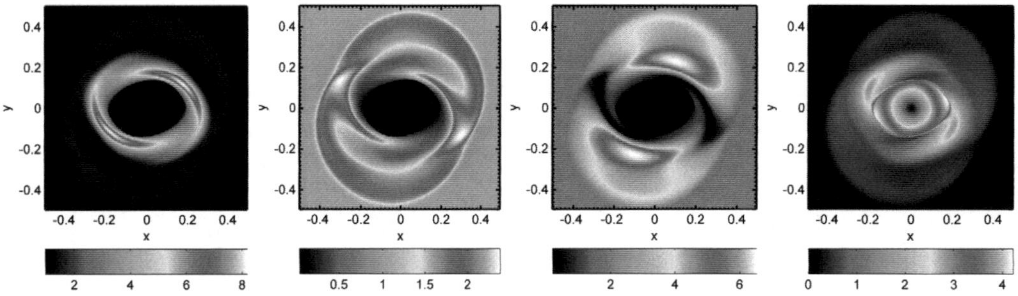

**Figure 6.** The rotor problem. The maps of density, pressure, magnetic pressure and Mach number are shown from left to right correspondingly.

waves with 1.09 and 1.07 Mach numbers, two rotational and one contact discontinuities [19].

The initial distribution is: $\rho = 1.08$, $P = 0.95$, $v_x = 1.2$, $v_y = 0.01$, $v_z = 0.5$, $B_x = 4/\sqrt{4\pi}$, $B_y = 3.6/\sqrt{4\pi}$, $B_z = 2/\sqrt{4\pi}$ at $x < 0$, and $\rho = 1$, $P = 1$, $v_x = v_y = v_z = 0$, $B_x = 4/\sqrt{4\pi}$, $B_y = 4/\sqrt{4\pi}$, $B_z = 2/\sqrt{4\pi}$ at $x > 0$. One can see a good agreement between results obtained both in our code and in the Athena code. However the peaks of $v_y$ and $B_y$ in our simulation are smoother than these obtained in the Athena code [18].

### *3.6. The rotor problem*

This test demonstrates fast rotation of the cylinder in the non-moving medium with homogeneous magnetic field. Initially there is a disc in the center of computational domain: $\rho = 10$, $v_x = -v_0 y/r_0$ and $v_y = v_0 x/r_0$ at $r < r_0 = 0.1$; $\rho = 1 + 9f(r)$, $v_x = -v_0 f(r) y/r_0$, $v_y = v_0 f(r) x/r_0$ at $r_0 < r < r_1 = 0.2$, and $\rho = 1$ at $r > r_1$; $P = 1$ and $B_x = 5/\sqrt{4\pi}$ are constant in the whole computational domain. Here the function $f(r) = (r_1 - r)/(r_1 - r_0)$ is the linear interpolation of the gas density and velocity. This configuration is strongly unstable, because the centrifugal forces are not balanced. A rotating gas is redistributed gradually within the computational domain capturing the stationary gas. Under these conditions the magnetic

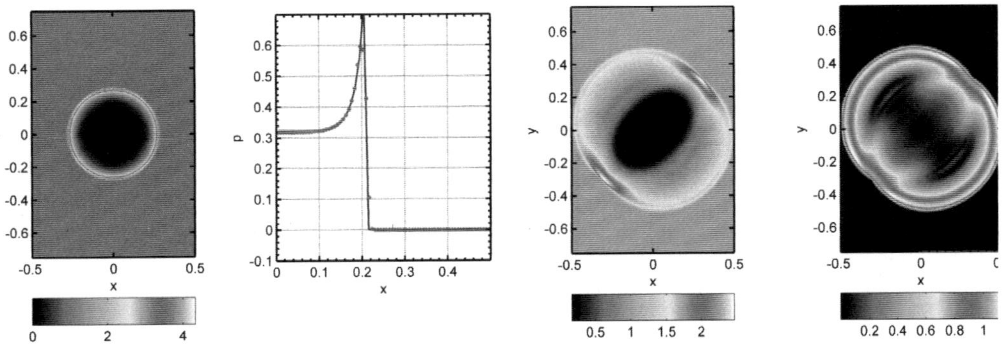

**Figure 7.** The Sedov-Taylor blast-wave. The 2D density map for pure gas dynamics (left panel) and the radial pressure profile at $t = 0.1$ for our simulations (red dots in left middle panel) compared with the exact solution (blue line). The 2D density and magnetic pressure maps at $t = 0.1$ for magneto-gas dynamics (two right panels).

field keeps the rotating material in the flattened form (Figure 6).

### *3.7. The Sedov-Taylor blast-wave*

One of the most well-known self-similar solution describes a strong shock wave originated from the point explosion. For zeroed magnetic field a numerical solution can be compared with the exact solution. In 2D we simulate a strong explosion with the following parameters: $\rho = 1$, $P = 10^{-5}$ are set in the whole computational domain, $P = \pi^2 r_0^2/(\gamma - 1)$ at $r < r_0$. We set $r = 0.01$ and take a computational domain $[-0.5, 0.5] \times [-0.75, 0.75]$ and number of cells $512 \times 768$. Figure 7 shows the 2D density map (left panel) and the radial pressure profile for our simulations (red dots in left middle panel) compared with the exact solution (blue line). One can see a good agreement between numerical data and analytic curve.

We consider a blast wave in a magnetized medium. At $t = 0$ we set $B_x = B_y = 1/\sqrt{8\pi}$. Despite of the symmetric initial distribution the shock front expands larger along the lines of the magnetic field.

### *3.8. The Kelvin-Helmholtz instability*

Shear instabilities are very common in astrophysical objects, so that numerical algorithms should reproduce such inabilities well. Here we simulate the evolution of two gas layers moving with different velocities and separated initially by a contact discontinuity: $\rho = 1$, $v_x = -0.5$ at $|y| < 0.25$ and $\rho = 2$, $v_x = 0.5$ otherwise, and $P = 2.5$ everywhere. We add a random perturbation of the velocity with the amplitude 0.001. The grid size is $512 \times 512$.

Figure 8 shows the evolution of gas density for pure gas dynamics (upper row of panels) and magneto-gas dynamics (bottom row of panels). One can note the turbulent flow formation due to shear instability at early times, which produces large scale vortexes at further nonlinear stage. The presence of magnetic field strongly changes the picture: a small scale perturbations disappear for initial longitudinal magnetic field $B_x = 0.5$ (see bottom line in Figure 8), whereas a large scale oscillation of a gas along magnetic field can be seen.

### *3.9. Spherical collapse of gravitating gas*

To test a gravity solver we use a standard 3D cosmological problem – the collapse of a gaseous sphere with initial density profile $\rho(r) = 1/r$. We set the cubic computational domain $(x, y, z) \in [-1, 1]$ with cell number $100^3$. Figure 9 shows the evolution of the density profile.

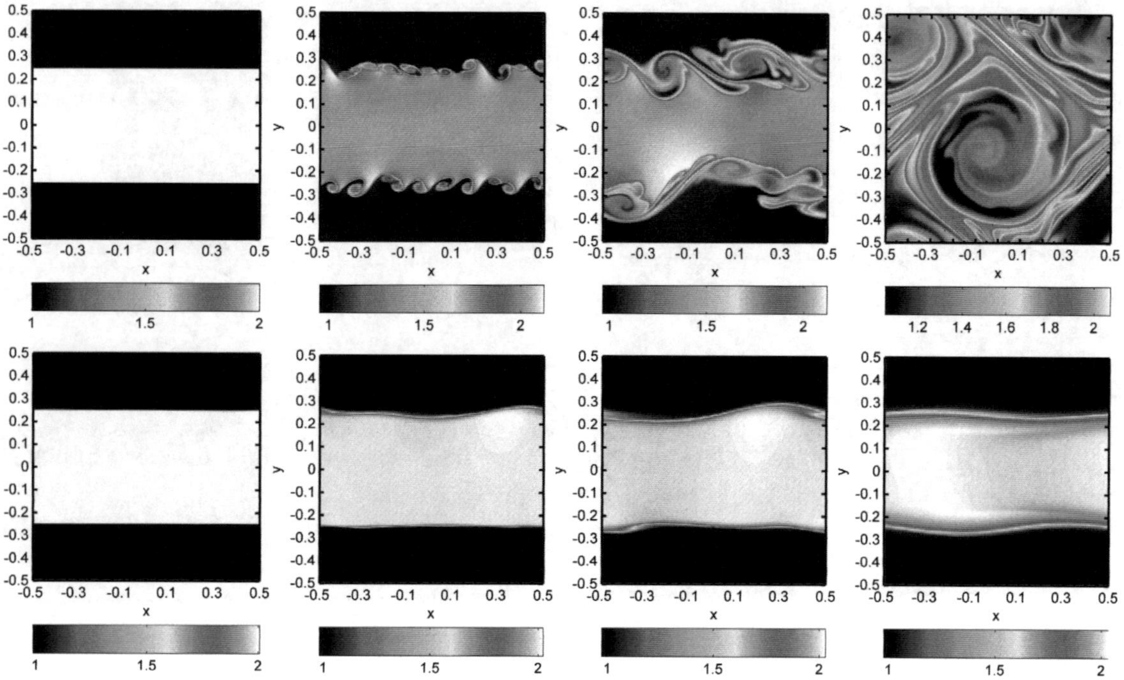


**Figure 8.** The Kelvin-Helmholtz instability. The gas density distributions at $t = 0, 0.1, 4, 12$ (from left to right) in the pure gas dynamics (top row of panels) and the same, but in the presence of magnetic field (bottom row of panels).

One can see that dense core forms due to homogeneous collapse and strong shock wave moves outwards (see right panel of the Figure).

### *3.10. Thermodynamical module*

To study the evolution of the interstellar medium in galaxies we develop a module of thermal processes. In this module we can switch between tabulated cooling/heating rates and calculation of rates for main chemical species in the interstellar medium. For the former we can use tables of cooling rates in the temperature range 10 K$< T < 10^8$ K calculated by [20, 21, 22]. In the latter we calculate the cooling and heating rates produced by specific emission processes and in the temperature range $T < 2 \times 10^4$ K, that is of a great importance for studying thermal state of the interstellar medium (see [23] for instance). We can investigate the thermal evolution of the interstellar medium for different abundances of heavy elements (metallicities). Our thermal block includes following cooling processes: cooling due to recombination and collisional excitation and free-free emission of hydrogen [24], molecular hydrogen cooling [25], cooling in the fine structure and metastable transitions of carbon, oxygen and silicon [26], energy transfer in collisions with the dust particles [27] and recombination cooling on the dust[28]. The heating rate takes into account photoelectric heating on the dust particles [28, 27], heating due to $H_2$ formation on dust and the $H_2$ photodissociation [29] and the ionization heating by cosmic rays [30]. For multi-level atoms the cooling rates are obtained from the level population equation assuming the optically thin steady-state regime (see e.g. [26]).

Figure 10 presents cooling rates in the low temperature ($T < 2 \times 10^4$ K) range adopted for our model of the interstellar medium.

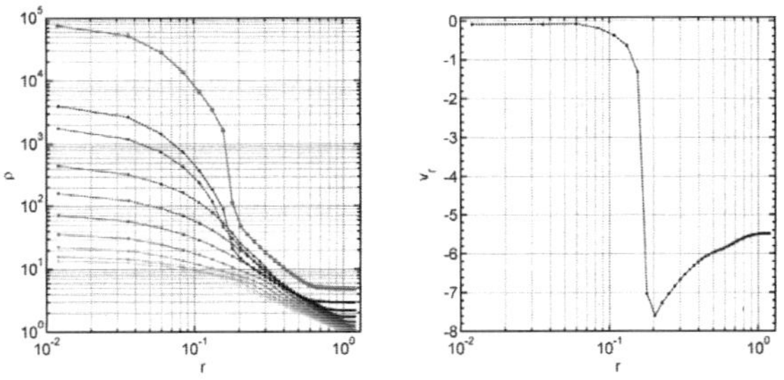

**Figure 9.** The density profile for $t = 0$ and $t = 1$ is depicted by cyan and black lines, respectively, the pressure profile for $t = 1$ is shown by red line (left panel). The velocity profile for $t = 1$ is shown at right panel.

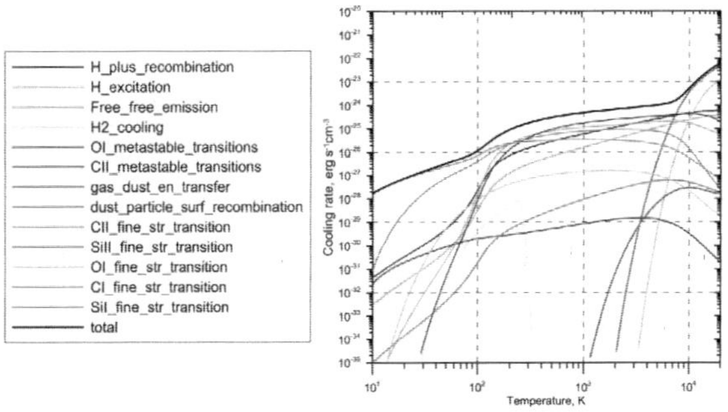


**Figure 10.** Cooling functions, $\Lambda/n_H^2$, for solar metallicity. The total cooling rate is depicted by black line.

*3.11. $H_2$ kinetics*

Molecular hydrogen ($H_2$) regulates star formation in galaxies [23], then to study the evolution of the interstellar medium we need to take into consideration the $H_2$ kinetics. Because of significant difference between dynamic and chemical timescales the $H_2$ kinetics is nonequilibrium and to get a source term in equation (2) we solve a system of ordinary differential equations for the following chemical species: H, $H^+$, $H_2$. Molecular hydrogen is formed on the surface of dust grains and dissociated by ultraviolet Lyman-Werner photons and cosmic rays. Because of strong absorption and scattering of ultraviolet photons in the neutral hydrogen and dust to mimic radiative transfer and $H_2$ self shielding we use a simplified approach introduced by [31] for calculation of neutral and molecular hydrogen column densities.

*3.12. Star formation implementation*

Stellar feedback effects can significantly influence on the galactic evolution. However, there is no consensus about the implementation of the star formation into gas dynamics, that is clearly demonstrated in [32]. Their results of the galaxy evolution simulations within the ΛCDM paradigm strongly depend on the star formation and feedback recipes: stellar mass, size, morphology and gas content of the galaxy at $z = 0$ vary significantly due to the different implementations of star formation and feedback. Despite this problem we include the star formation effects using an intuitive approach, that partly based on the well-known method offered in [33].

To form a star or in general stellar particle we need to find cells in a computational domain, which satisfy to our conditions for star formation. These conditions can be more or less sophisticated and sometimes may have intricate nature. Usually the following criteria for the stellar particle creation in a given cell are considered: the surface density should be greater than the threshold for star formation $\Sigma_t$ and simultaneously the temperature should be less than some critical level $T_t$. In the simulations presented below we have assumed $\Sigma_t > 10^4$ cm$^{-3}$ and $T_t \leq 10^2$ K. If these conditions are realised in a given cell, then we create a test stellar particle. The Salpeter initial mass function is assumed for each particle, which usually consists of $10^3 - 10^4$ stars. The initial velocity of this test particle is taken to be the same as its host cell. The kinematics of test particles are followed by the second-order integrator mentioned in Section 2.2. We take into account several feedback effects: stellar winds from massive stars, radiative pressure, supernova explosions, and stellar mass loss by low-mass stars [34]. Metals ejected by supernovae and lost by low-mass stellar population can strongly change chemical

**Figure 11.** The Galactic disc evolution. First three columns: stellar (top) and gaseous (bottom) distributions at $t = 0, 300, 600$ Myr. Right pictures: radial variations of the module of magnetic field (top) and vectors of magnetic field in the disc plane.

composition and thermal evolution of a gas. Because of local character of enrichment process the re-distribution (mixing) of metals is of great importance for further star formation in galaxies [35, 36, 37, 38].

## 4. Simulation of the multicomponent galactic disc

Here we present an example of the gravitationally unstable stellar-gaseous galactic disc evolution. In this simulation we take into account magnetic field and radiation processes. The stellar disc is simulated using $N$-body method with number of particles $N = 10^7$. The number of cells for gas dynamics equals to $1024 \times 1024 \times 128$. We start our simulation from the equilibrium state of the galactic gaseous disc within the fixed potential of the dark matter halo for Toomre's parameter $Q_T = 1.2$. A construction of initial equilibrium configuration for stellar-gaseous disc is described by [39] in detail. We assume that the galactic disc consists of stellar component and gas with temperature $10^4$ K in the middle plane. The magnetic field is set as the toroidal configuration with $B = 6$ $\mu$G.

Figure 11 presents the result of this simulation. The distribution of stellar particles (top row of three left panels) and gas surface density (bottom row of three left panels) are shown at time moments $t = 0; 300; 600$ Myr. One can see how the gravitational instability drives the formation of spiral pattern, which is clearly seen in both stellar and gaseous components. At $t = 300$ Myr it is easily found narrow and dense galactic shocks (bottom row of panels). Because of relatively small initial value $Q_T$ a flocculent structure is rapidly developed in the spiral arms [40]. The spatial structure of a gas becomes very complex due to numerous shear flows, thermal and gravitational instabilities, magnetic field pressure influence as well.

One can see the formation of the large number of dense gaseous complexes with number density $\sim 300 - 1000$ cm$^{-3}$. These complexes are mostly located in galactic spiral arms. This result is almost independent on numerical resolution and slightly dependent on the initial magnetic field. The increase of numerical resolution leads to formation of clouds on smaller scales. Such clouds are expected to give birth of stars.

Figure 11(upper right panel) shows the structure of magnetic field at galactic disc at $t = 600$ Myr. The amplitude of magnetic field decreases with radius from $|B| = 8\mu G$ at 4 kpc down to $|B| = 4\mu G$ at the outer part of the disc. This is in a good agreement as with observations [41] and with cosmological simulations [42]. The vector map (right bottom panel in Figure 11) represents the chaotic and regular velocity components. The regular field corresponds to the rotation of a gas and large-scale spiral structure in the disc.

## 5. Conclusion

In this paper we have described our three-dimensional numerical code for multi-component simulation of the galactic evolution. This code has been mainly developed to study the evolution of disc galaxies taking into account generation of spiral structure, physics of interstellar medium, formation of clouds and stars. Our code includes the following ingredients: $N$-body dynamics, ideal magneto-gas dynamics, self gravity of gaseous and stellar components, cooling and heating processes, star formation, chemical kinetics and multi-species gaseous and particle (for dust grains) advection. We present several tests for our code and show that our code have passed the tests with a resonable accuracy. It should be noted that our code is parallelized using the MPI library. We apply our code to study the large scale dynamics of galactic discs in context of formation and evolution of galactic spiral structure [39], molecular clouds formation [43] and disc-to-halo interactions [44].

## 6. Acknowledgments

The numerical simulations have been performed on the supercomputer 'Chebyshev' at the Research Computing Center (Moscow State University) and MVS100k at the Joint Supercomputer Center (Russian Academy of Sciences). This work was partially supported by the Russian Foundation of the Basic Research (grants 12-02-31452, 13-02-90767, 12-02-92704, 12-02-00685, 13-01-97062). SAK and EOV thanks to the Dmitry Zimin's "Dynasty" foundation for the financial support.

## References

[1] Kennicutt R C and Evans N J 2012 *Annual Review of Astronomy & Astrophysics* **50** 531–608
[2] Agertz O, Moore B, Stadel J, Potter D, Miniati F, Read J, Mayer L, Gawryszczak A, Kravtsov A, Nordlund Å, Pearce F, Quilis V, Rudd D, Springel V, Stone J, Tasker E, Teyssier R, Wadsley J and Walder R 2007 *Monthly Notices of the Royal Astronomical Society* **380** 963–978
[3] Tasker E J, Brunino R, Mitchell N L, Michielsen D, Hopton S, Pearce F R, Bryan G L and Theuns T 2008 *Monthly Notices of the Royal Astronomical Society* **390** 1267–1281
[4] Heß S and Springel V 2012 *Monthly Notices of the Royal Astronomical Society* **426** 3112–3134
[5] McNally C P, Lyra W and Passy J C 2012 *Astrophysical Journal Supplement Series* **201** 18
[6] Elmegreen B G and Scalo J 2004 *Annual Review of Astronomy & Astrophysics* **42** 211–273
[7] Draine B 2010 *Physics of the Interstellar and Intergalactic Medium* Princeton Series in Astrophysics (Princeton University Press) ISBN 9781400839087
[8] Yee K 1966 *IEEE Transactions on Antennas and Propagation* **14** 302–307
[9] Evans C R and Hawley J F 1988 *Astrophysical Journal, Part 1* **332** 659–677
[10] Miyoshi T and Kusano K 2005 *Journal of Computational Physics* **208** 315–344
[11] Harten A 1983 *Journal of Computational Physics* **49** 357
[12] Khoperskov A V, Zasov A V and Tyurina N V 2001 *Astronomy Reports* **45** 180–194
[13] Barnes J and Hut P 1986 *Nature* **324** 446–449
[14] Shu C W and Osher S 1989 *Journal of Computational Physics* **83** 32
[15] Bisikalo D, Kaygorodov P, Ionov D, Shematovich V, Lammer H and Fossati L 2013 *Astrophysical Journal* **764** 19
[16] Roediger E and Brüggen M 2007 *Monthly Notices of the Royal Astronomical Society* **380** 1399–1408
[17] Brio M and Wu C C 1988 *Journal of Computational Physics* **75** 400–422

[18] Stone J M, Gardiner T A, Teuben P, Hawley J F and Simon J B 2008 *Astrophysical Journal Supplement Series* **178** 137–177
[19] Ryu D and Jones T W 1995 *Astrophysical Journal* **442** 228–258
[20] Sutherland R S and Dopita M A 1993 *Astrophysical Journal Supplement Series* **88** 253–327
[21] Vasiliev E O 2011 *Monthly Notices of the Royal Astronomical Society* **414** 3145–3157
[22] Vasiliev E O 2013 *Monthly Notices of the Royal Astronomical Society* **431** 638–647
[23] Spitzer L 1978 *Physical processes in the interstellar medium*
[24] Cen R 1992 *Astrophysical Journal Supplement Series* **78** 341–364
[25] Galli D and Palla F 1998 *Astronomy and Astrophysics* **335** 403–420
[26] Hollenbach D and McKee C F 1989 *Astrophysical Journal* **342** 306–336
[27] Wolfire M G, McKee C F, Hollenbach D and Tielens A G G M 2003 *Astrophysical Journal* **587** 278–311
[28] Bakes E L O and Tielens A G G M 1994 *Astrophysical Journal* **427** 822–838
[29] Hollenbach D and McKee C F 1979 *Astrophysical Journal Supplement Series* **41** 555–592
[30] Goldsmith P F and Langer W D 1978 *Astrophysical Journal* **222** 881–895
[31] Dobbs C L 2008 *Monthly Notices of the Royal Astronomical Society* **391** 844–858
[32] Scannapieco C, Wadepuhl M, Parry O H, Navarro J F, Jenkins A, Springel V, Teyssier R, Carlson E, Couchman H M P, Crain R A, Dalla Vecchia C, Frenk C S, Kobayashi C, Monaco P, Murante G, Okamoto T, Quinn T, Schaye J, Stinson G S, Theuns T, Wadsley J, White S D M and Woods R 2012 *Monthly Notices of the Royal Astronomical Society* **423** 1726–1749
[33] Yepes G, Kates R, Khokhlov A and Klypin A 1997 *Monthly Notices of the Royal Astronomical Society* **284** 235–256
[34] Agertz O, Kravtsov A V, Leitner S N and Gnedin N Y 2013 *Astrophysical Journal* **770** 25
[35] Vasiliev E O, Dedikov S Y and Shchekinov Y A 2009 *Astrophysical Bulletin* **64** 317–324
[36] Vasiliev E O, Vorobyov E I, Matvienko E E, Razoumov A O and Shchekinov Y A 2012 *Astronomy Reports* **56** 895–914
[37] Dedikov S Y and Shchekinov Y A 2004 *Astronomy Reports* **48** 9–20
[38] de Avillez M A and Mac Low M M 2002 *Astrophysical Journal* **581** 1047–1060
[39] Khoperskov S A, Khoperskov A V, Khrykin I S, Korchagin V I, Casetti-Dinescu D I, Girard T, van Altena W and Maitra D 2012 *Monthly Notices of the Royal Astronomical Society* **427** 1983–1993
[40] Khoperskov A V, Just A, Korchagin V I and Jalali M A 2007 *Astronomy and Astrophysics* **473** 31–40
[41] Basu A and Roy S 2013 *Monthly Notices of the Royal Astronomical Society* **433** 1675–1686
[42] Pakmor R and Springel V 2013 *Monthly Notices of the Royal Astronomical Society* **432** 176–193
[43] Khoperskov S A, Vasiliev E O, Sobolev A M and Khoperskov A V 2013 *Monthly Notices of the Royal Astronomical Society* **428** 2311–2320
[44] Khoperskov A V, Khoperskov S A, Zasov A V, Bizyaev D V and Khrapov S S 2013 *Monthly Notices of the Royal Astronomical Society* **431** 1230–1239

# A new paradigm for petascale Monte Carlo simulation: Replica exchange Wang–Landau sampling

**Ying Wai Li[1,2], Thomas Vogel[1,3], Thomas Wüst[4], and David P. Landau[1]**

1 Center for Simulational Physics, The University of Georgia, Athens, GA 30602, USA
2 National Center for Computational Sciences, Oak Ridge National Laboratory, Oak Ridge, TN 37831, USA
3 Theoretical Division, Los Alamos National Laboratory, Los Alamos, NM 87545, USA
4 Swiss Federal Research Institute WSL, Zürcherstrasse 111, CH-8903 Birmensdorf, Switzerland

E-mail: yingwai.li@mailaps.org

**Abstract.** We introduce a generic, parallel Wang–Landau method that is naturally suited to implementation on massively parallel, petaflop supercomputers. The approach introduces a replica-exchange framework in which densities of states for overlapping sub-windows in energy space are determined iteratively by traditional Wang–Landau sampling. The advantages and general applicability of the method are demonstrated for several distinct systems that possess discrete or continuous degrees of freedom, including those with complex free energy landscapes and topological constraints.

## 1. Introduction

One of the great challenges accompanying recent developments in high performance computing hardware is the question of how to make truly efficient use of the huge numbers of cores found in today's "cutting edge" machines. There have been concomitant developments in methodology in computational statistical physics, one of the most recent of them being a parallel implementation of the very successful importance sampling Monte Carlo based approach termed "Wang–Landau sampling" [1]. In the original Wang–Landau sampling, the *a priori* unknown density of states $g(E)$ of a system is determined iteratively by performing a random walk in energy space ($E$) and sampling configurations with probability $1/g(E)$ (i.e. with a "flat histogram") [2, 3, 4]. Many studies have shown that this procedure is very powerful for studying a number of diverse problems including those with complex free energy landscapes because it circumvents the very long time scales encountered near phase transitions or at low temperatures. The method also facilitates the direct calculation of thermodynamic quantities, including the free energy, at any temperature from a *single* simulation. (In fact, random walks in the space of more than one variable allow thermodynamic information to be determined as a function of more than one thermodynamic field from a single simulation.) Wang–Landau sampling is also a generally applicable Monte Carlo procedure with only a few adjustable parameters, and it has been applied successfully to a wide range of simulation models, some with discrete degrees of freedom and some involve degrees of freedom that are continuous (see [5, 6, 7, 8] for examples). While modifications to the initial implementation have been proposed, e.g. by optimizing the "modification factor–flatness criterion" scheme [9, 10, 11], by introducing more efficient Monte Carlo trial moves [12, 13, 14], or "error correction" [15], the underlying simplicity has remained intact. Thus, the basic algorithm is an excellent candidate for parallel implementation.

A few simple attempts at parallelizing Wang–Landau sampling have been implemented, but these are useful only for a relatively small number of processors (cores). One early approach [3, 4] subdivided the total energy range into smaller sub-windows, each being sampled by an independent random walker. The total simulation time needed is then limited by the convergence of the slowest walker, but this can be tuned by an unequal partition of energy space. Nevertheless, "perfect" load balancing is impossible due to the *a priori* unknown nature of the complex free energy landscape. Furthermore, the individual energy sub-windows cannot be reduced arbitrarily in range because some regions of configurational space would then ultimately become inaccessible. In another approach to parallelization, multiple random walkers work simultaneously on the *same* density of states and histogram. Distributed memory (MPI [16]), shared memory (OpenMP [17], multi-thread [18]), and GPU [19] variants of this idea have been proposed; shared memory implementations have the advantage of not requiring periodic synchronization among the walkers and even allow for "data race" when updating $g(E)$ without noticeable loss in accuracy [17]. Although this second approach would appear to avoid the problems of the first approach, a recent, massively parallel implementation [19] found that correlations among the walkers could lead to a systematically underestimation of $g(E)$ in "difficult to access", low energy regions. The addition of a phenomenological bias to the modification factor alleviated this difficulty, the effective round-trip times of the individual walkers, however, are not improved through the use of this "ad hoc" method and the validity of the approach cannot be confirmed.

## 2. The new, parallel (replica exchange Wang–Landau) algorithm

Our new approach [1, 20] is a *generic*, parallel Wang–Landau scheme which combines the benefits of the original Wang–Landau (WL) sampling scheme with those of replica-exchange Monte Carlo [21, 22, 23]. Much of the success of the Wang–Landau algorithm has resulted from the combination of its simplicity and robustness, and our goal was to retain these qualities in the parallel algorithm. We begin by splitting up the total energy range into smaller, overlapping sub-windows. Sampling then proceeds in each energy sub-window by multiple, *independent* Wang–Landau walkers, each of which has its own instantaneous density of states and histogram. The key to this approach is that configurational, or replica, exchanges are allowed between walkers in overlapping energy sub-windows during the simulation. Consequently, each replica can travel back and forth over the entire energy space many times during a single simulation. The replica exchange move does not bias the overall procedure; thus the approach is applicable with any valid trial update or Wang–Landau convergence criterion (e.g., the "$1/t$ algorithm" [11]). Furthermore, the parallel algorithm does not impose any intrinsic limitation to the number of random walkers and the number of energy sub-windows that can grow with the size of the system. Therefore, the computational method should scale straightforwardly to many thousands of cores.

The standard Wang–Landau algorithm [2, 3] estimates the density of states, $g(E)$, using a single random walker in an energy range $[E_{\mathrm{min}}, E_{\mathrm{max}}]$. During the simulation, trial moves are accepted with a probability $P = \min\left[1, g(E_{\mathrm{old}})/g(E_{\mathrm{new}})\right]$, where $E_{\mathrm{old}}$ ($E_{\mathrm{new}}$) is the energy of the original (proposed) configuration. The estimation of $g(E)$ is continuously adjusted and improved using a modification factor $f$ (i.e. $g(E) \to f \times g(E)$) which starts with $f_0 > 1$ and progressively approaches unity as the simulation proceeds. A histogram, $H(E)$, keeps track of the number of visits to each energy $E$ during a given iteration. When $H(E)$ is sufficiently "flat", the next iteration begins with $H(E)$ reset to zero but keeping the estimate of $g(E)$ from the previous

Journal of Physics: Conference Series **510** (2014) 012012 doi:10.1088/1742-6596/510/1/012012

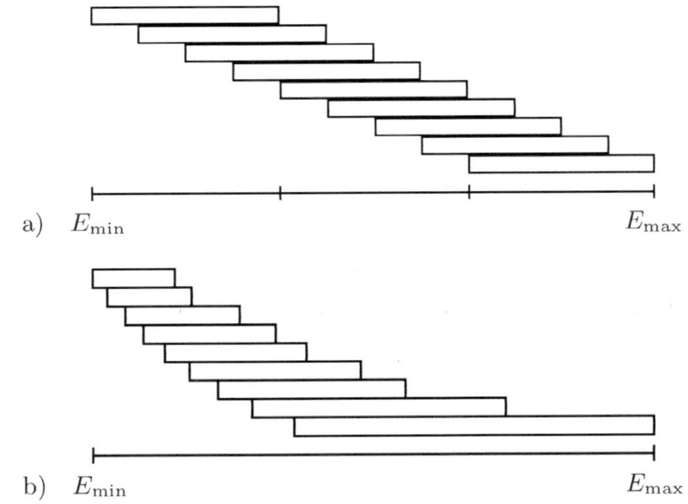


**Figure 1.** a) Subdivision of the global energy range into nine equal-size intervals with overlap $o = 75\%$. b) Run-time balanced subdivision with overlap to the higher energy interval $o \geq 75\%$. Multiple Wang–Landau walkers can be employed within each interval.

iteration, and $f$ reduced by some predefined rule (e.g. $f \to \sqrt{f}$). The simulation ends when $f$ reaches a sufficiently small value $f_{\text{final}}$ at which point the accuracy of $g(E)$ is proportional to $\sqrt{f}_{\text{final}}$ for sufficiently flat $H(E)$ [9].

In our parallel Wang–Landau framework, the global energy range is first split into $h$ overlapping sub-windows, and $m$ random walkers are used to sample each sub-window. The extent of the overlap $o$ should strike a balance between fast convergence of $g(E)$ and a reasonable exchange acceptance rate. While an overlap of $o \approx 75\%$ works well, excellent results can be obtained with other values [20]. (Different types of partitioning of the global energy into sub-windows may be used; but, in all cases, configurational exchange cannot occur without overlap of adjacent sub-windows, see Fig. 1 for examples.) Within an energy sub-window, each random walker performs standard Wang–Landau sampling. After a predetermined number of Monte Carlo steps, a "replica exchange" is proposed between two random walkers, $i$ and $j$, where walker $i$ chooses swap partner $j$ from a neighboring sub-window at random. For the sake of simplicity and generality in our discussion, we do not sort out walkers which are currently not in the overlap region before drawing the pairs. Let $X$ and $Y$ be the configurations that the random walkers $i$ and $j$ are "carrying" before the exchange; $E(X)$ and $E(Y)$ be their energies, respectively. From the detailed balance condition the acceptance probability $P_{\text{acc}}$ for the exchange of configurations $X$ and $Y$ between walkers $i$ and $j$ is:

$$P_{\text{acc}} = \min\left[1, \frac{g_i(E(X))}{g_i(E(Y))} \frac{g_j(E(Y))}{g_j(E(X))}\right] \tag{1}$$

where $g_i(E(X))$ is the instantaneous estimator for the density of states of walker $i$ at energy $E(X)$, cf. [24]. Note that if either of the random walkers has an energy that lies outside the range of the sub-windows of the other, the replica exchange cannot take place. In that case, we just disregard the exchange and carry out another number of MC steps until the next replica exchange is proposed. This combination of sampling trials defines the new parallel algorithm.

An important new feature of our formalism is that each random walker has its own $g(E)$ and $H(E)$ which are updated independently. Also, since every walker has to satisfy the Wang–Landau flatness criterion *independently* at each iteration, the systematic errors found in [19] are avoided. When all random walkers within an energy sub-window have attained flat histograms, their estimates for $g(E)$ are averaged and then redistributed among themselves before simultaneously proceeding to the next iteration. This procedure reduces the error during the simulation with $\sqrt{m}$ [20], i.e. as for uncorrelated WL simulations. Furthermore, increasing $m$ can improve the convergence of the Wang–Landau sampling by reducing the risk of statistical outliers in $g(E)$ resulting in slowing down subsequent iterations. Alternatively, it allows us in principle, to use a weaker flatness criterion [20], which is in the spirit of a concurrently proposed idea of merging histograms in multicanonical simulations [25].

The parallel simulation ends when the modification factors in all the energy intervals have reached $f_{\text{final}}$ and the $h \times m$ pieces of $g(E)$ fragments with overlapping energy sub-windows are then used to construct a single $g(E)$ over the entire energy range.

## 3. Data analysis and production run

To connect two pieces of consecutive, overlapping density of states fragments, say $g_i(E)$ and $g_j(E)$, we first calculate the inverse microcanonical temperatures, $\beta_i(E)$ and $\beta_j(E)$ by: $\beta(E) = \mathrm{d}\log[g(E)]/\mathrm{d}E$. The joining point, $E_{\text{join}}$, is determined as the point where $\Delta\beta = |\beta_i(E) - \beta_j(E)|$ vanishes or is the smallest. This procedure avoids discontinuities in the derivatives of the final density of states, which can lead to artificial peaks in derived functions like the heat capacity, for example. Next, $g_j(E)$ has to be rescaled using the value $g_i(E_{\text{join}})$ as the reference point to yield a correct ratio for different energy levels: $g_j(E) \to g_j(E)(g_i(E_{\text{join}})/g_j(E_{\text{join}}))$. Finally, a joined relative density of states can be obtained by:

$$g(E) = \begin{cases} g_i(E) & \text{if } E < E_{\text{join}} \\ g_j(E) & \text{if } E \geq E_{\text{join}} \,. \end{cases} \tag{2}$$

Here, we have assumed $j > i$ for Eq. (2) but this does not need to be the case for the previous rescaling procedure to work properly. See Fig. 2 for an illustration of the procedure described so far. Since there are $m$ pieces of density of states in each energy window, it is possible to compute the statistical errors by standard resampling techniques, e.g. jackknife or bootstrap methods, which can be found in standard textbooks. A more comprehensive discussion on this topic as applied to our data analysis scheme is presented in Ref. [20].

Very often one is not only interested in obtaining a one-dimensional density of states, but a two-dimensional (or even higher dimensional) density of states, $g(E,Q)$, where $Q$ is a physical quantity. Such joint densities of states can be particularly useful for the calculation of thermodynamic properties of order parameters, for instance. One way of doing this is to perform a two-dimensional Wang–Landau sampling as proposed in the original treatment [2, 3, 4]; a more efficient way is to carry out a two-dimensional multicanonical production run using $1/g(E)$, obtained from a one-dimensional Wang–Landau sampling, as the simulation weight and construct $g(E,Q)$ from measurements of $E$ and $Q$ (see, for example, Ref. [26] for details).

In any case, one is confronted with the necessity of constructing the entire two-dimensional, joint density of states from fragments of $g(E,Q)$. We found that a direct generalization from the aforementioned one-dimensional scheme with a slight modification is able to yield satisfactory outcomes [1]: let $g_i(E,Q)$ and $g_j(E,Q)$ be the two pieces of 2D density of states to be merged. We first calculate the normal vectors of the surfaces $S(E,Q) = \log\, g(E,Q)$:

$$\hat{N} = \begin{pmatrix} \partial S(E,Q)/\partial E \\ \partial S(E,Q)/\partial Q \\ -1 \end{pmatrix}, \tag{3}$$

[1] The construction of a higher-dimensional density of states, in principle, could follow the same fundamental scheme as far as the calculation of the normal vector (Eq. 3) is properly generalized to higher dimensions.

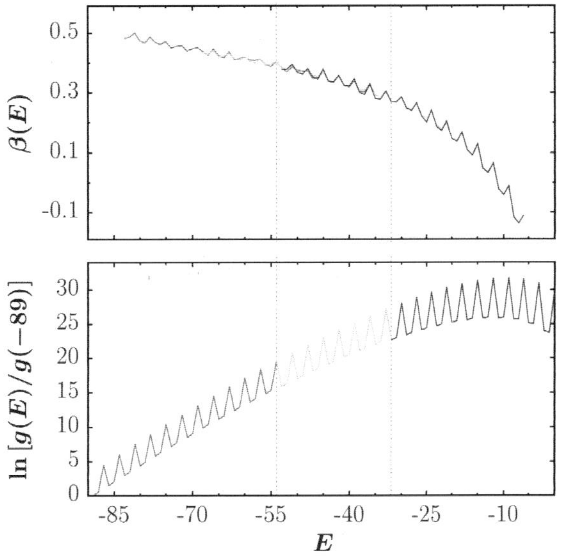


**Figure 2.** (color) Joining pieces of $g(E)$ for the HP 67mer adsorption on a weak attractive surface (Complete data shown in Fig. 5 a). (Top) First derivatives of raw DOS pieces from the three highest-energy windows. We applied a 5-point stencil with step-width $\Delta E = 3$. Derivatives coincide best at $E = -54$ (red vs. green curve) and at $E = -32$ (green vs. blue curve; marked by vertical dotted lines). (Bottom) Pieces are connected at these points to obtain overall DOS.

from which the unit normal vectors $\hat{n}$ are calculated at all points $(E, Q)$. The joining position can then be determined as the point where these vectors best coincide, i.e., where $\hat{n}_i \cdot \hat{n}_j$ is maximal. One of the $g(E, Q)$ fragments is again rescaled in the same way as in the 1D case, and finally a joined relative density of states is determined by:

$$g(E,Q) = \begin{cases} g_i(E,Q) & \text{if } (E,Q) \text{ is present only in window } i \\ g_j(E,Q) & \text{if } (E,Q) \text{ is present only in window } j \\ (g_i(E,Q) + g_j(E,Q))/2 & \text{if } (E,Q) \text{ is present in both windows.} \end{cases} \tag{4}$$

This procedure is used for the analysis of our example models in Sect. 5.

## 4. Performance measures and scaling

In order to assess the generality and performance of this parallel Wang–Landau scheme, we applied it to multiple, fundamentally different models.[2] As for most new computational schemes in statistical physics, we first tested our framework on the "fruit fly" of statistical physics, the Ising model. We simulated the 2D Ising model with system sizes as large as $256^2$ using more than 2000 cores. While single-walker Wang–Landau sampling typically takes more than a week to converge, the parallel scheme easily finishes within a few hours, with the same accuracy as the serial scheme. The deviations from the exact results are always of the same order as the statistical errors, which are $< 0.01\%$ in the peak region of the density of states.

To generally quantify the efficiency of the parallel WL scheme, we define the speed-up, $s_o(h, m)$, as the number of Monte Carlo steps taken by the slowest parallel WL walker

[2] We emphasize again, however, that there is nothing in our framework that restricts its use to these models!

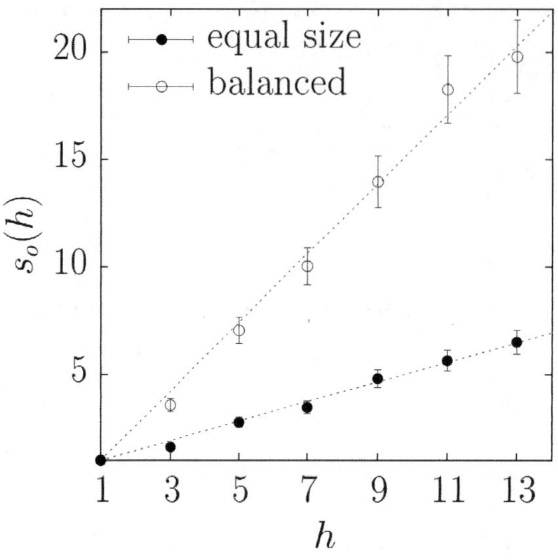


**Figure 3.** a) Dependence of the speed-up $s_o(h,m)$ on the number of energy windows and overlap $o = 75\%$ (filled symbols, cf. Fig. 1 a) and using a run-time balanced energy splitting (open symbols, cf. Fig. 1 b). Here, the calculation of the speed-up is based on the MC steps (MCS) needed to complete the first WL iteration. Measurements are performed using the lipid bilayer system, see text for details.

($N_o^{\text{parallel}}(h,m)$), as compared to that taken by a single walker ($N^{\text{single}}$):

$$s_o(h,m) = \frac{N^{\text{single}}}{N_o^{\text{parallel}}(h,m)}. \tag{5}$$

We measure this number for $h \lesssim 20$ in simulations of the lipid system to be introduced below. As shown in Fig. 3, we find strong scaling, i.e. the speed-up scales *linearly* with $h$ for a fixed number $m$ for both energy splittings as shown in Fig. 1. While the equal-size energy range splitting (Fig. 1 a) is the most basic approach, the run-time balanced energy splitting (Fig. 1 b) is chosen such that walkers in different energy sub-windows complete the first WL iteration after the same number of MC sweeps (within statistical fluctuations). As the growth behavior of WL histograms is in principle known [9], such an energy splitting can be estimated by analyzing the first-iteration histogram from a short pre-run with equal-size energy intervals.[3] We have shown in Ref. [1] that our method can also achieve weak scaling properties. With the capability of achieving both strong and weak scaling (i.e., by increasing the number of computing cores one can get results faster for the same system and/or simulate larger systems in the same period of time, respectively), our formalism becomes a promising candidate for large-scale applications.

## 5. Opening new vistas

To show how this parallel framework can allow us to examine previously unapproachable problems, we also applied the method to two very distinct and particularly challenging molecular systems: a coarse-grained continuum model for the self assembly of amphiphilic molecules (lipids) in explicit water and a discrete model for the surface adsorption of lattice proteins. In the first model, amphiphilic molecules, each of which is composed of a polar (P) head and two hydrophobic (H) tail monomers (P–H–H), are surrounded by solvent particles (W). The

[3] Of course, the energy-window setup can be further adapted as the simulation proceeds.

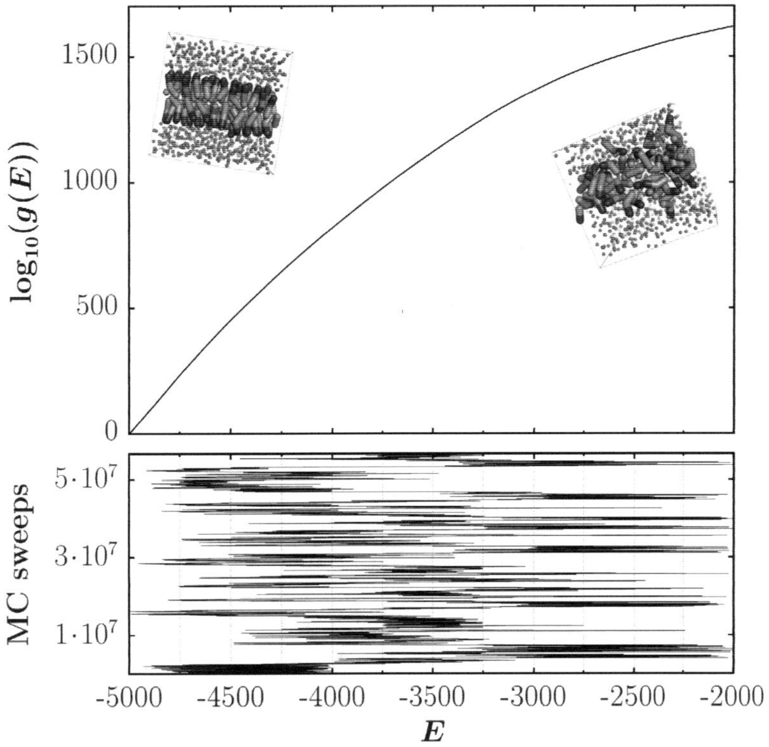


**Figure 4.** (color; Top) Logarithm of the density of states of the amphiphilic system containing 75 lipid molecules and a total of 1000 particles obtained by our parallel Wang–Landau scheme with the setup shown in Fig. 1 a. The pictures show a conformation where lipid molecules assemble and form a single cluster ($E \approx -2100$) and a low-energy bilayer configuration ($E \approx -4800$). (Bottom) Path of one single replica (out of 27 contributing to the above density of states) through energy space. Replica exchanges between walkers are proposed every $10^4$ sweeps (data also shown with that resolution), with acceptance rates between 30 and 55 %. Grid lines correspond to the borders of the individual energy windows.

interactions between H and W molecules, as well as those between H and P molecules, are purely repulsive. All other interactions between non-bonded particles are of Lennard-Jones type; bonded molecules are connected by a FENE potential, see Refs. [29, 30] for similar models. The second model uses the hydrophobic-polar (HP) lattice model [31] for protein surface adsorption. Here a protein is represented by a self-avoiding walk consisting of H and P monomers placed on a simple cubic lattice with an attractive substrate. For recent simulational results on this model and computational details see [26, 32].

Both models pose technical challenges due to high energy and/or configurational barriers. Simulations of either setup is impossible for all practical purposes using the traditional, single walker Wang–Landau method due to unreasonable resource demands. (Standard Metropolis sampling would be *many* orders of magnitude too slow.) The first model (the amphiphilic system) consists of 75 lipid molecules (each composed of three particles) and 775 solution particles with a continuous energy domain. The density of states $g(E)$ over an energy range covering the lipid bilayer formation spans more than 1600 orders of magnitude, which makes precise low temperature properties extremely difficult to estimate. Only with the parallel algorithm, it is possible to clearly observe and study the lipid bilayer formation and different bilayer phases in this system. See Fig. 4 for example data and Refs. [27, 28] for all simulation details.

The second model (the HP lattice protein) under consideration consists of 67 monomers [33] interacting with a weakly attractive surface, for which the total energy of the system can be calculated as:

$$E = -(n_{HH}\varepsilon_{HH} + n_{SH}\varepsilon_{SH} + n_{SP}\varepsilon_{SP}), \tag{6}$$

where $n_{HH}$, $n_{SH}$ and $n_{SP}$ are the number of hydrophobic interactions, surface-H interactions, and surface-P interactions, respectively; $\varepsilon_{HH}$, $\varepsilon_{SH}$ and $\varepsilon_{SP}$ are the corresponding energy scales. The discrete energy levels and the unequal interaction strengths between the H monomers and with the surface result in an unusual, sawtooth-like density of states, as shown in Fig. 5 a. This, combined with other obstacles such as the first-order like structural transitions in the system and the peculiar form of the density of states near the ground state, makes convergence for the entire energy range extremely time-consuming using only a single walker. Simulating smaller energy sub-windows individually as proposed by earlier studies was also not successful for this kind of system, due to the fact that too-small energy window would destroy ergodicity in the simulation. Consequently, some regions in the configurational space are unreachable, resulting in an incompletely simulated density of states.[4]

However, we successfully simulated this model by our parallel Wang–Landau algorithm within a reasonable time frame. The density of states over the entire energy range and the thermodynamic properties of a few structural properties are shown in Fig. 5. From $g(E)$, the average energy $\langle E \rangle$ and the heat capacity $C_V$, both shown in Fig. 5 b, can be calculated:

$$\langle E \rangle = Z^{-1} \sum_E E g(E) e^{-E/k_B T}, \tag{7}$$

$$C_V = \frac{\langle E^2 \rangle - \langle E \rangle^2}{k_B T^2}, \tag{8}$$

where $Z = \sum_E g(E) e^{-E/k_B T}$ is the partition function at temperature $T$ with Boltzmann factor $k_B$.

To identify precisely which transitions are responsible for the peak at $T \approx 0.9$ and the shoulder across $T \approx 1.0$–$2.0$ in $C_V$, we compare it with the thermodynamics of the number of contacts between different particle species, i.e., the contributions to the total energy given in Eq. (6), and the radius of gyration, $R_g^2 = \frac{1}{N}\sum_{i=1}^{N}(\vec{r}_i - \vec{r}_{cm})^2$, where $\vec{r}_{cm}$ is the center of mass of the configuration, $\vec{r}_i$ is the position of monomer $i$, and $N$ is the chain length. These results are shown in Fig. 5 c and d and were obtained through the joint density of states estimation as discussed earlier, where mean values are calculated via:

$$\langle Q \rangle = Z_Q^{-1} \sum_{E,Q} Q\, g(E,Q)\, e^{-E/k_B T}, \tag{9}$$

using the corresponding partition sum $Z_Q$.

From the derivative of the number of H–H contacts, $d\langle n_{HH}\rangle/dT$, we can see that the hydrophobic core formation occurs mainly at $T \approx 0.9$ and mildly at $T \approx 1.7$. By looking at the number of surface–H contacts, $d\langle n_{SH}\rangle/dT$, it is clear that adsorption is initiated by the attraction of the hydrophobic surface at $T \approx 1.1$. The polar monomers, although not attracted by the surface, are dragged to come in contact with the surface besides forming a polar shell of the protein. This causes a negative peak in $d\langle n_{SP}\rangle/dT$ at a lower temperature of $T \approx 0.9$. But recall that at the same temperature, the transition to ground state (hydrophobic core formation) is also taking place, which draws some H monomers from the surface to the core, resulting in a negative trough (visually, a peak) in $d\langle n_{SH}\rangle/dT$.

[4] This is not to be confused with the energy gaps shown in Fig. 5 a, as those are *real* missing energy levels for this system.

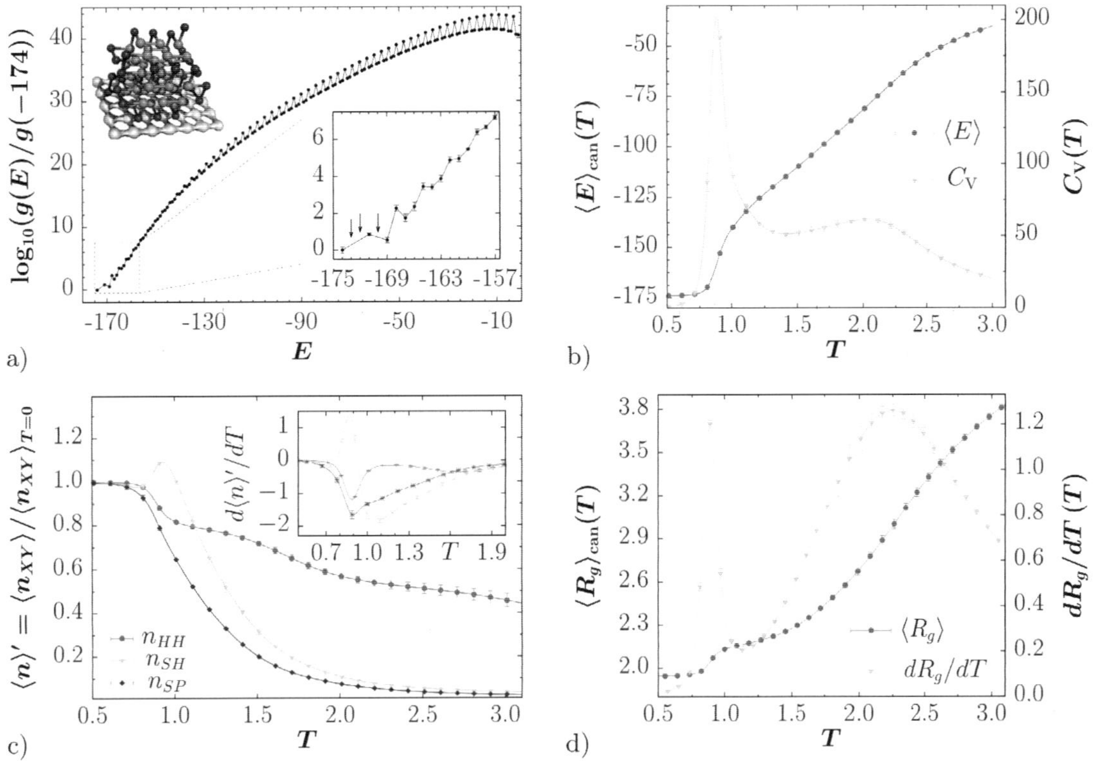


**Figure 5.** (color; a) Density of states of the lattice HP 67mer, where only H-monomers are attracted by the substrate. The H–H interaction is three times stronger than the surface attraction ($\varepsilon_{HH} = 3$, $\varepsilon_{SH} = 1$ and $\varepsilon_{SP} = 0$), leading to the unusual sawtooth like shape. The inset shows the error bars on the enlarged low-energy data. Note the two energy gaps, i.e. no conformations exist with $E = -173$, $-172$, and $-170$ (see arrows). The picture shows an adsorbed HP protein with energy $E = -174$. (b) The corresponding canonical mean energy and heat capacity; (c) number of single H–H interactions and surface contacts; and (d) radius of gyration.

The radius of gyration and its derivative, shown in Fig. 5 d, give us extra information about the structural changes of the system. The peak at $T \approx 2.2$ in $d\langle R_g\rangle/dT$ signals the $\theta$-transition[5], where an extended coil collapses into a globular structure. This confirms the above observation on $d\langle n_{HH}\rangle/dT$, as it is necessary to first (i.e., at a higher temperature) bring the monomers closer together before the mild hydrophobic core formation can take place at $T \approx 1.7$. Another peak found at $T \approx 0.9$ is a more obvious reinforcement of our previous observation. During the transition to the ground state, there is a competition between the major hydrophobic core formation (which tends to decrease $R_g$ significantly) and adsorption (which tends to increase $R_g$ slightly), causing a large fluctuation in $R_g$. Since both the hydrophobic core formation and adsorption take place at nearly the same temperature, a single, pronounced peak is observed in $C_V$ eventually. This is a typical Category III transition according to the scheme defined in Ref. [26].

Using the lipid system as an example and considering a much smaller global energy range accessible for single-walker simulations, we measure the speed-up defined by Eq. (5). The slope of the speed-up in completing the first iteration is $\approx 0.5$ for the equal size energy splitting and $\approx 1.6$ for the run-time balanced energy splitting (cf. Fig. 3), which is particularly remarkable

[5] For a review of the $\theta$-transition in lattice-polymer models, see, for example, Ref. [34].

as this indicates that the speed-up is *larger* than the number of processors used. For the HP protein, even a basic set-up of equal-size energy splitting with only a single walker per energy interval yields a speed-up of $s_{o=75\%}(h = 9) \approx 20$ compared to single walker Wang–Landau simulations. Again, we get a speed-up larger than the number of processors.[6] It is conceivable that the speed-up factor is "mysteriously" larger than the number of processors since the sum of entries needed to create flat histograms in all small energy windows can well be, and in the case of the balanced energy range splitting indeed is (data not shown), smaller than the number needed for a flat histogram on the whole energy space. Furthermore, our parallel scheme combines the advantages of two levels of parallelism which both contribute to the acceleration: first, as just mentioned, each walker only needs to attain a flat histogram in a smaller energy window; second, the replica exchange process can revitalize walkers from trapped states, thus shortening the time spent on redundant sampling of rare events. It also avoids an erroneous bias in $g(E)$ due to potential ergodicity breaking, as replicas can access the *entire* conformational space by walking through all energy windows (a typical time-series of a replica performing round trips in the full energy range of the lipid system is shown at the bottom in Fig. 4). A more detailed analysis and discussion can be found in Ref. [20].

## 6. Summary and Outlook

In summary, we introduced a generic, parallel framework for generalized ensemble Wang–Landau simulations making use of energy range splitting, replica exchange, and multiple random walkers. The method is simple, general, and leads to significant advantages over the traditional, serial algorithm. In our complete formulation, we consider multiple Wang–Landau walkers in independent parallelization directions and show that both strong and weak scaling can be achieved. With the ability to produce highly accurate results, and proven scalability up to $\approx$ 2000 cores without introducing any bias, we have demonstrated that our parallel scheme has the potential for extremely large scale parallel Monte Carlo simulations. Since the framework is complementary to other technical parallelization strategies, it is further extendible in a straightforward manner. This facilitates efficient simulations of larger and more complex systems, and thus provides a basis for many applications on petaflop machines and beyond.

## Acknowledgments

We thank M. Eisenbach and E.A. Zubova for constructive discussions. This work was supported by the National Science Foundation under Grants DMR-0810223 and OCI-0904685. Y.W. Li was partly sponsored by the Office of Advanced Scientific Computing Research; U.S. Department of Energy. Part of the work was performed at the Oak Ridge Leadership Computing Facility at ORNL, which is managed by UT-Battelle, LLC under Contract No. De-AC05-00OR22725. Supercomputer time was also provided by TACC under XSEDE grant PHY130009. Assigned: LA-UR-13-28353.

## References

[1] Vogel T, Li Y W, Wüst T, and Landau D P 2013 *Phys. Rev. Lett.* **110** 210603
[2] Wang F and Landau D P 2001 *Phys. Rev. Lett.* **86** 2050
[3] Wang F and Landau D P 2001 *Phys. Rev.* E **64** 056101
[4] Landau D P, Tsai S-H, and Exler M 2004 *Am. J. Phys.* **72** 1294
[5] Rathore N and de Pablo J J 2002 *J. Chem. Phys.* **116** 7225
[6] Alder S, Trebst S, Hartmann A K, and Troyer M 2004 *J. Stat. Mech.* **2004** P07008

[6] Note that we consider the effect of our parallel scheme as a *whole*, compared to the original, single-walker Wang–Landau method. That is, there are multiple factors contributing to the speed-up as defined here. In a purely technical definition, where algorithmic and methodological influences are absent, speed-ups larger than the number of used processors are obviously unreachable.

[7] Taylor M P, Paul W, and Binder K 2009 *J. Chem. Phys.* **131** 114907
[8] Langfeld K, Lucini B, and Rago A 2012 *Phys. Rev. Lett.* **109** 111601
[9] Zhou C and Bhatt R N 2005 *Phys. Rev.* E **72** 025701(R)
[10] Zhou C, Schulthess T C, Torbrügge S, and Landau D P 2006 *Phys. Rev. Lett.* **96** 120201
[11] Belardinelli R and Pereyra V 2007 *Phys. Rev.* E **75** 046701
[12] Yamaguchi C and Kawashima N 2002 *Phys. Rev.* E **65** 056710
[13] Wu Y, Körner M, Colonna-Romano L, Trebst S, Gould H, Machta J, and Troyer M 2005 *Phys. Rev.* E **72** 046704
[14] Wüst T and Landau D P 2009 *Phys. Rev. Lett.* **102** 178101
[15] Lee H K, Okabe Y, and Landau D P 2006 *Comput. Phys. Commun.* **175** 36
[16] Khan M O, Kennedy G, and Chan D Y C 2005 *J. Comput. Chem.* **26** 72
[17] Zhan L 2008 *Comput. Phys. Commun.* **179** 339
[18] Wang Z, Wang L, and He X 2013 *Soft Matter* **9** 3106
[19] Yin J and Landau D P 2012 *Comput. Phys. Commun.* **183** 1568
[20] Vogel T, Li Y W, Wüst T, and Landau D P (*in preparation*)
[21] Geyer C J 1991 *Computing Science and Statistics: Proceedings of the 23rd Symposium on the Interface* ed Keramidas E M (Fairfax Station, VA: Interface Foundation) p 156
[22] Marinari E and Parisi G 1992 *Europhys. Lett.* **19** 451
[23] Hukushima K and Nemoto K 1996 *J. Phys. Soc. Jpn.* **65** 1604
[24] Nogawa T, Ito N, and Watanabe H 2011 *Phys. Rev.* E **84** 061107 (note the misprint in the corresponding equation)
[25] Zierenberg J, Marenz M, and Janke W 2013 *Comput. Phys. Commun.* **184** 1155
[26] Li Y W, Wüst T, and Landau D P 2013 *Phys. Rev.* E **87** 012706
[27] Gai L, Vogel T, Maerzke K A, Iacovella C R, Landau D P, Cummings P T, and McCabe C 2013 *J. Chem. Phys.* **139** 054505
[28] Vogel T, Li Y W, Wüst T, and Landau D P (*submitted to J. Phys.: Conf. Ser.*)
[29] Goetz R and Lipowsky R 1998 *J. Chem. Phys.* **108** 7397
[30] Fujiwara S, Itoh T, Hashimoto M, and Horiuchi R 2009 *J. Chem. Phys.* **130** 144901
[31] Dill K A 1985 *Biochemistry* **24** 1501
[32] Li Y W, Wüst T, and Landau D P 2011 *Comput. Phys. Commun.* **182** 1896
[33] Yue K and Dill K A 1995 *Proc. Natl. Acad. Sci. USA* **92** 146
[34] Vogel T, Bachmann M, and Janke W 2007 *Phys. Rev.* E **76** 061803

25th IUPAP Conference on Computational Physics (CCP2013) IOP Publishing
Journal of Physics: Conference Series **510** (2014) 012013 doi:10.1088/1742-6596/510/1/012013

# Dynamic connectivity algorithms for Monte Carlo simulations of the random-cluster model

**Eren Metin Elçi and Martin Weigel**

Applied Mathematics Research Centre, Coventry University, Coventry, CV1 5FB, United Kingdom and
Institut für Physik, Johannes Gutenberg-Universität Mainz, Staudinger Weg 7, D-55099 Mainz, Germany

E-mail: elcie@uni.coventry.ac.uk, Martin.Weigel@coventry.ac.uk

**Abstract.** We review Sweeny's algorithm for Monte Carlo simulations of the random cluster model. Straightforward implementations suffer from the problem of computational critical slowing down, where the computational effort per edge operation scales with a power of the system size. By using a tailored dynamic connectivity algorithm we are able to perform all operations with a poly-logarithmic computational effort. This approach is shown to be efficient in keeping online connectivity information and is of use for a number of applications also beyond cluster-update simulations, for instance in monitoring droplet shape transitions. As the handling of the relevant data structures is non-trivial, we provide a Python module with a full implementation for future reference.

## 1. Introduction

The cluster-update algorithm introduced for simulations of the Potts model by Swendsen and Wang in 1987 [1] has been a spectacular success, reducing the effect of critical slowing down by many orders of magnitude for the system sizes typically considered in computer simulation studies. A number of generalizations, including an algorithm for continuous-spin systems and the single-cluster variant [2] as well as more general frameworks for cluster updates [3, 4], have been suggested following the initial work of Ref. [1]. The single bond update introduced by Sweeny [5] several years before Swendsen's and Wang's work is considerably less well known. This is mostly due to difficulties in its efficient implementation in a computer code, which is significantly more involved than for the Swendsen-Wang algorithm. In deciding about switching the state of a given bond from inactive to active or *vice versa*, one must know the consequences of the move for the connectivity properties of the ensemble of clusters, i.e., whether two previously disjoint clusters will become connected or an existing cluster is broken up by the move or, instead, the number of clusters will stay unaffected. If implemented naively, these connectivity queries require a number of steps which is asymptotically close to proportional to the number of spins, such that the resulting *computational critical slowing down* outweighs the benefit of the reduced autocorrelation times of the updating scheme. Even though it was recently shown that the decorrelation effect of the single-bond approach is asymptotically *stronger* than that of the Swendsen-Wang approach [6, 7], this strength can only be played once the computational critical slowing down is brought under control. Here, we use a poly-logarithmic dynamic connectivity algorithm as recently suggested in the computer science literature [8, 9, 10] to perform bond

Journal of Physics: Conference Series **510** (2014) 012013 doi:10.1088/1742-6596/510/1/012013

insertion and removal operations as well as connectivity checks in run-times only logarithmically growing with system size, thus removing the problem of algorithmic slowing down. As the mechanism as well as the underlying data structures for these methods are not widely known in the statistical physics community, we here use the opportunity to present a detailed description of the approach. For the convenience of the reader, we also provide a Python class implementing these codes, which can be used for simulations of the random-cluster model or rather easily adapted to different problems where dynamic connectivity information is required.

## 2. The random-cluster model and Sweeny's algorithm

We consider the random-cluster model (RCM) [11] which is a generalization of the bond percolation problem introducing a correlation between sites and bonds. It is linked to the $q$-state Potts model through the Fortuin-Kasteleyn transformation [12, 13, 14], generalizing the Potts model to arbitrary real $q > 0$. Special cases include regular, uncorrelated bond percolation ($q = 1$) as well as the Ising model ($q = 2$). To define the RCM, consider a graph $\mathcal{G} \equiv (V, E)$ with vertex set $V$, ($|V| \equiv N$), and edge set $E$, ($|E| \equiv M$). We associate an occupation variable $\omega(e) \in \{0, 1\}$ with every edge $e \in E$. We say that $e$ is *active* if $\omega(e) = 1$ and *inactive* otherwise. The state space $\Omega$ of the RCM corresponds to the space of all (spanning[1]) sub-graphs,

$$\Omega \equiv \{\mathcal{A} = (V, A) | A \subseteq E\} = \{0, 1\}^M. \tag{1}$$

A configuration is thus represented as $\vec{\omega} \equiv [\omega(e_1), \omega(e_2), ..., \omega(e_M)] \in \Omega$ and corresponds uniquely to a sub-graph $\mathcal{A}(\vec{\omega}) \equiv (V, A(\vec{\omega})) \subseteq \mathcal{G}$ with

$$A(\vec{\omega}) = \{e \in E | \omega(e) = 1\} \subseteq E. \tag{2}$$

The probability associated with a configuration $\vec{\omega} \in \Omega$ is given by the RCM probability density function (PDF)

$$\mu(p, q, \vec{\omega}) \equiv \frac{1}{Z(p, q)} \left[\prod_{e \in E} p^{\omega(e)} (1-p)^{1-\omega(e)}\right] q^{k(\vec{\omega})}, \tag{3}$$

$$Z(p, q) \equiv \sum_{\vec{\omega} \in \Omega} \left[\prod_{e \in E} p^{\omega(e)} (1-p)^{1-\omega(e)}\right] q^{k(\vec{\omega})}, \tag{4}$$

where $k(\vec{\omega})$ is the number of connected components (clusters) and $Z(p, q)$ denotes the RCM partition function. More generally, as a function of the parameters $p \in [0, 1]$, the density of active edges, and $q \in (0, \infty)$, the cluster number weight, these expressions define a family of PDFs. It is worthwhile to mention a number of limiting cases. For $q \to 1$ Eq. (3) factorizes and corresponds to independent bond percolation with $Z(p) \to 1$. In the limit of $q \to 0$ with fixed ratio $w = v/q$, on the other hand, it corresponds to bond percolation with local probability $w/(1 + w)$ and the condition of cycle-free graphs. Taking $w \to \infty$ or in the limit of $q \to 0$ and $v/q^\sigma$ constant for $0 < \sigma < 1$ we obtain the ensemble of uniform spanning trees for connected $G$. Naturally, in the latter two limits every edge in a configuration is a bridge.

Sweeny's algorithm [5] is a local bond updating algorithm directly implementing the configurational weight (3). We first consider its formulation for the limiting case $q \to 1$ of independent bond percolation. For an update move, randomly choose an edge $e \in E$ with uniform probability and propose a flip of its state from inactive to active or *vice versa*. Move acceptance can be implemented with any scheme satisfying detailed balance, for instance the Metropolis acceptance ratio $\min(v^{\Delta w}, 1)$ where $v \equiv p/(1-p)$ and $\Delta \omega = \pm 1$ for insertions and

[1] A subgraph $\mathcal{A} \subseteq G$ is spanning if it contains all vertices of $G$.

deletions of edges, respectively. This dynamical process is described by the following master equation:

$$\mathbb{P}(\vec{\omega}, t+1) = (1 - r(\vec{\omega}))\mathbb{P}(\vec{\omega}, t) + \sum_{\vec{\omega}' \neq \vec{\omega}} W(\vec{\omega}' \to \vec{\omega})\mathbb{P}(\vec{\omega'}, t), \tag{5}$$

where $r(\vec{\omega}) = \sum_{\vec{\omega}' \neq \vec{\omega}} W(\vec{\omega} \to \vec{\omega'})$ ensures proper normalization of $\mathbb{P}(\vec{\omega}, t+1)$, given the normalization of $\mathbb{P}(\vec{\omega}, t)$ . The Metropolis transition rates are then given by

$$W(\vec{\omega} \to \vec{\omega}') \equiv \frac{1}{M} \sum_{m=1}^{M} W_m(\vec{\omega} \to \vec{\omega}'), \tag{6}$$

$$W_m(\vec{\omega} \to \vec{\omega}') \equiv \min(1, v^{\Delta w}) \prod_{l \neq m} \left[\delta_{\omega_l, \omega_l'}\right]. \tag{7}$$

Eq. (6) expresses the uniform random selection of an edge and the corresponding edge dependent transition rate $W_m$ is defined in Eq. (7). The product of Kronecker deltas ensures the single-bond update mechanism, i.e., that only one edge per step is changed. From here, generalization to arbitrary $q \in (0, \infty)$ is straightforward, leading to a modified transition rate

$$W_m(\vec{\omega} \to \vec{\omega}') \equiv \min(1, q^{\Delta k} v^{\Delta \omega}) \prod_{l \neq m} \left[\delta_{\omega_l, \omega_l'}\right]. \tag{8}$$

We note that this Metropolis update is more efficient than a heat-bath variant for any value of $q$ apart from $q = 1$, where both rules coincide. Clearly, for $q \neq 1$, to compute the acceptance probability $W_m$ of a given trial move one must find $\Delta k$, the change in connected components (clusters) induced by the move. This quantity, equivalent to the question of whether the edge $e$ is a *bridge*, is highly non-local. Determining it involves deciding whether there exists at least one alternative path of active edges connecting the incident vertices $x$ and $y$ that does not cross $e = (x, y)$.

## 3. The connectivity problem

The dynamic connectivity problem is the task of performing efficient connectivity queries to decide whether two vertices $x$ and $y$ are in the same ($x \leftrightarrow y$) or different ($x \nleftrightarrow y$) connected components for a dynamically evolving graph, i.e., mixing connectivity queries with edge deletions and insertions. For a static graph, such information can be acquired in asymptotically constant time after a single decomposition, for instance using the Hoshen-Kopelman algorithm [15]. Under a sequence of edge insertions (but no deletions), it is still possible to perform all operations, insertions and connectivity queries, in practically constant time using a so-called union-and-find (UF) data structure combined with path-compression and tree-balance heuristics [16]. This fact has been used to implement a very efficient algorithm for the (uncorrelated) percolation problem [17]. An implementation of Sweeny's algorithm, however, requires insertions as well as deletions to ensure balance. Hence, we need to be able to remove edges without the need to rebuild the data structure from scratch.

This goal can be reached using a number of different techniques. Building on the favorable behavior of the UF method under edge insertions and connectivity queries, the data structure can be updated under the removal of an external edge (bridge) by performing breadth-first searches (BFSs) through the components connected to the two ends $x$ and $y$ of the edge $e = (x, y)$. Alternatively, one might try to do without any underlying data structure, answering each connectivity query through a separate graph search in breadth-first manner. In both cases, the process can be considerably sped up by replacing the BFSs by *interleaved* traversals alternating between vertices on the two sides of the initial edge and terminating the process as soon as one

Table 1: Asymptotic run-time scaling at criticality of the elementary operations of insertion or deletion of internal or external edges, respectively, using sequential breadth-first search (SBFS), interleaved BFS (IBFS), union-and-find (UF) or the fully dynamic connectivity algorithm (DC) as a function of the linear system size $L$.

| operation | SBFS | IBFS | UF | DC |
|---|---|---|---|---|
| internal insertion | $L^{d_F - x_2}$ | $L^{d_F - x_2}$ | const. | $\log^2 L$ |
| external insertion | $L^{\gamma/\nu}$ | $L^{d_F - x_2}$ | const. | $\log^2 L$ |
| internal deletion | $L^{d_F - x_2}$ | $L^{d_F - x_2}$ | $L^{d_F - x_2}$ | $\log^2 L$ |
| external deletion | $L^{\gamma/\nu}$ | $L^{d_F - x_2}$ | $L^{\gamma/\nu}$ | $\log^2 L$ |
| dominant | $L^{\gamma/\nu}$ | $L^{d_F - x_2}$ | $L^{\gamma/\nu}$ | $\log^2 L$ |

of the two searches comes to an end [18, 19, 7]. As, at criticality of the model, the sizes of the two cluster fragments in case of a bridge bond turn out to be very uneven on average, this seemingly innocent trick leads to dramatic run-time improvements [7]. The asymptotic run-time behavior of insertion and deletion steps for internal and external edges and the algorithms based on BFS or UF data structures is summarized in Table 1 for the case of simulations on the square lattice of edge length $L$. We expect the same bounds with the corresponding exponents to hold for general critical hypercubic lattices. Here, $\gamma/\nu$ denotes the finite-size scaling exponent of the susceptibility and $d_F - x_2$ is a geometric exponent related to the two-arm crossing behavior of clusters [19]. We note that $d_F - x_2 < \gamma/\nu$ for $0 < q \leq 4$ in two dimensions. Asymptotically, it is the most expensive operation which dominates the run-time of the algorithm and, as a consequence, it turns out that (for the square lattice) a simple BFS with interleaving is more efficient than the approach based on union-and-find, cf. Table 1.

In any case, the implementations discussed so far feature a computational effort for a sweep of bond updates that scales faster than linearly with the system size, thus entailing some computational critical slowing down. It is found in Ref. [7] that for most choices of $q$, this effect appears to asymptotically destroy any advantage of a faster decorrelation of configurations by the Sweeny algorithm as compared to the Swendsen-Wang method. An alternative technique based on more complicated data structures allows to perform any mix of edge insertions, deletions and connectivity queries in *poly-logarithmic* run-time per operation [8, 9, 10]. Poly-logarithmic here denotes polynomials of powers of the logarithm of the independent variable, e.g., system size $L$, of the form

$$f(L) = \sum_{k=0}^{K} \alpha_k \log^k L. \tag{9}$$

Here the base of the logarithm is not important and changes only the coefficients. In the following all logarithms are with respect to base 2. Given the observation of generally faster decorrelation of configurations by the Sweeny algorithm in the sense of smaller dynamical critical exponents [6, 7], the use of such (genuinely) dynamic connectivity algorithms (DC) allows for an asymptotically more efficient simulation of the critical random-cluster model [7]. In the following, we discuss the basic ideas and some details of the algorithm employed here.

### *3.1. Trees, Forests and Euler tours*

The DC algorithm is based on the observation that for a given sub-graph $\mathcal{A}$ it is possible to construct a spanning forest $\mathcal{F}(\mathcal{A})$ which is defined by the following properties:

- $x \leftrightarrow y$ in $\mathcal{F}(\mathcal{A})$ if and only if $x \leftrightarrow y$ in $\mathcal{A}$
- there exists exactly one path for every pair $x$, $y$ with $x \leftrightarrow y$

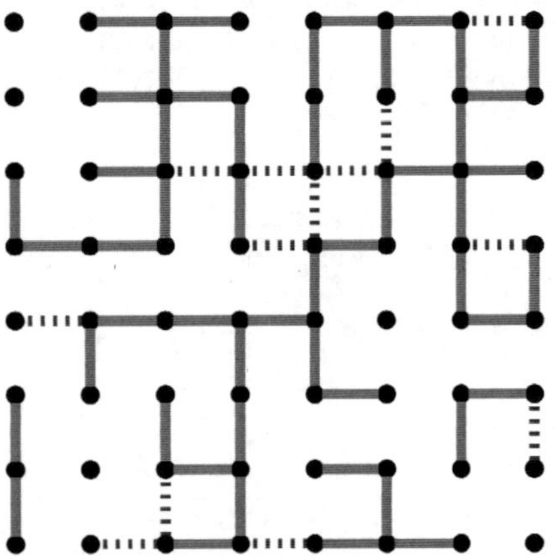

(a) Level 0.

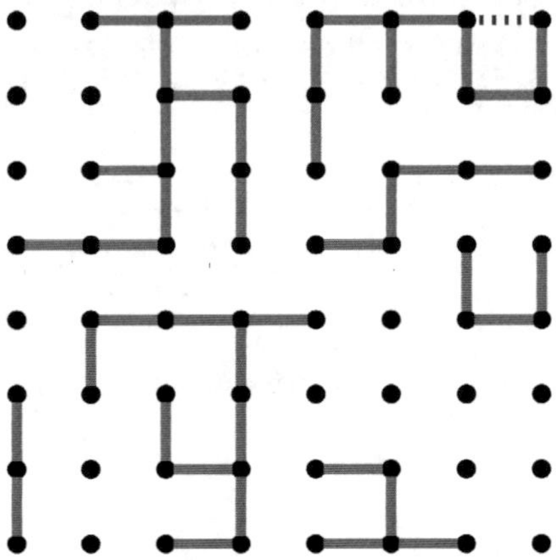

(b) Level 1.

Figure 1: A sub-graph $\mathcal{A}$ of a $8 \times 8$ square lattice. The solid (red) lines correspond to edges in one spanning forest $\mathcal{F}(\mathcal{A})$ and the dashed (blue) lines are additional edges not in the current spanning forest. Note that the graph has periodic boundary conditions and hence there are edges wrapping around the lattice horizontally and vertically. For the sake of clarity these are not shown.

In other words a spanning forest of a graph associates a spanning tree to every component, an example is given in Fig. 1a (solid lines only). One advantage of $\mathcal{F}(\mathcal{A})$ is that it has fewer edges than $\mathcal{A}$, but represents the same connectivity information. For the sub-graph $\mathcal{A}$, the distinction of *tree edges* $e \in \mathcal{F}(\mathcal{A})$ and *non-tree edges* $e \in \mathcal{A} \backslash \mathcal{F}(\mathcal{A})$ allows for a cheap determination of $\Delta k$. For the case of deleting an edge in $e \notin \mathcal{F}(A)$ we know that there is an alternative path connecting the adjacent vertices, namely the path in $\mathcal{F}(\mathcal{A})$, so this edge was part of a cycle and we conclude $\Delta k = 0$. If we insert an edge whose adjacent vertices are already connected in $\mathcal{F}(\mathcal{A})$ then we come to the same conclusion.

If, on the other hand, we want to insert a tree edge, i.e., an edge with adjacent vertices $x$, $y$ not yet connected, we observe that because $x$ and $y$ belong to separate spanning trees before the insertion of $e$, the new spanning subgraph obtained by linking $x$ and $y$ via $e$ is still a spanning tree. Hence the only modification on the spanning forest for the insertion of a tree edge is the amalgamation of two trees. This can be done in $\mathcal{O}(\log L)$ steps by using the following idea of Ref. [20] which also supports the deletion of bridges. For a given component $\mathcal{C}$ in $\mathcal{A}$ we transform the corresponding tree $\mathcal{T}(\mathcal{C})$ in $\mathcal{F}(\mathcal{A})$ into a directed circuit by replacing every edge $(x, y)$ by two directed edges (arcs) $[x, y]$ and $[y, x]$ and every vertex by a loop $[x, x]$. Figure 2 illustrates how to translate edge insertions or deletions into/from the spanning forest to modifications on the directed circuits. Deleting an edge from a tree splits it into two trees. The directed circuit therefore splits into two circuits corresponding to the two trees. When inserting a tree edge, i.e., merging two trees, we join the circuit by the two arcs, corresponding to $e$, at the vertices incident to $e$.

By storing the directed circuits for every component in so called Euler tour sequences (ETS), [8], all necessary manipulations on the directed circuits can be done in $\mathcal{O}(\log L)$ operations if we store each ETS in a separate balanced search tree such as, for instance, a red-black, AVL, or B tree [21]. Alternatively one can use self-adjusting binary search trees, so called splay trees [22]. In this case the bound is amortized, i.e., averaged over the complete sequence of operations. Due to a somewhat simpler implementation and the fact that a Monte Carlo simulation usually consists of millions of operations in random order naturally leading to amortization, we concentrated on

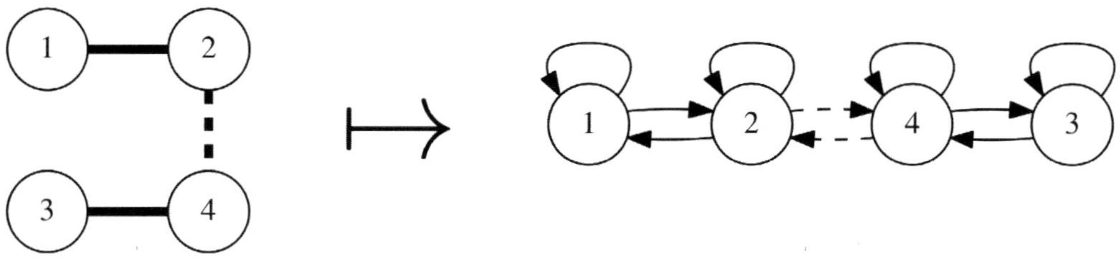


Figure 2: Mapping of a component to a directed circuit with loops. The dashed line is an edge to be deleted. The dashed arrows will be removed to update the directed circuit.

the splay-tree approach. Based on the ETS representation we can translate connectivity queries into checking the underlying search tree roots for equivalence. An in-depth discussion of the exact manipulations on the representing ETS is beyond the scope of this article and we refer the interested reader to the literature [8]. Here we restrict ourselves to considering the deletion of a bridge as an example. The initial graph is the one illustrated in the left panel of Fig. 2. One corresponding ETS sequence of the linearized directed cycle in the right panel of Fig. 2 is the following:

$$\mathcal{E} = [1,1] \to [1,2] \to [2,2] \to [2,4] \to [4,4] \to [4,3] \to [3,3] \to [3,4] \to [4,2] \to [2,1]. \quad (10)$$

Suppose now that we delete edge $(2,4)$ from the tree (cf. the dashed line in Fig. 2). This will result in a split of the component $\mathcal{C} \equiv \{1,2,3,4\}$ into two parts $\mathcal{C}_1 \equiv \{1,2\}$ and $\mathcal{C}_2 \equiv \{3,4\}$. In this case the deletion of edge $(2,4)$ translates into a cut of the original sequence $\mathcal{E}$ at the two arcs $[2,4]$ and $[4,2]$. The sequence $\mathcal{E}_2$ of arcs between these two edges corresponds to $\mathcal{C}_2$ and the concatenation of the remaining sequences without the two arcs corresponding to $(2,4)$ results in $\mathcal{E}_1$ representing $\mathcal{C}_1$:

$$\mathcal{E}_1 = [1,1] \to [1,2] \to [2,2] \to [2,1], \quad (11)$$

$$\mathcal{E}_2 = [4,4] \to [4,3] \to [3,3] \to [3,4]. \quad (12)$$

In summary, we see that by mapping every component in $\mathcal{A}$ to a tree in $\mathcal{F}(\mathcal{A})$ and every such tree to a directed circuit which we store in an ETS we are able to perform edge insertions/deletions into/from $\mathcal{F}(A)$ as well as connectivity queries with an amortised $\mathcal{O}(\log L)$ computational effort.

### *3.2. Edge hierarchy*

The remaining operation not implemented efficiently by the provisions discussed so far is the deletion of edges from $\mathcal{F}(\mathcal{A})$ which are not bridges, i.e., for which a replacement edge exists outside of the spanning forest. The DC algorithm first executes the tree splitting as in the case of a bridge deletion. Additionally, however, it checks for a reconnecting edge in the set of non-tree edges. If such an edge is found, it concludes $\Delta k = 0$ and merges the two temporary trees as indicated above by using the located non-tree edge, which hence now becomes a tree edge. If, on the other hand, no re-connecting edge is found, no additional work is necessary as the initially considered edge is a bridge. To speed up the search for replacement edges, we limit it to the smaller of the two parts $\mathcal{C}_1$ and $\mathcal{C}_2$ as all potential replacement edges must be incident to both components. To allow for efficient searches for non-tree edges incident to a given component using the ETS representation, the search-tree data structures are augmented such that the loop arc for every vertex stores an adjacency list of non-tree edges (vertices) incident to it. Further, every node in the underlying search tree representing the ETS carries a flag indicating if any non-tree edge is available in the sub-tree [which is a sub-tree of the search tree and not of $\mathcal{F}(\mathcal{A})$].

Journal of Physics: Conference Series **510** (2014) 012013 doi:10.1088/1742-6596/510/1/012013

This allows for a search of replacement edges using the Euler tour and ensures that any non-tree edge can be accessed in $\mathcal{O}(\log L)$ time.

It turns out that exploiting this observation is, in general, beneficial but not sufficient to ensure the amortized time complexity bound indicated for the DC algorithm in general. Suppose that the graph consists of a giant homogeneous component with $M = \alpha N$ and $0 < \alpha \leq 1$ and the edge deletion results, temporarily setting aside the question of possible replacement edges, in two trees with $\alpha_1 N$ and $(\alpha - \alpha_1)N$ incident edges, respectively, where $0 < \alpha_1 \leq \alpha/2$. Then the computational effort caused by scanning all possible non-tree edges is clearly $\mathcal{O}(N)$. In amortizing onto the insertions performed to build up this component, every such non-tree edge carries a weight of $\mathcal{O}(\log L)$. If this case occurs sufficiently frequently, it will be impossible to bound the amortized cost per operation. This problem is ultimately solved in the DC algorithm by the introduction of an edge hierarchy. The intuitive idea is to use the expense of a replacement edge search following a deletion to reduce the cost of future operations on this edge. This is done in such a way as to separate dense from sparse clusters and more central edges from those in the periphery of clusters. By amortizing the cost of non-tree edge scans and level increases over edge insertions it follows that one can reduce the run-time for graph manipulations to an amortized $\mathcal{O}(\log^2 L)$ and $\mathcal{O}(\log L)$ for connectivity queries [9]. Each time an incident non-tree edge is checked and found unsuitable for reconnecting the previously split cluster we promote it to be in a higher level. If we do this many times for a dense component we will be able to find incident non-tree edges very quickly in a higher level. These ideas are achieved in the DC algorithm by associating a level function to each $e \in E$,

$$0 \leq \ell(e) \leq \ell_{\max} \equiv \lfloor \log N \rfloor. \tag{13}$$

Based on this level function, one then constructs sub-graphs $\mathcal{A}_i \subseteq \mathcal{A}$ with the property

$$\mathcal{A}_i \equiv \{e \in \mathcal{A} | l(e) \geq i\}. \tag{14}$$

This induces a hierarchy of sub-graphs:

$$\mathcal{A}_{\ell_{\max}} \subseteq \cdots \subseteq \mathcal{A}_1 \subseteq \mathcal{A}_0 \equiv \mathcal{A}. \tag{15}$$

As described above, for every sub-graph we construct a spanning forest $\mathcal{F}_i \equiv \mathcal{F}(\mathcal{A}_i)$. Clearly the same hierarchy holds for the family of spanning forests. In other words the edges in level $i-1$ connect components/trees of level $i$. If an edge has to be inserted into $\mathcal{A}$, then it is associated to a level $l(e) = 0$ and hence it is in $\mathcal{A}_0$. To achieve an efficient search for replacement edges, the algorithm adapts the level of edges after deletions of tree edges in a way which preserves the following two invariants [9]:

(i) The maximal number of vertices in a component in level $i$ is $\lfloor N/2^i \rfloor$.

(ii) Any possible replacement edge for a previously deleted edge $e$ with level $l$ has level $\leq l$.

Trivially, both invariants are fulfilled when all edges have level 0. We now have to specify how exactly the idea of keeping important edges at low levels and unimportant ones at higher levels is implemented. To do this, suppose we deleted an edge from $\mathcal{F}_i$, i.e., at level $i$, and temporarily have $\mathcal{T} \to \mathcal{T}_1 + \mathcal{T}_2$ where (say) $\mathcal{T}_1$ is the smaller of the two, i.e., it has less vertices. Because of invariant (i) it follows that we are allowed to move the tree $\mathcal{T}_1$ (which is now at most half the size of $\mathcal{T}$) to level $i+1$ by increasing the level of all tree edges of $\mathcal{T}_1$ by one. After that we start to search for a replacement edge in the set of non-tree edges stored in the ETS of $\mathcal{T}_1$ in level $i$ where it also remains because of the fact that $\mathcal{F}_{i+1} \subseteq \mathcal{F}_i$. For every scanned non-tree edge we have two options:

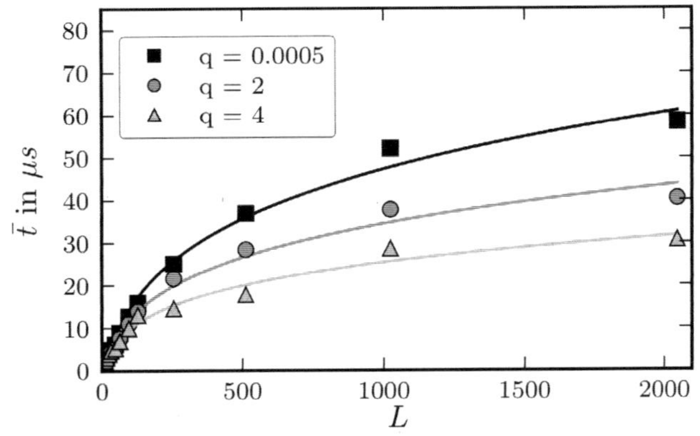


Figure 3: Average run-time for Sweeny's algorithm for the RCM for different values of $q$ at the critical point $v_c = \sqrt{q}$ on the 2D square lattice with periodic boundary conditions.

- It does not reconnect $\mathcal{T}_1$ and $\mathcal{T}_2$ and has therefore both ends incident to $\mathcal{T}_1$. In this case, we increase the level of this edge $i \to i + 1$. This implements the idea of moving unimportant edges in "dense" components to higher levels.
- It does reconnect and hence we re-insert it at level 0.

If we have not found a replacement edge at level $i$ we continue at level $i-1$. The search terminates after unsuccessfully completing the search at level 0 or when a replacement edge was found. In the first case it follows $\mathcal{C} \to \mathcal{C}_1 + \mathcal{C}_2$ whereas in the second case $\mathcal{C}$ remains unchanged.

Implementing this replacement-edge search following any tree-edge deletion introduces an upward flow of edges in the hierarchy of graphs and the level of an edge in the current graph never decreases. Focusing on a single edge, we see that it is sequentially moved into levels of smaller cluster size and hence the cost of future operations on this edge is reduced. Taking this into account it follows that the insertion of an edge has a cost of $\mathcal{O}(\log L)$ for inserting at level 0 plus it also "carries" the cost of all possible $\mathcal{O}(\log L)$ level increases with cost of $\mathcal{O}(\log L)$ each resulting in $\mathcal{O}(\log^2 L)$ amortized per insertion. Deletions on the other hand imply a split of cost $\mathcal{O}(\log L)$ in $\mathcal{O}(\log L)$ levels. In case of an existing replacement edge another contribution of $\mathcal{O}(\log L)$ caused by an insertion at level 0 is added. The contribution of moving tree edges to higher levels and searching for replacement edges (moving non-tree edges up) is already paid for by the sequence of previous insertions (amortization). The only missing contribution is the $\mathcal{O}(\log L)$ effort for obtaining the next replacement edge in an ETS. In total, deletions hence have an amortized computational cost of $\mathcal{O}(\log^2 L)$.

### *3.3. Performance and optimizations*

We tested the performance of the current DC implementation in the context of Sweeny's algorithm in comparison to the simpler approaches based on breadth-first search and union-and-find strategies. While the algorithm discussed here allows all operations to be performed in poly-logarithmic time, due to the complicated data structures the constants are relatively large. Our results show consistency with the poly-logarithmic run-time bounds derived. It appears, however, that very large system sizes are required to clearly see the superior asymptotic performance of the DC algorithm as compared to the BFS and UF implementations. For details see the more elaborate discussion in Ref. [7]. As an example, Fig. 3 shows the average run-time per edge operation as a function of the system size for three different choices of parameters.

Apart from run-time considerations, the implementation has a rather significant space complexity. Since we maintain $\mathcal{O}(\log L)$ overlapping forests over the $L^2$ vertices, the space complexity is $\mathcal{O}(L^2 \log L)$. A heuristic suggested in Ref. [10] to decrease memory consumption is a truncation of higher edge levels as these are, for the inputs or graphs considered in our application, sparsely populated. We checked the impact on our implementation by comparing run-times and memory consumptions for a truncation $\ell_{\mathrm{max}} \to \ell_{\mathrm{max}}/2$. We did not see any significant change in the run-time. On the other hand we observed a reduction of almost a factor of two in the memory consumption. This conforms to our observation that during the course of a simulation almost no edges reached levels beyond $\approx 10$ for system sizes $L \lesssim 1024$ where the actual maximal level according to Eq. (13) is $\ell_{\mathrm{max}} = 20$.

Likewise, a number of further optimizations or heuristics are conceivable to improve the typical run-time behavior. This includes a sampling of nearby edges when looking for a replacement edge before actually descending into the edge level hierarchy [10]. A number of such heuristics and experimental comparisons of fully and partially dynamics connectivity algorithms has been discussed in the recent literature, see Refs. [23, 24, 25]. A full exploration of these possibilities towards an optimal implementation of the DC class of algorithms for the purpose of the Sweeny update is beyond the scope of the current article and forms a promising direction for future extensions of the present work.

## 4. Sweeny Python class

We provide a Python class [26] encompassing four different implementations of Sweeny's algorithm based on:

- sequential breadth-first searches (SBFS)
- interleaved breadth-first searches (IBFS)
- union-and-find with interleaved breadth-first searches (UF)
- poly-logarithmic dynamic connectivities as discussed here (DC)

The package is built on top of a C library and it is therefore possible to use the library in a stand-alone compiled binary. The necessary source code is also provided. For more details see the related project documentation [26]. The source code is published under the MIT license [27]. Here we give a basic usage example, which simulates the RCM with $q = 2$ (the Ising model) at $v_c = \sqrt{2}$, using an equilibration time of 1000 sweeps, a simulation length of 10000 sweeps, and random number seed 1234567 using the DC implementation:

Listing 1: Example usage of Sweeny class.

```
from sweeny import Sweeny
sy = Sweeny(q=2.,l=64,beta=np.log(1. + np.sqrt(2.)),coupl=1.,
cutoff=1000,tslength=10000,rngseed=1234567,impl='dc')
sy.simulate()
```

In order to extract an estimate, say, of the Binder cumulant $R = \langle \mathcal{S}_4 \rangle / \langle \mathcal{S}_2^2 \rangle$ we need to retrieve the time series for $\mathcal{S}_4$ and $\mathcal{S}_2$,

Listing 2: Retrieving time series.

```
sec_cs_moment= sy.ts_sec_cs_moment
four_cs_moment = sy.ts_four_cs_moment
sec_cs_moment *= sec_cs_moment
binder_cummulant = four_cs_moment.mean()/sec_cs_moment.mean()
```

Once an instance of the Sweeny class is created, it is easy to switch the algorithm and parameters as follows:

Listing 3: Switching algorithm and parameters.

```
sy.init_sim(q=1.3,l=64,beta=np.log(1.+np.sqrt(1.3.)),coupl=1.,
cutoff=5000,tslength=50000,rngseed=7434,impl='ibfs')
```

## 5. Conclusions

We have shown how to implement Sweeny's algorithm using a poly-logarithmic dynamic connectivity method and we described the related algorithmic aspects in some detail. We hope that the availability of the source code and detailed explanations help to bridge the gap between the computer science literature on the topic of dynamic connectivity problems and the physics literature related to MC simulations of the RCM, specifically in the regime $q < 1$.

The availability of an efficient dynamic connectivity algorithm opens up a number of opportunities for further research. This includes studies of the tricritical value $q_c(d)$ where the phase transition of the random-cluster model becomes discontinuous for dimensions $d > 2$ [28, 18] as well as the nature of the ferromagnetic-paramagnetic transition for $q \to 0$ and $d > 2$ [29].

## Acknowledgments

E.M.E. would like to thank P. Mac Carron for carefully reading the manuscript.

## References

[1] Swendsen R H and Wang J S 1987 *Phys. Rev. Lett.* **58** 86–88
[2] Wolff U 1989 *Phys. Rev. Lett.* **62** 361–364
[3] Edwards R G and Sokal A D 1988 *Phys. Rev. D* **38** 2009
[4] Kandel D and Domany E 1991 *Phys. Rev. B* **43** 8539–8548
[5] Sweeny M 1983 *Phys. Rev. B* **27** 4445
[6] Deng Y, Garoni T M and Sokal A D 2007 *Phys. Rev. Lett.* **98** 230602
[7] Elçi E M and Weigel M 2013 *Phys. Rev. E* **88** 033303
[8] Henzinger M R and King V 1999 *J. ACM* **46** 502–516
[9] Holm J, de Lichtenberg K and Thorup M 2001 *J. ACM* **48** 723–760
[10] Iyer R, Karger D, Rahul H and Thorup M 2001 *J. Exp. Algorithmics* **6** 4
[11] Grimmett G 2006 *The random-cluster model* (Berlin: Springer)
[12] Fortuin C M and Kasteleyn P W 1972 *Physica* **57** 536–564
[13] Fortuin C M 1972 *Physica* **58** 393–418
[14] Fortuin C M 1972 *Physica* **59** 545–570
[15] Hoshen J and Kopelman R 1976 *Phys. Rev. B* **14** 3438–3445
[16] Cormen T H, Leiserson C E, Rivest R L and Stein C 2009 *Introduction to Algorithms* 3rd ed (Cambridge, MA: MIT Press)
[17] Newman M E J and Ziff R M 2001 *Phys. Rev. E* **64** 016706
[18] Weigel M 2010 *Physics Procedia* **3** 1499–1513
[19] Deng Y, Zhang W, Garoni T M, Sokal A D and Sportiello A 2010 *Phys. Rev. E* **81** 020102
[20] Tarjan R E 1997 *Math. Program.* **78**
[21] Knuth D E 1998 *The Art of Computer Programming: Sorting and Searching v. 3: The Classic Work Newly Updated and Revised* 2nd ed (Addison Wesley)
[22] Sleator D D and Tarjan R E 1985 *J. ACM* **32** 652–686
[23] Zaroliagis C D 2002 *Experimental Algorithmics* (*Lecture Notes in Computer Science* vol 2547) ed Fleischer R, Moret B and Schmidt E (Springer Berlin Heidelberg) pp 229–278
[24] Demetrescu C and Italiano G F 2006 *ACM Trans. Algorithms* **2** 578–601
[25] Demetrescu C, Eppstein D, Galil Z and Italiano G F 2010 ed Atallah M J and Blanton M (Chapman & Hall/CRC), chapter *Dynamic graph algorithms*, p 9
[26] Elçi E M Sweeny Python module at Github, `https://github.com/ernmeel/sweeny`
[27] MIT License, `http://opensource.org/licenses/MIT`
[28] Hartmann A K 2005 *Phys. Rev. Lett.* **94** 050601
[29] Deng Y, Garoni T M and Sokal A D 2007 *Phys. Rev. Lett.* **98** 030602

Journal of Physics: Conference Series **510** (2014) 012014 doi:10.1088/1742-6596/510/1/012014

# Non-equilibrium critical dynamics of pure and diluted 2D $XY$-model

**P V Prudnikov†, I S Popov‡**

Department of Theoretical Physics, Omsk State University, Omsk 644077, Russia

E-mail: †`prudnikp@univer.omsk.su`, ‡`diphosgen@mail.ru`

**Abstract.** Comprehensive study of the ageing and the violation of the fluctuation-dissipation theorem in a pure and diluted two-dimensional $XY$-model were carried out. Two-dimensional $XY$-model tend to a particular type of phase transition is called a phase transition *Berezinskii-Kosterlitz-Thouless* associated with the change of the correlation properties of the system.

## 1. Introduction

In the past few years, systems with slow dynamics have attracted considerable theoretical and experimental interest [1–3]. Ageing phenomena are observed during this everlasting non-equilibrium evolution. This is due to the predicted and observed ageing at a slow evolution of non-equilibrium systems and violations of the fluctuation-dissipation theorem [1, 4]. Known examples of such systems with abnormal slow dynamics are complex disordered systems like glasses: dipole, metal and spin glasses [5–7]. However, these features of the non-equilibrium behaviour, as shown by analytical and numerical studies [8], can also be observed in systems near the critical temperature as critical dynamics of such systems is characterized an abnormally large relaxation times. The two-dimensional $XY$-model demonstrates abnormal slow dynamics, but the main distinguishing feature of the 2D $XY$-model is that it shows an anomalous behaviour is not only close to the temperature phase transition *Berezinskii-Kosterlitz-Thouless* $T_{\mathrm{BKT}}$, but $T < T_{\mathrm{BKT}}$ [9, 10]. Ageing is a phenomenon of growth of the relaxation time system with the increase of the *age* of the material ie older samples respond more slowly [1, 4, 11]. The *age* of the system $t_w$ or also waiting time, being the time elapsed since the preparation of the system. An important aspect of ageing systems is that the equilibrium fluctuationdissipation theorem (FDT) does not hold.

Relevance of the study of two-dimensional $XY$-model due to the extensive a number of physical systems whose behaviour can be described by this model. There are examples of such systems: monolayer magnetic films of transition metals such as Co and Ni, on a nonmagnetic substrate, such as Cu [12]; an important class of planar magnetics [12, 13]; two-dimensional crystals [13]; superconducting thin films [13]; two-dimensional Bose liquid films of superfluid He [10, 13]; Josephson junction array and the array of SFS contacts [13, 14].

Most real-world systems contain structural defects that can have a significant influence on the critical behaviour of the system. According to the Harris criterion [15] the presence of structural defects can be significant if critical exponent of the heat capacity $\alpha$ is positive. Thus, it is predicted that the presence of structural defects does not affect to two-dimensional $XY$-model behaviour near the critical temperature $T_{\mathrm{BKT}}$. However, in low-temperature phase for

$T < T_{\rm BKT}$, as shown by analytical and numerical study of the equilibrium properties of the model [16–18], the presence of defects lead to changing of the equilibrium characteristics for the correlation features and their concentration dependence. However, the dynamics of structurally disordered $XY$-dimensional model was still not investigated carefully.

In this paper the non-equilibrium critical dynamics of pure and diluted $2D$ $XY$-model were investigated by Monte Carlo methods in whole low-temperatures *Berezinskii-Kosterlitz-Thouless* phase.

## 2. Model and methods

The $XY$-model is a example of the general lattice models with n-component parameter, with $n = 2$. Phase transition in such systems associated with the changing of the asymptotic behaviour of the correlation functions from an exponential decay in high-temperature phase to power decay in low-temperature phase.

The two-dimensional $XY$ model is defined by the Hamiltonian [9]:

$$H = -J \sum_{<i,j>}^{N} \vec{S}_i \vec{S}_j \tag{1}$$

where $J > 0$ is the exchange integral, $\vec{S}_i$ - classical planar spin, associated with the $i$-th node of two-dimensional lattice, $N = L^2$.

The Hamiltonian for disordered model has the form:

$$H = -J \sum_{<i,j>}^{N} p_i p_j \vec{S}_i \vec{S}_j \tag{2}$$

where $p_i$ are random variables: $p_i = 1$, if the $i$-th node is the spin, and $p_i = 0$, if the $i$-th node is a non-magnetic atom.

Dynamics of the two-dimensional $XY$-model is provided by two contributions: the spin waves and topological excitations such as vortex and antivortex (Fig. 1).

In this paper, for the study of non-equilibrium critical behaviour of two-dimensional $XY$-model was used Metropolis algorithm [19]. The dynamics of one-spin-flips Metropolis corresponds to dissipative processes which described by dynamic model $A$ in Halperin-Hohenberg classification [20])

$$\frac{\partial \phi(x,t)}{\partial t} = -\Gamma_0 \frac{\delta H}{\delta \phi(x,t)} + \xi(x,t), \tag{3}$$

where $\phi$ – order parameter, $\Gamma_0$ – kinetic factor, $\xi(x,t)$ – gaussian random force. The dynamic correlation functions are determined by the solution of this equation.

## 3. Ageing effects

The ageing effects in non-equilibrium critical relaxation of $2D$ $XY$-model were studied from low-temperature initial state with $T = 0$, $m_0 = 1$ and from high-temperature initial state with $T \gg T_{\rm BKT}$, $m_0 \ll 1$. The values of ageing times were chosen $t_w = 100$, 500 and 1000 MCS/s. The initial state with $m_0 \ll 1$ is prepared by the dynamic evolution of the system from $T_0 = 3.0 \gg T_{\rm BKT}$. The value of initial magnetization was $m_0 = 0.001$ with accuracy $\Delta m_0 = 0.01 m_0$.

The time dependencies of autocorrelation function were presented in Fig. 2. The data clearly demonstrates presence of two scaling regimes with power dependence of autocorrelation function:

$$A(t,t_w) = \left\langle \frac{1}{N} \sum_i \vec{S}_i(t) \vec{S}_i(t_w) \right\rangle - \left\langle \frac{1}{N} \sum_i \vec{S}_i(t) \right\rangle \left\langle \frac{1}{N} \sum_i \vec{S}_i(t_w) \right\rangle. \tag{4}$$

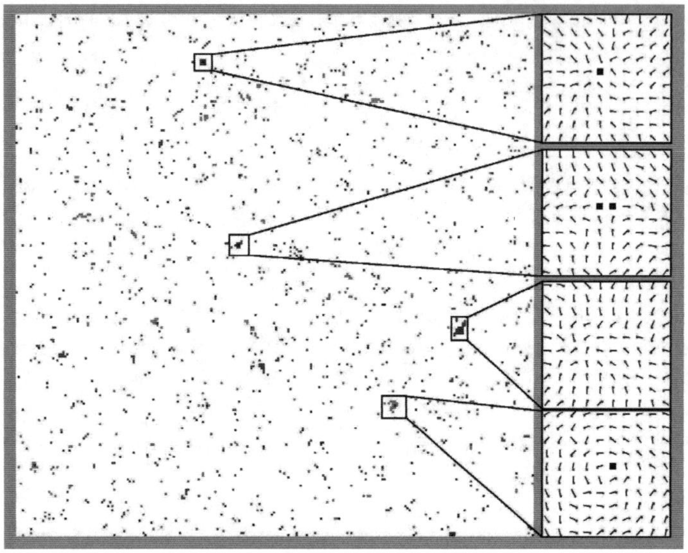

**Figure 1.** Spin system configuration for $p = 0.995$, $L = 256$, $T = 0.3$ at moment of time $t = 10000$ MCS/s after starting from the initial high-temperature state with $m_0 \ll 1$. Vortex excitations are designated by red points, antivortex excitations – blue points, structural defects – black points

To characterize these time intervals it have been introduced exponent time dependence of the autocorrelation function:

$$A(t, t_w) = (t - t_w)^{-\Delta_A}. \tag{5}$$

The obtained values of exponents $\Delta_A$ are presented for system with low-temperature and high-temperature initial state in Table. 1. It was shown that critical slow down and ageing effects becomes stronger with increasing $t_w$ for $T < T_{\mathrm{BKT}}$.

The scaling behaviour of autocorrelation function can be defined by form [22]

$$A(t, t_w) = \frac{1}{(t - t_w)^{\eta(T)/2}} \Phi\left(\frac{\xi(t)}{\xi(t_w)}\right). \tag{6}$$

where $\xi \sim t/\ln(t)$ is the correlation length, $\eta$ – spatial exponent insuring the best collapse of data, $\Phi$ is the scaling function which behaviour is presented in the insets of Fig. 2. $\Phi(x) = 1$ for small values of $x$. Long-time decay of $\Phi(x)$ is well described by a power law, i.e. $\Phi(x) \sim x^{-\lambda}$ for large $x$. The calculated values of the exponents $\lambda$ are presented in Table. 2. The value $\lambda = 0.5384(9)$ for $T = 0.3$ is in good agreement with value $\lambda = 0.54$ obtained in [22].

Presented in Fig. 2 the behaviour of $A(t, t_w)$ demonstrate that for time regime $t - t_w \sim t_w$ the influence of vortex excitations is irrelevant and the main contribution to the relaxation dynamics has only the spin-wave nature. In the time regime $t - t_w \gg t_w$ the role of vortex interaction have been increased and become significant.

It has revealed significant differences in the non-equilibrium evolution from the different initial states. These differences can be explained by role of vortex interaction. In system which started from high-temperature state the initial value of vortex-antivortex excitations is significant and vortex-antivortex interaction produce drastically effect on non-equilibrium dynamics. When system started from low-temperature state it needs energy for formation of vortex and antivortex excitations and the role of vortex interaction is not so significant.

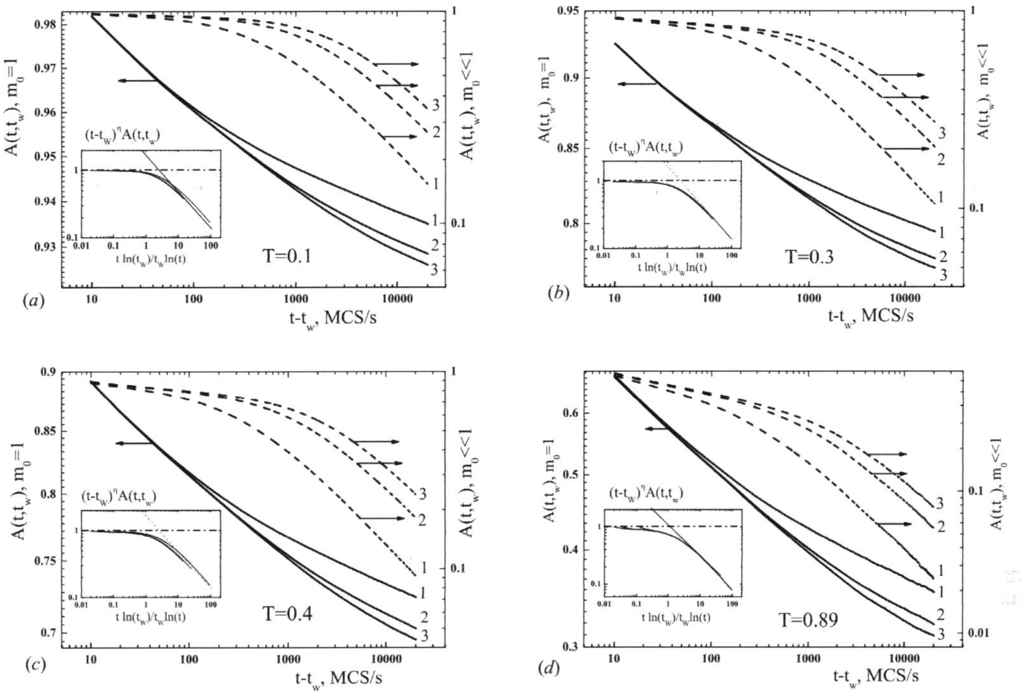


**Figure 2.** Time dependencies of the autocorrelation function $A(t, t_w)$ as a function $t - t_w$ for different non-equilibrium initial states and values of $t_w$: 1 – $t_w = 100$, 2 – $t_w = 500$, 3 – $t_w = 1000$ MCS/s. The relaxation curves were presented for low-temperature phase $T < T_{\text{BKT}}$ states for $T = 0.1$ $(a)$, $T = 0.3$ $(b)$, $T = 0.4$ $(c)$, $T = 0.89$ $(d)$. The solid lines are correspond to low-temperature initial state, the dashed lines and scaling (insets) – high-temperature initial state

## 4. The influence of structural disorder on the non-equilibrium critical behaviour

The presence of structural disorder and non-magnetic impurities breaks the interaction of vortices and antivortices. The vortices are pinneded 1 at the defects [25].

To estimate the $T_{\text{BKT}}(p)$ dependence on spin concentration $p$ for disordered systems was calculated temperature dependence of the ratio $R$

$$R = \frac{[\langle C(L/2)\rangle]}{[\langle C(L/4)\rangle]}, \tag{7}$$

where $C(L)$ is the correlation function

$$C(r) = \left\langle p_i p_{i+r} \vec{S}_i \vec{S}_{i+r} \right\rangle, \tag{8}$$

the angle brackets stand for an statistical average over MC realizations and the square brackets stand for additional average over disorder configurations.

Temperature dependencies of $R(L)$ were investigated for spin concentrations $p = 1.0$ (Fig. 3a), $p = 0.9$, $p = 0.8$ and $p = 0.7$ (Fig. 3b) for systems with linear sizes $L = 16, 32, 48$. The averaging was carried out on over 10000 MCS/s and 50 samples with different disorder configurations with 10 MC runs for each sample. The error bars for spin concentration $p = 1.0$ (Fig. 3a), 0.9 and 0.8 are smaller than the size of the symbols. The significant value of error bars for $p = 0.7$ (Fig. 3b) can be explained by the breakdown of the phase transition.

**Table 1.** The values of exponent $\Delta_A$ for different initial states

| $T/J$ | $t_w = 100$ | | $t_w = 500$ | | $t_w = 1000$ | |
|---|---|---|---|---|---|---|
| | low-temperature initial state with $m_0 = 1$ | | | | | |
| | [0;60] | [1000;10000] | [0;60] | [1000;10000] | [0;100] | [10000;20000] |
| 0.2 | 0.0185(4) | 0.0091(1) | 0.0197(3) | 0.0093(1) | 0.0190(3) | 0.0093(1) |
| 0.4 | 0.0379(8) | 0.0193(1) | 0.0400(6) | 0.0203(1) | 0.0389(5) | 0.0206(1) |
| 0.6 | 0.0603(10) | 0.0313(1) | 0.0635(9) | 0.0322(1) | 0.0620(6) | 0.0356(1) |
| 0.8 | 0.0903(12) | 0.0477(8) | 0.0948(12) | 0.0483(1) | 0.0931(8) | 0.0534(1) |
| 0.89 | 0.1112(15) | 0.0623(9) | 0.1176(40) | 0.0649(2) | 0.1164(9) | 0.0597(2) |
| | high-temperature initial state with $m_0 \ll 1$ | | | | | |
| | [0;60] | [1000;20000] | [0;60] | [5000;20000] | [0;60] | [1000;20000] |
| 0.2 | 0.041(2) | 0.461(3) | 0.026(5) | 0.457(8) | 0.024(6) | 0.41(5) |
| 0.4 | 0.064(1) | 0.484(1) | 0.050(9) | 0.461(8) | 0.046(4) | 0.42(7) |
| 0.6 | 0.096(1) | 0.499(3) | 0.078(1) | 0.479(7) | 0.071(6) | 0.44(3) |
| 0.8 | 0.148(1) | 0.514(9) | 0.117(3) | 0.489(8) | 0.106(5) | 0.44(4) |
| 0.89 | 0.184(1) | 0.612(3) | 0.146(9) | 0.614(7) | 0.135(7) | 0.56(4) |

**Table 2.** The values of exponents $\lambda$

| $T/J$ | 0.1 | 0.2 | 0.3 | 0.4 | 0.5 | 0.6 | 0.7 | 0.8 | 0.89 |
|---|---|---|---|---|---|---|---|---|---|
| $\lambda$ | 0.5216(3) | 0.5428(5) | 0.5384(9) | 0.5267(4) | 0.5188(9) | 0.5200(8) | 0.4983(8) | 0.4812(9) | 0.6151(3) |

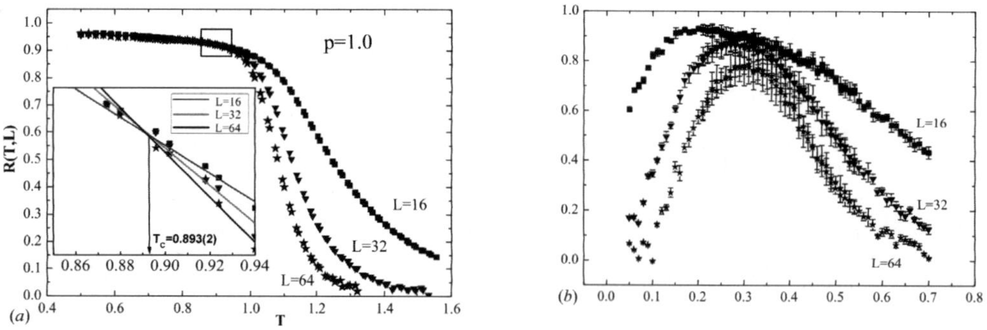


**Figure 3.** Temperature dependencies of ratio $R(L)$ for the spin concentration $p = 1.0$ ($a$) and $p = 0.7$ ($b$) for different $L$

The MC obtained values of critical temperature are $T_{\rm BKT}(p = 1.0) = 0.893(2)$, $T_{\rm BKT}(p = 0.9) = 0.679(7)$, $T_{\rm BKT}(p = 0.8) = 0.485(4)$. It can be seen that the presence of disorder considerably reduces the temperature of the phase transition. The value of $T_{\rm BKT}(p = 1.0) = 0.893(2)$ for pure system is in a good agreement with value $T_{\rm BKT} = 0.894(5)$ [26]. For $p = 0.7$

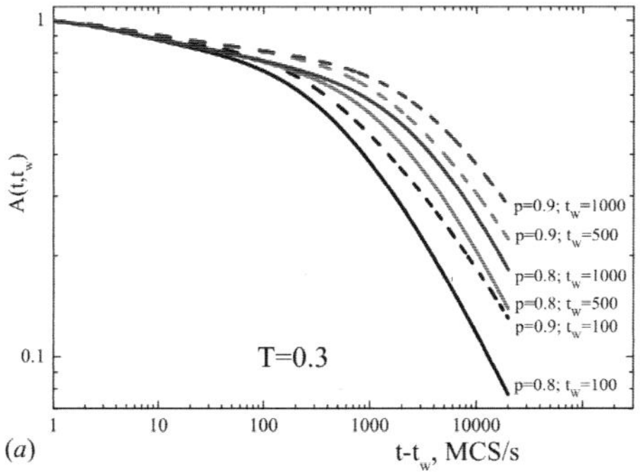


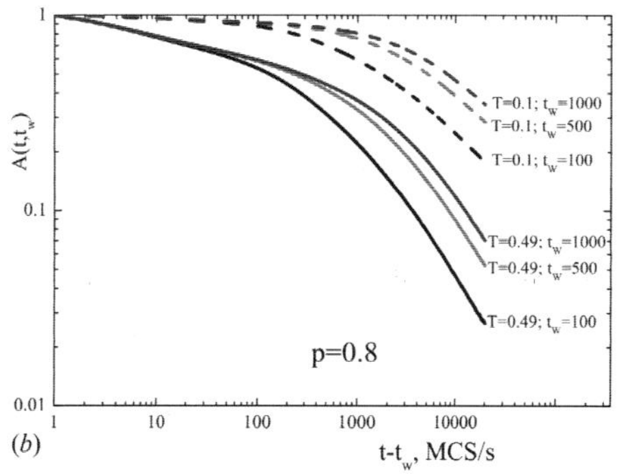


**Figure 4.** Time dependence of autocorrelation function $A(t, t_w)$ from $t - t_w$ for different $p$

the intersection wasn't find and this behaviour can be associated with the breakdown of the *Berezinskii-Kosterlitz-Thouless* phase transition.

Feature of non-equilibrium behaviour of systems with slow dynamics is a violation of the translational invariance of time due to the long-term effect of non-equilibrium initial states of such systems. This is manifested primarily in two-time characteristics of the system, such as the correlation function and the response function. When these functions are non-equilibrium processes depend on two variables, the temporary nature of $t$ and $t_w$ ( when $t > t_w$,) and not only on the difference but also on each separately. Moreover, this dependence is maintained for sufficiently long times observation $t$.

The ageing effects were investigated for disordered systems with $p = 0.8$ and $p = 0.9$ started from high-temperature initial state. The two-time autocorrelation function is defined for disordered systems by the following expression:

$$A(t, t_w) = \left[\left\langle \frac{1}{pN} \sum_i p_i \vec{S}_i(t) \vec{S}_i(t_w) \right\rangle\right] - \left[\left\langle \frac{1}{pN} \sum_i p_i \vec{S}_i(t) \right\rangle\right] \left[\left\langle \frac{1}{pN} \sum_i p_i \vec{S}_i(t_w) \right\rangle\right]. \quad (9)$$

The two-time dependencies of the autocorrelation function for different $p$ are presented in Fig. 4. The data is clerly shows that the presence of disorder lead to increasing of the system relaxation from initial non-equilibrium high-temperature state.

## 5. Violation of the fluctuation-dissipation theorem

Another slow dynamics effect at out-of-equilibrium stage is the violation of the fluctuation-dissipation theorem (FDT) [27]. FDT suggests relation between autocorrelation function $A(t, t_w)$ and response function $R(t, t_w)$ on some external field

$$R(t, t_w) = \frac{X(t, t_w)}{T} \frac{\partial A(t, t_w)}{\partial t_w}, \quad (10)$$

where $X(t, t_w)$ is so called fluctuation-dissipation ratio (FDR), and one assumes $t > t_w$. According to FDT $X(t, t_w) = 1$ in equilibrium. The asymptotic value of the FDR

$$X^{\infty} = \lim_{t_w \to \infty} \lim_{t \to \infty} X(t, t_w) \quad (11)$$

can be used as a new universal critical characteristic of a systems with slow dynamics.

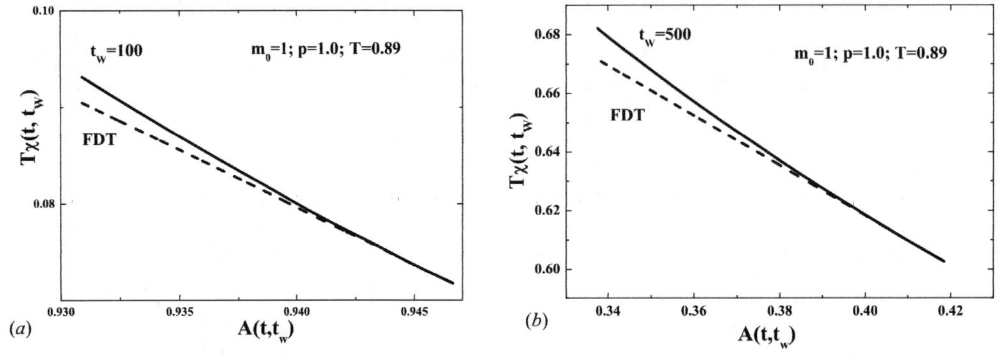


**Figure 5.** FDR plot for $T = 0.89$: $t_w = 100$ MCS/s ($a$) and $t_w = 500$ MCS/s ($b$). The dashed line corresponds to the $X(t, t_w) = 1$

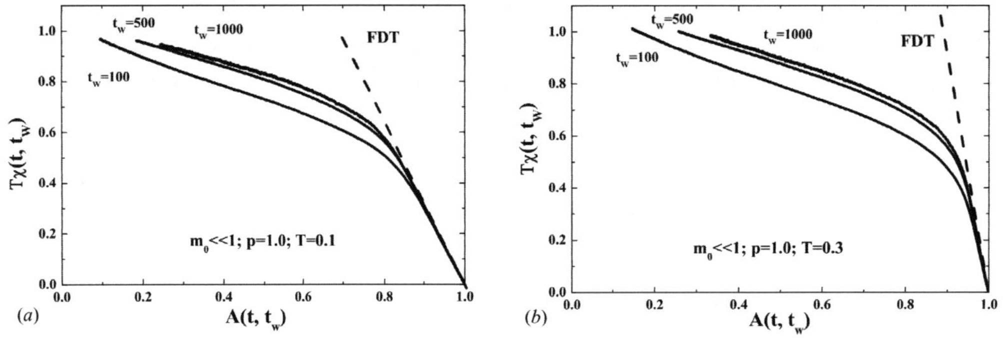


**Figure 6.** FDR plot for $T = 0.1$ ($a$) and $T = 0.3$ ($b$). The dashed line corresponds to the $X(t, t_w) = 1$

FDR can be estimated with application of small magnetic field after waiting time $t_w$. In this case the response function $R(t, t_w)$ can not be calculated directly and more useful quantity is the integrated response function [28].

$$\chi(t, t_w) = T \int_{t_w}^{t} dt' R(t, t'). \tag{12}$$

In the large time limit $\chi(A) = \int_A^1 X(q) dq$ and the FDR can be defined as

$$X(t, t_w) = - \lim_{A \to 0} \frac{\partial T\chi(t, t_w)}{\partial A(t, t_w)}. \tag{13}$$

$\chi(t, t_w)$ for the two-dimensional $XY$-model can be calculated based on the following relationship:

$$\chi(t, t_w) = \frac{1}{L^2 h^2} \sum_i \overline{< \vec{h_i}(t_w) \vec{S}(t) >}, \tag{14}$$

where the random magnetic field $h$ is characterized by two components which are independently drawn from a bimodal distribution $\pm h$. The overline means an average over the realizations

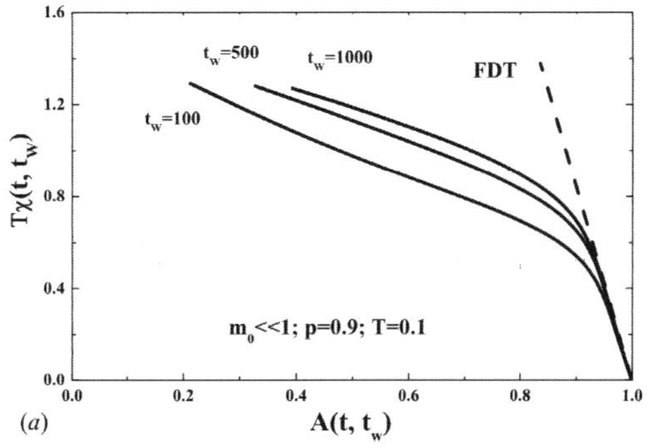


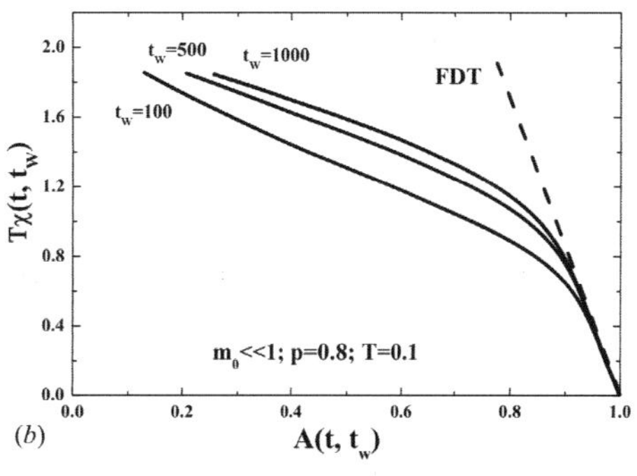


**Figure 7.** FDR plot for disordered system: $T = 0.1$; $p = 0.9$ $(a)$ and $p = 0.8$ $(b)$. The dashed line corresponds to the $X(t, t_w) = 1$

of the random magnetic field. We checked that the value $h = 0.04$ was small enough for the response to be linear.

In Fig. 5 we present for initial state with $m_0 = 1$ the obtained parametric $T\chi(t, t_w)$ versus $A(t, t_w)$ plot for $T_{BKT} = 0.89$ and for different waiting times $t_w$. The dashed line represents $X(t, t_w) = 1$, while the solid line represents $X(t, t_w) > 1$, which means that the susceptibility is larger than its equilibrium value and it takes place the violation of FDT. To obtain the FDR on the base of relation (13), we analysed the dependencies $T\chi(t, t_w)$ from $A(t, t_w)$, found the slopes of curves for different $t_w$, and then made extrapolation $t_w \to \infty$. The asymptotic value of FDR for system started from low-temperature initial state with $m_0 = 1$ is $X_\infty = 2.49(13)$.

The same procedure was used for describing FDT violation in non-equilibrium behaviour of the system started from high-temperature initial state. The MC obtained data is presented in Fig. 6. The asymptotic value of FDR for system started from initial state with $m_0 \ll 1$ $X_\infty = 0.47(3)$.

The violation of FDT in disordered 2D $XY$-model were discovered. We present in Fig. 7 the FDR plot for disordered system with different $p$. It clear demonstrates that the presence of disorder lead to increasing of FDT violation.

## 6. Conclusion

In this paper we have investigated the non-equilibrium effects in critical behaviour of pure and structural disordered 2D $XY$-model. It has revealed significant differences in the non-equilibrium evolution from the different initial states. These differences can be explained by role of vortex interaction. In system which started from high-temperature state the initial value of vortex-antivortex excitations is significant and vortex-antivortex interaction produce drastically effect on non-equilibrium dynamics. When system started from low-temperature state it needs energy for formation of vortex and antivortex excitations and the role of vortex interaction is not so significant.

The violation of FDT for different initial non-equilibrium states have been discovered. The final values of the asymptotical fluctuation-dissipation ratio are $X_\infty(m_0 \ll 1) = 0.47(3)$, $X_\infty(m_0 = 1) = 2.49(13)$.

The presence of structural disorder lead to increasing of the system relaxation and violation of FDT. It have been explained by vortex pinning on defects.

## Acknowledgments
The simulations were supported by the Supercomputing Center of Lomonosov Moscow State University and Joint Supercomputer Center of the Russian Academy of Sciences.

## References
[1] Afzal N and Pleimling M 2013 *Phys. Rev.* B **87** 012114
[2] Picone A and Henkel M 2004 *Nucl. Phys.* B **688** 217
[3] Schehr G and Paul R 2005 *Phys. Rev.* E **72** 016105
[4] Calabrese P and Gambassi A 2005 *J. Phys.* A. **38** R133
[5] Vincent E, Hammann J, Ocio M, Bouchaud J-P and Cugliandolo L F 1997 *Slow dynamics and ageing in spin glasses* ed Rubi M *Complex behaviour of glassy systems* Lecture Notes in Physics (Heidelberg, Springer) 492
[6] Vincent E 2007 *Ageing, rejuvenation and memory: the example of spin-glasses* ed Henkel M, Pleimling M, Sanctuary R *Ageing and the glass transition* Lecture Notes in Physics (Heidelberg, Springer) 716
[7] Dupuis V, Bert F, Bouchaud J-P, Hammann J, Ladieu F, Parker D and Vincent E 2005 *Pramana Journal of Physics* **64** 1109
[8] Pleming M and Gambassi A 2005 *Phys. Rev.* B **71** 180401(R)
[9] Berezinskii V L 1970 *JETP* **59** 907
[10] Kosterlitz J M and Thouless D J 1973 *J. Phys.* C **6** 1181
[11] Lei X W and Zheng B 2007 *Phys. Rev.* E **75** 040104
[12] Vaz C A F, Bland J A C and Lauhoff G 2008 *Reports on Progress in Physics* **71** 056501
[13] Korshunov S E 2006 *Physics-Uspekhi* **176** 233
[14] Zhukovskii E A, Ismagilov A M and Kopaev Yu V 1992 *JETP* **101** 231
[15] Harris A B 1974 *J. Phys.* C **7** 1671
[16] Berche B, Farinas-Sanchez A I, Holovatch Yu and Paredes R 2003 *Eur. Phys. J.* B **36** 91
[17] Kapikranian O, Berche B and Holovatch Yu 2007 *Eur. Phys. J.* B **56** 93
[18] Kapikranian O, Berche B and Holovatch Yu 2007 *Phys. Lett.* A. **366** 150
[19] Metropolis N, Rosenbluth A W, Rosenbluth M N, Teller A H and Teller E 1953 *J. Chem. Phys.* **21** 1087
[20] Hohenberg P C and Halperin B I 1977 *Rev. Modern Phys.* **49** 435
[21] Henkel M and Pleimling M 2010 *Non-equilibrium Phase Transitions. Volume 2: Ageing and dynamical scaling far from equilibrium* Theoretical and Mathematical Physics (Heidelberg, Springer) p 544
[22] Berthier L, Holdsworth P C W and Sellitto M 2001 *J. Phys.* A **34** 1805
[23] Binder K and Landau D P 1976 *Phys. Rev.* B **13** 1140
[24] Berche B 2003 *J. Phys.* A Math Gen **36** 585
[25] Pereira A R, Mól L A S, Leonel S A, Coura P Z and Costa B V 2003 *Phys. Rev.* B **68** 132409
[26] Gupta R and Baillie C F 1992 *Phys. Rev.* B **45** 2883
[27] Calabrese P and Gambassi A 2002 *Phys. Rev.* E **65** 066120
[28] Ricci-Tersenghi F 2003 *Phys. Rev.* E, **68** 065104
[29] Mayer P, Bissig H, Berthier L, Cipelletti L, Garrahan J P, Sollich P and Trappe V 2004 *Phys. Rev. Lett.* **93** 115701
[30] Redner S and Krapivsky P L 1998 *J. Phys.* A Math Gen **31** 9229

Journal of Physics: Conference Series **510** (2014) 012015 doi:10.1088/1742-6596/510/1/012015

# Ageing properties of three-dimensional pure and site-diluted Ising ferromagnets

**Vladimir V. Prudnikov**[†], **Pavel V. Prudnikov**[‡] and **Evgeny A. Pospelov**[§]

Department of Theoretical Physics, Omsk State University, Pr. Mira 55A, Omsk 644077, Russia

E-mail: [†]prudnikv@univer.omsk.su, [‡]prudnikp@univer.omsk.su, [§]posevg@yandex.ru

**Abstract.** Ageing properties of the three-dimensional pure and diluted Ising models are investigated for case with quenching from a high temperature initial state to their critical points. The two-times autocorrelation and response functions for the systems with spin concentrations $p = 1.0, 0.8$, and 0.6 are computed and analysed. The fluctuation-dissipation theorem violation for considered 3D Ising models are shown with the asymptotic values of the fluctuation-dissipation ratio $X^{\infty}(p) < 1/2$. The asymptotic values of universal fluctuation-dissipation ratio $X^{\infty}$ were measured , and it was shown that $X^{\infty}_{\text{disorder}}(p < 1) > X^{\infty}_{\text{pure}}(p = 1)$.

## 1. Introduction

The behaviour of systems with abnormally slow dynamics stays one of the main problem in theoretical physics [1, 2]. It demonstrates a wide range of interesting phenomena such as critical slowing down, ageing, and violation of the fluctuation-dissipation theorem [3]. Originally, these features were founded in complex spin glass systems [4] but different studies of these phenomena showed that the main properties of slow dynamics can be observed in simpler systems with second-order phase transition near critical point, because the evolution of such systems is characterized by large relaxation times.

According to dynamical scaling the relaxation time diverges as $t_{\text{rel}} \sim |T - T_c|^{-z\nu}$ in the vicinity of critical point $T_c$, where $z$ is the dynamic exponent, $\nu$ is the exponent for correlation length. Therefore a system does not achieve an equilibrium state at critical point while relaxation process. During out-of-equilibrium stage of dynamics for $t \ll t_{\text{rel}}$ ageing phenomena occur with two-time dependencies of correlation and response functions on characteristic time variables as waiting time $t_w$ and time of observation $t - t_w$ with $t > t_w$.

Another slow dynamics effect at out-of-equilibrium stage is the violation of the fluctuation-dissipation theorem (FDT) [5]. FDT suggests relation between correlation function $C(t, t_w)$ and response function $R(t, t_w)$ on some external field

$$R(t, t_w) = \frac{X(t, t_w)}{T} \frac{\partial C(t, t_w)}{\partial t_w}, \tag{1}$$

where $X(t, t_w)$ is so called fluctuation-dissipation ratio (FDR), and one assumes $t > t_w$. According to FDT $X(t, t_w) = 1$ in equilibrium. The asymptotic value of the FDR

$$X^{\infty} = \lim_{t_w \to \infty} \lim_{t \to \infty} X(t, t_w) \tag{2}$$

can be used as a new universal critical characteristic of a systems with slow dynamics with $X^\infty \neq 1$ as a signal of an asymptotic non-equilibrium dynamics.

Ageing effects were studied at critical point for 2D Ising model [3, 6] and 3D XY model [7] by Monte Carlo methods. The theoretical description results of the different models was reviewed in [5]. In the paper [8] the off-equilibrium fluctuation-dissipation relation for some disordered systems (diluted ferromagnets, random field Ising model and four-dimensional spin glasses) has been investigated for temperatures far from the critical point with $T < T_c$. The value of asymptotical FDR $X^\infty = 0$ was obtained for these systems. It has been shown that no evidences of replica symmetry breaking have been found in the diluted ferromagnetic model. Non-equilibrium dynamics of the 2D random-bond Ising model has been discovered in [9] by short-dynamics method

In this report we present the numerical investigation of the ageing properties and the fluctuation-dissipation ratio in ferromagnets described by three-dimensional Ising model with point-like defects at critical point.

## 2. Model and methods

The Hamiltonian of disordered ferromagnetic Ising model with local magnetic field $h_i$ is

$$H = -J \sum_{<i,j>} p_i p_j S_i S_j - \sum_i h_i S_i, \tag{3}$$

where $J > 0$ is the short-range exchange interaction between spins $S_i$ fixed at the lattice sites, and assuming values of $S_i = \pm 1$. Nonmagnetic impurity atoms form empty sites. In this case, occupation numbers $p_i$ assume the value 0 or 1, and $p_i$ equal to 1 if site contains spin and 0 otherwise. We considered the cubic lattice with periodic boundary conditions. Let us denote $N = pL^3$ is the number of spins where $L$ is the linear size of the lattice and $p$ is the spin concentration.

For magnetic systems the conjugate response function $R(t, t_w)$ with applied external field $h$ at time $t_w$ is defined by

$$R(t, t_w) = \frac{1}{V} \int d^d x \frac{\delta \langle S(x,t) \rangle}{\delta h(x, t_w)} |_{h=0} \,. \tag{4}$$

There are two ways to compute the FDR. The first way is connected with application of small magnetic field after waiting time $t_w$. In this case the response function $R(t, t_w)$ can not be calculated directly and more useful quantity is the integrated response function [11].

$$\chi(t, t_w) = T \int_{t_w}^{t} dt' R(t, t'). \tag{5}$$

In the large time limit $\chi(C) = \int_C^1 X(q) dq$ and the FDR can be defined as

$$X(t, t_w) = - \lim_{C \to 0} \frac{\partial(T\chi(t, t_w))}{\partial C(t, t_w)}. \tag{6}$$

Another way to calculate the FDR is the use of heat-bath dynamics that allows to obtain the response function directly. The derivative of the heat-bath probability transition is continuous and this fact allows to calculate the response function without taking into account an external magnetic field. We have applied the method suggested by Chatelain [10, 11]. Let us describe the details.

Heat bath dynamics is one of the realization of Glauber dynamics in Monte Carlo simulations [13]. The probability transition of single spin update $S_i \to S_i'$ reads

$$W_{sp}(S_i \to S_i') = \frac{\exp(-\beta H(S_i'))}{\sum_{S_j} \exp(-\beta H(S_j))}, \tag{7}$$

with $S_j$ summed over all possible states of spin $S_i$ before flipping, $\beta = 1/T$ is the inverse temperature. Since there are only two possible spin's states in Ising model $S_j = \pm 1$, the probability transition is

$$W_{sp}(S_i \to S_i^{'}) = \frac{\exp(-\beta H(S_i^{'}))}{\exp(\beta H(S_i)) + \exp(-\beta H(S_i))}. \tag{8}$$

Let us denote by $\varphi(\{S\}, t)$ the probability to obtain system with spin's configuration $\{S\}$ at time t. The dynamic evolution in critical point is Markov process and the master equation can be written as

$$\varphi(\{S\}, t + \Delta t) = (1 - \Delta t)\varphi(\{S\}, t) + \Delta t \sum_{\{S'\}} W(\{S'\} \to \{S\}, t)\varphi(\{S'\}, t). \tag{9}$$

$W(\{S'\} \to \{S\}, t)$ is the probability transition from state $\{S'\}$ to $\{S\}$ at time $t$ with normalization condition $\sum_{\{S'\}} W(\{S'\} \to \{S\}, t) = 1$. The conditional probability $\varphi(\{S\}, t|\{S'\}, t_w)$ satisfies the master equation too

$$\begin{aligned} \varphi(\{S\}, t + \Delta t|\{S'\}, t_w) = (1 - \Delta t)\varphi(\{S\}, t|\{S'\}, t_w) + \\ \Delta t \sum_{\{S''\}} W(\{S''\} \to \{S\}, t)\varphi(\{S''\}, t, \{S'\}, t_w). \end{aligned} \tag{10}$$

The conditional probability is defined by the Bayes theorem

$$\varphi(\{S\}, t) = \sum_{\{S'\}} \varphi(\{S\}, t|\{S'\}, t_w)\varphi(\{S'\}, t_w). \tag{11}$$

The probability transition from $\{S_i\}$ to $\{S_i^{'}\}$ reads (include (7))

$$W(\{S_i\} \to \{S_i^{'}\}, t) = \frac{1}{N}\sum_{k=1}^{N} W_k(\{S\} \to \{S^{'}\}) = \frac{1}{N}\sum_{k=1}^{N}\Big[\prod_{l \neq k} \delta_{S_l S_l^{'}}\Big] W_{sp}(S_k \to S_k^{'}). \tag{12}$$

The probability $W_k$ includes only one spin-flip update $S_k \to S_k^{'}$. Note that in simulations time unit is Monte Carlo step on spin (mcs/s) that defines $N = pL^3$ discrete spin updates. The product of deltas ensures that during spin flip only the one updates.

Using the master equation (9) and Bayes theorem the average of spin $\langle S_j \rangle$ at time $t > t_w$ can be written as

$$\begin{aligned} \langle S_j \rangle &= \sum_{\{S\}} S_j \varphi(\{S\}, t) = \sum_{\{S\},\{S'\}} S_j \varphi(\{S\}, t|\{S'\}, t_w + \Delta t)\varphi(\{S'\}, t_w + \Delta t) \\ &= \sum_{\{S\},\{S'\}} S_j \varphi(\{S\}, t|\{S'\}, t_w + \Delta t)\Big[(1 - \Delta t)\varphi(\{S'\}, t_w) + \\ &\qquad \Delta t \sum_{\{S''\}} W_k(\{S''\} \to \{S^{'}\})\Big]. \end{aligned} \tag{13}$$

In order to obtain the response function one needs to calculate derivative $\frac{\partial \langle S_j(t) \rangle}{\partial h_i}$ and takes the limit $h_i \to 0$. In (13) only $W_k$ depends on $h_i$ through Hamiltonian (3) and the derivative reads

$$\begin{aligned} \Big[\frac{\partial \langle S_j(t) \rangle}{\partial h_i}\Big]_{h_i \to 0} = \Delta t \delta_{k,i} \sum_{\{S\},\{S'\},\{S''\}} S_j \varphi(\{S\}, t|\{S'\}, t_w + \Delta t \\ \times \Big[\frac{\partial W_i}{\partial h_i}(\{S''\} \to \{S^{'}\})\Big]_{h_i \to 0}. \end{aligned} \tag{14}$$

The derivative defines the integrated response function when infinitesimal external field $h_i$ is applied at time range $[t_w, t_w + \Delta t]$

$$\chi_{ji}(t, [t_w, t_w + \Delta t]) = T \frac{\partial \big\langle S_j(t) \big\rangle}{\partial h_i}. \tag{15}$$

Taking derivative of probability transition at(14), one obtains

$$\left[\frac{\partial W_i}{\partial h_i}(\{S''\} \to \{S'\})\right]_{h_i \to 0} = \beta W_i(\{S''\} \to \{S'\})\left[S_i' - S_i^W\right], \tag{16}$$

where $S_i^W = \tanh(\beta J \sum_{m \neq i} S_m')$.

Using master equation (9) and taking into account (16), the integrated response function reads

$$\begin{aligned}\chi_{ji}(t, [t_w, t_w + \Delta t]) = \delta_{k,i} \sum_{\{S\},\{S'\}} & S_j \varphi(\{S\}, t | \{S'\}, t_w + \Delta t)\left[S_i' - S_i^W\right] \\ & \times \left[\varphi(\{S'\}, t_w + \Delta t) - (1 - \Delta t)\varphi(\{S'\}, t_w)\right]. \end{aligned} \tag{17}$$

The response function $R_{ji}$ is connected with the integrated response function by relation

$$\chi_{ji}(t, [t_w, t_w + \Delta t]) = T \int_{t_w}^{t_w + \Delta t} R_{ji}(t, s) ds = T R_{ji}(t, t_w) \Delta t + O(\Delta t^2). \tag{18}$$

As note above, the time is a discrete quantity in Monte Carlo simulations. Assuming $\Delta t = 1$ at (18), we obtain

$$\chi_{ji}(t, [t_w, t_w + \Delta t]) = \delta_{k,i} \Big\langle S_j(t) \left[S_i(t_w + 1) - S_i^W(t_w + 1)\right]\Big\rangle. \tag{19}$$

The final relation for calculation of the response function in Monte Carlo simulations reads

$$R(t, t_w) = \frac{1}{N} \sum_{i=1}^{N} R_{ii} = \frac{1}{N} \beta \sum_{i=1}^{N} \Big\langle S_i(t) \left[S_i(t_w + 1) - S_i^W(t_w + 1)\right]\Big\rangle, \tag{20}$$

where $R(t, t_w)$ is averaged on all spin flips during one Monte Carlo step.

The two-time autocorrelation function reads

$$C(t, t_w) = \Big\langle \frac{1}{N} \sum_{i=1}^{N} S_i(t) S_i(t_w) \Big\rangle, \tag{21}$$

and the autocorrelation function derivative by waiting time $t_w$ can be calculated by

$$\frac{\partial}{\partial t_w} C(t, t_w) = \frac{1}{N} \sum_{i=1}^{N} \Big\langle S_i(t) \left[S_i(t_w + 1) - S_i(t_w)\right]\Big\rangle. \tag{22}$$

Using definition (1) and relations (20) - (22), the fluctuation-dissipation ratio can be defined as

$$X(t, t_w) = \frac{T R(t, t_w)}{\frac{\partial}{\partial t_w} C(t, t_w)} = \frac{\sum_{i=1}^{N} \Big\langle S_i(t) \left[S_i(t_w + 1) - S_i^W(t_w + 1)\right]\Big\rangle}{\sum_{i=1}^{N} \Big\langle S_i(t) \left[S_i(t_w + 1) - S_i(t_w)\right]\Big\rangle}. \tag{23}$$

*2.1. Simulations with probing magnetic field*

In this part of investigations we computed time dependence of the autocorrelation function and integrated response function using the Metropolis algorithm. The simulations have been performed on cubic lattices with linear size $L = 128$, with periodic boundary conditions. The considered spin systems were quenched in the critical point from an infinite-temperature initial state. The averages are performed with the use of 5000 samples characterized by different independent configurations of defects. The two-time autocorrelation function was calculated via relation

$$C(t,t_w) = \left[ \left\langle \frac{1}{pL^3} \sum_{i=1}^{pL^3} p_i S_i(t) S_i(t_w) \right\rangle \right], \tag{24}$$

where the angle brackets stand for an average over initial configurations and realizations of the thermal noise, the square brackets are for averaging over the different impurity configurations. The two-time integrated response function $\chi(t,t_w)$ was computed by the application of a random magnetic field with small amplitude $h = 0.01$ in order to avoid nonlinear effects via:

$$\chi(t,t_w) = \overline{\left[ \left\langle \frac{1}{h^2 pL^3} \sum_{i=1}^{pL^3} h_i(t_w) S_i(t) \right\rangle \right]}, \tag{25}$$

where the line stands for an average over the random field realizations. The autocorrelation functions time dependencies are plotted in Figure 1 for different values of waiting time $t_w$ and spin concentrations $p$.

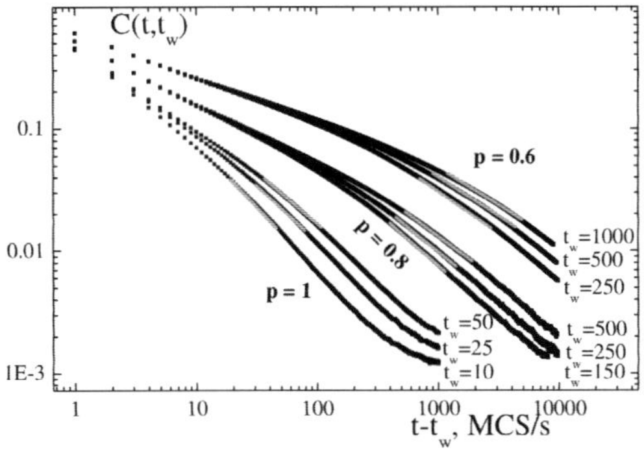


**Figure 1.** Autocorrelation plot for different $p$ and $t_w$. The error bars are smaller than the size of the symbols.

We can distinguish three stages of the relaxation for each system. The first stage is observed for small time separation $t - t_w \ll t_w$ where ageing does not exist and the dynamic evolution of the autocorrelation functions exhibits a stationary part and does not depend on waiting time. The second regime is realized for times $t - t_w \sim t_w \gg 1$ where the correlation functions at different waiting times do not superpose and are characterized by different slopes for each $t_w$. It is the ageing stage of relaxation. At long time separations with $t - t_w \gg t_w \gg 1$ the correlation functions decay as power law

$$C(t,t_w) \sim (t/t_w)^{-c_a}, \quad c_a = \frac{d}{z} - \theta', \tag{26}$$

where exponent $c_a$ is the same which describes time dependence of the autocorrelation function in the short-time regime of non-equilibrium behaviour with well-known initial slip exponent $\theta'$ for the magnetization ($M(t) \sim t^{\theta'}$) [12, 15]. At this stage ageing effects are not developed as well.

At the ageing regime (dashed lines in Figure 1) we approximated the autocorrelation function time-decay as power law $C(t, t_w) \sim (t - t_w)^{-b}$ with the slope exponent $b$ for estimation of correlations falling. Obtained values of $b$ are presented in the table 1 which show that decay of the correlation functions is slower with increasing of waiting time $t_w$. Also we point out that the presence of defects and increasing of their concentration leads to ageing gain.

**Table 1.** The values of the autocorrelation slope exponent for spin concentrations $p = 1.0$, $p = 0.8$ and 0.6.

| $t_w$ | $b$ | $t_w$ | $b$ | |
|---|---|---|---|---|
| | $p = 1.0$ | | $p = 0.8$ | $p = 0.6$ |
| 10 | 1.048(18) | 50 | 0.938(34) | 0.746(32) |
| 25 | 1.023(14) | 250 | 0.739(40) | 0.604(45) |
| 50 | 0.934(11) | 500 | 0.644(25) | 0.531(40) |
| 75 | 0.879(12) | 1000 | 0.569(30) | 0.467(36) |

For stage with $t - t_w \sim t_w \gg 1$ theory of ageing phenomena predicts for the correlation function following two-time scaling dependence $C(t, t_w) \sim t_w^{-2\beta/\nu z} F_c(t/t_w)$. The scaling function $F_c(t/t_w)$ should depends on the universality class of the system. We have checked the collapse of the different waiting time curves. The obtained data are plotted in Figure 2 which demonstrate a quite good collapse of the curves for different waiting times.

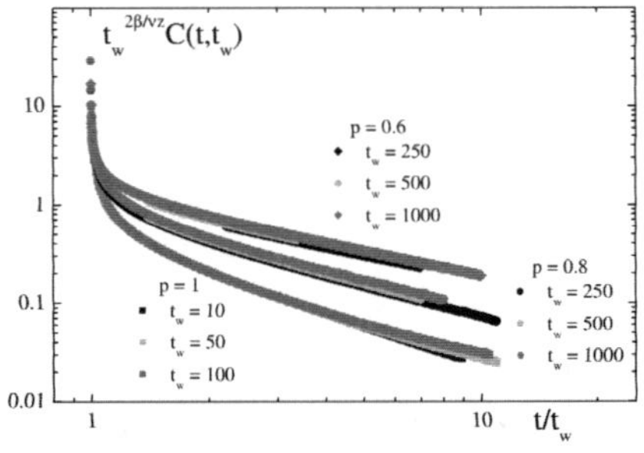


**Figure 2.** Scaling of autocorrelation function $C(t, t_w)$ for different spin concentrations $p$. The error bars are smaller than the symbols.

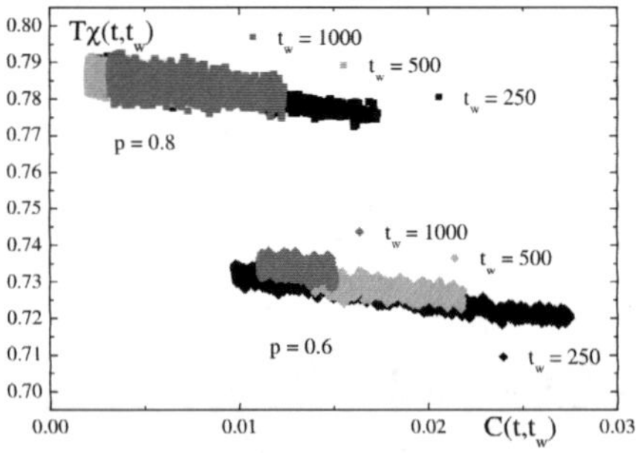


**Figure 3.** $T\chi(t, t_w)$ versus $C(t, t_w)$ plot for $p = 0.8$ and 0.6. The error bars are smaller than the size of the symbols.

For well-separated times with $t - t_w \gg t_w \gg 1$ the scaling function $F_c(t/t_w)$ is characterized by power law time dependence $F_c(t/t_w) \sim A_c(t/t_w)^{-c_a}$ with exponent $c_a$ from (26). From our data we obtained the following values of $c_a$ for systems with different spin concentrations: $c_a(p = 1.0) = 1.333(40), c_a(p = 0.8) = 1.237(22)$, and $c_a(p = 0.6) = 0.982(30)$. We must mention that these values of $c_a$ for pure and weakly diluted Ising models with $p = 0.8$ agree rather well with $c_a = 1.36(1)$ from paper [16] and with $c_a = 1.242(10)$ from paper [15] computed by the short-time dynamics method.

To obtain the FDR on the base of relation (6), we analysed the dependencies $T\chi(t, t_w)$ from $C(t, t_w)$, found the slopes of curves for different $t_w$, and then made extrapolation $t_w \to \infty$. The results are plotted in Figure 3 for diluted systems and presented in table 2.

**Table 2.** The values of the FDR $X^\infty$ for spin concentrations $p = 1.0$, $p = 0.8$, and 0.6.

| $t_w$ | $X^\infty$ | | |
|---|---|---|---|
| | $p = 1.0$ | $p = 0.8$ | $p = 0.6$ |
| 10 | 0.586(24) | | |
| 25 | 0.460(22) | | |
| 50 | 0.437(26) | | |
| 250 | | 0.708(16) | 0.726(13) |
| 500 | | 0.553(17) | 0.583(14) |
| 1000 | | 0.494(17) | 0.519(29) |
| $\to \infty$ | 0.391(12) | 0.419(11) | 0.443(10) |

The obtained values of the fluctuation-dissipation ratio for systems with different spin concentrations will be analysed below in this paper. In next subsection the results of the calculations of the two-time response function $R(t, t_w)$ are presented.

### *2.2. Heat-bath dynamics without application of external magnetic field*

The use of the heat-bath dynamics method allows us to compute and analyse the response function directly. In this case we have simulated the dynamics of the systems starting from the high-temperature initial state with magnetization values $m_0(p = 1) = 0.02$, $m_0(p = 0.8) = 0.01$, and $m_0(p = 0.6) = 0.005$. The response function was calculated using the relation (20). The obtained data were averaged over 90000 statistical realization in case of pure Ising model with $p = 1$ and over 6000 independent impurities configurations for site-diluted Ising models. Note, that each disordered configuration was averaged over 15 statistical realizations of initial state with $m_0 \ll 1$. The ageing properties of the response function are showed in Figure 4.

One can see that the response function has demonstrated the similar to the autocorrelation function time behaviour (see section 2.1). The early stage of critical relaxation $t - t_w \ll t_w$ is characterized by lack of the memory effects, and only at times $t - t_w \sim t_w$ the ageing is developed. On the analogy of the autocorrelation function, theoretical prediction for scaling form of the response function $R(t, t_w)$ is

$$R(t, t_w) \sim t_w^{-1-2\beta/\nu z} F_R(t/t_w). \tag{27}$$

To check this scaling dependency of the $R(t, t_w)$ we plotted $t^{(1+2\beta/\nu z} R(t, t_w)$ versus $(t - t_w)/t_w$. The obtained data are presented in Figure 5. The response function also demonstrates the collapse of curves for different waiting times into single curve with universal scaling dependency. Note that systems with different spin concentrations are characterized by different curves for two-time dependencies of the response function.

The scaling function $F_R(t/t_w)$ has power law time dependency $F_R(t/t_w) \sim A_R(t/t_w)^{-c_r}$ at times $t \gg t_w \gg 1$. At this stage ageing properties has not appeared, and $c_r$ is determined by the same exponents as $c_a$ in asymptotic dependency of the scaling function $F_c(t/t_w)$. So, $c_r = \frac{d}{z} - \theta'$.

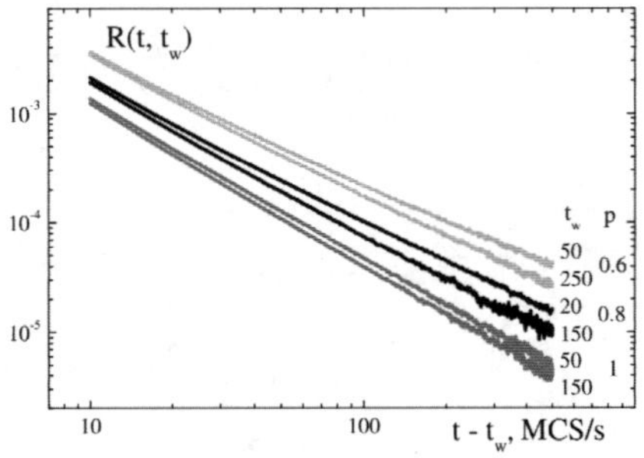


**Figure 4.** The response function plot $R(t,t_w)$ versus waiting time $(t-t_w)$ for different $p$. The error bars are smaller than the size of the symbols.

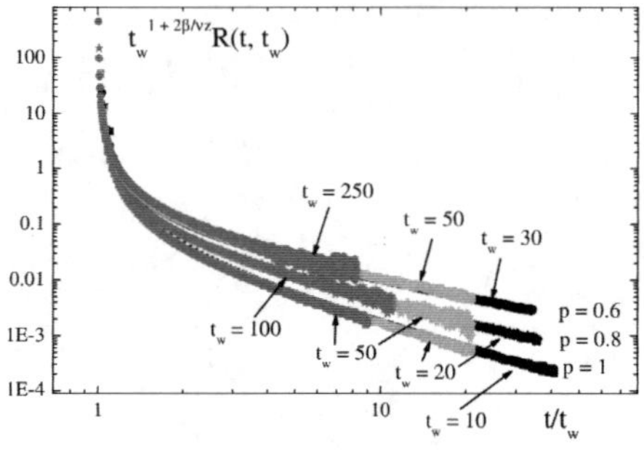


**Figure 5.** Scaling of the response function $R(t,t_w)$. The error bars are smaller than the size of the symbols.

**Table 3.** The values of exponents $c_r$ and $c_a$ .

| | $p=1.0$ | $p=0.8$ | $p=0.6$ |
|---|---|---|---|
| $c_a$ | 1.333(40) | 1.237(22) | 0.982(30) |
| $c_r$ | 1.357(18) | 1.251(22) | 0.950(8) |
| Jaster, et al. [16] | 1.362(19) | | |
| Prudnikov, et al. [15] | | 1.242(10) | |

We collected the values of $c_r$ and $c_a$ from section 2.1 in table 3 and compared their with values which have been obtained by the short-time dynamics method in other works. The exponents demonstrate a good agreement with short-time dynamics results in both pure and weak diluted systems cases.

Using expression (23), we have computed the fluctuation-dissipation ratio. The obtained data are plotted in Figure 6 for different spin concentrations. The plot of the $X(t,t_w)$ evolution consists of two different stages. It is expected that at times $t-t_w \ll t_w$ ageing properties have not appeared yet, and the $X(t,t_w)$ will be close to 1. We used time interval from $t-t_w \sim t_w$ to $t-t_w \gg t_w$ to obtain the asymptotic values of $X(t_w,p)$ for different waiting times.

In order to get the asymptotic values $X^\infty(p)$ (2) we have calculated $X(t_w,p)$ from plot in Figure 6 in the limit $t_w/(t-t_w) \to 0$, i.e. $t-t_w \gg t_w$. Using the obtained values of $X(t_w,p)$, we made extrapolation $t_w \to \infty$ to gain the asymptotic FDR $X^\infty(p)$. The results are showed in table 4 and in Figure 7.

The final values of the fluctuation-dissipation ratio are $X^\infty(p=1)=0.380(13), X^\infty(p=0.8)=0.413(11)$, and $X^\infty(p=0.8)=0.446(8)$. In the table 4 we include the FDR values obtained from analysis of the integrated response function (lower line), see section 2.1. The obtained the FDR values agree quite well with each other for both methods.

### *2.3. Conclusions*

We have investigated the non-equilibrium behaviour of pure and disordered three dimensional ferromagnetic Ising models quenched from high-temperature initial state to critical point.

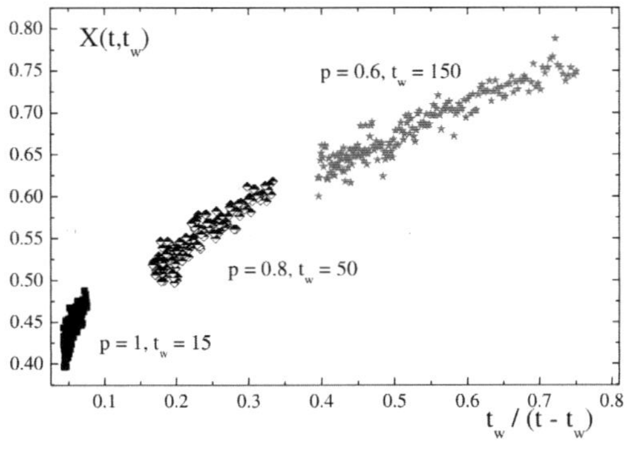


**Figure 6.** FD ratio $X(t, t_w)$ plot for different $p$.

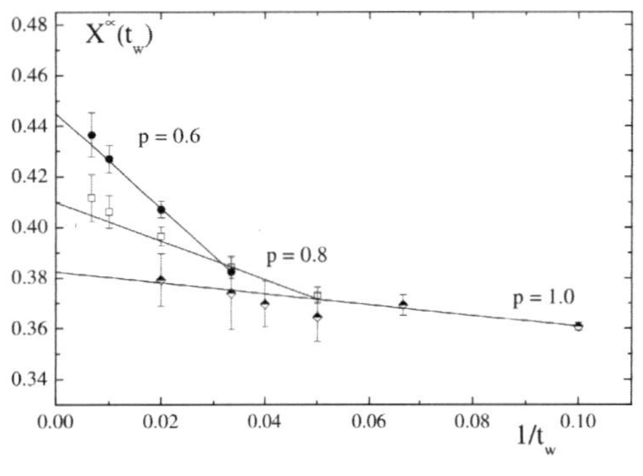


**Figure 7.** Extrapolation $X^{\infty}(t_w, p)$ in the limit $1/t_w \to 0$.

**Table 4.** Values of the FDR $X^{\infty}$ for spin concentrations $p = 1.0$, $p = 0.8$, and 0.6.

| $t_w$ | $X^{\infty}$ | | |
|---|---|---|---|
| | $p = 1.0$ | $p = 0.8$ | $p = 0.6$ |
| 10 | 0.361(2) | | |
| 15 | 0.371(4) | | |
| 20 | 0.365(10) | 0.373(9) | |
| 25 | 0.369(9) | | |
| 30 | 0.374(14) | 0.384(5) | 0.382(1) |
| 50 | 0.379(10) | 0.397(4) | 0.407(3) |
| 100 | | 0.406(6) | 0.427(6) |
| 150 | | 0.412(9) | 0.437(9) |
| $\to \infty$ | 0.380(13) | 0.413(11) | 0.446(8) |
| $\to \infty$, $\chi(t, t_w) vs C(t, t_w)$ | 0.391(12) | 0.419(11) | 0.443(10) |

Investigation of influence of structural disorder on ageing properties and violations of the equilibrium fluctuation-dissipation theorem in these systems is carried out.

On the base of analysis of the two-time autocorrelation $C(t, t_w)$ and response $R(t, t_w)$ functions were demonstrated the ageing effects. It has been shown that the presence of defects leads to ageing gain. The scaling relations $C(t, t_w) \sim t_w^{-a} F_C(t/t_w)$ and $R(t, t_w) \sim t_w^{-1-a} F_R(t/t_w)$ with $a = (d - 2 + \eta)/z = 2\beta/\nu z$ were checked.

For regime with well-separated times $t - t_w \gg t_w \gg 1$ it was shown that the scaling functions $F_c(t/t_w)$ and $F_R(t/t_w)$ are characterized by power law time dependencies and the obtained values of the exponents $c_a$ and $c_r$ demonstrate a good agreement with relation $c_a = c_r = d/z - \theta'$ for each spin concentration $p$ and with results of the short-time dynamics method (see the table 3).

The asymptotic values of the universal fluctuation-dissipation ratio were computed for these systems, and it has been proved the violation of the fluctuation-dissipation theorem for critical non-equilibrium behaviour of the three-dimensional diluted Ising models with $X^{\infty}(p) < 1/2$. So, the obtained values of $X^{\infty}(p)$ for pure and diluted systems equal $X^{\infty}(p = 1) = 0.380(13), X^{\infty}(p = 0.8) = 0.413(11)$, and $X^{\infty}(p = 0.8) = 0.446(8)$. From

Journal of Physics: Conference Series **510** (2014) 012015 doi:10.1088/1742-6596/510/1/012015

these results of our numerical investigations we can conclude that the non-equilibrium critical dynamics of the three-dimensional diluted Ising model is characterized by new FDR with $X^{\infty}_{disorder}(p < 1) > X^{\infty}_{pure}(p = 1)$, and values of the FDR are increased with growth of defect concentrations.

Renormalization-group investigations of non-equilibrium critical behaviour in pure Ising model and weakly dilute Random Ising Model (RIM) for purely relaxational dynamics were carried out in papers [17] and [18]. The asymptotic values of the FDR $X^{\infty}_{pure} = 0.429(6)$ and $X^{\infty}_{weak-diluted} \simeq 0.416$ were calculated for these models with the use of $\varepsilon$-expansion method. In the article [18] the authors has written that it is too complicated to define the influence of impurities based only on results $\varepsilon$-expansion in nearest orders of theory. They believe that calculations in higher orders of theory can help to clarify this complex question. The results of our study gave the answer on this question.

## Acknowledgments

The reported study was supported in part by Ministry of Education and Science of Russia through project No. 2.3046.2011 and by grant MU-4/2013 of the Omsk State University for young scientists.

The simulations were supported by the Supercomputing Center of Lomonosov Moscow State University and Joint Supercomputer Center of the Russian Academy of Sciences.

## References

[1] Henkel M and Pleimling M 2010 (Springer, Heidelberg) vol 2
[2] Afzal N and Pleimling M 2013 *Phys. Rev.* E **87** 012114
[3] Calabrese P, Gambassi A, and Krzakala F 2006 *J. Stat. Mech.* , **6** 2
[4] Berthier L and Kurchan J 2013 *Nature Phys.* **9** 310
[5] Calabrese P and Gambassi A 2005 *J. Phys.* A **38** R133
[6] Krzakala F and Ricci-Tersenghi F 2006 *J. Phys.: Conf. Ser.* **40** 42-49
[7] Abriet S and Karevski D 2004 *Eur. Phys. J.* B **41** 79
[8] Parisi G, Ricci-Tersenghi F and Ruiz-Lorenzo J J 1999 *Eur. Phys. J.* B **11** 317
[9] Luo H J, Schülke L and Zheng B 2001 *Phys. Rev.* E **64** 036123
[10] Chatelain C 2003 *J. Phys.* A **36** 10739
[11] Ricci-Tersenghi F 2003 *Phys. Rev.* E, **68** 065104
[12] Janssen H K, Schaub B, and Schmittmann B 1989 *Z. Phys.* B **73** 539
[13] Janke W 2008 (Springer Berlin Heidelberg) vol 739 pp 79-140
[14] Prudnikov V, Prudnikov P, Vakilov A and Krinitsyn A 2007 *JETP* **105** 371
[15] Prudnikov V, Prudnikov P, Krinitsyn A, Vakilov A, Pospelov E and Rychkov M 2010 *Phys. Rev.* E **81** 011130
[16] Jaster A, Mainville J, Scheulke L and Zheng B 1999 *J. Phys.* A **32** 1395
[17] Calabrese P and Gambassi A 2002 *Phys. Rev.* E **66** 066101
[18] Calabrese P and Gambassi A 2002 *Phys. Rev.* B **66** 212407

Journal of Physics: Conference Series **510** (2014) 012016 doi:10.1088/1742-6596/510/1/012016

# Melting Scenario of the Two-Dimensional Core-Softened System: First-Order or Continuous Transition?

**D.E. Dudalov, Yu.D. Fomin, E.N. Tsiok, and V.N. Ryzhov**

Institute for High Pressure Physics RAS, 142190 Kaluzhskoe shosse, 14, Troitsk, Moscow, Russia

E-mail: `ryzhov@hppi.troitsk.ru`

**Abstract.** We present a computer simulation study of the phase behavior of two-dimensional classical particles repelling each other through an isotropic core-softened potential. As in the analogous three dimensional case, a reentrant-melting transition occurs upon compression for not too high pressures. However, in two dimensions in the low density part of the phase diagram melting is a continuous two-stage transition, with an intermediate hexatic phase. All available evidence supports the Kosterlitz-Thouless-Halperin-Nelson-Young (KTHNY) scenario for this melting transition. On the other hand, at high density part of the phase diagram one first-order transition takes place. We expect that such a phenomenology can be checked in confined monolayers of charge-stabilized colloids with a softened core and water confined between two hydrophobic plates.

Despite the long history of investigations, the melting transition of most materials in two dimensions is not well understood, because theories explaining the transition on a microscopic scale are not available. Furthermore, the mechanism of melting depends on the details of the interactions between the particles forming the crystal lattice. Moreover, the defects reducing the translational order of the crystal are expected to play a mayor role. In their pioneering works, Halperin, Nelson, and Young [1, 2, 3], using the Kosterlitz-Thouless ideas [4], proposed the scenario of two-dimensional melting which is fundamentally different from the melting scenario of conventional three-dimensional systems. It has been shown that the transition between a crystal and an isotropic liquid can occur by means of two continuous transitions which correspond to dissociation of bound dislocation and disclination pairs, respectively. The low-temperature solid phase is characterized by quasi-long-range translational order and long-range bond-orientational order. Dislocations unbinding at some temperature $T_m$ leads to a phase with short-range translational order, but with quasi-long-range bond-orientational order. This intermediate phase is called a hexatic phase. The shear modulus of the hexatic phase $\mu$ is equal to zero. The properties of the hexatic phase are similar to those of a nematic liquid crystal, except that triangular lattices melt into a phase with persistent sixfold, rather than twofold order. Paired disclinations in the hexatic phase ultimately unbind themselves, driving a second transition at a higher temperature $T_i$ into an isotropic liquid.

This theory has strong support from experiments with electrons on helium [5, 6, 7] and computer simulations of the 2D electron systems [8, 9]. An experimental confirmation for the KTHNY theory for crystal melting in 2D has been found with the colloidal model system with repulsive magnetic dipole-dipole interaction [10, 11, 12]. However, a conventional first-order transition between a two-dimensional solid and an isotropic liquid is also a possibility (see, for example, [13, 14, 15, 16, 17]).

It should be noted that the KTHNY theory is phenomenological and seems universal. It is not clear from this theory whether the melting scenario depends on the shape of an intermolecular potential. Actually, the natural way to analyze this dependence is to use computer simulations. However, simulations are not reliable enough in the case of two-dimensional melting: it is interesting to note that the similar simulation methods have led to contradictory conclusions even when applied to the same systems [18, 19, 20, 21, 22, 23, 24, 25, 26, 27, 28, 29, 30, 31, 32, 33, 34, 35]. The problems are understandable since correlation times and lengths (translational and orientational) can be extremely long near the phase transition. Particularly severe is the situation for hard-core particles, where enormous samples and huge simulation times are required in order to discriminate between the various transition scenarios, while the state sampling may be less demanding for systems of soft particles whose steric constraints are less pronounced. A lot of efforts were made on computational studies of two-dimensional melting of hard-core potential systems including hard disks or Lennard-Jones potentials. Simulation results on these systems tend to favor a first-order transition scenario for melting, although some conflicting results also exist. In spite of all these efforts, a satisfactory answer has not been obtained yet for one of the most important questions in two-dimensional melting, which is as follows: what condition determines the existence of a hexatic phase and the nature of the melting transition? It seems natural to relate this behavior with the range and the softness of the potential [33, 34, 35].

It is well known that some liquids demonstrate anomalous behavior in some regions of thermodynamic parameters. The most common and well known example is water. The water phase diagrams have regions where a thermal expansion coefficient is negative (density anomaly), self-diffusivity increases upon compression (diffusion anomaly), and the structural order of the system decreases with increasing pressure (structural anomaly). Later on it was discovered that many other substances also demonstrate similar behavior. Some typical examples are silica, silicon, phosphorus, and many others. It is reasonable to relate this kind of behavior to the orientational anisotropy of the potentials, however, a number of studies demonstrate water-like anomalies in fluids that interact through spherically symmetric core-softening potentials with two length scales. A lot of different core-softened potentials were introduced [36, 37].

In this work we study a two-dimensional system with a potential with a hard core followed by a smooth shoulder. In three dimensions it does exhibit the density, the diffusion, and the structural anomalies observed in water. We present a simulation study of two-dimensional melting transition in the purely repulsive core-softened system introduced in our previous publications [38, 39, 40, 41, 42, 43, 44, 45, 46, 47].

The general form of the potential is written as

$$U(r) = \varepsilon \left(\frac{\sigma}{r}\right)^{14} + \frac{1}{2}\varepsilon\left(1 - \tanh(k_1\{r - \sigma_1\})\right). \tag{1}$$

Here $k_1 = 10.0$, and $\sigma_1 = 1.35$. The potential (1) is plotted in Fig. 1. It should be noted that potential (1) is very similar to the Fermi-Jagla potential suggested recently in Ref. [48], where the Fermi distribution function is used instead of the hyperbolic tangent in Eq. (1) in order to describe the smoothed step [46].

In the remainder of this paper we use the dimensionless quantities: $\tilde{\mathbf{r}} \equiv \mathbf{r}/\sigma$, $\tilde{P} \equiv P\sigma^2/\varepsilon$, $\tilde{V} \equiv V/N\sigma^2 \equiv 1/\tilde{\rho}, \tilde{T} \equiv k_BT/\varepsilon$. Since we will use only these reduced variables, the tildes will be omitted.

In three dimensions particles interacting through a purely repulsive potential given by Eq. (1) exhibit reentrant melting, a maximum melting temperature, superfragile glass behavior, and

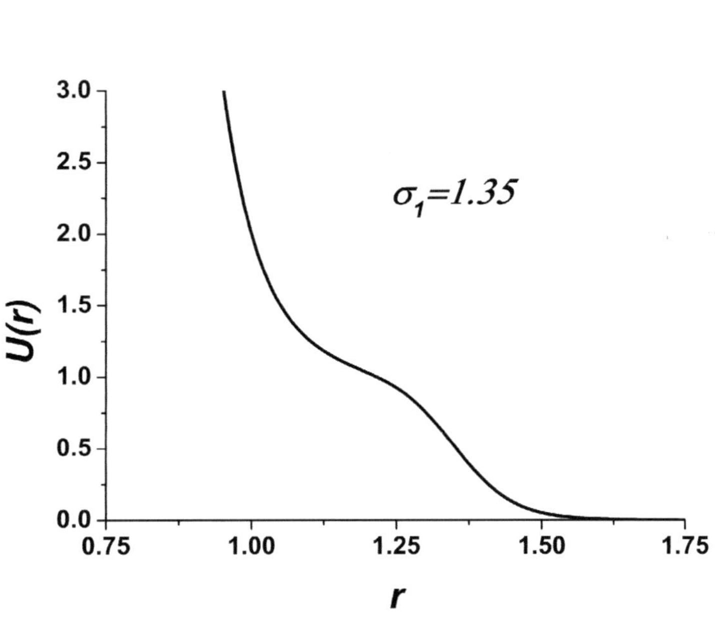


**Figure 1.** The potential (1) with $\sigma_1 = 1.35$.

thermodynamic anomalies similar to the ones found in water and silica [38, 39, 40, 41, 42, 43, 44, 45, 46, 47].

As it was discussed before [49, 50], there are two characteristic temperatures for the melting transition in two dimensions: the dislocation unbinding temperature $T_m$ and the first-order transition temperature $T_{MF}$. The modulus of the order parameter vanishes at the temperature $T_{MF}$ which can be obtained from the double-tangent construction for the free energies of liquid and solid phases. There are two possibilities [49, 50]: 1: $T_m < T_{MF}$. In this case the system melts via two continuous transitions of the Kosterlitz-Thouless type with the unbinding of dislocation pairs. 2: $T_m > T_{MF}$. The system melts via a first-order transition because of the existence of third-order terms in the Landau expansion as in the ordinary three-dimensional case [49, 50]. The phase diagram corresponding to $T_{MF}$, gives the limit of the thermodynamic stability of the solid phase. In order to conclude whether the melting occurs through the KTHNY scenario, the additional analysis is necessary.

We simulate the system in $NVT$ and $NVE$ ensembles using the molecular dynamics (LAMMPS package [52]). The number of particles in the simulation varied between 3200 and 102400. In order to find the transition points we carry out the free energy calculations for different phases and construct a common tangent to them. For the purely repulsive potentials we computed the free energy of the liquid by integrating the equation of state along an isotherm [51]: $\frac{F(\rho)-F_{id}(\rho)}{Nk_BT} = \frac{1}{k_BT}\int_0^\rho \frac{P(\rho')-\rho' k_BT}{\rho'^2}d\rho'$. Free energies of different crystal phases were determined by the method of coupling to the Einstein crystal [51]. The phase diagram calculated in this way corresponds to the first-order transition scenario.

In Fig. 2, we plot the phase diagram of the system in $\rho - T$ and $P - T$ coordinates. There is a clear maximum in the melting curve at low densities. The phase diagram consists of two isostructural triangular crystal domains (T) corresponding to close packing of the small and large disks separated by a structural phase transition and square lattice (S). Similar phase diagram in three dimensions was discussed in details in our previous publications [38, 39]. It is important to note that there is a region of the phase diagram where we have not found any stable crystal

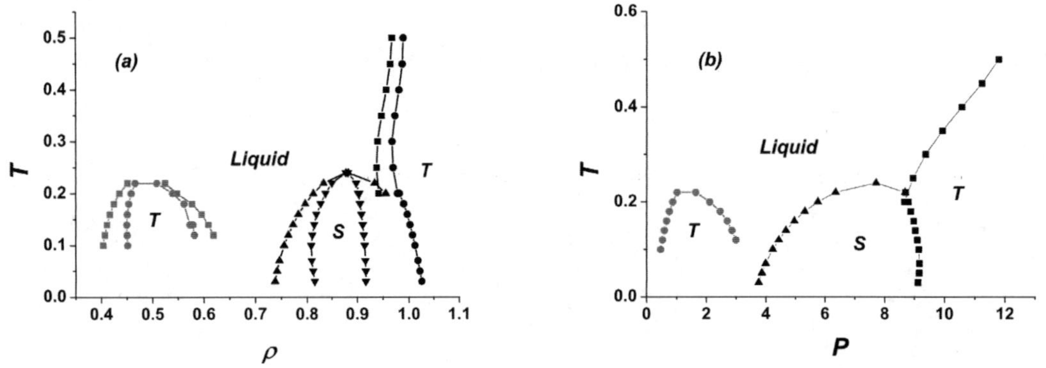


**Figure 2.** (a) Phase diagram of the system with the potential (1) in $\rho - T$ plane, where the Triangular (T) and Square (S) phases are shown. (b) Phase diagram of the same system in the $P - T$ plane.

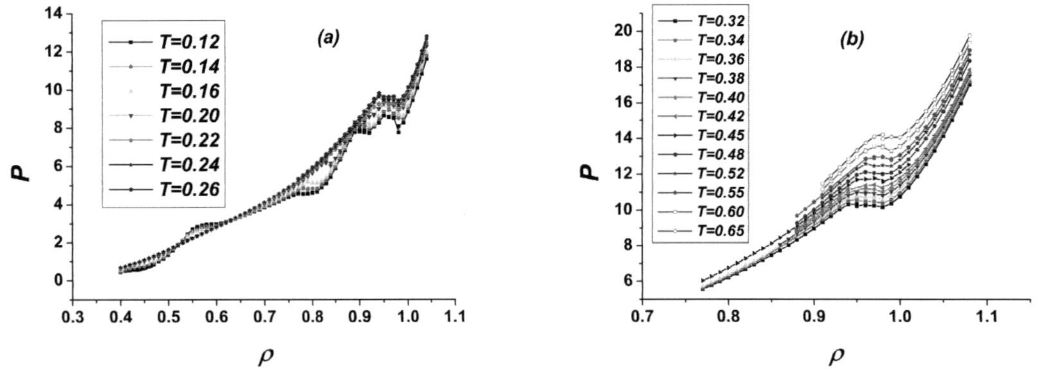


**Figure 3.** The low-temperature (a) and high-temperature (b) sets of isotherms.

phase. The results of three dimensional simulations [38, 46] suggest that a glass transition can occur in this region.

To disentangle first-order from continuous melting, we used the criteria described in the Ref. [31]. In Fig. 3 we present the low-temperature (Fig. 3(a)) and high-temperature (Fig. 3(b)) sets of isotherms. One can see that at low temperatures there are four regions on the isotherms corresponding to the phase transitions (see Fig. 2(a)), the low density ones being smooth as in the case of liquid-hexatic-solid transition [31] and the high densities part containing the Van der Waals loops characteristic of the first order phase transition. At high temperatures (see Fig. 2(b)) there is only one liquid-triangular lattice first-order transition. From Fig. 3 one can guess that the melting of the low-density and high-density parts of the phase diagram occurs with different scenarios: at low densities the KTHNY scenario is probable, while the high density phase melts through the first-order phase transition. As we are going to show in the following, the intermediate region between the solid and the (normal) fluid can be qualified as hexatic.

To confirm this guess, let us consider two different order parameters (OP), which are separately sensitive to the overall translational and orientational order, with their respective correlation functions. The translational OP is taken to be

$$\psi_T = \frac{1}{N}\left\langle \left| \sum_i e^{i\mathbf{G}\mathbf{r}_i} \right| \right\rangle, \tag{2}$$

where the sum is over the particle labels and $\mathbf{G}$ is any first shell reciprocal-lattice vector of the crystal. From its very definition, it follows that $\psi_T$ is sizeable only in a solid that is oriented in a way consistent with the length and direction of $\mathbf{G}$. Hence, $\psi_T$ is only measured on heating, where memory of the original crystal orientation is preserved as long as the system is large and remains solid. A sharp drop of $\psi_T$ signals the melting of the solid into a fluid, be it hexatic or normal.

At regular intervals during the simulation, we identify the $n_c(i)$ nearest neighbors (NN) of each particle $i$, together with the orientation $\theta_{NN}$ of each neighbor bond with respect to a reference axis. Whence, the orientational OP follows as

$$\psi_6 = \frac{1}{N}\left\langle \left| \sum_i \frac{1}{n_c(i)} \sum_{NN(i)} e^{6i\theta_{NN}} \right| \right\rangle = \frac{1}{N}\left\langle \left| \sum_i \Psi_6(\mathbf{r}_i) \right| \right\rangle. \tag{3}$$

The local bond-angular OP $\Psi_6(\mathbf{r}_i)$ enters the definition of the orientational correlation function (OCF):

$$G_6(r) = \rho^{-2}\left\langle \sum_{i,j}{}' \delta(\mathbf{r}_i - \mathbf{R})\delta(\mathbf{r}_i - \mathbf{R}')\Psi_6(\mathbf{r}_i)\Psi_6^*(\mathbf{r}_j) \right\rangle, \tag{4}$$

where the prime over the sum excludes $i = j$, and $r = |\mathbf{R} - \mathbf{R}'|$. The KTHNY theory predicts an algebraic $r^{-\eta(T)}$ large-distance decay of the OCF in the hexatic phase, which should be contrasted with the exponential asymptotic vanishing of angular correlations in a normal fluid. Another prediction of the theory is $\eta = 1/4$ at the hexatic-to-normal fluid transition point.

In Fig. 4, we represent the orientational order parameter (OOP) as a function of density for a set of temperatures. We see, that at the low density part of the phase diagram OOP behaves smoothly while at high densities one can see the abrupt change of OOP. This kind of behavior suggests again that the melting at low densities is continuous in accordance with the KTHNY scenario, and at high densities melting transition is of the first order.

In Fig. 5, we plot the two OPs for $\rho = 0.56$ as a function of temperature (an analogous behavior was observed for all the other densities). We see that $\psi_T$ vanishes at a slightly smaller temperature than $\psi_6$, which implies that the hexatic phase is confined to an narrow $T$ interval. In Fig. 6, the phase transition lines of the solid-hexatic and hexatic-liquid transitions are shown in comparison with the solid-liquid transition line (see Fig. 2). One can see that the transitions are mainly inside the solid region, obtained in the framework of the free-energy calculations. This fact also supports the idea that the melting in this region occurs through two continuous transitions. It is necessary to note, that in the case of the conventional first-order phase transition, the density change at the melting line maximum is equal to zero. We see, that the hexatic phase becomes narrower in the vicinity of the maximum, however, our calculations can not answer whether the width of hexatic region tends to zero at the maximum point. The errors in calculation of the OOP $\psi_6$ are less than 1%, while the errors of the translational order parameter $\psi_T$ do not exceed 5%.

A more direct evidence of the hexatic phase emerges from the large-distance behavior of the OCF. We plot this function in Fig. 7 at various densities across the hexatic phase for $T = 0.12$. It appears that the OCF decays algebraically in a $\rho$ region of limited extent, which roughly corresponds to the middle of the bridging region in Fig. 6 .

The similar analysis was made for the melting of the square lattice region of the phase diagram, and it was shown that the square lattice melts through the first-order phase transition.

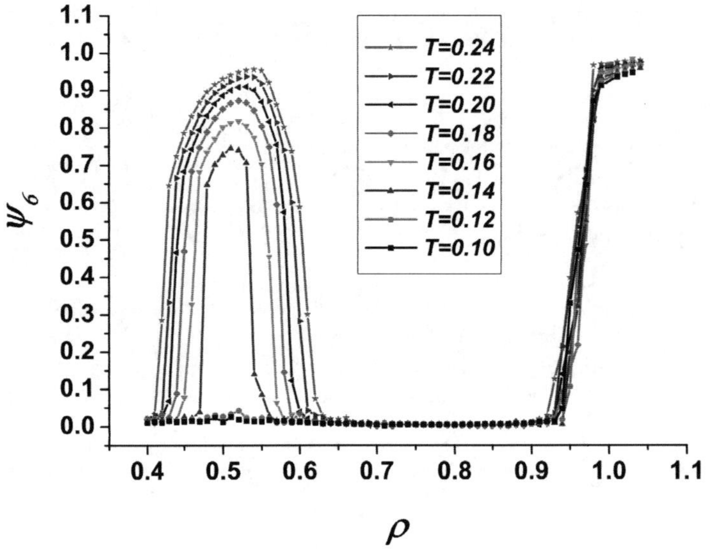


**Figure 4.** Orientational order parameter as a function of density for different temperatures.

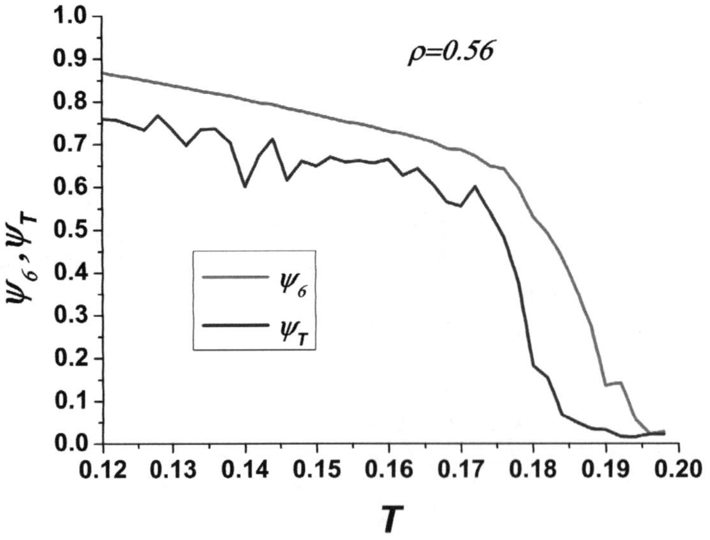


**Figure 5.** OPs $\psi_T$ and $\psi_6$ as functions of temperature for $\rho = 0.56$. It is clearly the narrow hexatic phase.

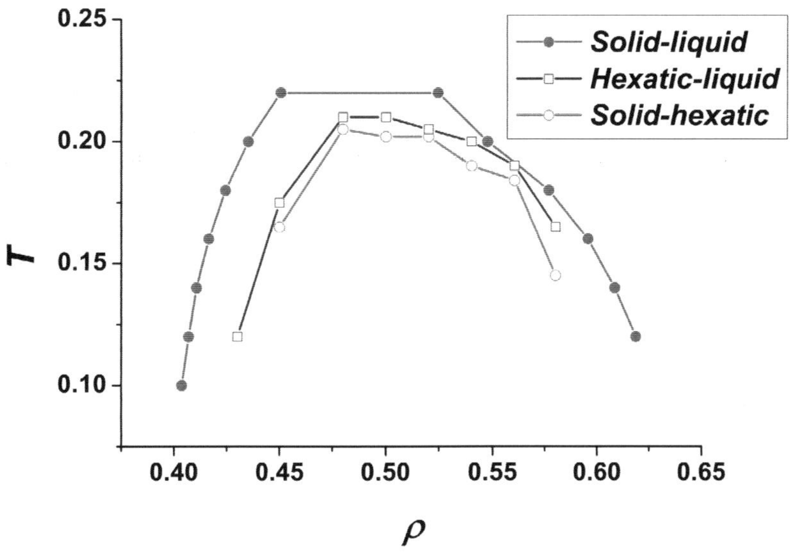


**Figure 6.** The low-density part of the phase diagram (Fig. 2(a)) along with the lines of solid-hexatic and hexatic-liquid transitions.

It should be noted, that the scaling analysis made in accordance with the algorithm in Refs. [21, 31] also supports the melting scenario described above. For the OOP we used a system of 102400 particles which was divided in subboxes. The subbox size parameter $M_b$ is equal to the number of subboxes along the edge of the total system and varies in our simulations from 1 to 16. As expected (see [21, 31]), the bond-orientational order parameter does not change in the ordered region while it increases with increasing the number of the subboxes in the liquid phase. At the same time, we observe an increase of OOP susceptibility without the change of the locations of the peaks maxima.

In conclusion, we have provided the first unambiguous evidence of the occurrence of two-stage continuous reentrant melting via a hexatic phase in the 2D core-softening model at low densities, while at high densities the melting occurs through the conventional first-order phase transition. We have validated a number of KTHNY predictions. This kind of behavior can be understood from the consideration of the potential (1) (see Fig. 1). It is widely believed that the 2D melting transition scenario corresponds to the KTHNY one for the softer potentials, however, the systems with hard potentials melt through the first-order transition. The behavior of the system described by the potential (1) is determined by the soft long-range part of the potential at low densities. At the same time, the hard core of the potential plays the main role at the high densities. It seems that this is the reason of the observed peculiarities of the phase diagram. The thermodynamic anomalies which does exist in this system will be considered in the forthcoming publications [53]. The present discovery of reentrant-hexatic behavior in the core-softening potential is relevant for many soft-matter systems. For instance, one can engineer colloidal particles interacting through a temperature modulated softened repulsion, which will likely exhibit reentrant melting in a range of packing fractions well below the density at which hard-core crystallization occurs. These results may be also useful for the qualitative understanding the behavior of confined monolayers of charge-stabilized colloids with a softened core and water confined between two hydrophobic plates [54, 55, 56].

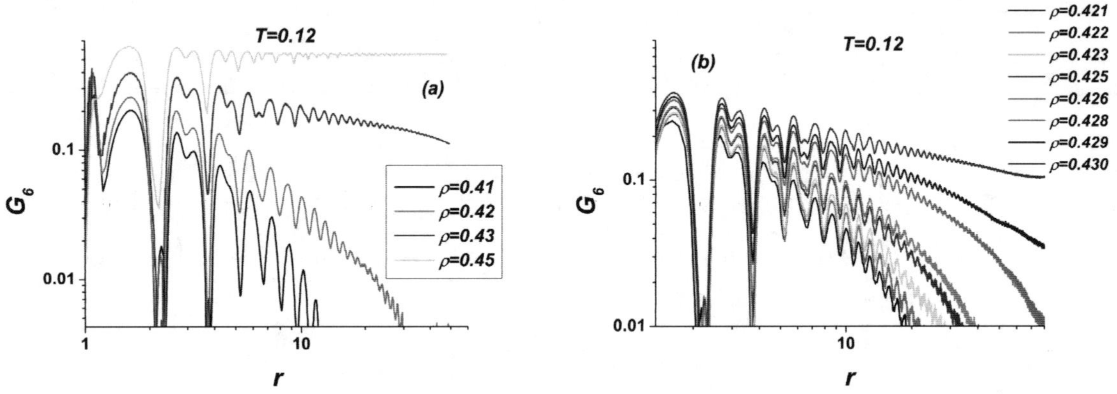


**Figure 7.** Log-log plots of the orientational correlation function $G_6(r)$ at selected densities across the hexatic region for $T = 0.12$. (a) Upon increasing $\rho$ from 0.41 to 0.45 there is a qualitative change in the large-distance behavior of $G_6(r)$, from constant (solid) to power-law decay (hexatic fluid), up to exponential decay (normal fluid). (b) Note that, consistently with the KTHNY theory, the decay exponent $\eta$ is less than 1/4 for $\rho > 0.43$.

We are grateful to S. M. Stishov, V. V. Brazhkin, and E.E. Tareyeva for stimulating discussions. Yu.F. and E.T. also thank the Russian Scientific Center Kurchatov Institute and Joint Supercomputing Center of the Russian Academy of Science for computational facilities. The work was supported in part by the Russian Foundation for Basic Research (Grants No 14-02-00451, 13-02-12008, 13-02-00579, and 13-02-00913) and the Ministry of Education and Science of Russian Federation (project MK-2099.2013.2).

[1] Nelson DR and Halperin BI 1978 *Phys. Rev. Lett.* **41** 121
[2] Nelson DR and Halperin BI 1979 *Phys. Rev.* B **19** 2457
[3] Young AP 1979 *Phys. Rev.* B **19** 1855
[4] Kosterlitz M and Thouless DJ 1973 *J. Phys.* C **6** 1181
[5] Grimes CC and Adams G 1979 *Phys. Rev. Lett.* **42** 795
[6] Mehrotra R, Guenin BM, and Dahm AJ 1982 *Phys. Rev. Lett.* **48** 641
[7] Stan MA and Dahm AJ 1989 *Phys. Rev.* B **40** 8995
[8] Morf RH 1979 *Phys. Rev. Lett.* **43** 931
[9] Strandburg KJ 1988 *Rev. Mod. Phys.* **60** 161
[10] Gasser Urs, Eisenmann C, Maret G, and Keim P 2010 *ChemPhysChem* **11** 963
[11] Zanh K and Maret G 2000 *Phys. Rev. Lett.* **85** 3656
[12] Keim P, Maret G, and von Grunberg HH 2007 *Phys. Rev.* E **75** 031402
[13] Chui ST 1983 *Phys. Rev.* B **28** 178
[14] Kleinert H 1983 *Phys. Lett.* A **95** 381
[15] Janke W and Kleinert H 1990 *Phys. Rev.* B **41** 6848
[16] Ryzhov VN 1991 *THEORETICAL AND MATHEMATICAL PHYSICS* **88** 990 (DOI: 10.1007/BF01027701)
[17] Ryzhov VN 1991 *ZHURNAL EKSPERIMENTALNOI I TEORETICHESKOI FIZIKI* **100**, 1627
[18] Alder BJ and Wainright TE 1962 *Phys. Rev.* **127**, 359
[19] Hoover WG and Ree FH 1968 *J. Chem. Phys.* **49** 3609
[20] Lee J and Strandburg KJ 1992 *Phys. Rev.* B **46** 11190
[21] Weber H, Marx D and Binder K 1995 *Phys. Rev.* B **51** 14636
[22] Mak CH 2006 *Phys. Rev.* E **73** 065104
[23] Jaster A 1999 *Phys. Rev.* E **59** 2594
[24] A. Jaster, Europhys. Lett., **42**, 277 (1998).
[25] Bagchi K, Andersen HC, and Swope W 1996 *Phys. Rev. Lett.* **76** 255
[26] Lozovik YE and Farztdinov VM 1985 *Solid State Commun.* **54** 725
[27] Ryzhov VN and Tareyeva EE 1995 *Phys. Rev.* B **51** 8789

[28] Ryzhov VN and Tareyeva EE 1995 *ZHURNAL EKSPERIMENTALNOI I TEORETICHESKOI FIZIKI* **108**, 2044
[29] Bernard EP and Krauth W 2011 *Phys. Rev. Lett.* **107** 155704
[30] Engel M, Anderson JA, Glotzer SC, Isobe M, Bernard EP, and Krauth W 2013 *Phys. Rev.* E **87** 042134
[31] Binder K, Sengupta S, and Nielaba P 2002 *J. Phys.: Condens. Matter* **14** 2323
[32] Bladon P and Frenkel D 1995 *Phys. Rev. Lett.* **74** 2519
[33] Lee SI and Lee SJ 2008 Phys. Rev. E **78** 041504
[34] Prestipino S, Saija F, and Giaquinta PV 2011 *Phys. Rev. Lett.* **106** 235701
[35] Prestipino S, Saija F, and Giaquinta PV 2012 *J. Chem. Phys.* **137** 104503
[36] Buldyrev SV, Malescio G, Angell CA, Giovambattista N, Prestipino S, Saija F, Stanley HE, and Xu L 2009 *J. Phys.: Condens. Matter* **21** 504106
[37] Vilaseca P and Franzese G 2011 *Journal of Non-Crystalline Solids* **357** 419
[38] Fomin YD, Gribova NV, Ryzhov VN, Stishov SM, and Frenkel D 2008 *J. Chem. Phys.* **129** 064512
[39] Gribova NV, Fomin YD, Frenkel Daan, and Ryzhov VN 2009 *Phys. Rev.* E **79** 051202
[40] Fomin YD, Ryzhov VN, Gribova NV 2010 *Phys. Rev.* E **81** 061201
[41] Fomin YD, Tsiok EN, and Ryzhov VN 2011 *J. Chem. Phys.* **134** 044523
[42] Fomin YD and Ryzhov VN 2011 *Phys. Lett.* A **375** 2181
[43] Fomin YD, Tsiok EN, and Ryzhov VN 2011 *J. Chem. Phys.* **135** 124512
[44] Fomin YD, Tsiok EN, and Ryzhov VN 2011 *J. Chem. Phys.* **135** 234502
[45] Fomin YD, Tsiok EN, and Ryzhov VN 2013 *European Physical Journal - Special Topics* **216** 165
[46] Ryltsev RE, Chtchelkatchev NM, and Ryzhov VN 2013 *Phys. Rev. Lett.* **110** 025701
[47] Fomin YD, Tsiok EN, and Ryzhov VN 2013 *Phys. Rev.* E **87** 042122
[48] Abraham JA, Buldyrev SV, and Giovambattista N 2011 *J. Phys. Chem.* B **115**, 14229
[49] Ryzhov V N and Tareyeva E E 2002 *Physica* A **314** 396
[50] Ryzhov V N and Tareyeva E E 2002 *THEORETICAL AND MATHEMATICAL PHYSICS* **130** 101 (DOI: 10.1023/A:1013884616321)
[51] Frenkel Daan and Smit Berend 2002 *Understanding molecular simulation (From Algorithms to Applications), 2nd Edition* (Academic Press)
[52] http://lammps.sandia.gov/
[53] D.E. Dudalov, Yu.D. Fomin, E.N. Tsiok, and V.N. Ryzhov, arXiv:1311.7534.
[54] Rice SA 2009 *Chem. Phys. Lett.* **479** 1
[55] Krott LB and Barbosa MC 2013 *J. Chem. Phys.* **138** 084505
[56] Almudallal AM, Buldyrev SV, and Saika-Voivod I 2012 *J. Chem. Phys.* **137** 034507

Journal of Physics: Conference Series **510** (2014) 012017 doi:10.1088/1742-6596/510/1/012017

# Application of the parallel multicanonical method to lattice gas condensation

**Johannes Zierenberg, Micha Wiedenmann and Wolfhard Janke**
Institut für Theoretische Physik, Universität Leipzig, Postfach 100 920, D-04009 Leipzig, Germany

E-mail: zierenberg@itp.uni-leipzig.de

**Abstract.** We present the speedup from a novel parallel implementation of the multicanonical method on the example of a lattice gas in two and three dimensions. In this approach, all cores perform independent equilibrium runs with identical weights, collecting their sampled histograms after each iteration in order to estimate consecutive weights. The weights are then redistributed to all cores. These steps are repeated until the weights are converged. This procedure benefits from a minimum of communication while distributing the necessary amount of statistics efficiently.

Using this method allows us to study a broad temperature range for a variety of large and complex systems. Here, a gas is modeled as particles on the lattice, which interact only with their nearest neighbors. For a fixed density this model is equivalent to the Ising model with fixed magnetization. We compare our results to an analytic prediction for equilibrium droplet formation, confirming that a single macroscopic droplet forms only above a critical density.

## 1. Introduction

Despite the large amount of existing literature, condensation remains under investigation up to today by means of analytical treatment, numerical studies, and experiments [1, 2, 3, 4, 5, 6, 7, 8]. In addition to the increasing interest in non-equilibrium properties and droplet size distributions, questions regarding the equilibrium properties remain open as well. For example, Biskup et al. [3, 4] showed that in equilibrium an over-saturated gas system is either in the evaporated phase or in the condensed phase consisting of a single macroscopic droplet. The probability for droplets of intermediate size is negligibly small. This result was proven for the two-dimensional Ising model and numerically verified [7, 8]. We set out to test this analytical result also for the three-dimensional case of a lattice gas. Due to the nucleation barrier, we applied a parallel implementation [9] (for related methods see [10, 11, 12]) of multicanonical simulations [13, 14, 15] to circumvent a possible hysteresis, driving the system back and forth between the condensed and evaporated phase.

We start by describing the main method and the theoretical arguments together with the model. This is followed by a comparison to micromagnetic simulations of the equivalent Ising model in order to verify the correctness. We discuss the speedup with increasing parallelization and compare the simulation effort with the independent micromagnetic simulation. Afterwards, the results in two and three dimensions are presented and discussed, followed by some concluding remarks.

Journal of Physics: Conference Series **510** (2014) 012017 doi:10.1088/1742-6596/510/1/012017

## 2. Method

The multicanonical method [13, 14, 15] enables the sampling of a broad parameter space in a single simulation by modifying the acceptance probability according to a simple idea. The Boltzmann factor is replaced, or sometimes extended, by a weight function with the purpose to increase the probability to reach states that are suppressed otherwise. This can be achieved for a variety of order parameters leading to multimagnetic, multibondic and many more realizations but can best be explained for the case of the internal energy $E$. Formally, this can be described by writing the partition function in terms of the density of states $\Omega(E)$,

$$Z_{\text{can}} = \sum_{E} \Omega(E) e^{-\beta E} \to \sum_{E} \Omega(E) W(E), \tag{1}$$

where we have already transformed the Boltzmann factor into the weight function $W(E)$. Analogous to the Metropolis algorithm, new states ($E_n$) are generated from old states ($E_o$) and accepted with

$$\min\left(1, \frac{W(E_n)}{W(E_o)}\right).$$

If we were to know $\Omega(E)$, we could sample the full energy range with a flat distribution choosing $W(E) = 1/\Omega(E)$. However, this is in general not the case such that $W(E)$ has to be obtained in an iterative way. This can be simple or complex; it can modify the weights on the fly or rely on equilibrium distributions in each iteration. A simple example would be to perform equilibrium simulations with initial weights $W^i$ and estimate the consecutive weights as $W^{i+1}(E) = W^i(E)/H^i(E)$. Hence, if a state occurs more often the weight is reduced more strongly.

We applied a parallel version of this multicanonical method which has been shown to exhibit an ideal scaling for the Ising model and a very good scaling for the $q = 8$ Potts model [9]. This parallel implementation relies on equilibrium iterations, distributing the sampling of the distribution $H^i(E)$ on $p$ independent processes. Since the weights are not modified within a single iteration, each process returns a contribution to the same distribution and they can simply be summed up:

$$H^i(E) = \sum_{j=1}^{p} H^i_j(E) \tag{2}$$

Afterwards, the consecutive weights are calculated from the previous weights together with the total histogram according to whatever rule one may choose, in our case by a recursive scheme [15].

Altogether, this leads to an efficient parallelization due to a very limited amount of communication. Below, we will show the applicability of the method to a general problem, here the condensation of a lattice gas in two and three dimensions. As in real life and other than in [9], we will not pay attention to the optimization of each degree of parallelization but want to demonstrate the value of this parallelization to modern computational statistical physics.

## 3. Condensation of a lattice gas

A general theory of equilibrium droplet condensation was proposed by Biskup et al. [3, 4], with a rigorous analytic solution for the two-dimensional Ising spin model. In principle they describe the interplay of entropy maximization by vacuum fluctuations with energy minimization by forming a droplet. They showed that the probability for intermediate-size droplets vanishes with increasing system size. The free energy consists of a contribution from the vacuum fluctuation of excess particles $\delta N$

$$F_{\text{fluc}} = \frac{(\delta N)^2}{2\hat{\kappa} V}, \tag{3}$$

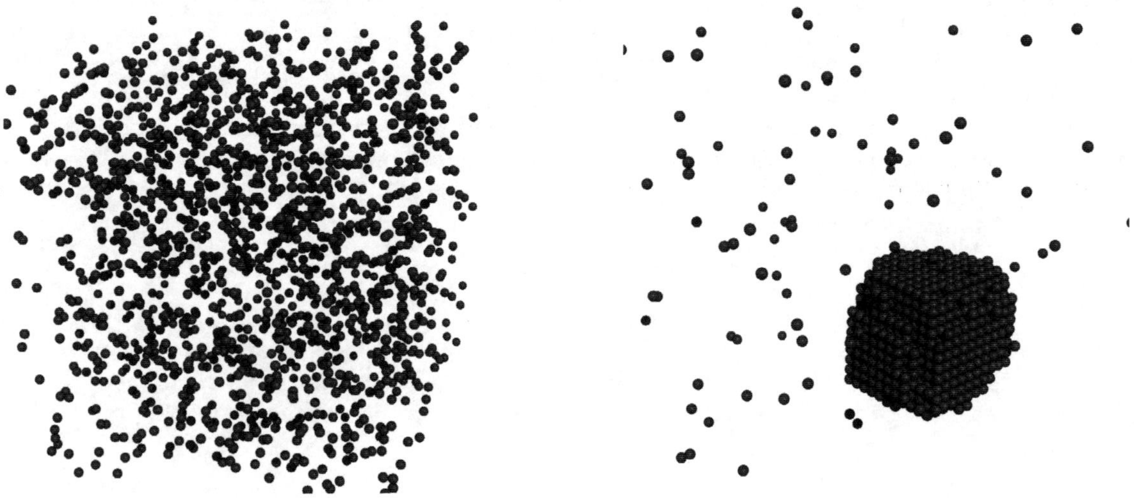

**Figure 1.** Example of (left) fluctuations of particle excess and (right) a single condensate in three dimensions.

with the isothermal compressibility $\hat{\kappa} = \beta\kappa = \beta\left\langle (N - \langle N \rangle)^2 \right\rangle / V$ and a contribution from the single largest droplet of size $V_D$

$$F_{\text{drop}} = \tau_W (V_D)^{\frac{d-1}{d}}, \tag{4}$$

where $\tau_W$ is the surface free energy of a (Wulff shaped) droplet of unit volume (see Fig. 1).

This was translated into the language of the Ising spin model in [3, 7]. Here, we will concentrate on the formulation of a lattice gas, where each lattice site is either occupied by a particle or empty, i.e. $n_i \in \{0, 1\}$. For a fixed number of particles this model is equivalent to the Ising model with spins $s_i \in \{-1, 1\}$, at fixed total magnetization, when $s_i = 2n_i - 1$:

$$\mathcal{H}^{\text{gas}} = -J \sum_{<i,j>} n_i n_j = \frac{1}{4}\mathcal{H}^{\text{Ising}} + c(N) \tag{5}$$

Thus, for fixed coupling constants, a scaling of the temperature by $T^{\text{Ising}} = T^I = 4T$ and the addition of a system-size dependent constant transforms the energy scale of an Ising model into the corresponding one of a lattice gas. All other observables are then mapped accordingly by scaling the temperature.

The contributions to the free energy can as well be formulated in the language of a lattice gas. Consider a temperature-dependent vacuum or background density $\rho_0 = N_0/V$, which may be identified with the spontaneous magnetization in the Ising model via

$$m_0 = 1 - 2\rho_0. \tag{6}$$

Adding particles to the system leads to a total particle excess $\delta N = N - N_0$. According to Biskup et al. this excess $\delta N$ may be decomposed into the particle excess inside the droplet $\delta N_D$ and the remaining particle excess in the fluctuating phase $\delta N_F$. For the infinite-size system and from the equivalence to the Ising model follows that the gas density is given by $\rho_G = \rho_0$ and the liquid density by $\rho_L = 1 - \rho_0$. Hence, in a droplet of size $V_D$ we expect an excess of particles $\delta N_D = (\rho_L - \rho_G)V_D$. Introducing the fraction of excess particles inside a droplet

$\lambda = \delta N_D/\delta N$, allows us to rewrite the decomposed particle excess as fractions $\delta N_D = \lambda\delta N$ and $\delta N_F = (1-\lambda)\delta N$. Altogether the total free energy $F = F_{\text{drop}} + F_{\text{fluc}}$ then reads

$$F = \left(\tau_W \left(\frac{\lambda\delta N}{1-2\rho_0}\right)^{\frac{d-1}{d}} + \frac{(1-\lambda)^2(\delta N)^2}{2\hat{\kappa}V}\right). \tag{7}$$

Rewriting the particle excess in terms of a volume

$$V_L = \frac{N - N_0}{1-2\rho_0} \tag{8}$$

and identifying $\lambda = V_D/V_L$ leads to

$$\begin{aligned} F &= \left(\tau_W(\lambda V_L)^{\frac{d-1}{d}} + \frac{(1-2\rho_0)^2}{2\hat{\kappa}V}V_L^2(1-\lambda)^2\right) \\ &= \tau_W V_L^{\frac{d-1}{d}}\left(\lambda^{\frac{d-1}{d}} + \Delta(1-\lambda)^2\right) \end{aligned} \tag{9}$$

with the dimensionless parameter

$$\Delta = \frac{(1-2\rho_0)^2}{2\hat{\kappa}\tau_W}\frac{V_L^{(d+1)/d}}{V}. \tag{10}$$

While $\hat{\kappa}$ has the same value as $\chi$ in the Ising model, $\tau_W$ has to be rescaled such that $\tau_W^I = 4\tau_W$. At fixed temperature, the parameters $\rho_0, \hat{\kappa}, \tau_W$ are constant. Minimizing eq. (9) leads to the infinite-size solution for the equilibrium volume fraction $\lambda = \lambda(\Delta)$ (see [3, 4]). Biskup et al. showed that below a critical value $\Delta_c^{2D} \approx 0.9186$ or $\Delta_c^{3D} \approx 0.8399$ there are merely gas fluctuations, while above there exists one macroscopic droplet together with background fluctuations. At $\Delta_c$ the mass of the largest droplet jumps to a fraction $\lambda_c^{2D} = 2/3$ or $\lambda_c^{3D} = 1/2$ of the total particle excess. Droplets of intermediate size have negligible probability.

In order to compare to the analytic predictions, we have to measure the average size of the largest droplet at different densities but fixed temperature. Because of the increasing correlation times in large systems, we applied the multicanonical method, driving the system between the evaporated and condensed phase, and reweighted to the relevant temperatures afterwards. The energy ranges were obtained through short parallel tempering [16] runs in the beginning of each simulation.

## 4. Results

In order to assess the parallel algorithm, we performed two independent simulations of equivalent but different models, namely the Ising spin model and the lattice gas. While the lattice gas was simulated with the parallel multicanonical algorithm explained above, the Ising model was simulated with a Metropolis algorithm at fixed magnetization, also referred to as micromagnetic simulation. Both approaches needed independent simulation runs for each measurement point. We chose the inverse temperature in the Ising language to be $\beta^I = 0.369$, which corresponds to a temperature $T = 0.6995$ in the particle picture. Figure 2 shows very good agreement between the two completely independent and different methods.

A direct comparison of simulation times is rather difficult. For one, the multicanonical method provides additional information for various temperatures, and this comes at a price: A time-consuming weight iteration, which is not contributing to the production run. Whether or not the overhead of the initial weight iteration is justified, will depend on the desired size of the statistical error. Furthermore, the approaches will scale differently with system size, which we

Journal of Physics: Conference Series **510** (2014) 012017 doi:10.1088/1742-6596/510/1/012017

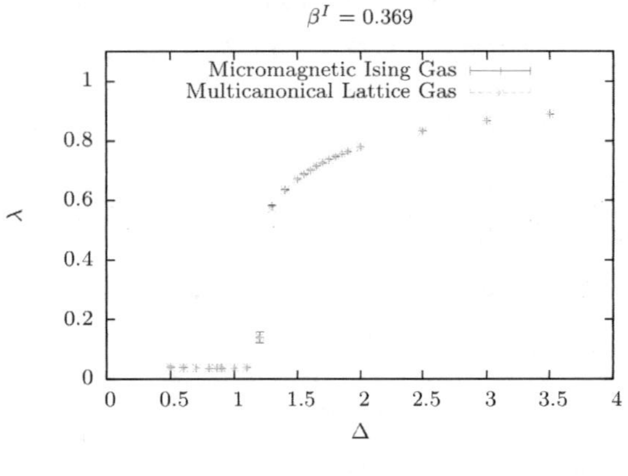


**Figure 2.** Comparison of simulation data from micromagnetic Metropolis simulations of the Ising model and parallel multicanonical simulations of the lattice gas in three dimension. Here, the volume is fixed at $L = 30$ while the number of spins/particles are modified.

do not want to address at this point. For the plot in Fig. 2, the average simulation time per data point of both methods was about half an hour on common 2.5GHz cores, but the multicanonical simulation used 36 cores instead of one in the case of the micromagnetic simulation. However, with increasing system size the free-energy barrier between the evaporated and the condensed phase increases [17], and the necessary computation time for the single-core micromagnetic simulation exceeds the limits of practicability.

Next, we discuss the speedup of the parallel method with increasing degree of parallelization. As mentioned above, we intentionally did not pay attention to optimization of the simulation parameters. Rather we want to demonstrate the applicability of the parallelization by choosing appropriate parameters from single-process simulations. Hence, we fixed the total number of sweeps per iteration to 25600 such that each process performed $25600/p$ sweeps. A single sweep consists of $L^d$ updates. The number of particles was chosen to be $N = 1000$ and the energy range adapted accordingly. The simulation was defined to be converged if a total of 30 "tunneling events" were counted, meaning that the simulations had traveled at least 30 times from the maximal energy to the minimal energy or vice versa.

When we want to estimate the speedup, we need to consider statistical averages. This is due to the fact that the weight iteration relies on the sampled data such that the number of iterations until convergence cannot be predicted and may, moreover, vary depending on the initial seed of the random number generator. To this end, we performed 32 simulations for each degree of parallelization. The effect is best seen in the average number of iterations until convergence (Fig. 3 (a)). An increasing number of independent Markov chains improves each estimate of the distribution belonging to the current weights. This optimizes the estimation for the successive weights and reduces the total amount of necessary number of iterations. Of course, this is in general not true on all scales [9], for example when the number of sweeps per iteration becomes too small or if the optimal number of sweeps is more complex than the assumed $1/p$ behavior.

The speedup is defined in terms of the average time until convergence; it is given by the ratio of a single-core to a $p$-core simulation:

$$S_p = \frac{\bar{t}_1}{\bar{t}_p}. \tag{11}$$

For a fair comparison, both times were obtained with the same program. Because the communication between processes occurs only at the end of every iteration, the speedup should be ideal with a linear behavior of slope one, i.e., doubling the amount of processes should speed up the time by a factor 2. We can see in Fig. 3 (b) that this is indeed true for the case of three dimensions and even better for two dimensions. This may be explained by the reduced number of iterations, due to the independence of Markov chains.

Having demonstrated the correctness of our model and the efficiency of our method, we want

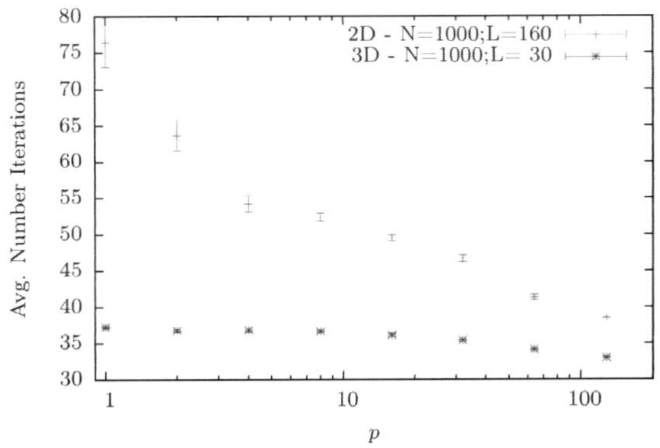


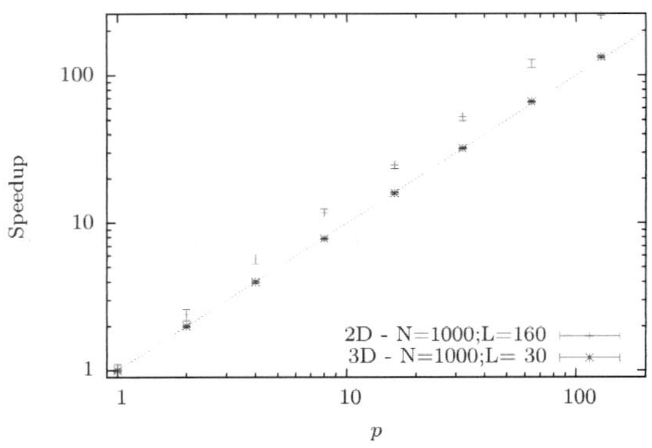


**Figure 3.** (a) Average number of iterations until convergence. (b) Average time speedup. The dotted line is a linear function of slope 1. Each average consists of 32 parallel multicanonical simulations.

to apply it to the problem of condensation in a lattice gas. Figure 4 shows the results for selected temperatures in two and tree dimensions with up to 24 cores in each simulation. The temperature in two dimension was chosen to correspond to the equivalent Ising temperature $T^I = 1.5$ used in Ref. [7], allowing us to use the same parameters. In three dimension we chose a temperature $T^I = 2$ below the roughening transition $T^I_R \approx 2.4537$ [18], where low-temperature approximations hold. This allowed us to use a low-temperature series expansion for the interface tension: $\sigma = 1.9072\ldots$ [19]. Furthermore, below the roughening transition the condensate tends to form a cube, such that the surface free energy may be assumed as $\tau^I_W \approx 6\sigma$. This is the result for the Ising model and has to be rescaled by the same factor as the temperature, hence $\tau_W \approx 6\sigma/4$ (the same argument applies to the two-dimensional case). The remaining free parameters, $\rho_0 = (1 - m_0)/2$ and $\hat{\kappa} = \chi$, were obtained by standard Metropolis simulations of the Ising model on very large lattices. An overview of the involved parameters is given in Table 1.

Opposite to Ref. [7], we always fixed the number of particles and varied the density by modifying the volume of the system. We can see in Fig. 4 that this choice leads to a similar behavior in two dimensions as in Ref. [7], where the predicted $\Delta_c$ separates the evaporated from the condensed phase quite well. Of course, we see finite-size effects as reported before, but the qualitative picture is satisfying. In three dimensions, on the other hand, we observe a strong deviation of small systems from the predicted critical value of the parameter $\Delta$. This becomes weaker but is still present for the largest systems investigated. Specifically, the system needs to go to larger $\Delta$ or, equivalently, particle density in order to condensate. With increasing system

**Table 1.** Converted parameters for the lattice gas system in two [8] and three dimensions, where for the latter case $\rho_0 = (1 - m_0)/2$ and $\hat{\kappa} = \chi$ are obtained from Metropolis simulations of the Ising model at various system sizes and $\tau_W \approx 6\sigma/4$ with $\sigma$ from a low-temperature expansion [19].

| Parameter | 2D | 3D |
|---|---|---|
| $T$ | 0.375 | 0.500 |
| $\rho_0$ | 0.00675 | 0.002739(1) |
| $\hat{\kappa}$ | 0.02708 | 0.00608(1) |
| $\tau_W$ | 1.06125 | $\approx 2.861$ |

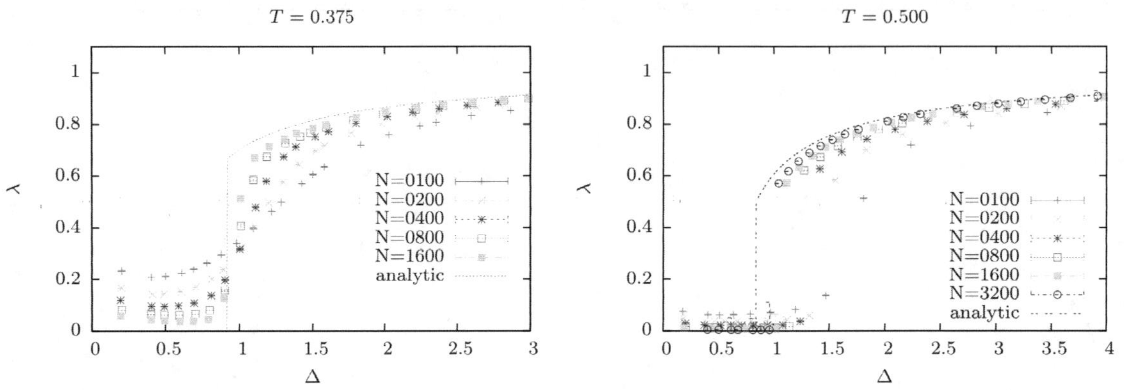


**Figure 4.** Condensation of a lattice gas in two (left) and three (right) dimensions. The temperature is chosen, such that the two-dimensional case corresponds to the Ising model at $T^I = 1.5$ in Refs. [7, 8] and the three-dimensional case is below the roughening transition. The differently colored curves belong to fixed number of particles $N$, while the density is varied by changing $L$. Each data point corresponds to an independent simulation with up to 24 cores.

size, the critical value of $\Delta$ decreases, while the shape of the curves seems to converge to the predicted functional shape. This may be a finite-size effect that is more pronounced in three dimensions, which calls for more detailed studies in future work. In both cases, we do not have any free parameters left.

## 5. Conclusion

We have demonstrated the applicability of an efficient but simple parallelization of the multicanonical method to the case of lattice gas condensation in two and three dimensions. In these cases, the speedup was shown to scale ideally (or even better) due to the fact that independent Markov chains improve the sampling of equilibrium distributions.

Applying this method to the problem of condensation, we could reproduce the results from [7] in two dimensions and present new results in three dimensions which indicate large finite-size effects. In the latter case, condensation systematically occurred at densities higher than predicted. While the deviation decreases with system size, the available data does not allow for quantitative predictions due to the low resolution at the critical density caused by varying the volume. This has still to be investigated in more detail.

## Acknowledgments

We would like to thank Martin Marenz for the joint development of a Monte Carlo Simulation Framework. This work has been partially supported by the Leipzig Graduate School of Excellence GSC185 "BuildMoNa", the Deutsch-Französische Hochschule DFH-UFA (grant CDFA-02-07) and the Sächsische DFG-Forschergruppe FOR877. The project was funded by the European Union and the Free State of Saxony.

## References

[1] Binder K and Kalos M H 1980 *J. Stat. Phys.* **22** 363
[2] Pleimling M and Selke W 2000 *J. Phys.* A **33**
[3] Biskup M, Chayes L and Kotecký R 2002 *Europhys. Lett.* **60** 32
[4] Biskup M, Chayes L and Kotecký R 2003 *J. Stat. Phys.* **116** 175
[5] Neuhaus T and Hager J S 2003 *J. Stat. Phys.* **113** 47

[6] Binder K 2003 *Physica* A **319** 99
[7] Nußbaumer A, Bittner E, Neuhaus T and Janke W 2006 *Europhys. Lett.* **75** 716
[8] Nußbaumer A, Bittner E and Janke W 2008 *Phys. Rev.* E **77** 041109
[9] Zierenberg J, Marenz M and Janke W 2013 *Comput. Phys. Comm.* **184** 1155
[10] Sugihara T, Higo J and Nakamura H 2009 *J. Phys. Soc. Jpn.* **78** 074003
[11] Slavin V V 2010 *Low Temp. Phys.* **36** 243
[12] Ghazisaeidi A, Vacondio F and Rusch L A 2010 *J. Lightwave Technol.* **28** 79
[13] Berg B A and Neuhaus T 1991 *Phys. Lett.* B **267** 249
[14] Berg B A and Neuhaus T 1992 *Phys. Rev. Lett.* **68** 9
[15] Janke W 1998 *Physica* A **254** 164
[16] Hukushima K and Nemoto K 1996 *J. Phys. Soc. Japan* **65** 1604
[17] Nußbaumer A, Bittner E and Janke W 2010 *Prog. Theor. Phys. Suppl.* **184** 400
[18] Hasenbusch M, Meyer S and Pütz M 1996 *J. Stat. Phys.* **85** 383
[19] Shaw L J and Fisher M E 1989 *Phys. Rev.* A **39** 2189

# Non-equilibrium critical behaviour of ultrathin magnetic and metamagnetic Ising films

**P V Prudnikov, A S Elin, M A Medvedeva**

Department of Theoretical Physics, Omsk State University, Omsk 644077, Russia

E-mail: prudnikp@univer.omsk.su

**Abstract.** Monte-Carlo simulations have used to observe both equilibrium and non-equilibrium critical behaviour of Ising thin films with simple cubic lattice structure. The thickness dependency of critical temperature was obtained for metamagnetic thin films. By exploring the short-time scaling dynamics, we have found critical exponents $z$ and $\theta'$ for ferromagnetic thin film. Aging effects were discovered for non-equilibrium behaviour of thin ferromagnetic films.

## 1. Introduction

Magnetic multilayer films provide convenient model systems for studying the physics of antiferromagnetic films and surfaces. The new physical phenomena can be observed in thin films and some states can be realized, which are not observed in bulk samples. Possible application field of ultrathin magnetic films range from spintronics devices [1] to high-density recording technology [2, 3].

It has been theoretically suggested that the highest areal density advantage for Heat assisted magnetic recording (HAMR) can be achieved when the maximum heating temperature is closed to or exceed the Curie temperature [4]. HAMR is the most likely candidate technology to achieve magnetic recording densities 1 TBit/in$^2$ [5, 6]. In this sense, it is important to understand the evolution of magnetization in ultrathin magnetic films with temperature, especially at temperatures close to or above the Curie temperature.

Artificial antiferromagnets, formed of nanostructured superlattices that are coupled antiferromagnetically, have been the focus of many recent studies, due to their high potential for technological applications [7]. Artificial antiferromagnets are heterostructures composed of ferromagnetic layers that are coupled antiferromagnetically via spacers. This structure yields a high level of control over both the intra- and interlayer interactions, allowing for a tailoring of the physical properties.

Examples include [Co/Pt]/Ru, where ferromagnetic Co/Pt multilayers are periodically separated by Ru layers [8, 9]. [Co/Pt]/Ru, with strong perpendicular anisotropy have been called synthetic metamagnets [10], as they can exhibit a regime with an antiferromagnetic phase at low external magnetic fields and a paramagnetic phase at large fields. Metamagnetic systems are systems which present both ferromagnetic and antiferromagnetic couplings simultaneously. Metamagnetic materials are of great interest because it is possible to induce novel kinds of critical behaviour by forcing competition between these couplings, especially by applying a magnetic field. Critical behaviour in metamagnetic films can be differ from bulk behaviour. Both new intermediate phase and critical end point exist for metamagnetic films (see the phase diagram from [11]) unlike bulk metamagnetics.

In this paper we study the equilibrium and non-equilibrium properties of Ising magnetic and metamgnetic films using Monte Carlo simulations by short-time dynamic method [12, 13].

The paper is organized in the following manner. The Ising metamagnetic model and the Monte Carlo simulation scheme are discussed in the next section (Section 2). The numerical results are given in Section 3. The paper ends with concluding remarks and summary in Section 4.

## 2. Model and simulation technique

In this study we consider the Ising hamiltonian

$$H = -J_{xy} \sum_{<i,j>} S_i S_j - J_z \sum_{(i,j)} S_i S_j - h \sum_i S_i, \tag{1}$$

where spins $S_i = \pm 1$, $J_{xy} > 0$ and $J_z$ are intralayer and interlayer coupling strengths, respectively; $J_z > 0$ for ferromagnetic film and $J_z < 0$ for metamagnetic film; $h$ is external magnetic field. The sums over $\langle i,j \rangle$ are the sums over nearest-neighbor pairs in the plane of film and $(i,j)$ – along the $z$ direction. The interaction is nearest-neighbor short-ranged in-plane and out-of-plane direction. Periodic and free boundary conditions are used for the in-plane and out-of-plane directions, respectively.

The simulations are carried out for systems of size $L \times L \times N$, where $L$ is linear size of layer and $N$ is number of layers. Metropolis algorithm was used for updating spin configurations.

The order parameter for ferromagnetic films ($J_z > 0$) can be defined as magnetization

$$m = \left\langle \frac{1}{N_s} \sum_i^{N_s} S_i \right\rangle, \tag{2}$$

where $N_s = NL^2$ is the number of spins, angle brackets denote the statistical averaging. The order parameter for metamagnetic films is the staggered magnetization

$$m_{stg} = |m_1 - m_2|, \tag{3}$$

where $m_1, m_2$ are the magnetization of odd and even layers, respectively.

The critical temperature of metamagnetic films $T_c$ is located via fourth-order cumulant $U_4(L,T)$, defined as

$$U_4(L,T) = \frac{1}{2}\left[3 - \frac{\langle m^4(L,T)\rangle}{\langle m^2(L,T)\rangle^2}\right], \tag{4}$$

where $T$ is temperature in $J/k_B$, the angle brackets denote the statistical averaging, fourth-order cumulant $U_4(L,T)$ is equilibrium characteristic of metamagnetic materials.

The cumulant $U_4(L,T)$ has a scaling form

$$U_4(L,T) = u\left[L^{1/\nu}(T - T_c)\right]. \tag{5}$$

The scaling dependence of the cumulant makes it possible to determine the critical temperature $T_c$ from the coordinate of the intersection points of the curves specifying the temperature dependence $U_4(L,T)$ for different $L$.

For investigation non-equilibrium properties of critical behaviour of magnetic materials we were considered time dependencies of magnetization $m(t)$ and second-order Binder cumulant $U_2(t)$, where time $t$ is measured in Monte-Carlo steps per spin (MCS/s). According to the argument of Janssen et al. [15] obtained with the RG method and $\varepsilon$-expansion, one may expect a generalized scaling relation for the time dependence of $k$-th moment the magnetization

$$m^{(k)}(t,\tau,L,m_0) = b^{-k\beta/\nu} m^{(k)}(b^{-z}t, b^{1/\nu}\tau, b^{-1}L, b^{x_0}m_0), \tag{6}$$

is realized after a time scale $t_{\rm mic}$ which is large enough in a microscopic sense but still very small in a macroscopic sense. In (6) $b$ is a spatial rescaling factor, $\beta$ and $\nu$ are the well-known static critical exponents, $z$ is the dynamic exponent, the new independent exponent $x_0$ is the scaling dimension of the initial magnetization $m_0$ and $\tau = (T - T_c)/T_c$ is the reduced temperature.

When system evolving from a completely ordered state ($m_0 = 1$) magnetization have a scaling form $m(t) \sim t^{-\beta/z\nu}$. The cumulant $U_2(t)$ is defined on the basis of the magnetization and its second moment like

$$U_2(t) = \left[\frac{\langle m^{(2)}(t)\rangle}{\langle m(t)\rangle^2} - 1\right] \sim t^{d/z}. \tag{7}$$

When system evolving from high-temperature state ($m_0 \ll 1$) we can choose the scaling factor $b = t^{1/z}$ and applying the scaling form (6) for $k = 1$ to the small quantity $t^{x_0/z}m_0$, one obtains

$$m(t,\tau,m_0) \sim m_0 t^{\theta'} F(t^{1/\nu z}\tau, t^{x_0/z}m_0) = m_0 t^{\theta'}\left(1 + a t^{1/\nu z}\tau\right) + O(\tau^2, m_0^2), \tag{8}$$

where $\theta' = (x_0 - \beta/\nu)/z$ has been introduced. For $\tau = 0$ and small enough $t$ and $m_0$, the scaling dependence for magnetization (8) takes the form $m(t) \sim t^{\theta'}$.

## 3. Results and discussion

### *3.1. Equilibrium critical behaviour*

For research equilibrium critical behaviour we considered $L = 64, 96, 128$ and $N = 2, 4, 6, 8, 10, 14, 20$. To ensure equilibrium we wait at least $10^4$ Monte-Carlo steps per spin (MCS/s) from its initial state before taking any measurements. Averages were calculated using $10^5$ MCS/s for each sample and 100 samples with temperature's shift equaling 0.001.

The temperature dependencies of staggered magnetization $m_{\rm stg}$ and staggered susceptibility $\chi_{\rm stg} = \left(\langle m^2_{\rm stg}\rangle - \langle m_{\rm stg}\rangle^2\right)/T$ for metamagnetics films were presented in Fig. 1.

From calculating fourth-order cumulant $U_4(L,T)$ (Fig. 2) we have obtained critical temperature's values, which presented in Table 1. These values demonstrate a good agreement with results of [16] for $J_z = 1$. The values of critical temperature $T_c$ evolve from 2D to 3D values with increasing film thickness $N$.

Whereas the phase diagram of the bulk system exhibits only one phase transition between the antiferromagnetic and paramagnetic phases, the phase diagram of thin Ising metamagnets includes an additional intermediate phase where one of the surface layers has aligned itself with the direction of the applied magnetic field. Transition from antiferromagnetic phase to intermediate phase ends at a critical end point with $T_{\rm cr} = 2.50(5)$ and $h_{\rm cr} = 0.98(1)$ [11] for $N = 8$. In Fig. 3 it is presented metamagnetic films behaviour in the magnetic field. For $T > T_{\rm cr}$ it demonstrates the transition from antiferromagnetic to paramagnetic phase only, but $T < T_{\rm cr}$ it demonstrates the discontinuous transitions from antiferromagnetic to intermediate and to paramagnetic phase.

### *3.2. Non-equilibrium properties*

For study non-equilibrium critical behaviour we have considered $L = 32$ and $N = 2$ and 4. We consider evolution of thin ferromagnetic film both from high-temperature ($m_0 \ll 1$) and from low-temperature ($m_0 = 1$) initial states. Starting from initial configurations, the system was

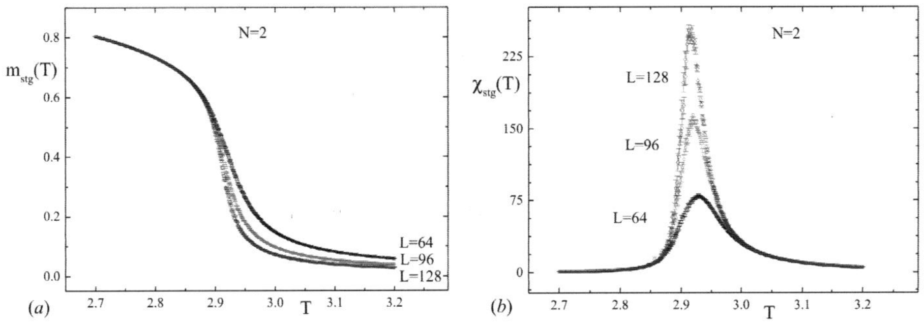


**Figure 1.** Temperature dependencies of staggered magnetization $m_{\text{stg}}$ $(a)$ and staggered susceptibility $\chi_{\text{stg}}$ $(b)$ for $N = 2$, $J_z = -0.5$ and different $L$

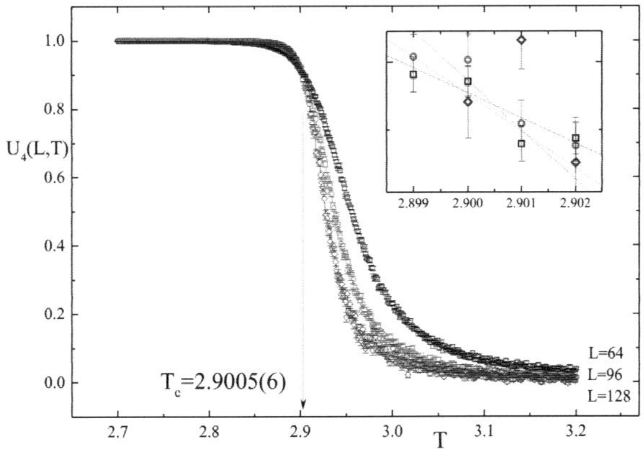


**Figure 2.** Binder's cumulants $U_4(L,T)$ as function of $T$ for $N = 2$, $J_z = -0.5$ and different $L$

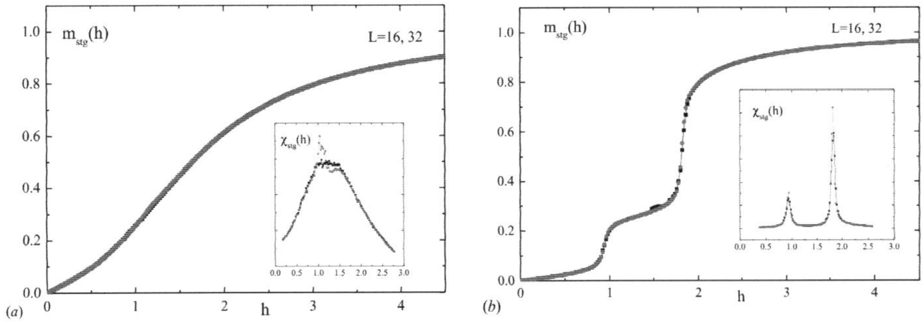


**Figure 3.** Field dependencies of staggered magnetization $m_{\text{stg}}$ and staggered susceptibility $\chi_{\text{stg}}$ (inset) for $T = 3.2 > T_{\text{cr}}$ $(a)$ and $T = 2.5 < T_{\text{cr}}$ $(b)$ for $N = 8$, $J_z = -0.5$

updated with Metropolis algorithm at the critical temperatures $T_c = 3.2076(4)$ for $N = 2$ and $T_c = 3.8701(3)$ for $N = 4$ [16]. In both cases simulations have been performed up to $t = 10000$ MCS/s.

We measured the time evolution of the magnetization $m(t)$, the second moment $m^{(2)}(t)$, which also allow to calculate the time-dependent cumulant $U_2(t)$. In Fig. 4 the magnetization is plotted for high-temperature initial state (Fig. 4a) and low-temperature initial state (Fig. 4b). The increasing of the magnetization $m(t)$ from a high-temperature initial magnetization $m_0 \ll 1$ at

**Table 1.** The critical temperatures $T_c(J_z, N)$ determined for magnetic films with different number of layers $N$ and different interlayer coupling strength $J_z$.

| N | $T_c$ | | |
|---|---|---|---|
| | $J_z = -1$ | $J_z = -0.5$ | $J_z = 1$ [16] |
| 2 | 3.2062(12) | 2.9005(11) | 3.2076(4) |
| 4 | 3.8691(10) | 3.3120(7) | 3.8701(3) |
| 6 | 4.1161(11) | 3.4607(9) | 4.1179(3) |
| 8 | 4.256(17) | 3.560(14) | 4.2409(2) |
| 10 | 4.344(50) | 3.585(24) | 4.3117(3) |
| 14 | 4.412(20) | 3.636(27) | — |
| 20 | 4.481(16) | 3.672(19) | 4.4381(2) |

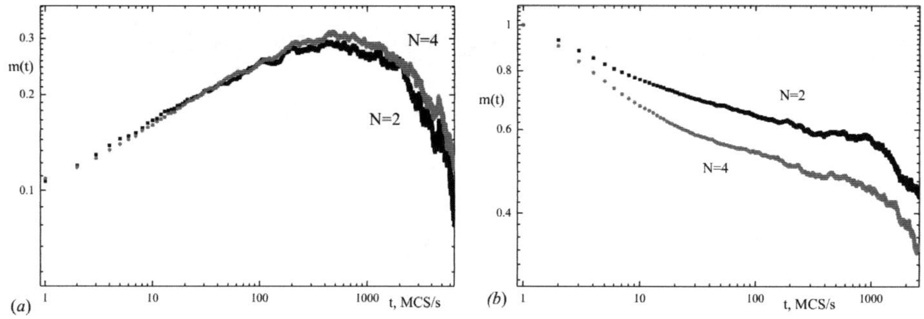


**Figure 4.** The relaxation of magnetization $m(t)$ from non-equilibrium initial $m_0 \ll 1$ $(a)$ and $m_0 = 1$ $(b)$ states for $N = 2$: $T_c = 3.2076$ and $N = 4$: $T_c = 3.8701$

short-time regime $t < t_{\rm cr} \simeq m_0^{-1/(\theta'+\beta/z\nu)}$ have form $m(t) \sim t^{\theta'}$. The initial rise of magnetization is changed to the well-known decay $m(t) \sim t^{-\beta/z\nu}$ for $t \gg t_{\rm cr}$. From slope of magnetization at high-temperature initial state on $t \in [0; 100]$ we can estimate the value of critical non-equilibrium exponent $\theta'$. From the slope of time dependence of the cumulant $U_2(t)$ on $t \in [0; 100]$ we can estimate the value of critical exponent $z$. The calculated values of $\theta'$ and $z$ are presented in Table 2. A comparison of the obtained values with the results for 2D Ising model in various works show the very good agreement ie thin ferromagnetic film belong universality class non-equilibrium behaviour of 2D Ising model.

For study ageing effects from high-temperature initial state we calculated time dependencies of two-time autocorrelation function

$$C(t, t_w) = \left\langle \frac{1}{N_s} \sum_i S_i(t) S_i(t_w) \right\rangle - \left\langle \frac{1}{N_s} \sum_i S_i(t) \right\rangle \left\langle \frac{1}{N_s} \sum_i S_i(t_w) \right\rangle, \quad (9)$$

where $t_w$ ("waiting time") is time which characterizes the time elapsed since the preparation of the sample prior to measurement of its quantities.

Time dependencies of autocorrelation function $C(t, t_w)$ for $t_w = 10, 20, 30, 50$ MCS/s are presented in Fig. 5 for $N = 2$ and 4. The resulting curves presented in Figs. 4 and 5 have been obtained by averaging over 1000 samples.

Autocorrelation function have scaling form

$$C(t, t_w) \sim t_w^{-2\beta/\nu z} F_C(t/t_w). \quad (10)$$

**Table 2.** Values of the critical exponents $z$ and $\theta'$.

| System | Method | $z$ | $\theta'$ |
|---|---|---|---|
| Present report | MC | 1.98(31) | 0.188(30) |
| | | 2D Ising | |
| Okano et al, 1997 [17] | MC | | 0.191(1) |
| Prudnikov et al, 2006 [18] | RG | 2.0842(39) | |
| Wang, Hu, 1997 [19] | MC | 2.166(7) | |
| Nightingale and Blöte, 2000 [20] | MC | 2:1667(5) | |
| Prudnikov, Markov, 1995 [21] | MC | 2.24(7) | |
| Prudnikov, Vakilov, 1992 [22] | RG | 2.277 | |
| Kalle, 1984 [23] | MC | 2.14(2) | |
| Jan et al., 1983 [24] | MC | 2.12(6) | |

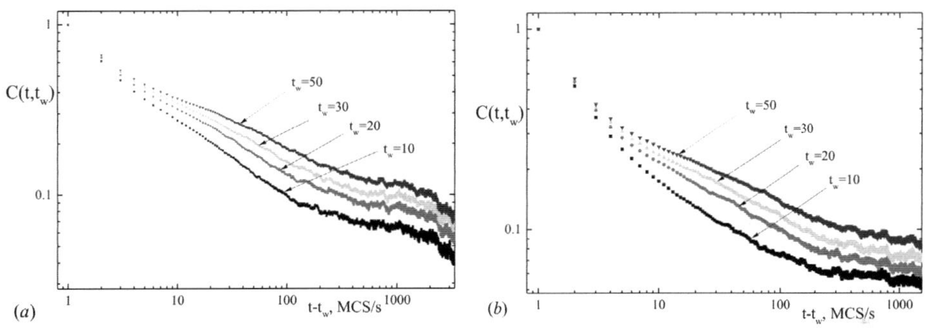


**Figure 5.** Time dependencies of autocorrelation function $C(t; t_w)$ for $N = 2$ (a) and $N = 4$ (b) for different $t_w$ ($m_0 \ll 1$)

Data presented in Fig. 5 clear demonstrates the presence of the two characteristics regimes: quasi-equilibrium regime at times $(t - t_w) \ll t_w$ and non-equilibrium regime at times $(t - t_w) \gg t_w$. At times $(t - t_w) \sim t_w$ there is a crossover regime with the dependence of correlation characteristics of the waiting time.

At quasi-equilibrium regime autocorrelation function have a scaling form $C(t) \sim t^{-c_a}$, where $c_a = \gamma/z\nu$. As we can see on Fig. 5 the increasing of system age $t_w$ lead to decreasing of value $c_a$ and consequently lead to amplification of critical slow-down and ageing effects.

## 4. Conclusions

We have studied the critical behaviour of Ising thin films on simple cubic lattice using extensive Monte Carlo simulations. We have found critical temperatures $T_c$ of metamagnetic films for different thickness $N$ and different coupling strength $J_z$. The film's $T_c$ evolve from 2D to 3D values with increasing film thickness.

We have researched non-equilibrium behaviour of ferromagnetic thin film by short-time dynamics method. The obtained values of critical exponents $z$ and $\theta'$ demonstrates that ultrathin ferromagnetic films with $N < 4$ belong to class of universal critical non-equilibrium behaviour of the 2D Ising model. Aging effects were discovered for non-equilibrium regime $(t - t_w) \gg t_w$.

## Acknowledgments
The reported study was supported in part by Ministry of Education and Science of Russia through project No. 2.3046.2011 and by grant MU-2/2013 of Omsk State University for young scientists.

The simulations were supported by the Supercomputing Center of Lomonosov Moscow State University and Joint Supercomputer Center of the Russian Academy of Sciences.

## References
[1] Mangin S, Ravelosona D, Katine J A *et al* 2006 *Nat. Mater.* **5** 210
[2] Fullerton E E, Margulies D T, Supper N *et al* 2003 *IEEE Trans. Magn.* **39** 639
[3] Chappert C, Fert A and Nguyen van Dau F 2007 *Nat. Mater.* **6** 813
[4] Lyberatos A and Guslienko K Y 2003 *J. Appl. Phys.* **94** 1119
[5] Rottmayer R E, Batra S, Buechel D *et al* 2006 *IEEE Trans. Magn.* **42** 2417
[6] Seigler M A, Challener W A, Gage E *et al* 2008 *IEEE Trans. Magn.* **44** 119
[7] Mukherjee T, Sahoo S, Skomski R *et al* 2009 *Phys. Rev.* B **79** 144406
[8] Hellwig O, Kirk T L, Kortright J B *et al* 2003 *Nat. Mater.* **2** 112
[9] Hellwig O, Berger A, Kortright J B, and Fullerton E E 2007 *J. Magn. Magn. Mater.* **319** 13
[10] Rössler U K and Bogdanov A N 2004 *J. Magn. Magn. Mater.* **269** L287
[11] Chou Y-L and Pleimling M 2011 *Phys.Rev.* B **84** 134422
[12] Albano E V, Bab M A, Baglietto G *et al* 2011 *Rep. Prog. Phys.* **74** 026501
[13] Prudnikov V, Prudnikov P, Krinitsyn A, Vakilov A, Pospelov E and Rychkov M 2010 *Phys. Rev.* E **81** 011130
[14] Santos M and Figueiredo W 2000 *Phys. Rev.* E **62** 1799
[15] Janssen H K, Schaub B and Schmittmann B 1989 *Z. Phys.* B **73** 539
[16] Laosiritaworn Y, Poulter J and Staunton J B 2004 *Phys. Rev.* B **70** 104413-1 (*Preprint* cond-mat/0307291v1)
[17] Okano K, Schülke L, Yamagishi K and Zheng B 1997 *Nucl. Phys.* B **485** 727
[18] Krinitsyn A S, Prudnikov V V and Prudnikov P V 2006 *Theor. Math. Phys.* **147** 561
[19] Wang F-G and Hu C-K 1997 *Phys. Rev.* E **56** 2310
[20] Nightingale M P and Blöte H W 2000 *Phys. Rev.* B **62** 1089
[21] Prudnikov V V and Markov O N 1995 *Europhys. Lett.* **29** 245
[22] Prudnikov V V and Vakilov A N 1992 *JETP* **101** 990
[23] Kalle C 1984 *J. Phys.* A **17** 801
[24] Jan N, Leo Moseley L and Stauffer D 1983 *J. Stat. Phys.* **33** 1

# Monte Carlo renormalization group of dilute 3D Ising dynamics

**V V Prudnikov, A N Vakilov, and S A Zolotarev**

Dept. of Theoretical Physics, Omsk State University, Omsk 644077, Russia

E-mail: prudnikv@univer.omsk.su

**Abstract.** Simulation of critical relaxation of the magnetization in a three-dimensional Ising model diluted by quenched nonmagnetic impurity atoms is carried out. The cubic systems with linear size $L = 144$ and spin concentrations $p = 0.95, 0.8, 0.6$, and 0.5 are studied. The dynamical critical exponent $z$ is determined by the Monte Carlo method combined with the dynamical renormalization group method. The obtained results are: z(p): $z(0.95) = 2.19 \pm 0.07$, $z(0.8) = 2.29 \pm 0.06$, $z(0.6) = 2.72 \pm 0.08$, and $z(0.5) = 2.75 \pm 0.08$.

## 1. Introduction

The investigation of critical behavior of disordered systems remains one of the main problems in condensed-matter physics and excites a great interest because all real solids contain structural defects [1, 2]. The structural disorder breaks the translational symmetry of the crystal and thus greatly complicates the theoretical description of the material. The influence of disorder is particularly important near the critical point where behavior of a system is characterized by anomalous large response on any even weak perturbation. The description of such systems requires the development of special analytical and numerical methods. The effects produced by weak quenched disorder on critical phenomena have been studied for many years [3–8]. According to the Harris criterion [3], the disorder affects the critical behavior only if $\alpha > 0$, the specific-heat exponent of the pure system, is positive. In this case, a new universal critical behavior, with new critical exponents, is established. In contrast, when $\alpha < 0$, the disorder appears to be irrelevant for the critical behavior. Only systems whose effective Hamiltonian near the critical point is isomorphic to the Ising model satisfy this criterion.A large number of publications is devoted to the study of the critical behavior of diluted Ising-like magnets by the renormalization-group (RG) methods, the numerical Monte-Carlo methods, and experimentally (for a review, see [2, 9–11]). The ideas about replica symmetry breaking in the systems with quenched disorder were presented in Refs. [12, 13]. A refined RG analysis of the problem has shown the stability of the critical behavior of weakly disordered three dimensional systems with respect to the replica symmetry breaking effects [14]. All obtained results confirm the existence of a new universal class of the critical behavior, which is formed by diluted Ising-like systems. However, it remains unclear whether the asymptotic values of critical exponents are independent of the rate of dilution of the system, how the crossover effects change these values, and whether two or more regimes of the critical behavior exist for weakly and strongly disordered systems. These questions are the subjects of heated discussions [2, 15] and extensive Monte Carlo simulations for site-diluted [16–18] and bond-diluted [19, 20] three-dimensional Ising models.

At present paper, a computer simulations of the critical dynamics of the diluted 3D Ising model by the Monte-Carlo method are considered for samples with spin concentrations $p = 0.95, 0.8, 0.6$, and 0.5. There is reason to believe that the influence of quenched impurities on the critical dynamics will be seen more clearly than in the equilibrium state, because of the specific conservation laws.

## 2. DESCRIPTION OF THE MODEL AND METHODS

We have considered the following Hamiltonian for the 3D site-diluted ferromagnetic Ising model defined in a cubic lattice of linear size L with periodic boundary conditions

$$H = -J \sum_{i,j} S_i S_j p_i p_j, \tag{1}$$

where the sum is extended to the nearest neighbors, $J > 0$ is the short-range exchange interaction between spins $S_i$ fixed at the lattice sites, and assuming values of $\pm 1$. Nonmagnetic impurity atoms form empty sites. In this case, occupation numbers $p_i$ assume the value 0 or 1 and are described by the distribution function

$$P(p_i) = (1-p)\delta(p_i) + p\delta(1-p_i) \tag{2}$$

with $p = 1 - c$, where $c$ is the concentration of the impurity atoms. In this paper, we have investigated systems with the spin concentrations $p = 0.95, 0.8, 0.6$, and 0.5. We have considered the cubic lattices with linear size $L = 144$. The Metropolis algorithm has been used in simulations. We consider only the dynamic evolution of systems described by the model A in the classification of Hohenberg and Halperin [21]. Ising-model dynamics is customarily described by the conditional probability function $P_s(t) \equiv P(\{S\}, t)$, which satisfied the Glauber's kinetic equation

$$\frac{dP_s}{dt} = -P_s(t) \sum_{s'} W(S \to S') + \sum_{s'} W(S' \to S) P_{s'}(t), \tag{3}$$

where $W(S \to S')$ is the transition rate from a microscopic state given by the spin configuration $\{S\}$ to a state with the configuration $\{S'\}$. In order that the Markov process described by Eq. (2) converge to the equilibrium state of a Gibbs ensemble with $P_s = \exp(-E_s/kT))$, the detailed-balancing condition $W(S \to S')P_s = W(S' \to S)P_s$ is sufficient but not necessary. This freedom of choice gives $W(S- > S')$ which is not determined uniquely. The function $W$ is usually chosen in the form of the Metropolis's function

$$W(S \to S') = \begin{cases} \exp(-\Delta E_{ss'}/kT), & \Delta E_{ss'} > 0 \\ 1, & \Delta E_{ss'} \leq 0 \end{cases} \tag{4}$$

or the Glauber's function

$$W(S \to S') = \exp(-\Delta E_{ss'}/kT)/[1 + \exp(-\Delta E_{ss'}/kT], \tag{5}$$

where $E_{ss'} = E_{s'} - E_s$ .

The relation $< A(t) >= \sum_s A_s P_s(t)$ determines the dynamical evolution of the quantity $A_s$, by means of the function $P_s(t)$ which is the solution of Eq.(3). Metropolis's algorithm, which consists of choosing randomly the spin $S_i$, and flipping the spin with probability determined by the function $W$ in Eq.(4), makes is possible to implement directly the Ising-model dynamics with relaxation of the magnetization $m_b(t) = \sum_i^N S_i/N$ to the equilibrium value determined by the thermostat temperature $T$. The time scale $t$ can be associated with the scale $\{S\}$ of successive configurations, assuming that $N$ system sites are chosen randomly per unit time. One time unit corresponds to one Monte Carlo step per spin. In simulation of the critical dynamics the initial state of the system is chosen when all spins are parallel ($m_b = 1$) and the temperature of the system is equal to the critical temperature. The critical temperature $T_c$ for dilute magnetic materials is a function of the impurity concentration $c_{imp}$ decreasing with increasing $c_{imp}$ and vanishing at the threshold concentration $c_{imp} = 1 - p_c$. For a cubic lattice of Ising spins $p_c \simeq 0.31$ and $T_c(p)$ are equal to: $T_c(0.95) \simeq 4.26267$, $T_c(0.8) \simeq 3.49948$, $T_c(0.6) \simeq 2.42413$, and $T_c(0.5) \simeq 1.84509$ [17] in the units $J/k$. We have used here the Monte Carlo method combined with the dynamical renormalization group method [22], to determine the dynamic exponent $z$ characterizing the critical increase in the relaxation time of the system $t_{rel} \sim |T - T_c|^{-z\nu}$. For this, the system was partitioned into blocks, where a block $b^d$ of neighboring spins was replaced by a single spin whose direction is determined by the direction of most spins in the block. The redefined spin system forms a new lattice with magnetization $m_b$. Let the magnetization of the initial lattice relax to some $m_1$ over a time $t_1$, and let the redefined system reach the same value $m_1$ over the time $t_b$. Then by using two systems with block size $b$ and $b'$ and determining the relaxation times $t_b$ and $t_{b'}$ of the block magnetizations $m_b$ and $m_{b'}$ to the same value $m_1$, the dynamic exponent $z$ can be determined from the relation

$$t_b/t_{b'} = (b/b')^z \tag{6}$$

or

$$z = \ln(t_b/t_{b'})/\ln(b/b') \tag{7}$$

in the limit of sufficiently large $b$ and $b' \to \infty$.

## 3. MEASUREMENTS OF THE DYNAMICAL CRITICAL EXPONENT z FOR 3D SITE-DILUTED ISING MODEL

We applied this algorithm to dilute systems with linear size $L = 144$ and spin concentrations presented above. The size of the system made it possible to partition it into blocks with sizes $b = 2, 3, 4, 6, 8, 9, 12, 16, 18, 24, 36, 48$, and 72. The procedure of block partitioning of the initial spin and impurity configurations was implemented on the basis of the criterion of spin connectivity. Thus a $b^d$-dimensional block was considered to be a spin block and replaced by an effective spin oriented in a direction determined by the direction of most spins in the block if the block contained a spin cluster connecting both opposite faces of the block. Otherwise, the block was considered to be an impurity block and replaced by an empty site in the renormalized lattice. A relaxation simulating procedure consisting of 10000 Monte Carlo steps per spin was performed for each system with 1000 runs with different impurity configurations over which the function $m_b(t)$ was averaged.

We demonstrate in fig.1 a)-d) the plots of the initial and renormalized magnetizations $m(t)$, as functions of time, for systems with spin concentrations $p = 0.95, 0.8, 0.6$, and 0.5 averaged over different impurity configurations. The relation (7) can be employed in order to determine the values of the exponent $z$ independently. However, the power-law character found for the relaxation of the magnetization at the critical temperature enabled us to employ, in contrast to [22] and [23], a different and, we believe, better-founded procedure for processing the curves for the renormalized magnetizations $m_b(t)$, as in our paper [24]. Thus the $m_b(t)$ curves plotted in a double logarithmic scale were approximated by the straight lines $\lg m_b = k_b \lg t + n_b$, by the least squares method in intervals $\triangle m_b$, corresponding best to a power-law variation of $m_b(t)$. Then, the coefficients $k_b$ were averaged and the average value $k_{av}$ gives possibility to determine the parameters $n_b$ of the straight lines $\lg m_b = k_{av} \lg t + n_b$ by extending the lines through the point of intersection with $\lg m_b = k_b \lg t + n_b$ at the center of the intervals $\triangle m_b$. As a result, the formula (7) for $z$ becomes as

$$z = (\bar{n}_{b'} - \bar{n}_b)/[k_{av} \lg(b/b')] \tag{8}$$

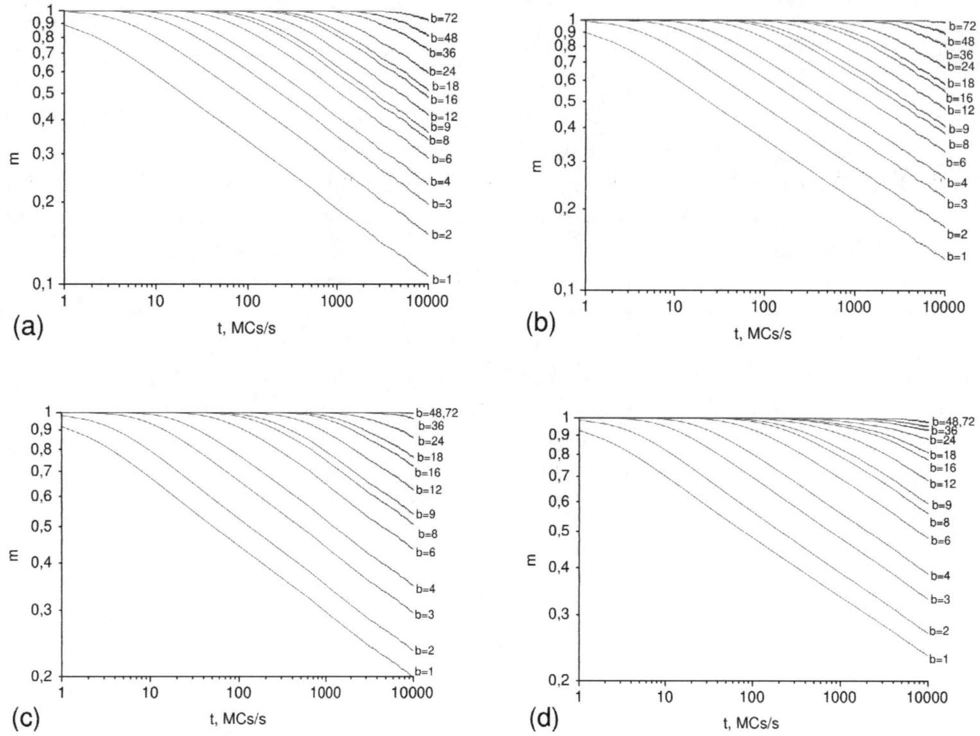


**Figure 1.** Initial $m_1$ and renormalized $m_b$ magnetizations as functions of time for the dilute Ising model with spin concentrations $p = 0.95$ (a), 0.8 (b), 0.6 (c), 0.5 (d)

**Table 1.** Values of the dynamic exponent $z$ for systems with different spin concentrations $p$

| $b$ | $p = 0.5$ | $p = 0.6$ | $p = 0.8$ | $p = 0.95$ |
|---|---|---|---|---|
| 8 | 2.638(50) | 2.554(45) | 2.283(35) | 2.246(25) |
| 9 | 2.659(50) | 2.571(50) | 2.285(35) | 2.240(30) |
| 12 | 2.689(60) | 2.604(60) | 2.287(40) | 2.226(30) |
| 16 | 2.699(60) | 2.636(60) | 2.288(40) | 2.213(35) |
| 18 | 2.690(65) | 2.655(65) | 2.290(45) | 2.210(35) |

Sets of values of the exponent $z$ corresponding to different values of $b$ with $b' = 1$ were obtained using the relation (8) (Table 1). For impurity systems the renormalization-group-transformation procedure reaches the proven asymptote of $m_b$, as a function of the block-partition parameter $b$, at larger values of $b$ than in the case of a pure system. So, for blocks with small values $b$ the presence of defects leeds both to loss of some initial paths of spin connectivity and appearance of a new paths of spin connectivity which are absent in the original lattice. This effect becomes less likely with increasing $b$. On the other hand, the changes of the renormalized magnetization $m_b$ are decreased with time for large values of $b$ and an errors for $m_b$ and therefore for the determination of $z$ are increased. For this reason, we selected for the analysis the values of the exponent $z$ corresponding to $b = 8, 9, 12, 16$, and 18.

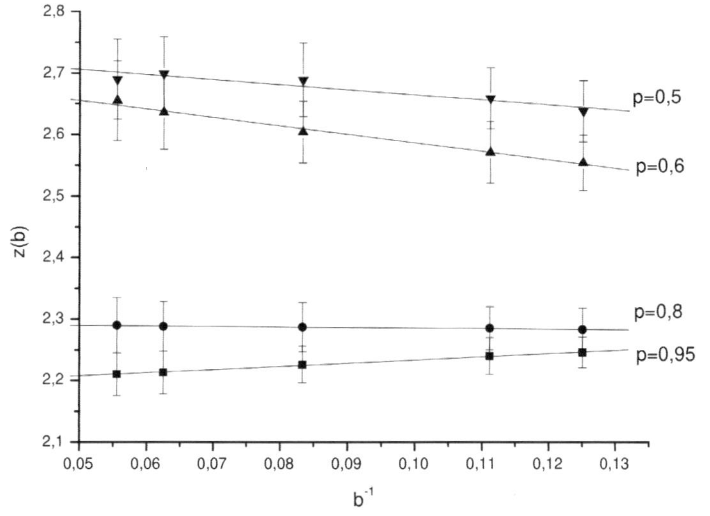


**Figure 2.** Dependence of dynamic exponent $z$ on $b$ for different spin concentrations $p$

The obtained dependence of $z$ on $b$ (fig. 2) made possible to realize the typical extrapolation for renormalization-group transformation to the case $b \to \infty$, assuming in concordance with [23] that

$$z_b = z_{b=\infty} + const \cdot b^{-1}. \tag{9}$$

The following results were obtained for considered impurity systems: $z(0.95) = 2.19 \pm 0.05$, $z(0.8) = 2.29 \pm 0.06$, $z(0.6) = 2.72 \pm 0.08$, and $z(0.5) = 2.75 \pm 0.08$. Hence it is clear that the value of the dynamical exponent $z$ for $p = 0.6$ is virtually identical to the value for $p = 0.5$, while for $p = 0.95$ and 0.8 they are in agreement with each other only to within the statistical uncertainties in their values.

## 4. ANALYSIS OF RESULTS AND CONCLUSIONS

We now compare the simulation results with other results of Monte Carlo simulations (MC), application of field-theoretical method with fixed-dimension $d = 3$ expansion (FTM), and experimental (EXP) investigations (Table 2). Values of $z$ from paper [27] with EXP results agree rather well with our results only for weakly diluted systems with $p = 0.95$, while a noticeable difference of the results is observed for strongly disordered systems. Starting from the universality concept for critical behavior of diluted Ising systems and that the asymptotic value of $z$ is independent of the dilution degree, the author in [28] obtained the asymptotic value $z = 2.41$ using the effective values of the exponent listed in Table II. The off-equilibrium critical dynamics of the 3D Ising model with the spin concentration varying in a wide range was analyzed in [29]. Assuming that $\gamma/(\nu z)$ and $\omega/z$ exponents are dilution independent, the authors obtained the asymptotic value of $z = 2.62(7)$ taking into account the leading corrections to the scaling dependence for the dynamical susceptibility. In this case, the value of the exponent $\omega = 0.50(13)$ obtained in [29] is strongly inconsistent with $\omega = 0.25(10)$ from the field theory calculations [31] and not so well agreement with $\omega = 0.37(6)$ from Monte Carlo results computed in [16]. In addition, so small errors for the dynamical critical exponent $z = 2.62(7)$ couldn't explain given by authors so large errors for non-universal parameters $A(p)$ and $B(p)$ for strongly diluted samples

**Table 2.** Values of the obtained critical exponent $z$ and comparison to other results of Monte Carlo simulations (MC), field-theoretical method with fixed-dimension $d = 3$ expansion (FTM), and experimental (EXP) investigations

| | 0.5 | 0.6 | 0.8 | 0.95 |
|---|---|---|---|---|
| Rosov et al., 1992 [27] EXP | | | 2.18(10) | |
| Prudnikov et al.., 1992 [24] MC | 2.65(12) | 2.58(9) | 2.20(7) | 2.19(7) |
| Heuer, 1993 [28] MC | | 2.93(3) | 2.38(1) | 2.16(1) |
| Parisi et. al., 1999 [29] MC | 2.62(7) | 2.62(7) | 2.62(7) | 2.62(7) |
| Hasenbusch et. al., 2007 [30] MC | | 2.35(2) | 2.35(2) | |
| Prudnikov et al., 1992 [25] FTM | | | 2.237(34) | |
| Prudnikov et al., 2006 [26] FTM | | | 2.1792(13) | |
| Prudnikov et al., 2010 [18] MC | | | 2.208(22) | 2.185(17) |
| present paper | 2.75(8) | 2.72(8) | 2.29(6) | 2.19(5) |

in comparison with lightly diluted samples. It must be inversely from physics of phenomenon. In [30] the purely relaxational dynamics (model A) at criticality in three-dimensional disordered Ising systems whose static critical behaviour belongs to the randomly diluted Ising universality class was studied. Authors considered the site-diluted and bond-diluted Ising models, and the $\pm$J Ising model along the paramagnetic-ferromagnetic transition line. They performed Monte Carlo simulations at the critical point using the Metropolis algorithm and studied the dynamic behaviour in equilibrium at various values of the disorder parameter. The results provide an evidence of the existence of a unique model-A dynamic universality class which describes the relaxational critical dynamics in all considered models. In particular, the analysis of the size dependence of suitably defined autocorrelation times at the critical point provides the estimate $z = 2.35(2)$ for the universal dynamic critical exponent. Authors also study in [30] the off-equilibrium relaxational dynamics following a quench from $T = \infty$ to $T_c$. In agreement with the field-theory scenario, the analysis of the off-equilibrium dynamic critical behaviour gave an estimate of $z$ that is consistent with the equilibrium estimate $z = 2.35(2)$.

In [25] we gave a field-theoretical description of the critical dynamics of dilute magnetic materials directly for the three-dimensional case. In the two-loop approximation, using the Pade-Borel summation technique, we obtained the critical exponent $z(p) = 2.237$, valid for impurity concentrations much less than the spin-percolation threshold. A similar calculation performed in the three-loop approximation for diluted Ising system [26] with the use of different methods of summation gave the value $z = 2.1792(13)$. The comparison of these FTM results with our present MC results and results in [18, 24] shows that they are in good agreement for weakly diluted systems with $p = 0.95$ and 0.8 and with experimental value of $z = 2.18(10)$ obtained in [27] under investigations of the dynamic critical behavior of weakly diluted Ising-like magnet $Fe_pZn_{1-p}F_2$ with $p = 0.9$. The simulation results give a much higher values of the dynamic exponent $z$ for spin concentrations $p = 0.6$ and 0.5. We attribute this to the fact that for a cubic lattice of Ising spins with $p \leq p_c^{(imp)} \simeq 0.69$ the impurities form a connecting cluster, which for $T < T_c$, coexists with a connecting spin cluster right up to the spin percolation $p_c = 1 - p_c^{(imp)}$. As a result, the spin correlation length in the region $p_c \leq p \leq p_c^{(imp)}$ is not the only scale determining the behavior of the system near the critical temperature $T_c(p)$). The character of impurity scattering of long-wavelength fluctuations of the magnetization also changes. By analogy with [32,33] and the works of one of us, [34,35], where the influence of the correlation of the impurities and extended structural defects on the critical properties of disordered systems was investigated, there are grounds for believing that in the region $p_c \leq p \leq p_c^{(imp)}$ the existence of an extended impurity structure results in a change in the Harris criterion for the effect of quenched point defects. For this reason, the change in sign of the heat-capacity exponent $\alpha$ (from positive to negative) at a transition from pure to impurity critical behavior in Ising-like magnetic materials does not limit the new type of critical behavior determined by extended impurity structure.

## Acknowledgments

The reported study was supported in part by Ministry of Education and Science of Russia through project No. 2.3046.2011 and by grant MU-3/2013 of Omsk State University for young scientists.

The simulations were supported by the Supercomputing Center of Lomonosov Moscow State University and Joint Supercomputer Center of the Russian Academy of Sciences.

## References

[1] Prudnikov P V and Prudnikov V V 2009 *Phys. Rev.* B **80** 024115
[2] Folk R, Holovatch Yu, and Yavorskii T 2003 *Phys. Usp.* **46** 169
[3] Harris A B 1974 *J. Phys.* C **7** 1671
[4] Harris A B and Lubensky T C 1974 *Phys. Rev. Lett.* **33** 1540
[5] Khmelnitski D E 1975 *Sov. Phys. JETP* **41** 981
[6] Prudnikov V V, Belim S V, Ivanov A V, Osintsev E V, and Fedorenko A A 1998 *JETP* **87** 527
[7] Prudnikov V V, Prudnikov P V, and Fedorenko A A 1999 *JETP* **89** 325
[8] Prudnikov V V, Prudnikov P V, Kalashnikov I A, and Rychkov M V 2010 *JETP* **110** 253
[9] Selke W, Shchur L N, and Talapov A L 1994 *Annual Reviews of Computational Physics* vol 1, ed D. Stauffer World Scientific, (Singapore:World Scientific) p 17
[10] Cardy J L, 1996 *Scaling and Renormalization in Statistical Physics* Cambridge University Press, (Cambridge: Cambridge University Press)
[11] Pelissetto A and Vicari E 2002 *Phys. Rep.* **368** 549
[12] Dotsenko V S, Harris A B, Sherrington D, and Stinchcombe R B 1995 *J. Phys. A* **28** 3093
[13] Dotsenko V S and Feldman D E 1995 *J. Phys.* A **28** 5183
[14] Prudnikov V V, Prudnikov P V, and Fedorenko A A 2001 *Phys. Rev.* B **63** 184201; Prudnikov V V and Prudnikov P V 2002 *Sov. Phys. JETP* **95** 550
[15] Berche B, Berche P E, Chatelain C, and Janke W 2005 *Condens. Matter Phys.***8** 47
[16] Ballesteros H G, Fernandez L A, Martin-Mayor V, Muoz Sudupe A, Parisi G, and Ruiz-Lorenzo J J 1998 *Phys. Rev.* B **58** 2740
[17] Prudnikov V V, Prudnikov P V, Vakilov A N, and Krinitsyn A S 2007 *Sov. Phys. JETP* **105** 371
[18] Prudnikov V V, Prudnikov P V, Krinitsyn A S, Vakilov A N ,Pospelov E A, and Rychkov M V 2010 *Phys. Rev.* E **81** 011130
[19] Berche P E, Chatelain C, Berche B, and Janke W 2004 *Eur. Phys. J.* B **38** 463
[20] Fytas N G and Theodorakis P E 2010 *Phys. Rev.* E **82** 062101
[21] Hohenberg P C and Halperin B I 1977 *Rev. Mod. Phys* **49** 435
[22] Jan N, Moseley L L, and Stauffer D 1983 *J. Stat. Phys.* **33** 1
[23] Kalle C 1984 *J. Phys.* A **17** L801
[24] Prudnikov V V and Vakilov A N 1992 *JETP Lett.* **55** 741;Prudnikov V V and Vakilov A N 1993 *Sov. Phys. JETP 103* 962
[25] Prudnikov V V and Vakilov A N 1992 *JETP* **74** 990
[26] Krinitsyn A S, Prudnikov V V, and Prudnikov P V 2006 *Teor. Mat. Fiz.* **147** 559
[27] Rosov N, Hohenemser C, and Eibschutz M 1992 *Phys. Rev.* B **46** 3452
[28] Heuer H-O 1993 *J. Phys.* A **26** L333; Heuer H-O 1993 *J. Phys.* A **26** L341
[29] Parisi G, Ricci-Tersenghi F, and Ruiz-Lorenzo J J 1999 *Phys. Rev.* E **60** 5198
[30] Hasenbusch1 M, Pelissetto A, and Vicari E 2007 *J. Stat. Mech.: Theor. Exp.* **11** 11009
[31] Pelissetto A and Vicari E 2000 *Phys. Rev.* B **62** 6393
[32] Dorogovtsev S N 1981 *Sov. Phys.JETP* **53** 1070
[33] Weinrib A and Halperin B I 1983 *Phys. Rev.* B **27** 413
[34] Prudnikov V V 1983 *J. Phys.* C **16** 3685
[35] Prudnikov V V and Lawrie I D 1984 *J. Phys.* C **17** 165

Journal of Physics: Conference Series **510** (2014) 012020 doi:10.1088/1742-6596/510/1/012020

# Ageing and non-equilibrium critical phenomena in Monte Carlo simulations

**V V Prudnikov**[†], **P V Prudnikov**[‡]

Department of Theoretical Physics, Omsk State University, 55A Pr. Mira, Omsk 644077, Russia

E-mail: [†]prudnikv@univer.omsk.su, [‡]prudnikp@univer.omsk.su

**Abstract.** It is considered the non-equilibrium critical evolution of statistical systems which displays some features, such as ageing and violation of the fluctuation-dissipation theorem. We review here some theoretical results of computations that have been obtained in recent years for universal quantities, such as the exponents determining the scaling behaviour of dynamic response and correlation functions and the fluctuation-dissipation ratio, associated with the non-equilibrium critical dynamics, with particular focus on the 3D pure and diluted Ising models with Glauber dynamics. We analyse an influence of critical fluctuations, different non-equilibrium initial states and presence of nonmagnetic impurities in spin systems on two-time dependence of correlation and response functions on characteristic time variables as waiting time $t_w$ and time of observation $t - t_w$ with $t > t_w$. We discuss the obtained values of non-equilibrium exponents for autocorrelation and response functions and values of the universal long-time limit of the fluctuation-dissipation ratio $X^\infty$. Analysis of simulation results show that the insertion of disorder leads to new values of $X^\infty$ with $X^\infty_{\text{diluted}} > X^\infty_{\text{pure}}$.

## 1. Introduction

The collective behaviour of statistical systems close to critical points is characterized by an extremely slow dynamics which, in the thermodynamic limit, doesn't give them to relax to an equilibrium state after a change in some thermodynamic parameters. The non-equilibrium evolution in this case displays some of the features typically observed in glassy materials, such as ageing and violation of the fluctuation-dissipation theorem (FDT) [1–5]. It can be detected through determination of dynamic susceptibilities and correlation functions of the order parameter, the scaling behaviour of which is characterized by universal exponents, scaling functions, and amplitude ratios. This universality allows one to calculate these quantities in suitable simplified models [6, 7] and Monte Carlo methods are a natural way for this analysis.

We consider here some of the theoretical results of computations that have been obtained in recent years for universal quantities, such as the fluctuation-dissipation ratio, associated with the non-equilibrium critical dynamics, with particular focus on original results of authors for the 3D pure and diluted Ising models with Glauber dynamics.

The investigation of critical behaviour of disordered systems remains one of the main problems in condensed-matter physics and excites a great interest, because all real solids contain structural defects. The structural disorder breaks the translational symmetry of the crystal and thus greatly complicates the theoretical description of the material. The influence of disorder is particularly important near the critical point where behaviour of a system is characterized by anomalous large response on any even weak perturbation. The description of such systems requires the development of special analytical and numerical methods.

The effects produced by weak quenched disorder on critical phenomena have been studied for many years [8–14]. According to the Harris criterion [8] the disorder affects the critical behaviour only if $\alpha$, the specific heat exponent of the pure system, is positive. In this case a new universal critical behaviour, with new critical exponents, is established. In contrast, when $\alpha < 0$, the disorder appears to be irrelevant for the critical behaviour. Only systems whose effective Hamiltonian near the critical point is isomorphic to the Ising model satisfy this criterion ($\alpha > 0$) for drastic change of universality class of the critical behaviour with introduction to system of whatever concentration of short-range correlated quenched defects [15].

A large number of publications is devoted to the study of the critical behaviour of diluted Ising-like magnets by the renormalization-group (RG) methods, the numerical Monte Carlo methods, and experimentally (for a review see [16]). The ideas about replica symmetry breaking in the systems with quenched disorder were presented in Refs. [17, 18]. A refined RG analysis of the problem has shown the stability of the critical behaviour of weakly disordered three-dimensional systems with respect to the replica symmetry breaking effects [19, 20]. All obtained results confirm the existence of a new universal class of the critical behaviour, which is formed by diluted Ising-like systems. However, it remains unclear, whether the asymptotic values of critical exponents are independent of the rate of dilution of the system, how the crossover effects change these values, and whether two or more regimes of the critical behaviour exist for weakly and strongly disordered systems. These questions are the subjects of heated discussions [21, 22].

In the present paper, we numerically investigate the non-equilibrium critical dynamics with a non-conserved order parameter (model A) [23] in the 3D pure and site-diluted Ising systems with spin concentrations $p = 0.8$ and 0.6.

## 2. Non-equilibrium critical dynamics, main peculiarities and characteristics

Statistical systems with slow dynamics have recently attracted considerable theoretical and experimental interest, in view of the rich scenario of phenomena they display: dramatic slowing down of relaxation processes, hysteresis, memory effects, etc. After a perturbation a system with slow dynamics does not achieve equilibrium even after long times and its dynamics is not invariant either under time translations or under time reversal, as it should be in thermal equilibrium. During this never-ending relaxation ageing occurs: two-time quantities such as response and correlation functions depend on characteristic time variables as waiting time $t_w$ and time of observation $t - t_w$ with $t > t_w$ not via $t - t_w$ only. Important, that decays for these quantities as functions of time of observation $t - t_w$ are slower for larger waiting times $t_w$.

At variance with one-time quantities like the order parameter converging to asymptotic values in the long-time limit, two-time quantities clearly demonstrate the ageing. Ageing was known to occur in such disordered and complex systems as glassy materials [1–3] and only in the last ten years has attention been focused on simpler systems such as critical ones, whose universal features can be rather easily investigated by using different methods and which might provide insight into more general cases [24–26].

Consider a system with a critical point at temperature $T_c$, order parameter $S(x,t)$, and prepare it in some initial configuration which might corresponds to an equilibrium state at a given temperature $T_0$. At time $t = 0$ bring the system in contact with a thermal bath with temperature $T_s$ not equal $T_0$. The relaxation process is expected to be characterized by some equilibration time $t_{\text{eq}}(T_s)$ such that for $t \gg t_{\text{eq}}(T_s)$ equilibrium is attained and the dynamics is stationary and invariant under time reversal, whereas for $0 < t \ll t_{\text{eq}}(T_s)$ the evolution depends on the specific initial condition. Upon approaching the critical point $T_s = T_c$ the equilibration time diverges $t_{\text{eq}} \sim |T - T_c|^{-z\nu}$, where $z$ is the dynamic critical exponent, $\nu$ is the exponent for correlation length, and therefore equilibrium is never achieved. To monitor the time evolution we

Journal of Physics: Conference Series **510** (2014) 012020 doi:10.1088/1742-6596/510/1/012020

consider the average order parameter $M(t) = \frac{1}{V}\int d^d x \, \langle S(x,t)\rangle$, the time-dependent correlation function of the order parameter

$$C(t, t_w) = \frac{1}{V}\int d^d x[\langle S(x,t)S(0,t_w)\rangle - \langle S(x,t)\rangle \langle S(0,t_w)\rangle], \tag{1}$$

where $< ... >$ stands for the mean over the stochastic dynamics, and the linear response (susceptibility) $R_x(t, t_w)$ to a small external field, applied at time $t_w$, which is defined by relation

$$R(t, t_w) = \frac{1}{V}\int d^d x \left.\frac{\delta \langle S(x,t)\rangle}{\delta h(x, t_w)}\right|_{h=0}. \tag{2}$$

Causality implies $R(t, t_w > t) = 0$. According to the general picture of the relaxation process, one expects that for $t > t_w \gg t_{\text{eq}}(T_s)$, $C(t, t_w) = C^{\text{eq}}(t - t_w)$ and $R(t, t_w) = R^{\text{eq}}(t - t_w)$, where $C^{\text{eq}}$ and $R^{\text{eq}}$ are the corresponding equilibrium quantities, related by the fluctuation-dissipation theorem (FDT)

$$R^{\text{eq}}(t) = -\frac{1}{T_s}\frac{dC^{\text{eq}}(t)}{dt}. \tag{3}$$

The FDT suggests the definition of the so-called fluctuation-dissipation ratio (FDR)

$$X(t, t_w) = \frac{T_s R(t, t_w)}{\partial_{t_w} C(t, t_w)} \tag{4}$$

with $t > t_w$. For $t > t_w \gg t_{\text{eq}}(T_s)$ the FDT yields $X(t, t_w) = 1$. This is not generically true in the ageing regime. The asymptotic value of the FDR

$$X^\infty = \lim_{t_w\to\infty}\lim_{t\to\infty} X(t, t_w) \tag{5}$$

is a very useful quantity in the description of systems with slow dynamics, since $X^\infty = 1$ whenever the ageing evolution is interrupted and the system crosses over to equilibrium dynamics, i.e. $t_{\text{eq}}(T_s) < \infty$. Conversely, $X^\infty \neq 1$ is a signal of an asymptotic non-equilibrium dynamics. Moreover, $X^\infty$ can be used to define an effective non-equilibrium temperature $T_{\text{eff}} = T/X^\infty$, which might have some features of the temperature of an equilibrium system, e.g. controlling the direction of heat flows and acting as a criterion for thermalization.

What is known in general about $X^\infty$? As a consequence of the fluctuation-dissipation theorem $X^\infty(T > T_c) = 1$. On the other hand, on the basis of general scaling arguments for the phase ordering regime [27], it has been shown that $X^\infty(T < T_c) = 0$. These results are expected to be actually independent of the specific system and of its microscopic details. For $T = T_c$ there are no general arguments constraining the value of $X^\infty$ and therefore it has to be determined for each specific model. In table 1 we report some of the values that has been found either by exact solutions or by means of Monte Carlo (MC) simulations (a more complete table can be found in [24]). Clearly, $X^\infty(T = T_c)$ depends on the model and on the space dimensionality $d$. Nevertheless it has been argued on the basis of scaling arguments [27, 29] that $X^\infty(T = T_c)$ should be a universal quantity associated with the critical dynamics.

At present time, it is well known that two-time dependence for autocorrelation and response functions for systems starting from high-temperature initial state with $m_0 = 0$ (or $m_0 \ll 1$) satisfies the following scaling forms:

$$\begin{aligned} C(t, t_w) &= A_C(t - t_w)^{a+1-d/z}\,(t/t_w)^{\theta-1}\, f_C(t_w/t), \\ R(t, t_w) &= A_R(t - t_w)^{a-d/z}\,(t/t_w)^{\theta}\, f_R(t_w/t), \end{aligned} \tag{6}$$

**Table 1.** Values of the fluctuation-dissipation ratio $X^\infty$ for some systems for high-temperature initial state with $m_0 \ll 1$

| Model | | | $T < T_c$ | $T = T_c$ | $T > T_c$ |
|---|---|---|---|---|---|
| Free Gaussian Field | [28] | Exact | – | 1/2 | 1 |
| dD Spherical Model | [29] | Exact | 0 | $1 - 2/d$ | 1 |
| 1D Ising Model | [30] | Exact | – | 1/2 | 1 |
| 2D Ising Model | [29] | MC | | 0.26(1) | |
| | [31] | MC | | 0.340(5) | |
| | [32] | MC | | 0.330(5) | |
| 2D 3-states Potts Model | [32] | MC | | 0.406(1) | |
| 2D 4-states Potts Model | [32] | MC | | 0.459(8) | |
| 3D Ising Model | [29] | MC | 0 | $\approx 0.40$ | |
| 3D XY Model | [33] | MC | 0 | 0.43 | |

where $f_C(t_w/t)$ and $f_R(t_w/t)$ are finite for $t_w \to 0$, $a = (2-\eta-z)/z$, $\theta = \theta' - (2-z-\eta)/z$, $\theta'$ is the initial slip exponent [34]. $A_R$ and $A_C$ are non-universal amplitudes which are fixed by the condition $f_{R,C}(0) = 1$. With this normalization $f_{R,C}$ are universal. From these scaling forms the universality of $X^\infty$ follows as an amplitude ratio $X^\infty = A_R/[(1-\theta)A_C]$ [24, 27, 29].

A remarkable property of non-equilibrium relaxation is the increase of magnetization $M(t) \sim t^{\theta'}$ from a non-zero initial magnetization $m_0 \ll 1$ at short times $t < t_{\rm cr} \sim m_0^{-1/(\theta'+\beta/z\nu)}$ (an example of $M(t)$ behaviour is given in Fig.7). The initial rise of magnetization is changed to the well known decay $M(t) \sim t^{-\beta/z\nu}$ for $t \gg t_{\rm cr}$. The critical exponents $\theta$ and $\theta'$ depend on the dynamic universality class [23] and have been calculated by the RG method for a number of dynamic models such as the model with a non-conserved order parameter [22, 34–36] (model A), the model with an order parameter coupled to a conserved density [37] (model C), and the models with reversible mode coupling [38] (models E, F, G, and J).

It can be distinguished three stages of the non-equilibrium relaxation process. The first quasi-equilibrium stage is observed for small time separation $t - t_w \ll t_w$ with $t_w \gg 1$ where ageing does not exist and the dynamic evolution of the correlation and response functions exhibits a stationary part and does not depend on waiting time with $C = C(t-t_w)$ and $R = R(t-t_w)$. The second ageing regime is realized for times $t - t_w \sim t_w \gg 1$ where the correlation and response functions are derived by relations

$$C(t, t_w) \sim t_w^{-2\beta/(\nu z)} \hat{F}_c(t/t_w), \quad R(t, t_w) \sim t_w^{-2\beta/(\nu z)-1} \hat{F}_R(t/t_w), \tag{7}$$

and therefore at different waiting times do not superpose and are characterized by different slopes for each $t_w$ (in (7) it was used the relation $2\beta/(\nu z) = d/z - a - 1$). At long time separations with $t - t_w \gg t_w \gg 1$ the scaling functions in (7) decay as power law

$$\hat{F}_c(t/t_w) \sim (t/t_w)^{-c_a}, \quad \hat{F}_R(t/t_w) \sim (t/t_w)^{-c_r}, \tag{8}$$

where exponent $c_a = d/z - \theta'$ is the same which describes time dependence of the autocorrelation function in the short-time regime of non-equilibrium behaviour [22, 39, 40]. At this short-time dynamics stage ageing effects are not developed too. Scaling analysis of response function $R(t, t_w)$ behaviour in short-time dynamics regime predicts that $c_r = c_a$.

Renormalization-group investigations of non-equilibrium critical behaviour in $d$-dimensional pure systems with $n$-component order parameter and weakly dilute Random Ising Model (RIM) for purely dissipative dynamics (model A) were carried out in papers [41] and [42]. The asymptotic values of the FDR $X^\infty$ were calculated with the use of $\varepsilon$-expansion method ($\varepsilon = 4-d$) in two-loop approximation for pure systems [41]

$$\frac{(X^\infty)^{-1}}{2} = 1 + \frac{n+2}{4(n+8)}\varepsilon + \varepsilon^2 \frac{n+2}{(n+8)^2}\left[\frac{n+2}{8} + \frac{3(3n+14)}{4(n+8)}\right] + O(\varepsilon^3), \tag{9}$$

and in one-loop approximation for diluted Ising model [42]

$$X^\infty = \frac{1}{2} - \frac{1}{4}\sqrt{\frac{6\varepsilon}{53}} + O(\varepsilon). \tag{10}$$

It was obtained $X^\infty_{\rm 3DIs} = 0.429(6)$ for the three-dimensional Ising ($\varepsilon = 1, n = 1$) model, $X^\infty_{\rm 3DXY} = 0.416(8)$ for XY ($\varepsilon = 1, n = 2$) model, and $X^\infty_{\rm 2DIs} = 0.30(5)$ for the two-dimensional Ising ($\varepsilon = 2, n = 1$) model which are in good agreement with Monte Carlo results for these models given in Table 1. For 3D diluted Ising model (that is the only physical relevant case) it was obtained value $X^\infty_{\rm 3DRIM} \simeq 0.416$. To this order it is not clear whether disorder really changes $X^\infty$ in a sensible way or not. In any case this could not be safely stated from low-order computations since the $\sqrt{\varepsilon}$-expansion is known to be not well-behaved at $d = 3$ [16, 20, 43]. Results of Monte Carlo investigations of non-equilibrium critical dynamics for the 3D diluted Ising model will be presented in this article below.

If initial state of system is characterized by magnetization $m_0 \neq 0$ the renormalization-group analysis of non-equilibrium dynamics for model A predicts that the correlation $C(t, t_w)$ and response $R(t, t_w)$ functions display the following scaling behaviours after a quench to $T = T_c$ [34, 44]

$$\begin{aligned} C(t,t_w) &= a_C(t-t_w)^{a+1-d/z}\,(t/t_w)^{\theta-1}\,F_C(t_w/t, t/t_m), \\ R(t,t_w) &= a_R(t-t_w)^{a-d/z}\,(t/t_w)^{\theta}\,F_R(t_w/t, t/t_m), \end{aligned} \tag{11}$$

where modification of these scaling relations in comparison with (6) is connected with new time scale $t_m$ set by the initial value of the magnetization $m_0$ and which displays a universal dependence on it

$$t_m = B_m m_0^{-1/\kappa}, \tag{12}$$

where the universal scaling exponent $\kappa > 0$ is given, in terms of static and dynamic equilibrium and non-equilibrium exponents, by $\kappa = \theta + a + \beta/(\nu z)$.

The two-time quantities $C(t, t_w)$ and $R(t, t_w)$ are homogeneous functions of the three time scales $t - t_w$, $t_w$, and $t_m$. In particular, when $t_w < t \ll t_m$, which is always for case with $m_0 = 0$, the scaling form of $C$ and $R$ becomes as in equation (6) with $F_{C,R}(x, 0) = f_{C,R}(x)$. In the opposite case with large times compared to $t_m$, i.e. $t_m \ll t_w < t$ the scaling form of $C(t, t_w)$ and $R(t, t_w)$ becomes [44]:

$$\begin{aligned} C(t,t_w) &= \bar{a}_C(t-t_w)^{a+1-d/z}\,(t/t_w)^{\bar\theta-1}\,\bar{F}_C(t_w/t), \\ R(t,t_w) &= \bar{a}_R(t-t_w)^{a-d/z}\,(t/t_w)^{\bar\theta}\,\bar{F}_R(t_w/t), \end{aligned} \tag{13}$$

where new exponent $\bar\theta = -\beta\delta/(\nu z) = -(1+a+\beta/(\nu z))$ and $\bar{F}_{C,R}$ are universal scaling functions related to the large-$y$ behaviour of $F_{C,R}(x, y)$. In the ageing regime which is realized for times $t - t_w \sim t_w \gg t_m$ the correlation and response functions are derived by relations

$$C(t,t_w) \sim t_w^{-2\beta/(\nu z)}\tilde{F}_C(t/t_w), \quad R(t,t_w) \sim t_w^{-2\beta/(\nu z)-1}\tilde{F}_R(t/t_w), \tag{14}$$

with the scaling functions $\tilde{F}_{C,R}(t/t_w)$ which decay at long time separations limit with $t - t_w \gg t_w \gg t_m$ as power law

$$\tilde{F}_{C,R}(t/t_w) \sim (t/t_w)^{-\phi}, \tag{15}$$

where exponent $\phi = d/z - a + \beta\delta/(\nu z)$.

In paper [44] it was investigated the non-equilibrium behaviour of the $d$-dimensional Ising model with purely dissipative dynamics during its critical relaxation from a magnetized initial state. The universal scaling forms of the two-time response and correlation functions of the magnetization were derived within the field-theoretical approach and the associated scaling functions were computed in first order of the $\varepsilon$-expansion. It was shown that ageing behaviour is clearly displayed and the asymptotical universal fluctuation-dissipation ratio is characterized by relation

$$X^{\infty} = \frac{4}{5} - \left(\frac{73}{600} - \frac{\pi^2}{100}\right)\varepsilon + O(\varepsilon^2), \tag{16}$$

which gives $X^{\infty}_{3DIs} \simeq 0.78$ for the three-dimensional Ising ($\varepsilon = 1, n = 1$) model and $X^{\infty}_{2DIs} \simeq 0.75$ for the two-dimensional Ising ($\varepsilon = 2, n = 1$) model. The obtained results were confirmed by Monte Carlo simulations of the two-dimensional Ising model with Glauber dynamics, from which it was found $X^{\infty}_{MC} = 0.73(1)$ [44].

## 3. Monte Carlo simulations of the 3D pure Ising models with relaxation from high-temperature and low-temperature initial states

The one from simplest non-trivial models in which ageing occurs is the Ising model in $d$-dimensions evolving with a purely dissipative dynamics after a quench to the critical point. Its Hamiltonian on an hypercubic lattice is given by

$$H = -J \sum_{<i,j>} S_i S_j, \tag{17}$$

where $J > 0$ is the short-range exchange interaction between spins $S_i$ fixed at the lattice sites, and assuming values of $S_i = \pm 1$. We performed our Monte Carlo simulations using the heat bath updating rule [45], simulating a large Ising spin system on cubic lattice with linear size $L = 128$ with periodic boundary conditions. It were computed the magnetization $M(t)$ and the two-time autocorrelation function $C(t, t_w)$

$$M(t) = \left\langle \frac{1}{L^3} \sum_{i=1}^{L^3} S_i(t) \right\rangle, \quad C(t, t_w) = \left\langle \frac{1}{L^3} \sum_{i=1}^{L^3} S_i(t) S_i(t_w) \right\rangle - M(t)M(t_w), \tag{18}$$

where the angle brackets stand for an average over initial configurations and MC realizations. The averaging of $C(t, t_w)$ is carried out on 3000 MC runs for every $t_w$.

For case when we have simulated the dynamics of the systems starting from the high-temperature initial state with magnetization value $m_0 = 0.02$, the response function and fluctuation-dissipation ratio were calculated using the relations [32]

$$R(t, t_w) = \frac{1}{TL^3} \sum_{i=1}^{L^3} \Big\langle S_i(t) \big[ S_i(t_w + 1) - S_i^W(t_w + 1) \big] \Big\rangle, \tag{19}$$

where $S_i^W = \tanh(J \sum_{m \neq i} S_m / T)$ and

$$X(t, t_w) = \frac{TR(t, t_w)}{\frac{\partial}{\partial t_w} C(t, t_w)} = \frac{\sum_{i=1}^{N} \Big\langle S_i(t) \big[ S_i(t_w + 1) - S_i^W(t_w + 1) \big] \Big\rangle}{\sum_{i=1}^{N} \Big\langle S_i(t) \big[ S_i(t_w + 1) - S_i(t_w) \big] \Big\rangle}. \tag{20}$$

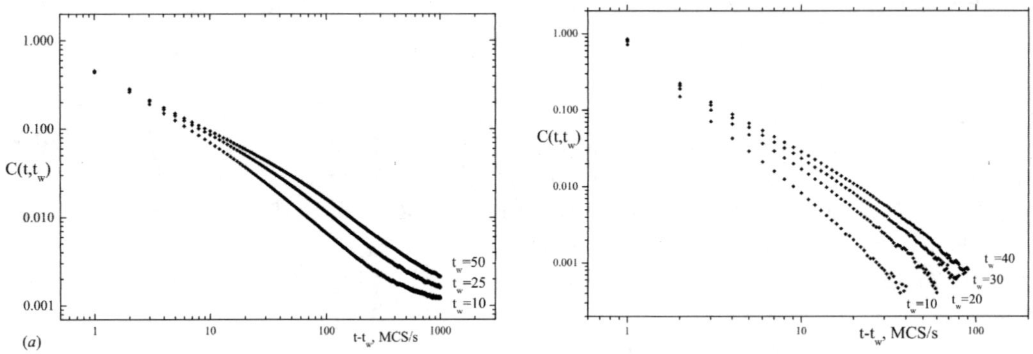


**Figure 1.** Time dependence of correlation function $C(t, t_w)$ as a function $t - t_w$ for different initial non-equilibrium states $(a)$ $m_0 \ll 1$; $(b)$ $m_0 = 1$

The averaging of $R(t, t_w)$ and $X(t, t_w)$ is carried out on 90000 MC runs for every $t_w$.

For case when we have simulated the dynamics of the systems starting from the low-temperature initial state with magnetization value $m_0 = 1$ we calculated the integrated response function [46]:

$$\chi(t, t_w) = T \int_{t_w}^{t} dt' R(t, t') \tag{21}$$

with response function determined by (2). In the large time limit $\chi(C) = \int_C^1 X(q)dq$ and the fluctuation-dissipation ratio can be defined as

$$X(t, t_w) = -\lim_{C \to 0} \frac{\partial(T\chi(t, t_w))}{\partial C(t, t_w)}. \tag{22}$$

The MC obtained time dependencies of the autocorrelation function $C(t, t_w)$ and the response function $R(t, t_w)$ from observation time $t - t_w$ for different initial non-equilibrium states and $t_w$ values are presented in Fig.1 and 2. These functions demonstrate the ageing effects that is the slowing down of time correlations and decreasing of response with increasing of system age $t_w$.

In the ageing regime the time dependence of correlation functions is characterized by scaling relations (7) and (14). These scaling relations are rather well displayed by our data, as shown in Fig. 3. For time intervals with $(t - t_w)/t_w \gg 1$ the values of exponents $c_a = 1.333(40)$ and $c_r = 1.357(18)$ were determined which demonstrate a good agreement with each other and with value $c_r = 1.362(19)$ [39] obtained by short-time dynamics method.

For completely ordered initial state with $m_0 = 1$ the short-time dynamics regime for autocorrelation function $C(t, t_w)$ (Fig. 4) is characterized by the exponent $\phi$ (15). We obtained the value $\phi = 2.76(6)$, which is in a good agreement with theoretical value $\phi = 1 + d/z + \beta/(\nu z) = 2.737(8)$ calculated with exponents $\beta = 0.325(1)$, $\nu = 0.630(1)$ [47], $z = 2.024(6)$ [48].

The obtained data for fluctuation-dissipation ratio for case with $m \ll 1$ are presented in Fig. 5. The asymptotical value of FDR $X^\infty = 0.380(13)$ was obtained by linear extrapolation of data for $X(t_w/(t - t_w))$ to limit $t_w/(t - t_w) \to 0$.

For completely ordered initial state with $m_0 = 1$ the asymptotical value of the FDR $X^\infty = 0.77(6)$ was obtained as realization of limit dependence $T\chi(C)$ (Fig. 6) in accordance with relation (22). This value of $X^\infty$ is in excellent agreement with theoretical field value $X^\infty \simeq 0.78$ [44].

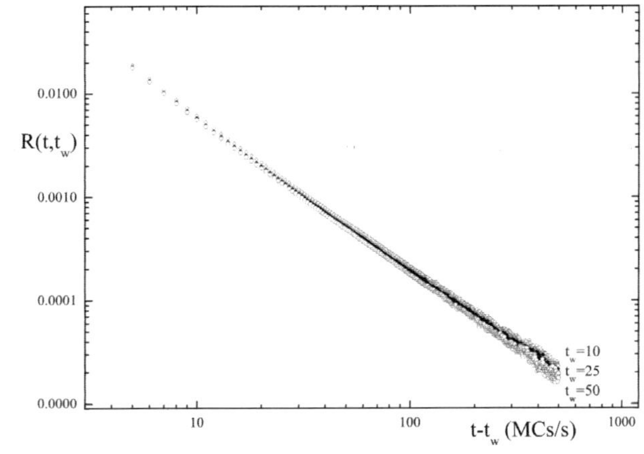


**Figure 2.** Time dependence of response function $R(t, t_w)$ as a function $t - t_w$ for high-temperature initial state with $m_0 \ll 1$ and for different values of $t_w$

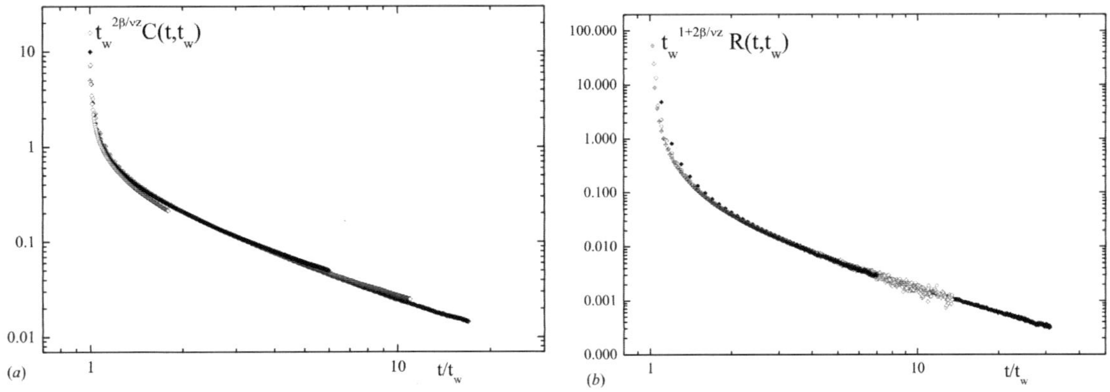


**Figure 3.** Scailing collapse of correlation $C(t, t_w)$ $(a)$ and response $R(t, t_w)$ $(b)$ functions for high-temperature initial state with $m_0 \ll 1$

## 4. Monte Carlo simulations of the 3D diluted Ising models with relaxation from high-temperature initial state

The Hamiltonian of diluted by nonmagnetic impurity atoms ferromagnetic Ising model is given by

$$H = -J \sum_{<i,j>} p_i p_j S_i S_j, \tag{23}$$

where random occupation numbers $p_i$ take the values 0 or 1, and $p_i$ equal to 1 if site of lattice contains spin and 0 otherwise. We considered the cubic lattice with periodic boundary conditions. Let us denote, $N_s = pL^3$ is the number of spins in cubic lattice with linear size $L$ and $p$ is the spin concentration.

We performed our Monte Carlo simulations also with the use of the heat bath updating rule for Ising spin system with the spin concentrations $p = 0.8$ and $0.6$ on cubic lattice with $L = 128$.

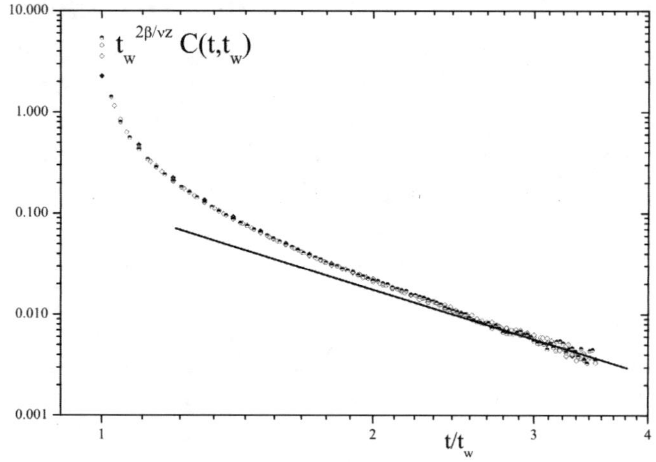


**Figure 4.** Scailing collapse of correlation function $C(t, t_w)$ for low-temperature initial state with $m_0 = 1$

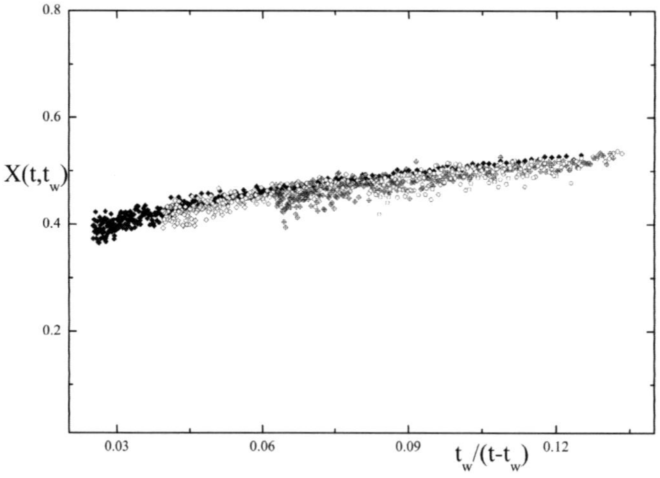


**Figure 5.** Fluctuation-dissipation ratio $X(t, t_w)$ as function $t_w/(t - t_w)$ for $t - t_w \gg t_w$

It were computed the magnetization and the two-time autocorrelation function

$$M(t) = \left[\left\langle \frac{1}{N_s} \sum_{i=1}^{N_s} p_i S_i(t) \right\rangle\right], \quad C(t, t_w) = \left[\left\langle \frac{1}{N_s} \sum_{i=1}^{N_s} p_i S_i(t) S_i(t_w) \right\rangle\right], \tag{24}$$

where the square brackets stand for additional average over disorder configurations. The averaging of $C(t, t_w)$ was carried out on over 1000 samples with different disorder configurations with 15 MC runs for each sample.

In this part of investigations we have simulated the non-equilibrium critical dynamics of the systems starting from the high-temperature initial state only with magnetization value $m_0 = 0.01$ for $p = 0.8$ and $m_0 = 0.005$ for $p = 0.6$. The response function and fluctuation-dissipation ratio

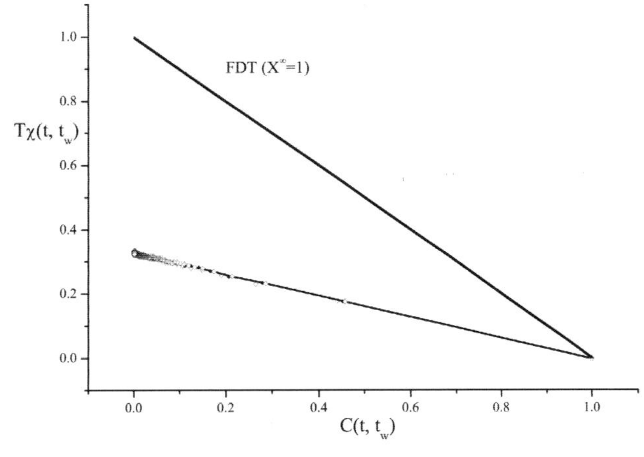


**Figure 6.** Dependence of $T\chi(t, t_w)$ as a function of $C(t, t_w)$ for different values of $t_w$ with demonstration of its deviation from line with $X^{\infty} = 1$

were calculated using the relations

$$R(t, t_w) = \frac{1}{T}\Big[\frac{1}{N_s}\sum_{i=1}^{N_s}\Big\langle S_i(t)[S_i(t_w+1) - S_i^W(t_w+1)]\Big\rangle\Big], \tag{25}$$

$$X(t, t_w) = \frac{\Big[\sum_{i=1}^{N_s}\Big\langle S_i(t)[S_i(t_w+1) - S_i^W(t_w+1)]\Big\rangle\Big]}{\Big[\sum_{i=1}^{N}\Big\langle S_i(t)[S_i(t_w+1) - S_i(t_w)]\Big\rangle\Big]}, \tag{26}$$

where $S_i^W = \tanh(J\sum_{m\neq i} S_m/T)$. The averaging of $R(t, t_w)$ and $X(t, t_w)$ was carried out on over 5000 samples with different disorder configurations with 15 runs for each sample.

We present in Fig. 7 the time evolution of the magnetization $M(t)$ from different initial states with magnetization $m_0 = 0.02$ at $T_c = 4.5114(1)$ for pure Ising model, $m_0 = 0.01$ at $T_c = 3.4995(2)$ for samples with spin concentration $p = 0.8$, and $m_0 = 0.005$ at $T_c = 2.4241(1)$ for samples with $p = 0.6$ [21, 22] which are characterized by power-law dependence in form $M(t) \sim t^{\theta'}$ in initial stage of evolution. The time scale of a critical initial increase of the magnetization is $t_{\rm cr} \sim m_0^{-1/(\theta'+\beta/z\nu)}$. The initial rise of magnetization is changed to the well-known decay $m(t) \sim t^{-\beta/z\nu}$ for $t \gg t_{\rm cr}$. However, in the limit of $m_0 \to 0$, the time scale $t_{\rm cr}$ goes to infinity. The times with $t < t_{\rm cr}$ can be considered as time intervals of non-equilibrium behaviour of considered systems. The values of exponent $\theta'$ calculated with the use of corrections to scaling are following: $\theta' = 0.106(4)$ for $p = 1$, $\theta' = 0.127(16)$ for $p = 0.8$, and $\theta' = 0.186(39)$ for $p = 0.6$ [22, 49, 50].

The time dependence of autocorrelation $C(t, t_w)$ and response $R(t, t_w)$ functions is plotted in Fig. 8 for different values of waiting time $t_w$ and spin concentrations $p$. These curves demonstrate that the ageing effects are increased with growth of defect concentrations.

To check the scaling dependence of $C(t, t_w)$ and $R(t, t_w)$ given by relations (7) we plotted $t^{2\beta/(\nu z)}C(t, t_w)$ and $t^{1+2\beta/(\nu z)}R(t, t_w)$ versus $t/t_w$ with the use of exponent values: $z = 2.191(21)$, $2\beta/\nu = 1.016(32)$ for $p = 0.8$ [22] and $z = 2.663(30)$, $2\beta/\nu = 0.924(80)$ for $p = 0.6$ [49, 50]. The obtained data are presented in Fig. 9. These functions demonstrate the collapse of curves for different $t_w$ with fixed spin concentration $p$ into single curve with universal scaling dependence. Note, that systems with different $p$ are characterized by different scaling functions $F_{C,R}(t, t_w, p)$.

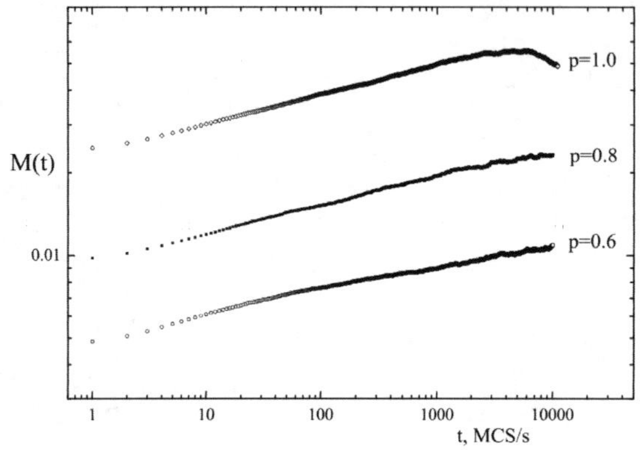


**Figure 7.** Time dependence of magnetization $M(t)$ for different spin concentrations $p$

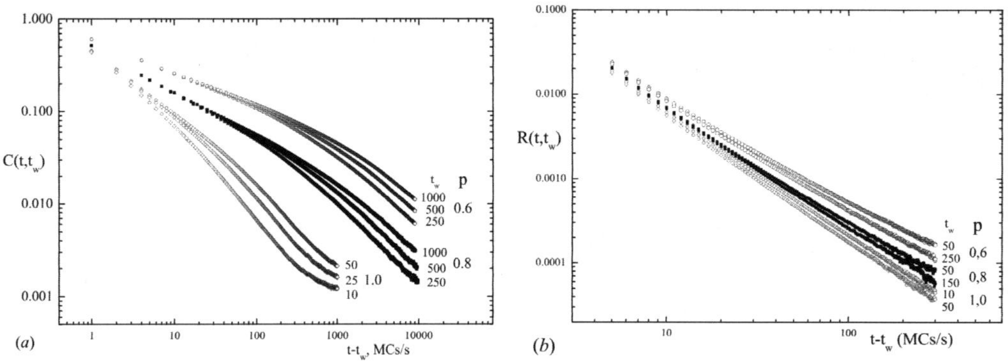


**Figure 8.** Time dependence of correlation $C(t, t_w)$ $(a)$ and response $R(t, t_w)$ $(b)$ functions from $t - t_w$ for different spin concentrations $p$

The scaling functions $F_{C,R}(t, t_w, p)$ have the power-law dependence (8) for evolution stage with long time separations $t - t_w \gg t_w \gg 1$ which are characterized by exponents $c_a$ and $c_r$. We determined values of these exponents: $c_a = 1.237(22)$, $c_r = 1.251(22)$ for $p = 0.8$ and $c_a = 0.982(30)$, $c_r = 0.950(8)$ for $p = 0.6$. They demonstrate very good agreement with each other for fixed spin concentration $p$, but differences between values $c_a$ and $c_r$ for different $p$ and for case of pure Ising model with $c_a = 1.333(40)$ and $c_r = 1.357(18)$ exceed the statistical errors of their determination. These values of $c_a$ are in a good agreement with values $c_a = 1.242(10)$ for weakly diluted system with $p = 0.8$ and $c_a = 0.941(21)$ for strongly diluted system with $p = 0.6$ obtained in [22, 49, 50] by short-time dynamics method, but are in a poor agreement with value $c_a = 1.05(3)$ from paper [51] obtained as independent on $p$ characteristics of non-equilibrium behaviour of the autocorrelation function. Reasons for this discrepancy have been discussed in detail in our paper [22].

At later investigations we have computed the fluctuation-dissipation ratio in compliance with relation (25). The obtained data are plotted in Fig. 10 for different spin concentrations $p$. For analysis we used time interval from $t - t_w \sim t_w$ to $t - t_w \gg t_w$. In contrast to pure Ising model the data of $X(t, t_w)$ for the diluted Ising model are characterized by an explicit dependence from $t_w$. At the beginning, we calculated the asymptotical values of $X(t_w, p)$ for different $t_w$ from plot in Fig. 10 in the limit $t_w/(t - t_w) \to 0$, and then, using the obtained values of $X(t_w, p)$, we made extrapolation $1/t_w \to 0$ to gain the asymptotical fluctuation-dissipation ratios $X^{\infty}(p)$.

Journal of Physics: Conference Series **510** (2014) 012020 doi:10.1088/1742-6596/510/1/012020

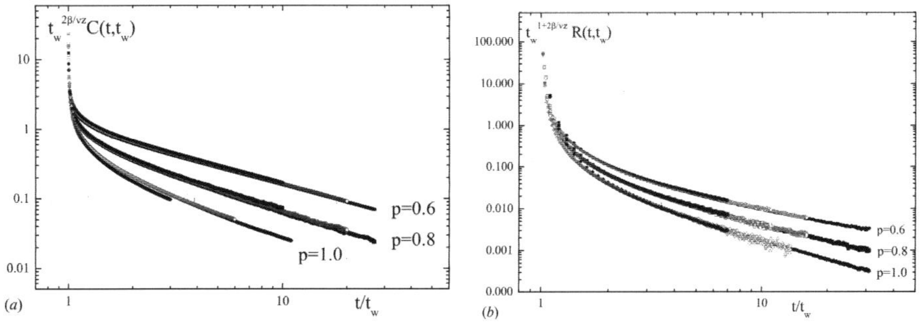


**Figure 9.** Scailing collaps of correlation $C(t, t_w)$ $(a)$ and response $R(t, t_w)$ $(b)$ functions for different spin concentrations $p$

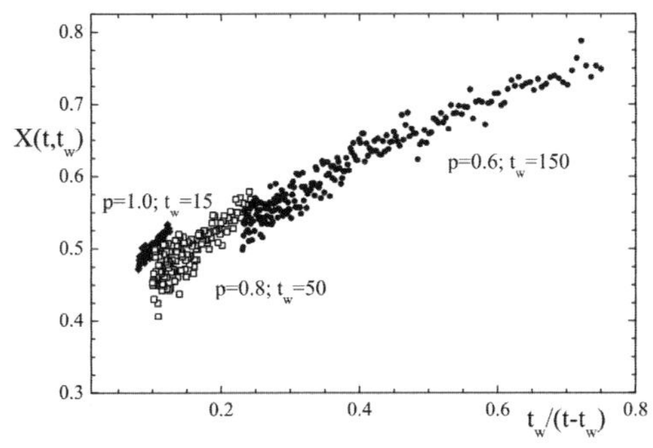


**Figure 10.** Dependence FDR $X(t, t_w)$ as function of $t_w/(t-t_w)$ for $t - t_w \gg t_w$ and for different $p$

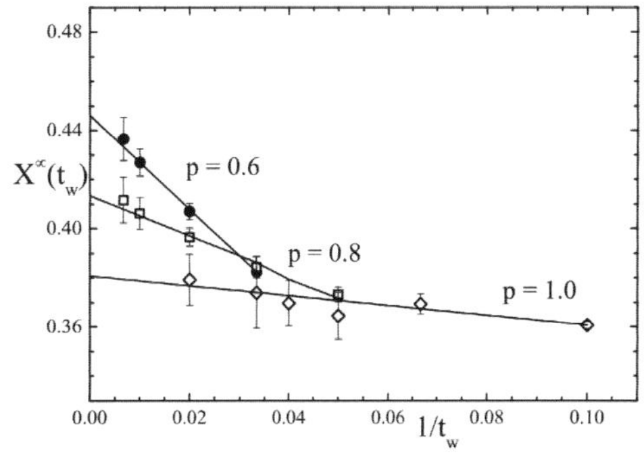


**Figure 11.** Dependence FDR $X(t_w, p)$ as function of $1/t_w$ for different $p$. Asymptotical value of $X^\infty(p)$ can be obtained in limit $1/t_w \to 0$

The results are presented in Fig. 11. The final values of the asymptotical fluctuation-dissipation ratio are $X^\infty(p = 1) = 0.380(13)$, $X^\infty(p = 0.8) = 0.413(11)$, and $X^\infty(p = 0.6) = 0.446(8)$. We are giving in table 2 the values for critical universal characteristics like exponents $c_a$, $c_r$ and the asymptotical fluctuation-dissipation ratio $X^\infty$, obtained at present paper for different spin concentrations $p = 1.0$, 0.8, and 0.6, and values of the critical exponents $2\beta/\nu$, $z$, and $\theta'$ for these spin concentrations which were used in procedure of $c_a$ and $c_r$ determination and were calculated by Monte Carlo simulations of non-equilibrium critical dynamics in Refs. [22, 39, 49, 50].

## 5. Conclusions

In this paper we have reviewed some theoretical results of computations in the range of non-equilibrium phenomena that have been obtained in recent years for universal quantities, such as the exponents determining the scaling behaviour of dynamic response and correlation functions and the fluctuation-dissipation ratio, associated with the non-equilibrium critical dynamics. In original part, we have considered the results of our MC simulations for the 3D pure and diluted Ising models with Glauber dynamics. The influence of critical fluctuations, different non-equilibrium initial states and site-quenched disorder on two-time dependence of the correlation and response functions at non-equilibrium critical regime were investigated for these models.

Analysis of the time dependencies for the autocorrelation and response functions at ageing

**Table 2.** Values of obtained critical exponents and the fluctuation-dissipation ratio $X^\infty$ for different spin concentrations $p$

| $p$ | $2\beta/\nu z$ | $z$ | $\theta'$ | $c_a$ | $c_r$ | $X^\infty$ |
|---|---|---|---|---|---|---|
| 1.0 | 1.040(18) [22] | 2.034(16) [22] | 0.106(4) [22] | 1.333(40) | 1.357(18) | 0.380(13) |
| | 1.034(4) [39] | 2.042(6) [39] | 0.108(2) [39] | | | |
| 0.8 | 1.016(32) [22] | 2.191(21) [22] | 0.127(16) [22] | 1.237(22) | 1.251(22) | 0.413(11) |
| 0.6 | 0.924(80) [49,50] | 2.663(30) [49,50] | 0.186(39) [49,50] | 0.982(30) | 0.950(8) | 0.446(8) |

regime of evolution showed that the ageing effects are increased with growth of defect concentrations. The obtained values of the autocorrelation critical exponents $c_a$ and the fluctuation-dissipation ratio $X^\infty$ demonstrate that these systems belong to different classes of non-equilibrium critical behaviour.

Analysis of simulation results show that the insertion of disorder leads to new universal FDRs with $X^\infty_{\text{strong diluted}} > X^\infty_{\text{weak diluted}} > X^\infty_{\text{pure}}$.

## Acknowledgments
The reported study was supported in part by Ministry of Education and Science of Russia through project No. 2.3046.2011.

The simulations were supported by the Supercomputing Center of Lomonosov Moscow State University and Joint Supercomputer Center of the Russian Academy of Sciences.

## References
[1] Berthier L and Kurchan J 2013 *Nature Phys.* **9** 310
[2] Vincent E, Hammann J, Ocio M, et al. 1997 *Lect. Notes Phys.* **492** 184
[3] Bouchaud J P, Cugliandolo L F, Kurchan J, and Mezard M 1998 *Spin Glasses and Random Fields* Directions in Condensed Matter Physics ed Young A P (Singapore: World Scientific) vol 12 p 443
[4] Cugliandolo L F 2003 *Slow relaxation and nonequilibrium dynamics in condensed matter* Les Houches, Ecole dEte de Physique Theorique ed Barrat J-L et al.(Berlin: Springer) vol 77 p 371
[5] Henkel M and Pleimling M 2010 *Non-equilibrium Phase Transitions. Volume 2: Ageing and dynamical scaling far from equilibrium* Theoretical and Mathematical Physics (Heidelberg, Springer) p 544
[6] Afzal N and Pleimling M 2013 *Phys. Rev.* E **87** 012114
[7] Parisi G, Ricci-Tersenghi F, and Ruiz-Lorenzo J J 1999 *Eur. Phys. J.* B **11** 317
[8] Harris A B 1974 *J. Phys.* C **7** 1671
[9] Khmelnitskii D E 1976 *Sov. Phys. JETP* **41** 981
[10] Grinstein G and Luther A 1976 *Phys. Rev.* B **13** 1329
[11] Prudnikov V V and Vakilov A N 1992 *JETP Lett.* **55** 741; *Sov. Phys. JETP* **103** 962
[12] Prudnikov V V and Vakilov A N 1993 *JETP* **76** 469
[13] Prudnikov V V, Belim S V, Ivanov A V, Osintsev E V, and Fedorenko A A 1998 *JETP* **87** 527
[14] Yin J Q, Zheng B, Prudnikov V V, and Trimper S 2006 *Eur. Phys. J.* B **49** 195
[15] Prudnikov V V, Prudnikov P V, and Fedorenko A A 1999 *J. Phys. A: Math. Gen.* **32** 8587; 2000 *Phys. Rev.* B **62** 8777
[16] Pelissetto A and Vicari E *Phys. Rep.* **368** 549
[17] Dotsenko V S, Harris A B, Sherrington D, and Stinchcombe R B 1995 *J. Phys. A: Math. Gen.* **28** 3093
[18] Dotsenko V S and Feldman D E 1995 *J. Phys. A: Math. Gen.* **28** 5183
[19] Prudnikov V V, Prudnikov P V, and Fedorenko A A 2001 *Phys. Rev.* B **63** 184201
[20] Prudnikov V V and Prudnikov P V 2002 *JETP* **95** 550
[21] Prudnikov V V, Prudnikov P V, Vakilov A N and Krinitsyn A S 2007 *JETP* **105** 371
[22] Prudnikov V V, Prudnikov P V, Krinitsyn A S, Vakilov A N, Pospelov E A and Rychkov M V 2010 *Phys. Rev.* E **81** 011130
[23] Hohenberg P C and Halperin B I 1977 *Rev. Mod. Phys.* **49** 435

[24] Calabrese P and Gambassi A 2005 *J. Phys. A: Math. Gen.* **38** R133
[25] Berthier L, Holdsworth P C W, and Sellitto M 2001 *J. Phys. A: Math. Gen.* **34** 1805
[26] Gambassi A 2008 *Eur. Phys. J.* B **64** 379
[27] Godréche C and Luck J M 2002 *J. Phys: Condens. Matter* **14** 1589
[28] Cugliandolo L F, Kurchan J, Parisi G 1994 *J.Phys.(France)* **4** 1641
[29] Godréche C and Luck J M 2000 *J. Phys. A: Math. Gen.* **33** 9141
[30] Lippiello E and Zannetti M 2000 *Phys. Rev.* E **61** 3369
[31] Mayer P, Berthier L, Garrahan J P, and Sollich P 2003 *Phys. Rev.* E **68** 016116
[32] Chatelain C 2003 *J. Phys. A: Math. Gen.* **36** 10739; Chatelain C 2004 *J. Stat. Mech.: Theor. Exp.* P06006
[33] Abriet S and Karevski D 2004 *Eur. Phys. J.* B **41** 79
[34] Janssen H K, Schaub B, and Schmittmann B 1989 *Z. Phys.* B **73** 539
[35] Prudnikov V V, Prudnikov P V, Kalashnikov I A, and Tsirkin S S 2008 *Sov. Phys. JETP* **106** 1095
[36] Oerding K and Janssen H K 1995 *J. Phys. A: Math. Gen.* **28** 4271
[37] Oerding K and Janssen H K 1993 *J. Phys. A: Math. Gen.* **26** 3369
[38] Oerding K and Janssen H K 1993 *J. Phys. A: Math. Gen.* **26** 5295
[39] Jaster A, Mainville J, Schulke L, and Zheng B 1999 *J. Phys. A: Math. Gen.* **32** 1395
[40] Prudnikov V V, Prudnikov P V, Zheng B, Dorofeev S V, and Kolesnikov V Yu 2007 *Prog. Theor. Phys* **117** 973
[41] Calabrese P and Gambassi A 2002 *Phys. Rev.* E **66** 066101
[42] Calabrese P and Gambassi A 2002 *Phys. Rev.* B **66** 212407
[43] Folk R, Holovatch Yu and Yavorskii T 2003 *Phys. Usp.* **46** 169
[44] Calabrese P, Gambassi A, and Krzakala F 2006 *J. Stat. Mech.* **6** 2
[45] Janke W 2007 *Ageing and the Glass Transition* Lect. Notes. Phys. ed M Henkel, M Pleimling, and R Sanctuary (Berlin Heidelberg: Springer) chapter 5 p 207-260
[46] Ricci-Tersenghi F 2003 *Phys. Rev.* E **68** 065104
[47] Guida R and Zinn-Justin J 1998 *J. Phys. A: Math. Gen.* **31** 8103
[48] Prudnikov V, Prudnikov P and Krinitsyn A 2006 *Theor. Math. Phys.* **147** 559
[49] Prudnikov V V, Prudnikov P V, Vakilov A N, Krinitsyn A S, and Rychkov M V 2009 *Proceedings of workshop on computational technogies in natural sciences*, Comp. Phys. ed R R Nazirov (Moscow: KDU publ.) vol 1 p 240-263 (in Russian)
[50] Prudnikov V V, Prudnikov P V and Vakilov A N 2013 Description of nonequilibrium critical behaviour in disordered systems by the theoretical methods (Moscow: Fizmatlit) (in Russian)
[51] Schehr G and Paul R 2005 *Phys. Rev.* E. **72** 016105

# Imitation Monte Carlo methods for problems of the Boltzmann equation with small Knudsen numbers, parallelizing algorithms with splitting

**A I Khisamutdinov**[1,2], **N N Velker**[2]

[1] Novosibirsk State University, 2 Pirogova Str., Novosibirsk, Russia
[2] A.A. Trofimuk Institute of Petroleum Geology and Geophysics SB RAS, 3 Akademika Koptyuga Prosp., Novosibirsk, Russia

E-mail: `KhisamutdinovAI@ipgg.sbras.ru`

**Abstract.** The talk examines a system of pairwise interaction particles, which models a rarefied gas in accordance with the nonlinear Boltzmann equation, the master equations of Markov evolution of this system and corresponding numerical Monte Carlo methods. Selection of some optimal method for simulation of rarefied gas dynamics depends on the spatial size of the gas flow domain. For problems with the Knudsen number $Kn$ of order unity "imitation", or "continuous time", Monte Carlo methods ([2]) are quite adequate and competitive. However if $Kn \leq 0.1$ ( the large sizes) , excessive punctuality, namely, the need to see all the pairs of particles in the latter, leads to a significant increase in computational cost( complexity). We are interested in to construct the optimal methods for Boltzmann equation problems with large enough spatial sizes of the flow. Speaking of the optimal, we mean that we are talking about algorithms for parallel computation to be implemented on high-performance multi-processor computers. The characteristic property of large systems is the weak dependence of sub-parts of each other at a sufficiently small time intervals. This property is taken into account in the approximate methods using various splittings of operator of corresponding master equations. In the paper, we develop the approximate method based on the splitting of the operator of master equations system "over groups of particles" ([7]). The essence of the method is that the system of particles is divided into spatial subparts which are modeled independently for small intervals of time, using the precise"imitation" method. The type of splitting used is different from other well-known type "over collisions and displacements", which is an attribute of the known Direct simulation Monte Carlo methods. The second attribute of the last ones is the grid of the "interaction cells", which is completely absent in the imitation methods. The main of talk is parallelization of the imitation algorithms with splitting using the MPI library. New constructed algorithms are applied to solve the problems: on propagation of the temperature discontinuity and on plane Poiseuille flow in the field of external forces. In particular, on the basis of numerical solutions, comparative estimates of the computational cost are given for all algorithms under consideration.

## 1. Introduction

The paper examines a system of pairwise interaction particles to model a rarefied gas in accordance with the nonlinear smoothed Boltzmann equation, the master equations of Markov evolution of this system and corresponding numerical Monte Carlo methods. Selection of an optimal simulation method of rarefied gas dynamics depends on the spatial size of the gas flow domain. "Imitation" or "continuous time" Monte Carlo methods ([1, 2, 3, 4]) are quite adequate and competitive for problems with the Knudsen number $Kn \simeq 1$. However if $Kn < 0.1$ (the large sizes) , excessive punctuality, namely, a need to see all the pairs of particles in the latter, leads to a significant increase of the computational cost( complexity). In this study we have been interested in developing optimal solution methods for the Boltzmann equations with spatial sizes of the flow that are large enough. Saying optimal, one has meant the algorithms for parallel computation should be implemented on high-performance multi-processor computers. The characteristic property of large systems is the weak dependence of its sub-parts on each other at a sufficiently small time intervals. This property is considered in the approximate methods using various splittings for the operator of the corresponding master equations. In the paper we have presented the approximate method based on the splitting of a master- equation- system operator "over groups of particles" ([5, 6, 7]). The essence of the method is that the system of particles is divided into spatial subparts which are modeled "quasi"- independently for small intervals of time, using the precise imitation method. The type of splitting used is different from another well-known type "over collisions and displacements" being an attribute of the known direct simulation Monte Carlo methods ([8, 9]). The second attribute of the last ones is a grid of "interaction cells" that is completely absent in the imitation methods. The main issue this paper addresses to is parallelization of the imitation algorithms with splitting performed using MPI library. The new constructed algorithms have been applied to solve problems: on propagation of the temperature discontinuity and on the plane Poiseuille flow in external-force field both for cases with small Kn. In the paper we have also given some estimates of the computational cost for all the algorithms considered which are in agreement with numerical results.

## 2. Common construction and notations

Evolution of rarefied gas is simulated in a bounded connected domain $(V)$, $(V) \subset R^3$, with the piecewise smooth boundary $(\partial V)$. Translatory movement and pairwise elastic collisions are simulated for the gas particles. It is assumed that the boundary has been split into two nonintersecting subdomains $(\partial V_1)$ and $(\partial V_2)$. Particles may fly in and fly out of the surface $(\partial V_1)$ and reflect back to $(V)$ from $(\partial V_2)$ with account for the given boundary conditions. Particles flying out of $(V)$ do not return.

Here are some common designations:
$(\bar{V}) = (V) \cup (\partial V)$, $(G) = (V) \times R^3$, $(\bar{G}) = (\bar{V}) \times R^3$; where $\|V\|$ is the volume of domain $(V)$; $\mathcal{N} = \mathcal{N}_t$ is the random number of particles in $(\bar{V})$ at the moment of time $t$, $t \geq 0$, $0 \leq \mathcal{N}_t < \infty$, $N$ is realization of $\mathcal{N}$;
$(r^i, v^i), i = 1, \cdots, N$, are the particles' radius vectors and velocities;
$x^i = (r^i, v^i)$, $\mathcal{R} = (r^1, \cdots, r^N)$, $\mathcal{V} = (v^1, \cdots, v^N)$, $\mathcal{X} \equiv \mathcal{X}^{(N)} = (\mathcal{R}, \mathcal{V}) = (x^1, \cdots, x^N)$;
$\mathcal{P}_0 \equiv \{\mathcal{N}_t, \mathcal{X}_t\}_{t=0}^{\infty}$ is the Markov jump process of gas-particle evolution with states in $\mathcal{S}$; $\mathcal{S} = \cup_{N=0}^{\infty} \mathcal{S}_N$; $\mathcal{S}_N = N \times (\bar{G})^N$; $\mathcal{S}_0$ - is the state when the domain $(\bar{V})$ does not contain particles; $\mu_t^{(N)}(\mathcal{X}) = \mu_t^N(\mathcal{X}) = \mu_t^{(N)}(x^1, \ldots, x^N)$ are the phase densities of gas at the moment of time t defined on $\mathcal{S}$;
$p^{(N)}(t) \equiv p^N(t) = \int\limits_{(\bar{G})^N} \mu_t^{(N)}(x^1, \ldots, x^N) dx^1 \ldots dx^N$; $p^{(0)}(t) \equiv p^0(t)$ is the probability that the domain $(\bar{V})$ does not contain particles at the moment of time $t$; $f(t, x)$ is the density of an average number of particles in a random process at the moment of time $t$; $\overline{N}_0 = \mathcal{M}(\mathcal{N}_0) = \int f(0, x) dx = \bar{n}_0 \cdot \|V\|$; we note that the density $f(t, x)$ is an approximation to the solution of the smoothed Boltzmann equation; $\mathcal{E}_{t^*}(x)$ is the given function measurable in $(\bar{G})$, $t^*$ is some given moment of time, $0 < t^* < \infty$; so that all considered functionals can be represented as $\mathcal{I}_{t^*} = \int f(t^*, x) \cdot \mathcal{E}_{t^*}(x) dx$.

Note, that if the integration domain is not specified, the integration is performed over $(\bar{G})$.

## 3. Imitation methods

### *3.1. Algorithms of Imitation methods*

Imitation, or "continuous time", Monte Carlo method to model rarefied gas evolution has been proposed in ([1, 2]). In this method, gas evolution is considered as predetermined process $\mathcal{P}_0$ with the following jumps: changing of velocities of a pair of particles due scattering; changing of the velocity and coordinates during flying out of a particle on the boundary ($\partial V_2$); changing of the number of particles in domain (V) as a result of flying out or flying in of a particle through the boundary ($\partial V_1$). The algorithm implementing this process was proposed, for example, in ([4]), and it will not be presented below. It should be noted only, that this algorithm has two distinctive features. First, it uses the well-known method of "fictitious collisions" for pair interactions modeling (see [10, 11]). Second, pair scattering frequency at arbitrary time depends on the relative velocity of the particles and the differences of their coordinates. In this paper one will refer to this algorithm as $A_0$.

Approximate simulation method, based on splitting the operator, corresponding to the exact master equations, "over groups of particles" has been proposed in [5, 6]. This method is based on the idea stated in the introduction: if we divide the particle system into multiple spatial subparts, then particles from the different sub-parts will rarely interact with each other at short time intervals. This feature virtually allows to model evolution of gas in the sub-parts almost independently from each other. Simulation inside an arbitrary splitting intervalis carried out in two stages. At the first stage, gas evolution is modeled independently in each of the $K_{sub}$ groups, and modeling within a group is carried out the way is done for the process $\mathcal{P}_0$. At the second stage, one performs correction for interaction of the particles that belong to different groups. The algorithm, implementing this method, has been presented in [7]. Its description is given below, and what will follow is the way of how to parallelize it.

(i) The time axis is split into intervals $[t_{d-1}, t_d)$ , $d = 1, 2, ..., M$. All the particles in the system are split into the $K_{sub}$ groups. Simulation of the evolution of rarefied gas inside each $d$-th interval is performed in two stages.
   (a) At the first stage, the evolution of gas is simulated independently in each group. Simulation is performed based on algorithm, implementing the process $\mathcal{P}_0$.
   (b) The Simulation is considered as complete when the condition $t' \geq t_d$ for the current moment of time$t'$ is satisfied.

(ii) After the first stage is complete, the particles are redistributed by the new $K_{sub}$ groups. At the same time, the distribution of particles over old groups at the first stage is stored in the memory. Then, one starts the second stage of the simulation.
   (a) Value $t_{d-1}$ is selected as the initial (zero) jump at the $d$-th step, while all the coordinates and velocities of the particles are fixed at this moment of time in the same way as they were at the moment of time $t_d$ at the end of the first stage.
   (b) Gas evolution is modeled in each of the new groups independently as in the algorithm that implements the process $\mathcal{P}_0$, but with account for two features:
      1. state of the system does not change between jumps (that means, that the particles remain stationary)
      2. only one type of jump (pair scattering) is considered, and the scattering may be physical only for the particles that have belonged to different groups at the first stage.
   (c) The second-stage simulation is considered complete when the condition $t' \geq t_d$ is satisfied for the current moment of time $t'$. At the same time, the distribution of particles by groups performed at the beginning of the second stage is used for the implementation of the first stage at the next $(d+1)$-th step.

(iii) The trajectory is interrupted when the M-th step is complete.

Journal of Physics: Conference Series **510** (2014) 012021 doi:10.1088/1742-6596/510/1/012021

### *3.2. Parallelization of the algorithm $A_1$*

The structure of the algorithm $A_1$ causes its natural parallelization. Its basic idea is that the simulation of particles evolution in each group at the first and second stage of an arbitrary step of splitting by time should be carried out on different processors of a multiprocessor device. One has used MPI library for the code implementation of this parallelism. The algorithm $A_1^{Par}$ for parallel computation based on the algorithm $A_1$ is described below.

(i) The number of groups $K_{sub}$ is chosen to be equal to the number of parallel processes. Paragraph 1 of the algorithm $A_1$ is implemented in parallel for all the processes. In each $i$-th process, $i = 1, .., K_{sub}$, the initial spatial energy and angular distribution has been given only for the particles that belong only to the i-th group.

(ii) Paragraphs 1.1 and 1.2 of the algorithm $A_1$ is implemented simultaneously in all processes. It should be noted, that since only the i-th group has been defined in the i-th process, it is equivalent to the simultaneous modeling of gas evolution in all the $K_{sub}$ groups.

(iii) The particles, defined in each process, are sorted into two subgroups at the end of the first phase. The first subgroup includes the particles, which by the end of the first stage still belong to the group,corresponding to the process. The second subgroup includes the particles that have fallen into any of the other groups or taken off the system. For example, if a particle's belongs to the $i$-th group means it is located in some $i$-th spatial subdomain of the system. The division into subgroups in i-th group has been performed as follows: the particle belongs to the first subgroup, if it have not left the i-th subdomain at the end of the first stage; the particle belongs to the second subgroup, when it has flown out of the i-th subdomain.

(iv) Numbers and phase coordinates of the particle that belong to the second subgroup in each process, is forwarded to the processes of the respective groups to which they belong. Afterwards, they are included in the lists of new groups of particles, and the information about them in the old group is erased. Thus there has been an automatic redistribution of particles on the new $K_{sub}$ groups, referred to the paragraph 2 of the algorithm $A_1$. Doing so, one should keep in mind, how the particles have been distributed in the old groups at the first stage.

(v) Paragraphs 2.1-2.3 of the algorithm $A_1$ is implemented in parallel in each of the processes.

(vi) Paragraph 3 of the algorithm $A_1$ is implemented in parallel in each of the processes.

## 4. Results of numerical modelling and analysis of computation costs

### *4.1. Problem of temperature discontinuity propagation*

Algorithm $A_1^{Par}$ has been applied to model a problem of temperature discontinuity propagation. Studying this issue has been conducted and reported in a number of papers, however, using of the algorithm $A_1^{Par}$ has allowed us to consider the problem for the sufficiently small values of $Kn$. In the problem monoatomic gas is placed in a halfspace $z > 0$ bounded by the plane $z = 0$. Diffuse reflection with temperature $T_0$ is realized on the plane. The gas is in equilibrium with the temperature $T_0$ and (numerical) density $n_0$ until to time t = 0 ; at $t = 0$, the temperature of the plane $z = 0$, changes abruptly from $T_0$ to $2 \cdot T_0$ causing temperature discontinuity. Its propagation is studied further.

In the numerical simulation the evolution of the gas is regarded in a rectangular parallelepiped $(V) : 0 \leq z \leq L, \quad -d_x \leq x \leq d_x, \quad -d_y \leq y \leq d_y$. The periodic conditions are set on the parallel to the axis $OZ$ faces. Diffuse reflection is modeled on the face $z = 0$, with a temperature of $2 \cdot T_0$. Departure and entry of particles in accordance with the equilibrium conditions of $z = \infty$ are modeled on the face $z = L$. Scattering of particles pairs is modeled following the law of hard spheres, and the scattering cross section is constant and equal to $\sigma$. Dimensionless variables

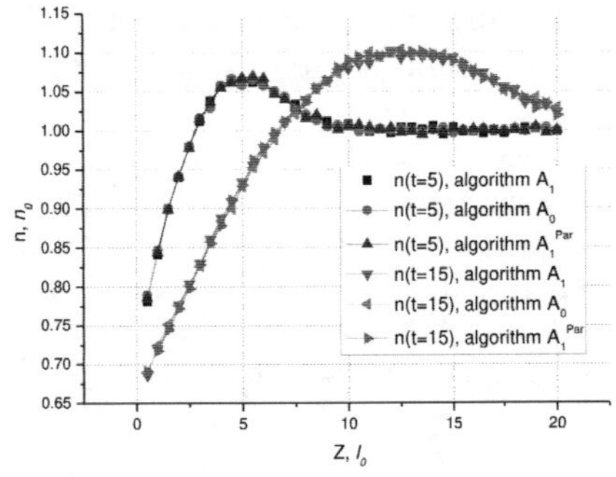


**Figure 1.** Density profiles. The temperature discontinuity propagation problem, $L = 20$.

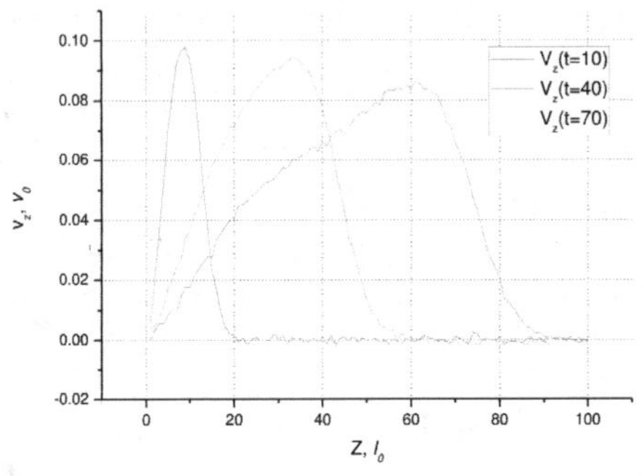


**Figure 2.** Velocity $v_z$ component profiles. The temperature discontinuity propagation problem, $L = 100$.

are used in the problem: $[L_0] = \frac{1}{\sqrt{2} \cdot n_0 \sigma}$ is the length unit ( mean free path of equilibrium gas); $[v_0] = \sqrt{2\frac{kT_0}{m}}$ is the velocity unit; $[E_0] = kT_0$ is the energy unit. Here $n_0$ is the initial density of particles, $T_0$ is the initial temperature of the equilibrium gas, $m$ is the mass of a gas particle, $k$ - the Boltzmann constant. The parameters defining the lateral dimensions of the parallelepiped $(V)$ in terms of dimensionless quantities are: $d_x = d_y = 0.5$. Any two particles with the numbers $k$ and $l$ are considered interacting if the condition $|z_k - z_l| < 0.1$ is satisfied.

Modeling of temperature discontinuity propagation problem for $L = 20$ $(Kn = 0.05)$ has been carried out to compare the algorithms $A_1$, $A_1^{Par}$ and $A_0$. The initial density of particles is equal to 100. Figure 1. shows the calculated density profiles at times $t = 5$ and $t = 15$. The simulation has implemented six thousand trajectories. Standard statistical error of the calculated density is not more then 0.5%. From the figure you can see that the results of simulation are in agreement with each other within the statistical error specified. The same good agreement presents for graphs displaying such physical quantities as the longitudinal velocity, longitudinal and transverse components of temperature and heat flux.

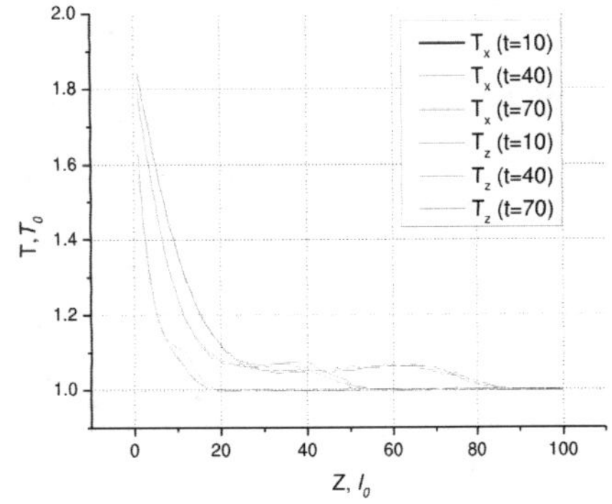


**Figure 3.** Temperature components profiles. The temperature discontinuity propagation problem, $L = 100$.

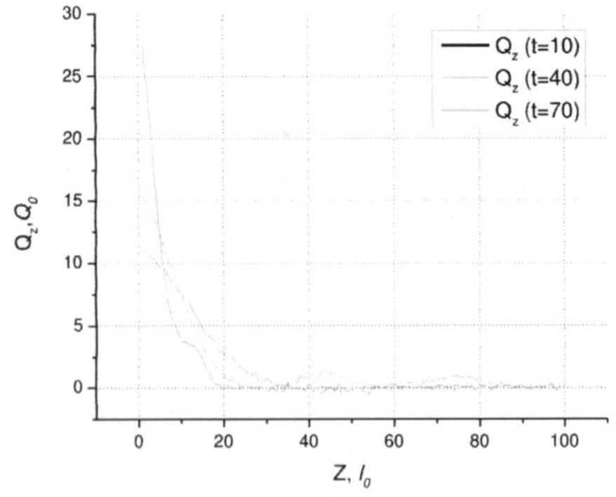


**Figure 4.** Heat flux $Q_z$ component profiles. The temperature discontinuity propagation problem, $L = 100$.

The further series of calculations presented have been obtained for the same problem at $Kn = 0.01$. For the modeling the algorithm $A_1^{Par}$ has been obtained. Fig. 2, 3 and 4 show the average velocity (particle flow), the temperature components and the heat flux at various time instants. In total, the simulation has implemented six thousand trajectories. Standard statistical error has been not more then 2.2% for the velocity, 3.7% - for the temperature, 9% - for the heat flow.

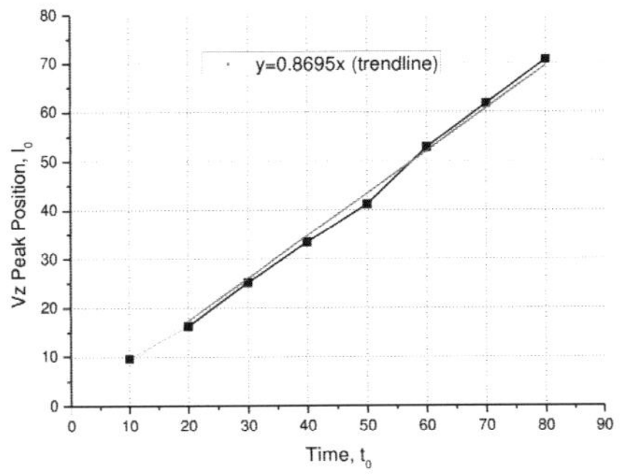


**Figure 5.** The position of $v_z$ maximum at different time instants.

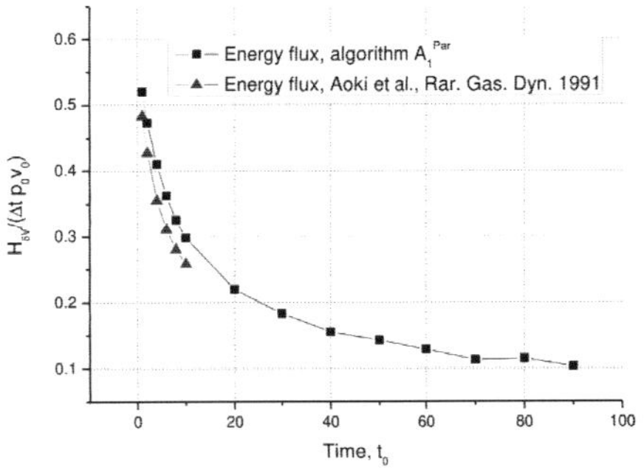


**Figure 6.** Energy flux through the surface $z = 0$. The temperature discontinuity propagation problem, $L = 100$.

These results show that the perturbation caused by the wall's temperature jump propagates with approximately constant speed. It can be best seen from the average velocity component $v_z$ behavior, demonstraited in Fig. 2. Fig. 5 shows the position of the $v_z$ maximum at different time instants. The figure shows that it is changing linearly with time, and its velocity is approximately $v_0 = 0,87$. The fig. 5 shows that in the region of the front temperature komponets' inversion is present, indicating the non-uniform redistribution of the internal energy between the longitudinal and transverse degrees of freedom. The heat flux profiles, shown in Fig. 4 have prominent features near the heated wall and in the vicinity of the disturbance front. The flux $Q_z$ is decreasing with time, asymptotically tending toward zero, near the plane $z = 0$. The flow has a characteristic hump, in the vicinity of the front disturbance, moving with approximately constant velocity. The humps height is decreasing with time, while it is spreading. Fig. 6 demonstrates the time dependence of energy flux through the surface $z = 0$, computed in ([12]) using the BGK for times from the interval $[1, 10]$ and using the algorithm of $A_1^{Par}$ for times from the interval $[1, 100]$. It is clear From Fig. 6 that, as well as the heat flow, the flow of energy through the surface $z = 0$ is asymptotically approaching zero. The difference between the results obtained with BGK and $A_1^{Par}$ can be explained by the fact that BGK is an approximate method in contrast to the Monte Carlo ones.

### *4.2. Comparison and estimating of the algorothms computation costs*

Fig. 7 shows the ratio of the computation costs for the $A_0$ and $A_1$ algorithms and for the $A_0$ and $A_1^{Par}$ algorithms depending on the number of splitting groups. The data are presented for the temperature discontinuity propagation problem with a typical size $L = 100$. The figure demonstrates that the algorithms $A_1$ and $A_1^{Par}$ have been more beneficial in terms of computation costs if compared with the algorithm $A_0$. In the case of $A_1$ the simulation is performed using a single processor. In the case when the $A_1^{Par}$ is applied the number of processors involved in the simulation is equal to $K_{sub}$. Saying computation costs one means product of the algorithm implementation time and the dispersion of the corresponding method. In accord with the theoretical ideas the numerical experiments have showed that all three methods under consideration have close dispersions, so the ratio of computational costs is equivalent to the ratio of implementation times.

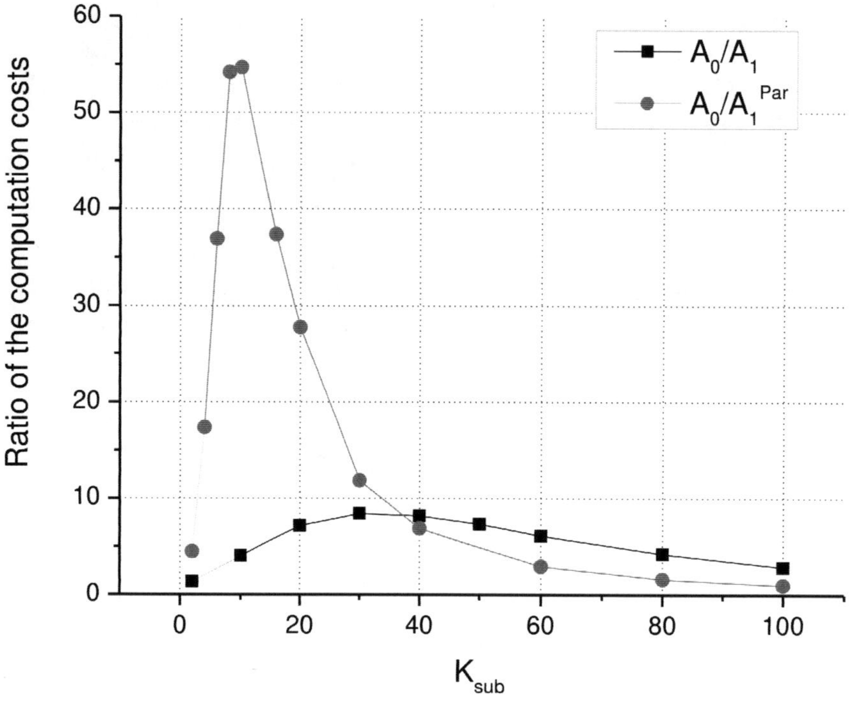


**Figure 7.** The ratio of the computation costs for the $A_0$ and $A_1$ algorithms and for the $A_0$ and $A_1^{Par}$ algorithms as a functions of the splitting groups number $K_{sub}$. The temperature discontinuity propagation problem, $L = 100$

Fig. 7 shows that the two presented curves have maxima, and the maximum for the algorithm $A_1^{Par}$ has been reached for a smaller number of splitting groups. The maximum value for the algorithm $A_1^{Par}$ exceeds the maximum value for the algorithm $A_1$.

Asymptotic by $n_0$ estimates of the $A_1$ and $A_1^{Par}$ implementation time are given below and they are an explanation of the form of numerical estimates obtained above. Such an estimate has been obtained for the algorithm $A_1$ in [5, 7]:

$$T_{A_1} = C \cdot \frac{n_0^2}{K_{sub}} \cdot (1 + O(\frac{1}{n_0})) \tag{1}$$

This estimate can be detailed, highlighting explicitly the member responsible for the redistribution of particles on the new $K_{sub}$ groups between the first and second stage of the simulation.

$$T_{A_1} = C \cdot \frac{n_0^2}{K_{sub}} \cdot (1 + O(\frac{1}{n_0})) + C_1 \cdot K_{sub} \cdot n_0 \tag{2}$$

Since the simulation of the evolution of the particles within the groups in the algorithm $A_1^{Par}$ is implemented in parallel , rather than sequentially, as in the algorithm of $A_1$, then implementation of the first and second stages is reduced to $K_{sub}$ times in this case. The estimate also contains a term that describes the exchange of information between the $K_{sub}$ parallel processes between the first and second stage of the simulation. Each of $K_{sub}$ parallel processes sends information about particles flied out from the corresponding group to other $K_{sub} - 1$ processes. The number of such particles is proportional to $n_0$. Therefore the estimate for the $T_{A_1^{Par_1}}$ should be written as:

$$T_{A_1^{Par_1}} = C \cdot \frac{n_0^2}{K_{sub}^2} \cdot (1 + O(\frac{1}{n_0})) + C_2 \cdot K_{sub}^2 \cdot n_0 \tag{3}$$

It should be noted that constants $C$, $C_1$ and $C_2$ in (2) and (3) depend on the properties of multiprocessor system used for simulation. Data exchange in the algorithms $A_1$ and $A_1^{Par}$ has been implemented in a such way, that $C_1 > C_2$ (it is possible, that $C_1 >> C_2$ ). The formulas (2) and (3) are asimptotic expressions of the $A_1$ and $A_1^{Par}$ realization times for large $n_0$ and for small and large values of $K_{sub}$. If one neglects the remainder term $O(\frac{1}{n_0})$ in (2) and (3) and considers $T_{A_1}$ and $T_{A_1^{Par}}$ as functions of $K_{sub}$, approximate estimates of the positions of the minima on $K_{sub}$ can be obtained:

$$K_{sub}^{A_1} = \sqrt{\frac{C \cdot n_0}{C_1}}, K_{sub}^{A_1^{Par}} = \sqrt[4]{\frac{C \cdot n_0}{C_2}} \tag{4}$$

The minimum values of the expressions $T_{A_1}$ and $T_{A_1^{Par}}$ are, respectively:

$$(T_{A_1})_{min} = 2 \cdot n_0 \cdot (C \cdot C_1 \cdot n_0)^{\frac{1}{2}}, (T_{A_1^{Par}})_{min} = 2 \cdot n_0 \cdot (C \cdot C_2 \cdot n_0)^{\frac{1}{2}} \tag{5}$$

As follows from (4), $K_{sub}^{A_1} > K_{sub}^{A_1^{Par}}$ under the condition $n_0 > \frac{C_1^2}{C \cdot C_2}$. This condition is valid for sufficiently large $n_0$. Hence, the ratio of the implementation time minima should be written as:

$$\frac{(T_{A_1^{Par}})_{min}}{(T_{A_1})_{min}} = \frac{C_2^{\frac{1}{2}}}{C_1^{\frac{1}{2}}} \tag{6}$$

The conclusions of the above analysis demonstrate good agreement with the curves in fig. 7. The similar computation costs behavior has been also got for the problem on plane Poiseuille flow in the field of external forces. In general, it has been obtained that the algorithm $A_1^{Par}$ has good enough gain in the computational cost.

Algorithm $A_1^{Par}$ does not contain any restrictions on the dimension of the problems. Volumes of the sub-parts in the method should not be small. Therefore, we suppose that the estimates and the conclusions retain their validity for the spacially multi-dimensional cases.

[1] Khisamutdinov A I 1985 Ob imitatsionnom metode Monte-Karlo dlya modelirovaniya dinamiki razrezhennykh gazov (On Imitation Method of Monte-Carlo for Simulation of Dynamics of Rarefied Gases) *Preprint* VTs SO AN SSSR 599.

[2] Khisamutdinov A I 1988 Imitatsionnoe statisticheskoe modelirovanie kineticheskogo uravneniya razrezhennykh gazov *Dokl. Akad. Nauk SSSR* **302** no.1 75-79.

[3] Khisamutdinov A I 2005 On some properties of Markov processes and Monte Carlo methods for inhomogeneous Boltzmann equation *Russ. J. Numer. Anal. Math. Modelling* **20** no.2 131-60.

[4] Khisamutdinov A I 2004 On development of Continuous Time Monte Carlo methods for problems of Boltzmann equation with external forces *Transport. Theor. Stat. Phys.* **33** No.1 69-89.

[5] Khisamutdinov A I 1989 Priblizhennyi metod statisticheskogo modelirovaniya, ispolzuyushchii slabuyu zavisimost' podchastei (Approximated Method of Static Simulation Using Weak Dependence of Subparts) *Vopr. Korrekt. Zad. Anal. IM SO AN SSSR* 154-60.

[6] Khisamutdinov A I 2000 On connection between "Continuous time" and "Direct simulation" Monte Carlo methods for Boltzmann equation and on some new approximate methods *Monte Carlo Meth. and Appl.* **6** No.4 323-40.

[7] Khisamutdinov A I and Velker N N 2012 On Reduction of Computational Cost of Imitation Monte Carlo Algorithms for Modeling Rarefied Gas Flows *Math. Mod. and Comp. Simul.* **4** No.2 187-202.

[8] Bird G A 1994 *Molecular gas dynamics and the Direct Simulation of Gas Flows* (Oxford: Clarendon Press).

[9] Belotserkovskii O M and Yanitskii V E 1975 The statistical method of particles in cells for the solution of problems of the dynamics of a rarefied gas *Zh. Vychisl. Mat. Mat. Fiz.* **15** No.5 1195-08 No.6 1553-67

[10] Koura K 1986 Null-collision technique in the direct-simulation Monte Carlo method *Phys. Fluids.* **29**, No.11 3509-11.

[11] Khisamutdinov A I 2007 Statisticheskoe modelirovanie odnogo tipa par sluchainykh velichin s ispolzovaniem fiktivnykh skachkov (Statistic Simulation of Single Type of Pairs of Arbitrary Quantities Using Fictive Jumps) *ZhVM i MF*, **47** No.1 162-73.

[12] Aoki K, Sone Y, Nishino K and Sugimoto H 1991 Numerical Analysis of unsteady motion of a rarefied gas caused by sudden changes of wall temperature with special interest in the propagation of a discontinuity in the velocity distribution function *Raref. Gas Dynam.* 222-31.

Journal of Physics: Conference Series **510** (2014) 012022 doi:10.1088/1742-6596/510/1/012022

# Bond orientation properties in lipid molecules of membranes: molecular dynamics simulations

**Alexander L. Rabinovich**[a], **Alexander P. Lyubartsev**[b]

[a] Institute of Biology, Karelian Research Center of Russian Academy of Sciences, 11 Pushkinskaya str., Petrozavodsk 185910, Russian Federation

[b] Division of Physical Chemistry, Department of Material and Environmental Chemistry, Stockholm University, SE-10691 Stockholm, Sweden

E-mail: [a] rabinov@krc.karelia.ru; [b] alexander.lyubartsev@mmk.su.se

**Abstract.** Atomistic molecular dynamics simulations have been carried out for 16 different fully hydrated phosphatidylcholine lipid bilayers, having 16 or 18 carbon atoms in fully saturated $sn-1$ chain and from 18 to 22 carbon atoms in $sn-2$ chain with different degree of unsaturation, with the purpose to investigate the effect of unsaturation on physical properties of lipid bilayers. Special attention has been paid to profiles of C-C and C-H bond order parameters of lipid molecules and the orientational fluctuations of these bond vectors. It was shown that the study of anisotropy degree of bond orientations probability distributions allows distinguishing extended regions with different types of angular fluctuations of bonds in a membrane formed by lipid molecules with unsaturated chains.

## 1. Introduction

One of the most important molecular systems in nature is a biomembrane. Biomembranes play a role of barriers separating the interior of a cell from the outside environment and are in control of the movement of various compounds into and out of cells. Biomembranes, built mostly by lipid molecules arranged in a bilayer, are very complex heterogeneous systems, which are characterized by anisotropy of structural and physical properties, knowledge of which is a key element of our general understanding of membrane functioning. A typical biological membrane contains many species of lipids, phosphatidylcholine (PC) being the most abundant one. The commonly occurring fatty acid (FA) chains of lipids may contain 1 – 6 carbon-carbon double bonds of the *cis*-configuration in different positions. In most cases, at least half of the FA chains are unsaturated. Unsaturated FA tails of lipids are of great importance for the structure and function of animal and plant membranes [1–16]. The double bonds of polyunsaturated (PU) chains are, as a rule, methylene-interrupted. Throughout this paper, the shorthand nomenclature, i.e., the notation of $M : k(n-j)cis$ for describing the structure of each hydrocarbon chain of lipids will be used. Here the number $M$ specifies the total number of carbon atoms in the chain, the number $k$ specifies the number of double bonds, $j$ denotes the number of carbons between the chain terminal $CH_3$ group and the nearest double bond, assuming that all the other double bonds are methylene-interrupted, whereas *cis* refers to the conformation around the double bonds. For brevity, the fragment $(n-j)cis$ in the notation is frequently omitted.

Although the history of study of FA composition of different membranes in relation to the tentative role of lipid unsaturation is comparatively long [1–16], full understanding of the effects of lipid unsaturation on various physical properties of membranes at the molecular level, affecting their functioning, is not yet achieved. In order to investigate the role of lipid tails unsaturation we have carried out molecular dynamics (MD) simulations of 16 hydrated liquid crystalline phase PC bilayers composed of phospholipids with *sn*-1 saturated and *sn*-2 unsaturated chains. The following PC molecules were considered in the simulations:

1-stearoyl-2-oleoyl-*sn*-glycero-3-PC (18:0/18:1(n-9)cis PC),
1-stearoyl-2-linoleoyl-*sn*-glycero-3-PC (18:0/18:2(n-6)cis PC),
1-stearoyl-2-linolenoyl-*sn*-glycero-3-PC (18:0/18:3(n-3)cis PC),
1-stearoyl-2-octadecatetraenoyl-*sn*-glycero-3-PC (18:0/18:4(n-3)cis PC),
1-stearoyl-2-octadecapentaenoyl-*sn*-glycero-3-PC (18:0/18:5(n-3)cis PC),
1-stearoyl-2-arahidonoyl-*sn*-glycero-3-PC (18:0/20:4(n-6)cis PC),
1-stearoyl-2-eicosapentaenoyl-*sn*-glycero-3-PC (18:0/20:5(n-3)cis PC),
1-stearoyl-2-docosahexaenoyl-*sn*-glycero-3-PC (18:0/22:6(n-3)cis PC),
1-palmitoyl-2-oleoyl-*sn*-glycero-3-PC (16:0/18:1(n-9)cis PC),
1-palmitoyl-2-linoleoyl-*sn*-glycero-3-PC (16:0/18:2(n-6)cis PC),
1-palmitoyl-2-linolenoyl-*sn*-glycero-3-PC (16:0/18:3(n-3)cis PC),
1-palmitoyl-2-octadecatetraenoyl-*sn*-glycero-3-PC (16:0/18:4(n-3)cis PC),
1-palmitoyl-2-octadecapentaenoyl-*sn*-glycero-3-PC (16:0/18:5(n-3)cis PC),
1-palmitoyl-2-arahidonoyl-*sn*-glycero-3-PC (16:0/20:4(n-6)cis PC),
1-palmitoyl-2-eicosapentaenoyl-*sn*-glycero-3-PC (16:0/20:5(n-3)cis PC),
1-palmitoyl-2-docosahexaenoyl-*sn*-glycero-3-PC (16:0/22:6(n-3)cis PC).

## 2. Models and methods

Each simulated bilayer was composed of 128 lipids of one of 16 studied types, with 64 lipids in each leaflet. All hydrogen atoms were explicitly included in the computations. Each bilayer was hydrated by 3840 water molecules, which corresponds to condition of full hydration with 30 water molecules per lipid. The system was put into rectangular periodic cell, with the Z-axis parallel to the bilayer normal. The size of the box was varying during the simulations under condition of semianizotropic $NPT$-ensemble with two degrees of freedom: one in Z-direction and another in XY-direction, so that the box sizes in X and Y direction were equal at each time moment. The pressure was set to 1 bar and the temperature to 303 K.

The force field for lipids was based on the CHARMM27 parameter set for lipids [17] with modifications described in paper [18], which include scaling of the 1-4 electrostatic interactions by factor 0.83, and recalculated charges of the lipid headgroup on the bases of *ab-initio* computations. For bilayer composed of 14:0/14:0 PC lipids, modifications introduced in paper [18] provide perfect agreement with experimental data for the area per lipid, as well as with the X-ray structure factor and NMR order parameters. For lipids considered in this work, we use the same partial charges as in ref. [18] in the lipid headgroup including esters, while for tails we used charges adopted from the original CHARMM27 force field, with scaling of 1-4 electrostatic interactions by factor 0.83. All intramolecular bond and angle parameters, as well as Lennard-Jones interactions, were taken as in the original CHARMM27 force field [17]. Water molecules were described by the flexible SPC model [19]. Use of this water model, in connection with modified CHARMM27 force field, was verified in paper [18].

The double time step algorithm [20] was used to treat separately fast forces (covalent bonds, angles, torsions, collision Lennard-Jones forces within 5 $\AA$ distance) with time step 0.25 fs, and longer range forces with time step 2.5 fs. The long-range electrostatic interactions were treated by the Ewald summation method [21]. The reciprocal part of the Ewald sum was cut at the condition that the remaining terms do not contribute more than on 0.0001 level of the total value. The $\alpha$ parameter of Ewald sum was set to $\alpha = 2.6r_{cut}$, and cutoff distance $r_{cut} = 14$ Å was optimized for computational performance according to ref. [22]. The dispersion correction from the Lennard-Jones interactions outside the cutoff distance was included into pressure [21].

In the starting configuration, the lipids were set parallel to each other, organized in a regular manner in two layers, and water molecules were distributed outside the bilayer. The systems were simulated 1 ns under constant volume and then 1 ns under constant pressure and isotropic cell fluctuations. The obtained configurations were considered as starting points for longer simulations with independent cell fluctuations in Z and XY directions. The time reversible Nose-Hoover constant temperature – constant-pressure algorithm [23] was implemented, with the thermostat and barostat relaxation time 30 fs and 1 ps respectively. All the systems were simulated during 100 ns. The first 20 ns of the simulations was considered as equilibration and disregarded in further analysis. Atom coordinates were saved each 1 ps in trajectories for further analysis. The simulation software was MDynaMix package v.5.2 [24].

## 3. Results and discussion

### *3.1. Area per lipid*

The average area per lipid $A_{pl}$ is one of the most fundamental parameters of a lipid bilayer. The area monitoring in computer simulations is one of the most common ways to determine whether the bilayer system has reached equilibrium. Moreover, the average area per lipid defined in zero-tension computer simulations is a parameter which is most often used to define the quality of the force field, though more reliable validation of a force field can be done by comparison of simulated and experimental structure factors of the system [25].

From the observation of the time evolution of the simulated areas per lipid and calculations of block averages a conclusion was made that 20 ns of equilibration time is enough for all considered bilayer systems. Average values of the area per lipid computed during the 80 ns production stage, as well as compressibility modulus of the bilayers computed from the fluctuations of the total areas as $K_A = k_B TA/\langle(A - \langle A \rangle)^2$ ($A$ being the total area of the bilayer patch) are listed in Table 1. Some of available experimental average lipid areas are also given in Table 1. We should note that there is a wide spread of experimental data concerning some of the bilayers. Many experimental structural methods were reviewed in works [26,27], and considerable quantitative uncertainty was demonstrated in structural results for lipid bilayers, including the value of area $A_{pl}$ in the fluid phase.

A rather weak trend of the simulation and experimental results is that the average area is slightly increasing with the degree of unsaturation. Still, bilayers with highest degree of unsaturation, 16:0/22:6 PC and 18:0/22:6 PC show somewhat smaller $A_{pl}$ than those of the nearest less unsaturated bilayers. The reason for the decrease of $A_{pl}$ values of molecules 16:0/22:6 PC and 18:0/22:6 PC can possibly become more clear after analysis of packing arrangement of different chains. It is also worth to note the tendency that bilayers with higher average area per lipid have higher relative area fluctuations which render to their lower compressibility modulus.

### *3.2. Bond order parameters*

The traditional measure for the orientational mobility of C-H bonds in lipid bilayers is the C-H bond order parameter which is available from $^2H$ NMR experiments [31–33]; it is defined as:

$$S_{CH} = (3\langle\cos^2\theta_{CH}\rangle - 1)/2 \tag{1}$$

where $\theta_{CH}$ is the angle between the C-H bond vector (C-D bond in $^2H$ NMR) and the bilayer normal. The order parameter can change from $S_{min} = -0.5$ (the bonds are always oriented parallel to the bilayer plane) to $S_{max} = 1$ (the bonds are always parallel to the bilayer normal). Strictly speaking, it is impossible to establish the sign of the C-D bond order parameter from

Journal of Physics: Conference Series **510** (2014) 012022 doi:10.1088/1742-6596/510/1/012022

**Table 1.** Average areas per lipid, $A_{pl}$, and bilayer compressibility modulus, $K_A$, obtained for mixed-chain liquid-crystalline phase unsaturated phosphatidylcholine bilayers by MD simulations of the present work; $T = 303$ K

| lipid | $A_{pl}$ $(\pm\Delta)^a$, $nm^2$, sim | $A_{pl}$, $nm^2$, exp | $K_A$, mN/m |
|---|---|---|---|
| 16:0/18:1(n-9)cis PC | 0.634 (±0.002) | 0.63 [28][b] | 458 |
| 16:0/18:2(n-6)cis PC | 0.636 (±0.003) | 0.66 [28][b] | 349 |
| 16:0/18:3(n-3)cis PC | 0.632 (±0.002) | | 359 |
| 16:0/18:4(n-3)cis PC | 0.647 (±0.003) | | 312 |
| 16:0/18:5(n-3)cis PC | 0.646 (±0.004) | | 274 |
| 16:0/20:4(n-6)cis PC | 0.652 (±0.003) | 0.68 [28][b] | 339 |
| 16:0/20:5(n-3)cis PC | 0.656 (±0.003) | | 337 |
| 16:0/22:6(n-3)cis PC | 0.643 (±0.002) | 0.70 [28][b] | 447 |
| 18:0/18:1(n-9)cis PC | 0.627 (±0.002) | 0.614 [29][c]; 0.666 [30][d] | 440 |
| 18:0/18:2(n-6)cis PC | 0.625 (±0.002) | 0.673 [30][d] | 432 |
| 18:0/18:3(n-3)cis PC | 0.637 (±0.002) | 0.666 [30][d] | 366 |
| 18:0/18:4(n-3)cis PC | 0.646 (±0.003) | | 381 |
| 18:0/18:5(n-3)cis PC | 0.653 (±0.003) | | 339 |
| 18:0/20:4(n-6)cis PC | 0.646 (±0.003) | 0.706 [30][d] | 428 |
| 18:0/20:5(n-3)cis PC | 0.653 (±0.002) | 0.691 [30][d] | 387 |
| 18:0/22:6(n-3)cis PC | 0.637 (±0.004) | 0.692 [29][c]; 0.716 [30][d] | 323 |

[a] statistical error $\Delta$ for 20-100 ns is evaluated from the variance of 10 ns block averages;
[b] Langmuir film balance, 297K; [c] $^2$H NMR and X-ray; [d] $^2$H and $^{31}$P NMR

the $^2$H NMR experiments, that is why experimental values are usually given as absolute values $|S_{CD}|$ [31]. Evidently, computer simulations can provide sign of the order parameters directly.

The C-H bond order parameters, calculated in our MD simulations, are presented in Fig. 1, 2 in the form of $-S_{CH}$, both for saturated and unsaturated chains. The plots demonstrate general features of C-H order parameters which agree with results from previous experimental (NMR) and computer simulation works. For the saturated chains the order parameter profile exhibits a small drop at the 3-rd carbon position, then reaches a plateau, and then drops gradually to the chain end. Behaviour of unsaturated chains is clearly different. Already a chain with a single double bond (18:1) has a clear drop in the order parameter for carbons at and near position of the double bond. The region of small order parameters is increasing with increase of the number of double bonds in the tail, and for polyunsaturated chains with $4-6$ double bonds most of order parameters are below 0.1 and often even below 0.05. A clear tendency is that the more unsaturated *sn*-2 chain the smaller average of the absolute values of order parameters $|S_{CH}|$.

Another important result of our MD simulations is that C-H bond order parameters of both signs (plus and minus) are not uncommon, they coexist in polyunsaturated chains of PC lipids. Would the corresponding $^2$H NMR-data be available, they could be seen in these ranges as positive $|S_{CD}|$ values, as it is indicated by dashed lines in Figs. 1, 2. Generally speaking, this fact gives raise to a question how the correct signs of the bond order parameters of a chain of complicated chemical structure (e.g., a polyunsaturated chain) can be established when deuterium NMR data ($|S_{CD}|$) only are available; a computer modelling can possibly help in such cases.

Computer simulations allow to determine order parameters for other bonds for which experimental data are unavailable. The $S_{CC}$ values of the consecutive carbon-carbon bonds can give additional insight into ordering properties of lipid chains. The $S_{CC}$ bond order parameters calculated in our MD simulations are presented in Figs. 3, 4).

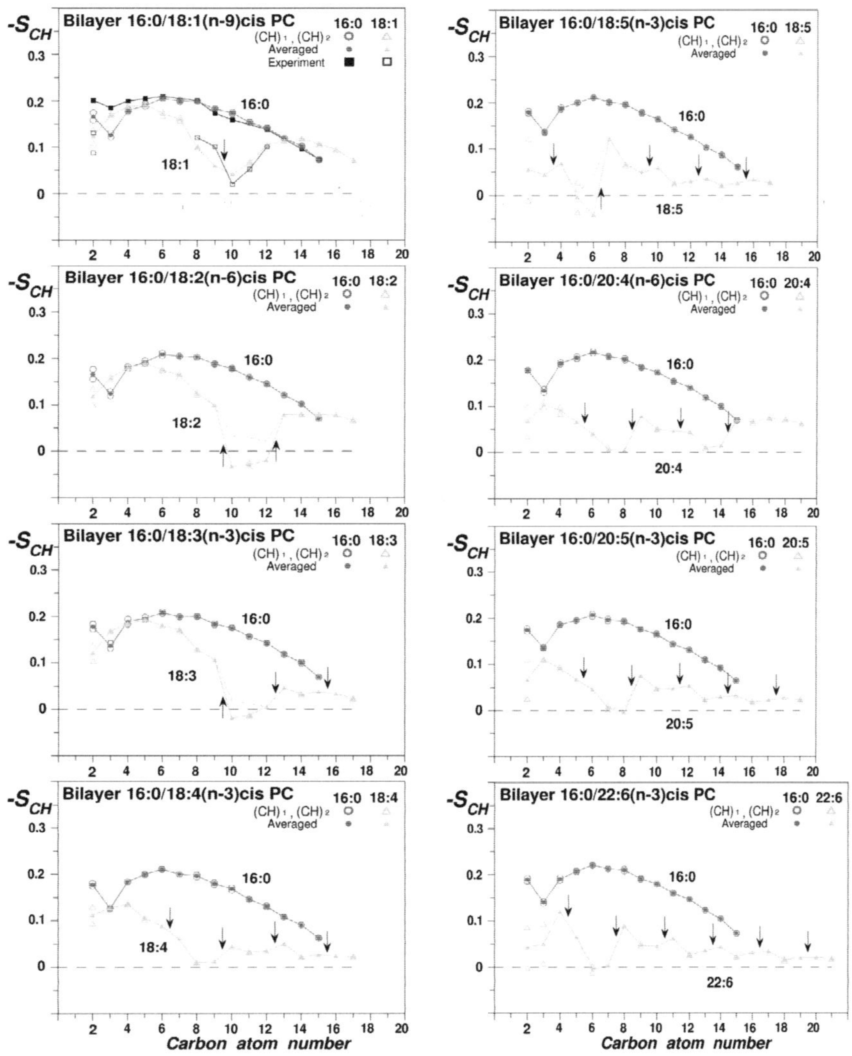


**Figure 1.** Order parameter $-S_{CH}$ as a function of the carbon atom number of the lipid hydrocarbon chains of PC bilayers with *sn*-1 chain 16:0 as determined from the MD simulations. The arrows indicate the double bonds. Dashed line shows the ranges where simulated MD-data (solid lines) and possible $^2$H NMR-data could be different due to the sign uncertainty of NMR data.

The $S_{CC}$ values in all the saturated and unsaturated chains are seen to vary non-monotonously. All the $S_{CC}$ profiles are zigzag-shaped and exhibit a trend of decreasing values toward the chain end. A difference in the behaviour of saturated and unsaturated tails is also well seen. The zigzag-shaped $S_{CC}$ profiles of the saturated chains are characterized by a systematic, strictly alternating rise and fall of successive $S_{CC}$ values ("odd-even effect", [34]). The order parameter $S_{CC}$ profiles of unsaturated chains are also zigzag-shaped, but differ significantly in size and shape from those in the saturated chains. The $S_{CC}$ order parameters of single C-C bonds next to the double bonds C=C in all unsaturated *sn*-2 chains are significantly, several times, lower than that of the double bond C=C and of the corresponding single bond C-C in the saturated chains *sn*-1. The substitution of double C=C bonds for a single C-C bonds affects

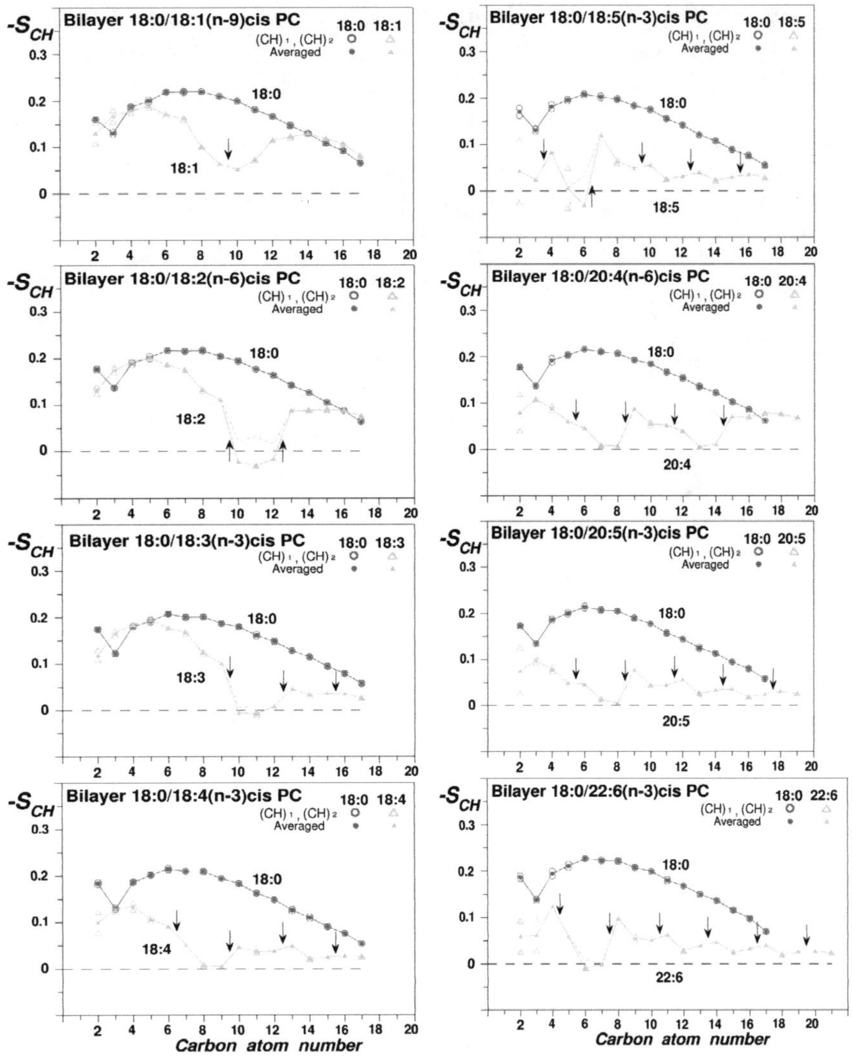


**Figure 2.** Order parameter $-S_{CH}$ as a function of the carbon atom number of the lipid hydrocarbon chains in PC bilayers with *sn*-1 chain 18:0 as determined from the MD simulations. The arrows indicate the double bonds. Dashed line shows the ranges where simulated MD-data (solid lines) and possible $^2$H NMR-data could be different due to the sign uncertainty of NMR data.

the C-C order parameters resulting in a sharp decline in the $S_{CC}$ of the single bonds adjacent to the double bonds, and in an increase in the $S_{CC}$ of the double bonds.

### *3.3. Bond orientation distribution functions and anizotropy coefficient*

The order parameters for primarily C-H, but also for C-C bonds are routinely computed in many works on computer simulations of lipid membranes. They however do not provide full information about the ordering and orientation of the bonds. The order parameter defined by eq. 1 is an inherently integral quantity depending both on ordering of the bonds and their orientation. In a general case (except extreme cases $S = -0.5$ and $S = 1$) it is not possible to distinguish between these two contribution. For example, order parameter zero can

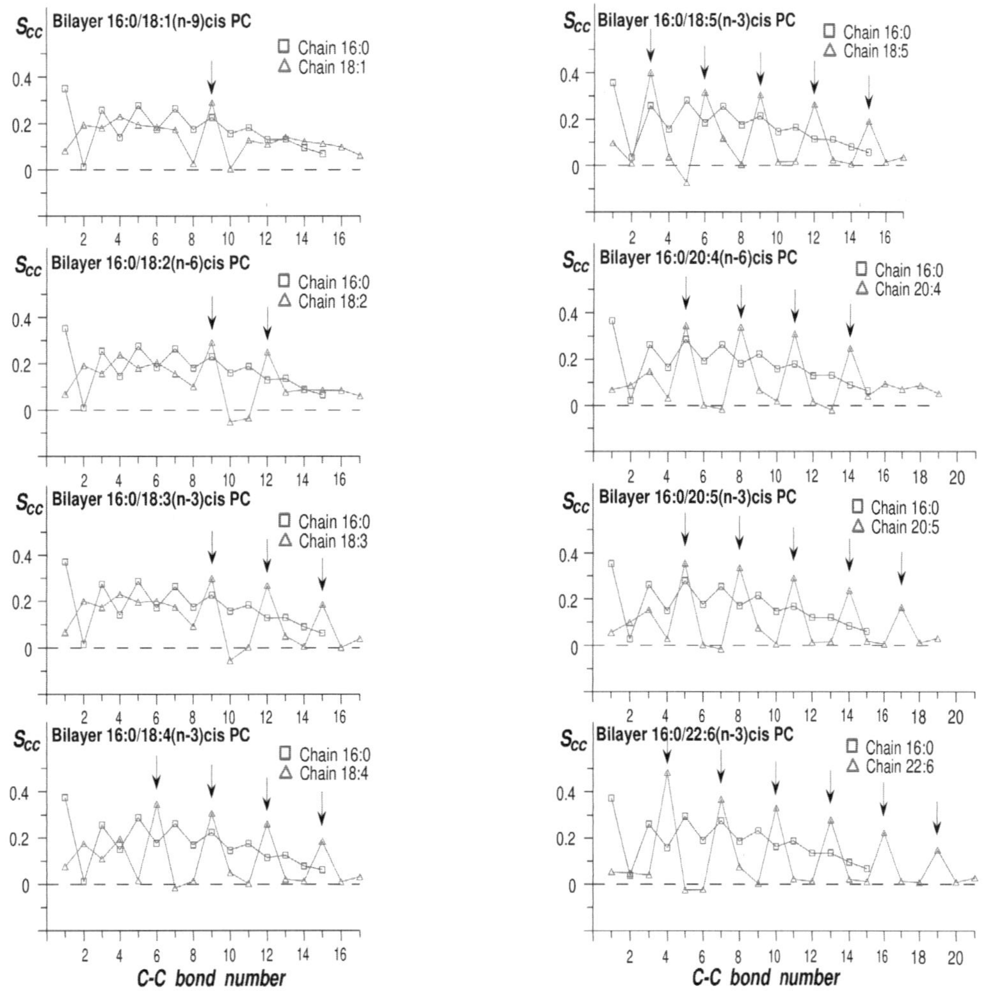


**Figure 3.** Order parameter $S_{CC}$ as a function of the $C-C$ bond number of the lipid hydrocarbon chains in PC bilayers with $sn$-1 chain 16:0 as determined from the MD simulations. The arrows indicate the double bonds.

be observed both in the case of completely isotropic distribution of the bond direction, and for perfectly ordered bonds in the direction which forms with the bilayer normal the so-called "magic angle" for which $3 \cdot Cos^2\theta - 1 = 0$. Evidently, for distinguishing effects of "inherent ordering" and orientation one needs to know the distribution of angle between the bond and the bilayer normal [35].

In order to compute the orientational bond distribution, the range $0 \leq \theta \leq \pi$ was divided into 60 equal bins, and histograms of hits into each bin $n_i$ were collected. From these histograms, the angular distributions were determined as $\rho(\theta) = n_i/n_{tot}$, where $n_{tot}$ is the total number of frames in the MD-trajectory multiplied by the number of lipids. For further analysis it is appropriate to divide the obtained distributions by $\sin\theta$ accounting for the solid angle area corresponding to the polar angle $\theta$, and consider functions $\rho(\theta)/\sin\theta$. For the completely isotropic distribution of bonds $\rho(\theta)/\sin\theta$ function is constant.

We have computed such distributions for all C-H and C-C bonds in the tails of the simulated

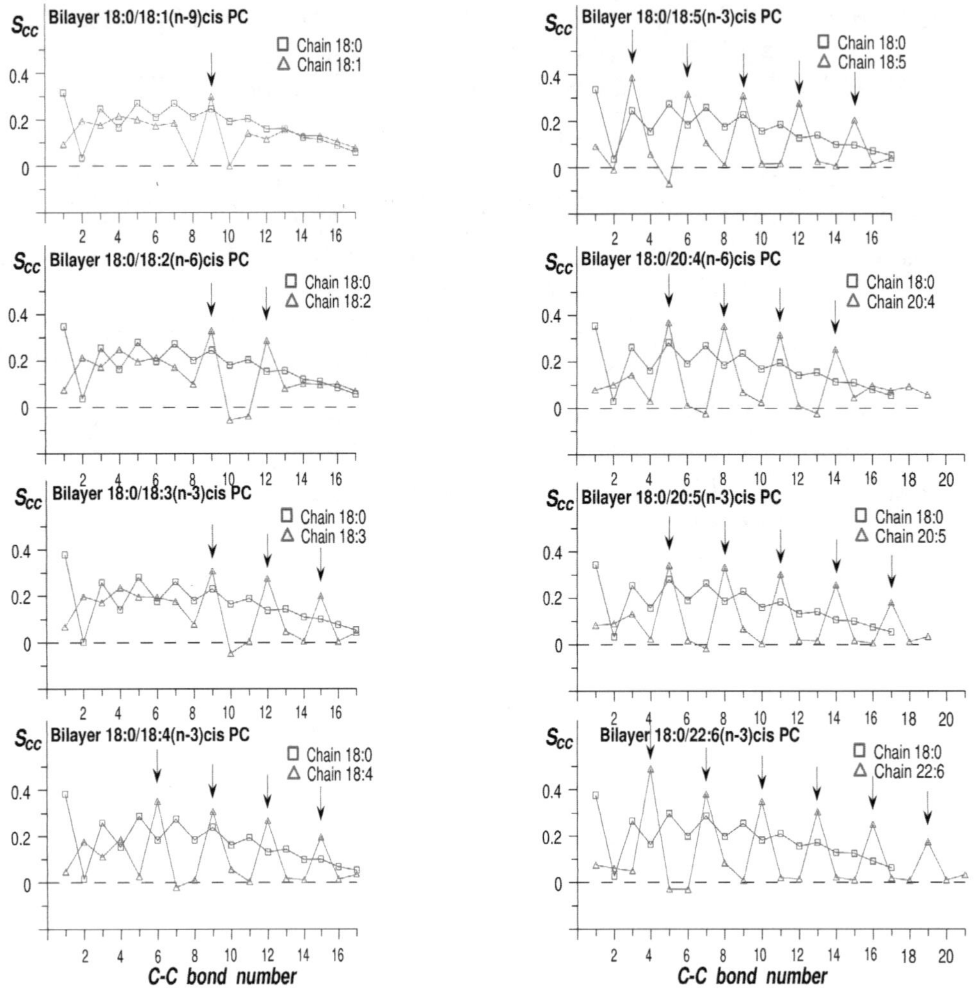


**Figure 4.** Order parameter $S_{CC}$ as a function of the $C-C$ bond number of the lipid hydrocarbon chains in PC bilayers with *sn*-1 chain 18:0 as determined from the MD simulations. The arrows indicate the double bonds.

bilayers. These distributions provide a very detailed picture of bond ordering in lipid bilayers. The obtained distributions were of varying types, differing by a sharpness of the maximum, asymmetry, presence of plateau or other features, being dependent on the type of the bond, presence of double bonds in the chain, position along the chain. In order to characterize different bonds on equal footing, we propose to introduce a new parameter called "anisotropy coefficient" which describes relative sharpness of the distribution for each bond:

$$D_A = \frac{[\rho(\theta)/\sin\theta]_{max} - [\rho(\theta)/\sin\theta]_{min}}{[\rho(\theta)/\sin\theta]_M} \qquad (2)$$

where $[\rho(\theta)/\sin\theta]_{max}$ is the maximum value of $\rho(\theta)/\sin\theta$ for the given bond, $[\rho(\theta)/\sin\theta]_{min}$ is its minimum value while $[\rho(\theta)/\sin\theta]_M$ is the maximum value of all $[\rho(\theta)/\sin\theta]_{max}$ taken over all bonds of the given type in the both chains of the lipid. Evidently, $0 \leq D_A \leq 1$, and zero values of $D_A$ corresponds to a uniform (isotropic) distribution with $[\rho(\theta)/\sin\theta]_{max} = [\rho(\theta)/\sin\theta]_{min}$.

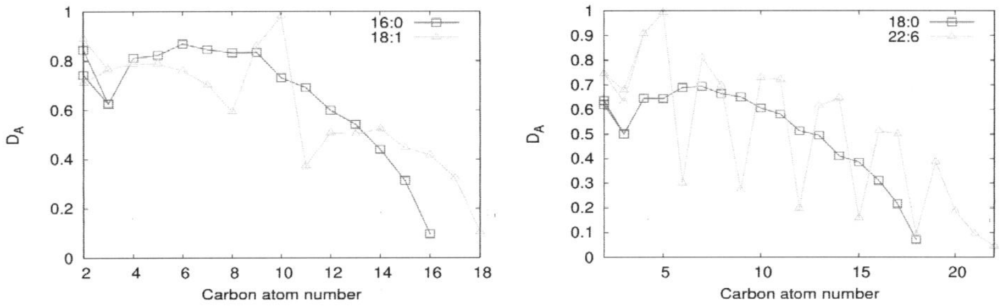


**Figure 5.** Anisotropy coefficients for C-H bonds of 16:0/18:1(n-9)cis PC bilayer (a) and of 18:0/22:6(n-3)cis PC bilayer (b)

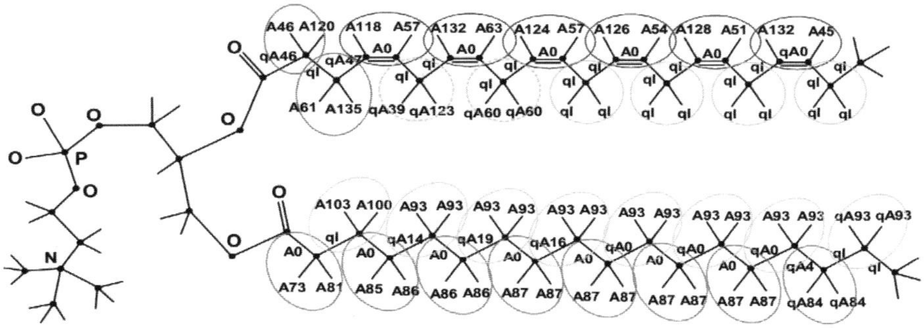


**Figure 6.** Characterization of orientational ordering of C-H and C-C bonds in a 18:0/22:6(n-3)cis PC bilayer. Lines of same colour show bonds with similar ordering behaviour.

The introduced anisotropy coefficient describes how well the bonds are orientationally ordered (relative to a bond with the maximum ordering) irrespectively to the ordering direction. The $D_A$ coefficients for C-H bonds of 16:0/18:1(n-9)cis PC and 18:0/22:6(n-3)cis PC bilayers are plotted in Fig 5.

The profiles of the anisotropy coefficient for unsaturated lipid chains in Fig. 5 are clearly different from the respective order parameter profiles in Fig 1 and 2. While C-H order parameters at carbons of double bonds are small, the anisotropy coefficients of the same bonds are rather high, indicating that the small C-H order parameter at the double bonds is essentially *orientational* and not *ordering* effect.

We can also (somewhat arbitrary) to call orientational distribution of a bond quasi-isotropic if its anisotropy coefficient $D_A < 0.25$ and denote by letter $qI$. Bonds with $0.25 < D_A < 0.368$ we shall call quasi-anisotropic ($qA$) and with $D_A > 0.368$ anisotropic ($A$). The number 0.368 in the above definition corresponds to the inverse value of $e$ number ($1/e \approx 0.368$). Quasi-anisotropic and anisotropic bonds have typically a preferred orientation (at which the distribution reaches maximum), therefore for such bonds we complement notation $qA$ or $A$ by an index expressing the preferred angle in degrees. An example of characterization of all C-H and C-C bonds in tails of 18:0/22:6(n-3)cis PC bilayer according to this scheme is given in Fig. 6. We beleive that such combined classification of bonds creates possibilities for deeper insight into ordering and orientational effects in the membrane interior.

## 4. Conclusions

We presented here results of molecular dynamics simulations of 16 lipid bilayers differing by degree of tail unsaturation and the number of carbons in the lipid tails. The analysis was concentrated on the orientational ordering behaviour of different lipids. Besides of computations of the conventional "order parameter" we suggested to characterize the bonds according to a newly introduced "anisotropy coefficient" computed from the angular distributions between the bond and the bilayer normal. We have demonstrated that such analysis allows one to get deeper insight into ordering and orientational behaviour of various bonds and to separate these two effects. In the forthcoming publications, we are planning to present a detailed analysis of orientational order of all 16 simulated bilayers according to the scheme presented here, and discuss possible biological implications.

## Acknowledgments

This work has been supported by Swedish Institute Visby programme (grant 00675/2009), Vetenskapsrådet (grant 621-2007-5256), Russian Foundation for Basic Research (grant 10-03-00201), by grant No. 1642.2012.4 for leading Russian research schools. The authors thank Swedish National Infrastructure for Computing (SNIC) for granting computer facilities.

## References

[1] Tinoco J 1982 *Progr. Lipid Res.* **21** 1 – 45
[2] Brenner R R 1984 *Progr. Lipid Res.* **23** 69 – 96
[3] Stubbs C D and Smith A D 1984 *Biochim. Biophys. Acta* **779** 89 – 137
[4] Bell M V, Henderson R J and Sargent J R 1986 *Comp. Biochem. Physiol.* **83B** 711 – 719
[5] Murphy D J 1986 *Biochim. Biophys. Acta* **864** 33 – 94
[6] Neuringer M, Anderson G J and Connor W E 1988 *Ann. Rev. Nutr.* **8** 517 – 541
[7] Quinn P J, Joo F and Vigh L 1989 *Progr. Biophys. & Molec. Biol.* **53** 71 – 103
[8] Mitchell D C, Gawrisch K, Litman B J and Jr N S 1998 *Biochemical Society Trans.* **26** 365 – 370
[9] Gawrisch K, Eldho N V and Holte L L 2003 *Lipids* **38** 445 – 452
[10] Stillwell W and Wassall S R 2003 *Chem. Phys. Lipids* **126** 1 – 27
[11] Rabinovich A L, Ripatti P O, Balabaev N K and Leermakers F A M 2003 *Phys. Rev. E* **67** 011909
[12] Valentine R C and Valentine D L 2004 *Prog. Lipid Res.* **43** 383 – 402
[13] Feller S E and Gawrisch K 2005 *Curr. Opin. Struct. Biol.* **15** 416 – 422
[14] Gawrisch K, Soubias O and Mihailescu M 2008 *Prostagl., Leukotr. & Essent. Fatty Acids* **79** 131 – 134
[15] Stillwell W 2008 *Chem. Phys. Lipids* **153** 1 – 2
[16] Wassall S R and Stillwell W 2009 *Biochim. Biophys. Acta* **1788** 24 – 32
[17] Feller S E and A D MacKerell J 2000 *J. Phys. Chem. B* **104** 7510 – 7515
[18] Högberg C J, Nikitin A M and Lyubartsev A P 2008 *J. Comp. Chem.* **29** 2359 – 2369
[19] Toukan K and Rahman A 1985 *Phys. Rev. B* **31** 2643 – 2648
[20] Tuckerman M, Berne B J and Martyna G J 1992 *J. Chem. Phys.* **97** 1990 – 2001
[21] Allen M P and Tildesley D J 1987 *Computer simulations of liquids* 2nd ed (Oxford: Clarendon)
[22] Fincham D 1994 *Mol. Simulations* **13** 1 – 9
[23] Martyna G J, Tuckerman M E, Tobias D J and Klein M L 1996 *Mol. Phys.* **87** 1117 – 1157
[24] Lyubartsev A P and Laaksonen A 2000 *Comput. Phys. Commun.* **128** 565 – 589
[25] Benz R W, Castro-Román F, Tobias D J and White S H 2005 *Biophys. J.* **88** 805 – 817
[26] Nagle J F and Tristram-Nagle S 2000 *Biochim. Biophys. Acta* **1469** 159 – 195
[27] Nagle J F and Tristram-Nagle S 2000 *Curr. Opin. Struct. Biol.* **10** 474 – 480
[28] Smaby J M, Momsen M M, Brockman H L and Brown R E 1997 *Biophys. J.* **73** 1492 – 1505
[29] Köenig B W, Strey H H and Gawrisch K 1997 *Biophys. J.* **73** 1954 – 1966
[30] Gawrisch K and Holte L L 1996 *Chem. Phys. Lipids* **81** 105 – 116
[31] Seelig J 1977 *Q. Rev. Biophys.* **10** 353 – 418
[32] Seelig A and Seelig J 1980 *Q. Rev. Biophys.* **13** 19 – 61
[33] Davis J H, Bloom M, Butler K W and Smith I C P 1980 *Biochim. Biophys. Acta* **597** 477 – 491
[34] van der Ploeg P and Berendsen H J C 1982 *J. Chem. Phys.* **76** 3271 – 3276
[35] Schmidt-Rohr K and Hong M 1996 *J. Phys. Chem.* **100** 3861 – 3867

# Intramolecular structures in a single copolymer chain consisting of flexible and semiflexible blocks: Monte Carlo simulation of a lattice model

**Julia A. Martemyanova, Victor A. Ivanov**
Faculty of Physics, Moscow State University, Moscow 119991, Russia

**Wolfgang Paul**
Institut für Physik, Martin-Luther-Universität Halle-Wittenberg, 06099 Halle (Saale), Germany

**Abstract.** We study conformational properties of a single multiblock copolymer chain consisting of flexible and semiflexible blocks. Monomer units of different blocks are equivalent in the sense of the volume interaction potential, but the intramolecular bending potential between successive bonds along the chain is different. We consider a single flexible-semiflexible regular multiblock copolymer chain with equal content of flexible and semiflexible units and vary the length of the blocks and the stiffness parameter. We perform flat histogram type Monte Carlo simulations based on the Wang-Landau approach and employ the bond fluctuation lattice model. We present here our data on different non-trivial globular morphologies which we have obtained in our model for different values of the block length and the stiffness parameter. We demonstrate that the collapse can occur in one or in two stages depending on the values of both these parameters and discuss the role of the inhomogeneity of intraglobular distributions of monomer units of both flexible and semiflexible blocks. For short block length and/or large stiffness the collapse occurs in two stages, because it goes through intermediate (meta-)stable structures, like a dumbbell shaped conformation. In such conformations the semiflexible blocks form a cylinder-like core, and the flexible blocks form two domains at both ends of such a cylinder. For long block length and/or small stiffness the collapse occurs in one stage, and in typical conformations the flexible blocks form a spherical core of a globule while the semiflexible blocks are located on the surface and wrap around this core.

## 1. Introduction

The collapse transition of a single flexible or semiflexible homopolymer macromolecule has been studied in detail by means of theoretical methods [1–9], computer simulations [10–18], as well as experiments [19–21]. Upon decreasing the solvent quality a single flexible macromolecule undergoes a transition into a structureless compact globule [22–24], while the stiff-chain (semiflexible) homopolymers collapse into structured globules, for example, into toroidal or cylindrical structures [6–8,14,23,25–27]. The complete diagram of states for a single semiflexible homopolymer macromolecule was investigated in Refs. [18,23,28,29]. The stability regions of the coil, the isotropic liquid globule, the crystalline isotropic globule, and the toroidal and cylindrical globule have been found. In these studies of homopolymer chains the stiffness along the chain was assumed to be homogeneous.

Such a model with uniform stiffness describes real macromolecules, but there are many examples where it is important to study the behavior of macromolecules which are heterogeneous in terms of rigidity. One such example are polymers for organic electronics [30]. Also, many biopolymers consist of flexible (coiled) and rigid parts (e.g., $\alpha$-helices, $\beta$-sheets) and form a stable spatial structure, which is also achievable kinetically [31]. Such conformations of polymers can be studied using a model with rigid and flexible chain parts. Previously, this model was used to simulate intramolecular structures in Refs. [33, 34], but in those studies the volume interaction parameter of flexible and rigid blocks was also different. However, a systematic study of the phase diagram of a polymer chain consisting of flexible and rigid blocks, which are identical in terms of volume interaction, still has to be performed. Such a study could show the general trends of the behavior of macromolecules with non-uniform stiffness distribution along the chain and possible structures which can be formed in various conditions. In this contribution we have started a systematic investigation of the behavior of the chain of length 64 monomer units consisting of rigid and flexible parts. We have performed Monte Carlo simulations on a cubic lattice using the bond fluctuation model. The length of the rigid and flexible blocks was equal to each other, and the composition was 1:1. We have studied macromolecules with block length from 4 to 32 monomer units and considered different values of stiffness of the rigid (semiflexible) block.

## 2. Model and Simulation Technique

In our simulation we have used the standard bond fluctuation model [35]. In this model each monomeric unit of a chain of length $N$ is represented as an elementary cube on the simple cubic lattice, blocking all 8 sites at the corners of this cube from further occupation. The lattice spacing $a \equiv 1$ is taken as the unit of length. The length of the bonds between monomeric units can take the values $2, \sqrt{5}, \sqrt{6}, 3$, and $\sqrt{10}$. The set of bonds consists of 108 different bond vectors, and 87 different angles between subsequent bonds can occur. We have studied a single copolymer chain composed of an alternating sequence of flexible (A) and semiflexible (B) blocks of monomer units. The total length of the chain was equal to 64 monomer units. The block length was varied from 4 to 32, i.e., half of the chain length (diblock copolymer). To model the chain stiffness, we introduced a bending potential depending on the angle $\theta$ between subsequent bond vectors $l_i$ and $l_{i+1}$,

$$U_{bending}(\theta) = \varepsilon_\alpha(\cos\theta + 1), \tag{1}$$

where $\varepsilon_\alpha = \varepsilon_{\alpha A} = 0$ for the A monomer units, and $\varepsilon_\alpha = \varepsilon_{\alpha B}$ for B monomer units. The stiffness parameter $\varepsilon_{\alpha B}$ is varied from 4 to 16 (in units of $k_B T$) with step 4 for long blocks in the copolymer.

To model the quality of the solvent, we introduce an attractive square well potential between all monomer units (A and B monomers are equivalent in the sense of the volume interaction potential) which are less than $\sqrt{6}$ lattice units apart in space:

$$U_{contact}(r) = \begin{cases} -\varepsilon, & r = 2, \sqrt{5}, \sqrt{6}, \\ 0, & \text{otherwise}, \end{cases} \tag{2}$$

where $\varepsilon \equiv 1$ in units of $k_B T$. This potential $U_{contact}$ was applied also between monomer units adjacent along the chain, and there was no potential on the bond length between these monomer units.

We have performed our simulations in acubic box of size $150 \times 150 \times 150$ with periodic boundary conditions. To change the conformation of the chain we used local moves of monomer units as well as rearrangements of large parts of the chain by means of the configurational bias (CB) algorithm [36]. Additionally, we have implemented the Wang-Landau (WL) sampling technique [37] to estimate the density of states $g(E)$. Then, for all moves, the acceptance probability $p_{acc}$ was equal to

$$p_{acc} = \min\{1, g(E_{old}/g(E_{new})\} \tag{3}$$

where $E_{old}$ and $E_{new}$ are the total energies of the old and the new conformations, respectively.

For CB moves, the following procedure was used. A monomeric unit $i^*$ and the part of the chain (from the first monomer unit of the chain until unit $i^*$ or from the unit $i^*$ until the end of the chain) was chosen at random, and the selected part of the chain was removed. For such a trial move, one starts from the end and counts the number $k_i^{(old)}$ of free (unoccupied) lattice sites where the considered monomer $i$ could potentially be located (respecting bond constraints). Then, the next monomer is treated analogously, and so forth, until $i^*$ is reached. Then, we attempt to regrow the removed part of the chain in a new conformation, choosing at every step of the growth attempt the position of the next monomeric unit $i$ at random out of the $k_i^{(new)}$ possible lattice sites that comply with excluded volume and bond length constraints. At the end of this regrowth trial, we calculate weights of the old and the new conformations

$$w_{new} = \prod_{i=i^*+1}^{N} k_i^{(new)} \qquad w_{old} = \prod_{i=i^*+1}^{N} k_i^{(old)}. \tag{4}$$

Finally, the trial move is accepted with probability

$$p_{acc}(CB) = \min\{1, [g(E_{old})/g(E_{new})](w_{new}/w_{old})\}. \tag{5}$$

We divide the whole interval of the total energy, $E$, in sub-intervals (windows) of width about 200. We start the simulation in each window with $g(E) \equiv 1$ for all values of $E$. We accumulate the histogram, $H(E)$, of visits to energy $E$ and simultaneously update the density of states $g(E)$ (multiplying it by a factor $f$) until the histogram $H(E)$ becomes flat within the pre-defined deviation (error bar), which is the so called flatness parameter. The length of such an iteration of the WL algorithm is not fixed. Then, we update the factor $f$ (we start with the initial value $f = e$ and take the square root of the modification factor $f$ after each iteration) and the flatness parameter (we start from the flatness parameter equal to 0.2 and then multiply it by 0.95 after every iteration of the WL algorithm) and put the histogram $H(E)$ to zero. We stop the iteration procedure after $f$ reaches the value 1.0000001 and perform the productive simulation run with the obtained density of states $g(E)$ in order to accumulate the histograms of observable parameters. Then, the density of states $g(E)$ is used to calculate the temperature dependencies of average values of the observable parameters, i.e., the gyration radius, the total energy and different energy terms, parameters of the globular shape and intraglobular orientational ordering of segments, etc., as well as their histograms in a wide region of temperatures using the following re-weighting formulas:

$$\langle E \rangle (T) = \frac{1}{Z(T)} \sum_E E g(E) e^{-E/k_B T} \tag{6}$$

$$Z(T) = \sum_E g(E) e^{-E/k_B T} \tag{7}$$

$$\left\langle R_G^2 \right\rangle (T) = \frac{1}{Z(T)} \sum_E \overline{R}_G^2(E) g(E) e^{-E/k_B T}, \tag{8}$$

where $\overline{R}_G^2(E) = \sum_{conf,E} R_G^2 / \sum_{conf,E} 1$, i.e., the average gyration radius over conformations with given value of the total energy $E$ calculated using the histogram file accumulated in the course of the productive simulation run.

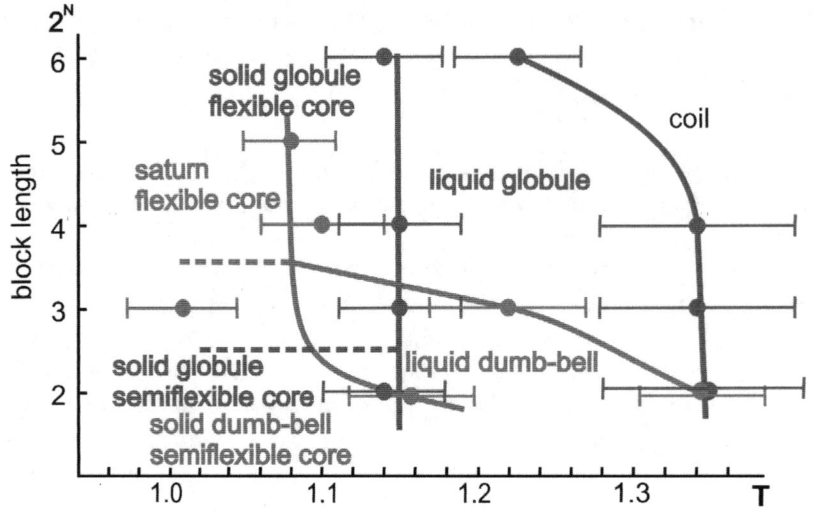


**Figure 1.** Diagram of states for the chain length 64 monomer units in varables block length (presented as power of 2) vs. temperature. Red colour is for $\varepsilon_{\alpha B} = 4$, blue color is for $\varepsilon_{\alpha B} = 8$.

The intraglobular orientational ordering of bond vectors can be characterized by orientational order parameters which are the eigenvalues $\eta_1$, $\eta_2$, $\eta_3$ of the tensor

$$Q_{\alpha\beta} = \frac{1}{N-1} \sum_{i=1}^{N-1} \frac{1}{2} \left( 3 \left\langle e_i^\alpha e_i^\beta \right\rangle - \delta_{\alpha\beta} \right), \tag{9}$$

where $e_i^\alpha$ is the $\alpha$-th component of the unit vector $\vec{e}_i \equiv \vec{l}_i / |\vec{l}_i|$ along the bond $\vec{l}_i$ connecting monomers $i$ and $i+1$.

For this model the phase diagram for flexible chains has been determined using the Wang-Landau simulation method [38], and the state diagram of semiflexible chains has been determined by means of expanded ensemble simulation techniques [23].

## 3. Results and discussion

We have obtained a preliminary diagram of states (fig. 1) for the flexible-semiflexible block copolymer chain in variables block length vs. temperature. The block length is presented as a power of 2. The points with error bars indicate the simulation data while the solid and dashed lines are guides for the eye. The diagram of states for $\varepsilon_{\alpha B} = 4$ is shown in red colour, and that one for $\varepsilon_{\alpha B} = 8$ is shown in blue colour. For the flexible homopolymer chain the temperature of the liquid-solid globule transition is equal to $T \approx 1.18$, as has been calculated in our previous work [23].

For the high temperature we observe, of course, coil conformations for both values of the stiffness parameter of semiflxible blocks. Upon decreasing temperature, the coil–globule transition occurs first, and it is followed by the transition between liquid globule and solid globule at lower temperatures. One can see the difference between the chains with different stiffness of semiflexible blocks ($\varepsilon_{\alpha B} = 4$ and $\varepsilon_{\alpha B} = 8$). The most obvious difference is that there is only a single transition between the coil and globule states for the chains with $\varepsilon_{\alpha B} = 8$ and large block length.

Now let us discuss the structure of the solid globule which appears to be quite different for the chains with short and long block length for both values of the stiffness parameter. For a more flexible chain (with $\varepsilon_{\alpha B} = 4$), the solid globule has approximately spherical shape but for short block length the globular core is formed by semiflexible blocks (while the flexible blocks form the shell) and for long block length (more than 8 monomer units in one block) the core is formed by flexible blocks (while the semiflexible blocks form the shell). In case of $\varepsilon_{\alpha B} = 8$, for the long block length only one transition occurs (like for the large stiffness of a hompolymer chain). There is a transition from coil to "saturn" like globule [39] with a flexible core (like for the case of $\varepsilon_{\alpha B} = 4$ and long blocks). With decreasing the block length we observed a two stage transition: first, the transition from coil to liquid dumbbell globule, and second, the transition to solid dumbbell globule which has a semiflexible core. The location of the transition between these two different types of intraglobular monomer packing (semiflexible core vs. flexible core) is indicated by the horizontal dashed line for both values of the stiffness parameter.

In fig. 2 we present some conformations for the chain of length 64 monomer units for different block length and different values of the stiffness parameter, and transition temperatures are indicated in this figure as well. Semiflexible blocks are shown using red colour, the flexible ones using green colour. Transition temperatures were defined from maxima in the temperature dependence of the full energy fluctuations. The temperature dependence of the contact energy looks very similar to that of the full energy, but the positions of the fluctuation maxima are slightly different. In fig. 2a) wee see a spherical globule with semiflexible core and in figures b) and c) the core is composed by flexible blocks of monomer units.

The criteria for determining the structure of conformations were the contact energy and the gyration radius (of the whole chain and of different blocks separately). In fig. 3 and fig. 4 the temperature dependence of the contact energy is shown. In this figure, the black lines are for the total contact energy, the red ones for contacts between monomer units of flexible blocks, the green ones for contacts between monomer units of semiflexible blocks and the blue ones for contacts between monomer units of flexible and semiflexible blocks (the estimated error in these curves is less than one percent). As one can see, with increasing the block length from 4 monomer units (fig. 3) to 8 monomer units (fig. 4) the red and green lines change their positions. For short blocks the monomer units of semiflexible blocks are more compact (have more contacts) while for longer blocks the flexible blocks are more compact (have more contacts). This confirms that the globule has a semiflexible core for short blocks and a flexible core for longer blocks. The same situation is observed for chains with $\varepsilon_{\alpha B} = 8$. For short blocks we find dumbbell conformations (fig. 2 d) and e)) where semiflexible blocks form a cylinder-like globule as a core, and flexible ones form two spherical globules at both ends of the cylinder-like core. For long blocks (fig. 2 f) and g)) we see "saturn"-like morphologies with a flexible core while the semiflexible blocks are located on the surface and wrap around this core.

The collapse occurs in two stages for small stiffness and for short blocks for larger stiffness. In fig. 5 wee see two maxima for the fluctuations of the total contact energy (black curve), indicating two stages of the collapse. The right maximum corresponds to the coil – liquid globule transition and the left one to the liquid – solid globule transition. Earlier, we have shown [23] that, similarly, for homopolymer macromolecules with large stiffness parameter the collapse transition occurs as a single transition while for smaller stiffness parameter it occurs in two stages.

## 4. Conclusions

We have studied conformational properties of a single multiblock copolymer chain consisting of flexible and semiflexible blocks and obtained different non-trivial globular morphologies for different values of the block length and the stiffness parameter. Our conlcusion is that the collapse can occur in one or in two stages depending on the values of both these parameters. For short block length and/or large stiffness the collapse occurs in two stages, because it goes through intermediate (meta-)stable structures, like a dumbbell shaped conformation. In such conformations the semiflexible blocks form a cylinder-like core, and the flexible blocks form two

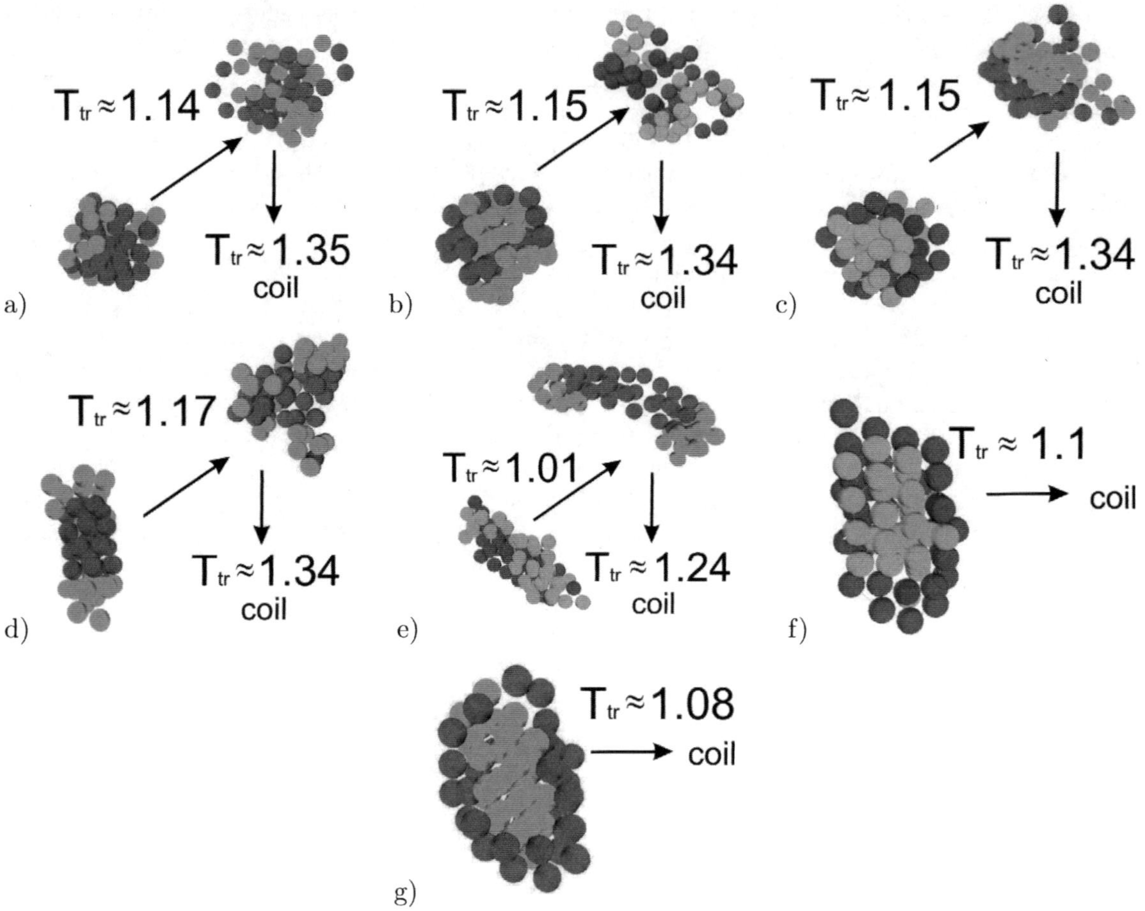


**Figure 2.** Conformations observed for the chain length 64 monomer units for semiflexible blocks (shown in red colour) with $\varepsilon_{\alpha B} = 4$: a) block length equal to 4 monomer units; b) block length equal to 8 monomer units; c) block length equal to 16 monomer units; for semiflexible blocks with $\varepsilon_{\alpha B} = 8$: d) block length equal to 4 monomer units; e) block length equal to 8 monomer units (dumbbell like conformation); f) block length equal to 16 monomer units ("saturn"-like conformation); g) block length equal to 32 monomer units ("saturn"-like conformation).

domains at both ends of such a cylinder. For long block length and/or small stiffness the collapse occurs in one stage, and in typical conformations the flexible blocks form the spherical core of a globule while the semiflexible blocks are located on the surface and wrap around this core ("saturn"-like conformations). Keeping the stiffness parameter of the semiflexible blocks to be constant and varying the block length we can observe the intraglobular structural transition from dumbbell to "saturn"-like conformations. To check whether these conclusions will be also valid for longer chains, we are now studying the chains consisting of 256 monomer units.

**Acknowledgments**
The financial support from DFG (project PA 473/10-1) and RFBR (grants 13-03-91334-NNIO-a and 12-03-31254-mol) is highly appreciated. We also thank the Supercomputer center of Moscow State University for providing computer time on "Chebyshev" supercomputer.

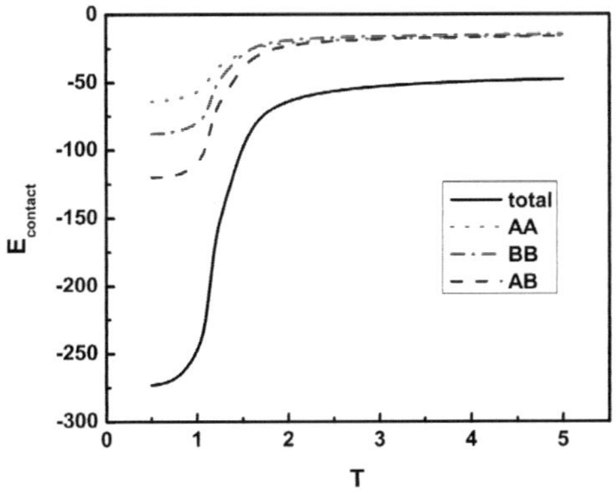


**Figure 3.** Contact energy (total contacts, AA-, BB- and AB-contacts) for the chain of length 64 monomer units consisting of alternating blocks of 4 monomer units, $\varepsilon_{\alpha B} = 4$.

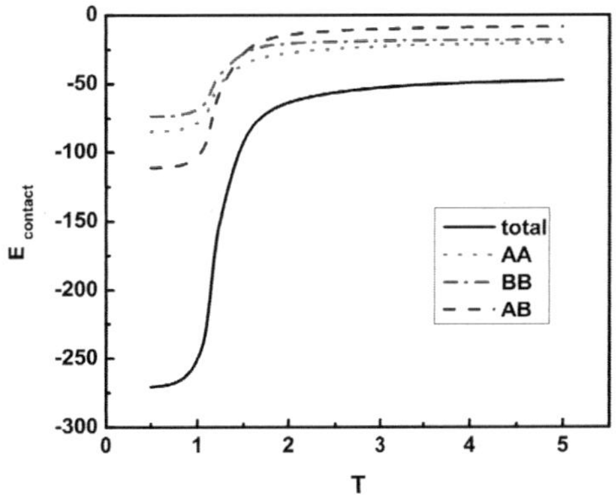


**Figure 4.** Contact energy (total contacts, AA-, BB- and AB-contacts) for the chain of length 64 monomer units consisting of alternating blocks of 8 monomer units, $\varepsilon_{\alpha B} = 4$.

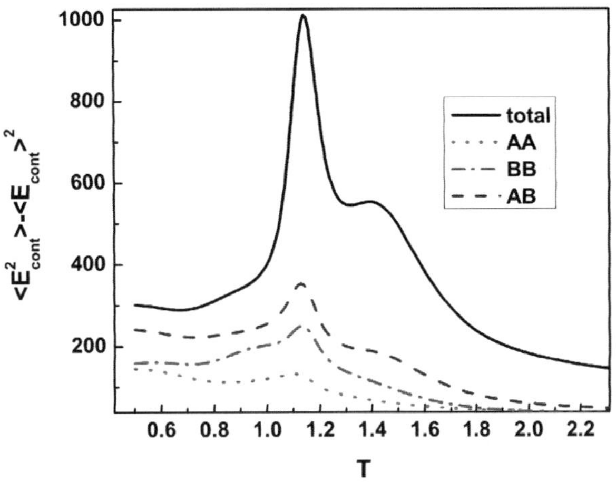


**Figure 5.** Fluctuations of contact energy (total contacts, AA-, BB- and AB-contacts) for the chain of 64 monomer units consisting of blocks of 4 monomer units, $\varepsilon_{\alpha B} = 4$.

## References

[1] Grosberg A Yu , Khokhlov A R 1997 *Statistical physics of macromolecules* AIP
[2] Lifshitz I M, Grosberg A Yu and Khokhlov A R 1978 *Rev. Mod. Phys.* **50** 683
[3] Grosberg A Yu, Kuznetsov D V 1992 *Macromolecules* **25** 1970; 1980; 1991; 1996
[4] Grosberg A Yu, Khokhlov A R 1981 *Adv. Polym. Sci.* **41** 53
[5] Grosberg A Yu, Zhestkov A V 1986 *J. Biomol. Struct. Dyn.* **3** 859
[6] Yoshikawa K, Takahashi M, Vasilevskaya V V, Khokhlov A R 1996 *Phys. Rev. Lett.* **76** 3029
[7] Vasilevskaya V V, Khokhlov A R, Kidoaki S, Yoshikawa K 1997 *Biopolymers* **41** 51
[8] Ubbink J, Odijk T 1995 *Biophys. J.* **68** 54
[9] Kuznetsov Yu A, Timoshenko E G, Dawson K A 1996 *J. Chem. Phys.* **105** 7116
[10] Elyashevich A M and Skvortsov A M 1971 *Mol. Biol.* **5** 204
[11] Birshtein T M, Sariban A A and Skvortsov A M, 1975 *Polym. Sci. U.S.S.R.* **17** 1962; 1976 **18** 1978; 1976 **18** 2734
[12] Birshtein T M, Skvortsov A M , Sariban A A 1977 *Macromolecules* **10** 202
[13] Khalatur P G 1979 *Polym. Sci. U.S.S.R.* **21** 2687; ibid. 1980 **22** 2050; ibid. 1980 **22** 2226
[14] Ivanov V A, Paul W, Binder K 1998 *J. Chem. Phys.* **109** 5659
[15] Rampf F, Paul W, Binder K 2005 *Europhys. Lett.* **70** 628; Rampf F, Binder K, Paul W 2006 *J. Polym. Sci. Part B: Polym. Phys.* **44** 2542
[16] Paul W, Rampf F, Strauch T, Binder K 2007 *Macromol. Symp.* **252** 1; Paul W, Strauch T, Rampf F, Binder K 2007 *Phys. Rev. E.* **75** 060801
[17] Siretskiy A, Elvingson C, Vorontsov - Velyaminov P, Khan M O 2011 *Phys. Rev. E* **84** 016702
[18] Ivanov V A, Stukan M R, Vasilevskaya V V, et al. 2000 *Macromol. Theory Simul.* **9** 488
[19] Wu C, Wang X 1998 *Phys. Rev. Lett.* **80** 4092
[20] Maurstad G, Stokke B T 2004 *Biopolymers* **74** 199
[21] Danielsen S, Varum K M, Stokke B T 2004 *Biomacromolecules* **5** 928
[22] Paul W, Mueller M 2001 *J. Chem. Phys.***115** 630
[23] Martemyanova J A, Stukan M R, Ivanov V A, Mller M, Paul W and Binder K 2005 *J. Chem. Phys.* **122** 174907
[24] Zhou Y, Karpulus M, Wichert J M, Hall C K 1997 *J. Chem. Phys.* **107** 10691
[25] Noguchi H, Yoshikawa K 1998 *J. Chem. Phys.* **109** 5070
[26] Cooke I R, Williams D R M 2004 *Physica A* **339** 45
[27] Sakaue T, Yoshikawa K 2002 *J. Chem. Phys.* **117** 6323
[28] Stukan M R, Ivanov V A , Grosberg A Yu, et al. 2003 *J. Chem. Phys.* **118** 3392
[29] Paul W, Mueller M, Binder K, et al. 2005 *In: Computer Simulations of Liquid Crystals and Polymers* ed by P Pasini, C Zannoni, S Zumer. (NATO Science Series II: Mathematics, Physics and Chemistry) **177** (Dordrecht/Boston/London: Kluwer Academic Publishers) p 171
[30] Topham P D, Parnell A J, Hiorns R C 2011 *J.Polym. Sci. Part B: Polym. Phys.* **49** 5636
[31] Petsko G A, Ringe D 2009 *Protein Structure and Function* (Oxford: Oxford University Press)
[32] Rothemund P W K 2006 *Nature* **440** 297
[33] Cooke I R, Williams D R M 2004 *Macromolecules* **37** 5778; Hernandez-Zapata E, Cooke I R, Williams D R M 2004 *Physica A* **339** 40
[34] Neratova I V, Komarov P V, Pavlov A S, Ivanov V A 2011 *Russ. Chem. Bull. Int. Ed.* **60** 229
[35] Carmesin I, Kremer K 1988 *Macromolecules* **21** 2819; Deutsch H - P, Binder K 1991 *J. Chem. Phys.* **94** 2294; Paul W, Binder K, Heermann D W and Kremer K 1991 *J. Phys. II* **1** 37; 1991 *J. Chem. Phys.* **95** 7726
[36] Smit B 1995 *Mol. Phys.* **85** 153; Wilding N B, Müller M, Binder K 1996 *J.Chem.Phys.* **105** 802
[37] Wang F, Landau D P 2001 *Phys. Rev. Lett.* **86** 2050; Wang F, Landau D P 2001 *Phys. Rev. E* **64** 056101
[38] Rampf F, Binder K and Paul W 2006 *J. Polym. Sci. Part B: Polym. Phys.* **44** 2542
[39] Yoshinaga N, Yoshikawa K 2007 *J. Chem. Phys.* **127** 044902

Journal of Physics: Conference Series **510** (2014) 012024 doi:10.1088/1742-6596/510/1/012024

# Monte Carlo simulation of critical properties of ultrathin anisotropic Heisenberg films

**M A Medvedeva**[1], **P V Prudnikov**[2]

Department of Theoretical Physics, Omsk State University, Omsk 644077, Russia

E-mail: [1]`mmed@mail.ru`, [2]`prudnikp@univer.omsk.su`

**Abstract.** The crossover from two-dimensional monolayer to three-dimensional system in ultrathin magnetic films is studied using a Monte Carlo technique. Ultrathin magnetic films is modeled as an anisotropic Heisenberg model. Finite-size scaling of the critical exponents as a function of film thickness is investigated. It was found that the magnetic behaviour changes from two-dimensional to three-dimensional universality class with increasing film thickness.

## 1. Introduction

The magnetic behaviour of ultrathin films has become of great technological importance due to the applications in magnetic storage devices [1]. Magnetic order in ultrathin ferromagnetic films is very complex due to a strong influence of the shape and the magnetocrystalline anisotropies of the sample. In the past 20 years, a considerable amount of experimental results on different aspects of magnetism in ultrathin films has appeared [2]. Nevertheless it is difficult to reach general conclusions even in seemingly basic things such as the kind of magnetic order at low temperatures. In view of this complexity, theoretical work on simplified models and computer simulations are essential for rationalizing and guiding new experimental work.

The dimensional crossover of magnetic properties from two-dimensional ($2D$) to three-dimensional ($3D$) character in magnetic multilayers has currently attracted much interest as a result of both technological and fundamental importance. Of particular interest is the critical behaviour of magnetic thin films for which the dimensionality $d$ is not well established. It is interesting to consider how magnetic properties such as the magnetization $m$, magnetic susceptibility $\chi$, and critical temperature $T_c$ depend on the thickness of the film.

Critical temperature $T_c$ in multilayered systems are known to change from $2D$ to $3D$ values with increasing numbers of layers. Magnetic films, however, should belong to a $2D$ universality class owing to the correlation length being constrained by the film thickness and allowed to expand only in the in-plane direction. This is not apparent from experimental studies of thin films of Ni [3] which provide evidence of a dimensional crossover of the critical exponents $\beta$ from $2D$ to $3D$.

For magnetic systems, the spin dimensionality as well as the spatial extension determines the universality class, giving rise to a myriad of ordering phenomena on different length scales. Furthermore, there are transition regions not represented by any universality class with corresponding critical exponents, but representing something in between. For example, the thickness dependence of the critical exponents of thin magnetic films exhibit such a transition, in which the exponents are continuously varying with the thickness of the layers, from typical

Journal of Physics: Conference Series **510** (2014) 012024 doi:10.1088/1742-6596/510/1/012024

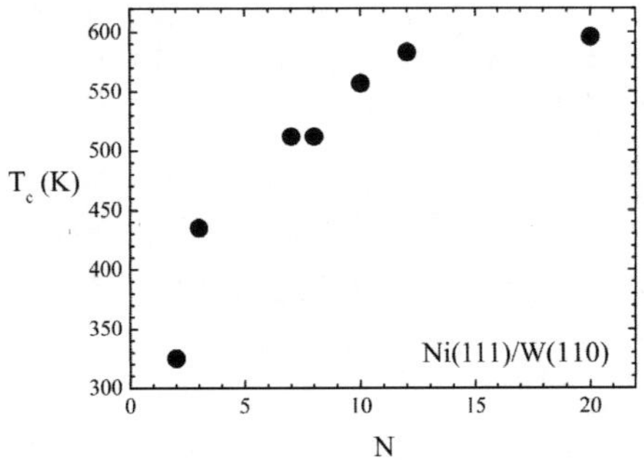


**Figure 1.** Dependence of critical temperature from film thickness [3]

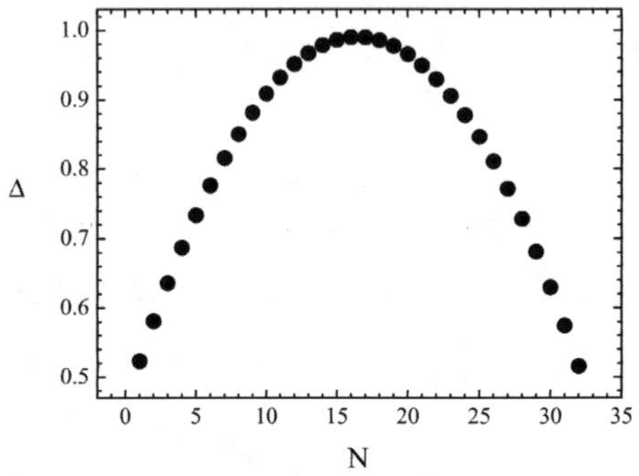


**Figure 2.** Dependence of anisotropy constant from film thickness

two-dimensional (2D) Ising ($\beta = 0.125$) to three-dimensional (3D) Heisenberg ($\beta = 0.364(4)$ [4]) behaviour.

## 2. Models and methods

We have performed Monte Carlo simulations on the anisotropic Heisenberg model [5] with Hamiltonian for ultrathin films on a square lattice of side $L$ as follows:

$$H = -J\sum_{ij}[(1-\Delta)(S_i^x S_i^x + S_i^y S_i^y) + S_i^z S_i^z], \tag{1}$$

where $\mathbf{S}_i = (S_i^x, S_i^y, S_i^z)$ is a unit vector in the direction of the classical magnetic moment at lattice site $i$, the sum is extended over nearest-neighbor pairs on the cubic lattice, $J > 0$ being the exchange constant, and $\Delta$ characterizes the amount of anisotropy. $\Delta = 0$ is the isotropic Heisenberg case, $\Delta = 1$ the Ising case. In order to study the critical properties of anisotropic Heisenberg magnets we thus made Monte Carlo calculations, studying cubic lattice with periodic and free boundary conditions are used for the in-plane and out-plane directions, respectively. The anisotropy constant for different sizes of the film was chosen from experimental studies of thin films of Ni(111)/W(110) [3]. The thickness dependence of critical temperatures presented in figure 1. The anisotropy constant was chosen proportional to the critical temperature for different film thicknesses. The resulting dependence was shown in figure 2.

The simulation are carried out for simple cubic films of size $N_s = L \times L \times N$ where $L \times L$ represents the number of sites (spins) in each layer of the film and $N$ is the number of layers. We consider temperature interval $T = 0.01 \div 5.01\ J/k_B$ with step $T_{\text{step}} = 0.02$. The spin configurations of the films are updated using the Swendsen-Wang cluster algorithm [6]. The spin system is divided into clusters. The bonds between sites in cluster are created with probability $1 - \exp\left[-2J(\mathbf{S}_i\mathbf{r})(\mathbf{S}_j\mathbf{r})/T\right]$ only if the condition $(\mathbf{S}_i\mathbf{r})(\mathbf{S}_j\mathbf{r}) > 0$ is true, where $\mathbf{r}$ is random unit vector. After that each cluster is flipped with probability 1/2. We always used a completely ordered ferromagnet as a starting configuration. We measured the total magnetization

$$m = \left\langle \frac{1}{N_s}\left[\left(\sum_i^{N_s} S_i^x\right)^2 + \left(\sum_i^{N_s} S_i^y\right)^2 + \left(\sum_i^{N_s} S_i^z\right)^2\right]^{1/2} \right\rangle, \tag{2}$$

out-plane magnetization

$$m_z = \left\langle \frac{1}{N_s}\sum_i^{N_s} S_i^z \right\rangle, \tag{3}$$

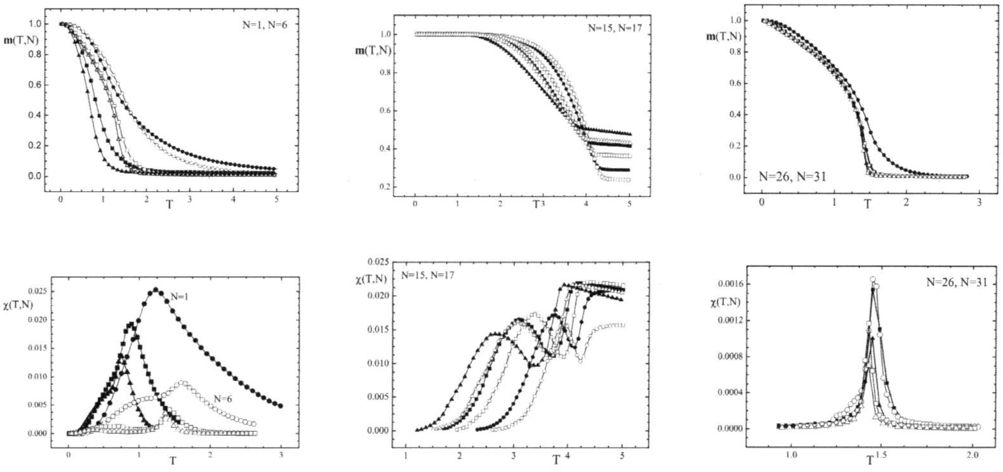


**Figure 3.** Temperature dependence of magnetization and susceptibility of thin films with different size $N = 1 \div 31$

the in-plane magnetization

$$m_{\parallel} = \left\langle \frac{1}{N_s} \left[ \left( \sum_i^{N_s} S_i^{\mathrm{x}} \right)^2 + \left( \sum_i^{N_s} S_i^{\mathrm{y}} \right)^2 \right]^{1/2} \right\rangle, \tag{4}$$

where angle brackets denote the statistical averaging. Temperature dependence of magnetization and susceptibility for different groups presented on figure 3. These curves were obtained by averaging over 3000 samples for each size of films. Temperature dependence of the susceptibility $\chi_m \sim [\langle m^2 \rangle] - [\langle m \rangle]^2$ was calculated for different lattice size in order to estimate the critical temperature $T_c$. The position of the susceptibility maximum allowed to estimate range of values of the critical temperature. Temperature dependencies of the forth order Binder cumulant were calculated in order to clarify the critical temperatures. The scaling dependence of the cumulant makes it possible to determine the critical temperature $T_c$ from the coordinate of the intersection points of the curves specifying the temperature dependence $U_4(L,T)$ for different $L$.

In early experiments on Fe/Cu(100) films [7, 8], it were observed a spin reorientation transition (SRT) from a region with perpendicular magnetization to one with in-plane magnetization. SRT transitions in films has actively been studied both theoretically [9, 10] and experimentally [11, 12], recently.

We measured orientational order parameter $O_z$ [13, 14]

$$O_\alpha = \left\langle \left| \frac{n_h^\alpha - n_v^\alpha}{n_h^\alpha + n_v^\alpha} \right| \right\rangle, \tag{5}$$

where angle brackets denote the statistical averaging, $\alpha \in \{x, y, z\}$, $n_h$ and $n_v$ are the number of horizontal and vertical pairs of nearest neighbor spins with antialigned perpendicular component,

$$\begin{aligned} n_h^\alpha &= \sum_{\mathbf{r}} \Big\{ 1 - \mathrm{sgn}\,[S^\alpha(r_x, r_y), S^\alpha(r_x + 1, r_y)] \Big\}, \\ n_v^\alpha &= \sum_{\mathbf{r}} \Big\{ 1 - \mathrm{sgn}\,[S^\alpha(r_x, r_y), S^\alpha(r_x, r_y + 1)] \Big\}. \end{aligned} \tag{6}$$

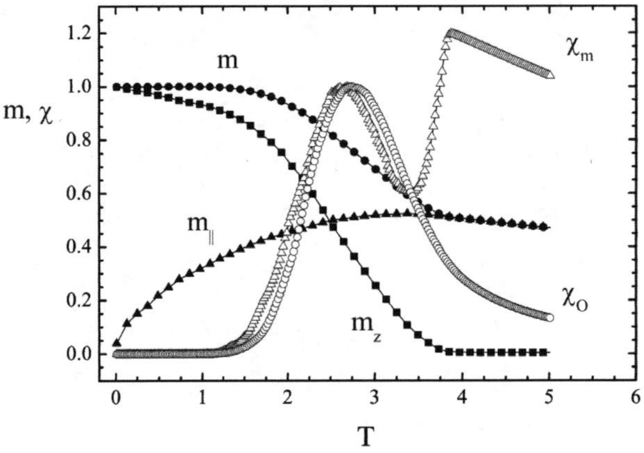


**Figure 4.** Temperature dependence of $m(T)$, $m_z(T)$, $m_{||}(T)$ and $\chi_m(T)$, $\chi_O(T)$ for films with size $N = 15$ $L = 64$

It was observed a transition from an out-of-plane ordering at low temperatures to an in-plane configuration as described by the magnetization behaviour shown in figure 4. We show $m(T)$, $m_{||}(T)$, $m_z(T)$ and susceptibility $\chi_m(T)$, $\chi_O(T) \sim [\langle O_z^2 \rangle] - [\langle O_z \rangle]^2$ in the same figure for comparison. The susceptibility $\chi_m(T)$ curve presents two peaks see figure 4. The peak at low temperature is pronounced and is centered in the temperature in which occurs the rapid decrease of the out-of-plane magnetization $T = 2.71$. The susceptibility $\chi_O(T)$ has a maximum at the same temperature, that is why first peak is corresponds to spin reorientation transition from ferromagnetic phase to planar ferromagnetic. The second peak appears at $T = 3.91$ and it corresponds to the second order phase transition from a ferromagnetic state to a paramagnetic. In the temperature $T = 1.75$ the out-of-plane magnetization starts to decrease and becomes zero

**Table 1.** Dependence of critical temperature $T_c(N)$ from film thickness

| $N$ | $SRT$ | $FM-PM$ | $N$ | $SRT$ | $FM-PM$ |
|---|---|---|---|---|---|
| 2 | - | 1.03 | 13 | 1.87 | 1.87 |
| 3 | - | 1.15 | 14 | 2.45 | 3.41 |
| 4 | - | 1.25 | 15 | 2.71 | 3.91 |
| 5 | - | 1.31 | 16 | 2.89 | 4.15 |
| 6 | - | 1.35 | 17 | 2.87 | 4.09 |
| 7 | - | 1.39 | 21 | 1.17 | 1.61 |
| 8 | 0 | 1.45 | 22 | 0.77 | 1.55 |
| 9 | 0.65 | 1.49 | 23 | 0 | 1.46 |
| 10 | 0.91 | 1.57 | 26 | - | 1.43 |
| 11 | 1.13 | 1.61 | 30 | - | 1.43 |
| 12 | 1.43 | 1.63 | 31 | - | 1.43 |

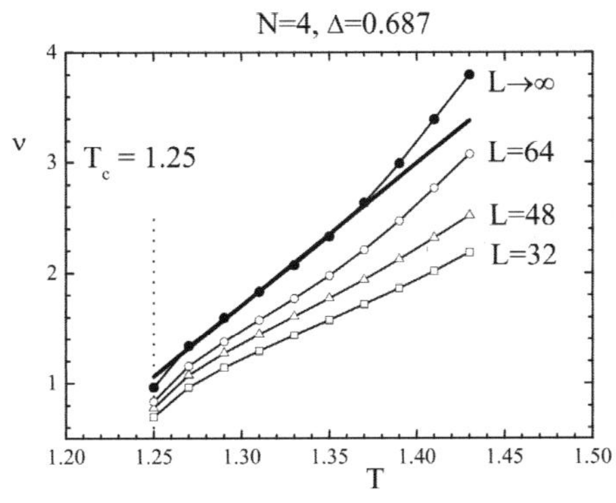


**Figure 5.** Temperature dependence critical exponents $\nu(T, L)$

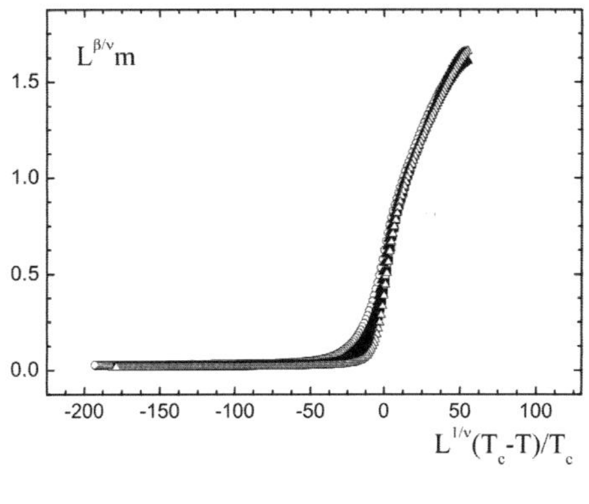


**Figure 6.** Dependence $f(m) = L^{\beta/\nu}m$ as a function of $L^{1/\nu}(T_c - T)/T_c$

at $T = 3.89$. In this temperature interval the state with the biased magnetization is possible too.

The thickness dependence of critical temperature is presented in table 1 for film with thickness from $N = 2$ to $N = 31$. The FM-PM – is the second order phase transition from ferromagnetic phase to paramagnetic, SRT – spin reorientation transition from ferromagnetic phase to planar ferromagnetic. In the experimental [15] and theoretical [9] studies devoted to the study of single-layer magnetic materials, it is predicted that the orientation of the spin transition is a weak first-order transition.

We consider finite size scaling form for film geometry [16] to find how $m$ and $\chi$ scale with the size $L$ and thickness $N$ of the systems and use this to extract the effective critical exponents from our results. The basic finite-size scaling ansatz [17] rests on an assumption that only a single correlation length $\xi$ is needed to describe the critical properties of thin films. Hence the empirical scaling forms for $m$ and $\chi$ at some fixed $N$'s can be written as

$$\begin{aligned} \langle m(T,N)\rangle &= L^{-\beta/\nu}\tilde{m}(L^{1/\nu}\tau, N) \\ \chi(T,N) &= L^{\gamma/\nu}\tilde{\chi}(L^{1/\nu}\tau, N) \end{aligned} \quad (7)$$

where $\gamma$, $\beta$, and $\nu$ are the effective critical exponents associated with $\chi$, $m$, and $\xi$, respectively. For monolayer system with $N = 1$, the effective exponents are the critical exponents for the $2D$ system. The functions $\tilde{\chi}$ and $\tilde{m}$ are scaling functions for a given $N$ and $\tau = T/T_c - 1$ is the reduced temperature. We consider the $f(m) = L^{\beta/\nu}m$ as a scaling function of $L^{1/\nu}(T_c - T)/T_c$. We calculate these scaling functions for thin film with thickness from 2 to 5 layers with $L = 32$, 48, 64. These scaling functions for a range of $L$ collapsed onto a single curve with the correct critical temperature and effective critical exponents. The results are shown in figure 6.

The effective exponent $1/\nu$ can be extracted from the derivative of the cumulant $U_4$ with respect to $L$ at $T_c$ owing to its variation with system size as $L^{1/\nu}$. The scaling dependence of the cumulant allowed to calculate critical exponents $\nu$.

$$\frac{dU_4}{dT} \sim L^{1/\nu} \quad (8)$$

We calculated the critical exponent for different temperatures above critical temperature and for different linear sizes of lattice (figure 5). The effective value of $\nu$ can be obtained only in

**Table 2.** Critical exponents $\nu$, $\beta$, $\gamma$ and effective dimension $d_{\text{eff}}$

| $N$ | $\nu$ | $\beta$ | $\gamma$ | $d_{\text{eff}}$ |
|---|---|---|---|---|
| 2 | 1.021(18) | 0.126(8) | 1.816(69) | 2.046(59) |
| 3 | 1.011(27) | 0.128(8) | 1.770(94) | 2.007(125) |
| 4 | 1.018(14) | 0.126(9) | 1.713(112) | 2.007(101) |
| 5 | 0.972(59) | 0.129(9) | 1.609(150) | 1.992(98) |

thermodynamic limit. These values were calculated using linear approximation of linear size $L$ to infinity and linear approximation of temperature to critical [18]. The exponents for films with thickness from $N = 2$ to $N = 6$ are $\nu(N = 2) = 1.021(18)$, $\nu(N = 3) = 1.011(27)$, $\nu(N = 4) = 1.018(14)$, $\nu(N = 5) = 0.972(59)$, $\nu(N = 6) = 0.974(62)$. These values are very close to the 2D Ising values $\nu(\text{2D Ising}) = 1$.

Note that if equation (7) correctly encapsulates the nature of magnetic critical behaviour in films, we can extract the effective exponents $\beta/\nu$ and $\gamma/\nu$ from the slopes of the log-log plots of $m$ or $\chi$ against $L$ at $T_c$. Based on the hyperscaling relation $\gamma/\nu + 2\beta/\nu = d$, it is possible to consider the effective dimensionality $d_{\text{eff}}$. For $N/L \ll 1$, $2D$-like behaviour is expected; i.e., $d_{\text{eff}}$ should stay close to 2.

We perform the calculation of the effective dimension $d_{\text{eff}} = \gamma/\nu + 2\beta/\nu$. Using expressions

$$m \sim L^{-\beta/\nu}, \qquad \chi \sim L^{\gamma/\nu}, \tag{9}$$

we have found that, in case $N = 2-5$, $d_{\text{eff}}$ has a value of 2 within error bars as shown in table 2 as expected. This confirms the $2D$ universality in thin films and ensures the possibility of using equation (7) to describe the critical behaviour of the thin films.

The exponents $\beta$ can be obtained from the temperature dependence of the magnetization near the critical point $m \sim (T_c - T)^\beta$ for different thickness of the magnetic film (table 3). The crossover from two-dimensional Ising model to three-dimensional Heisenberg model with increasing film thickness was found from this dependence.

Anisotropy constant $\Delta$ is close to 1 for thin films with thickness from 15 to 17. This case corresponds to Ising model. For these films we can select two temperature intervals corresponding to the $2D$ and $3D$ Ising critical behaviour. In first temperature interval near the critical point exponents $\beta(L = 48) = 0.123(7)$, $\beta(L = 64) = 0.124(5)$ for $N = 15$ (figure

**Table 3.** Critical exponents $\beta$ of magnetization for different film thickness

| $N$ | $\beta$ | $N$ | $\beta$ | $N$ | $\beta$ |
|---|---|---|---|---|---|
| 2 | 0.126(9) | 9 | 0.294(14) | 18 | 0.338(2) |
| 3 | 0.128(8) | 10 | 0.310(13) | 21 | 0.344(2) |
| 4 | 0.126(9) | 12 | 0.299(11) | 22 | 0.352(6) |
| 5 | 0.129(9) | 13 | 0.314(5) | 25 | 0.343(4) |
| 6 | 0.170(11) | 15 | 0.324(7) | 26 | 0.358(1) |
| 7 | 0.184(5) | 16 | 0.329(8) | 30 | 0.370(2) |
| 8 | 0.286(11) | 17 | 0.343(10) | 31 | 0.368(2) |

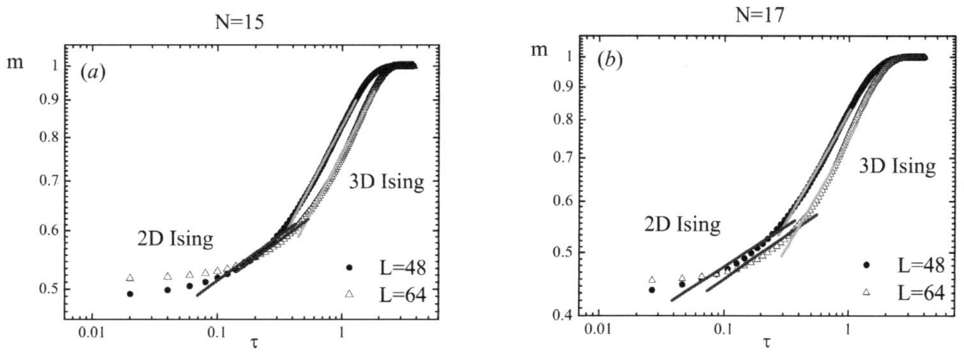


**Figure 7.** Dependence of magnetization from reduced temperature $\tau = (T_c - T)/T_c$

7($a$)) and $\beta(L = 48) = 0.126(8)$, $\beta(L = 64) = 0.127(7)$ for $N = 17$ (figure 7($b$)) are close to $2D$ Ising value of exponent $\beta = 0.125$. Critical exponents are $\beta(L = 48) = 0.316(2)$, $\beta(L = 64) = 0.319(3)$ for $N = 15$ and $\beta(L = 48) = 0.336(2)$, $\beta(L = 64) = 0.337(3)$ for $N = 17$ in the second temperature interval. This corresponds to $3D$ Ising critical behaviour $\beta = 0.325$[19]. This is because the correlation length is increased in all directions when approaching the critical temperature and demonstrate the $3D$ critical behaviour, reaching a boundary film continues to increase only in-plane direction of the film and shows critical behaviour of $2D$ system.

## 3. Conclusion

We have studied the magnetic behaviour of Heisenberg thin films in simple cubic structures with periodic and free boundary conditions for the in-plane and out-plane directions, respectively using extensive Monte Carlo simulations.

For thin films with $N = 9..22$ spin-reorientation transition was discovered. For these films the temperature of SRT have been determine from peak of susceptibility $\chi_O(T)$. In these system we consider the SRT from out-plane to in-plane ferromagnetic phase and second order phase transition from in-plane ferromagnetic to paramagnetic phase.

We have found the dimensional crossover of $m$ from $2D$ to $3D$ like with increasing film thickness. We have examined the critical regime of these systems in detail and extracted effective critical exponents based on a finite-size scaling method for films. For thin films with $N \leqslant 5$ the exponents are essentially the same as $2D$ and from this it can be implied that thin films fall into the $2D$ class. For thickness $N \geqslant 15$ the exponent $\beta$ move to $3D$ values as expected. The thickness dependence of critical exponents demonstrate the crossover from $2D$ Ising to $3D$ Ising and then to $3D$ Heisenberg behaviour.

## Acknowledgments

The reported study was supported by grant MU-2/2013 of the Omsk State University for young scientists and by the Supercomputing Center of Lomonosov Moscow State University and Joint Supercomputer Center of the Russian Academy of Sciences.

## References

[1] Chappert C, Fert A, F Nguyen van Dau 2007 *Nature Mater* **6** 813
[2] Vaz C A F, Bland J A C, Lauhoff G 2008 *Rep. Prog. Phys.* **71** 056501
[3] Li Y, Baberschke K 1992 *Phys. Rev. Lett.* **68** 1208

[4] Chen K, Ferrenberg A M, Landau D P 1993 *Phys. Rev.* B **48** 3249
[5] Binder K, Landau D P 1976 *Phys. Rev.* B **13** N 3 1140
[6] Swendsen R H, Wang J S 1987 *Phys. Rev. Lett.* **58** 86
[7] Pappas D P, Kamper K P, Hopster H 1990 *Phys. Rev. Lett.* **64** 3179
[8] Allenspach R, Bischof A 1992 *Phys. Rev. Lett.* **69** 3385
[9] Carubelli M, Billoni O V, Pighín S A, *et al.* 2008 *Phys. Rev.* B **77** 134417
[10] Rogelio Díaz-Méndez, Mulet R 2012 *J. Magn. Magn. Mater.* **324** 128
[11] Cai W, Muraishi S, Shi J, Nakamura Y, Liu W, Yu R 2012 *Appl. Phys.* A **109** 69
[12] Bordel C, Juraszek J, Cooke D W, *et al.* 2012 *Phys. Rev. Lett.* **109** 117201
[13] Booth I, MacIsaac A B, Whitehead J P, Dé Bell K 1995 *Phys. Rev. Lett.* **75** 950
[14] Ambrose M C, Stamps R L 2013 *Phys. Rev.* B **87** 184417
[15] MacIsaak A B, Dé Bell K, Whitehead J P 1998 *Phys. Rev. Lett.* **80** 616
[16] Dillmann O, Janke W, Muller M and Binder K 2001 *J. Chem. Phys.* **114** 5853
[17] Privman V 1990 *Finite Size Scaling and Numerical Simulation of Statistical Systems*, ed Privman V (Singapore: World Scientific)
[18] Prudnikov P V, Medvedeva M A 2012 *Prog. Theor. Phys.* **127** N 3 369
[19] Prudnikov V V, Prudnikov P V, Krinitsyn A S, *et al.* 2010 *Phys. Rev.* E **81** 011130

# The Fluctuation-Dissipation Theorem of Colloidal Particle's energy on 2D Periodic Substrates: A Monte Carlo Study of thermal noise-like fluctuation and diffusion like Brownian motion

**Amin Najafi**

*Department of physics, Islamic Azad University, (Hamedan Branch) Hamedan, 734, Iran*

Najafi.physic@gmail.com

**Abstract**. Using the Monte Carlo simulations, we have calculated mean-square fluctuations in statistical mechanics, such as those for colloids energy configuration are set on square 2D periodic substrates interacting via a long range screened Coulomb potential on any specific and fixed substrate. Random fluctuations with small deviations from the state of thermodynamic equilibrium arise from the granular structure of them and appear as thermal diffusion with Gaussian distribution structure as well. The variations are showing linear form of the Fluctuation-Dissipation Theorem on the energy of particles constitutive a canonical ensemble with continuous diffusion process of colloidal particle systems. The noise-like variation of the energy per particle and the order parameter versus the Brownian displacement of sum of large number of random steps of particles at low temperatures phase are presenting a markovian process on colloidal particles configuration, too.

## 1. Introduction

More recently, colloidal assemblies interacting with 2D periodic substrates have been investigated [1,2, 3, 5, 6,7]. Simulations of colloids on 2D periodic substrates have considered the case where the number of colloids is an integer multiple of the number of potential minima [2] There are also a two-stage melting transition for the trimer state at a filling of three colloids per minima for the square substrate [2]. Colloidal particle assemblies in two dimensions (2D) are an ideal system to study ordering and melting, since quantities such as diffusion, dislocation dynamics, and local ordering can be directly observed, in which is typically not the case for atomic and molecular systems [4]. More recently, colloidal assemblies interacting with 2D periodic substrates have been investigated as well [2, 3, 5, 6].

## 2. The correlation function and the fluctuation-Dissipation Theorem

The correlation function [7],[8] contains important information about a phase transition. There is an order-parameter density $m(r)$, such that the order parameter can be written as:

$$M = \left\langle \int d^3r\, m(r) \right\rangle \qquad (1)$$

Where <> denotes ensemble average. The correlation function is defined as:

$$\Gamma(r) = \langle m(r)m(0)\rangle - \langle m(r)\rangle\langle m(0)\rangle \qquad (2)$$

We can rewrite the order parameter more explicitly in the form:

$$\frac{M}{V} = \frac{1}{V}\int d^3r \frac{Tr[m(r)e^{-H/kT}]}{Tre^{-H/kT}} = \frac{Tr[m(0)e^{-H/kT}]}{Tre^{-H/kT}} \qquad (3)$$

The last relation is obtained under the assumption that the system is translationally invariant. Assume that the magnetic field H is coupled linearly to the order parameter:

$$H = H_0 - H\int d^3r\, m(r) \tag{4}$$

Where $H_0$ is the Hamiltonian for H=0, Differentiating (3) with respect to H, we obtain:

$$\chi = \frac{1}{kT}\int d^3r\, [\langle m(r)m(0)\rangle - \langle m(0)\rangle^2] \tag{5}$$

Comparison with the definition of the correlation function gives:

$$\chi = \frac{1}{kT}\int d^3r\Gamma(r) \tag{6}$$

By Monte Carlo simulations, we have calculated mean-square fluctuations for colloids energy configuration are set on square 2D periodic substrates interacting via a long range screened Coulomb potential on any specific and fixed substrate. The correlation function for ensemble average of colloids energy is defined as:

$$\Gamma(r) = \langle E(r)E(0)\rangle - \langle E(r)\rangle\langle E(0)\rangle \tag{7}$$

And then the heat capacity is defined as:

$$Cv = \frac{1}{kT}\int d^3r\, [\langle E(r)E(0)\rangle - \langle E(0)\rangle^2] \tag{8}$$

and comparison with definition of the correlation function gives:

$$Cv = \frac{1}{kT}\int d^3r\Gamma(r) \tag{9}$$

That is a special case of fluctuation- dissipation theorem with relation between the heat capacity and the energy fluctuations of colloids interacting via a long range screened Coulomb potential on square 2D periodic substrates. Here, we have calculated this theorem for any specific and fixed substrate of colloidal configurations separately.

## 3. Simulation of diffusion of colloidal particle and results

### *3.1. Model of simulation*

Colloidal ordering and melting on 2D periodic square and triangular substrates using Langevin simulations have been studied by C. Reichhardt [1][11][12].They also studied colloidal ordering and disordering on two-dimensional periodic substrates where the number of colloids per substrate minima is two or three by Langevin simulations [11,12].Here we use the Monte Carlo simulation and metropolis algorithm for survey of particles' motion and obtain the statistical quantities and colloidal particle configuration. $10^6$ Loops of Monte Carlo are used in this experiment as the number of particles's motion. We consider a 2D system of N=150 colloidal particles interacting with a repulsive Yukawa potential in the presence of a substrate. We perform a Monte Carlo simulation of a 2D system with periodic boundary conditions in the x and y directions. There are N colloids interacting with a 2D periodic substrate which has $N_{min}$ ,minima. We focus on the case $\frac{N}{N_{min}} = 1$.

The overdamped equation of motion for a single colloid i is:

$$\frac{dr_i}{dt} = f_i + f_s + f_T, \quad (10)$$

Where,

$$f_i = \sum_{j \neq i}^{N} \nabla_i V(r_{ij}) \quad (11)$$

Is the interaction force from the other colloids, which we take to be a Yukawa or screened Coulomb form:

$$V(r_{ij}) = (Q^2 / |r_i - r_j|)\exp(-\kappa |r_i - r_j|) \quad (12)$$

The colloid-colloid interaction is a long range Yukawa or screened Coulomb potential. Here Q is the charge of the particles, which we set to 1.0, $1/\kappa$ is the screening length, and $r_{ij}$ is the position of particle ( i, j). The system length is measured in units of the lattice constant $a_0$ and we take the screening length $\frac{1}{\kappa} = \frac{a_0}{10}$ .We consider a square periodic

$$V_s = \frac{-a_0 A}{2\pi}[\cos(\frac{2\pi x}{a_0})\hat{x} + \cos(\frac{2\pi y}{a_0})\hat{y}] \quad (13)$$

substrate with a lattice constant $a_0$ and strength A.

In this work we keep $a_0$ fixed and vary A. The thermal force $f_T$ is a randomly fluctuating force from random kicks. The results of our simulation has been obtained on the (table1) and (table2) and 'figures1', 'figure2','figure3','figure4' and 'figure5' as well.

### *3.2. Simulation of diffusion of colloidal particle configuration*

In the project we verify the diffusion law in Brownian motion of colloidal particle interacting Yukawa potential on 2D periodic substrate as well. 'figure 1a','figure1b','figure1c' and 'figure1d' show different stages of the stochastic diffusion process in low& (critical)&high phases for substrate strength(A=1) with topological defects[13] such as dislocation and declination[14,15,16,17,18].In 'figures2',the Specific heat of the colloidal particle configurations vs T has been done in low temperatures phase. The variation versus temperature is performed for substrate strength (A=3) as well. The specific heat has two Gaussian-like peaks at low critical temperature phase. Two successive peaks on the variation of heat capacity of colloids strongly indicate to two consecutive stage of diffusion in low critical temperatures phase. Actually curves of 'figures2','figure3'represent data from 'figures1' with Gaussian-like distribution. the Brownian displacement of diffusion colloids in 'figures2' and 'figure3', in which is a sum of a large number of random steps, has a Gaussian distribution. Obtained results in (table1), confirm the diffusion law and the fluctuation-Dissipation theorem for colloidal particle's energy interacting with Yukawa potential on 2D periodic substrate numerically too. In this here, we have obtained results of diffusion of colloidal particle for different phase consist of low temperature phases, critical phase and high temperature phase for different potential strength of colloidal particles configuration. Two-step process of continuous diffusion occurs at critical temperatures phase: 'figures1b','figure1c'. In

Journal of Physics: Conference Series **510** (2014) 012025 doi:10.1088/1742-6596/510/1/012025

very low temperatures phase particles are completely positionally and orientationally ordered: 'figure1a'. After a two-step process of continuous diffusion of colloids during intermediate critical temperature phase'figure1a''figure1b', configurations of particles are completely disordered in high temperatures phase: 'figure1d'.

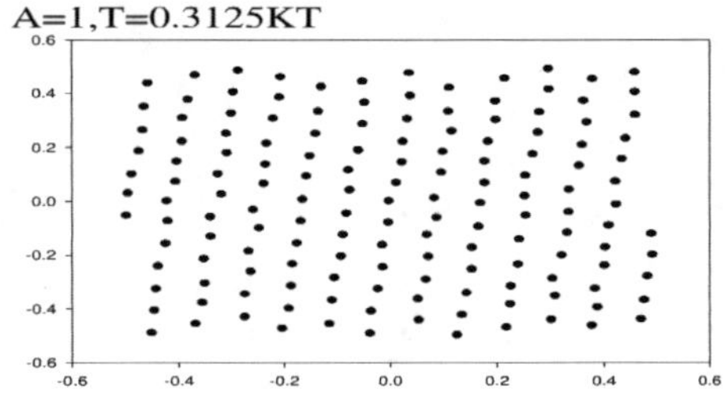


(1a): Very low temperature phase(ordered phase for A=1)

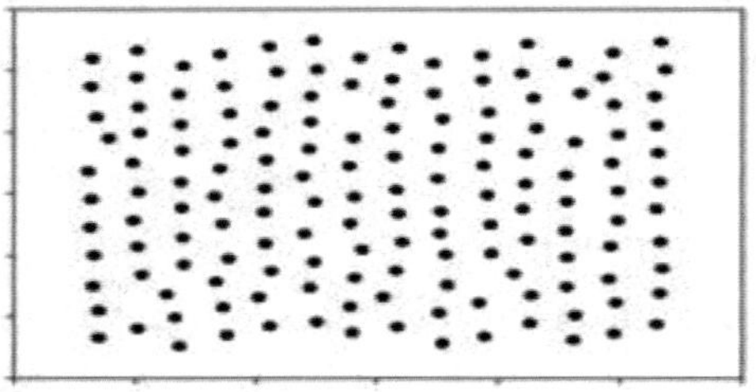

(1b):First diffusion process: first intermediate(critical) phase, (A=1 )

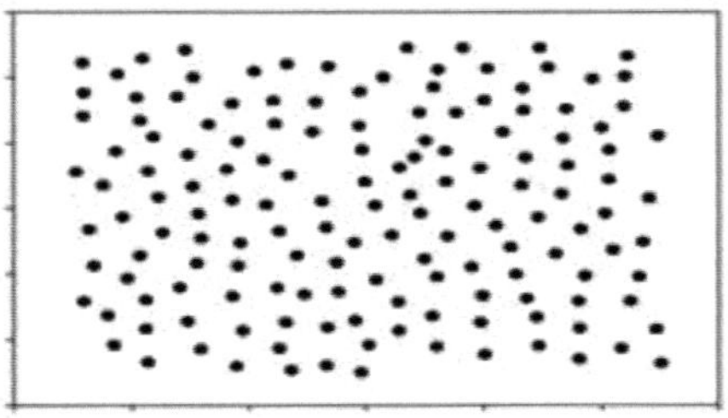

(1c):Second diffusion :second intermediate(critical) phase(A=1))

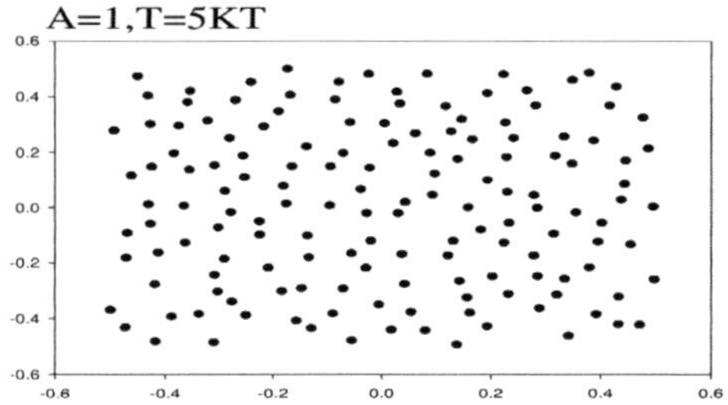


(1d):High temperature phase: colloids are diffused)

**Figure1.** Continuous diffusion process of the colloid configurations (black dots) for a square 2D periodic substrate with A = 1,N=150(The number of colloids). One colloid per potential minima with the colloids forming a square commensurate lattice, $N = N_{\min}$ : (1a) at very low temperature phase(ordered phase)(T=0.3125KT) ( $\beta = \frac{1}{T}$ with $(K = 1)$).(1b) First intermediate critical phase (low temperature), where first diffusion occurs(T~0.91KT)positional order is lost but orientational is preserved.(1c) Second intermediate critical phase (low temperature), where second diffusion occurs(T~0.96KT) orientational order is lost ,too.(1d)high temperature phase: T=5KT,colloids are diffused completely.

## 4. The numerically result of the Fluctuation-Dissipation Theorem of Colloidal Particle's energy

Thermodynamic quantities are supposed to be constant when the system is in thermal equilibrium. If we measure them with high precision, however, we will notice that they undergo small fluctuations. For example, the pressure a gas exerts on a wall fluctuates because of the randomness of atomic impacts. The internal energy of the gas fluctuates because it exchanges energy with the environment via atomic collisions. These fluctuations arise from the granular structure of matter and appear as thermal noise. We have calculated the mean-square fluctuations in statistical mechanics, such as those for the system's energy. We can usually ignore them because they are vanishingly small in the thermodynamic limit. We surveyed the relation of fluctuation-dissipation of energy to form:

$$\beta(\overline{E^2} - \overline{E}^2) = TC_V \qquad (14)$$

We'll investigate the actual range of the system's energy, in which is likely to be diffusion. In order to, we must write the average energy of canonical ensembles as follow:

$$U \equiv \langle E \rangle = \frac{\sum_r E_r \exp(-\beta E_r)}{\sum_r \exp(-\beta E_r)} \quad (15)$$

By assume the energy $E_r$ as the constant quantity; we will take the derivative U with respect to $\beta$

$$\text{As result: } \frac{\delta U}{\delta \beta} = -\frac{\sum_r E^2{}_r \exp(-\beta E_r)}{\sum_r \exp(-\beta E_r)} + \frac{\left[\sum_r E_r \exp(-\beta E_r)\right]^2}{\left[\sum_r \exp(-\beta E_r)\right]^2} = -\langle E^2 \rangle + \langle E \rangle^2 \quad (16)$$

$$\text{Thus we have: } \quad \langle (\Delta E)^2 \rangle \equiv \langle E^2 \rangle + \langle E \rangle^2 = -\left(\frac{\delta U}{\delta \beta}\right) = kT^2 \left(\frac{\delta U}{\delta T}\right) = kT^2 C_v \quad (17)$$

Using the relation above, for fluctuation of the relative root mean square of E, we will have:

$$\frac{\sqrt{\langle (\Delta E)^2 \rangle}}{\langle E \rangle} = \frac{\sqrt{kT^2 C_v}}{U} \quad (18)$$

With $O(N^{-1/2})$, that N is the number of particles in the system. By expansion of the logarithm of the probability density P(E) around the most probable of energy value , $E^* \approx U$ ,will have:

$$\ln\left[e^{-\beta E} g(E)\right] = \left(-\beta U \frac{S}{K}\right) + \frac{1}{2}\frac{\partial^2}{\partial E^2} \ln\left\{e^{-\beta E} g(E)\right\}|_{E=U} (E-U)^2 + \ldots$$

$$= -\beta(U - TS) - \frac{1}{2kT^2 C_v}(E-U)^2 + \ldots$$

(19)

As result, will have:

$$P(E) \alpha e^{-\beta E} g(E) \simeq e^{-\beta(U-TS)} \exp\left\{-\frac{(E-U)^2}{2kT^2 C_v}\right\} = e^{-\beta(\Delta E)} \exp\left\{-\frac{(\Delta E)^2}{2kT^2 C_v}\right\} \quad (20)$$

This function is represented a Gaussian distribution at E, with average value U and variance $\sqrt{kT^2 C_v}$ .For quantity $\frac{\Delta E}{U} = \frac{\sqrt{kT^2 C_v}}{U}$, the distribution is also Gaussian that its average value is unit with variance $\sqrt{kT^2 C_v} / U$ actually $O(N^{-1/2})$ .As for limit $N \to \infty$ ,we will approach to the Dirac delta function.

In (table1), we have such result for our colloidal configuration. For example, for A=0 and $\beta$=1 will have: $U = \langle E \rangle$ =212.54, $\sqrt{\langle (\Delta E)^2 \rangle} = \Delta E$ =0.351511024, $\frac{\sqrt{\langle (\Delta E)^2 \rangle}}{\langle E \rangle} = \frac{\Delta E}{\langle E \rangle} = \frac{\sqrt{kT^2 C_v}}{U}$ =0.001653858$\ll 1$

For A=1 and $\beta$=1, will have:

$U = \langle E \rangle$=211.05, $\sqrt{\langle (\Delta E)^2 \rangle} = \Delta E$ =0.348883935, $\frac{\sqrt{\langle (\Delta E)^2 \rangle}}{\langle E \rangle} = \frac{\Delta E}{\langle E \rangle} = \frac{\sqrt{kT^2 C_v}}{U}$ =0.001653087$\ll 1$

We easily see that, $U = \langle E \rangle$ for A=0 $> U = \langle E \rangle$ for A=1.

More results are obtained in Table1 completely. With more study of results obtained of (tables1) numerically, we will obtain such $O(N^{-1/2})$ for topological Defect Colloidal Particles on square 2D Periodic Substrates interacting via a long range screened Coulomb potential. Finally, by survey of partition function of canonical ensemble of this system, will have:

$$Q_N(V,T) \simeq e^{-\beta(U-TS)} \int_0^{\infty} \frac{-(E-U)^2}{2kT^2C_v} dE = e^{-\beta(U-TS)} \sqrt{2kT^2C_v} \int_0^{\infty} e^{-x^2} dx = e^{-\beta(U-TS)} \sqrt{2kT^2C_v} \quad (21)$$

Therefore: $$-kT \ln Q_N(V,T) \equiv F \approx (U-TS) - \frac{1}{2} kT \ln(2\pi kT^2 C_v) \Rightarrow F \approx U - S \quad (22)$$

The last sentence is order of lnN, whereas the others are order of N. Therefore, we can ignore the last sentence with respect the others. As the result, we obtain a thermodynamic sentence according to the formulation of canonical ensembles.The following equation is used to calculate the order parameter's variation of particles:

$$\overline{m} = \cos\left(\frac{2\pi}{A}\sum_{i=1}^{N} x_i + \frac{2\pi}{A}\sum_{i=1}^{N} y_i\right)$$
$$M = \frac{\overline{m}}{N} \quad (23)$$

These formulas being applied to average magnetization of all particle and will be then calculated the order parameter of any particle.

**Table1.** By looking at the tables, we can easily such relation between the heat capacity and energy fluctuation of colloids (for different substrates: A=0,1),(N=150:number of colloids) duration temperature interval: T~ 0.9KT to T~ 1 , are actually the critical temperatures interval.

| T | E A=0 | $\Gamma(r)$ A=0 | $C_v$ (A=0) | M A=0 | $\sqrt{\langle(\Delta E^2)\rangle}$ | $\frac{\Delta E}{\langle E \rangle}$ (A=0) | E A=1 | $\Gamma(r)$ (A=1) | $C_v$ (A=1) | M (A=1) | $\sqrt{\langle(\Delta E^2)\rangle}$ | $\frac{\Delta E}{\langle E \rangle}$ (A=1) |
|---|---|---|---|---|---|---|---|---|---|---|---|---|
| 1.00 | 212.54 | 0.1235 | 0.01527 | 0.8698 | 0.351511024 | 0.001653858 | 211.05 | 0.12172 | 0.014817 | 0.8771 | 0.348883935 | 0.001653087 |
| 0.99 | 212.51 | 0.1079 | 0.01178 | 0.8729 | 0.328603104 | 0.001546295 | 211.03 | 0.10628 | 0.011409 | 0.8796 | 0.326006135 | 0.001544833 |
| 0.98 | 212.50 | 0.10967 | 0.01227 | 0.8885 | 0.331164612 | 0.001558422 | 211.01 | 0.11201 | 0.012801 | 0.8768 | 0.334678951 | 0.001586081 |
| 0.97 | 212.45 | 0.17419 | 0.03127 | 0.8730 | 0.417360755 | 0.001964513 | 210.97 | 0.16725 | 0.028828 | 0.8840 | 0.408962101 | 0.001938485 |
| 0.96 | 212.46 | 0.11091 | 0.01281 | 0.8777 | 0.33303153 | 0.001567502 | 210.85 | 0.27592 | 0.079256 | 0.8776 | 0.525280877 | 0.002491254 |
| 0.95 | 212.44 | 0.11585 | 0.01411 | 0.8747 | 0.340367449 | 0.001602182 | 210.96 | 0.1093 | 0.012562 | 0.8802 | 0.330605505 | 0.001567148 |
| 0.94 | 211.85 | 0.30929 | 0.10161 | 0.8749 | 0.556138472 | 0.002625152 | 210.94 | 0.106 | 0.011935 | 0.8756 | 0.325576412 | 0.001543455 |
| 0.93 | 212.31 | 0.26160 | 0.07342 | 0.8700 | 0.511468474 | 0.002409064 | 210.91 | 0.1149 | 0.014165 | 0.8791 | 0.338969025 | 0.001607174 |
| 0.92 | 211.97 | 0.37456 | 0.15204 | 0.8683 | 0.612013072 | 0.002887263 | 210.4 | 0.36164 | 0.14173 | 0.8787 | 0.601365114 | 0.002858199 |
| 0.91 | 211.61 | 0.13240 | 0.01919 | 0.8754 | 0.363868108 | 0.001719522 | 210.41 | 0.38153 | 0.15935 | 0.8707 | 0.617681148 | 0.002935607 |
| 0.908 | 211.60 | 0.14175 | 0.02222 | 0.8786 | 0.376497012 | 0.001779286 | 210.1 | 0.13422 | 0.019918 | 0.8783 | 0.366360478 | 0.001743743 |
| 0.90 | 212.54 | 0.12634 | 0.01783 | 0.8761 | 0.355443385 | 0.001680107 | 211.05 | 0.12172 | 0.014817 | 0.8771 | 0.348883935 | 0.001653087 |

Here, we have calculated the ensemble average of colloid's energy. The correlation function for ensemble average of colloid's energy colloids and the mean-square fluctuations of colloid energy during temperature interval between T ~ 0.8KT to T ~ 1KT (critical interval) is calculated for substrates strength (A=0,1). By examining of the numerically results obtained of (tables1), the special case of "fluctuation- dissipation theorem" for energy of colloidal particle configurations will be satisfied as well. We can readily write relation between the heat capacity and the energy fluctuation of our colloidal particle configuration according to 'equations (14-20)'.

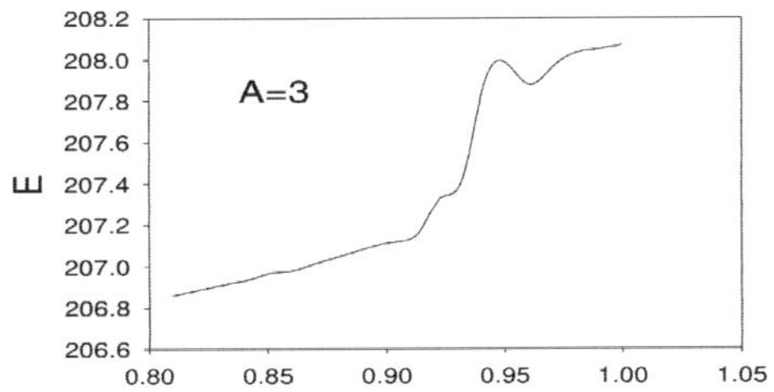


(2a)

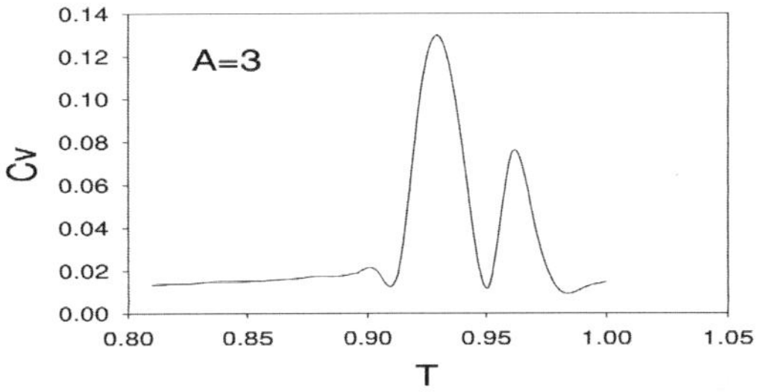


( 2b)

**Figure2**. The graph of the energy and the Specific heat fluctuation of the colloidal particle configurations vs T in low temperatures phase for A=3, (N=150). Two successive peaks on the variation of them strongly indicate resembles and present that of KT-like of first order phase transition, the theory of Kosterlitz, Thouless, Halperin, Nelson and Young (***KTHNY)*** defect-mediated topological phase transitions.[14-18].

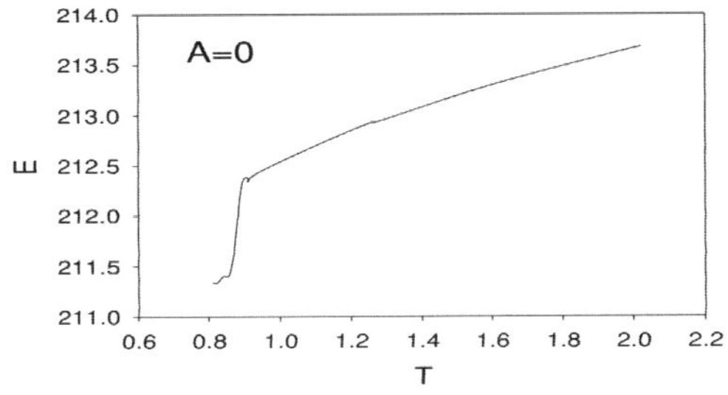


(3a)

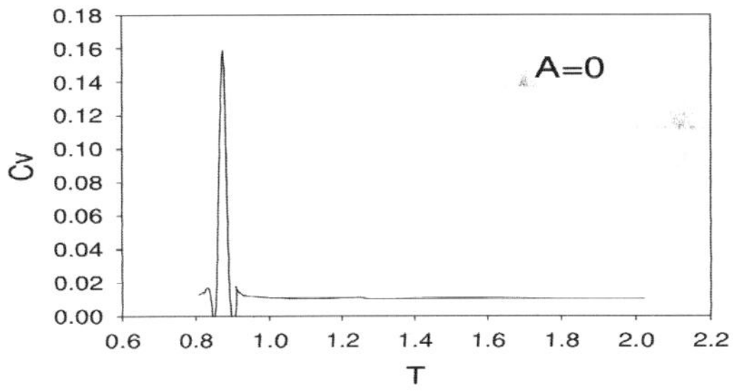


(3b)

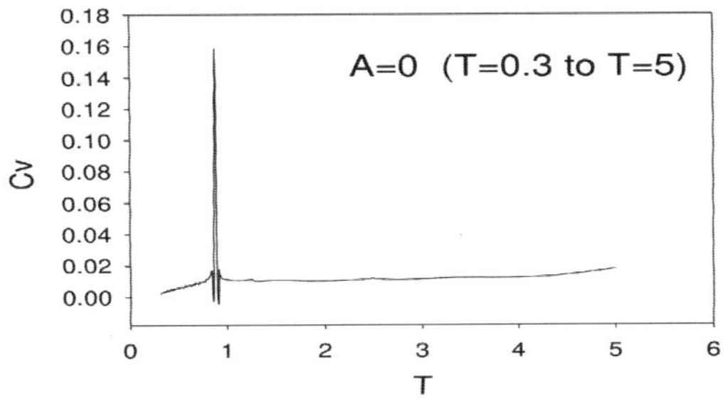


(3d)

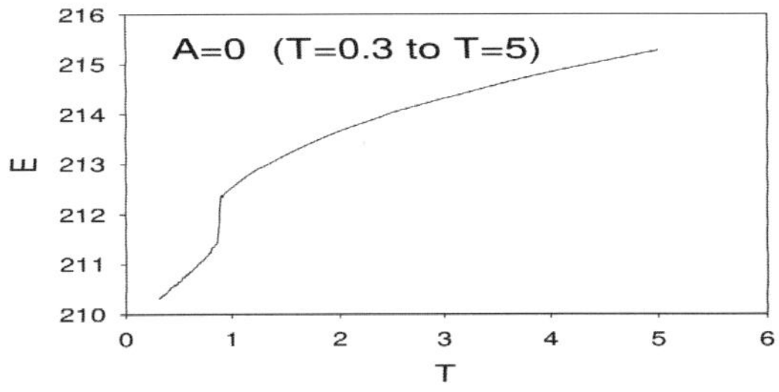


(3e)

**Figure3**. The graph of the Specific heat of the colloidal particle configurations between (T=0.8KT to T=2KT),(N=150).Here both low temperature and high temperature phase are considered. Two successive phase transition (Tc1~0.9224KT, Tc2~0.94148KT for A=0[17,18]) occurred over low critical temperatures (T ~ 0.8KT toT ~ 1KT) confirming of Gaussian distribution structure. But after T=1KT ,the fluctuations of Cv have mild and fairly constant trend over this range of temperature (T ~ 1KT to T ~ 2KT) in comparison with the previous low temperature phase(T ~ 0.8KT to T=1KT). In high temperature phase (T ~ 1KT-T ~ 2KT), configurations have no kind of ordering and topological structures of defects (disclination and dislocation[14-18]) have been totally destroyed and the diffusion process has been done completely. 'figure3a' refer to ensemble average of colloid's energy vs T. 'figure3b' indicate to fluctuation of heat capacity vs temperature with Gaussian distribution structure indicating linear form of the Fluctuation-Dissipation Theorem. 'figure 3d', 'figure3e' refer to variation of energy and heat capacity of colloidal configuration at all temperature regions(low, critical and high temperature: T ~ 0.3 to T ~ 5) for A=0.

Random fluctuations with small deviations from the state of thermodynamic equilibrium arise from the granular structure of them and appear as thermal diffusion with Gaussian distribution structure, are showing linear form of the Fluctuation-Dissipation Theorem constitutive a canonical ensemble. The data obtained of (tables1) is completely similar to 'equations (14-20)', with Gaussian distribution structure; all are confirming of linear form of the Fluctuation-Dissipation Theorem constitutive a canonical ensemble.

According to 'figure 4', there are relatively well comparisons between various substrate strength of colloid energy configuration versus temperature at low temperatures phase pointing to critical low temperature points.

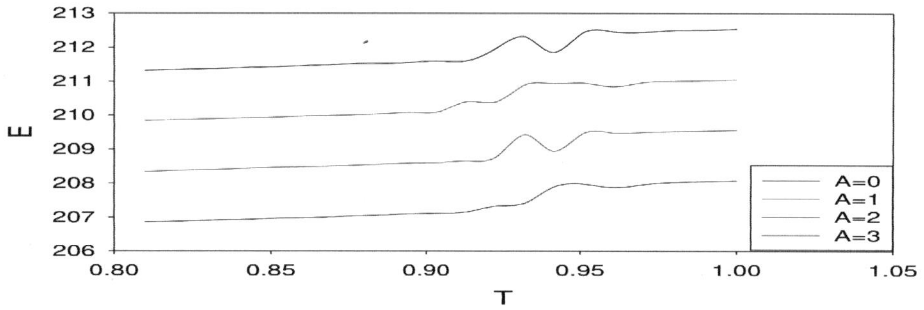


**Figure4.** There is relatively well comparison between different substrate strength of the colloids's energy versus temperature at low temperatures phase (N=150). As can be seen, as A increases from zero to A=3, the energy decrease reversely for all the temperatures given in low temperature phase. Actually with increasing of substrate strength, the energy of colloidal particle configuration monotonically decreases. The calculated energy of colloidal particles on (table1) for A=0 and A=1, also confirm the results of 'figure4' as well.

## 5. Simulation of noise-like fluctuation of colloids

In this here, according to the number of motion of each particle in system of colloidal configuration the variation of order parameter is characterized with 'equation (23)' at the end of particles motion.

Actually, the above equation is showing the phase relationship between different particles on the critical temperatures phase (T ~ 0.8KT-T ~ 1KT) with different substrates strength. The variation of the order parameter versus the number of particle's motion is calculated at the definite critical temperatures [17, 18] with various substrates strength. We perform $10^6$ Loops of Monte Carlo as the number of particles' motion in this experiment. In (table2), for example: A=1, T ~ 1, the energy per particle is calculated for 1000000 loops (the number of particle's motion) separately. The variations of order parameter versus the number of particle's motion (Markov Points) all are confirming of **noise-like fluctuation and Brownian motion** of colloidal particles as well. The fluctuations arise from the granular structure of matter and appear as thermal noise 'figures 5'. There is the large number of displacement in the coordinate of Brownian particles 'figures 5'.If we now want to measure of the energy and the order parameter, as obtained in 'figures5' for different temperatures, we will finally get a Gaussian distribution of energy performed in 'figures 2', 'figure3'and(table1) as well.

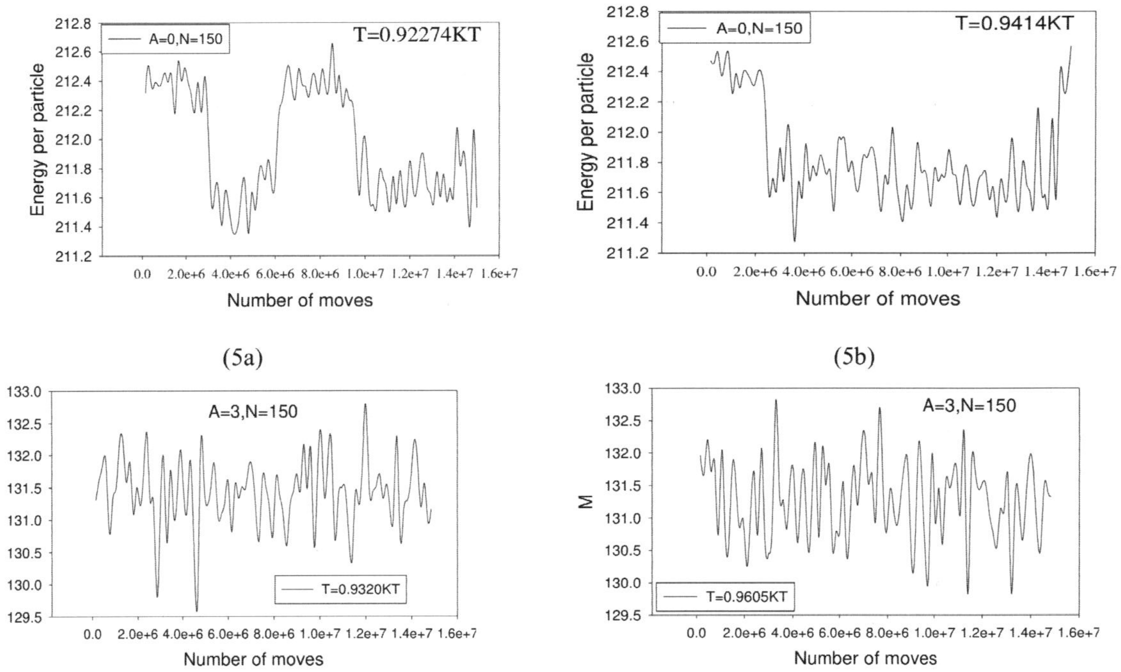


(5a) (5b)

**(5c )** **(5d)**

**Figure5**. The potential Energy per particle, and the order parameter versus the number of particles's motion in the critical temperatures interval (critical temperatures T ~ 0.8KT to T ~ 1KT) confirming of **noise-like fluctuation and diffusion like Brownian motion** of colloidal particles. The variation of energy per particle versus the number of moving of particles: For A=0,N=150: (5a) at T ~ 0.92274KT. (5b) The variation of the order parameter per particle versus the number of moving of particles for A=3,N=150: (5c) at T ~ 0.9320KT. (5d) at T ~ 0.9605KT.

**Table2.** The energy per particle is calculated for 1000000the number of particle's motion as loops of Monte Carlo Simulation separately (A=1,T=1 ,N =150) .

| The number of motion | Energy per particle | The number of motion | Energy per particle | The number of motion | Energy per particle | The number of motion | Energy per particle | The number of motion | Energy per particle |
|---|---|---|---|---|---|---|---|---|---|
| 100000 | 211.0299 | 300000 | 211.0417 | 500000 | 211.0875 | 700000 | 211.037 | 900000 | 211.206 |
| 200000 | 211.1988 | 400000 | 210.9702 | 600000 | 211.1137 | 800000 | 211.003 | 1000000 | 211.209 |

## 6. Conclusion

Using the Monte Carlo simulations , we have calculated mean-square fluctuations in statistical mechanics, such as those for colloids energy configuration are set on square 2D periodic substrates interacting via a long range screened coulomb potential for different substrate. **Random fluctuations with small deviations from the state of thermodynamic equilibrium arise from the granular (colloidal) structure of them and appear as thermal diffusion with a Gaussian distribution structure. The obtained results in the paper are showing linear form of the Fluctuation-Dissipation Theorem on the particle's energy constitutive a canonical ensemble with continuous diffusion process of colloidal particles system**. The numbers on the (table1) write the numerically results of the Fluctuation-Dissipation Theorem of Colloidal Particle's energy interacting with Yukawa potential on 2D periodic substrate. The observed results of both graph of colloidal particles configuration 'figure2' and the graph of the variation of statistical quantities of them 'figure3' at three phase of low, critical and high temperatures represent diffusion process of colloids completely [17,18] . We can easily that, the continuous process of phase transition of colloidal particles happened in the low temperatures phase 'figure3'. A good comparison is done between different substrates strength of the energy configuration of colloids versus temperature at low temperatures phase 'figure4'. As can be seen, as A increases from zero to A=3, the energy decrease all the temperatures given in low phase reversely. With the increasing substrate strength the colloidal particle's energy will monotonically decreases. The colloidal particles's energy for A=0, A=1 on the (table1) are also confirming the issue. The variations of energy per particle were calculated according to the random Brownian motion of colloidal particles (or loops of Monte Carlo Simulation (table2)) as well. The obtained results on the (table2) are actually consistent with results of 'figure5a', 'figure5b', too. **The variations of the order parameter versus the number of particle's motion 'figures5c', 'figure5d' are also confirming of noise-like fluctuation of colloidal particles as well.**

## 7. References

[1] Reichhardt C and Olson C J 2002 *Phys. Rev. Lett.* **88** 248301.
[2] Brunner M and Bechinger C 2002 *Phys. Rev. Lett.* **88** 248302.
[3] Murray C A and Winkle D H V 1987 *Phys. Rev. Lett.* 58 1200 Zahn K, Lenke R and Maret G 1999 *Phys. Rev. Lett.* 82 2721 Pertsinidis A and Ling X S 2001 *Nature (London)* 413 147.
[4] Mangold K, Leiderer P and Bechinger C 2003 *Phys. Rev. Lett.* **90** 158302.
[5] Korda P T, Spalding G C and Grier D G 2002 *Phys. Rev. B* 66 024504 Korda P T, Taylor M B and Grier D G 2002 *Phys. Rev. Lett.* **89** 128301.
[6] M. Brunner and C. Bechinger, *Phys. Rev. Lett.* 88, 248302 (2002); K. Mangold, P. Leiderer and C. Bechinger, ibid. 90, 158302 (2003).
[7] Prof.Matthias Troyer; Computational PhysicsII-ETH Zurich,SS 2004.
[8] W.Janke:*Monte Carlo in classical statistical Methods physic*,Lect.notes phys 739,79-140(2008),Speringer-verlag Berlin Heidelberg 2008.
[9] Reichhardt C, Olson C J, Scalettar R T and Zim´anyi G T 2001 *Phys. Rev. B* **64** 144509.
[10] M Mikulis, C. J. Olson Reichhardt, C. Reichhardt, R T Scalettar and G T Zim´anyi, *J. Phys: Condens. Matter* **16**, 7909 (2004).
[11] C. Reichhardt, C.J. Olson Reichhardt, *Europhys. Lett.* **68**, 303 (2004).
[12] A. Libál, C. Reichhardt and C. J. Olson Reichhardt1, *Phys. Rev. E* **75**, 011403 (2007).
[13] D.R. Nelson, *Defects and Geometry in Condensed Matter Physics (Cambridge Univ. Press, Cambridge, 2002).*
[14] Kosterlitz J M and Thouless D J 1973 *J. Phys. C: Solid State Phys.* 6 1181.
[15] Nelson D R and Halperin B I 1979 *Phys. Rev. B* 19 2457–84.
[16] Halperin B I and Nelson D R 1978 *Phys. Rev. Lett.* 41 121–4.
[17] Najafi Amin et al, *Proceeding of 2nd IPM Conference on Soft Matter, Biological and Statistical Physics, School of Physics, IPM, Tehran* (2013).
[18] Najafi Amin, *Proceeding of XXV IUPAP International Conference on Statistical Physics. STATPHYS25, Seoul, Korea* (2013).

Journal of Physics: Conference Series **510** (2014) 012026 doi:10.1088/1742-6596/510/1/012026

# Frustrations and phase transitions in the Ising model on square lattice

**A K Murtazaev[1], M K Ramazanov[1] and F A Kassan-Ogly[2]**
[1]Institute of Physics, Dagestan Scientific Center, Russian Academy of Sciences, 367003 Makhachkala, Russia
[2]Institute of Metal Physics, Ural Division, Russian Academy of Sciences, ul. S. Kovalevskoi 18, Ekaterinburg, 620990, Russia

E-mail: sheikh77@mail.ru

**Abstract.** The phase transitions in two-dimensional antiferromagnetic Ising model are studied on a square lattice by taking the interactions of next-nearest neighbors into account. The model is investigated on basis of the replica Monte Carlo algorithm and the histogrammic analysis of data. The diagram of the critical temperature dependence on an interaction value of next-nearest neighbors is plotted. The studied model reveals the phase transition of second order and the frustration points depending on relative interaction values and magnetic field. A effect, heat capacity splitting near the frustration points, is found.

## Introduction

The modern physics of condensed state uses different lattice models for quantitative description of phase transitions (PTs) and critical properties (CP). Theoretical methods used on simple lattice models could solve very limited number of tasks. One of such models is a two-dimensional Ising model. The Ising model with interaction of nearest neighbors has been well studied by different methods and approaches [1-4]. The same model on a square lattice with ferromagnetic interactions of first and second neighbors has been accurately solved. However the taking into account of antiferromagnetic interactions of second nearest neighbors in classical two-dimensional Ising model is attended by quenching of the main state and appearing of different phases and phase transitions. Furthermore, accounting for interactions of next-nearest neighbors can influence on a critical behavior of the model, in particular, the various anomalies of critical properties appear [5].

First renormalization group calculations and numerical simulation by the Monte Carlo (MC) method for two-dimentional antiferromagnetic Ising model on a square lattice with interactions of next-nearest neighbors were carried out in works [6, 7] by the end of the 1970s. Authors supposed the presence of a phase transition of second order, estimated the temperature of the phase transition, and calculated critical exponents.

In works [8-10], authors reported the second order PTs in antiferromagnetic Ising model on a square lattice with interactions of next-nearest neighbors. Also this model could possess "anomalous" critical exponents. Additionally, the dependence of critical exponents on correlation $J_{NNN} / J_{NN}$ ($J_{NN}$ and $J_{NNN}$ denote the constants of exchange interactions of nearest and next-nearest neighbors, correspondingly) was determined. But the scenario of a continuous phase transition was doubted after calculations by the mean-field theory which recognized the phase transition of first order [11].

According to works [12-14] the first order PTs were observed for Ising model on a square lattice with ferromagnetic interactions of nearest neighbors and with antiferromagnetic interactions of next-nearest neighbors in value intervals $J_{NNN} / J_{NN} = 0.5 \div 1.2$ in the system. In our previous work [15] we estimated the antiferromagnetic Ising model on a square lattice with taking into account the interactions of next-nearest neighbors for $J_{NNN} / J_{NN} = 1$ and determined the second order phase transition.
The main goal of the present work is to provide the accurate determination of a PT order of antiferromagnetic Ising model on a square lattice with taking into account next-nearest neighbors within value intervals $0.1 \leq J_{NNN} / J_{NN} < 1$ on basis of the replica Monte Carlo algorithm using a reliable and verified scheme and a unified technique.

**Model and method**

The antiferromagnetic Ising model on a square lattice taking into account interactions of second nearest neighbors is described by Hamiltonian [15]:

$$\mathrm{H} = -J_{NN}\sum_{\langle i,j\rangle}(S_i \cdot S_j) - J_{NNN}\sum_{\langle i,j\rangle}(S_i \cdot S_j) - h\sum_i s_i \qquad (1)$$

where $S_{i,j}=\pm 1$ is the Ising spin. The first term in the Equation (1) accounts for the exchange interaction of nearest neighbors by the value of $J_{NN}<0$, and the second term considers next-nearest neighbors by $J_{NNN}<0$, $h$ is the external magnetic field.
The investigation of phase transitions in frustrated spin systems by traditional theoretical, experimental, and numerical methods faces with complex problems. This is connected with the inherent in such models problems of multiple valleys of local energy minima. Such systems can be rigorously and sequentially studied by the Monte Carlo methods on basis of microscopic Hamiltonians [16-19] but the standard Monte Carlo methods fail to solve these problems. Therefore, in recent time many new versions of Monte Carlo algorithms are developed for solution of these problems. The most powerful and effective algorithms for estimation of PTs and CPs in frustrated systems proved to be the replica Monte Carlo algorithms [20].
By now the replica Monte Carlo algorithms and the finite-size scaling theory become the basic tools for the investigation of critical properties of such complex systems.
To analyze a nature of phase transitions and the peculiarities of thermal characteristics near the critical point is used a histogram method [21]. The histogram analysis allows to estimate the reliability and accuracy of data derived from Binder cumulants calculations by Monte Carlo method and also to determine other important parameters [22].
The histogram method uses a random walk in the energy space and allows to detect accurate estimations for energy state densities $g(U)$. The probability of transition from one state into another is determined by the formula

$$P(U \to U') = \min[g(U)/g(U'),1] \qquad (2)$$

where $U$ and $U'$ denote the energies before and after spin flip.
The calculations were carried out for the systems with periodic boundary conditions and linear sizes $L\times L = N$, $L = 20\div 150$. The exchange interaction correlation of next-nearest neighbors and nearest neighbors changed in value intervals $0.1 \leq J_{NNN} / J_{NN} < 1$. The system was brought into a thermodynamic equilibrium state by cut-off $\tau_0 = 4\times 10^5$ MC step/spin, which is several times larger than the nonequilibrium part, where $\tau_0$ denotes the length of nonequilibrium part of Markovian chain. The thermodynamic properties were averaged along the Markovian chain of length up to $\tau = 100\tau_0$ MC step/spin.

**Simulation results**

The temperature behavior of heat capacity and susceptibility was determined by the expressions [23]:

$$C = (NK^2)\left(\langle U^2\rangle - \langle U\rangle^2\right), \quad (3)$$

$$\chi = \begin{cases} (NK)\left(\langle m^2\rangle - \langle m\rangle^2\right), & T < T_N \\ (NK)\langle m^2\rangle, & T \geq T_N \end{cases} \quad (4)$$

where $K = J / k_B T$, $N$ is the number of particles, $U$ is the internal energy, $m$ is the order parameter. The order parameter $m$ was derived from expressions [15]:

$$m_\lambda = \frac{4}{N}\sum_{i\in\lambda} S_i, \text{ where } \lambda = 1,2,3,4, \quad (5)$$

$$m^a = [m_1 + m_2 - (m_3 + m_4)]/4, \quad (6)$$

$$m^b = [m_1 + m_4 - (m_2 + m_3)]/4, \quad (7)$$

$$m = \sqrt{(m^a)^2 + (m^b)^2}, \quad (8)$$

where $m_1$, $m_2$, $m_3$, $m_4$ are the order parameters of sublattices.

Figures 1 and 2 present the temperature dependences of the heat capacity and susceptibility obtained at $L$=80 for different values $J_{NNN} / J_{NN}$ (here and elsewhere a statistic error does not exceed the sizes of symbols used for plotting the curves). Note that decrease of $J_{NNN} / J_{NN}$ in intervals $1 \geq J_{NNN} / J_{NN} \geq 0.6$ is accompanied with maxima shift towards lower temperatures and simultaneously the absolute maxima of heat capacity and susceptibility increases. We interpreted such behavior by the reason that when increasing the interactions of next-nearest neighbors, the contribution of interaction energy increases in modulus what enhances the rigidity of the system, and correspondingly the phase transition temperature arises. The increase in the absolute values of maxima is accounted for competing of nearest neighbors and next-nearest neighbors.

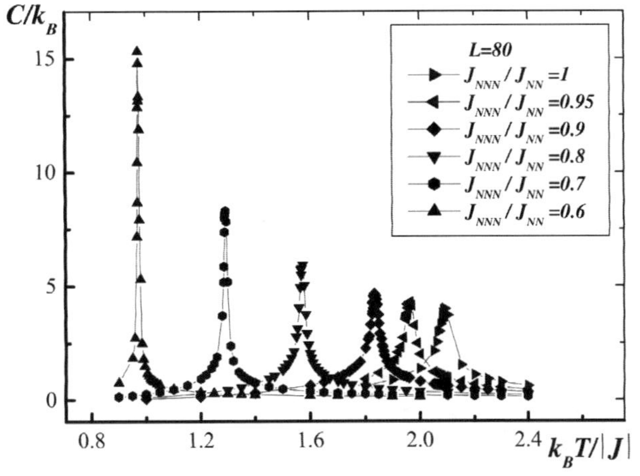


**Figure 1.** The dependence of heat capacity $C/k_B$ on the temperature $k_BT/|J|$ for different $J_{NNN} / J_{NN}$.

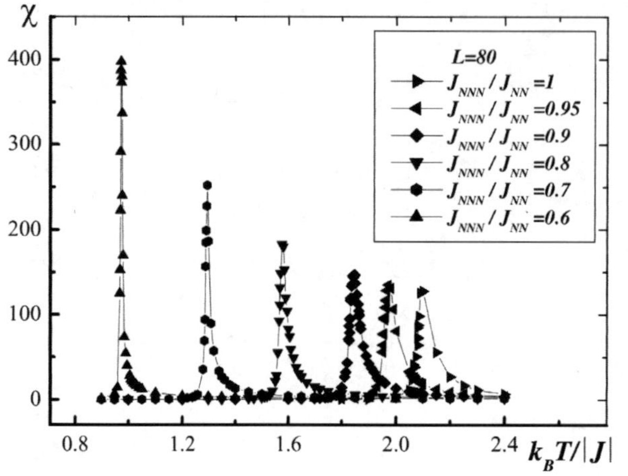


**Figure 2.** The dependence of susceptibility $\chi$ on the temperature $k_BT/|J|$ for different $J_{NNN} / J_{NN}$.

Figures 3 and 4 show the temperature curves of the heat capacity and susceptibility in intervals $0.1 \leq J_{NNN} / J_{NN} \leq 0.4$. It is notable that well-marked maxima are observed in all temperature curves of heat capacity $C$ and susceptibility $\chi$ for all r near the critical temperature. We observe opposite behavior in the case when $0.1 \leq J_{NNN} / J_{NN} \leq 0.4$. The phase transition temperature shifts towards higher temperatures at the decrease of r from 0.4 to 0.1.
We determined the critical temperatures $T_N$ using the method of forth-order Binder cumulants [24, 25]:

$$U_L = 1 - \frac{\langle m^4 \rangle_L}{3 \langle m^2 \rangle_L^2}, \tag{9}$$

$$V_L = 1 - \frac{\langle U^4 \rangle_L}{3 \langle U^2 \rangle_L^2}, \tag{10}$$

where $U_L$ is the cumulant in magnetization, $V_L$ is the cumulant in energy.

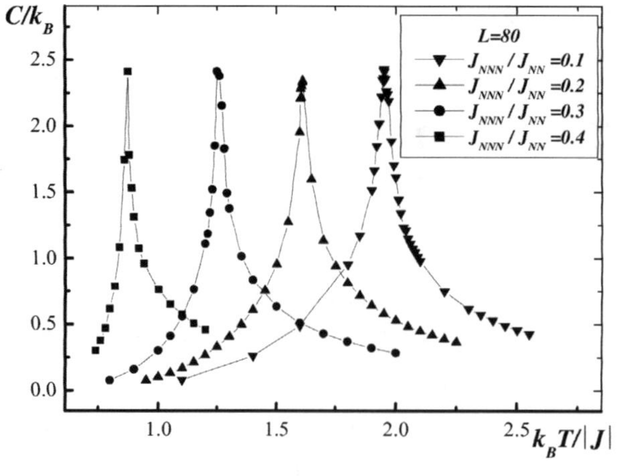


**Figure 3.** The dependence of heat capacity $C/k_B$ on the temperature $k_BT/|J|$ for different $J_{NNN} / J_{NN}$.

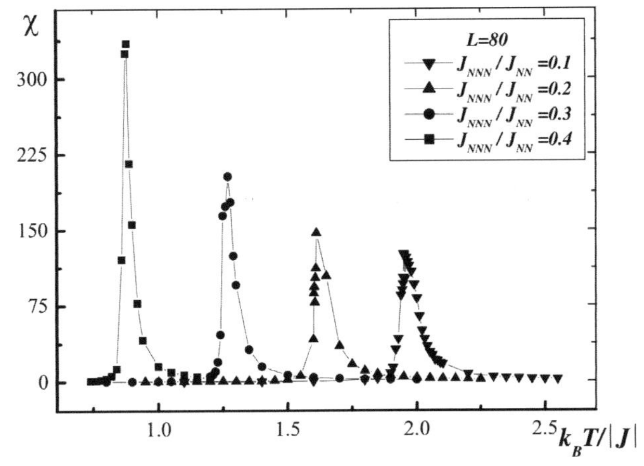


**Figure 4.** The dependence of susceptibility $\chi$ on the temperature $k_BT/|J|$ for different $J_{NNN} / J_{NN}$.

From Expressions (9) and (10) the critical temperature can be detected to a high accuracy. It should be noted that Binder cumulants provide good testing of PTs type in the system. It is common knowledge that the temperature curves of Binder cumulants $U_L$ have a well-marked cross point at second order PTs [25].

Figures 5 presents the $U_L$ temperature curve at $J_{NNN} / J_{NN}$ =0.7 for different values of $L$. This figure shows the accuracy in the detection of critical temperature. As is evident from insert, the well-marked cross point observed in the critical region (here and elsewhere the temperature is expressed in the unite of $|J|/k_B$) indicates the second order PTs. The critical temperatures for other $J_{NNN} / J_{NN}$ values were determined in a similar way.

The temperature dependence of cumulants in energy $V_L$ at $J_{NNN} / J_{NN}$ =0.7 for different $L$ is shown in Figures 6. As is obvious from the figure, the minimum magnitude of $V_L$ disappears at increase in $L$ and the value of $V_L$ tends to 2/3 at $T \rightarrow 0$ what also is typical for second order PTs. By a diagram form in Figures 6 one can predict a type of the phase transition in the system. A detail description is reported in our previous work [24].

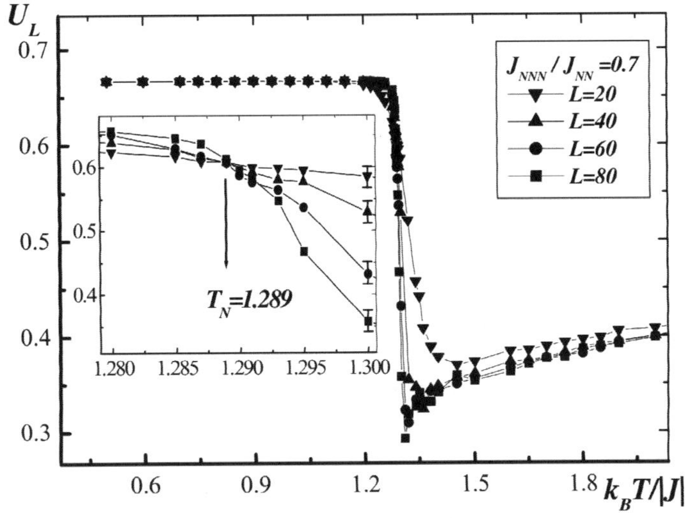


**Figure 5.** The dependences of Binder cumulant $U_L$ on the temperature $k_BT/|J|$ for $J_{NNN} / J_{NN}$ =0.7.

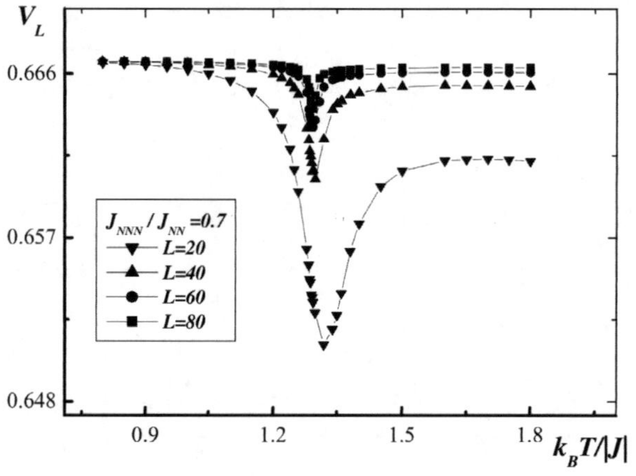


**Figure 6.** The dependences of Binder cumulant $V_L$ on the temperature $k_BT/|J|$ for $J_{NNN}/J_{NN}$=0.7.

Figures 7 presents the dependence of critical temperature on an interaction value of next-nearest neighbors. As is evident from the diagram, the different phases cross at a point $J_{NNN}/J_{NN} = 0.5$: 1) ferromagnetic, 2) paramagnetic, and 3) super antiferromagnetic lines. The figure demonstrates that the critical temperature $T_N = 0$ and phase transition are lack for the value $J_{NNN}/J_{NN} = 0.5$. This comes from the similarity in the interactions of nearest neighbors and next-nearest neighbors at $J_{NNN}/J_{NN} = 0.5$ and the whole system frustrates. It disorders the system and results in a disappearance of phase transition. It can be inferred by Figures 8 where the heat capacity temperature curve at $J_{NNN}/J_{NN} = 0.5$ for big size lattices (150) has not an abrupt jump but changes smoothly. A similar behavior of the heat capacity temperature curve is observed for $J_{NNN}/J_{NN}$=0.49 and $J_{NNN}/J_{NN}$=0.495 which are below the value $J_{NNN}/J_{NN} = 0.5$. For two-dimensional case in all frustration points, the temperature dependence of the heat capacity is of smooth maximum form. When moving away from the frustration point, a sharp lambda-shaped peak appears in the heat capacity and the smooth maximum is slightly decreasing (heat capacity splitting). At further moving away from the frustration point the smooth maximum vanishes and the sharp peak remains only. This is demonstrated in Figures 9. Also the figure exhibits the well-marked maxima near the critical temperature for values $J_{NNN}/J_{NN} = 0.505$ and $J_{NNN}/J_{NN} = 0.51$, which are above the value $J_{NNN}/J_{NN} = 0.5$.

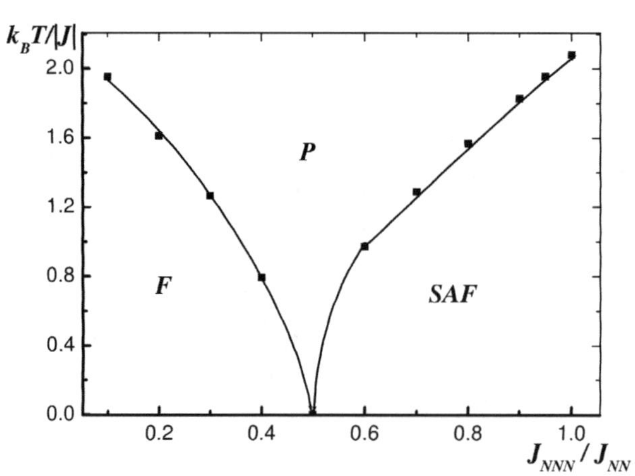


**Figure 7.** The phase diagram of dependence critical temperature on the interaction value of next-nearest neighbors, where ***F*** – ferromagnetic phase, ***P*** – paramagnetic phase, ***SAF*** – super antiferromagnetic phase.

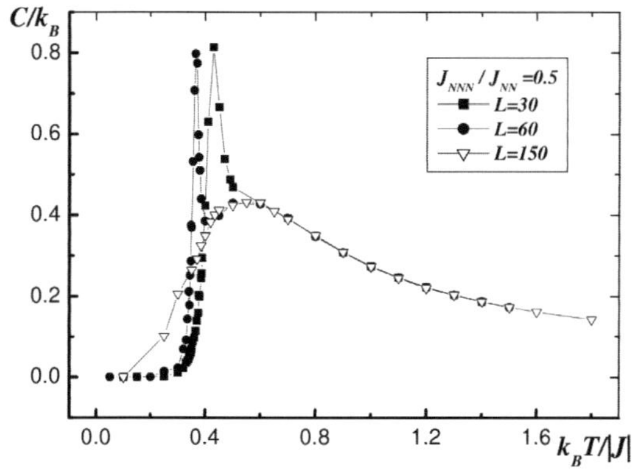


**Figure 8.** The dependence of heat capacity $C/k_B$ on the temperature $k_BT/|J|$ for $J_{NNN} / J_{NN} = 0.5$.

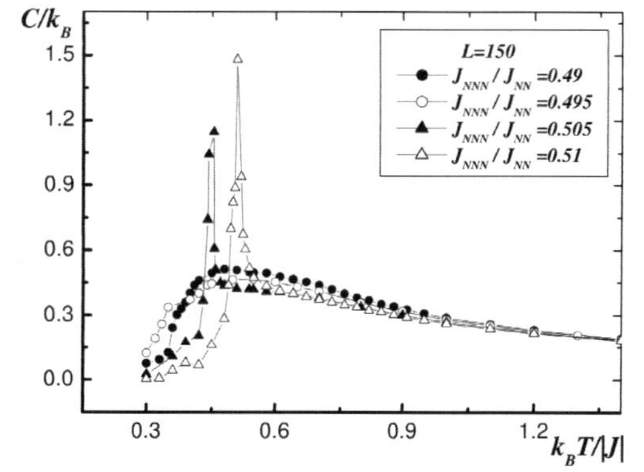


**Figure 9.** The dependence of heat capacity $C/k_B$ on the temperature $k_BT/|J|$ for $L = 150$.

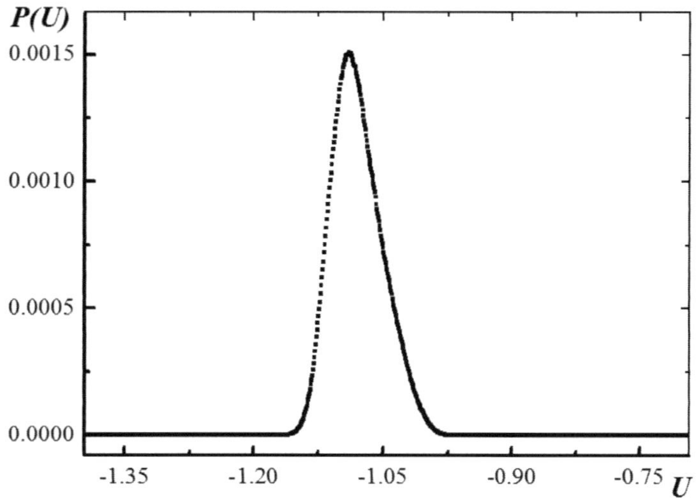


**Figure 10.** The histogram of energy distribution for $J_{NNN} / J_{NN}$ =0.7.

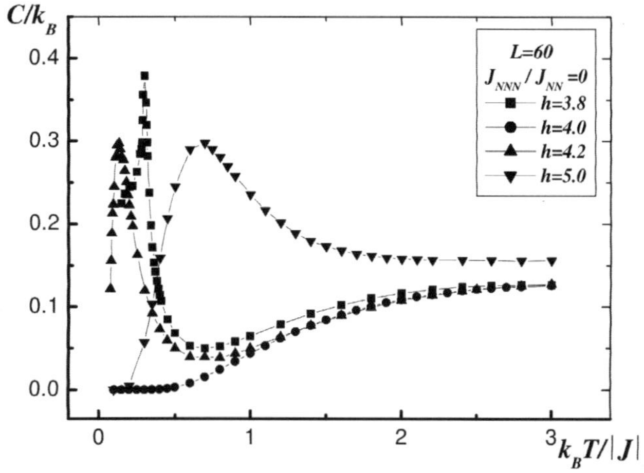


**Figure 11.** The dependence of heat capacity $C/k_B$ on the temperature $k_BT/|J|$ for different $h$.

In work [26], authors reported the phase diagram for phase transition temperature dependence $T_N$ on the ratio of exchange parameters $J_{NNN} / J_{NN}$ for the similar model. They determined that the second order phase transition occurred in value intervals $J_{NNN} / J_{NN} < 0.5$ and $J_{NNN} / J_{NN} \geq 0.948$ and the first order phase transition happened in intervals $0.5 < J_{NNN} / J_{NN} < 0.948$. According to our investigations the second order phase transition took place in intervals $0.5 < J_{NNN} / J_{NN} \leq 1$. This was confirmed by the histogram analysis.

The histogram of energy distribution for $J_{NNN} / J_{NN} = 0.7$ is presented in Figures 10. The diagram is plotted near the critical point for a lattice with $L = 150$. In the Figure, a maximum typical for second order phase transitions is observed [24]. Similar histograms were plotted for other values of $J_{NNN} / J_{NN}$. According to obtained data the second order phase transitions are observed in all long interval $0.1 \leq J_{NNN} / J_{NN} \leq 1$, except when $J_{NNN} / J_{NN} = 0.5$ where the phase transition fails to take place. The point $J_{NNN} / J_{NN} = 0.5$ is the frustration point.

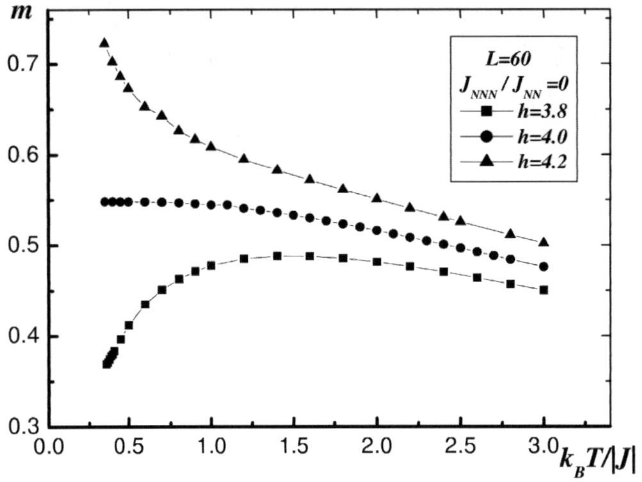


**Figure 12.** The dependence of order parameter $m$ on the temperature $k_BT/|J|$ for different $h$.

Figures 11 and 12 present the temperature dependence plots of heat capacity and order parameter on the external magnetic field for a case when accounting the interactions of nearest neighbors only. The plots exhibit the anomalous behavior in the heat capacity and order parameter dependences for the field value $h = 4.0$. This comes from the fact that when the value of external magnetic field is 4.0, the system is frustrated even without considering the interactions of next-nearest neighbors.

## Conclusion

The phase transitions in two-dimensional antiferromagnetic Ising model with taking into account of the interactions of next-nearest neighbors are studied using the exact and efficient replica Monte Carlo algorithm. A behavior of phase transitions for different correlations of $r$ of nearest neighbors and next-nearest neighbors is analyzed by the histogram and Binder cumulants methods. The phase diagram of critical temperature dependences on the interaction value of next-nearest neighbors is plotted. The second order phase transition is shown to happened for all values in intervals $0.1 \leq J_{NNN} / J_{NN} < 1$ except $J_{NNN} / J_{NN} = 0.5$. The value $J_{NNN} / J_{NN} = 0.5$ is detected to be the frustration point for this model. It is shown that this model frustrated in the external field at $h$=4.0 even when taking into account the nearest neighbors only.

## Acknowledgments

The study was supported by The Ministry of education and science of Russia, project 14.B37.21.1092, by the Russian Foundation for Basic Research (project no. 13-02-00220, project no. 12-02-95604).

## References

[1] Barber M N 1979 *J. Phys.* A **12** 679
[2] Landau D P and Binder K 1985 *Phys. Rev.* B **31** 5946
[3] Plascak J A 1992 *Physica* A **183** 563
[4] Ising E 1925 *Z. Phys.* **31** 253
[5] Landau D P and Binder K 2000 *Monte Carlo Simulations in Statistical Physics* (Cambridge University Press, Cambridge)
[6] Nightingale M P 1977 *Phys. Lett.* A **59** 486
[7] Swendsen R H and Krinsky S 1979 *Phys. Rev. Lett.* **43** 177
[8] Nightingale M P 1977 *Phys. Lett.* A **59** 468
[9] Binder K and Landau D P 1980 *Phys. Rev.* B **21** 1941
[10] Minami K and Suzuki M 1994 *J. Phys.* A **27** 7301
[11] Moran-Lopez J L, Aguilera-Granja F and Sanchez J M 1993 *Phys. Rev.* B **48** 3519
[12] Moran-Lopez J L, Aguilera-Granja F and Sanchez J M 1994 *J. Phys.: Condens. Matter* **6** 9759
[13] Lopez-Sandoval E, Moran-Lopez J L and Aguilera-Granja F 1999 *Solid State Commun.* **112** 437
[14] Buzano C and Pretti M 1997 *Phys. Rev.* B **56** 636
[15] Murtazaev A K, Ramazanov M K and Badiev M K 2011 *Low Temp. Phys.* **37** 1001
[16] Kassan-Ogly F A, Menshenin V V, Filippov B N, Murtazaev A K, Ramazanov M K and Badiev M K 2011 *Solid State Phenomena* **168-169** 435
[17] Kassan-Ogly F A, Filippov B N, Murtazaev A K, Ramazanov M K and Badiev M K 2012 *JMMM* **324** 3418
[18] Murtazaev A K, Kamilov I K and Ramazanov M K 2006 *Low Temp. Phys.* **32** 241
[19] Murtazaev A K, Ramazanov M K and Badiev M K 2009 *Low Temp. Phys.* **35** 521
[20] Mitsutake A, Sugita Y and Okamoto Y 2001 *Biopolymers (Peptide Science)* **60** 96
[21] Wang F and Landau D P 2001 *Phys. Rev. Lett.* **86** 2015
[22] Ramazanov M K 2011 *JETP Lett.* **94** 311
[23] Binder K and Wang J - Sh 1989 *J. Stat. Phys.* **55** 87
[24] Murtazaev A K, Ramazanov M K and Badiev M K 2010 *Phys. Solid State* **52** 1673
[25] Binder K and Heermann D W 1988 *Monte Carlo Simulation in Statistical Physics* (Springer, Berlin)
[26] Anjos A R, Viana J R and Sousa J R 2008 *Phys. Lett.* A **372** 1180

# Optical Spectrum Analysis of Real-Time TDDFT Using the Maximum Entropy Method

**M. Toogoshi, M. Kato, S. S. Kano, and Y. Zempo**

Computer and Information Sciences, Hosei University, 3-7-2 Kajino, Koganei 184-8584 Japan

E-mail: zempo@hosei.ac.jp

**Abstract.** In the calculation of time-dependent density-functional theory in real time, we apply an external field to perturb the optimized electronic structure, and follow the time evolution of the dipole moment to calculate the oscillator strength distribution. We solve the time-dependent equation of motion, keeping track of the dipole moment as time-series data. We adopt Burg's maximum entropy method (MEM) to compute the spectrum of the oscillator strength, and apply this technique to several molecules. We find that MEM provides the oscillator strength distribution at high resolution even with a half of the evolution time of a simple FFT of the dynamic dipole moment. In this paper we show the effectiveness and efficiency of MEM in comparison with that of FFT. Not only the total number of time steps, but also the length of the autocorrelation, the lag, plays an important role in improving the resolution of the spectrum.

## 1. Introduction

Time-dependent density functional theory (TDDFT) is one of the most prominent and widely used methods for calculatiing excited states of medium-to-large molecules, and it is recognized as a powerful tool for studying electronic transitions of molecules [1, 2]. TDDFT is, in principle, capable of analyzing electronic behaviors if the exact local exchange-correlation (xc) functional is known. In practical calculations, for valence-excited states with the excitation energy well below the ionization potential, accurate results have been obtained for an approximate xc functional [2].

We used TDDFT to study the optical properties of several materials within this framework where the electron-hole interaction is at the LDA level. Though the excitonic effects may be involved by solving the Bethe-Salpeter equation for the electron-hole Green's function [3], we neglect them for the sake of simplicity.

We focused on the maximum entropy method (MEM) to derive the electronic transition spectrum from the time-dependent dipole moment given by TDDFT, and evaluated the efficiency of the method. In our TDDFT, we employed a spatial grid with a fixed domain to express the physical system and solved the equation of motion by the finite difference approach, i.e., the real-time and real-space technique [4], without using explicit bases such as plane waves or Gaussian. The time evolution of the dipole moment of the system was tracked by solving the time-dependent equation of motion.

To determine the effectiveness of the new method, we used MEM, instead of using the FFT of the dynamic dipole moment to calculate the absorption spectrum. For smaller molecules, we found MEM provided spectra of fairly high resolution with a relatively small number of time-series data. MEM is attractive because a significant portion of the computational time is spent for calculations of time-series data of dipole moment. The theory with our numerical details is given in the next section, and the observed characteristics of MEM follow along with our conclusions.

## 2. Theory and numerical details

### *2.1. Time-dependent density-functional theory*

In this section, we briefly describe our TDDFT calculation procedure [5]. The basis is the density functional theory (DFT) [6] with the local density approximation (LDA). The total energy of the ground state can be derived from the Kohn-Sham equation (KS) [7]. DFT is much less successful in describing optical responses and absorption spectra where electronic excited states are involved. However, this difficulty is, in principle, overcome by the extension of DFT to its time-dependent version, TDDFT, which was established by Runge and Gross [1].

In analogy to the time-independent case, the TDDFT equation of motion coupled with pseudopotentials is given by

$$\Big\{ -\frac{1}{2}\nabla^2 + V_{ion}^{ps}(\mathbf{r}) + V_H(\mathbf{r},t) + V_{XC}[\rho(\mathbf{r},t)] + V_{ext}(\mathbf{r},t)\Big\}\psi_j(\mathbf{r},t) = i\frac{\partial}{\partial t}\psi_j(\mathbf{r},t) , \tag{1}$$

where $V_{ion}^{ps}$ is the ionic pseudopotential, $V_H$ is the Hartree potential, and $V_{XC}$ is the xc potential. Since the exact time-dependent xc potential is not known, the originally non-local time-dependent xc potential is replaced with a time-independent local one. This is reasonable when the density varies slowly with time. This approximation allows the use of a standard local ground-state xc potential in the TDDFT framework. The Hartree and xc potentials can be determined from the electronic charge density, $\rho(\mathbf{r},t) = \sum_j |\psi_j(\mathbf{r},t)|^2$. The summation is over all the occupied states $j$. The Hartree potential is determined by $\nabla^2 V_H = -4\pi\rho$, and the usual local density approximation (LDA) is used as the xc potential $V_{XC}$ in our study. For the ionic potential, we employed the pseudopotential $V_{ion}^{ps}$ in the separable form so that only the valence electrons were considered [8]. Prior to the calculation of optical responses, we first solved the usual, time-independent formulation of the pseudopotential-DFT method [9] to obtain the optimized electronic structure [10]. We then applied an external field $V_{ext}$ to the system as a perturbation and followed the linear responses of the system in real time. The xc functional we adopted is originally for the electronic ground states. This method of TDDFT has been effectively used, in cases where the potential was time-dependent, to study the behaviors of electrons in oscillating electric and magnetic fields and hence excited-state reactions [11]. In our calculations, the real-time and real space technique was adopted in solving Eq. (1) by means of the finite difference approach [4]. The uniform grid was used in our study for simplicity.

The time-dependent wave function is given by $\psi(t) = \exp[-iHt]\tilde{\psi}(0)$, with the initial wave function $\tilde{\psi}$ at $t = 0$,

$$\tilde{\psi}|_{t=0} = e^{ik_z z}\psi(0), \tag{2}$$

where $H$ is the Hamiltonian of the system and $k$ is a small wave number of the external perturbation in the $z$ direction. In the linear response, the time-dependent polarizability is proportional to the dipole matrix element $\mu_\xi(t) = \langle\psi(t)|\xi|\psi(t)\rangle$, where $\xi$ represents $x$, $y$, and $z$. The polarizability $\alpha_\xi(\omega)$ is numerically calculated by the fast Fourier transform (FFT) of $\mu_\xi(t)$ and averaged as $\alpha = (\alpha_x + \alpha_y + \alpha_z)/3$. The oscillator strength distribution $S(\omega)$ is related to the imaginary part of the polarizability,

$$S(\omega) = \frac{2\omega}{\pi}\mathrm{Im}\alpha(\omega). \tag{3}$$

In our FFT calculation for $C_{60}$ as shown in Fig. 1, the grid spacing was $\Delta x = 0.3$ Å and the Kohn-Sham eigenvalues converged well. The total volume, $(N_x\Delta x)^3$, was chosen so that there was a certain amount of space between the edge and any atom and the absorption boundary condition was applicable. For the time evolution, the time step was $\Delta t = 0.001\ eV^{-1}$ and the total number of the steps $N$ was varied. The calculated oscillator strength distribution of $C_{60}$ for $N = 5000$ shows a clear separation between the absorption peaks in comparison with that for $N = 2000$. The width of each peak was inversely proportional to the total evolution time, $T = N\Delta t$. In the spectrum for $N = 5000$, the low energy absorption peaks at 3.5, 4.4, and 5.6 eV agreed well with the experimentally observed peaks at 3.8, 4.8, and 5.8 eV [10].

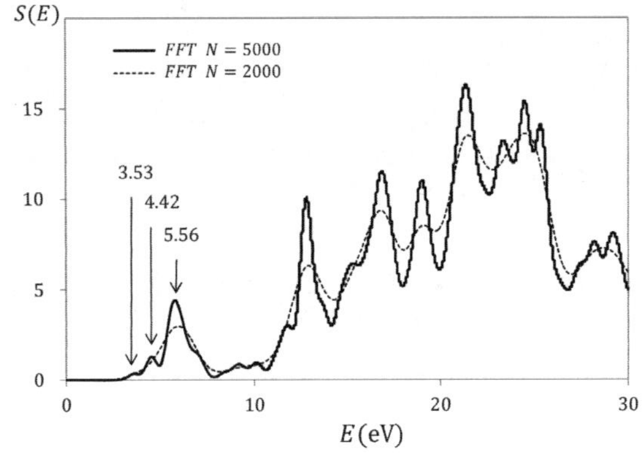


**Figure 1.** Absorption spectrum of $C_{60}$ obtained from the FFT of the dynamic dipole moment. The solid and dashed lines correspond to $N = 5000$ and 2000 steps, respectively.

The real-space algorithm offers the computational complexity of $O(N^2)$ and is attractive for the calculation of large systems. Also of potential interest is the intuitive feature of the method: the real-space approach allows us to grasp a clear physical image of the wave functions, which is not the case when the wave functions are expanded to the explicit basis sets. With the real-time approach, we can follow the evolution of the wave functions much more directly. These advantages helped our coding procedures.

### *2.2. Maximum entropy method*

As shown in the previous section, the resolution of the FFT absorption spectrum $S(\omega)$ is $\sim T^{-1}$, corresponding to the finite spectral bandwidth of the limited time series of $\mu(t)$, where $T = N\Delta t$ is the total evolution time. MEM, in contrast, estimates the full spectrum from the limited data based on information theory. The entropy in information theory has been recognized as a measure of uncertainty [12, 13]. Any inferences made from incomplete information should use with the probability distribution which maximizes the entropy under the constraints of available information [14]. The present study is based on Burg's method [15], where the dipole moment $\mu$ is assumed to be a random variable and the samples are given by TDDFT, $\mu_m = \mu(m\Delta t)$. The calculated autocorrelation $C_m$ at the time lag $m$ is

$$C_m = \frac{1}{N-|m|} \sum_{n=0}^{N-|m|-1} \mu_n \mu_{n+|m|}. \tag{4}$$

For $N \to \infty$, the Fourier transformation of the autocorrelation function is the power spectrum $P(\omega)$ [16] which is directly comparable to $S(\omega)$.

$$P(\omega)/\Delta t = \sum_{m=-\infty}^{\infty} C_m \exp(-i\omega m\Delta t) \tag{5}$$

$$C_m = \frac{1}{2\pi} \int_{-\pi/\Delta t}^{\pi/\Delta t} P(\omega) \exp(i\omega m\Delta t) d\omega. \tag{6}$$

Journal of Physics: Conference Series **510** (2014) 012027 doi:10.1088/1742-6596/510/1/012027

When each random variable $\mu_n$ obeys the Gaussian distribution,

$$h \propto \int_{-\pi/\Delta t}^{\pi/\Delta t} \log P(\omega) d\omega \tag{7}$$

is maximized, and thus $h$ is taken as the entropy. Conversely, under the constraints of Eq. (6) with the given value of Eq. (4), we find the value $\hat{P}(\omega)$ which maximizes Eq. (7). The solution is

$$\hat{P}(\omega)/\Delta t = \left| \frac{\beta}{1 + \sum_{k=1}^{M} a_k \exp(-ik\omega\Delta t)} \right|^2, \tag{8}$$

where $M(\leq N)$ is the maximum number of $|m|$ in Eq. (4). The parameters $a_k$ are the Lagrange multipliers which are the solution of the Yule-Walker equation

$$C_m = \sum_{k=1}^{M} a_k C_{m-k} + |\beta|^2 \delta_{m,0} \tag{9}$$

or equivalently, the linear Toeplitz matrix equation [16],

$$\begin{pmatrix} C_0 & C_1 & \cdots & C_M \\ C_1 & C_0 & \cdots & C_{M-1} \\ \vdots & \vdots & & \vdots \\ C_M & C_{M-1} & \cdots & C_0 \end{pmatrix} \begin{pmatrix} 1 \\ a_1 \\ \vdots \\ a_M \end{pmatrix} = \begin{pmatrix} |\beta|^2 \\ 0 \\ \vdots \\ 0 \end{pmatrix} \tag{10}$$

The Yule-Waker equation Eq. (9) is also derived by the autoregressive model of

$$\mu_m = -\sum_{k=1}^{M} a_k \mu_{m-k} + e_m, \tag{11}$$

where $e_m$'s are the white noise errors [17]. In this context, the important number for $M$ is

$$M_A = \min\{\text{FPE}(M)\}, \tag{12}$$

where

$$\text{FPE}(M) = \frac{N + M + 1}{N - M - 1} |\beta|^2. \tag{13}$$

FPE is Akaike's final prediction error [18, 19].

## 3. Optical spectra of MEM

In this section, we discuss the calculated absorption spectra of MEM compared with those of FFT.

### *3.1. Ethylene*

A simple molecule ethylene was chosen to confirm the efficiency of MEM. The molecular structure was based on the ground state. The calculation of the time evolution was the same as the real-time TDDFT procedure described in Sec. 2.1. Figure 2 (a) shows the dynamic dipole moment up to 20000 time steps. For comparison, the absorption spectra by FFT of $\mu(t)$ are shown in Fig. 2 (b), (c), (d) and (e) with the number of time steps $N$ = 1000, 2000, 10000, and 20000, respectively. The spectral resolution is higher for larger $N$.

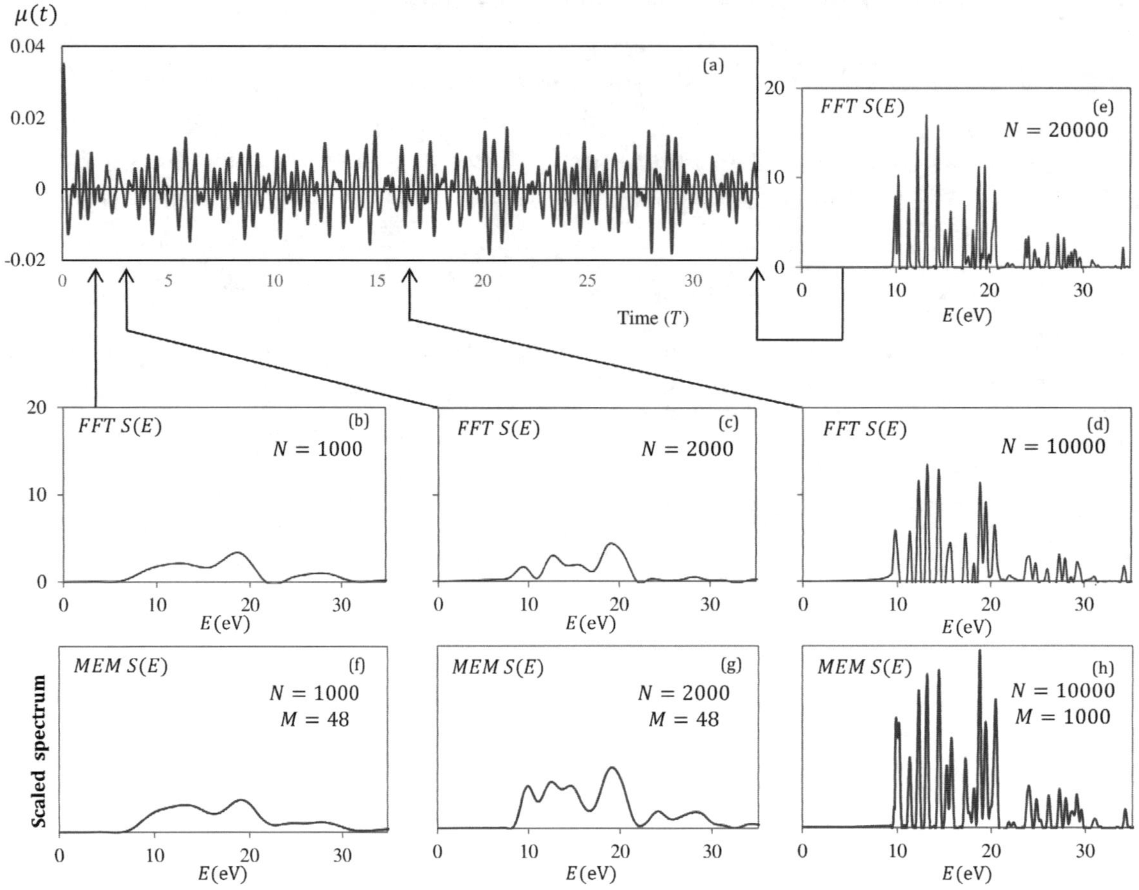


**Figure 2.** Ethylene absorption spectra. FFT of (a)The dynamic dipole moment provide the spectra calculated with (b) 1000, (c) 2000, (d) 10000 and (e) 20000 time steps, respectively. (f)(g)(h) show the results in MEM with the same number of time steps as in (b)(c)(d) with $M$ = 48, 48, and 38. The MEM data are scaled to allow comparison with those of FFT.

The results of MEM in Fig. 2 (f), (g) and (h) are similar to those of FFT. The resolution in Fig. 2 (h) is comparable with that in Fig. 2 (e) but the number of time steps is one half, which is the greatest advantage of MEM. MEM becomes more efficient if a suitable number $M$ is chosen for the lag of autocorrelation. In the cases Fig. 2 (f) and (g), we used Akaike's FPE ($M_A = 48$). In the case of Fig. 2 (h), $M_A$ was 38 but we adopted $M = 1000$, which was almost 25 times larger, to make the spectrum similar to that of FFT. Use of more data are in the autocorrelation calculation with larger lags and the estimation of lower-frequency components gives greater reliability. However, there are cases in which too large a value of $M$ provides unphysical results such as pseudo-peaks and splits.

### *3.2. Fluorene*

Poly(9,9-dialkyl-fluorene) and their substituted derivatives are expected to be the basic materials for blue emitting polymer diodes, and their electronic structures have been extensively studied [3, 20, 21, 22, 23]. We employed the oligomer of fluorene (FL) with $n$ = 8, and the results are shown in Fig. 3. Figure 3 (a) shows the dynamic dipole moment with the total time steps $N = 20000$. The spectrum of oligo-FL in Fig. 3 (b) was calculated by MEM with $N = 10000$ and $M = 1000$. As in the calculation of ethylene, the absorption spectrum is comparable with that of the FFT calculation in Fig. 3 (c), where FFT took $N = 20000$ steps to obtain the spectrum. The broad peak at 490 nm for MEM and FFT may correspond to the experimentally observed one at 348 nm for poly-FL. This discrepancy is due to the inherent problem of DFT [5]. In this MEM calculation, we employed $M = 1000$, which was almost forty times larger than Akaike's FPE ($M_A = 26$). We did not observe a significant change in the spectrum with further increased $M$.

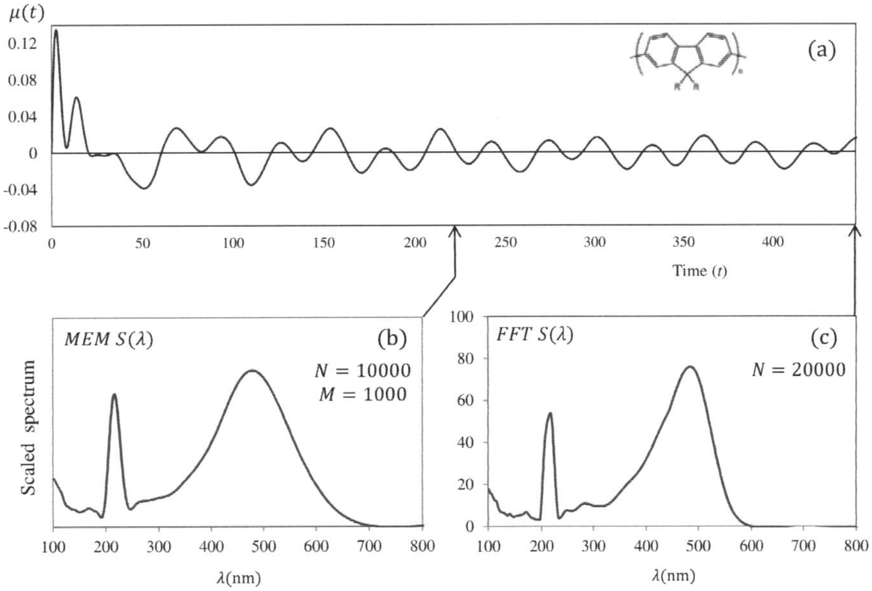


**Figure 3.** Optical absorption spectrum of oligo-$FL_8$. The dynamic dipole moment (a) is transformed into the spectrum obtained by MEM (b) and by FFT (c). The MEM result is scaled so that both intensities of the peaks look the same.

### *3.3. C60*

Figure 4 (a) shows the difference between the MEM and FFT spectra obtained in the same time steps $N = 5000$ for $C_{60}$. In this calculation of MEM, Akaike's FPE ($M_A = 166$) was employed. As noted in Section 2.1, the FFT and thus MEM spectra reasonably simulated the experimental absorption peaks at 3.8, 4.8, and 5.8 eV.

The MEM spectrum has a different intensity profile from that of FFT: the MEM intensities of the low-energy region increased and those of the high-energy region decreased in comparison with the FFT intensities. As shown in Fig. 4 (b), in the case of $N = 5000$ ($M_A = 166$), there are three peaks in the region from 3 to 7 eV, whereas in the case of $N = 3000$ ($M_A = 62$) and 4000 ($M_A = 62$), there are no separated peaks there.

## 4. Concluding remarks

We confirmed that MEM is an efficient method for obtaining absorption spectra from TDDFT. In our calculation, in comparison with the conventional FFT, only a half as many steps were

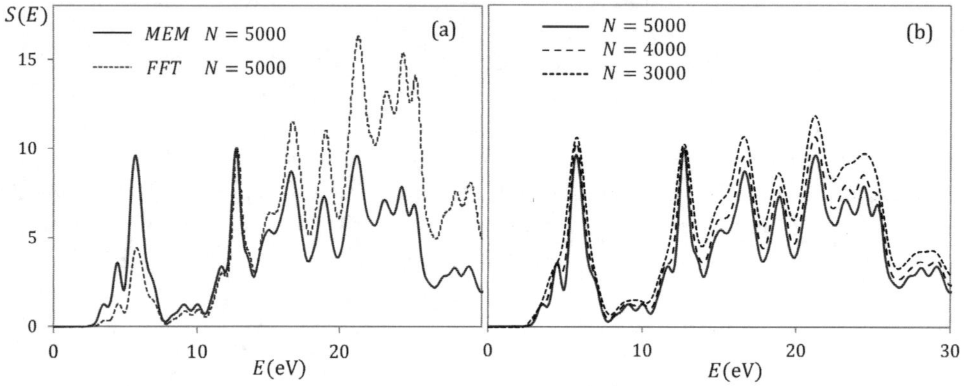


**Figure 4.** (a) Comparison of MEM (solid line) and FFT (dashed line) spectra of $C_{60}$. The number of time steps $N = 5000$ was the same for MEM and FFT. The value $M_A = 166$ was used in the MEM. (b) Comparison of MEM for $N = 3000$, 4000, and 5000 with $M_A = 62, 62$, and 166, respectively. All MEM results are scaled so that the peaks around 13 eV have the same intensities for FFT and MEM.

required to obtain the same spectral resolution, and the peak positions were as reliable as in FFT. Since the FFT or MEM procedure is carried out after the TDDFT calculation, the reduction in computational time afforded by MEM is a remarkable advantage, for example, in screening processes to find candidate materials from thousands of trial materials. The total number of steps $N$ and the lag of autocorrelation $M$ are quite sensitive to the spectrum.

## Acknowledgments
The authors would like to express their gratitude to Professor K. Yabana for providing his code as our base code and also for useful discussions and encouragement. This work was partially supported by JSPS Grants-in-Aid for Scientific Research (C) Grant number 25390158, Sumitomo Chemical Co., Ltd. and Simulatio Corporation.

## References
[1] Runge E and Gross E K U 1984 *Phys. Rev. Lett.* **52**, 997
[2] Dreuw A and Head-Gordon M 2005 *Chem. Rev.* **105**, 4009
[3] e.g. Hummer K, Puschnig P, Sagmeister S and Ambrosch-Draxl C 2006 *Mod. Phys. Lett.* B**20**, 261, and the references therein.
[4] Chelikowsky J, Troullier N, Wu K and Saad Y 1994 *Phys. Rev. B***50**, 11355
[5] Zempo Y, Akino, Ishida M, Ishitobi M and Kurita Y 2008 *J. Phys. Condens. Matter* **20**, 064231
[6] Hohenberg P and Kohn W 1964 *Phys. Rev.* **136**, B864
[7] Kohn W and Sham L. J 1965 *Phys. Rev.* **140**, A1133
[8] Troullier N and Martins J L 1991 *Phys. Rev.* B**43**, 1993
[9] e.g. Pickett W E 1989 *Comput. Phys. Reports* **9** , 115
[10] Yabana K and Bertisch G F 1996 *Phys. Rev.* B**54**, 4484
[11] Tateyama Y, Oyama N, Ohno T and Miyamoto Y 2006 *J. Chem. Phys.* **124**, 124507
[12] Shannon C 1948 *Bell Syst. Tech. J.*,**27**, 379 and 623
[13] Jaynes E T 1957 *Phys. Rev.* **106**, 620
[14] Jaynes E T 1982 *Proc. IEEE* **70**, 939
[15] Burg J P 1968 *Advanced Study Institute on Signal Processing*, 15-0, NATO, Erschede, Netherlands
[16] Haykin S and Kesler S 1979 *"Non-linear Methods of Spectral Analysis"*, ed S. Haykin, Springer (Berlin)
[17] Kauppinen J K , Maffat D F, Hollberg M R and Mantch H H 1991 *Appl. Spectrosc.* **45**, 411
[18] Akaike H 1969 *Annals of the Institute of Statistical Mathematics*, **21**, 243

[19] Akaike H 1970 *Annals of the Institute of Statistical Mathematics*, **22**, 203
[20] Yu W L, Cao Y, Pei J, Huang W and Heeger A J 1990 *Appl. Phys. Lett.* **75** 3270
[21] Wang J F, Feng J K, Ren A M, Liu X D, Ma Y G, Lu P and Zhang H X 2004 *Macromolecules* **37**, 3451
[22] Brière J F and Côté M 2004 *J. Chem. Phys.* B**108**, 3123
[23] Grice A W, Bradley D D C, Bernius, M T, Inbasekaran M, Woo EP and Wu W W 1998 *Appl. Phys. Lett.* **73**, 629

Journal of Physics: Conference Series **510** (2014) 012028 doi:10.1088/1742-6596/510/1/012028

# Pressure-induced semimetallic behavior of calcium from *ab initio* calculations

**M V Magnitskaya[1], N L Matsko[2], V S Baturin[2] and Yu A Uspenskii[2]**

[1] L.F. Vereshchagin Institute for High Pressure Physics, Russian Academy of Sciences, 142190 Troitsk, Moscow, Russia
[2] P.N. Lebedev Physical Institute, Russian Academy of Sciences, Leninskii prosp. 53, 119991 Moscow, Russia

E-mail: magnma@gmail.com

**Abstract.** A loss of metallic properties in fcc calcium under high pressure is studied *ab initio* using the density functional theory (DFT) and *GW* approximation. It is found that a more correct description of many-electron effects given by *GW* method does not provide significant changes in the behavior of electronic spectrum in comparison with DFT approach. We note that the obtained width of (pseudo)gap is highly sensitive to the **k**-point sampling used for density of states calculation. The analysis of fcc calcium's band structure at $p \sim 20$ GPa shows that the crossing of bands at the Fermi level is removed if the spin-orbit coupling is taken into account.

## 1. Introduction

Pressure-induced partial 'dielectrization' of electronic spectrum in the fcc phase of alkaline-earth metals Ca and Sr, as well as similar to them (isovalent and isostructural) rare-earth metal Yb, has long attracted attention of both experimental and theoretical physicists (see, e.g., review paper [1]). Experimentally, upon compression the electrical resistance $\rho$ of the three elements noticeably increases and reaches a maximum at a certain pressure value (not exceeding 15 GPa), the effect being most pronounced in case of Yb [2–4]. In addition, near this pressure value, a thermopower anomaly is observed for each of these elements [4–6]. Upon further compression, the re-entrance of metallic resistivity occurs at the point of structural transition to a bcc phase that remains metallic at all pressures. Notice that this pressure-induced transition occurs from a close-packed to more loosely-packed structure, in contrast, say, to alkali metals, where the opposite structural sequence is observed.

Of the three divalent elements mentioned, Ca exhibits the weakest resistance anomaly. Its resistance begins to increase at $p \sim 5$–6 GPa and reaches maximum values between the pressures of 12 and 19.5 GPa [3], the latter corresponds to the fcc→bcc transition [7]. In this pressure range, calcium behaves as a semimetal or narrow-band semiconductor, with a negative temperature coefficient of resistance from 77 to 300 K [3].

There are many *ab initio* calculations of band structure and phase transformations in Ca, however, only some of them [1, 8–13] touch upon the loss of metallicity in the fcc phase. Notice that each of these studies on Ca predicts different onset pressures for semimetallic behavior, which are, in turn, in rather poor agreement with experiment. The spread in values is rather wide — 4 to 19 GPa for metal–semimetal and 7 to 21 GPa for semimetal–metal transitions. In early publications [8–10], much attention has been paid to the choice of approximation for exchange and correlation. However, these

Journal of Physics: Conference Series **510** (2014) 012028 doi:10.1088/1742-6596/510/1/012028

efforts did not eliminate uncertainties concerning the pressure range where a semimetallic state of Ca occurs. It is also not completely clear, whether a real gap or a pseudogap exists in this pressure range. The situation looks more definite for Sr and Yb, where the energy gap, small as it is, has apparently been obtained in *ab initio* calculations [14–16].

In this paper, we report on electronic structure calculations of Ca in the range of stability of the fcc phase — from normal pressure $p$ = 1 atm to 20 GPa. The aim of this work is to find out the origin of (pseudo)gap and understand what parameters of calculation determine the pressure range of semimetallic behavior in Ca.

## 2. Calculation method

Our calculations were mainly performed in the framework of density functional theory (DFT) [17], with semilocal generalized gradient approximation (GGA) for exchange-correlation potential [18]. We employed the Vienna Ab-Initio Simulation Package (VASP) [19] and the projector augmented wave (PAW) pseudopotential method [20]. The Brillouin zone (BZ) was sampled with Monkhorst–Pack **k**-point meshes [21] and the valence electron configuration $3s^2p^64s^2$ was taken. The integration over BZ was carried out using the improved tetrahedron method [22]. The maximum number of **k**-points in the 1/48 irreducible BZ used for evaluating the electron density of states was equal to 3107 (the 50×50×50 sampling). The plane wave cutoff of 280 eV was chosen, which gives the total energy convergence of $10^{-6}$ eV. To adequately explore fine details of electronic spectra, the spin-orbit coupling was taken into account.

The DFT approach precisely describes crystal structures and total energies. The Kohn–Sham eigenvalues provide sufficiently accurate approximation for quasiparticle electronic spectra for most metals [17]. However, as is well known, the energy gaps are underestimated within DFT, which is related to the use of local exchange-correlation potentials. For this reason, we also employed the *GW* method [23], in which the quasiparticle spectra are obtained from the Dyson equation using the following simple approximation to the self-energy operator:

$$\Sigma^{GW}(\mathbf{r},\mathbf{r}',E)=\frac{\mathrm{i}}{2\pi}\int \mathrm{d}E' G(\mathbf{r},\mathbf{r}',E+E')W(\mathbf{r},\mathbf{r}',E') \quad . \tag{1}$$

Here, $G(\mathbf{r}, \mathbf{r}', E + E')$ is the single-electron Green function and $W(\mathbf{r}, \mathbf{r}', E')$ is the dynamically screened Coulomb interaction. The combination of two symbols $G$ and $W$ that appear in Eq. (1) is used as the method name. The *GW* method provides the dielectric gaps and entire quasiparticle spectra of solids in good agreement with experiment. The main shortcoming of this method is high computational cost, about 1000 times higher compared to DFT, which is due to exact calculation of $W(\mathbf{r}, \mathbf{r}', E')$. The self-energy $\Sigma^{GW}$ is non-local and energy dependent, which provides a more precise description of quasiparticle spectra.

As fully converged *GW* calculations are exceedingly demanding, we used the so-called $G_0W_0$ approximation which is known to be sufficiently accurate [24]. In this approximation, the DFT-GGA eigenvalues and eigenfunctions are used to construct a first guess to the Green function, $G_0$, and screened Coulomb interaction, $W_0$. Then the Dyson equation is solved with the self-energy $\Sigma^{G_0W_0}$. Thus, actually only the first iteration is done. Our test runs made up to the fourth iteration, $G_4W_4$, provided the results not too different from those obtained with $G_0W_0$ approximation. The *GW* calculations were made with the following parameters: maximum 145 **k**-points in the IBZ (the 16×16×16 grid), 96 bands involved in self-energy calculation, the energy cutoff of about 190 eV, and frequency grid spacing of 0.45 eV.

## 3. Results and discussion

Electronic structure of Ca was computed at several compressions $\xi = V/V_0$, where $V$ and $V_0$ are specific volumes at given pressure $p$ and normal pressure $p_0$, respectively. A value of pressure corresponding to a particular relative volume $V/V_0$ can be determined from the $T = 0$ equation of state, $p(V)$, measured

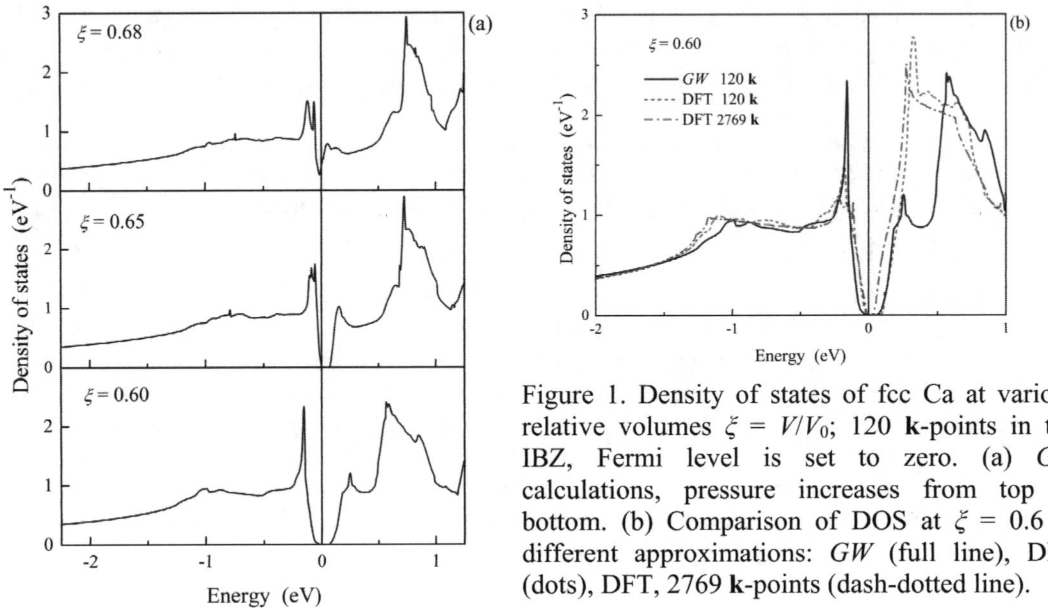


Figure 1. Density of states of fcc Ca at various relative volumes $\xi = V/V_0$; 120 **k**-points in the IBZ, Fermi level is set to zero. (a) *GW* calculations, pressure increases from top to bottom. (b) Comparison of DOS at $\xi = 0.6$ in different approximations: *GW* (full line), DFT (dots), DFT, 2769 **k**-points (dash-dotted line).

experimentally or calculated from first principles. The obtained VASP results well agree with our previous calculations [1] performed using the all-electron Augmented Plane Wave (APW) method. Hence, to evaluate the pressure, we used the equation of state obtained in [1]. The highest considered compression was $0.6V_0$, which corresponds to $p \sim 20$ GPa.

A non-zero density of states (DOS) at the Fermi level, $D(E_F)$, is a characteristic of metal. The DOS of fcc Ca calculated within *GW* approximation at three volumes is drawn in Figure 1a. On compression, $D(E_F)$ decreases and at $V/V_0 \sim 0.67$ becomes equal to zero. Upon further compression, $D(E_F)$ remains zero down to $0.6V_0$, where the bcc phase becomes energetically preferable over the fcc phase. As is seen, the gap width increases with pressure. In our DFT calculations made with the same grid of 120 **k**-points, a compression at which $D(E_F)$ vanishes is somewhat lower ($\sim 0.72V_0$). That is the change of exchange-correlation interaction somehow affects the electronic spectrum.

Figure 1b shows the comparison of DOS at $V/V_0 \sim 0.6$ for *GW* (full line) and DFT (dotted line) calculations. It is seen in the figure that there is a significant distinction between the two DOS curves in the conduction band region, while the valence band remains almost unaffected. The energy gap width also does not noticeably change as the *GW* approximation is applied.

These *GW* results are most probably not fully converged with respect to **k**-point sampling. However, increasing the **k**-point number $N_{\mathbf{k}}$ in *GW* calculations is computationally very expensive. The DFT calculations are far less demanding, so we made several test computations with considerably denser **k**-grids up to 3107 **k**-points in the IBZ. These tests demonstrate that with progressively increasing $N_{\mathbf{k}}$, the gap width is reduced and the pressure of metal-to-semimetal transition increases. The DOS for 2769 **k**-points is displayed in Figure 1b, together with the *GW* and DFT results for 120 **k**-points. The DOS profiles for 2769 (dash-dotted line) and 120 **k**-points (dotted line) are similar to each other. The main difference is that in the former case, the gap width is five times larger. As is seen, the **k**-point number $N_{\mathbf{k}}$ is more important for the gap width than the choice of approximation for exchange and correlation. Notice that the pressure of gap occurrence also significantly depends on this number.

The reason for such sensitivity of the gap width to the $N_{\mathbf{k}}$ value can be illustrated in the corresponding band structure plots. The results of our VASP calculations are presented in Figure 2, where the comparison of calcium's band structure at normal volume $V_0$ (a) and $V = 0.6V_0$ (b) is shown.

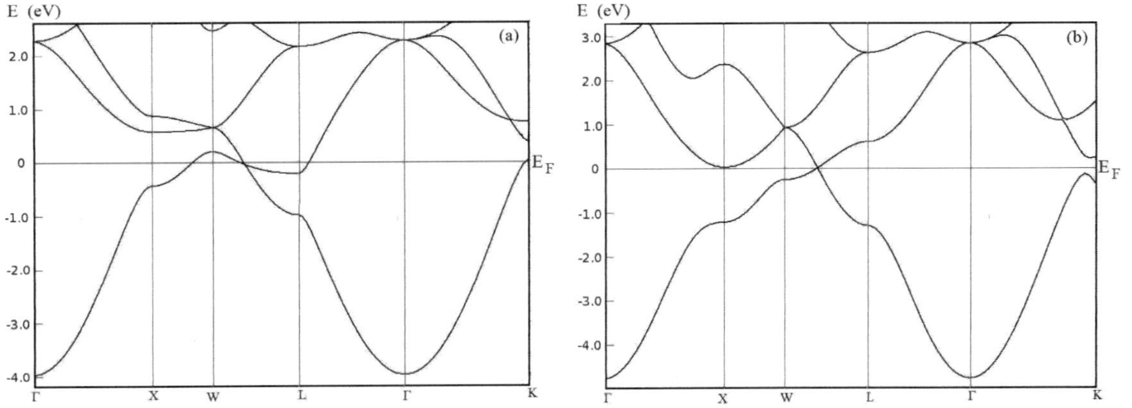


Figure 2. Band structure of fcc Ca along high-symmetry directions of BZ at normal volume $V_0$, $p \approx 0$ (a) and on compression to $V = 0.6V_0$, $p \sim 20$ GPa (b).

It is seen in the figure that the only place where the bands intersect the Fermi level $E_F$, is the point of crossing (accidental degeneracy) between the first and the second bands in the W–L line of BZ. That is, the $p = 0$ Fermi surface of Ca consisting of equal-sized hole and electron parts, almost disappears at $V/V_0 = 0.6$, which corresponds to a semimetallic state. Similar behavior of band structure upon compression has been found in other calculations of Ca [8–10], as well as Sr and Yb [14–16].

The existence of band crossing in the W–L line implies that the DOS at $E_F$ is very small but not zero. The origin of the gap seen in the DOS plots (Figure 1) is related to the way how the DOS is constructed by the corresponding algorithm. Namely, for any index $i$, the $i$th band at the point $\mathbf{k}_1$ is connected with the $i$th band at the neighbouring point $\mathbf{k}_2$. Band intersections are determined by compatibility relations for the symmetry groups of adjacent **k**-vectors. However, the procedure of DOS calculation does not support the analysis of wave function symmetry at all **k**-vectors, which is actually not important for DOS evaluation. Thus, the dependence of gap width on **k**-point number $N_\mathbf{k}$ is related to the accuracy of describing the band crossing. The larger $N_\mathbf{k}$, the narrower is the gap. Strictly speaking, the spectra shown in Figure 1 are characterized by a pseudogap with extremely low DOS values around the Fermi level. This picture is qualitatively the same within both DFT and *GW* approaches, so in *GW* calculations, one should also expect the gap narrowing with increasing $N_\mathbf{k}$. However we did not check this directly because of high computational cost involved.

It is natural to suppose that the existence of semimetallic (or semiconducting) state in fcc divalent metals can be due to the fact that the accidental degeneracy of the two lowest bands in W–L line is lifted by spin-orbit (SO) coupling. According to the relativistic *ab initio* calculations [14, 15] of heavier elements Sr and Yb (atomic numbers $Z$ = 38 and 70), the SO coupling does remove the degeneracy in the W–L direction. A maximum SO splitting $\Delta_{SO}$ of bands found for Sr and Yb is of ~0.1 and ~0.3 eV, respectively [14, 15]. For calcium with $Z = 20$, one should expect a several times smaller value of SO splitting (~0.02 eV), because $\Delta_{SO} \sim 1/Z^2$.

In paper [12] the effect of SO interaction on electronic structure of Ca has been considered using an assumed value of SO coupling constant. The paper's authors, however, refrain from exact estimates based on the fact the SO interaction for Ca is very weak. We performed relativistic PAW calculations of fcc Ca at $V = 0.6V_0$ on the mesh of 3107 **k**-points. It turned out that for the SO splitting to be reliably evaluated, a high **k**-space resolution corresponding to 400 points along the W–L direction is necessary. It should be noted that the calculation [12] of Ca band structure was made on the mesh of 89 **k**-points in the IBZ, with a larger number of 505 **k**-points used for DOS integration. Probably, it is a rather coarse **k**-point sampling that did not allow for exploring the effect of SO coupling in [12].

Journal of Physics: Conference Series **510** (2014) 012028 doi:10.1088/1742-6596/510/1/012028

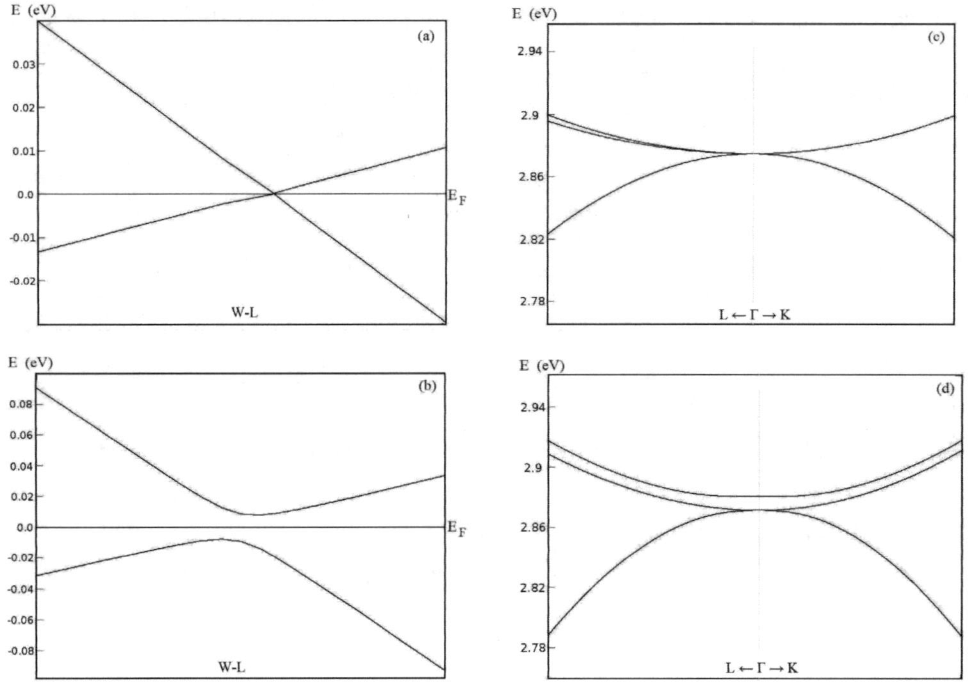


Figure 3. Band structure of fcc Ca at $V = 0.6V_0$ on an enlarged scale. (a, b) the first and the second bands in the W–L direction evaluated without and with SO coupling, respectively. (c, d) the triply degenerate $\Gamma_{25'}$ level at the $\Gamma$ point, without and with SO coupling, respectively. Energy is measured from the Fermi level $E_F$.

An enlarged view of the vicinity of $E_F$ along the W–L line calculated without and with SO coupling is shown in Figures 3a and 3b, respectively. As is seen, the SO coupling in Ca, small as it is, definitely removes the degeneracy of two lowest bands at the Fermi level $E_F$. For reference, the neighborhood of a triply degenerate $\Gamma_{25'}$ level is also presented, with and without SO coupling (Figure 3d and 3c, respectively). The threefold degeneracy of $\Gamma_{25'}$ level is seen to be lifted by SO interaction according to symmetry relations. The obtained value of SO splitting $\Delta_{SO}$ ~0.01 eV for the $\Gamma_{25'}$ level and ~0.015 eV for the levels near the crossing point. These values are close to the above estimate. It follows from the fcc lattice symmetry that the $\Gamma_{25'}$ level at the $\Gamma$ point is d-like ($t_{2g}$), while the two lowest bands in the W–L line that belongs to the hexagonal face of BZ are largely p-like. As is known, the higher orbital quantum number *l*, the less the amplitude of corresponding wave function near the core region, which mainly contributes to SO interaction. Therefore, it is only natural that SO splitting $\Delta_{SO}$ for $\Gamma_{25'}$ level is less than for the p-like levels in Figure 3b.

## 4. Conclusions

To conclude, we studied from first principles the changes that occur in the electronic properties of fcc Ca on compression down to the relative volume $V/V_0 = 0.6$ ($p \sim 20$ GPa). An analysis of factors and calculation parameters expected to influence the results was made. Our DFT calculations were made using the semilocal gradient approximation for the exchange-correlation potential. The electronic spectrum of Ca near $0.6V_0$ is observed to exhibit a pseudogap feature due to the crossing (accidental degeneracy) of two lowest bands at the Fermi level $E_F$. Application of the more exact non-local *GW* approximation significantly changes the conduction band shape, but hardly affects the valence band and pseudogap region, as compared to our DFT results.

The description of the pseudogap region is shown to be much more sensitive to the **k**-point number $N_{\mathbf{k}}$ used for DOS evaluation than to the choice of approximation for exchange and correlation. This is not surprising, because the larger $N_{\mathbf{k}}$, the more accurate is the description of the mentioned band crossing at $E_F$. In our DFT–GGA calculations, the **k**-point number was gradually increased up to 3107 **k**-points in the IBZ, while the *GW* calculations, because of their high computational cost, were made with considerably smaller $N_{\mathbf{k}}$. For the same reason, we investigated the effect of spin-orbit coupling also only within DFT. In this case the accidental degeneracy of bands is lifted, which gives rise to a rather small band splitting of 0.015 eV at the band crossing point. In principle, it is of interest to carry out similar GW calculations with increasing k-point number and SO coupling included, but in our opinion, this would not qualitatively change the obtained DFT results.

The calculated value of SO splitting is close to the limit of computational accuracy. Nevertheless such narrow gaps could affect the low-frequency response, and thereby be observable in optical experiments in the IR region. The described behavior of calcium's electronic spectrum upon compression should manifest itself in the pressure dependence of dc electrical resistivity, especially if one bears in mind that the application of high pressure produces numerous defects in a solid, with accompanying energy states within the gap. Notice that our calculations are performed at zero temperature. At room temperature, the existence of these states, with their number unknown, may lead to some discrepancies in behavior between a real material and perfect crystal. Anyway, the small energy gap should be thermally smeared out, because its estimated width is of the order of 100 K. In our opinion, it is this smeared gap which is responsible for the weak resistivity and thermopower anomalies experimentally observed in Ca at high pressure and room temperature.

**Acknowledgements**
The authors acknowledge funding from Russian Academy of Sciences, Russian Foundation for Basic Research (Nos. 12-02-31638, 12-02-31774, 13-02-00655 and 13-02-00913) and the Government of Russian Federation (No. 14.A12.31.0003). The numerical calculations are performed at the Joint Supercomputer Center of RAS.

**References**

[1] Maksimov E G, Magnitskaya M V and Fortov V E 2005 Non-simple behavior of simple metals at high pressure *Phys. Usp.* **48** 761–80
[2] McWhan D B, Rice T M and Schmidt P H 1969 *Phys. Rev.* **177** 1063–71
[3] Dunn K J and Bundy F P 1981 *Phys. Rev.* B **24** 1643–50
[4] Divakar C, Mohan M and Singh A K 1982 *Solid State Commun.* **41** 833–4
[5] Chandra Shekar N V, Meng J F, Polvani D A and Badding J V 2000 *Solid State Commun.* **116** 443–5
[6] Brazhkin V V, Tsiok O B and Magnitskaya M V 2013 *JETP Lett.* **97** 490–4
[7] Olijnyk H and Holzapfel W B 1984 *Phys. Lett.* A **100** 191–4
[8] Altmann S L, Harford A R and Blake R G 1971 *J. Phys.* F: *Metal Phys.* **1** 791–805
[9] McCaffrey J W, Anderson J R and Papaconstantopoulos D A 1973 *Phys. Rev.* B **7** 674–84
[10] Mickish D J, Kunz A B and Pantelides S T 1974 *Phys. Rev.* B **10** 1369–83
[11] Jan J P and Skriver H L 1981 *J. Phys.* F: *Metal Phys.* **11** 805
[12] Wang G M, Papaconstantopoulos D A and Blaisten-Barojas E 2003 *J. Phys. Chem. Solids* **64** 185
[13] Oganov A R, Ma Y, Xu Y, Errea I, Bergara A and Lyakhov A O 2010 *PNAS* **107**, 7646–51
[14] Johansen G and Mackintosh A R 1970 *Solid State Commun.* **8** 121–4
[15] Kubo Y 1987 *J. Phys.* F: *Metal Phys.* **17** 383
[16] Mutlu R H 1996 *Phys. Rev.* B **54** 16321–4
[17] Hohenberg P and Kohn W 1964 *Phys. Rev.* **136** B864
Kohn W and Sham L J 1965 *Phys. Rev.* **140** A1133
[18] Perdew J P, Burke K and Ernzerhof M 1996 *Phys. Rev. Lett.* **77** 3865

[19] Kresse G and Furthmüller J 1996 *Comp. Mater. Sci.* **6** 15–50
[20] Blöchl P E 1994 *Phys. Rev.* B **50** 17953–79
[21] Monkhorst H J and Pack J D 1979 *Phys. Rev.* B **13** 5188–92
[22] Blöchl P E, Jepsen O and Andersen O K 1994 *Phys. Rev.* B **49** 16223–33
[23] Hedin L and Lundqvist S 1970 Effects of electron-electron and electron-phonon interactions on the one-electron states of solids *Solid State Physics* (vol. 23) ed F Seitz, D Turnbull and H Ehrenreich (New York: Academic) pp 1–181
[24] Aryasetiawan F and Gunnarson O 1998 *Rep. Prog. Phys.* **61** 237

# Truly self-consistent solution of Kohn-Sham equations for extended systems with inhomogeneous electron gas

**A.Ya. Shul'man and D.V. Posvyanskii**

V.A. Kotel'nikov Institute of Radio Engineering and Electronics of the RAS
125009 Moscow, Russia

E-mail: ash@cplire.ru, posvyanskii@cplire.ru

**Abstract.** The density functional approach in the Kohn-Sham approximation is widely used to study properties of many-electron systems. Due to the nonlinearity of the Kohn-Sham equations, the general self-consistent solution method for infinite systems involves iterations with alternate solutions of the Poisson and Schrödinger equations. One of problems with such an approach is that the charge distribution, updated by solving the Schrödinger equation, may be incompatible with the boundary conditions of the Poisson equation for Coulomb potential. The resulting instability or divergence manifests itself most appreciably in the case of infinitely extended systems because the corresponding boundary-value problem becomes singular. In this work the stable iterative scheme for solving the Kohn-Sham equations for infinite systems with inhomogeneous electron gas is described based on eliminating the long-range character of the Coulomb interaction, which causes the tight coupling of the charge distribution with the boundary conditions. This algorithm has been previously successfully implemented in the calculation of work function and surface energy of simple metals in the jellium model. Here it is used to calculate the energy spectrum of quasi-two-dimensional electron gas in the accumulation layer at the semiconductor surface $n$-InAs. The electrons in such a structure occupy states that belong to both discrete and continuous parts of the energy spectrum. This causes the problems of convergence in the usually used approaches, which do not exist in our case. Because of the narrow bandgap of InAs, it is necessary to take the nonparabolicity of the conduction band into account; this is done by means of a new effective mass method. The calculated quasi-two-dimensional energy bands correspond well to experimental data measured by the angle resolved photoelectron spectroscopy technique.

## 1. Introduction

The density functional approach in the Kohn-Sham approximation is widely used to study properties of many-electron systems. The set of equations to describe the nonuniform electron gas of infinite system in the Kohn-Sham approximation should be satisfied by the self-consistent distribution of electron density $N(\mathbf{r})$ and Coulomb potential $U(\mathbf{r})$. Due to the nonlinearity of the Kohn-Sham equations, the general method of self-consistent solution involves iterations with the alternative solving of the Poisson and Schrödinger equations [1]. One of problems of such an approach is that the charge distribution renewed by means of the Schrödinger equation solution may be incompatible with boundary conditions on Poisson equation for Coulomb potential, as it has long been known since the articles of Bardeen [2] and Lang and Kohn [3]. The resulting instability or even divergence of iterations manifests itself most appreciably in the case of infinitely extended systems because the corresponding boundary-value problem becomes singular. It was shown in [1, 4], the published attempts to deal with this difficulty are reduced in fact to cancellation of iterative method declared initially, and its replacing somehow approximate scheme of calculation. In general, these schemes are usually semi-empirical and do not permit to evaluate the extent of deviation from the exact solution.

In [1] the iterative algorithm of solving the Kohn-Sham equations for unbounded systems with inhomogeneous electron gas has been formulated, which is based on eliminating the long-range character of Coulomb interaction being a cause of tight coupling of the charge distribution with boundary conditions. This algorithm was applied to a semi-infinite degenerate electron gas, limited infinitely high potential wall or self-consistent potential barrier [4]. The work function and surface energy of simple metals in the jellium model have been calculated and successfully compared with the published experimental and computational results [5, 6].

Here we present the results of applying our algorithm to the calculation of the energy spectrum of quasi-two-dimensional (2D) electron gas in accumulation layer on the semiconductor surface. In such a structure, the states electrons are occupied belong both discrete and continuous parts of the energy spectrum. Therefore, the high density of electrons that bound near the surface and degenerate electron gas in the bulk of semiconductor coexist simultaneously. The stable convergence of iterations was obtained even in this case, in which the self-consistency previously considered unattainable [7]. To assess the adequacy of the method at the real structures, we calculated the energy spectrum of the quasi-2D electron gas in the case of the accumulation layer at the surface of $n$-InAs. The calculated energy levels correspond well to experimental data measured by the angle resolved photoelectron spectroscopy (ARPES) method [8]. In addition, the difference has been estimated between self-consistent solutions in the Hartree approximation and with taking into account the exchange-correlation interaction.

## 2. Description of the computational scheme

In this section, we describe a theoretical framework of our analysis. The atomic system of units $|e| = m = \hbar = 1$ is used everywhere based on the charge $|e|$ and mass $m$ of the free electron, unless otherwise specified. Therein the unit of length is the Bohr radius $a_B$, and the unit of energy is Hartree $Ha = e^2/a_B$.

In the Kohn-Sham approximation, the gas of interacting electrons is described by the set of equations for single-particle wave function $\Psi_\varepsilon$

$$\frac{1}{2}\nabla^2\Psi_\varepsilon(\mathbf{r}) + (\varepsilon - U_{\text{eff}}(\mathbf{r}))\Psi_\varepsilon(\mathbf{r}) = 0. \tag{1}$$

Here $\varepsilon$ is the energy eigenvalue of single-particle state. The effective potential $U_{\text{eff}}(\mathbf{r})$ is the sum of Coulomb U and exchange-correlation $U_{\text{xc}}$ potentials

$$U_{\text{eff}}(\mathbf{r}) = U(\mathbf{r}) + U_{\text{xc}}(\mathbf{r}). \tag{2}$$

The Coulomb potential energy of electrons obeys Poisson equation and boundary conditions corresponding to the posed problem

$$\nabla^2 U = 4\pi(N_+(\mathbf{r}) - N(\mathbf{r})), \tag{3}$$

where $N_+(\mathbf{r})$ is the density of positive charge, $N(\mathbf{r})$ is the electron density.

The exchange-correlation potential $U_{\text{xc}}$ is taken in the local approximation of the density functional

$$U_{\text{xc}}(\mathbf{r}) = \left.\frac{d[N\varepsilon_{\text{xc}}(N)]}{dN}\right|_{N=N(\mathbf{r})}, \tag{4}$$

where $\epsilon_{\rm xc} = \epsilon_{\rm x} + \epsilon_{\rm c}$ is the sum of exchange and correlation energy per particle. The density of electrons is determined through the single-particle wave functions by the expression

$$N(\mathbf{r}) = 2 \int\limits_{\varepsilon \leq \varepsilon_{\rm F}} \mathfrak{D}(\varepsilon) |\Psi_\varepsilon|^2, \tag{5}$$

where factor 2 takes into account the spin degeneracy of single-particle states.

As it was pointed in Introduction, the boundary conditions to the Poisson equation impose certain requirements on the spatial distribution of electron density. The result of integration in Eq. (5) does not necessarily compatible with these requirements until the self-consistency is reached. However, the incompatibility problem of boundary conditions with the right-hand side of Poisson equation does not arise, if to present the total density of electrons as the sum

$$N(\mathbf{r}) = N_{\rm ind}\left[U_{\rm eff}(\mathbf{r})\right] + N_{\rm qu}(\mathbf{r}). \tag{6}$$

Here the induced charge density $N_{\rm ind}$ depends explicitly on the unknown potential. As a result, the long-range Coulomb interaction is replaced with the screened interaction of finite range. The induced density of electrons is related with the effective potential by an approximate local expression that is usual for the semiclassical degenerate Fermi gas

$$N_{\rm ind}(\mathbf{r}) = \frac{2^{3/2}}{3\pi^2}(\varepsilon_{\rm F} - U_{\rm eff}(\mathbf{r}))^{3/2}. \tag{7}$$

The function $N_{\rm qu}(\mathbf{r})$ in Eq. (6) determines a quantum correction to the semiclassical $N_{\rm ind}(\mathbf{r})$.

For greater transparency from here on we restrict ourselves to the approximation of the Hartree's self-consistent potential $U$. The full description of the algorithm with allowance for the exchange-correlation potential $U_{\rm xc}$ can be found in [4]. However all numerical results presented here were obtained by using the complete system of the Kohn-Sham equations including $U_{\rm xc}$.

The separation of the induced density transforms the Poisson equation (3) to the form

$$\nabla^2 U + 4\pi N_{\rm ind}(U) = 4\pi(N_+(\mathbf{r}) - N_{\rm qu}(\mathbf{r})). \tag{8}$$

The quantum correction $N_{\rm qu}(\mathbf{r})$ to the electron density introduced here by means of Eqs. (5) - (7) is defined as

$$N_{\rm qu}(\mathbf{r}) = 2 \int\limits_{\varepsilon \leq \varepsilon_{\rm F}} \mathfrak{D}\left\{\varepsilon\right\} |\Psi_\varepsilon(\mathbf{r})|^2 - N_{\rm ind}(\mathbf{r}). \tag{9}$$

The algorithm of consecutive steps $i = 0, 1, 2, ...$ for iterative solving of equations (1) and Eq. (8) can be now formulated as follows

1. One neglects small-scale quantum variations of electron density at the initial step of iterations

$$N^0_{\rm qu}(\mathbf{r}) = 0. \tag{10}$$

2. Solve the nonlinear Poisson equation with the known right-hand side and the imposed boundary conditions

$$\nabla^2 U^i(\mathbf{r}) + 4\pi N_{\rm ind}(U^i) = 4\pi\left[N_+(\mathbf{r}) - N^i_{\rm qu}(\mathbf{r})\right]. \tag{11}$$

3. Find the total electron density self-consistently with $U(\mathbf{r})$

$$N_s^i(\mathbf{r}) = N_{\rm ind}\left[U^i(\mathbf{r})\right] + N_{\rm qu}^i(\mathbf{r}). \tag{12}$$

4. Find the eigenfunctions of the single-particle Kohn-Sham Hamiltonian

$$\frac{1}{2}\nabla^2\Psi_\varepsilon^i(\mathbf{r}) + \left(\varepsilon - U_{\rm eff}^i(\mathbf{r})\right)\Psi_\varepsilon^i(\mathbf{r}) = 0. \tag{13}$$

5.Find anew the total electron density, using the obtained eigenvalues and eigenfunctions

$$N^i(\mathbf{r}) = 2\int_{\varepsilon\leq\varepsilon_{\rm F}} \mathfrak{D}(\varepsilon)|\Psi_\varepsilon^i|^2. \tag{14}$$

6. Compute the density $N_{\rm qu}(\mathbf{r})$ of quantum correction for the next step of iteration, using the found total electron density

$$N_{\rm qu}^{i+1}(\mathbf{r}) = N^i(\mathbf{r}) - N_{\rm ind}\left[U^i(\mathbf{r})\right]. \tag{15}$$

7. Return to solving the Poisson equation with the new function $N_{\rm qu}^{i+1}(\mathbf{r})$ in the right-hand side

$$i = 0, 1, 2, .....$$

The convergence criterion for this iteration process is the value of $\delta$, which measures the maximal difference between the electron density $N_s^i(\mathbf{r})$, Eq. (12), obtained after solving the Poisson equation, and the density $N^i(\mathbf{r})$, Eq. (14), found by means of the Schrödinger equation

$$\max|N_s^i(\mathbf{r}) - N^i(\mathbf{r})|/N(\infty) < \delta. \tag{16}$$

## 3. Work function of simple metals

The efficiency of the proposed algorithm has been successfully tested with the self-consistent calculations of the electronic properties of a number of typical unbounded systems [4]. From the results published out there we will present here only the part related to the calculation of the work function of simple metals in the jellium model.

In the approximation of a homogeneous positive background, the work function of simple metals was calculated in many papers. However the published data reveal sometimes not only quantitative, but also qualitative discrepancy. As a reason we believe the complete self-consistency of electron density and effective potential have not been reached in these works. The purpose of this part of the present work is to demonstrate the truly self-consistent solution to one of the classical problems of the theory of inhomogeneous electron gas, and appreciate the importance of self-consistency by comparing the results with those obtained by other methods in the same formulation of the problem. We have had no aim to improve the jellium model or the density functional theory for calculating the surface properties of simple metals. Our goal was to find an exact solution of the known problem in its conventional statement.

Let the positive background occupy the half-space $z > 0$, where 0 is coordinate at the metal-vacuum boundary. Following [4], we introduce dimensionless variables

$$\zeta = k_{\rm F}z, \qquad u(\zeta) = U(z)/\epsilon_{\rm F}^0, \qquad n(\zeta) = N(z)/N_+, \qquad R_s = \left(\frac{3}{4\pi N_+ a_B^3}\right)^{1/3}$$

where $k_{\rm F} = (3\pi^2 N_+)^{1/3}$ is the Fermi wave vector, $\epsilon_{\rm F}^0 = k_{\rm F}^2/2$ is the Fermi energy of non-interacting electrons.

The boundary condition for the Poisson equation were taken as

$$\left.\frac{du}{d\zeta}\right|_{\zeta=-\infty} = 0, \qquad \left.\frac{du}{d\zeta}\right|_{\zeta=\infty} = 0. \tag{17}$$

Hence, the neutral in a whole metal is considered. Wave functions $\Psi_\varepsilon$ of single-particle states exponentially damped at $\zeta \to -\infty$ for states with $\varepsilon \le \varepsilon_{\rm F}$. In the bulk of metal the eigenfunctions $\Psi_k$ have the asymptotic form

$$\Psi_k(\zeta) = \sqrt{\frac{2}{\pi}}\sin(k\zeta + \gamma_k), \tag{18}$$

where dimensionless quantum number $k = \sqrt{2\varepsilon}$ is normalized to $k_{\rm F}$. The eigenfunctions (18) are normalized to delta function $\delta(k - k')$.

The self-consistent electron density and effective potential have been obtained after convergence the offered iteration process at various values of Wigner-Seitz radius $0.3 \le R_s < 5$ [4]. To check the degree of self-consistency of the obtained solutions, we have used the exact strict relationship for the self-consistent electrostatic potential derived in [9]. It relates the equilibrium potential difference between the bulk of metal and the edge of the positive background with a derivative of the energy per electron with respect to the background density

$$U(0) - U(\infty) = N_+ \frac{d\epsilon}{dN_+} \equiv \tilde{\Delta}_{\rm BV}, \tag{19}$$

where $\epsilon = \frac{3}{5}\varepsilon_{\rm F}^0 + \varepsilon_{\rm xc}$ is the total mean energy per electron in the homogeneous electron gas. Contrary to the work [6] where the criterion (19) was explicitly used in the process of calculations, this relationship was not taken into account in our procedure. So, we can use Eq. (19) to verify if the calculated potential is actually self-consistent with the electron density. Good agreement between $\Delta = u(0) - u(\infty)$ and $\Delta_{\rm BV} = \tilde{\Delta}_{\rm BV}/\varepsilon_{\rm F}^0$ should prove there was obtained the truly self-consistent solution of Kohn-Sham equations.

**Table 1.** The $R_s$ dependence of the parameters $\Delta$, $\Delta_{\rm BV}$ and the calculated work function $W$.

| $R_s$ | $\Delta$ | $\Delta_{\rm BV}$ | $W$(meV) |
|---|---|---|---|
| 0.3 | 0.3751 | 0.3751 | 3.12 |
| 0.5 | 0.358 | 0.358 | 3.21 |
| 1.0 | 0.317 | 0.317 | 3.40 |
| 1.3 | 0.291 | 0.291 | 3.51 |
| 1.5 | 0.274 | 0.274 | 3.60 |
| 1.65 | 0.261 | 0.261 | 3.63 |
| 1.8 | 0.248 | 0.248 | 3.64 |
| 2.07 | 0.224 | 0.224 | 3.60 |
| 2.3 | 0.204 | 0.204 | 3.53 |
| 2.5 | 0.186 | 0.186 | 3.48 |
| 3.28 | 0.115 | 0.115 | 3.12 |
| 3.99 | 0.051 | 0.051 | 2.87 |
| 4.96 | -0.047 | -0.048 | 2.50 |

Table 1 presents the values of parameters $\Delta$ and $\Delta_{BV}$ at specified values of $R_s$. The agreement of both quantities up to three significant figures is easy to see.

Also in this table we included the values of the work function $W$ calculated for different values of $R_s$ from the found potential curves. A comparison of the values of work function calculated in [4] with published experimental and theoretical data is shown in Figure 1. Here the dependence $W(R_s)$ found in [4] (curve 3) is compared to similar ones published in the papers [5] and [6]. For $R_s > 2$ there is semi-quantitative agreement with those obtained in [5] (curve 1) and [6] (curve 2. One can see from the figure that both curves 2 and 3 have maxima at $R_s \approx 1.7$, however, at small $R_s$ there is a qualitative difference. According to [6], the work function should have had a local minimum at $R_s \approx 1$, which is not confirmed by our calculation. This local minimum has possibly resulted from poor convergency of iterations during the solution of Kohn-Sham equations at small $R_s$, which has been declared by authors of the article [6]

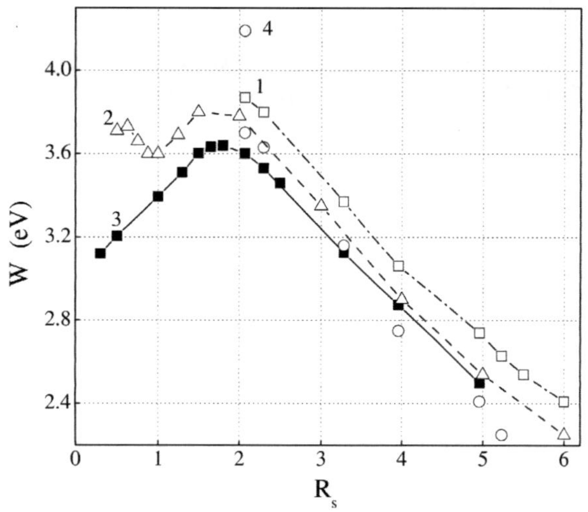


**Figure 1.** Work function in the homogeneous positive background model. Curves 1, 2 represent results of calculations by Lang and Kohn (1971) [5] and Perdue and Wang (1988) [6], respectively. Curve 3 (solid line) was calculated by the algorithm under discussion [4]. Circles 4 show experimental values of work function for simple metals. (See references in [4])

## 4. Energy spectrum of quasi-two-dimensional electron gas in accumulation layer at the semiconductor surface of $n$-InAs.

So far, we have had to deal with systems where only the states of continuous spectrum were occupied by electrons. Now let us consider the infinite systems, in which the localized states of the discrete spectrum coexist with delocalized states of the continuous spectrum. An example is the electron gas in the accumulation layer at the surface of semiconductor. As it was noted in the Sec. 3.1 of our paper [4], approximate calculation methods, that have been reported in the literature for such systems (see [10] page 463 and [7, 11]), do not lead to the self-consistent solution. Thus, it was natural to see whether our algorithm can be applied in such a problem.

First of all, it was necessary to make sure that the algorithm works for large values of $N_{qu}$, which, according to (9), have to originate from the large discrepancy between the number $N_{ind}$ of semiclassical and $N_{2D}$ quasi-two-dimensional electrons inside the quantum well. Second, the problems of this kind are typical of the physics of semiconductor nanostructures. Therefore, it was necessary to ensure that the self-consistent calculations in such a statement can be made.

One of such systems is the structure of $n$-InAs with an accumulation layer at the semiconductor surface. In these structures there is an increased concentration of charge carriers near the surface, the movement of which is limited in the direction normal to the surface. We denote this direction as $z$-axis. In the directions parallel to the surface ($x$-$y$ plane) carriers are free moving. Structures with an accumulation layer of electrons show a very large downward bending of the conduction band in the near-surface region (see Fig. 2). Hence the induced density, defined by (7), can reach sufficiently high values, which may vary widely from the true self-consistent density. What could break the convergence of the proposed iterative algorithm.

The first direct observation of surface-bound electronic states in structures based on $n$-InAs with accumulation layer was made by Tsui [12], using the tunneling spectroscopy method. However, such data are not very useful for comparison with our calculations, since the bias voltage on the structure varies during the measurement process and, accordingly, there is changing the electron energy spectrum. The development of the ARPES has enabled direct observation of these bound states and the two-dimensional electron dispersion law [8]. To make it easier to compare the results of calculations with measurements we have taken the numerical parameters of structures in accordance with the data in this paper.

Figure 2 illustrates the band diagram of structure with an accumulation layer at the semiconductor surface. It represents schematically the downward bending of the conduction band and positions of the size-quantized energy levels. The shaded area corresponds to the states of continuous spectrum, which are occupied by electrons of the degenerate gas.

Since the InAs is a narrow-bandgap semiconductor, it is important to take account for the nonparabolicity of the conduction band within calculations of energy spectrum of quasi-2D electron states. In the two-band approximation of the Kane's **kp**-theory [13, 14], the energy spectrum $\varepsilon(k)$ of narrow-bandgap semiconductors in the neighborhood of Γ-points of the Brillouin zone is determined by the equation

$$\varepsilon\left(1+\frac{\varepsilon}{E_g}\right)-\frac{k^2}{2m_0^*}=0, \tag{20}$$

where $E_g$ is the energy gap of the semiconductor, $m_0^*$ is the effective mass at the bottom of the conduction band, and $\mathbf{k}=(\mathbf{k}_{\|},k_z)$ is the quasi-wavevector of electrons. Two roots

$$\varepsilon_1(\mathbf{k})=\frac{E_g}{2}\left(\sqrt{1+\frac{2k^2}{m_0^*E_g}}-1\right),\ \varepsilon_2(\mathbf{k})=-\frac{E_g}{2}\left(\sqrt{1+\frac{2k^2}{m_0^*E_g}}+1\right) \tag{21}$$

of this equation describe the energy of electrons in the non-parabolic conduction band and that of light holes in the valence band, respectively. At $E_g\to\infty$ these expressions become the usual one-band quadratic spectra of energy for the electrons and light holes. If the periodic potential of the semiconductor crystal is perturbed by smooth non-periodic potential $U(\mathbf{r})$, the states of electrons and holes in this case are defined as eigenstates of the effective Hamiltonian and obey the effective mass equation. In the derivation of the effective Hamiltonian, either Wannier function expansion (see e.g. [15], Ch.6) or Luttinger-Kohn function expansion (see [16] and, e.g., [17], Ch.21) of the sought-for wave function is used.

In the one-band approximation, both methods lead to the same result. The effective mass equation can be obtained from the dispersion equation

$$\varepsilon-k^2/2m_0^*=0$$

by using the replacement rule

$$\varepsilon\to\varepsilon-U(\mathbf{r}),\ \mathbf{k}\to-\mathrm{i}\nabla_{\mathbf{r}} \tag{22}$$

and considering left-hand-side as the characteristic polynomial of the operator $\varepsilon_1(-\mathrm{i}\nabla_{\mathbf{r}})\equiv\hat{\mathbf{k}}^2/2m_0^*$ with the problem of eigenvalues and eigenfunctions in the form

$$[\varepsilon-U(\mathbf{r})-\varepsilon_1(-\mathrm{i}\nabla_{\mathbf{r}})]\psi_\varepsilon(\mathbf{r})=0.$$

Journal of Physics: Conference Series **510** (2014) 012029 doi:10.1088/1742-6596/510/1/012029

After we convert our Eq. (20) to the form

$$(\varepsilon - \varepsilon_2(\mathbf{k}))(\varepsilon - \varepsilon_1(\mathbf{k})) = 0, \tag{23}$$

we can apply the same replacement rule (22) to the two-band dispersion equation (20). This leads to a scalar equation of the effective mass theory[1] instead of matrix equation for two envelope functions, which are usually considered [18]. In explicit form we obtain the following approximate equation for the eigenfunctions

$$-\frac{1}{2m_0^*}\Delta\Psi_\varepsilon(\mathbf{r}) = (\varepsilon - U(\mathbf{r}))\left(1 + \frac{\varepsilon - U(\mathbf{r})}{E_g}\right)\Psi_\varepsilon(\mathbf{r}) \tag{24}$$

that can be seen directly from Eq. (20). In deriving this equation from Eq. (23), the commutator

$$[\hat{E}_1, U]\Psi_\varepsilon(\mathbf{r}) \simeq \frac{1}{m_0^* E_g}\frac{\partial}{\partial \mathbf{r}}\left(\frac{\partial U}{\partial \mathbf{r}}\Psi_\varepsilon(\mathbf{r})\right)$$

has been omitted. It is of the order of the square of the small parameter, $(L_g/L)^2$, where $L_g = 1/\sqrt{m_0^* E_g}$ and $L$ is the characteristic scale of $U(\mathbf{r})$. Thus, the applicability of the effective mass theory in the two-band case requires additional condition $L_g/L < 1$ for smoothness of the potential $U$, besides conventional $a/L << 1$, where $a$ is the lattice constant of semiconductor. In the case of InAs, $a \simeq 0.6$ nm and the ratio $L_g/a \leq 5$, so the quantum well of the accumulation layer with $L \geq 30$ nm should be good described in the effective mass approximation.

In order to find the boundary conditions for the solution of equation (24) one needs to know how that solution relates with the exact eigenstate $\Phi_\varepsilon(\mathbf{r})$ of the conduction band in the presence of the potential $U(\mathbf{r})$. The successive derivation of Eq. (24) is based on the expansion of $\Phi_\varepsilon(\mathbf{r})$ in terms of the Bloch functions $\phi_{c\mathbf{k}}(\mathbf{r})$ of the conduction band and gives rise to the expression

$$\Psi_\varepsilon(\mathbf{r}) = \int d\mathbf{r}'\Phi_\varepsilon(\mathbf{r}')w_c(\mathbf{r}' - \mathbf{x}),\ w_c(\mathbf{r}) = \int \frac{d\mathbf{k}}{(2\pi)^{3/2}}\phi_{c\mathbf{k}}(\mathbf{r}) \tag{25}$$

where $w_c(\mathbf{r})$ is the Wannier function of the conduction band. The expression (25) is simply the sliding smoothing of the function $\Phi_\varepsilon(\mathbf{r})$ with a window in the form of Wannier function. The characteristic scale of the latter is of order of the lattice constant of semiconductor. Therefore, the boundary and the normalizing conditions for the smoothed solution $\Psi_\varepsilon$ of (24) should be the same as those for the exact solution $\Phi_\varepsilon$.

The one-dimensional inhomogeneity of the accumulation layer allows one to write the three-dimensional wave function $\Psi(\mathbf{r})$ of electrons as the product of a plane wave that depends on the coordinate $\mathbf{r}_\parallel$ parallel to the surface, and the function $\psi_\varepsilon(z)$

$$\Psi_\varepsilon(\mathbf{r}) = \psi_{\varepsilon,\mathbf{k}_\parallel}(z)\frac{1}{2\pi}\exp(i\mathbf{k}_\parallel \mathbf{r}_\parallel). \tag{26}$$

Further we use the dimensionless units $n = N/N_+$, $\zeta = k_F z$, $\bar{k} = k/k_F$, $\epsilon = \varepsilon/\varepsilon_F^0$, $u_{\text{eff}} = U_{\text{eff}}/\varepsilon_F^0$ with the choice of parameters in the LDA of $U_{xc}$ as in [4]. The function $\psi(z)$ satisfies to the one-dimensional Schrödinger equation of the form

$$-\Delta^2\psi_{\epsilon,\bar{k}_\parallel}(\zeta) = (\epsilon - u_{\text{eff}}(\zeta))\left(1 + \frac{\epsilon - u_{\text{eff}}(\zeta)}{\epsilon_g} - \frac{k_\parallel^2}{k_F^2}\right)\psi_{\epsilon,\bar{k}_\parallel}(\zeta), \tag{27}$$

[1] A. Shul'man, to be published elsewhere

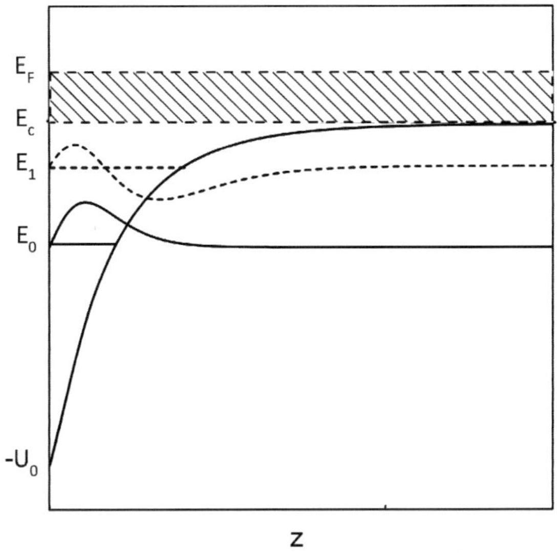


**Figure 2.** A downward band bending at $n$-InAs surface. Quantized states and their corresponding wave functions are shown.

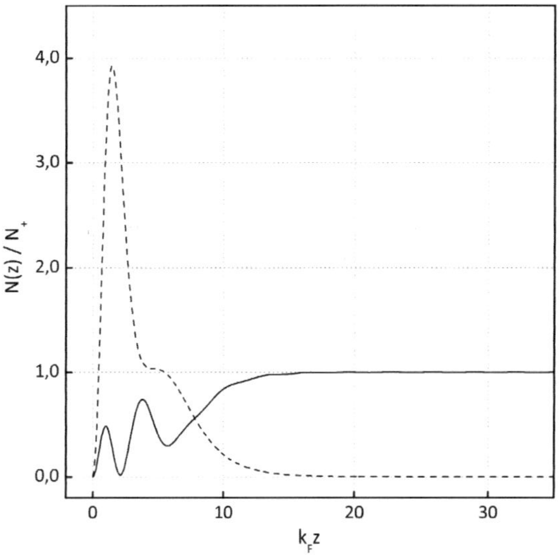


**Figure 3.** Self-consistent electron density of electrons localized in two size-quantized subbands (dashed line) and the density of delocalized electrons (solid line)

Boundary conditions for continuous states are

$$\psi_{\bar{k}_z,\bar{k}_\parallel}(0) = 0, \qquad \lim_{\zeta->\infty} \psi_{\bar{k}_z,\bar{k}_\parallel}(\zeta) = \sqrt{\frac{2}{\pi}} \sin(\bar{k}_z\zeta + \gamma_{\bar{k}_z,\bar{k}_\parallel}), \tag{28}$$

here

$$\bar{k}_\mathrm{z}^2 = \left[(\epsilon - u_\mathrm{eff}(\infty))\left(1 + \frac{\epsilon - u_\mathrm{eff}(\infty)}{\epsilon_g}\right) - \frac{k_\parallel^2}{k_\mathrm{F}^2}\right]. \tag{29}$$

For local states of the discrete spectrum we have

$$\psi_{\mathrm{j},\bar{k}_\parallel}(0) = 0, \qquad \lim_{\zeta->\infty} \psi_{\mathrm{j},\bar{k}_\parallel}(\zeta) = 0. \tag{30}$$

The effects of the nonparabolic conduction band should be included for solving Poisson equation as well. Eq. (8) can rewritten as

$$\frac{d^2u}{d\zeta^2} + c_n n_\mathrm{ind}(u) = c_n(1 - n_\mathrm{qu}(\zeta)), \tag{31}$$

where $c_n = (8/3\pi)(4/9\pi)^{1/3}R_s$ and induced density $n_\mathrm{ind}$ is given by

$$n_\mathrm{ind}(\zeta) \equiv \frac{\bar{k}_F^3(\zeta)}{3\pi^2} = \frac{2^{3/2}}{3\pi^2}\left[(\epsilon_\mathrm{F} - u_\mathrm{eff}(\zeta))\left(1 + \frac{\epsilon_\mathrm{F} - u_\mathrm{eff}(\zeta)}{\epsilon_g}\right)\right]^{3/2}. \tag{32}$$

Here the Fermi energy $\epsilon_\mathrm{F} \neq 1$ due to the $U_\mathrm{xc}$ contribution and nonparabolicity of InAs conduction band. Potential $u(\zeta)$ must satisfy the boundary condition

$$u(0) = -U_0/\varepsilon_\mathrm{F}^0 \qquad \left.\frac{du}{d\zeta}\right|_{\zeta=\infty} = 0.$$

Writing of the Poisson equation in the form of (31) corresponds to the condition that the ion charge forms the homogeneous positive background of density $N_+(z) = N_+$ for $z > 0$.

Self-consistent electron density and effective potential has been obtained by an iterative solution of the equations (27) and (31). The proposed algorithm has provided the convergence of the process. The total density of electrons was given by the expression

$$n(\zeta) = 6\int_0^1 d\bar{k}_z \int_0^{\sqrt{1-\bar{k}_z^2}} d\bar{k}_\parallel^2 |\psi_{\bar{k}_z,\bar{k}_\parallel}(\zeta)|^2 + \frac{1}{2\pi} \sum_{j=\text{occupied states}} \int_0^{\bar{k}^2_{\parallel\max}} d\bar{k}_\parallel^2 |\psi_{j,\bar{k}_\parallel}(\zeta)|^2, \quad (33)$$

where $k_{\parallel\max}$ is a root of equation $\varepsilon_j(k_\parallel^2) = \varepsilon_F$. The first term in (33) is the density of delocalized states $n_{3D}$ and the second one is the density of quasi-two-dimensional electrons $n_{2D}$.

Fig. 3 demonstrates the spatial dependence of $n_{3D}$ (solid curve) and $n_{2D}$ (dashed curve) for the case when there are two size-quantized bands in the quantum well. It can be seen, that in the region of the quantum well the density $n_{2D}$ of localized electrons, which is poorly approximated by the semiclassical expression (32), is much higher than the density of electrons in the states of the continuous spectrum. Nevertheless, the algorithm under study has shown the stable convergence of iterations.

Fig. 4 shows the energy dispersion of electronic state as a function of $k_\parallel$. We calculated the dispersion law $E_j(k)$ of quasi-two-dimensional subbands for the structure with the materials parameters presented in [8]. Circles in the Fig. 4 represent the measured energy of electrons for different values of $k_\parallel$. They were extracted from the insert in Fig. 2 of the article [8]. Authors of this work have assumed that clean surface of the samples corresponds to the condition of flat zones. In this case the measured band bending at the surface covered with some layer of Cs has been estimated as $U_0 = 400$ meV. However, the best match of our results with experimental dispersion curve was achieved with $U_0 = 435$ meV (see Fig. 4).

We believe the band bending $\Delta U_0 = 35$ meV for a clean surface could exist, but to be unobservable for two reasons: 1) the possible absence of a bound state in a shallow potential well, and 2) since this value is within the stated experimental error of order of 35 meV.

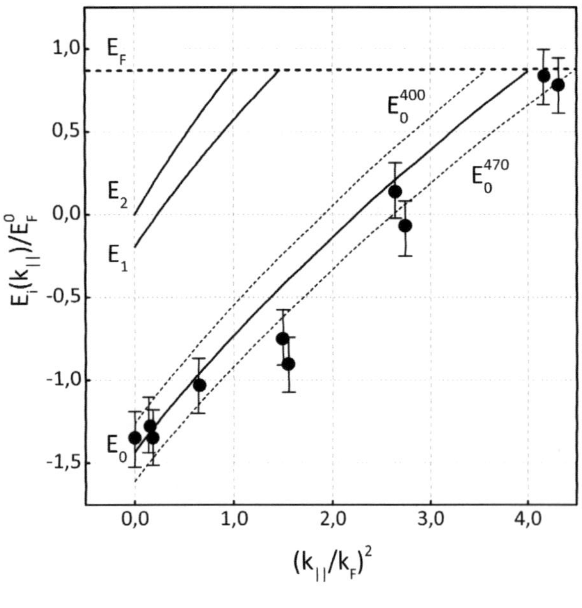


**Figure 4.** Dispersion curves of the size-quantized subbands for the band bending $U_0 = 435$ meV in the accumulation layer at the surface of n-InAs(solid lines). Circles show the experimental data for the ground subband measured in [8] by ARPES. The calculated dispersion laws for $U_0 = 400$ meV and $U_0 = 470$ meV (dashed lines) have been plotted to show the sensitivity of the results to the changes in the depth of quantum well.

In the end we recall that all presented results of calculations were obtained with account the exchange-correlation interaction. It turned out, however, that for the semiconductor structures with parameters specified in [8], contribution to the energy of size-quantized subbands due to $u_{\mathrm{xc}}$ is small and is approximately 3 meV.

**Acknowledgments**
This work was supported in part by grants of Russian Foundation for Basic Researches ##10-02-00869 and 13-02-1238.

## References

[1] Shul'man A Ya 2006 *J. Phys.: Conf. Ser.* **35**, 163
[2] Bardeen J 1936 *Phys. Rev.* **49**, 653
[3] Lang N D and Kohn W 1970 *Phys. Rev.* B **1**, 4555
[4] Posvyanskii D V and Shul'man A Ya 2009 *JETP* **109**, 145 [*Zh. Eksp. Teor. Fiz.* **136**, 169 (2009)]
[5] Lang N D and Kohn W 1971 *Phys. Rev.* B **3**, 1251
[6] Perdew J P and Wang Y 1988 *Phys. Rev.* B **38**, 12228
[7] Appelbaum J A and Baraff G A 1971 *Phys. Rev.* B **4**, 1246
[8] Aristov V Yu, Le Lay G, Zhilin V M, Indlekofer G, Grupp C, Taleb-Ibrahimi A and Soukiassian P 1999 *Phys. Rev.* B **60**, 7752
[9] Budd H F and Vannimenus J *Phys. Rev. Lett.* **31**, 1281, 1430 (Erratum)
[10] Ando T, Fowler A B and Stern F 1982 *Rev. Mod. Phys.* **54**, 437
[11] Baraff G A and Appelbaum J F 1972 *Phys. Rev.* B **5**, 475
[12] Tsui D C 1970 *Phys. Rev. Lett.* **24**, 303
[13] Kane E O 1957 *J. Phys. Chem. Solids* **1**, 249-261
[14] Abakumov V N, Perel V I, Yassievich I N 1991 *Nonradiative Recombination in Semiconductors* (Modern Problems in Condensed Matter Sciences vol. 33) ed V M Agranovich and A A Maradudin (Amsterdam: North-Holland) appendix 1
[15] Ziman J M 1972 *Principles of the Theory of Solids* 2nd edn (Cambridge: Cambridge Univ. Press)
[16] Luttinger J M and Kohn W 1955 *Phys. Rev.* **97**, 869-883
[17] Bir G L and Pikus G E 1974 *Symmetry and strain-induced effects in semiconductors* (Jerusalem: Keter Publishing House Ltd)
[18] Bastard G 1990 *Wave mechanics applied to semiconductor heterostructures* (Les Ulis: Les Editions de Physique)

# Effect of nitrogen impurity on the structural, mechanical and phonon properties of diamond from first-principle study

**T.A. Ivanova[1,2], B.N. Mavrin[1]**
[1] Institute of Spectroscopy of RAS, Fizicheskaya 5, Troitsk, Moscow 142190, Russia
[2] Technological Institute for Superhard and Novel Carbon Materials, Centralnaya 7a, Troitsk, Moscow 142190, Russia

E-mail: tas_88@mail.ru

**Abstract.** By means of density functional theory, we investigated physical properties of nitrogen-doped diamond and compared them with those of pure diamond. We showed that the nitrogen in substitution causes the strong lattice deformation near impurity, decreases elastic moduli, hardness and their anisotropy. We have confirmed experimental data that the face (111) is harder than face (100) in diamond. The nitrogen incorporation leads to an appearance of the resonant localized modes. The total and partial phonon densities of states, the mode localization, and the Raman and IR intensities in nitrogen-doped diamond were analyzed.

## Introduction

The incorporation of impurities into the crystal leads to the change in the interatomic forces in the vicinity of the impurity and a relaxation not only adjacent, but also more distant atoms from the impurity [1]. As a result of this disturbance the crystal translational symmetry is broken and localized vibrational modes in the crystal phonon structure may appear. When the masses of the atoms of the impurity and the crystal are close, as it is in the case of diamond doped with nitrogen, one can expect only the appearance of resonant localized modes insides the phonon density of states of the pure crystal. The resonance modes hybridize with the lattice vibrations, and their theoretical study requires taking into account a large number of atoms in the crystal lattice with an impurity [2]. Ab-initio quantum mechanical methods are the most accurate for the investigation of the resonant vibrational states in crystals with an impurity, as well as for calculations of the physical properties of materials [3].

The nitrogen atom is the simplest and most dominant impurity in natural diamonds and especially in diamonds synthesized with a high pressure and high temperature or by CVD it mainly substitutes the carbon atoms at concentrations <1021 atoms/cm$^3$ (diamond type Ib). The neutral nitrogen in substitutional position is a donor impurity in diamond with an ionization energy of ~ 1.7 eV. Even at low concentrations, impurities and defects play an important role in the physical properties of diamond. The experimental [4] and theoretical [5,6] studies showed that the substitution of the carbon atoms by nitrogen leads to the elongation of the one of four N-C bonds, that causes local symmetry $C_{3v}$ for a nitrogen atom. The reason for this nitrogen asymmetric position may be a Jahn-Teller instability [4] or the effects of local chemical bonds [7]. The previous ab initio calculations ([6] and refereces therein) concerned only the local structural distortion near nitrogen and electronic properties. The effect of nitrogen doping on the radial and bond angle distribution, phonon structure and mechanical properties are studied insufficiently in diamond. It is known that the IR absorption bands near 1130 [8] and 1344 [9] cm$^{-1}$ appear in nitrogen-doped diamond and their intensity is gained with doping.

In this paper, using first-principles methods in the approximation of density functional theory, we study the structural parameters, elastic constants, elastic moduli, anisotropy of the elastic properties and hardness, the phonon density of states and localization of vibrational excitations in neutral-nitrogen-doped diamond, and these data are compared with those of the diamond without impurities.

## 2. Computational details

First-principles calculations were performed in the plane wave basis approximation, using the ABINIT package [10]. The FHI98PP pseudopotentials with GGA (PBE) exchange-correlation functional (from the ABINIT site) were used in the structural and elastic calculations and LDA Troulier-Martin pseudopotentials [10] in the Raman and IR calculations. We expect core electron effects to bonding at doping. Therefore, we implemented the effectively all-electron calculations to include these effects, using the projector augmented wave method. We used cutoff energy of 20 Ha for the wave functions and 60 Ha for the charge density. During the relaxation of nitrogen-doped structures and calculating the phonon density we used a 8x8x8 Monkhorst-Pack sampling [11] and 6x6x6 at the elastic constants calculations. The convergence of computations was controlled by parameter $10^{-10}$ Ha for total electron energy and less than 0.01 eV/Å for forces. The structure of the nitrogen-doped diamond was represented by 64-atom cubic supercell (the 2x2x2 repetition of 8-atom elementary diamond cell), where one carbon atom was replaced by nitrogen ($NC_{63}$ cell). The distance between nitrogen atoms in the next supercells exceded 7 Å that allowed one not to consider an interaction between them. The supercell structure was relaxed with respect to the atom positions and lattice constants.

**Table 1.** Strains used to calculate elastic constants of cubic $NC_{63}$. Strains 2 and 3 are volume-conserving. $B_2^V$ and $B_2^\delta$ are quadratic coefficients in the polynomial approximation of $\Delta E(V)$ and $\Delta E(\delta)$, respectively.

| | Strains | Elastic constants |
|---|---|---|
| 1 | $\varepsilon_{xx} = \varepsilon_{yy} = \varepsilon_{zz} = \delta$ | $c_{11} + 2c_{12} = 3B_2^V V_0$ |
| 2 | $\varepsilon_{xx} = \delta,\ \varepsilon_{yy} = -\delta,\ \varepsilon_{zz} = \delta^2(1-\delta^2)^{-1}$ | $c_{11} - c_{12} = B_2^\delta / V_0$ |
| 3 | $\varepsilon_{xy} = \varepsilon_{yx} = \delta/2,\ \varepsilon_{zz} = \delta^2(4-\delta^2)^{-1}$ | $c_{44} = 2B_2^\delta / V_0$ |

For the calculating of elastic constants $c_{ij}$ we used the method of small strains $\varepsilon_{ij}$ imposed to the equilibrium structure with the cell volume $V_0$, that leaded to the change of total electron energy $E^{tot}$ after relaxation on size of elastic energy $\Delta E$ [12]:

$$\Delta E = E^{tot}(\varepsilon) - E^{tot}(0). \tag{1}$$

Cubic crystal has only three independent elastic constants of stiffness: $c_{11}$, $c_{12}$ and $c_{44}$ [13], which can be found using 3 types of strains with $\delta = -0.02 \div 0.02$ (Table 1). Elastic constants of compliance $s_{ij}$ are associated with elastic constants of stiffness $c_{ij}$ in cubic crystal by a ratio [13]:

$$\begin{gathered} s_{11} = (c_{11} + c_{12}) / s_0, \\ s_{12} = -c_{12} / s_0, \\ s_{44} = 1 / c_{44}, \\ s_0 = (c_{11} - c_{12})(c_{11} + 2c_{12}). \end{gathered} \tag{2}$$

In the isotropic approximation bulk modulus $B$, shear modulus $G$, Young's modulus $E$ and Poisson's ratio $\sigma$ are also defined by constants $c_{ij}$ [14]:

$$\begin{gathered} B=(c_{11}+2c_{12})/3, \\ G=(G_V+G_R)/2, \\ G_V=(c_{11}-c_{12}+3c_{44})/2, \\ G_R=5c_{44}(c_{11}-c_{12})/(4c_{44}+3(c_{11}-c_{12})), \\ E=9BG/(3B+G), \\ \sigma=(3B-2G)/(6B+2G). \end{gathered} \tag{3}$$

If we consider anisotropy of the cubic medium, the elastic moduli depend on direction [15]:

$$\begin{gathered} G_{lmn}^{-1}=s_{44}-4(s_{11}-s_{12}-s_{44}/2)\Phi, \\ E_{lmn}^{-1}=s_{11}-2(s_{11}-s_{12}-s_{44}/2)\Phi, \\ \sigma_{lmn}=-\frac{s_{12}+(s_{11}-s_{12}-s_{44}/2)\Phi}{s_{11}-2(s_{11}-s_{12}-s_{44}/2)\Phi}, \\ \Phi=l^2m^2+m^2n^2+l^2n^2, \end{gathered} \tag{4}$$

where $l$, $m$, $n$ are the directional cosines for direction $[lmn]$. The expression $\Phi$ in Eq. (4) is equal zero for the direction [100] and 1/3 for [111].

For the computation of phonon density of states in $NC_{63}$ we used the direct method of calculating interatomic force constants. In this method first of all we calculate forces arising between atoms by shifting each atom from equilibrium in direction of axes x, y and z by ± 0.02 Å. For the sake of minimization our calculations we used symmetrical properties of crystal $NC_{63}$.

The force constants matrix with the dimension 192x192 were constructed by dividing derived forces by the atom shifts, then the standard crystal lattice theory was applied for the phonon density of states (PDOS) computation using program PHON [16]. We calculated total PDOS and partial PDOS which reproduces separately the nitrogen and carbon atoms contribution to the nitrogen and carbon atoms contribution to the PDOS. Partial PDOS was calculated also for different groups of carbon atoms equidistant from nitrogen atom. The PDOS and partial PDOS calculations were based upon the atomic displacements in the normal modes and the results were broadened by 0.3 $cm^{-1}$.

## 3. Results and discussions

### *3.1. Structural properties*

After the relaxation nitrogen atom turned out to be displaced along the direction [111], occupying local position $C_{3v}$ and retaining the fourfold coordination, as it was showed in previous works [5,6]. In the $NC_{63}$ cell the carbon tetrahedrons adjacent to the nitrogen has been deformed. For the detailed investigation of local structure we calculated the radial and bond angle distributions in crystal $NC_{63}$.

The distribution of interatomic bondlengths in relaxed $NC_{63}$ crystal cell (Figure 1) shows that one N-$C_1$ bond has the length about 2 Å, while three others N-$C_3$ less than 1.5 Å. Most of the others C-C bonds length are groupped near 1.54 Å that is near the C-C length in pure diamond. But three $C_1$-$C_2$ bonds adjacent to the lengthy bond $C_1$-N become shorter than 1.5 Å. The lattice constant of cubic crystal $NC_{63}$ increases slightly in comparison with the doubled lattice constant of the diamond cell (7.159 and 7.09 Å, correspondingly). The lattice expansion at substitutional nitrogen doping was observed in experiments [17,18].

The bond angle distribution, that is the angles distribution between adjacent bonds for every atom shows (Figure 2) that average angle between bonds is close to the tetrahedral angle $\theta = 109.47^{\circ}$. Some deviations from $\theta$ are observed only for atoms, bonded to the nitrogen atom and to the nearest carbon atoms. From the radial (Figure1) and bond angle (Figure 2) distributions we can conclude that the substitutional nitrogen atom in diamond brought only local distortions in the diamond lattice.

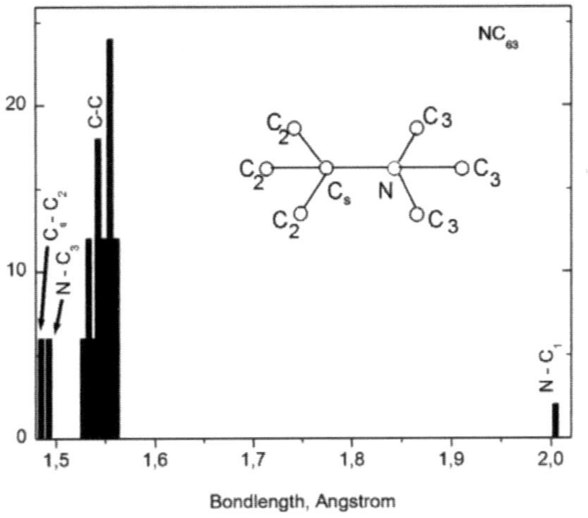


**Figure 1** The distribution of bonds depending on bondlengths in a relaxed crystal $NC_{63}$. Insert: a schematic arrangement of the nearest carbon atoms near nitrogen.

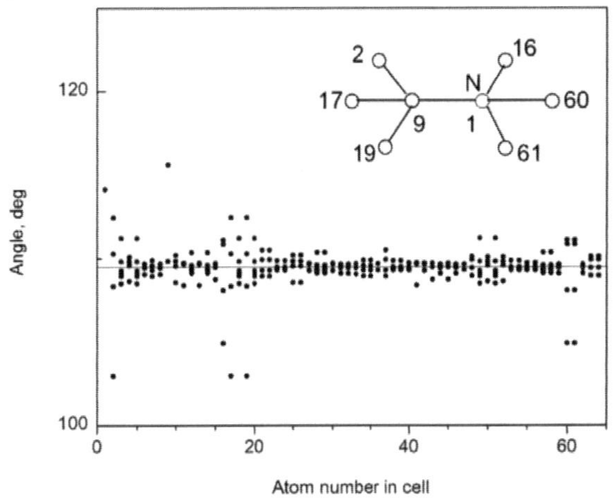


**Figure 2** The bond angle distribution depending on the distance from the origin of the $NC_{63}$ cell.

### 3.2. Mechanical properties

The calculated dependences $\Delta E^{tot}(V)$ and $\Delta E(\delta)$ allowed us to find the bulk modulus $B$ and elastic constants $c_{11}$, $c_{12}$ and $c_{44}$ for diamond and crystal $NC_{63}$. The calculated elastic constants and the elastic moduli in the isotropic approximation for diamond and crystal $NC_{63}$ are presented in Table 2. The theoretical values for diamond are compared with experimental data [19-21]. The elastic constants and moduli of crystal $NC_{63}$ were less than in diamond, that point to the lower stiffness and greater compressibility of nitrogen-doped diamond.

**Table 2.** Elastic constants and moduli in the isotropic approximation (in GPa) for diamond and $NC_{63}$.

| | | $c_{11}$ | $c_{12}$ | $c_{44}$ | $B$ | $G$ | $E$ | σ | $A$ | $k$ |
|---|---|---|---|---|---|---|---|---|---|---|
| Diamond | Calc. | 1022 | 151 | 595 | 467 | 543 | 1174 | 0.081 | 1.37 | 1.16 |
| | Exp. | 1078 | 126 | 577 | 444 | 535 | - | - | - | - |
| $NC_{63}$ | Calc. | 1026 | 134 | 532 | 432 | 496 | 1076 | 0.084 | 1.19 | 1.15 |

The distinction between the diamond and crystal $NC_{63}$ elastic properties is clearly visible using a graphical representation of a quasilinear states equation [22]:

$$\mathrm{Ln}[px^2/(3(1-x))] = \mathrm{Ln}Y = \mathrm{Ln}B + A_0(1-x), \\ x = (V/V_0)^{1/3},\ A_0 = 3(B'-1)/2, \quad (5)$$

where $p$ is the pressure and $B' = dB/dp$. The pressure $p$ was obtained after the cell relaxation at a given $x$. Figure 3 shows that the dependences Ln$Y$ are almost linear for both crystals. As far as Ln$Y$ = Ln$B$ when $x = 1$, it allows us to find bulk moduli (Table 2). The value Ln$Y$ at $x = 1$ for a crystal $NC_{63}$ has less valuc than for the diamond (Figure 3) and the bulk modulus is accordingly smaller. The inclination of the linear dependence *LnY* that is the coefficient $A_0$ in the Eq. (5), is a little bit more for $NC_{63}$, than for diamond, therefore the coefficient $B'$ is also larger (3.85 and 3.70, correspondingly).

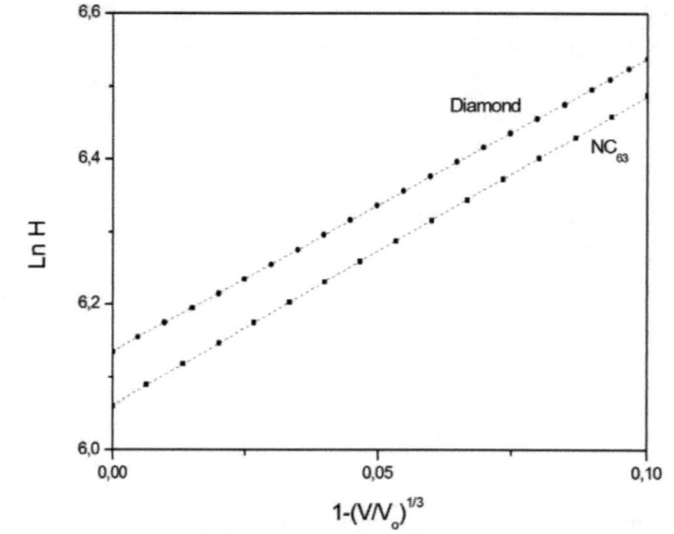


**Figure 3** The quasilinear equation of states for diamond and $NC_{63}$. The parameter 1-$(V/V_0)^{1/3}$ corresponds to a range of pressure of 0-250 GPA.

The material elasticity can be characterized by Pug's parameter [23]: $k = G/B$. Since this parameter is much greater than 0.5 in both crystals (Table 2), materials are brittle, although the parameter $k$ in the crystal $NC_{63}$ is smaller. The higher $NC_{63}$ elasticity is confirmed by greater value of Poisson's ratio $\sigma$, which shows the change in the transversal dimensions at the longitudinal elongation.

The elastic anisotropy of isotropic medium is usually estimated by Zener factor [24] $A = 2c_{44}(c_{11} - c_{12})^{-1}$. Table 2 shows that the factor $A$ decreases in the crystal $NC_{63}$ compared to diamond, that means that nitrogen-doped diamond becomes more isotropic.

A more detailed analysis of the elastic properties anisotropy can be carried out with the help of Eq. (4) and the results are given in Table 3. Anisotropic elastic moduli $G_{100}$ and $G_{111}$ (Table 3), exceed the isotropic value of $G$ (Table 2), that may indicate that isotropic approximation is insufficient for the study of elastic properties in both diamond and crystal $NC_{63}$. The anisotropy of elastic properties , i.e. relations $G_{111}/ G_{100}$ and $E_{111}/ E_{100}$ are less in crystal $NC_{63}$ than in diamond that confirms smaller anisotropy of elastic properties in crystal $NC_{63}$ in comparison with diamond.

**Table 3.** Elastic constants of compliance $s_{ij}$ (in $10^{-4}$ $GPa^{-1}$) and anisotropic elastic moduli (in GPa) for diamond and $NC_{63}$.

| | $s_{11}$ | $s_{11}$ | $s_{11}$ | $G_{100}$ | $G_{111}$ | $E_{100}$ | $E_{111}$ | $\sigma_{100}$ | $\sigma_{111}$ |
|---|---|---|---|---|---|---|---|---|---|
| Diamond | 10.172 | -1.309 | 16.806 | 595 | 787 | 1017 | 1231 | 0.128 | 0.035 |
| $NC_{63}$ | 10.054 | -1.161 | 18.801 | 532 | 610 | 1005 | 1131 | 0.115 | 0.063 |

Nowadays the hardness calculation is based on semi-empirical models ([25] and references there), from which we have used a model that assumes a correlation between the hardness $H$ and the elastic moduli [25,26]:

$$H = 2(G^3 / B^2)^{0.585} - 3 \quad (6)$$

This model has been satisfactorily described the hardness of various materials [25]. If in Eq. (6) we use the value $G$ in the isotropic approximation (Table 2), we find that hardness is 92 GPa for diamond and 86 GPa for crystal $NC_{63}$. Thus the hardness of nitrogen-doped diamond decreases. If we use the anisotropic elastic moduli values (Table 3), we find that the $H_{100}$ = 108 GPa and $H_{111}$ = 152 GPa for diamond and 98 and 132 GPa, correspondingly, for the crystal $NC_{63}$. The calculated hardness anisotropy in diamond correlates with the experimental data [27] and suggests that hardness of the face (111) is much larger than the face (100) hardness. The ratio $H_{111}/ H_{100}$ in diamond is higher than in the crystal $NC_{63}$, i.e. the hardness of nitrogen-doped diamond is not only reduced, but also have tendency of reduction the anisotropy.

### *3.3. Phonon density states and localization*

In the phonon density of states (PDOS) of the Crystal $NC_{63}$ (Figure 4), that it is similar to thediamond PDOS, the most intense band near 1220 $cm^{-1}$ is downshifted by 40 $cm^{-1}$ and all the bands are much wider. Since the band at 1220 $cm^{-1}$ is caused by the contribution of optical modes in diamond, its low-frequency shift in $NC_{63}$ indicates that the interatomic interaction between the carbon atoms weakens in nitrogen-doped diamond.

To investigate the nitrogen atom contribution and local carbon atoms contribution in the PDOS, we calculated the partial contributions of atoms in $NC_{63}$ cell (Figure 5). The inset in Figure 5 shows one eighth cell. The carbon atom 6 is the furthest from the nitrogen atom 1. The nitrogen atom forms with the carbon atom 3 a long bond (~ 2 Å), and a short bond with the atom 2. The nitrogen atom contribution (PDOS1 in Figure 5) is concentrated mainly near 800 and 1100 $cm^{-1.}$ The majority of carbon atoms contribution is maximum at 1220 $cm^{-1}$ (PDOS2, PDOS4-PDOS6), except the atom 3 contribution, which is removed from the nitrogen atom and forms three short bonds with nearest carbon atoms, that causes the appearance of band at 1329 $cm^{-1}$ (PDOS3). Thus, the vibrations with the nitrogen participation have frequencies within the diamond PDOS, i.e. the carbon atom replacement by nitrogen leads only to the emergence of the impurity resonant localized modes in the diamond phonon structure.

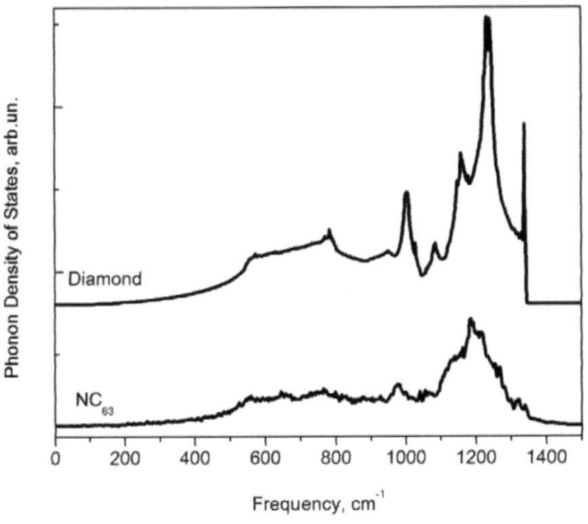


**Figure 4** The total PDOS in diamond and $NC_{63}$

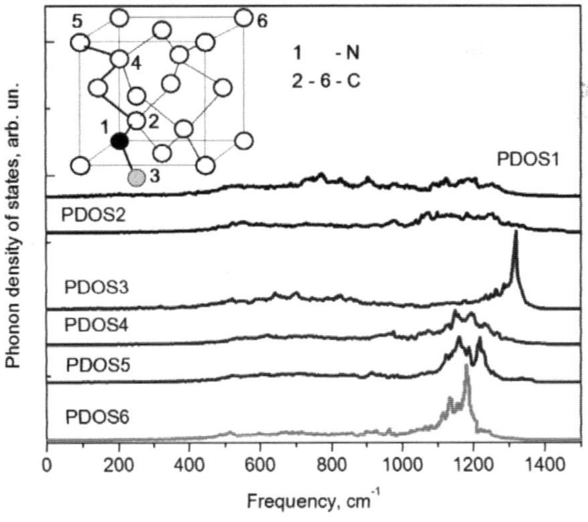


**Figure 5** The partial PDOS of atoms in $NC_{63}$: PDOS1 is the partial PDOS of nitrogen, PDOS2-PDOS6 are the partial PDOS of carbon atoms 2-6 that are shown in inset. Inset: one eighth cell $NC_{63}$.

As far as in optical experiments (Raman scattering and IR absorption) the spectrum intensity is determined mainly by modes in the centre of Brillouin zone, we investigated PDOS of modes with $\boldsymbol{q}$ = 0. In particular, the local vibrational density of nitrogen atom (LPDOS) in $NC_{63}$ at $\boldsymbol{q}$ = 0 can be found as [28]

$$LPDOS(\omega) = g(\omega)\left|\boldsymbol{e}(N,\omega)\right|^2, \qquad (7)$$

where $g(\omega)$ is the total PDOS and $\boldsymbol{e}(N, \omega)$ is the nitrogen displacement vector in the normalized eigenvector of the mode at the ω frequency at $\boldsymbol{q} = 0$. Figure 6 shows that intensity maxima in LPDOS are in the same frequency region as in the partial PDOS1 (Figure 5), calculated for the contribution of the nitrogen atom to PDOS from all Brillouin zone. The displacements of the nitrogen and the nearest carbon atoms are shown in Figure 7 (inset) for the mode 800 cm-1 with the intensive LPDOS. It is seen that in this mode the atom of nitrogen moves in an antiphase with the closest atoms of carbon. The displacements amplitudes for the other carbon atoms in this mode are much smaller.

The eigenvectors at $\boldsymbol{q} = 0$ allowed us to investigate the mode localization in $NC_{63}$. For this one often calculate the inverse participation ratio (*IPR*), being the next sum over all cell atoms $n$ [29]:

$$IPR(\omega) = \Sigma[\boldsymbol{e}_i(N,\omega)\cdot\boldsymbol{e}_i(N,\omega)]^2, \qquad (8)$$

$IPR(\omega) = 1$ for a fully localized mode and $IPR(\omega) = 1/n$ for delocalized modes. In Figure 7 the majority of the modes have $IPR(\omega) \approx 2/n$ and the most localized are acoustic mode at 330 cm-1 and the optical mode at 1300 cm-1 with $IPR(\omega) \approx 6/n$. The modes at 600, 800 and 1100 $cm^{-1}$ with the greatest nitrogen participation also have values ($IPR(\omega) \approx 4/n$), that exceed IPR of the majority of modes. Hence, impurity modes in nitrogen-doped diamond are weakly localized because of the strong hybridization with lattice modes.

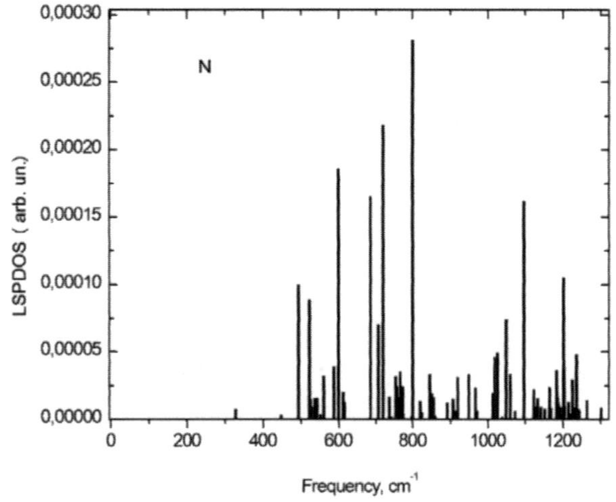


**Figure 6** The local vibrational density of nitrogen in $NC_{63}$ at $\boldsymbol{q} = 0$.

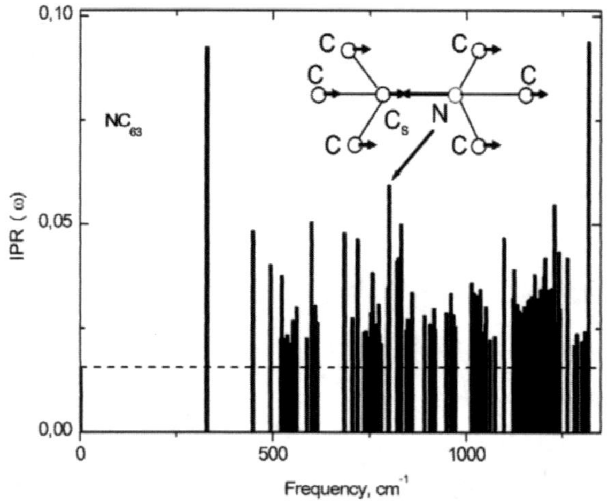


**Figure 7** The inverse participation ratio IPR for modes in $NC_{63}$ depending on frequency. Inset: displacements of the nitrogen and the nearest carbon atoms in mode at 800 $cm^{-1}$.

*3.4. Raman and IR spectra*

We have calculated the Raman and IR spectra for the crystal $NC_{63}$ in LDA approximation. Previously the relaxation with this pseudo-potential was carried out. The nonresonant Raman spectrum intensity is proportional to the differential scattering cross section, which is determined by the square of the components in Raman tensor. The Raman tensor components were calculated in the Placzek approximation [30]. In the scattering geometry which is defined by tensor component $\alpha_{zz}$, more intensive are the bands that are below 900 $cm^{-1}$, whereas in the crossed geometry $\alpha_{xz}$ intense bands are concentrated near 1300 $cm^{-1}$ (Figure 8). The nitrogen atom contribution is maximum in the intensity of bands marked by asterisks in Figure 8.

The infrared absorption intensity is proportional to the square of the tensor components of the effective Born charge and eigenvectors [31]. The Born effective charge tensor was determined by the derivative of the forces acting upon the atom to the applied electric field [32]. The nitrogen atom has the largest contribution to the IR intensity for the bands near 500 and 1120 $cm^{-1}$ (Figure 8). Previously [8] the band near 1130 $cm^{-1}$ was observed in the IR spectrum and its intensity depended on the nitrogen concentration. In the calculated IR spectrum are seen the intense bands in the 1310-1330 $cm^{-1}$ region with a small nitrogen contribution. Their intensity is mainly caused by the shortened $C_2$-$C_1$ bonds adjacent to the lengthy bond $C_1$-N (Figs. 1 and 5). These bands may be assigned to the local vibrational modes of carbons near nitrogen. If to consider that frequencies are underestimated in our calculations by ~ 1 %, it is possible assume that the observed IR band at 1344 $cm^{-1}$ [9] is one of these modes.

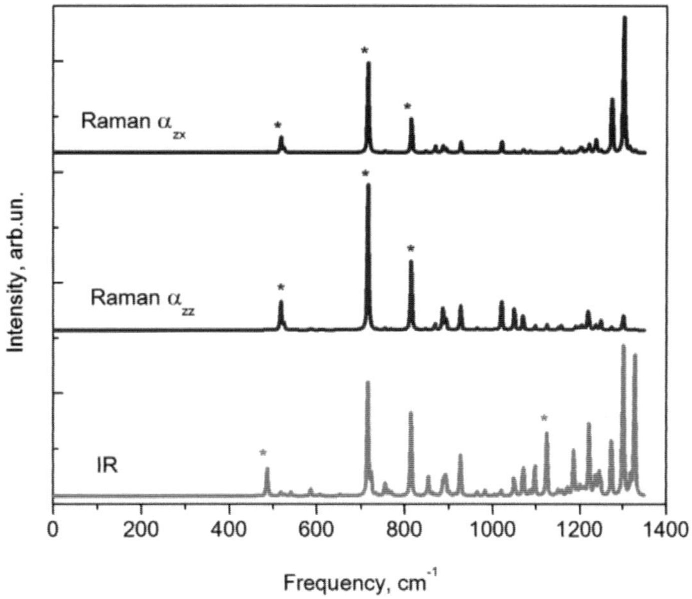


**Figure 8** The Raman ($\alpha_{zz}$, $\alpha_{xz}$) and IR intensities of modes in $NC_{63}$. The modes with the greatest contribution of nitrogen are designated by an asterisk.

## 4. Conclusion

Using first-principles calculations we investigated the structural, mechanical and phonon properties of the diamond doped with nitrogen and compared them with the properties of pure diamond. The calculated results for the pure diamond are in good agreement with experimental data and other ab-initio calculations.

When nitrogen is in a substitution position, the lattice constant of diamond slightly increases, the elastic constants and moduli reduces. In nitrogen-doped diamond the elastic anisotropy decreases, the crystal becomes less brittle. The investigation of elastic anisotropy has shown that isotropic approximation is nonsufficient for the study of elastic properties in both pure and nitrogen-doped diamond. The hardness of nitrogen-doped diamond decreases. The calculated hardness anisotropy of diamond confirmed the experimental data that the hardness of the face (111) is much higher than the hardness of the face (100). The nitrogen doping tends to reduce the hardness anisotropy.

The PDOS bands of diamond are shifted down at nitrogen doping that testifies to weakening of interatomic interaction in nitrogen-doped diamond. The study of partial PDOS has shown that the nitrogen contribution is seen only within the PDOS of pure diamond that causes an appearance of resonant localized modes. The modes with the greatest nitrogen participation are more localized in

Journal of Physics: Conference Series **510** (2014) 012030 doi:10.1088/1742-6596/510/1/012030

comparison with the majority modes. Unlike pure diamond the Raman tensor component $\alpha_{zz}$ is not equal to zero. The crossed component $\alpha_{xz}$, as expected, is largest for modes near 1300 $cm^{-1}$.

Our calculations are consistent with the following experimental data: a $C_{3v}$ local symmetry of nitrogen site in diamond, a lattice expansion in nitrogen-doped diamond, the frequencies of localized modes of carbons near nitrogen have the highest value in comparison to other modes, the noticeable contribution of nitrogen in IR intensity near 1100 $cm^{-1}$, the calculated elastic constants in diamond, hardness of the (111) face in diamond is higher than that of the (100) face.

Results of the carried-out calculations showed that, as masses of nitrogen and carbon differ a little, local modes of nitrogen have a resonant character, they are strongly hybridized with the diamond modes and they become poorly localized. On the other hand, a lattice distortion by nitrogen has the local character in nitrogen-doped diamond.

**Acknowledgements**
This work was supported by program "Physics of New Materials and Structures" of OFN RAS. The calculations were carried out at Interdepartmental Supercomputer Centre of RAS.

**References**

[1] Maradudin A A 1966 *Solid State Physics* **18** 273
[2] Petzke K 1999 *Phys. Rev.* B **60** 12726
[3] Baroni S, de Gironcoli S, Corso A D and Giannozzi P 2001 *Rev. Mod. Phys.* **73** 515
[4] Smith W V, Sorokin P P, Gelles I L and Lasher G J 1959 *Phys. Rev.* **115** 1546
[5] Kajihara S A, Antonelli A, Bernholc J and Car R 1991 *Phys. Rev. Lett.* **66** 2010
[6] Lombardi E B, Mainwood A, Osuch K and Reynhardt E C 2003 *J. Phys.: Condens. Matter* **15** 3135
[7] Bachelet G B, Baraff G A, and Shluter M 1981 *Phys. Rev.* B **24** 4736
[8] Collins A T 1980 *J. Phys.* C **13** 2641
[9] Collins A T and Woods G S 1982 *Phil. Mag.* **46** 77
[10] www.abinit.org
[11] Monkhorst H J and Pack J D 1976 *Phys. Rev.* B **13** 5188
[12] Wu Z, Chen X-J, Struzhkin V V, and Cohen R E 2005 *Phys. Rev* B **71** 214103
[13] Nye J F 1985, *Physical Properties of Crystals: Their Representation by Tensors and Matrices* Oxford Univ. Press.
[14] Wu Z-j, Zhao E-j, Xiang H-p, Hao X-f, Liu X-j and Meng J 2007 *Phys. Rev.* B **76** 054115
[15] Date E H F and Andrews K W 1969 *Brit. J. Appl. Phys. (J. Phys. D)* **2** Ser. 2 1373
[16] Alfè D 2009 *Computer Physics Communications* **180** 2622
[17] Kiflavi I, Mayer A E, Spear P M, Van Wyk J A and Woods GS 1994 *Phil. Mag.* B **69** 1141
[18] Boyd S R, Kiflavi I andWoods G S 1994 *Phil. Mag.* B **69** 1149
[19] McSkimin H J and Andreath J P 1972 *J. Appl. Phys.* **43** 2944
[20] Grimsditch M H and Ramdas A K 1975 *Phys. Rev.* B **11** 3139
[21] Zouboulis E S, Grimsditch M, Ramdas A K and Rodriguez S 1998 *Phys. Rev.* B **57** 2889
[22] Francisco E, Blanco M A and Sanjurio G 2001 *Phys. Rev.* B **63** 094107
[23] Pugh S F 1954 *Philos. Mag.* **45** 823
[24] Zener C 1948 *Elasticity and anelasticity of metals*, Chicago, University of Chicago Press
[25] Chen X-Q, Niu H, Li D and Li Y 2011 *arXiv*: 1102.4063v1 [cond-mat.mtrl-sci]
[26] Niu H, Wei P, Sun Y, Chen X-Q, Franchini C, Li D and Li Y 2011 *Appl. Phys. Lett.* **99** 031901
[27] Blank V, Popov M, Lvova N, Gogolinsky K and Reshetov V 1997 *J. Mater. Res.* **12** 3109
[28] Fultz B 2010 *Progress in Materials Science* **55** 247
[29] Atta-Fynn R and Biswas P 2009 *J. Phys.: Condens. Matter.* **21** 265801
[30] Sosso G S, Caravati S, Gatti C, Assoni S and Bernasconi M 2009 *J. Phys.: Condens. Matter* **21** 245401
[31] Giustino F and Pasquarello A 2008 *Phys. Rev.* B **78** 075307
[32] Gonze X and LeeC 1997 *Phys. Rev.* B **55** 10355

# Crystal equivalent temperature model in process of nonlinear conversion of laser radiation

**O A Ryabushkin, D V Myasnikov and A I Baranov**
Moscow Institute of Physics and Technology, Institutskiy per. 9, Dolgoprudnyy, 141700, Russia
Kotelnikov Institute of Radio-engineering and Electronics of RAS, Vvedensky Sq. 1, Fryazino, 141190, Russia

E-mail: roa228@mail.ru

**Abstract.** Theoretical modeling was performed for estimation of influence of thermal effects in nonlinear-optical crystals on process of second harmonic generation of laser radiation. Due to absorption of both pump and generated radiation the temperature distribution inside crystal is substantially nonuniform as in cross-section as along its length. Three dimensional heat conduction problem was solved considering two types of heat sources, conditioned by exponential decay and parabolic growth of laser radiation intensity along crystal length. Longitudinal temperature gradient inside crystal is detrimental for process of laser radiation conversion.

## 1. Introduction

Currently due to rapid growth of laser physics and technology one of the urgent goals is development of laser sources that can generate high-power radiation at wavelengths in broad spectrum range. As a rule, generation wavelengths of modern laser sources are specified by the active media and cannot be widely tuned. Promising solution of spectral broadening is conversion of laser radiation wavelength in nonlinear-optical crystals [1]. Nonlinear-optical processes include parametric oscillation as well as multiple harmonic generation of laser radiation. These processes are coherent i.e. phase difference or phase velocity relation between interacting electromagnetic waves plays crucial role. As refractive indices of the crystal and therefore phase velocities of waves to a considerable extent depend on crystal temperature the last one should be accurately measured and controlled for efficient radiation conversion. Recently noncontact impedance spectroscopy method has been proposed for precise temperature measurement of nonlinear-optical crystals interacting with laser radiation [2-4]. It is well known that all nonlinear-optical crystals are piezoelectrics. Each crystal sample has its own set of intrinsic vibration modes. Piezoelectric resonances are observed in nonlinear-optical crystal response to the applied radiofrequency electric field when frequency of excitement corresponds to any intrinsic mode frequency. It was demonstrated that with high accuracy piezoelectric resonance frequencies $Rf_n$ ($n$ – mode index) linearly depend on crystal uniform temperature $\theta$ with piezoelectric resonance thermal coefficients $K_n^{\text{prt}}$. When we consider interaction with laser radiation the crystal is heated due to optical absorption and commonly its temperature is nonuniformly distributed in correspondence with laser intensity profile. It was proved experimentally [2, 3] and theoretically [4] that for the last case nonuniform temperature distribution can be characterized using notion of crystal equivalent temperature. Crystal equivalent temperature is directly determined from measured resonance frequency shift $\Delta Rf_n(P)$ caused by laser irradiation of $P$ power taking into account $K_n^{\text{prt}}$ value of the resonance. Equivalent temperature concept has been already successfully implemented in well-known laser calorimetry technique for determination of optical absorption and heat transfer coefficients of nonlinear-optical crystals [5].

Nonlinear-optical crystal temperature measurement is of great interest directly in course of nonlinear-optical conversion of laser radiation [6-8]. In this case crystal is heated nonuniformly along propagation direction of laser radiation as far as optical absorption coefficients of pump and converted wavelengths can differ significantly. For theoretical interpretation of experimental results obtained in [6-8] it is necessary to elaborate model of acoustical vibration of bulk crystal sample under condition of nonuniform transverse and longitudinal crystal temperature distribution. First step to resolving of this problem is finding the solution of three dimensional heat conduction equation with spread heat source specified by the absorption of pump and generated laser radiation, which powers vary along propagation distance. Second harmonic generation (SHG) of single-mode laser radiation in nonlinear-optical crystal will be considered further.

## 2. Model of second harmonic generation in nonlinear-optical crystals

In order to find out temperature distribution inside nonlinear-optical crystal in course of SHG it is necessary to solve equations that describe nonlinear-optical SHG process. Dielectric crystal with nonzero second-order nonlinear-optical susceptibility is considered for this purpose. Nonlinear polarization $P_{\text{NL}}$ of dielectric occurs when light wave with electric field $E$ goes through it:

$$P_{\text{NL}} = \varepsilon_0 \chi^{(2)} : EE \,. \tag{1}$$

Here colon denotes tensor sum, $\varepsilon_0$ is permittivity of vacuum and $\chi^{(2)}$ is third rank tensor of second-order nonlinear-optical susceptibility. Overall electric field space-time dependence will be the sum result of first and second harmonic waves propagating along $x_3$ can be represented as

$$E(x_3,t) = \frac{1}{2}\mathrm{e}_1\left[A_1(x_3,t)\exp\left(i(\omega t - k_1 x_3)\right) + c.c.\right] + \frac{1}{2}\mathrm{e}_2\left[A_2(x_3,t)\exp\left(i(2\omega t - k_2 x_3)\right) + c.c.\right]. \tag{2}$$

Here index '1' and '2' refers to the first ($\omega$) and second ($2\omega$) harmonic respectively; '*c.c.*' means complex conjugation; $A$ denotes complex envelope; $\omega$ is frequency; $\mathrm{e}_i$ are unit vectors representing electric field polarization; $k_i$ are wave vectors that depend on the refractive indices $n_\omega$ and $n_{2\omega}$ that are complex values in general case:

$$k_1 = \frac{n_\omega \omega}{c}, \quad k_2 = \frac{n_{2\omega} 2\omega}{c}, \tag{3}$$

where $c$ is light velocity in vacuum. By Substituting (1) into (2) we can write in explicit form components of nonlinear polarization that oscillate at $\omega$ and $2\omega$ frequencies:

$$P_{\text{NL}} = \frac{1}{4}\chi_\omega^{(2)}:\mathrm{e}_1\mathrm{e}_1 A_1^2(x_3)\exp\left[i(2\omega t - 2k_1 x_3)\right] + \frac{1}{2}\chi_{2\omega}^{(2)}:\mathrm{e}_1\mathrm{e}_2 A_1^*(x_3)A_2(x_3)\exp\left[i(\omega t + k_1 x_3 - k_2 x_3)\right] + c.c., \tag{4}$$

Here '*' denotes complex conjugation; subscript of $\chi_\omega^{(2)}$ indicates on its dispersion properties.

Wave equation that describes evolution of electromagnetic wave in dielectric can be written as [9]

$$\nabla^2 E - \frac{1}{\varepsilon_0 c^2}\frac{\partial^2 D}{\partial t^2} = \frac{1}{\varepsilon_0 c^2}\frac{\partial^2 P_{\text{NL}}}{\partial t^2}, \tag{5}$$

Here $D = \varepsilon_0(1+\varepsilon)E$ is electric displacement vector where $\varepsilon$ is dielectric permittivity of the crystal. By substitution (2) and (4) in equation (5) we receive the following system:

$$\begin{cases} \left(\nabla^2 - \dfrac{1}{\varepsilon_0 c^2}\dfrac{\partial^2}{\partial t^2}\varepsilon :\right)\mathrm{e}_1 A_1(x_3)\exp\left[i(\omega t - kx_3)\right] = \dfrac{2}{\varepsilon_0 c^2}\dfrac{\partial^2}{\partial t^2}\chi_\omega : \mathrm{e}_1\mathrm{e}_2 A_1^*(x_3)A_2(x_3)\exp\left[i(\omega t + k_1 x_3 - k_2 x_3)\right] \\ \left(\nabla^2 - \dfrac{1}{\varepsilon_0 c^2}\dfrac{\partial^2}{\partial t^2}\varepsilon :\right)\mathrm{e}_2 A_2(x_3)\exp\left[i(2\omega t - k_2 x_3)\right] = \dfrac{4}{\varepsilon_0 c^2}\dfrac{\partial^2}{\partial t^2}\chi_{2\omega} : \mathrm{e}_1\mathrm{e}_1 A_1^2(x_3)\exp\left[i(2\omega t - 2k_1 x_3)\right]. \end{cases} \tag{6}$$

For simplification we assume that dielectric medium is low-loss and low-nonlinear and neglecting the second derivatives with respect to coordinates of complex envelopes (slow varying amplitude approximation). Then after performing differentiation in (6) we obtain following equation that describe evaluation of first and second harmonics:

$$\begin{cases} \dfrac{dA_1}{dx_3} + \alpha_1 A_1 = -i\sigma_1 A_1^* A_2 \exp(-i\Delta k x_3), \\ \dfrac{dA_2}{dx_3} + \alpha_2 A_2 = -i\sigma_2 A_1^2 \exp(i\Delta k x_3). \end{cases} , \qquad (7)$$

Here $\Delta k = k_2 - k_1$ is wave detuning and following designations are used:

$$\begin{array}{ll} \alpha_1 = k_1 \mathrm{e}_1 \dfrac{\mathrm{Im}(\varepsilon_\omega):\mathrm{e}_1}{2n_\omega^2}, & \alpha_2 = k_2 \mathrm{e}_2 \dfrac{\mathrm{Im}(\varepsilon_{2\omega}):\mathrm{e}_2}{2n_{2\omega}^2}, \\ \sigma_1 = \dfrac{1}{2} k_1 \mathrm{e}_1 \dfrac{\chi_\omega^{(2)}:\mathrm{e}_1\mathrm{e}_2}{n_\omega^2}, & \sigma_2 = \dfrac{1}{4} k_2 \mathrm{e}_2 \dfrac{\chi_{2\omega}^{(2)}:\mathrm{e}_1\mathrm{e}_1}{n_{2\omega}^2}. \end{array} , \qquad (8)$$

It can be concluded from the system (7) that SHG process has the highest efficiency when wave detuning $\Delta k$ is close to zero. As it was mentioned earlier crystal temperature is the key parameter that affects the $\Delta k$ value because refractive indices $n_\omega$ and $n_{2\omega}$ are temperature dependent. In those points of the dielectric media where $\Delta k$ is nonzero conversion efficiency decreases. Thus besides crystal average temperature it is also important to know temperature distribution inside crystal volume.

## 3. Three dimensional model of crystal heating

Stationary 3D heat conduction problem have to be solved with heat source, conditioned by absorption of first and second harmonic, in order to obtain 3D temperature distribution of nonlinear-optical crystal in process of SHG.

$$\begin{cases} \kappa_{\mathrm{cr}} \nabla^2 \theta_{\mathrm{cr}} + \alpha(\omega) I(x_1, x_2) f(x_3) = 0, \\ -\kappa_{\mathrm{cr}} \dfrac{\partial \theta_{\mathrm{cr}}}{\partial n}\Big|_{\partial\Gamma} = h^{\mathrm{T}} (\theta_{\mathrm{cr}} - \theta_{\mathrm{a}})\big|_{\partial\Gamma}. \end{cases} , \qquad (9)$$

Here $\kappa_{\mathrm{cr}}$ is thermal conductivity of the crystal, $\alpha(\omega)$ is optical absorption coefficient, which depends on frequency; $I(x_1, x_2)$ is radiation intensity transverse distribution; $\theta_{\mathrm{cr}}$ and $\theta_{\mathrm{a}}$ are temperatures of the crystal and air respectively; $n$ is the normal vector to the interface; $h^{\mathrm{T}}$ is the heat transfer coefficient; $|\partial\Gamma$ denotes values at the interface. Function $f(x_3)$ represents longitudinal dependence of radiation intensity along propagation direction.

Generally speaking systems of equations (7) and (9) are coupled because distributions of harmonic intensities change the crystal temperature distribution and the last one in turn affects the harmonic intensities through the wave detuning.

System (9) can be solved employing variational principle. It states that for equation type $\hat{L}u(x_1, x_2, x_3) = F(x_1, x_2, x_3)$, where $\hat{L}$ is differential operator, which is real, linear, self-conjugate and positive-definite, the solution in region D is equivalent to such function $u(x_1, x_2, x_3)$ that minimizes the following functional [10, 11]

$$\Phi(u) = \langle \hat{L}u, u \rangle - 2\langle u, F \rangle. \qquad (10)$$

In our case $\hat{L} = -\nabla^2$ and function $u(x_1, x_2, x_3)$ represents $\theta_{\mathrm{cr}}$ distribution. Triangular brackets denote the scalar product of square-integrable functions $\xi$ in region D:

$$\langle \xi_i, \xi_j \rangle = \iiint\limits_{\mathrm{D}} \xi_i(x_1, x_2, x_3) \xi_j(x_1, x_2, x_3) dx_1 dx_2 dx_3 . \qquad (11)$$

Rayleigh-Ritz method is popular and convenient approach for solving differential equations. For minimization of (10) we select basis functions $\{\psi\}$ that produce full set in D region. For simplicity of

Journal of Physics: Conference Series **510** (2014) 012031 doi:10.1088/1742-6596/510/1/012031

calculations we chose basis functions that satisfy boundary conditions of the given problem (9). Then unknown function $\theta_{cr}(x_1,x_2,x_3)$ is expanded using selected basis functions.

$$\theta_{cr}(x_1,x_2,x_3)=\theta_a+\sum_i a_i\psi_i \,. \tag{12}$$

Here $a_i$ are expansion coefficients. Air temperature $\theta_a$ is constant and for simplification of basis function type it can be treated separately.

Substituting (12) into (10) and using condition of minimum for function of several variables, which is equality to zero of function derivatives with respect to variables at the point of minimum, we obtain so-called Rayleigh system:

$$\left\langle -\nabla^2\psi_i,\psi_j\right\rangle a_j=\frac{\alpha(\omega)}{\kappa_{cr}}\left\langle I(x_1,x_2)f(x_3),\psi_i\right\rangle . \tag{13}$$

In case explicit forms of basis functions and distributed heat source are known then scalar products can be calculated. As follows equation (13) becomes system of linear equations with unknown expansion coefficients $a_j$. So that it can be solved using standard methods of linear algebra.

## 4. Experimental results

Experiment of crystal equivalent temperature measurement in course of SHG was conducted using nonlinear-optical periodically poled lithium niobate (PPLN). Crystal is a rectangular parallelepiped with dimensions 2.6×1×4.9 mm$^3$, polled along $x_2$. Block-scheme of experimental setup used for observation of piezoelectric resonances in crystals is introduced in Figure 1 (a). PPLN crystal is placed inside the capacitor, formed by two strip electrodes. Voltage from the radiofrequency (RF) generator is applied to this capacitor. Voltage frequency $f$ is swept while its low value amplitude is fixed. Crystal response is analyzed using lock-in amplifier by measuring voltage $U_R$ on the load resistor R, which is connected in series with the capacitor.

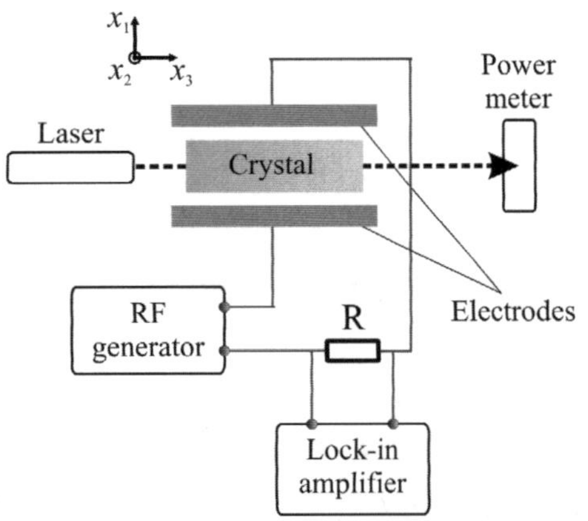


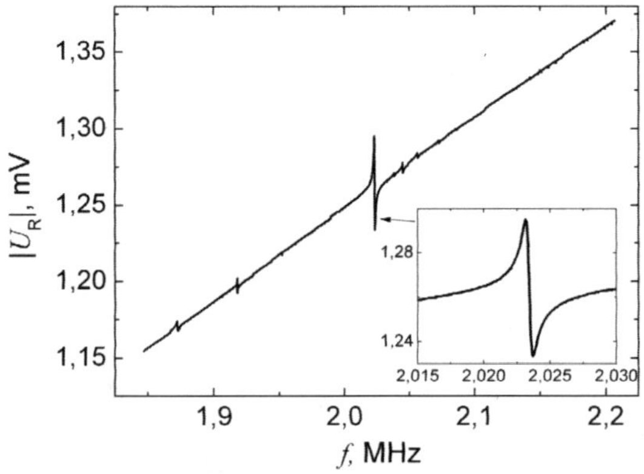


**Figure 1.** (a) Simplified block-scheme of experimental setup; (b) Frequency dependence of the voltage amplitude measured on load resistor R for PPLN crystal, inset shows dependence of the response near PPLN piezoelectric resonance.

Piezoelectric resonances are clearly observed in measured spectrum. Piezoelectric resonance frequencies $Rf_n$ are precisely measured due to high quality factors. Piezoelectric resonance thermal coefficients $K_n^{prt}$ are determined by measuring linear resonance frequency shift with temperature in crystal uniform heating calibration experiment [2, 3]:

$$\Delta Rf_n = K_n^{prt}\Delta\theta_{cr} \,. \tag{14}$$

When laser radiation goes through the crystal its nonuniform temperature distribution is characterized by *crystal equivalent temperature* $\Theta_{eq}$. This temperature is obtained directly by measuring resonance frequency shift, caused by influencing laser power, in respect to the frequency at crystal initial temperature $\theta^0_{cr}$ at zero power.

$$\Theta_{eq}(P_0) = \theta^0_{cr} + \frac{\Delta R f_n(P_0)}{K_n^{prt}} . \tag{15}$$

Experiment of SHG in PPLN crystal was performed using as a pump source the continuous wave polarized single-mode ytterbium fiber laser (λ=1064 nm) [7, 8]. Figure 2 a shows dependence of crystal equivalent temperature on pump power measured for two cases of pump polarization $e_1$ direction. When $e_1 \| x_1$ second harmonic is not generated and crystal is linearly heated with pump power with 1.35 K/W rate. For the $e_1 \| x_2$ case at first PPLN equivalent temperature linearly changes with pump power when second harmonic power $P_{sh}$ is low, i.e. wave detuning $\Delta k$ is far from zero. However in this case substantial nonlinear behavior of PPLN equivalent temperature is observed at phase matching when second harmonic generation efficiency is much higher.

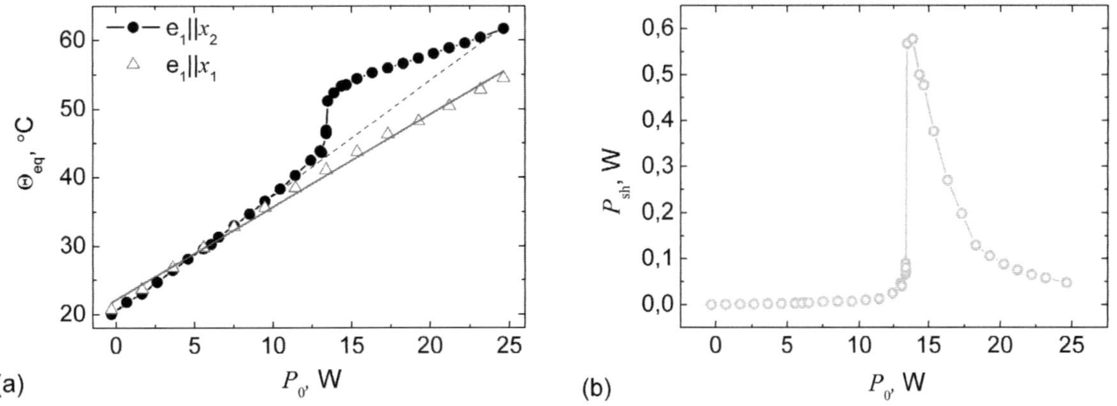


**Figure 2.** Dependencies on pump power of PPLN crystal equivalent temperature for two cases of pump polarization $e_1$ directions (a) and generated second harmonic power for the $e_1 \| x_2$ case (b).

## 5. Calculation results

For modeling temperature distribution of nonlinear-optical crystal, heated by laser radiation, the convective heat exchange at crystal ends and lateral surfaces should be taken as boundary conditions for solving heat conduction problem (9). This is due to experimental procedure of crystal equivalent temperature determination that supposes unclamped boundaries of the crystal.

For single-mode beam with radius $r$ and power $P_0$ laser radiation intensity has Gaussian profile of transverse distribution

$$I(x_1, x_2) = \frac{2P_0}{\pi r^2} \exp\left(-2\frac{x_1^2 + x_2^2}{r^2}\right), \tag{16}$$

Estimation of temperature effects will be considered for two different cases of longitudinal dependencies of laser intensity:

$$f_1(x_3) = \exp(-\alpha x_3), \quad f_2(x_3) = \left(\frac{x_3}{L_3}\right)^2 \exp(-\alpha x_3), \tag{17}$$

Here $L_3$ is crystal dimension along $x_3$, and $P_0$ is pump laser power. First function describes exponential decay of pump radiation in the absence of second harmonic and another one is the solution of (7) in assumptions of $\Delta k = 0$ and low conversion degree.

Also for simplicity we shall seek the solution of (9) in the cuboid crystal with equal dimensions along $x_1$ and $x_2$ axes, i.e. $L_1 = L_2 \equiv L$, $L_3$ dimension along $x_3$ direction and assume $\alpha_1 = \alpha_2 \equiv \alpha$. Taking into account boundary conditions and symmetry of the problem it is convenient to use following basis:

$$\psi_{mnl} = \left\{ \left[ \left( \frac{x_1}{L} \right)^{2m} + \frac{2m}{\mu L} - 1 \right] \left[ \left( \frac{x_2}{L} \right)^{2n} + \frac{2n}{\mu L} - 1 \right] + \left[ \left( \frac{x_1}{L} \right)^{2n} + \frac{2n}{\mu L} - 1 \right] \left[ \left( \frac{x_2}{L} \right)^{2m} + \frac{2m}{\mu L} - 1 \right] \right\} \cdot$$

$$\cdot \begin{bmatrix} \left( \frac{x_3}{L_3} \right)^l + \frac{x_3}{L_3} \frac{l - \mu L_3}{\mu L_3 - 1} & \text{for } l \text{ odd} \\ \left( \frac{x_3}{L_3} \right)^l + \frac{l}{\mu L} - 1 & \text{for } l \text{ even} \end{bmatrix} . \quad (18)$$

Here $\mu = h^{\mathrm{T}} / \kappa_{\mathrm{cr}}$ and total degree $2m+2n+l$ of each basis function does not exceed $N$. For calculations it was chosen $N$=12. Each polynomial in (18) satisfies boundary conditions at all crystal facets and due to symmetry doesn't depend on permutation of subscripts $m$ and $n$. Functions contain polynomials only with even degree of $x_1$ and $x_2$ due to the symmetry in respect to $x_1x_3$ and $x_2x_3$ planes (see figure 1).

System (13) was solved using following parameters of lithium niobate crystal: $L$=2 mm; $L_3$=10 mm; $\kappa_{cr}$=4.6 WK$^{-1}$m$^{-1}$; $r$=0.5 mm; $P_0$=3 W; $\alpha$=10$^{-2}$ cm$^{-1}$; $h^{\mathrm{T}}$=43 Wm$^{-2}$K$^{-1}$, $\theta_a$=0 °C. Calculated temperature distributions at lateral facet of the PPLN are shown in figure 3 for two cases given by (17).

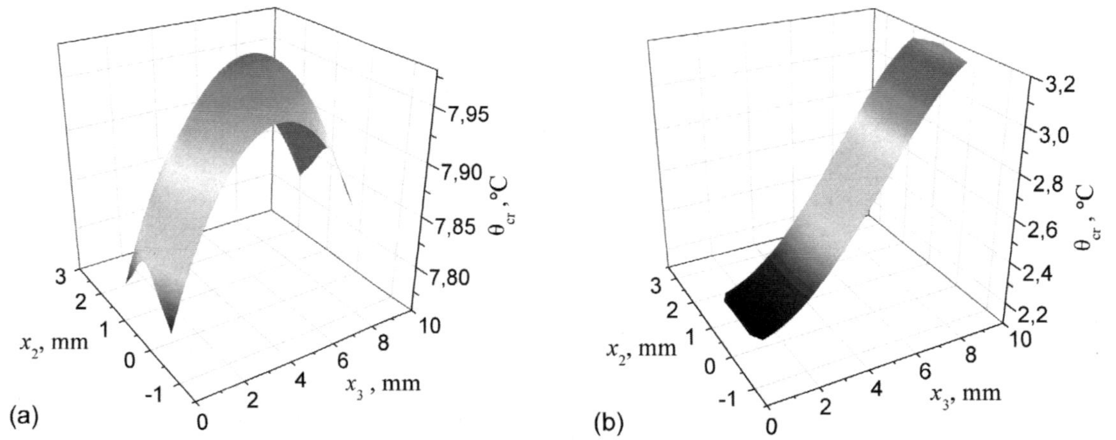


**Figure 3.** Calculated temperature distributions at lateral facet ($x_2x_3$) of the PPLN crystal for exponential $f_1(x_3)$ (a) and quadratic $f_2(x_3)$ (b) heat source distributions.

It can be seen that from figure 3 (a) that longitudinal temperature distribution is not monotonous for the case of intensity exponential decay. At both ends of the crystal the convective heat exchange occurs resulting in lower temperature at crystal ends in comparison with its centre as far as heat source is almost independent on $x_3$ coordinate. Here maximum temperature difference inside crystal is small (about 0.2 °C) in comparison with crystal average heating.

However if intensity growth is parabolic (see figure 3 (b)) the temperature field distribution differs from previous case. Maximum temperature difference in transverse direction is low. Still in longitudinal direction it is comparable with crystal average temperature rise. Calculations also reveal that in this case crystal temperature change has rather linear tendency along propagation direction of light ($x_3$) because thermal conductivity somehow flattens temperature distribution. It means that thermal effects influence on SHG process can be estimated using linear approximation of crystal longitudinal temperature change.

It should be notified that system (9) is linearly dependent on laser power $P_0$ and optical absorption coefficient $\alpha(\omega)$, which is assumed to be small enough. Thereby temperature distribution inside crystal linearly depends on variation of these parameters, i.e. absolute values of crystal temperature alter while relative temperature distribution remains unchanged.

Figure 4 shows first and second harmonic power distributions along crystal length as a result of numerical calculation of system (7). Figure 4 (a) represents ideal case of uniform crystal temperature distribution and wave detuning $\Delta k=0$.

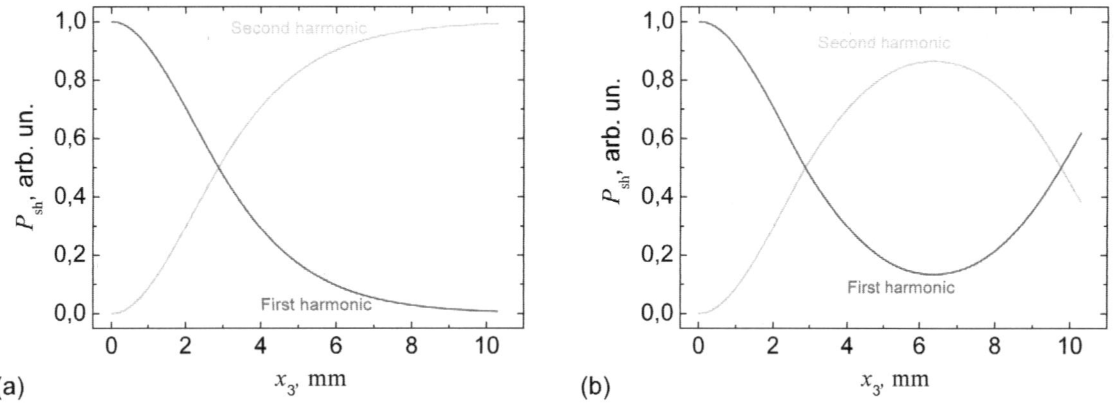


**Figure 4.** Calculated distributions of harmonics power along PPLN length for the case of uniform temperature distribution and zero wave detuning (a), for the case of linear temperature gradient 1°C/cm along crystal length (b).

Results, shown in figure 4 (b), are obtained when linear longitudinal temperature gradient inside crystal is considered. Here wave detuning difference from zero $\Delta k \neq 0$ is caused by temperature difference of 1°C between the input and output crystal facets. In the beginning conversion process is similar to the ideal case but at some crystal length the process of inverse energy conversion from second harmonic to the first one initiates due to $\Delta k$ distinction from zero. It is clear that second harmonic conversion efficiency strongly decreases longitudinal temperature gradient inside crystal is presented.

## 6. Discussion and conclusions

Three dimensional temperature distribution inside nonlinear-optical crystal was calculated for two different cases of laser intensity longitudinal dependence. First one is exponential decay and the second is quadratic rise of intensity along crystal length. Figure 3 (a) represents nonuniform crystal temperature distribution when second harmonic is not generated. In turn figure 3 (b) corresponds to crystal heating in course of second harmonic generation assuming uniform heating due to absorption of pump radiation. It is revealed that in case of absorption alone the crystal maximal temperature difference is much less than its average temperature change. However, in course of SHG the transverse temperature gradient of the crystal remains small. And yet longitudinal temperature difference is high enough and can be of the same order as crystal average temperature rise. This is important because even small temperature difference of 1°C along crystal length (10 mm) may substantially impede nonlinear-optical conversion process.

Two situations introduced in figure 3 (a) and (b) can be easily realized in experiment by variation of single parameter that is laser polarization direction. PPLN crystal equivalent temperature change with laser power measured for two orthogonal directions of pump polarization $e_1||x_1$ and $e_1||x_2$ (see figure 2) reveals instances with absence of SHG and presence of SHG respectively. Crystal equivalent temperature linearly depends on pump power when $e_1||x_1$ (no SHG) and calculated crystal temperature is almost uniform in this case. However in course of efficient SHG when $e_1||x_2$ the equivalent temperature nonlinearly changes with pump power and corresponding longitudinal (along $x_3$) temperature distribution is substantially nonuniform. Thus experimentally measured equivalent temperature of the crystal heated by laser radiation is some sort of average crystal temperature.

Evidently averaging depends on piezoelectric resonance selection. Model of crystal equivalent temperature for the case of uniform temperature distribution has been already introduced [4]. For accurate description of crystal real temperature averaging in course of SHG it is necessary to elaborate model of piezoelectric resonance frequency shift under strong longitudinal temperature gradient inside crystal. Besides problems of SHG (7) and crystal heating (9) should be solved conjointly. These are the goals of future investigation.

**Reference**

[1] Franken P A, Hill A E, Peters C W and Weinreich G 1961 Generation of optical harmonics *Phys. Rev. Lett.* **7** 118

[2] Konyashkin A V, Doronkin A V, Tyrtyshnyy V A and Ryabushkin O A 2009 A radio-frequency impedance spectroscope for studying interaction of high-power laser radiation with crystals *Instr. and Exp. Tech.* **52** 816

[3] Ryabushkin O A, Myasnikov D V, Konyashkin A V amd Tyrtyshnyy V A 2011 Equivalent temperature of nonlinear-optical crystals interacting with laser radiation *J. of Europ. Opt. Soc. – Rapid Publications* **6** 11032

[4] Ryabushkin O A and Myasnikov D V 2012 Experimental determination and the theoretical model of an equivalent temperature of nonlinear-optical crystals interacting with high-power laser radiation *Quantum Electronics* **42** 539

[5] Ryabushkin O A, Konyashkin A V, Myasnikov D V, Tyrtyshnyy V A and Vershinin O I 2013 Piezoelectric resonance calorimetry of nonlinear-optical crystals under laser irradiation *Proc. SPIE 8847 Photonic Fiber and Crystal Devices: Advances in Materials and Innovations in Device Applications VII* (San Diego California USA, 25 August 2013) 88470Q

[6] Ryabushkin O, Konyashkin A, Myasnikov D and Tyrtyshnyy V 2012 Piezoelectric resonance spectroscopy of crystals in process of laser frequency conversion *2012 IEEE International Ultrasonic Symposium Proceedings* (Dresden Germany 7–10 October 2012) pp 2191-2194

[7] Ryabushkin O A, Konyashkin A V, Myasnikov D V, Tyrtyshnyy V A and Baranov A I 2013 Equivalent temperature of nonlinear-optical crystals in process of laser frequency conversion *CLEO®/Europe - IQEC 2013: Conference Digest* (Munich Germany 12-16 May 2013) CE 8.5 WED

[8] Ryabushkin O A, Konyashkin A V, Myasnikov D V, Tyrtyshnyy V A and Baranov A I 2013 Acousto-resonance spectroscopy of nonlinear-optical crystals in course of laser frequency conversion *International Conference on Coherent and Nonlinear Optics (ICONO) Conference on Lasers, Applications, and Technologies (LAT)* (Moscow Russia 18–22 June 2013) LFH8

[9] Dmitriev V G and Tarasov L V 2004 *Applied nonlinear optics* (Moscow Russia: Fizmatlit)

[10] Sadiku M N O 2001 *Numerical techniques in electromagnetics* (2nd edition, Boca Raton Florida USA: CRS Press LLC)

[11] Tyn M-U Lokenath D 2007 *Linear Partial Differential Equations for Scientists and Engineers* (4th edition, Boston USA: Birkhauser)

Journal of Physics: Conference Series **510** (2014) 012032 doi:10.1088/1742-6596/510/1/012032

# Structural and electronic properties of small silicon clusters

**V S Baturin[1], S V Lepeshkin[1], M V Magnitskaya[2], N L Matsko[1] and Yu A Uspenskii[1]**

[1] P.N. Lebedev Physical Institute, Russian Academy of Sciences, Leninskii prosp. 53, 119991 Moscow, Russia

[2] L.F. Vereshchagin Institute for High Pressure Physics, Russian Academy of Sciences, 142190 Troitsk, Moscow, Russia

E-mail: baturin@lpi.ru

**Abstract.** The atomic structure and electronic spectrum of silicon nanoclusters (Si-ncs) $Si_7$, $Si_{10}$, $Si_{10}H_{16}$ and $Si_{10}H_{20}$ are calculated using the evolutionary algorithm with total energy computed within density functional theory and generalized gradient approximation (DFT-GGA). When analysing the low-energy structures, we pay significant attention to their symmetry and interatomic bond geometry. The candidate structures arising in the process of evolutionary algorithm convergence are also considered and classified by their topology and grouping near local energy minima. Possible ways to improve the convergence of evolutionary computation are discussed. Addressing qualitative criteria for the ground-state atomic structure of Si-ncs, we consider correlations between the density of electronic states and the total energetics of clusters in the ground state and low-energy-isomer configurations.

## 1. Introduction

It is well known that due to reduced dimensionality, nanoclusters demonstrate properties essentially different from those of bulk material. These differences, often very unusual, underlie a vast area of applications, both existing and potential. This area covers optics, nanoelectronics, nanophotonics, solar energetics, biology and medicine [1, 2, 3]. For example, silicon nanoclusters (Si-ncs), which are the point of interest in this article, show bright visible photoluminescence [4], whereas in bulk silicon the radiative transitions are suppressed due to indirect band gap.

The origins of these differences lie in the quantum confinement and the significant surface effects. The share of surface atoms grows as the cluster size decreases. Being on the surface, an atom forms less bonds compared to the 'inner' atoms. In turn, this affects the contribution of the atom to the total energy. Moreover, if a cluster is small enough ($\sim 1$ nm), the distinction between core and shell atoms becomes vague. So, when trying to make any conclusions regarding the cluster characteristics, one has to take into account its exact atomic structure.

The existing structural analysis technique has been developed aimed mostly to analyse crystal lattice. Therefore, when it comes to nanoscale objects, the experimental structure determining faces several difficulties [5]. The structures of very small clusters can be obtained using indirect methods such as DFT analysis of IR multiple-photon dissociation spectra, allowing to reconstruct structures for $N \leq 10$ ($N$ is a number of atoms) [6]. Yet, in general, the experimental data on

Journal of Physics: Conference Series **510** (2014) 012032 doi:10.1088/1742-6596/510/1/012032

cluster structures are rather scarce. Thus, at the present time theoretical predictions grow in importance.

The structure prediction is clearly an attractive field for many theorists. Not only for the prediction itself, but also because it implies methods of new materials design. At the dawn of the global structure optimization studies, the unbiased *ab initio* global minimum search seemed unrealizable [7]. So, different researches compared the energies of structures, that were built either with application of predetermined symmetries [8], or using 'cutting' from the bulk lattice, or applying some other tricks. In any case, these techniques contained prescriptions concerning the structure sought. With increasing computer performance many promising computational methods emerged, which involved direct energy calculation. Of course, they are not strictly unbiased, because otherwise they are reduced to random search. First of all, the dimensionality of search space ($d = 3N+3$, where $N$ is the number of atoms per unit cell for crystals, or $d = 3N$ for clusters) is reduced because these algorithms operate not with all possible atomic positions, but on the manifold of locally optimized (relaxed) structures. Even the random search method, combined with local optimization delivers correct solutions – though for small systems ($N < 8$ – 10) [9]. And second, these methods rule out absurd structures in advance, such as those with two atoms on the same spot. Except for these two points, these methods contained no other mandatory constraints. The advanced methods include simulated annealing [10], basin hopping [11], minima hopping [12], metadynamics [13] and evolutionary algorithms [14]. Many of these methods rely on the fact that low energy structures share some similarities i.e. they are close to each other in configurational space. This assumption, used by chemists for a long time, is now proved right for many real systems [15].

In this article we study geometry and electronic structure of small bare silicon and hydrogen passivated silicon nanoclusters. This choice was dictated by several reasons. First, as is mentioned above, silicon clusters exhibit properties that can be useful for future technological applications. Second, their structures are simple enough to study them systematically by means of *ab initio* methods. And third, despite the simplicity we can trace trends of cluster forming and reveal correlations between electronic and atomic structures. For the global structure optimization we chose the evolutionary algorithm as implemented in USPEX code. A set of blind tests showed this method to be one of the most effective [16]. Using topology analysis and intermediate evaluations of evolutionary algorithm we investigate cluster isomerism. Comparing isomers for several atomic compositions we explore dependence of electronic structure details on number and locations of atoms.

## 2. Methods

The problem of structure optimization is mathematically formulated as the search for the global minimum on a very irregular multidimensional surface determined by potential energy (potential energy surface, PES). Of course, there's no general recipe to solve this problem. All of the global optimization tools are essentially heuristic. The practice, however, shows that situation is more optimistic. Among others mentioned above, USPEX code developed by Oganov's group, shows very good results in structure prediction of crystalline materials, molecular crystals, clusters and surfaces [14, 21].

The underlying evolutionary algorithm of the USPEX code is based on the principles of natural selection. By applying evolutionary operations to the generations of structures and ruling out the least fit ones, we progressively obtain better and better species. The algorithm consecutive steps can be shortly described as follows. The first step is the choice of an adequate representation of the problem, i.e. a one-to-one correspondence between a point in the search space and a set of numbers. Then goes the initialization of the first generation – a set of points in the search space which satisfy some constraints. After that, the algorithm determines the fitness of each member of the population and chooses some of them as parents of the next generation.

Journal of Physics: Conference Series **510** (2014) 012032 doi:10.1088/1742-6596/510/1/012032

These parents undergo specially designed evolution operators that produce offspring. The next generation is evaluated and the procedure of new generation creation and evolution repeats. This loop is stopped once some halting criteria are achieved.

We would like to emphasize the importance of step 1 – the representation choice. The mathematical representation of the species, the evolution operators and the fitness of the species are basically what distinguishes one evolutionary algorithm from another. Unlike standard use of binary strings in genetic algorithms, the USPEX code has physically meaningful representation of both objects and operators. The code uses real numbers for the lattice vectors and atomic positions representation. On the one hand it increases the difficulty of the problem, but on the other hand it allows to construct physically meaningful evolution operators.

The evolutionary operators in USPEX can be divided into two major groups: heredity, that uses a few parent solutions to product one offspring solution, and mutation, where a single parent is used to build a single child. Heredity operators are the essential part of any evolutionary algorithm. They allow to utilize the information gathered during algorithm execution. Since the properties of the crystal are determined by the arrangement of atoms, the most physically meaningful way to implement heredity is to use a combination of pieces cut from two or more parent structures. In this case, the lattice of the child is in some way the average of its parent structures. As for mutation, here we have the diversity of operations including atomic permutation, random distortion, soft-mode mutation, etc.

Once the code has generated another set of the offspring structures, each structure is tested for correspondence with a number of constraints, and if it satisfies all of them, it undergoes a local optimization procedure. Constraints provide the feasibility of candidate structures, and include the following conditions. First, the distance between atoms cannot be less than some user-defined limit. Second, no lattice vector could be shorter than some threshold length, determined by e.g. the biggest atomic diameter. The third condition concerns the angles between lattice vectors. It guarantees that a unit cell is not too flat.

Local optimization means the relaxation of the candidate structures to a local minimum. This step allows to find the lowest free energy structures in the current generation. Candidate structures are relaxed using *ab initio* methods, thus, this part is the most computationally expensive. The USPEX code has a great flexibility concerning the choice of relaxation procedures. We used the generalized gradient approximation (GGA) [18] of density functional theory, as implemented in the Quantum Espresso code [17] for cluster structure relaxation. The pseudopotential is Troullier-Martins-type[19] norm-conserving pseudopotential with GGA exchange-correlation functional, generated by fhi98PP code [20]. Self-consistent electronic structure is calculated for given atomic positions of the candidate structure. Then atoms are moved to decrease system energy and electronic structure is calculated again. This procedure repeats until the energy convergence is achieved. USPEX code allows to improve relaxation parameters step by step: first relaxation of a given structure could be done roughly, then the precision of further relaxations is improved until required accuracy is achieved. Such strategy results in a notable performance improvement. For the numerical optimization during structure relaxation the BFGS algorithm (iterative quasi-Newton method) was implemented.

Usually for systems consisting of about tens of atoms and more, the total number of trial structures is of the order of $10^3$. Therefore, in order to save time, the maximum precision used in each relaxation is lowered and is often not sufficient for quantitative analysis. In this case it is worthwhile to pick out a number of lowest-lying topologically different structures obtained during the whole evolutionary computation and re-compute them with more precise accuracy. The refinement can even result in the interchange of the isomers, though only those very close in energy. The procedure of classification is described in the next section.

### 3. Structure classification and cluster geometry

In this section we explore the process of global search with the evolutionary algorithm and consider its intermediate and final results, which give rich information about the lowest energy (ground-state) structure and first isomer configuration of clusters. Examining the trial structures geometry may be useful to find possible routes to speed-up the search of the optimal cluster structures.

As we have noted above, not only the final result but also execution process of the algorithm might provide useful information. The subject of interest is the geometry and the total energy distribution of numerous trial structures, which lie energetically slightly above the ground state. Many of these structures have a similar arrangement of atoms and, probably, relate to the same local minimum of energy. Their occurrence in evolutionary searching is caused by imperfect atomic relaxation and a possible existence of soft modes in atomic displacements. The removal of such structures from the process of searching may save much computational time and resources. For this reason a proper classification of trial atomic structures is highly important for the search of the ground state atomic configuration.

Of course, the term 'similar arrangement of atoms' demands exact mathematical definition, and numerous measures have been proposed. Probably the most natural way is to classify the trial structures by their relationship with the representative isomer. It is well known, that the isomers are divided into two major groups: structural isomers and stereoisomers. The structural isomers are distinguished by bond topology, while stereoisomers differ only in spatial orientation of bonds. Structural isomers can be filtered using method based on graph theory [22]. Using this technique we sorted the total structural data array into groups, each of which corresponds to a distinct structural isomer.

The objects of our consideration are two bare Si-ncs ($Si_7$ and $Si_{10}$) and two Si-ncs passivated by hydrogen ($Si_{10}H_{16}$ and $Si_{10}H_{20}$). We chose $Si_7$ and $Si_{10}$ clusters for their structures have been studied in detail both experimentally and theoretically. Cluster $Si_{10}H_{16}$ is one of the widely used model systems for cluster-related studies, as its structure is a highly symmetrical result of cutting-from-bulk procedure. Cluster $Si_{10}H_{20}$ is notable for a high number of low energy isomers. In Figure 1 we present results of topology analysis applied to evolutionary procedure carried out for $Si_7$, $Si_{10}$, $Si_{10}H_{16}$ and $Si_{10}H_{20}$ clusters. The first thing we note here is that during its execution the algorithm tends to produce numerous structures close to the global minimum. Comparing results for $Si_7$ and $Si_{10}H_{20}$ we also see that the distance between structural isomers decreases when we go from smaller systems to larger ones. Most likely this is caused by the increasing of the total number of degrees of freedom. More total degrees of freedom means more 'soft' degrees of freedom, which provide small energy differences in the total energy.

The optimal (ground state) structures and the first structural isomers are shown in Figure 2. Our results, given by evolutionary algorithm for bare clusters are in perfect agreement with the experimental study [6]. The $Si_7$ and $Si_{10}$ have 5-fold and 3-fold symmetry axes respectively. It is important to note, that the coordination numbers of atoms in both of these clusters do not correspond to the common silicon valency of 4. This means that the bonding in these clusters differs from four $\sigma$-bonds and $sp^3$-hybridization inherent in the diamond crystal lattice. This fact affects the cluster stability and spectrum details. Below we discuss the relation between this difference and energetic characteristics.

The ground state structure of $Si_{10}H_{16}$ indeed coincides with a fragment of the diamond crystal lattice having broken bonds saturated by hydrogen. The structure has tetrahedral symmetry. Looking at its first isomer and the structures of $Si_{10}H_{20}$ we see that bigger systems are not necessarily highly symmetrical, which once again emphasizes that the global search should be unbiased.

The situation with stereoisomers is more complicated.Given two topologically identical structures we cannot tell for sure whether they are related to the same local minimum or to the different ones. Let us consider two pairs of $Si_{10}H_{20}$ structures corresponding to different topologies (Figure 3). After high-precision atomic relaxation, the structures A1 and A2 retain their different atomic arrangements. The situation with the structures B1 and B2 is opposed. We note that even at first glance the shape of B2 structure suggests the existence of a local energy maximum, but not a minimum. To check this assumption, we used a small structure distortion with subsequent high-precision relaxion. Indeed, B2 eventually relaxed to B1, which confirms our assumption. This shows, that B1 and B2 are related to one stereoisomer, while A1 and A2 are related to two different stereoisomers. However, to distinguish between two topologically identical structures in general case, one should trace several intermediate configurations between them to find whether these structures are separated by a potential barrier.

## 4. Correlation between the Kohn-Sham electron spectrum and cluster stability

The density of electronic states (DOS) $N(E)$ is the standard characteristic of electronic structure in a material. While DOS determines directly only one contribution to the total energy - the sum of one-electron energies, it may provide useful information on the total energetics of a material. For example, in metals a high value of $N(E_F)$ ($E_F$ is the Fermi energy) is an indication of possible low stability or even instability of the system with respect to lattice distortions or spin polarization. For this reason, when studying the most stable atomic structures of Si-ncs, we paid attention to their DOS in order not to miss possible correlations between the behavior of $N(E)$ and the total energetics of clusters.

Figures 4 – 7 show the calculated dependencies $N(E)$ (smoothed with the Gaussian of 0.1 eV width) for the ground state structures and the lowest-energy structural isomers in the clusters $Si_7$, $Si_{10}$, $Si_{10}H_{16}$, and $Si_{10}H_{20}$. One can see in these figures that the DOS of the ground state structure is similar to that of all presented isomers with the same chemical composition. Except for small DOS variations between the atomic configurations of the same composition, we did not find any regular distinctions in their $N(E)$ dependencies. At the same time, the comparison of bare Si-ncs ($Si_7$ and $Si_{10}$) with Si-ncs passivated by hydrogen ($Si_{10}H_{16}$ and $Si_{10}H_{20}$) reveals

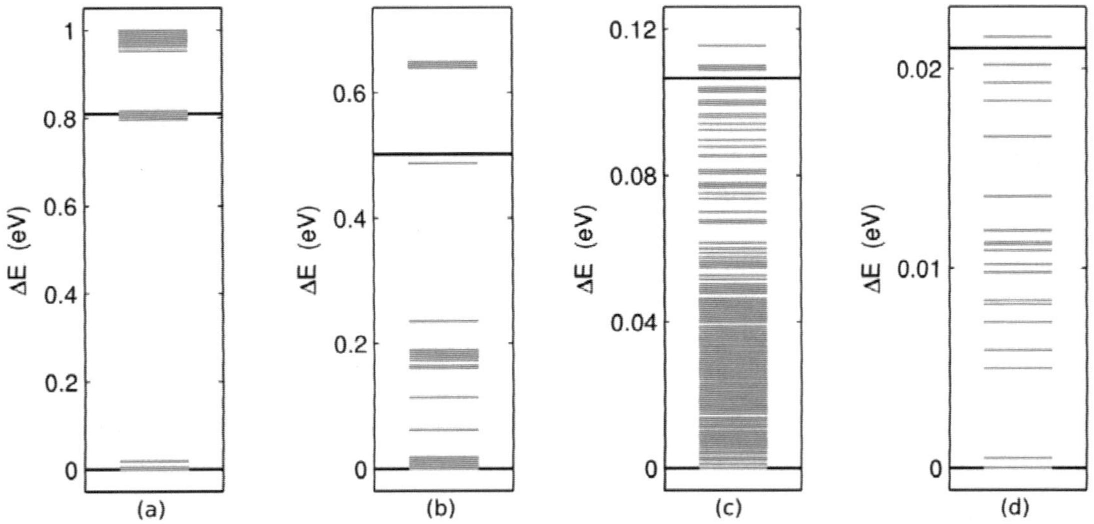


**Figure 1.** Energies of clusters (a) $Si_7$, (b) $Si_{10}$, (c) $Si_{10}H_{16}$ and (d) $Si_{10}H_{20}$: Energies of all candidate structures are shown by short green dashes, while ground state and first structural isomer structures are shown by long black lines. Notice that energy scales on the graphs are very different.

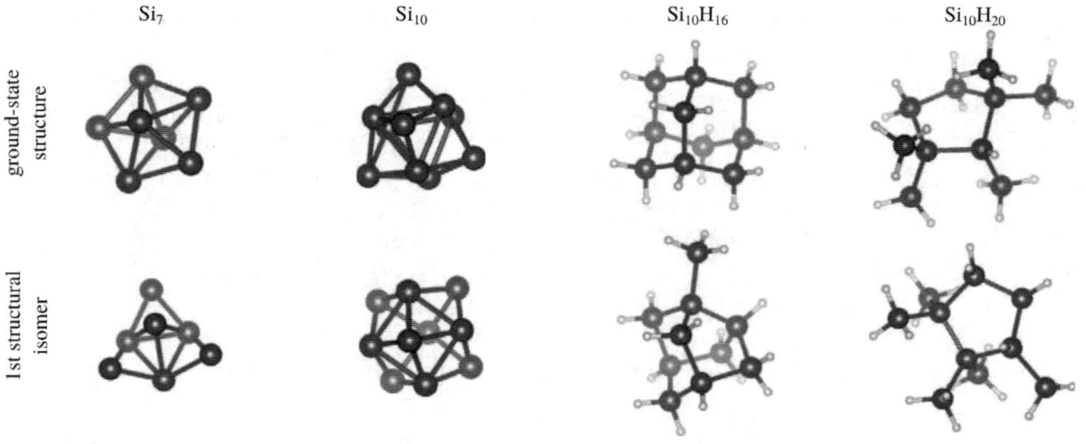


**Figure 2.** The ground-state and first structural isomers of $Si_7$, $Si_{10}$, $Si_{10}H_{16}$ and $Si_{10}H_{20}$.

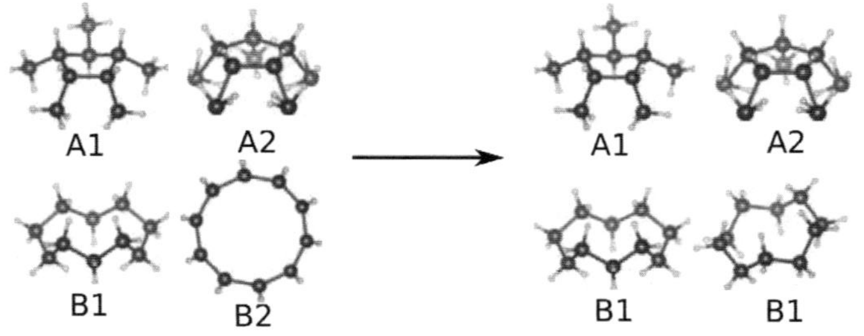


**Figure 3.** Two pairs of stereoisomers (A1, A2 and B1, B2) related to different topologies (A and B), before (left-hand) and after (right-hand) relaxation with preliminary small random distortion. B2 relaxes to B1, showing that the B2 structure corresponds to a local maximum of the total energy

evident differences in their DOS – the HOMO-LUMO gap is narrower and the valence band is wider in bare Si-ncs than in passivated ones. To quantify this observation using an integral characteristic appropriate for comparison, we defined the valence band width by the formula:

$$W_{\mathrm{val}} = 2\sqrt{3}\left\{\frac{\int \mathrm{d}E\theta(E_{\mathrm{F}}-E)N(E)(E-\overline{E})}{\int \mathrm{d}E\theta(E_{\mathrm{F}}-E)N(E)}\right\}^{1/2}, \quad (1)$$

where the factor $2\sqrt{3}$ is fitted to reproduce exactly a rectangular DOS width. In this equation

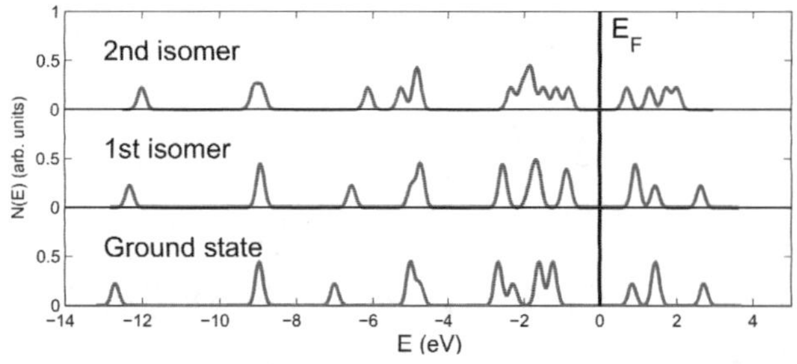


**Figure 4.** DOS of $Si_7$ cluster calculated for the lowest-energy structures.

the energy center of the valence band is:

$$\overline{E} = \frac{\int dE\theta(E_F - E)N(E)E}{\int dE\theta(E_F - E)N(E)}. \tag{2}$$

Table 1 presents the main parameters of electronic structure for the ground state geometry of the clusters $Si_7$, $Si_{10}$, $Si_{10}H_{16}$ and $Si_{10}H_{20}$ – the width of the HOMO-LUMO gap, the valence band width, the reaction energy (for clusters $Si_{10}H_{2m}$ with $m = 0, 8, 10$) given by the formula $\varepsilon_R = E_{tot}(Si_{10}H_{2m}) - E_{tot}(Si_{10}) - mE_{tot}(H_2)$, and the energy distance of the first low-energy structural isomer from the ground state of a cluster. For clusters $Si_{10}H_{16}$ and $Si_{10}H_{20}$ stabilized by hydrogen (energy gain about 6.2 eV), the HOMO-LUMO gap is wider by 2.2 – 2.6 eV and the valence band width is narrower by 2.6 – 3.0 eV than for the bare silicon clusters $Si_7$ and $Si_{10}$. It is noticeable that the energy difference between the first structural isomer and the ground state structure is large for bare silicon clusters (about 0.5 eV) and is only 0.01 – 0.1 eV for Si-ncs passivated by hydrogen. In general, our results indicate that the HOMO-LUMO gap width and the valence band width correlate well with reaction energy and can be used for the qualitative estimate of cluster stability.

**Table 1.** The main parameters of electronic structure (in eV) for the clusters $Si_7$, $Si_{10}$, $Si_{10}H_{16}$ and $Si_{10}H_{20}$ – the HOMO-LUMO gap $\Delta g$, the valence band width $W_{val}$, the reaction energy $\varepsilon_R$, and the energy difference between the first structural isomer and the ground state, $E_1 - E_0$.

| Cluster | $\Delta g$ | $W_{val}$ | $\varepsilon_R$ | $E_1 - E_0$ |
|---|---|---|---|---|
| $Si_7$ | 2.10 | 11.8 | – | 0.81 |
| $Si_{10}$ | 2.10 | 11.8 | 0 | 0.50 |
| $Si_{10}H_{16}$ | 4.70 | 9.2 | -6.21 | 0.11 |
| $Si_{10}H_{20}$ | 4.34 | 8.9 | -6.26 | 0.02 |

## 5. Discussion and conclusions

In this paper, we presented our calculations of atomic structure and electronic spectra in silicon nanoclusters $Si_7$, $Si_{10}$, $Si_{10}H_{16}$ and $Si_{10}H_{20}$. Analysing them allowed to reveal several features of the global search process applied to semiconductor clusters. In searching for least-energy structure, we ran over most promising atomic configurations delivered by the evolutionary algorithm. A general observation is that the evolutionary algorithm realized in the USPEX code gives good convergence of trial configurations to a ground-state atomic structure. The convergence behavior is not the same for different clusters, but depends on cluster characteristics. For the clusters $Si_7$, $Si_{10}$, $Si_{10}H_{16}$ with the low-lying-isomer energy well above the ground state,

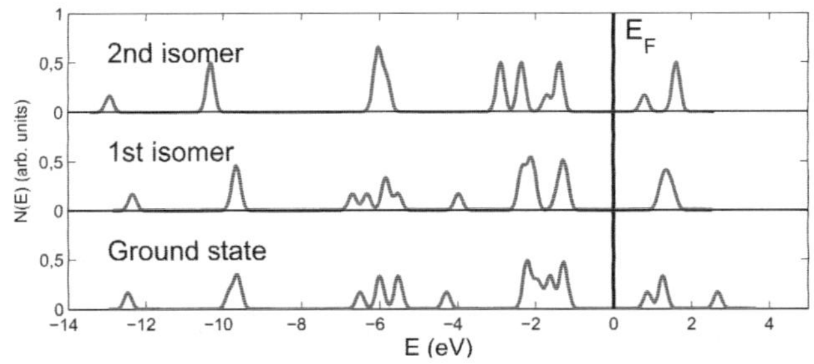


**Figure 5.** The same as in Figure 4, but for $Si_{10}$.

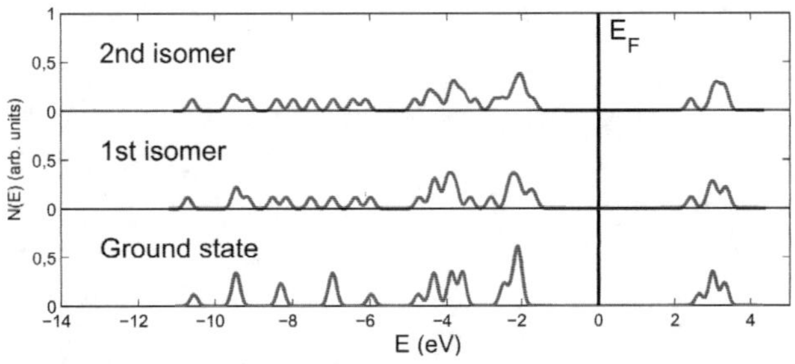


**Figure 6.** The same as in Figure 4, but for $Si_{10}H_{16}$.

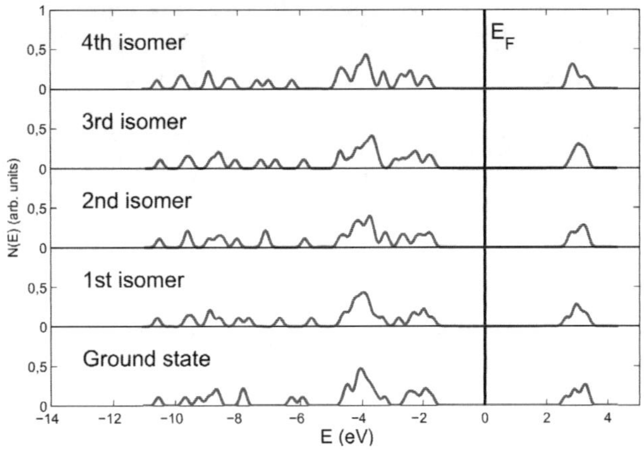


**Figure 7.** The same as in Figure 4, but for $Si_{10}H_{20}$.

the convergence is stable and fast. However, in the case of cluster $Si_{10}H_{20}$, which has several isomers separated from the ground state by $\Delta E_{\rm tot} < 0.1$ eV, the convergence behavior is less stable, with a three times larger number of iterations needed.

The analysis of atomic configurations arising in these iterations shows (see Section 3) that both structural and stereoisomers emerge among them. As the atomic relaxation is in many cases incomplete, a number of slightly perturbed structures close to local energy minima also arise in the course of computations as independent candidates. The calculation of these perturbed structures provides no helpful information and only increases the amount of computation. For this reason, the removal of these structures from computation could significantly accelerate searching for the ground state atomic structure. It should be noted that the perturbed structures can be easily distinguished from structural isomers by the comparison of bond topologies using the graph theory methods [22]. The distinction between perturbed structures and stereoisomers is not so evident and generally requires additional laborious calculations. In this connection, a challenging problem is to define a measure for quantifying the similarity between different atomic structures. The use of such measure in combination with the evolutionary algorithm will assist further progress in atomic structure prediction.

The density of electronic states is a useful characteristic which can be a qualitative indication of a low stability or even instability of an investigated material. Our analysis of electronic DOS for nanoclusters of varied composition shows that a wider HOMO-LUMO gap and a smaller valence band width generally correlate with the cluster stability and energy gain from formation of clusters. The comparison between electronic DOS of ground state and that of low-lying isomer structures in the clusters of the same composition shows, however, no noticeable correlations with atomic structure stability. Our results presented in Figures 4 – 7 and Table 1 suggest that DOS is a rather rough qualitative characteristic of cluster stability, which is sufficient for the detection of large reaction energy changes $\Delta\varepsilon_R \geq 1$ eV, but insensitive to small energy differences $\Delta E_{\mathrm{tot}} < 0.1$ eV between the low-energy isomers and ground-state structure, as well as between the low-energy isomers themselves.

## Acknowledgments

This work is supported in part by programmes of Russian Academy of Sciences, Russian Foundation for Basic Research (projects 13-02-00655, 13-02-00913, 12-02-31638, 12-02-31774 and 11-02-00615), Ministry of Education and Science of Russia and the Government of the Russian Federation (No. 14.A12.31.0003). The numerical calculations are performed at the Joint Supercomputer Center of RAS. Authors are grateful to Artem R. Oganov of Stony Brook University for insightful discussions.

## References

[1] 2007 *Nanosilicon* ed V Kumar (Amsterdam: Elsevier)
[2] 2009 *Silicon Nanophotonics* ed L Khriachtchev (Singapore: World Scientific)
[3] O'Farrel N, Houlton A and Horrocks B 2006 *Int. J. Nanomedicine* **4** 451–472
[4] Akcakir O, Therrien J, Belomoin G, Barry N, Gratton E and Nayfeh M 2000 *Appl. Phys. Lett.* **76** 1857
[5] Billinge S J, Levin I *Science* **316** 561
[6] Haertlet M, Lyon J T, Pieterjan C, de Haeck J, Lievens P and Fielicke A 2012 *J. Chem. Phys.* **136** 064301
[7] Maddox J 1988 *Nature* **335** 201
[8] Jug K, Schluff H-P, Kupka H and Iffert R 1988 *J. Comp. Chem* **9** 8 803–809
[9] Schmidt M U, Englert U 1996 *J. Chem. Soc.* **10** 2077–2082
[10] Pannetier J, Bassasalsina J, Rodriguez-Carvajal J and Caignaert V 1990 *Nature* **346** 343–345
[11] Wales D J, Doye J P K, *J. Phys. Chem.* **A101** 5111–5116
[12] Goedecker S 2004 *J. Chem. Phys.* **120** 9911–9917
[13] Martoňák R, Laio A and Parrinello M 2003 *Phys. Rev. Lett.* **90** 075503
[14] Oganov A R, Glass C W 2006 *J. Chem. Phys.* **124** 244704
[15] Umemoto K, Wentzovitch R M, Saito S and Miyake T 2010 *Phys. Rev. Lett.* **104** 125504
[16] Oganov A R, Schön J C, Jansen M, Woodley S M, Tipton W W and Henning R G 2011 *Modern Methods of Crystal Structure Prediction* ed A R Oganov (Weinheim: WILEY-VCH) pp 223–231
[17] Gianozzi P *et al* 2009 *J. Phys.: Condens. Matter* **21** 395502
[18] Perdew J P, Burke K and Wang Y 1996 *Phys. Rev.* B **54** 16533
[19] Troullier N and Martins J L 1991 *Phys. Rev.* B **43** 19932006
[20] Fuchs M and Scheffler M 1999 *Comput. Phys. Commun.* **119** 67-98
[21] Zhu Q, Li L, Oganov A R, Allen P B 2013 *Phys. Rev.* B **87** 195317
[22] Pietrucci F and Andreoni W 2011 *Phys. Rev. Lett.* **107** 085504

Journal of Physics: Conference Series **510** (2014) 012033 doi:10.1088/1742-6596/510/1/012033

# Concept of equivalent temperature of the nonlinear-optical crystal interacting with nonuniform laser radiation

**O A Ryabushkin[1,2], D V Myasnikov[1], A V Konyashkin[1,2], I S Ulyanov[1]**
1. Moscow Institute of Physics and Technology, Institutskiy per. 9, Dolgoprudnyy, 141700, Russia
2. Kotelnikov Institute of Radio-engineering and Electronics of RAS, Vvedensky Sq. 1, Fryazino, 141190, Russia

E-mail: roa228@mail.ru

**Abstract.** Experimental criterion was determined for application of equivalent temperature of nonlinear-optical crystal heated both uniformly and nonuniformly by laser radiation with known spatial distribution of intensity. Novel coefficients are introduced in laser physics that characterize thermal response of crystal to the laser radiation propagating through it. Theoretical model is proposed for calculation of these coefficients. Physical parameters of nonlinear-optical crystals were determined that have significant influence upon correspondence between calculated and measured values of introduced coefficients.

## 1. Introduction

Generation of radiation in solid state lasers is always accompanied by the conversion of some part of optical pump power into heat leading to the active medium temperature rise. In condition of continuous generation of radiation the thermodynamic characterization of laser medium is based on the exploitation of nonuniform and at the same time nonequilibrium temperature.

There is one important class of thermodynamic problems in laser physics that refers to the laser radiation frequency conversion in nonlinear-optical crystals. It is essential that efficient harmonics generation and parametric conversion of laser radiation is governed by phase matching conditions that are fulfilled only in certain temperature range of the crystal [1]. However during nonlinear-optical process it is typical that optical absorption coefficients of both pump radiation and converted one change with pump power. Moreover heat exchange coefficient between crystal and air also depends on crystal temperature. These facts lead to uncontrollable rise of crystal temperature with pump power and can result in irreversible crystal degradation (optical damage). Precise nonlinear-optical crystal temperature control is crucial in such circumstances.

Another class of thermodynamic problems in optics is connected with laser calorimetry [2]. At present laser calorimetry method is international standard for determination of optical absorption coefficients of solid optical media. This technique supposes irradiation of the test sample by low level optical power (usually several Watts) and the last one is not heated significantly provided that its optical absorption coefficient is not high.

Present work is dedicated to the elaboration of theoretical model together with experimental realization, based on acousto-resonant spectroscopy technique, of precise temperature measurement of the nonlinear-optical crystals nonuniformly heated by laser radiation [3-9]. Such fact that all nonlinear-optical crystals possess piezoelectric properties admits direct noncontact excitation and detection of acoustical vibrations in crystal using external radiofrequency (RF) electric field. When one of the intrinsic vibration modes frequencies $Rf_n$ of the crystal sample coincides with the probe electric field frequency $f$ the piezoelectric resonance occurs. Equivalent heating temperature during crystal interaction with laser radiation of power $P$ can be directly determined basing on experimentally measured piezoelectric resonance frequency shift $\Delta Rf_n(P)$ provided that resonance frequency dependence on temperature $Rf_n(\theta)$ for the case of crystal uniform heating was determined.

## 2. Equivalent temperature concept

As it was clarified propagation of laser radiation through the crystal results in its nonuniform heating due to optical absorption. For example single-mode laser radiation intensity has Gaussian distribution in cross section to the propagation direction. In this case crystal temperature distribution $\theta_{cr}(x_1, x_2)$ can be obtained by solving nonstationary heat conduction equation with appropriate boundary condition at the crystal-air interface. In assumption of weak optical absorption and crystal convective air cooling it can be written as follows

$$\begin{cases} \rho_{cr}c_{cr}\dfrac{\partial\theta_{cr}}{\partial t} = \kappa_{cr}\left(\dfrac{\partial^2\theta_{cr}^2}{\partial x^2} + \dfrac{\partial^2\theta_{cr}^2}{\partial y^2}\right) + \alpha(\lambda)I(x_1, x_2), \\ \rho_a c_a\dfrac{\partial\theta_a}{\partial t} = \kappa_a\left(\dfrac{\partial^2\theta_a^2}{\partial x^2} + \dfrac{\partial^2\theta_a^2}{\partial y^2}\right), \\ -\kappa_{cr}\dfrac{\partial\theta_{cr}}{\partial \mathbf{n}} = h^T\left(\theta_{cr}^b - \theta_a^b\right). \end{cases} \tag{1}$$

Here subscripts "cr" and "a" denote reference of values to the crystal and air respectively; $\rho$ and $c$ are density and specific heat capacity; $\kappa$ is thermal conductivity; $\alpha(\lambda)$ is crystal optical absorption coefficient, which is wavelength dependent; $I(x_1, x_2)$ is radiation intensity spatial distribution; $\theta_{cr}^b$ and $\theta_a^b$ are temperatures at the crystal-air interface; $\mathbf{n}$ is the normal vector to the interface; $h^T$ is the heat transfer coefficient. At the outer boundary air temperature is $\theta_0$. Due to relatively low crystal overheating its cooling, conditioned by mechanism of spontaneous radiation, is not considered.

Mathematically simplified solution of problem (1) is used in laser calorimetry technique, which is standardized method for measuring optical absorption and heat transfer coefficients [2]. In case laser radiation of power $P$ at $\lambda$ wavelength propagates along crystal length $L_3$ during time interval $0<t<t_1$ then crystal temperature change caused by optical absorption of laser radiation can be expressed as:

$$\theta_{cr}(t) - \theta_0 = \begin{cases} \dfrac{\alpha L_3 P\tau}{mc_{cr}}\left[1 - \exp(-t/\tau)\right], \ 0 < t < t_1 \\ \dfrac{\alpha L_3 P\tau}{mc_{cr}}\left[\exp(-t/\tau) - \exp(-(t-t_1)/\tau)\right], \ t > t_1. \end{cases} \tag{2}$$

Here $m$ is crystal mass. Equation (2) is obtained by solving combined nonstationary heat conduction equations (1) taking into account several simplifying assumptions such as that temperature distribution inside crystal volume is uniform during interaction with laser radiation and crystal overheating is relatively small in comparison with ambient air temperature. Characteristic time $\tau$ describes crystal temperature loss.

$$\tau = \frac{mc_{cr}}{h^T S} \tag{3}$$

where $S$ is overall area of crystal surface. Its value can be obtained by fitting of the experimental data of crystal heating ($0<t<t_1$) and cooling ($t>t_1$) by the function of the following type

$$\theta_{cr}(t) = \theta_a^b + \Delta\theta_{cr}\exp(-t/\tau) \tag{4}$$

In case crystal parameters and laser power $P$ are known then optical absorption $\alpha$ and heat transfer $h^T$ coefficients can be calculated using fitting parameters $\tau$ and $\Delta\theta_{cr}$ as follows

$$\alpha L_3 = \frac{mc_{cr}\Delta\theta_{cr}}{\tau P}; \quad h^{T} = \frac{mc_{cr}}{S\tau}. \tag{5}$$

Kinetics of the sample temperature is measured during heating by laser radiation and subsequent cooling after the laser is switched off. However, in this case sample temperature or to say more precisely air temperature in the vicinity of the sample surface is measured indirectly using external thermal detectors.

Equivalent temperature concept has been recently employed in calorimetry technique. In case of piezoelectric crystals its temperature is directly measured using piezoelectric resonance frequency temperature dependence. It was shown that in case of piezoelectric crystal uniform temperature $\theta_{cr}$ change its piezoelectric resonance frequencies $Rf_n$ in a first approximation linearly shift with temperature [3-9]

$$Rf_n(\theta_{cr}) = Rf_n(\theta_0) + K_n^{prt}\left(\theta_{cr} - \theta_0\right). \tag{6}$$

Here $K_n^{prt}$ is the piezoelectric thermal coefficient of the particular $n$-th vibration mode and $\theta_0$ is initial crystal temperature. Obviously crystal temperature $\theta_{cr}$, which in this case is equal to surrounding air temperature, can be determined if resonance frequency shift is known

$$\theta_{cr} = \theta_0 + \frac{Rf_n(\theta_{cr}) - Rf_n(\theta_0)}{K_n^{prt}}. \tag{7}$$

It was also demonstrated that in case of linear optical absorption the piezoelectric resonance frequencies linearly depend on laser power $P$

$$Rf_n(P) = Rf_n(0) + K_n^{pro} P. \tag{8}$$

Here $K_n^{pro}$ are piezoelectric resonance optical coefficients and $Rf_n(0)$ corresponds to resonance frequency at $P = 0$.

Nonuniform temperature of the crystal can be precisely determined using notion of *crystal equivalent temperature* $\Theta_{eq}$. Crystal equivalent temperature is expressed as follows

$$\Theta_{eq}(P) = \theta_0 + \Delta\Theta_{eq}(P). \tag{9}$$

Temperature $\theta_0$ corresponds to $P = 0$ and $\Delta\Theta_{eq}(P)$ is crystal equivalent heating temperature determined directly from the $n$-th piezoelectric resonance frequency shift

$$\Delta\Theta_{eq}(P) = \frac{Rf_n(P) - Rf_n(0)}{K_n^{prt}}. \tag{10}$$

In linear case of $Rf_n$ dependence on $P$ equation (10) transforms in

$$\Delta\Theta_{eq}(P) = \frac{K_n^{pro}}{K_n^{prt}} P = \beta P. \tag{11}$$

Here $\beta$ is piezoelectric resonance optothermal coefficient, which value depends mostly on crystal optical absorption $\alpha(\lambda)$, and heat exchange conditions.

True thermodynamic temperature of the crystal interacting with laser radiation can be expressed as

$$\theta_{cr}\left(x_1, x_2, x_3, P\right) = \theta_0 + \Delta\Theta_{eq}\left(P\right) + \delta\theta\left(x_1, x_2, x_3, P\right). \tag{12}$$

Our calculations revealed that always the crystal overall temperature rise greatly exceeds nonuniform part. It means that thermodynamic temperature of the crystal heated by laser radiation can be successfully replaced by equivalent temperature. In many interesting cases such as nonlinear-optical frequency conversion the crystal is heated nonlinearly with laser power. Still its equivalent temperature can be determined using equation (9) - (11) [8].

## 3. Experimental results

### *3.1. Experimental setup*

Simplified block-scheme of the experimental setup for the nonlinear-optical crystal's electrical impedance measurements during interaction with laser radiation is shown in figure 1 (a). Crystal sample is placed between strip metallic electrodes. Two electrodes form a capacitor and function as an antenna that excites crystal vibration and simultaneously detects its radiofrequency response. Voltage of the frequency $f$ from the radiofrequency generator is applied to the capacitor that is connected in series with the load resistor $R$. Radiofrequency field is a probe field and its power that affects the crystal is very low ($10^{-3}$ - $10^{-6}$ W). Lock-in amplifier is used for measuring complex voltage $U_R$ on $R$ that is proportional to the electrical current flowing through the crystal. Thermostat is used for the crystal temperature $\theta_{cr}$ stabilization.

Figure 1 (b) shows dependencies of the $U_R$ voltage amplitude on frequency $f$ measured for two mutually perpendicular orientations of quartz crystal in respect to the probe electric field $\mathbf{E}^{rf}$. Specimen of α-quartz is rectangular parallelepiped 2.9×2.9×11.2 mm$^3$ with polished facets. At certain frequencies $Rf_n$ piezoelectric resonances between the external electric field and crystal internal vibration modes are observed. Figure 1 (b) inset shows spectrum at piezoelectric resonance $Rf$ = 1.52768 MHz (at $\theta_{cr}$ = 18 °C) that was selected for further investigations.

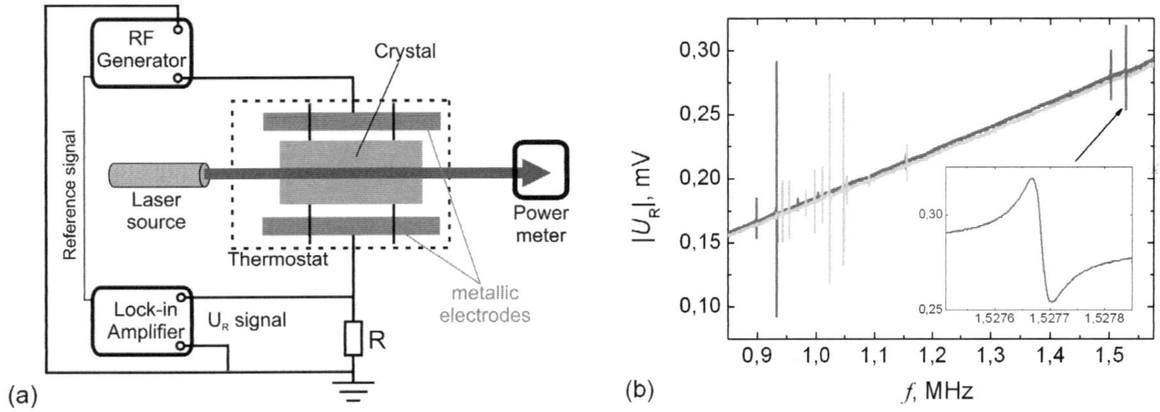


**Figure 1.** (a) Block-scheme of experimental setup; (b) Frequency dependencies of the $U_R$ voltage amplitude measured at $\theta_{cr}$=18 °C for two orthogonal orientations of quartz crystal in respect to electric field $\mathbf{E}^{rf}$.

### *3.2. Temperature calibration and crystal heating by laser radiation*

For investigation of the crystal interaction with laser radiation the temperature calibration of the selected piezoelectric resonance frequencies is made. For precise measurement of crystal equivalent temperature sensitive piezoelectric resonances, i.e. with high piezoelectric resonance thermal coefficient are preferable. Figure 2 (a) shows dependence of the selected resonance frequency on crystal temperature during its uniform heating. Linear approximation (6) gives piezoelectric resonance thermal coefficient $K^{prt}$ = -53.6 Hz/K of the selected mode.

Equivalent temperature of the crystal heated by laser radiation of power $P$ can be directly determined using relations (9) and (10) by measuring piezoelectric resonance frequency shift $Rf(P)$ of the calibrated piezoelectric resonance. Figure 2 (b) shows dependence of the selected resonance frequency on laser power. Laser source is continuous wave single-mode fiber laser that generates polarized radiation at wavelength $\lambda$ = 1064 nm. Collimated Gaussian beam of 1.4 mm diameter (beam quality factor M$^2$ is close to 1) was directed along crystal length $L_3$=11.2 mm. Values of $Rf(P)$ at each power $P$ were measured after reaching stationary temperature state of the quartz crystal with surrounding air.

Linear approximation (8) gives value of piezoelectric resonance optical coefficient $K^{pro}$ = -42.5 Hz/W of selected resonance. Then optothermal coefficient $\beta$, which is independent on resonance selection, can be obtained using (11). Its value is $\beta$ = 0.79 K/W.

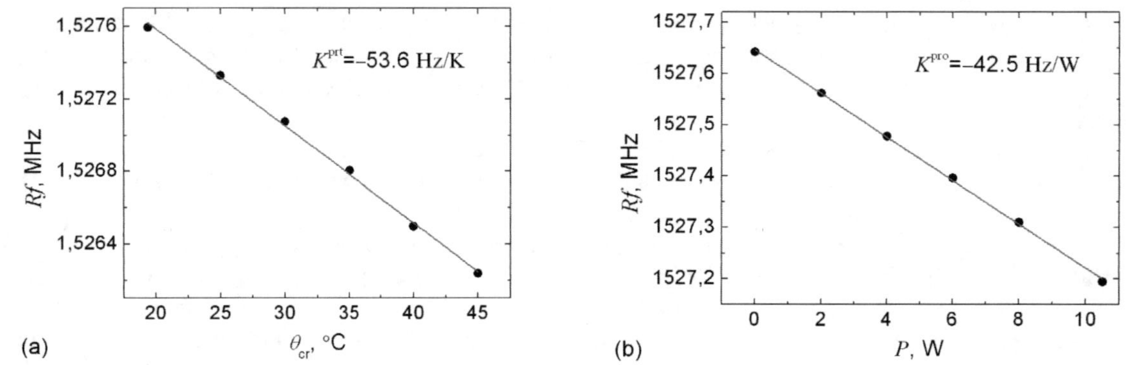


**Figure 2.** Piezoelectric resonance frequency dependencies on crystal temperature during its uniform heating (a) and on laser power (b).

*3.3. Crystal temperature kinetics measurement during and after laser irradiation*

Crystal piezoelectric resonance frequency kinetics is measured during interaction with laser radiation. Then both optical absorption and heat transfer coefficients can be obtained replacing $\theta_{cr}$ in (2) and (4) by the value of $\Theta_{eq}$ represented by (9). Figure 3 shows experimental results of our crystal heating by laser radiation of 10.5 W power at 1064 nm wavelength and subsequent cooling.

Exponential fittings of quartz crystal heating and cooling kinetics using equation (4) give characteristic temperature kinetics times $\tau_h = 53.18$ s and $\tau_c = 53.19$ s respectively. Corresponding values of optical absorption at $\lambda = 1064$ nm and heat transfer coefficients are calculated using equations (3) and (5): $\alpha = 2.4 \cdot 10^{-3}$ cm$^{-1}$ and $h^T = 23.5$ Wm$^{-2}$K$^{-1}$ respectively.

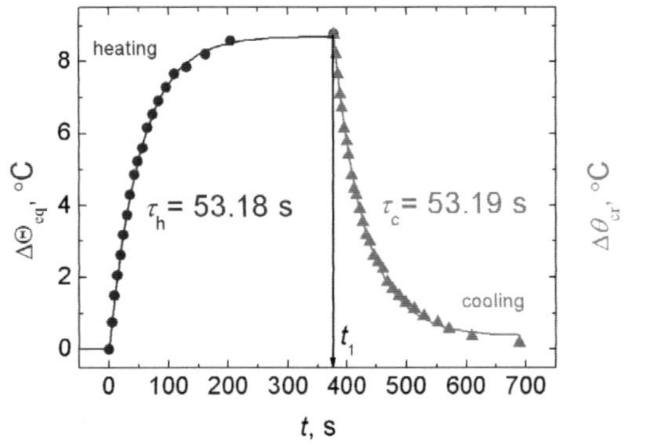


**Figure 3.** Quartz crystal temperature kinetics during its heating by laser radiation (10.5 W at 1064 nm) and subsequent cooling.

Experimental results reveal that both heating and cooling events gives almost equal values of characteristic temperature kinetic time ($\tau_h \approx \tau_c$). This is the main criteria for validity of equivalent temperature concept of the crystal nonuniformly heated by laser radiation. When laser radiation affects the crystal its temperature is nonuniform and can be calculated using equation (1). It is essential that after switching off laser radiation the temperature distribution inside crystal rapidly becomes uniform due to the high value of its thermal conductivity. It means that in the last case equivalent temperature of the crystal is identical to the crystal thermodynamic temperature.

## 4. Equivalent temperature model

*4.1. Stages of generalized model*

Modeling of equivalent temperature of nonlinear-optical crystal nonuniformly heated by laser radiation should contain following stages:

- Determination of crystal elastic and piezoelectric tensors temperature dependence;

- Measurement of crystal impedance (admittance) spectrum for finding piezoelectric resonances [3];
- Measurement of piezoelectric resonance thermal coefficients $K_i^{\mathrm{prt}}$ from crystal uniform heating experiment [3];
- Theoretical calculation of crystal internal vibration modes spectrum and identification with experimentally observed piezoelectric resonances [9];
- Measurement of piezoelectric resonance optical coefficients $K_i^{\mathrm{pro}}$ from crystal nonuniform laser heating experiment [3];
- Determination of optothermal coefficient $\beta$ for the crystal equivalent temperature $\Theta_{eq}(P)$ determination at given laser power $P$ [3];
- Measurement of laser radiation intensity distribution;
- Measurement of optical absorption $\alpha(\lambda)$ and heat transfer $h^{\mathrm{T}}$ coefficients [4, 6].
- Theoretical calculation of the nonuniform temperature distribution $\theta_{cr}(x_1, x_2, x_3, P)$ in crystal heated by laser radiation by solving heat conduction equation (1). Determination of $\theta_{cr}(x_1, x_2, x_3, P)$ maximum $\theta_{max}(P)$ and minimum $\theta_{min}(P)$ [5, 9];
- Theoretical calculation of crystal equivalent temperature $\Theta_{eq}(P)$ for the laser power $P$ at $\lambda$ wavelength in compliance with elastic and piezoelectric coefficients distribution that corresponds to calculated nonuniform temperature distribution [5, 9];
- Measured and calculated equivalent temperature should lie between calculated values $\theta_{min}(P)<\Theta_{eq}(P)<\theta_{max}(P)$ [5, 9].

### *4.2. Basic equations and boundary conditions*

The vibration of the sample is characterized by the vector of mechanical displacement $u_i(x_1,x_2,x_3)$ $i=\overline{1,3}$ and electric potential $\varphi(x_1,x_2,x_3)$. Equations describing motion of the sample can be written as

$$\begin{cases} \rho\ddot{u}_i = T_{ij,j} + F_i \\ D_{i,i} = 0 \\ T_{ij} = c_{ijkl}S_{kl} - d_{ijk}E_k \\ D_i = \varepsilon_{ij}E_j + d_{ijk}S_{jk} \end{cases}. \qquad (13)$$

Here point above the variable means time derivative and $G_{i,j} = \partial G_i/\partial x_j$ denotes spatial derivative; indices $i,j,k,l=\overline{1,3}$; $\rho$ is sample density; $T_{ij}$ and $S_{ij}=0.5\left(u_{i,j}+u_{j,i}\right)$ are stress and strain tensors respectively; $F_i$ are components of body force vector; $D_i$ and $E_i = -\varphi_{,i}$ are components of electric induction and electric field vectors respectively; $c_{ijkl}$, $d_{ijk}$, $\varepsilon_{ij}$ are elastic, piezoelectric and permittivity tensors respectively. Region occupied by the sample is denoted by $\Omega$, its boundary is denoted $\partial\Omega \equiv \Gamma$. In general system (13) should be supplemented by initial and boundary conditions. For our purposes only boundary conditions have to be considered. These are split into mechanical and electrical boundary conditions of the first or second kind. They can be represented as follows:

$$u_i(\mathbf{r})=\overline{u}_i, \mathbf{r}\in\Gamma_{\mathrm{u}};\quad T_{ij}(\mathbf{r})n_j=\overline{F}_i, \mathbf{r}\in\Gamma_{\mathrm{F}};\quad \varphi(\mathbf{r})=\overline{\varphi}, \mathbf{r}\in\Gamma_{\varphi};\quad D_i(\mathbf{r})n_i=\overline{D}, \mathbf{r}\in\Gamma_{\mathrm{D}}. \qquad (14)$$

Here $\mathbf{r}$ is the radius vector. Boundaries are $\Gamma=\Gamma_{\mathrm{u}}\cup\Gamma_{\mathrm{F}}$ and $\Gamma=\Gamma_{\varphi}\cup\Gamma_{\mathrm{D}}$. For the determination of resonance frequencies for bodies of different geometry the common approach is to use Lagrange-Hamilton variation principle.

### *4.3. Variation principle*

As was noted by many authors (e.g., see [10-12]) variational formulation is very useful tool for numerical solution of the problems with unspecified boundary conditions. Tiersten [10] proposed to use this method in such a way that no constraints are imposed on the variations. His idea was to add to the main functional the constraints multiplied by undetermined Lagrange multipliers. Following Tiersten one can obtained functional for piezoelectric medium in the following form:

$$J=\int_{t_0}^{t}\left[\iiint_{\Omega}\left(\frac{1}{2}\rho\dot{u}_i^2-\frac{1}{2}c_{ijkl}S_{ij}S_{kl}+\frac{1}{2}\varepsilon_{ij}E_iE_j+d_{ijk}S_{ij}E_k\right)d\Omega\right]dt$$
$$+\int_{t_0}^{t}\left[\iint_{\Gamma_F}\overline{F}_iu_idA-\iint_{\Gamma_D}\overline{D}_in_i\varphi dA+\iint_{\Gamma_u}T_{ij}n_j\left(u_i-\overline{u}_i\right)dA-\iint_{\Gamma_\varphi}D_in_i\left(\varphi-\overline{\varphi}\right)dA\right]dt. \quad (15)$$

Here $n$ is normal vector to the surface and $dA$ is area element. By finding stationary points of (15) using natural boundary conditions $\Gamma_u=\Gamma_\varphi=\varnothing,\ \overline{F}_i=\overline{D}=0$ we will obtain the set of eigenmodes of the sample. The process of obtaining stationary solutions is described, for example, by Demarest [12] for the case $\varepsilon_{ij}=d_{ijk}=0$. It may be divided into two stages: separation of time dependence and variation of spatial part of the functional. The latter stage was described by Tiersten and Holland [10, 11]. The most important conclusion concerning the time dependence is that it is harmonic with frequency $w$. So later on we consider only spatial part of the functional [9].

$$J_S=\iiint_{\Omega}\left(\rho w^2u_i^2-c_{ijkl}S_{ij}S_{kl}+\varepsilon_{ij}E_iE_j+2d_{ijk}S_{ij}E_k\right)d\Omega. \quad (16)$$

In order to solve numerically problem (16) the Rayleigh-Ritz method is usually used. In this method unknown functions of displacement and potential are expanded using some set of orthonormal basis functions $\{\psi_p\}$ of the total number $N_b$ ($p=\overline{1,N_b}$). It means that equations for eigenmodes can be obtained by substitution series expansions $u_i=C^{u}_{ip}\psi_p$ and $\varphi=C^{\varphi}_p\psi_p$ in Lagrangian (16) and subsequent differentiation with respect to each expansion coefficient $C$ [9]. Extremum of the functional (16) corresponds to the condition of equal to zero derivatives $\partial J_S(C)/\partial C=0$. In matrix form equations for eigenmodes can then be written as

$$\begin{cases}C^{\varphi}=-\mathrm{V}^{-1}\Pi^{\mathrm{T}}C^{u}\\ \left(\mathrm{Y}-\Pi\mathrm{V}^{-1}\Pi^{\mathrm{T}}\right)C^{u}=w^2\mathrm{W}C^{u}\end{cases}. \quad (17)$$

Here 'T' and '-1' designate transposed and inverse matrix respectively and $C^{u}=\left(C^{u}_1,C^{u}_2,C^{u}_3\right)^{\mathrm{T}}$. Matrices W, Y, Π and V are ($p=\overline{1,N_b}$, $q=\overline{1,N_b}$)

$$\mathrm{W}_{pq}=\iiint_{\Omega}\rho\psi_p\psi_qd\Omega,\quad \mathrm{Y}^{ik}_{pq}=\iiint_{\Omega}c_{ijkl}\frac{\partial\psi_p}{\partial x_j}\frac{\partial\psi_q}{\partial x_l}d\Omega,$$
$$\Pi^{i}_{pq}=\iiint_{\Omega}d_{ijl}\frac{\partial\psi_p}{\partial x_j}\frac{\partial\psi_q}{\partial x_l}d\Omega,\quad \mathrm{V}_{pq}=\iiint_{\Omega}\varepsilon_{jl}\frac{\partial\psi_p}{\partial x_j}\frac{\partial\psi_q}{\partial x_l}d\Omega. \quad (18)$$

After finding the eigenfrequencies $w_n$ and pertinent spatial distributions of the displacement and potential the piezoelectric resonance thermal coefficients $K^{\mathrm{prt}}_n=\frac{1}{2\pi}\frac{dw_n}{d\theta}$ can be calculated if temperature dependence of the elastic constants $c_{ijkl}(\theta)$ is known. Detailed calculation can be found elsewhere [6, 9]. As it was estimated the contribution to resonance frequency shift of other parameters (e.g. density $\rho$) variation with temperature is not determinative. It was also demonstrated that for common values of material piezoelectric and dielectric coefficients the contribution of piezoelectric effect to intrinsic vibration mode frequencies doesn't exceed 0.5% [13]. So for simplification we shall neglect $\Pi\mathrm{V}^{-1}\Pi^{\mathrm{T}}$ term in (17). Using perturbation theory the resonance frequency shift can be expressed as:

$$\frac{dw_n}{d\theta}=\frac{C^{u}_{ip,n}\dfrac{d\mathrm{Y}^{ik}_{pq}}{d\theta}C^{u}_{kq,n}}{2\rho w_n}. \quad (19)$$

During interaction with laser radiation the nonuniform temperature distribution inside the crystal leads to an additional shift of its eigenmode frequencies, because here unlike the uniform case elastic constants will have spatial dependence. If we consider small uniform temperature perturbation $d\theta$ then resulting elastic coefficients $c_{ijkl}$ can be represented by

$$c_{ijkl} = c^0_{ijkl} + \frac{dc_{ijkl}}{d\theta} d\theta . \tag{20}$$

Here $c^0_{ijkl}$ corresponds to the initial uniform temperature $\theta_0$. By substituting (20) into equation (18) for matrix Y we obtain using relation (19) the following expression for piezoelectric resonance thermal coefficients:

$$K^{\text{prt}}_n = \frac{\frac{dc_{ijkl}}{d\theta} C^{\text{u}}_{ip,n} B^{jl}_{pq} C^{\text{u}}_{kq,n}}{4\pi\rho w_n} . \tag{21}$$

Here $C^u_{ip,n}$ are expansion coefficients of mechanical displacement that corresponds to the $n$-th vibration mode. And matrix B is

$$\text{B}^{jl}_{pq} = \iiint_\Omega \frac{\partial \psi_p}{\partial x_j} \frac{\partial \psi_q}{\partial x_l} d\Omega . \tag{22}$$

## 5. Comparison of experimental and theoretical results

Eigenfrequencies of quartz in rectangular parallelepiped form with dimensions $L_1\times L_2\times L_3$ can be found as solution of eigenvalue problem (17) using Legendre polynomials $P_{p_{xi}}(x_i)$ as basis functions [9, 12]:

$$\psi_p(x_1, x_2, x_3) = \prod_{i=1}^{3} \left[ \left( \frac{2p_{xi}+1}{L_i} \right)^{1/2} P_{p_{xi}} \left( \frac{2x_i}{L_i} \right) \right] . \tag{23}$$

Such basis is convenient to use here because for such sample shape it forms orthogonal system:

$$\iiint_\Omega \psi_p \psi_q d\Omega = \delta_{p_{x1}q_{x1}} \delta_{p_{x2}q_{x2}} \delta_{p_{x3}q_{x3}} . \tag{24}$$

Here $p_{xi}$ is polynomial degree specified by $i=\overline{1,3}$ and $p=\overline{1,N_b}$ where $N_b$ is total number of basis functions. Vibration spectrum $w_n$ is calculated from the known sample dimensions, its mass, and a set of surmised elastic constants. The main problem is that number of calculated modes in narrow frequency range greatly exceeds number of experimentally observed ones. Special identification procedure should be performed for finding correct correspondence between certain measured mode and one of the calculated modes. Recently we have introduced novel simple methods of modes identification. First one relies on comparison of measured and calculated piezoelectric resonance thermal coefficients (for more details see [9]) and second one is based on comparison of resonance dimensional coefficients, i.e. mode frequency shift with the change of sample size [7]. These coefficients can be experimentally measured and also calculated using expression (21).

Figure 4 (a) shows calculated and measured resonance spectra of our quartz crystal (2.9×2.9×11.2 $\text{mm}^3$) in frequency range from 0.9 MHz to 1.5 MHz. Initial set of quartz crystal elastic constants together with its dependence on temperature were taken from literature [14]. By variation of elastic constants (about 1%) and its temperature derivatives (about 2%) values we have identified calculated mode ($Rf^{\text{calc}} = 1527700 \pm 100$ Hz at 18 °C; $\left(K^{\text{prt}}\right)^{\text{calc}} = -54 \pm 1$ Hz/K ) with experimentally measured mode ($Rf^{\text{exp}} = 1527681 \pm 1$ Hz at 18 °C, $\left(K^{\text{prt}}\right)^{\text{exp}} = -53.6 \pm 0.1$ Hz/K), which was selected for detailed investigations (see section 3). It can be seen that there are few experimental modes that do not match well enough with calculated ones. Disagreement can be attributed to the imperfection of crystal form and errors of sample sizes determination.

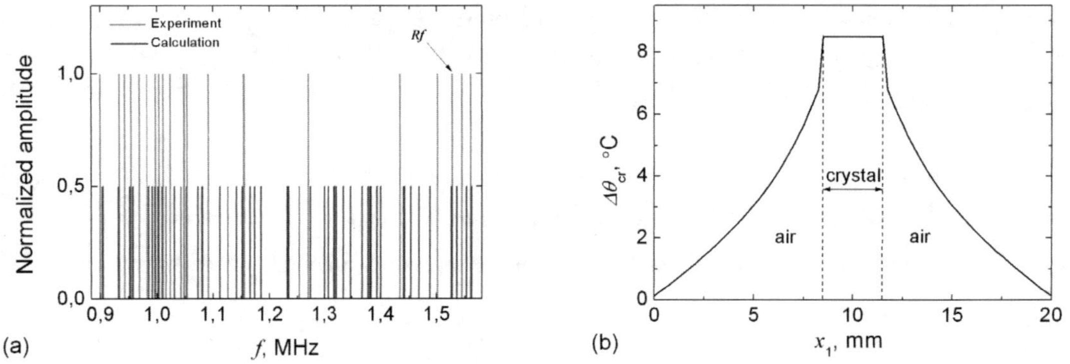


**Figure 4.** (a) Calculated (0.5 level) and measured (1 level) resonance spectra of quartz crystal (rectangular parallelepiped 2.9×2.9×11.2 mm$^3$). Arrow indicates resonance mode that was identified. (b). Calculated stationary temperature distribution in quartz crystal, irradiated by 10.5 W laser power, and ambient air in respect to the thermostat temperature [9].

Stationary temperature distribution inside crystal heated by laser radiation and surrounding air can be calculated using equation (1) with equal to zero time derivatives. Figure 4 (b) shows calculated stationary temperature distribution for our quartz crystal (2.9×2.9×11.2 mm$^3$) irradiated by laser radiation of power $P = 10.5$ W at $\lambda = 1064$ nm. Laser radiation intensity has Gaussian distribution with beam diameter $d = 1.4$ mm at $1/e^2$ level:

$$I\left(x_1, x_2\right) = \frac{2P}{\pi\left(d/2\right)^2} \exp\left(-2\frac{x_1^2 + x_2^2}{\left(d/2\right)^2}\right). \tag{25}$$

Values of optical absorption and heat transfer coefficients: $\alpha(\lambda) = 2.4 \cdot 10^{-3}$ cm$^{-1}$ and $h^T = 23.5$ Wm$^{-2}$K$^{-1}$ were measured in calorimetric experiment (see section 3.3). Other necessary physical parameters of quartz crystal and air were taken from literature.

Crystal overall temperature rise is about 8 °C and maximum difference of temperature inside crystal is of $\Delta\theta_{cr} = 10^{-2}$ °C order. Temperature gradient in the surrounding air is relatively high in comparison with crystal volume. Also there is a gap of about 1 °C between crystal and air temperatures at crystal interface. These results reveal that for characterization of crystal response to the propagating laser radiation it is necessary to use equivalent temperature concept.

Calculation of crystal equivalent temperature is more of academic rather than practical interest. From practical point of view it is necessary to calculate:

- Temperature distribution inside volume of the crystal heated by laser radiation with given spatial intensity distribution. Nonuniform temperature distribution is not measured expetrimentally
- Frequencies of internal vibration modes of the crystal when its temperature is uniform;
- Dependence of all vibration mode frequencies on uniform crystal temperature and comparison of calculated piezoelectric resonance thermal coefficients with measured ones.

Identification of crystal vibration modes gives opportunity to amend accuracy of experimental values of elastic constants together with its temperature derivatives. These are key parameters of nonlinear-optical crystal that specify its threshold values of optical fluence, which leads to optical damage of crystals.

Frequency shifts of resonances caused by absorption of laser radiation of power $P$ can be calculated, using e.g. perturbation theory. This frequency shift can be associated with the crystal uniform temperature change that gives the same shift value. Thus determined temperature change is called an equivalent heating temperature of the crystal. Implementation of perturbation theory gives the following expression for the crystal equivalent heating temperature [9]:

$$\Delta\Theta_{\text{eq}}(P)=\frac{\frac{\partial c_{ijkl}}{\partial\theta}\delta\theta_{rs}(P)C^{\text{u}}_{ip,n}Z^{jl}_{pqrs}C^{\text{u}}_{kq,n}}{\frac{\partial c_{ijkl}}{\partial\theta}C^{\text{u}}_{ip,n}\text{B}^{jl}_{pq}C^{\text{u}}_{kq,n}}, \tag{26}$$

Where matrix B is expressed by equation (22) and matrix Z is

$$\text{Z}^{jl}_{pqrs}=\iiint_{\Omega}\psi_r(x_1)\psi_s(x_2)\frac{\partial\psi_p}{\partial x_j}\frac{\partial\psi_q}{\partial x_l}d\Omega. \tag{27}$$

Expression (26) contains temperature derivatives of elastic constants. Values $\text{B}^{jl}_{pq}$, $\text{Z}^{jl}_{pqrs}$ depend only on sample geometry. Vector $C^{\text{u}}_{ip,n}$ characterizes given $n$-th resonance. Coefficients $\delta\theta_{rs}(P)$ characterize the temperature distribution of the irradiated crystal with transverse dimensions $L_1$ and $L_2$.

$$\delta\theta_{rs}=\int_{-\frac{L_1}{2}}^{\frac{L_1}{2}}\int_{-\frac{L_2}{2}}^{\frac{L_2}{2}}\delta\theta(x_1,x_2)\psi_r(x_1)\psi_s(x_2)dx_1dx_2\,. \tag{28}$$

Temperature dependence on the coordinate $x_3$ along the radiation propagation direction ($L_3$) is not considered due to low optical absorption. Crystal temperature perturbation $\delta\theta$, caused by interaction with laser power, can be expanded using our orthonormal basis functions $\{\psi_p\}$:

$$\delta\theta(x_1,x_2)=\sum_{r,s}\delta\theta_{rs}\psi_r(x_1)\psi_s(x_2), \tag{29}$$

Calculations performed for several resonance vibration modes of crystal in frequency range 20 kHz to 180 kHz revealed that difference between measured value of equivalent heating temperature and calculated one using equation (26) is less than 0.01% [9]. Higher accuracy cannot be achieved because of values of elastic constants and its temperature derivatives as well as sample sizes are determined with lower precision.

## 6. Main conclusions

Important applications of equivalent temperature concept is determination of optical absorption $\alpha(\lambda)$ and heat transfer coefficients $h^{\text{T}}$ in laser calorimetry technique [6]. Classical laser calorimetry, which is standardized method for optical absorption measurement, is based on finding of the correspondence between the solution of nonstationary heat conduction equation and experimentally measured crystal temperature kinetics when it is heated and cooled during and after laser irradiation respectively. Classical approach, which supposes indirect temperature measurement via external thermal detectors, lacks accuracy and crystal equivalent temperature kinetics should be measured instead. Despite that degree of crystal temperature nonuniformities in cases of heating and cooling differ from each other the characteristic kinetic times, measured in each case, have close values due to both low optical absorption and application of low-power laser radiation. However, when high-power radiation is involved especially in nonlinear-optical interactions the application of equivalent temperature concept plays crucial role. It has been already successfully used for determination of crystal temperature in course of nonlinear-optical frequency conversion of laser radiation. Second harmonic generation experiment using nonlinear-optical periodically polled lithium niobate (PPLN) crystal was recently performed [8]. It was observed that when second harmonic power is low the PPLN equivalent temperature linearly depend on pump laser power. However, when PPLN temperature is close to phase matching temperature and second harmonic is efficiently generated, the crystal equivalent temperature exhibit substantially nonlinear dependence on pump power.

## Reference

[1] Dmitriev V G and Tarasov L V 2004 *Applied nonlinear optics* (Moscow Russia: Fizmatlit)
[2] ISO11551 2003 Test method for absorptance of optical laser components *International Organization for Standartization* (Geneva Switzerland)

Journal of Physics: Conference Series **510** (2014) 012033 doi:10.1088/1742-6596/510/1/012033

[3] Ryabushkin O A, Konyashkin A V, Myasnikov D V, Tyrtyshnyy V A and Baranov A I 2013 Equivalent temperature of nonlinear-optical crystals in process of laser frequency conversion *CLEO®/Europe - IQEC 2013: Conference Digest* (Munich Germany 12-16 May 2013) CE 8.5 WED

[4] Ryabushkin O A, Myasnikov D V, Konyashkin A V and Vershinin O I 2013 Kinetics of equivalent temperature of nonlinear-optical crystals *CLEO®/Europe - IQEC 2013: Conference Digest* (Munich Germany, 12-16 May 2013) CE-P.18 TUE

[5] Ryabushkin O A, Konyashkin A V, Myasnikov D V, Tyrtyshnyy V A and Baranov A I 2013 Equivalent temperature of nonuniformly heated nonlinear-optical crystals in course of laser radiation frequency conversion *Proc. SPIE 8847 Photonic Fiber and Crystal Devices: Advances in Materials and Innovations in Device Applications VII* (San Diego California USA, 25 August 2013) 884711

[6] Ryabushkin O A, Konyashkin A V, Myasnikov D V, Tyrtyshnyy V A and Vershinin O I 2013 Piezoelectric resonance calorimetry of nonlinear-optical crystals under laser irradiation *Proc. SPIE 8847 Photonic Fiber and Crystal Devices: Advances in Materials and Innovations in Device Applications VII* (San Diego California USA, 25 August 2013) 88470Q

[7] Ryabushkin O A, Myasnikov D V and Konyashkin A V 2013 Novel method for identification of intrinsic vibration modes in piezoelectric crystals *XXV IUPAP Conference on Computational Physics, Book of Abstracts* (Moscow Russia, 20-24 August 2013) pp 89 (full paper in current proceedings)

[8] Ryabushkin O A, Myasnikov D V and Baranov A I 2013 The equivalent temperature model in process of nonlinear conversion of laser radiation *XXV IUPAP Conference on Computational Physics, Book of Abstracts* (Moscow Russia, 20-24 August 2013) pp 90 (full paper in current proceedings)

[9] Myasnikov D V 2011 Model of resonant interaction of radiofrequency electric field with piezoelectric crystals during its interaction with laser radiation *PhD Thesis, Moscow Institute of Physics and Technology, Dolgoprudnyy Moscow region, Russia* (*in Russian*)

[10] Tiersten H F 1968 Natural boundary and initial conditions from a modification of Hamilton's principle *J. Math. Phys.* **9** 1445

[11] Holland R 1968 Resonant properties of piezoelectric ceramic rectangular parallelepipeds *J. Acoust. Soc. Am.* **43** 988

[12] Demarest H H 1971 Cube-resonance method to determine the elastic constants of solids *J. Acoust. Soc. Am.* **49** 768

[13] Ohno I 1990 Rectangular parallelepiped resonance method for piezoelectric crystals and elastic constants of alpha-quartz *Phys. Chem. Minerals* **17** 371

[14] Ohno I, Harada K, Yoshitomi C 1990 Temperature variation of elastic constants of quartz across the α-β transition constants of alpha-quartz *Phys. Chem. Minerals* **33** 1

# Dynamics of Two-dimensional Electron Gas in Non-uniform magnetic field

**O G Balev[1] and I A Larkin[2,3]**

**Abstract.**
We have theoretically studied dynamics of the two-dimensional electron system (2DES) placed in a strong laterally non-uniform magnetic field, which appears due to ferromagnetic film on the top of heterostructure. It is shown that lateral inhomogeneity of a strong magnetic field allows itself "magnetic gradient" or special magnetic-edge magnetoplasmons. This mechanism is different from usual "density gradient" edge magnetoplasmons. We have solved self-consistently Poisson equation for non-uniform density distribution of the 2DES for realistic heterostructure together with hydrodynamic equation of 2D Fermi liquid. As a result eigen value problem has been obtained that corresponds to the motion of charge density wave perpendicular to magnetic gradient. It is shown that for non-monotonic distribution of magnetic field "magnetic gradient" magnetoplasmons may move in both directions. To solve eigen value problem we have compared two types of numerical approaches: first is grid method that diagonalizes large Hermitian matrix and second is semi-analytical approach that expand each eigen mode on the set of orthogonal functions.

[1] Departamento de Física, Universidade Federal do Amazonas, 69077-000, Manaus, AM, Brazil
[2] Department of Physics, Minho University, Braga 4710-057, Portugal
[3] Institute of Microelectronics Technology RAS, 142432 Chernogolovka, Russia

E-mail: ogbalev@ufam.edu.br
vaniala2000@yahoo.co.uk

Edge magnetoplasmons (EMPs) have been studied for 2DES subjected to a lateral confining potential and a strong homogeneos magnetic fiels [1, 2, 3, 4, 5]. The EMPs appear due to a strong change of the stationary local electron density at the edge of the channel that induces a strong modulation of the local magnetoconductivity [1, 3, 5] and can be charactarized also as "density gradient" EMPs [6]. Recently [6] the chiral modes in 2DES induced by the lateral inhomogeneity of a strong magnetic field and named magnetic-edge magnetoplasmons (MEMPs), or "magnetic gradient" edge magnetoplasmons, have been studied. They are localized in a vicinity of magnetic-edge, i.e., the region of magnetic field inhomogeneity. Here we present new results on MEMPs. In particular, for conditions when modulation of the magnetic field, within 2DES, is not very small in comparison with a finite external spatially homogeneous magnetic field; here purely symmetric or antisimmetric MEMPs will be absent.

We consider homogeneous 2DES, localized within $z = 0$ plane, that is embedded in GaAs based sample, with the dielectric constant $\varepsilon$, that occupies a half-space $z < d$. In addition, 2DES is subjected to a strong laterally inhomogeneos magnetic field $\mathbf{B}(y) = B(y)\hat{\mathbf{z}}$, which appears due to ferromagnetic semi-infinite film of a finite thickness $\eta d$. This film is located at $y < 0$, $d(1+\eta) > z > d$. Fig. 1 presents a model geometry under discussion; the x-axis points out of the figure plane. A finite external spatially homogeneous magnetic field $\mathbf{B}_{ext} = B_{ext}\hat{\mathbf{z}}$ is applied as well; $B_{ext} > 0$. We assume that the surface of heterostructure crystal, at $z = d$, has

Journal of Physics: Conference Series **510** (2014) 012034 doi:10.1088/1742-6596/510/1/012034

pined potential [7], or covered by a very thin (of the thichness $\ll d$) homogeneous nonmagnetic-metallic gate. In addition, we assume that at $z = 0$ plane the ions jellium background of the constant area density $n_I$ is located. Following Ref. [6], we assume that magnetic field $B(y)$ is a smooth function of $y$, with the characteristic scale $\Delta y$ and the static electron density is well approximated by $n_I$.

**Figure 1.** 2DES, at $z = 0$, is embedded in heterostructure, $z < d$. Potential at $z = d$ is pined. Ferromagnetic semi-infinite film ($y < 0$, $d < z < d(1+\eta)$) induces inhomogeneity of magnetic field within 2DES.

Following [6], for the low-frequency and the long-wavelength conditions, the current density induced by a wave can be calculated in the quasi-static approximation as

$$j_x(y) = \sigma_{xx}(y)E_x(y) - \sigma^0_{yx}(y)E_y(y), \quad j_y(y) = \sigma_{yy}(y)E_y(y) + \sigma^0_{yx}(y)E_x(y), \tag{1}$$

where we have suppressed the exponential factor $\exp[-i(\omega t - k_x x)]$ and common arguments $\omega$, $k_x$ in $j_\mu(y)$, $E_\mu(y)$. Dissipation is neglected, assuming a clean 2DES and sufficiently low temperatures $T$ [6].

For our setup, shown in Fig. 1, we assume that magnetic moment of the ferromagnetic semi-infinite film is constant, $\mathbf{M}_0 = M_0\hat{\mathbf{z}}$. Then readily it follows [8, 6] that

$$B(y) = B_{ext} - 2M_0\{\arctan(Y) - \arctan(Y/(1+\eta))\}, \tag{2}$$

where $Y = y/d$. We assume that $B_0 > 0$ and $M_0/B_{ext} \leq 1$.

**Figure 2.** Dimensionless phase velocity for six most fast symmetric MEMPs, for $2M_0/B_{ext} \ll 1$, as function of $\eta$. MEMPs of both positive and negative chirality are present.

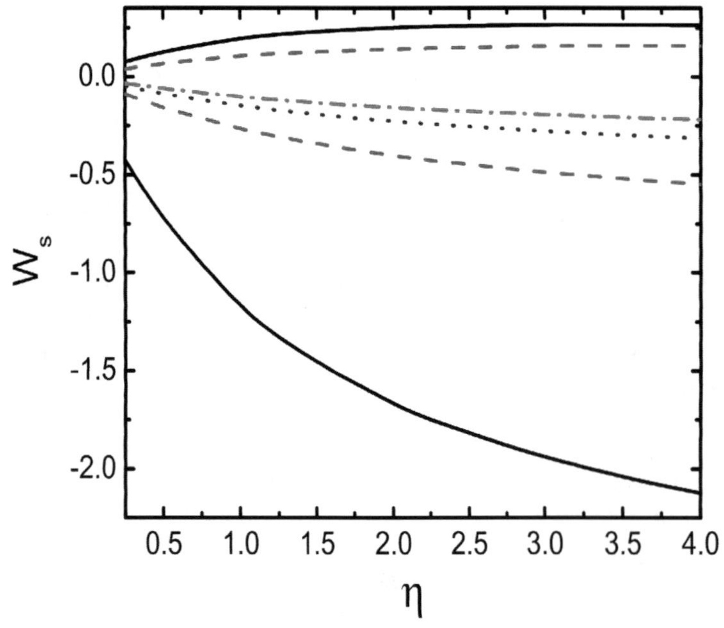

**Figure 3.** Dimensionless phase velocity for six most fast antisymmetric MEMPs, for $2M_0/B_{ext} \ll 1$, as function of $\eta$. MEMPs of both positive and negative chirality are present.

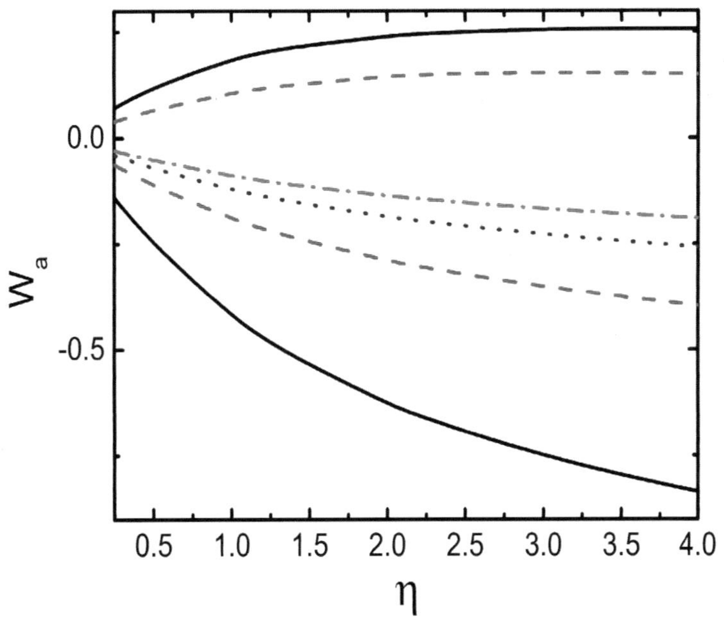


For $B(y)$ given by Eq. (2), using Eq. (1), the Poisson equation and the continuity equation, we obtain the integral equation for the wave charge density $\rho(\omega, k_x, Y)$ as

$$\frac{\omega}{k_x}\rho(\omega, k_x, Y) - \frac{2|e|cn_I}{\varepsilon B_0} g_0(Y) f_0(Y) \int_{-\infty}^{\infty} dY' \rho(\omega, k_x, Y') R_g^{(0)}(|Y - Y'|; k_x d) = 0, \quad (3)$$

where $B_0 = B_{ext}^2/(4M_0)$, exact dimensionless gradient $B(y)$ along $y$ is given by

$$g_0(Y) = \{\frac{(1+\eta)^{-1}}{1 + Y^2/(1+\eta)^2} - \frac{1}{1+Y^2}\}. \quad (4)$$

and prefactor

$$f_0(Y) = \{1 - \frac{2M_0}{B_{ext}}[\arctan(Y) - \arctan(\frac{Y}{1+\eta})]\}^{-2}. \quad (5)$$

In $R_g^{(0)}(|Y - Y'|; k_x d)$ we took the screening of the Coulomb potential by the equipotential surface at $z = d$ into account, namely:

$$R_g^{(0)}(|Y - Y'|; k_x d) = K_0(|k_x d||Y - Y'|) - K_0(|k_x d|\sqrt{(Y - Y')^2 + 4}), \quad (6)$$

where $K_0(x)$ is the modified Bessel function. We assume that $k_x d \ll 1$. Eqs. (3)-(6) show that at $k_x \to 0$ the magnetic edge waves have acoustic type of dispersion.

To solve equation (3) numerically, we use two methods. The first one is a grid approximation, where we use uniform grid with step $\Delta Y = 0.02$ with span $Y \in [-19, 19]$. As a result, equation (3) is approximated by large Hermitian matrix and eigen values of this matrix correspond to eigen values of equation (3). The drawback of this approach is that it is not an accurate approximation of equation (3) at $Y \to \pm\infty$.

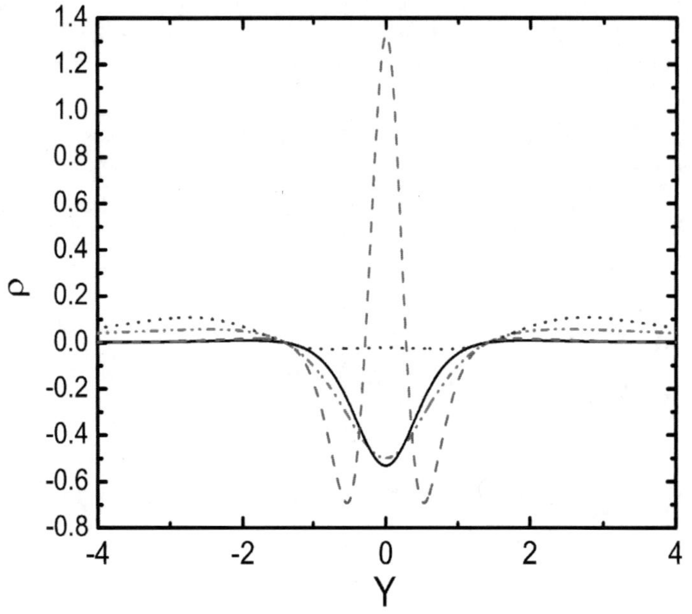


**Figure 4.** Spatial profiles, at $\eta = 1$, for three fastest symmetric MEMPs are plotted by the solid curve, for $W_s = -1.175255$, the dashed curve, for $W_s = -0.265200$, and the dotted curve, for $W_s = 0.196089$. The dot-dot-dashed curve plots $g_0(Y)$.

Another way to solve equation (3) numerically is to introduce a new variable $X = \frac{2}{\pi}\arctan(Y)$ that varies from $-1$ to $1$ while $Y$ changes from $-\infty$ to $\infty$. In Eq. (4) as function of $X$ the left hand side we will denote as $g_0(X)$, to simplify notations. First, we consider the limit $M_0/B_{ext} \ll 1$ where we approximate $1/B(Y)$ as

$$\frac{1}{B(y)} \approx \frac{1}{B_{ext}} - \frac{1}{2B_0}\{\arctan(Y/(1+\eta)) - \arctan(Y)\}, \tag{7}$$

in this approximation prefactor $f_0(Y) = 1$. In Ref. [6] it is shown that Eq. (3) at $f_0(Y) = 1$ has a complete set of solutions can be presented as the set of symmetric solutions and the set of antisymmetric solutions.

Then the symmetric solutions of Eq. (3) are given as [6]

$$\rho_s(\omega, k_x; X) = g_0(X)\sum_{k=0}^{\infty} a_k^s(\omega, k_x)\left[\cos(k\pi X) - \frac{1}{2}\delta_{k,0}\right], \tag{8}$$

and they are defined by the system of linear homogeneous equations

$$W a_n^s - \sum_{k=0}^{\infty} r_{n,k}^s a_k^s = 0, \tag{9}$$

where $n = 0, 1, 2, ...$; $\delta_{k,0}$ is the symbol Kroneckera, the matrix elements

$$r_{n,k}^s = \frac{\pi}{2}\int_{-1}^{1} dX \cos(n\pi X)\int_{-1}^{1} dX' R_g^{(0)}(|\tan(\frac{\pi}{2}X) - \tan(\frac{\pi}{2}X')|, |k_x d|)$$
$$\times\left[\frac{1+\eta}{1+\eta(2+\eta)\cos^2(\pi X'/2)} - 1\right]\left[\cos(k\pi X') - \frac{1}{2}\delta_{k,0}\right], \tag{10}$$

$W = \frac{\omega}{k_x}\frac{\varepsilon B_0}{2|e|cn_I}$ is the dimensionless wave velocity and its sign corresponds to a sign of the chirality or of the phase velocity. On Fig. 2 we plot dimensionless phase velocity $W_s$ as function of $\eta$ for six most fast symmetric MEMPs calculated from Eq. (9). In Fig. 2 it is seen that $|W_s|$ of the most fast MEMP with the negative chirality is much larger than $W_s$ of the most fast MEMP with the positive chirality.

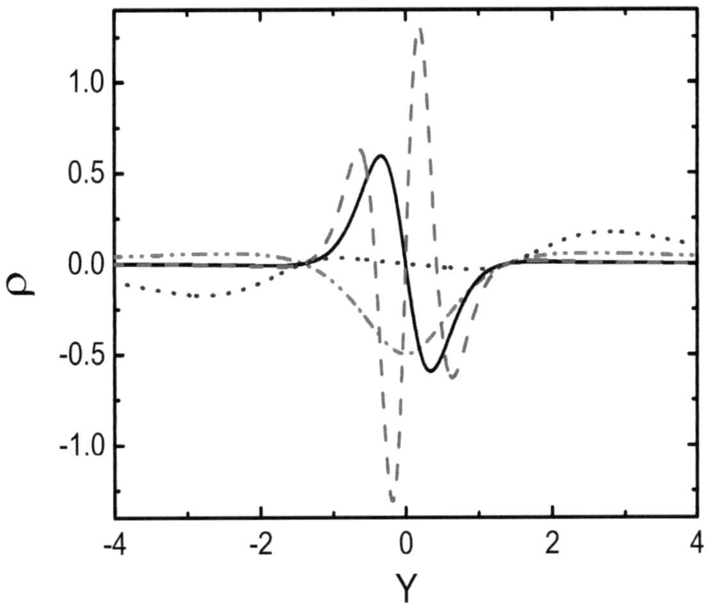


**Figure 5.** Spatial profiles, at $\eta = 1$, for three fastest antisymmetric MEMPs are plotted: by the solid curve, for $W_s = -0.418959$, the dashed curve, for $W_s = -0.189269$, and the dotted curve, for $W_s = 0.185154$. The dot-dot-dashed curve plots $g_0(Y)$.

The antisymmetric solutions of Eq. (3) are given as

$$\rho_a(\omega, k_x; X) = g_0(X) \sum_{k=1}^{\infty} a_k^a(\omega, k_x) \sin(k\pi X), \tag{11}$$

and they are defined by the system of linear homogeneous equations

$$W a_n^a - \sum_{k=1}^{\infty} r_{n,k}^a a_k^a = 0, \tag{12}$$

where $n = 1, 2, 3, ...$; the matrix elements

$$r_{n,k}^a = \frac{\pi}{2} \int_{-1}^{1} dX \sin(n\pi X) \int_{-1}^{1} dX' R_g^{(0)}(|\tan(\frac{\pi}{2}X) - \tan(\frac{\pi}{2}X')|, |k_x d|)$$
$$\times \left[\frac{1+\eta}{1+\eta(2+\eta)\cos^2(\pi X'/2)} - 1\right] \sin(k\pi X'). \tag{13}$$

Eq. (9) and Eq. (12) have infinite set of positive and negative eigen values, however, physically only a finite number of these modes, with largest $|W|$, meet assumed conditions as slower modes have the characteristic scale of the spatial structure along $y$ of the order of magnetic length or smaller. On Fig. 3 we plot dimensionless phase velocity $W_a$ as function of $\eta$ for six most fast antisymmetric MEMPs calculated from Eq. (12). In Fig. 3 it is seen that $|W_a|$ of the most fast MEMP with the negative chirality is essentially larger than $W_a$ of the most fast MEMP with the positive chirality. From Figs. 2, 3 it follows that the fastest mode is the fundamental symmetric MEMP which has negative chirality (as well as the sign of velocity) and only two nodes that coincide with the nodes of $g_0(Y)$. The latter is a common factor for obtained density profiles Eqs. (8), (11).

In Fig. 4, using Eqs. (8)-(9), and Fig. 5, by using Eqs. (11)-(12), we plot charge density profiles of three fastest symmetric and antisymmetric modes for $\eta = 1$; $g_0(Y)$ is also shown. It is seen that for the modes with the negative chirality the wave charge density is strongly localized within the region of negative $g_0(Y)$, i.e, $-1.75 \leq Y \leq 1.75$. In addition, in Figs. 4, 5 the charge density of the MEMPs with the positive chirality, is more localized within the region of positive $g_0(Y)$, at $|Y| > 1.75$.

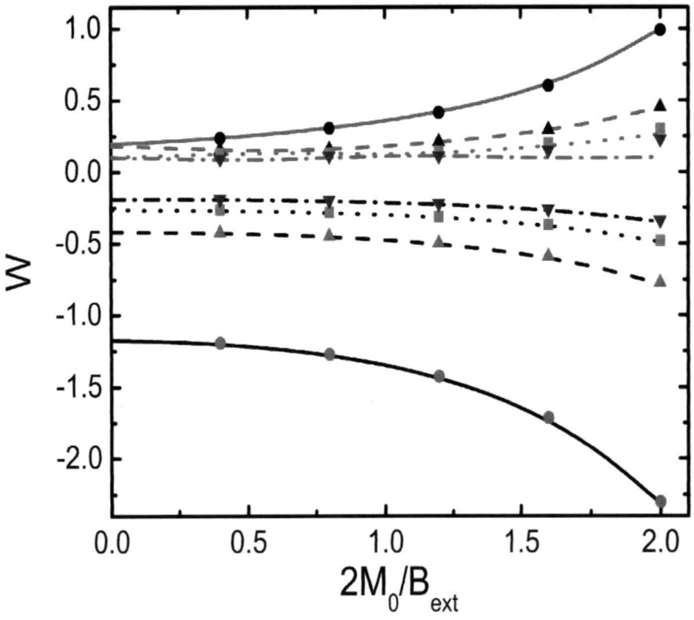


**Figure 6.** Dimensionless phase velocities for four most fast MEMPs with negative chirality and four most fast MEMPs with positive chirality as function of $2M_0/B_{ext}$, for $\eta = 1$, calculated from Eqs. (17). With full circles, triangles, squares, and upside-down triangles we show results for pertinent MEMPs calculated by another numerical approach, where equation (3) was approximated by the grid with step $\Delta Y = 0.02$ on finite domain [-19,19].

If exact form of $f_0(Y)$ is taken into account, Eq. (3) readily reduces to the integral equation for the wave charge density $\rho(\omega, k_x, X)$ as

$$W\rho(\omega, k_x, X) - g_0(X)f_0(X)\int_{-1}^{1}\frac{(\pi/2)dX'}{\cos^2(\pi X'/2)}\rho(\omega, k_x, X')R_g^{(0)}(|\tan(\frac{\pi}{2}X) - \tan(\frac{\pi}{2}X')|; k_x d) = 0, \tag{14}$$

where factor

$$f_0(X) = \{1 - \frac{2M_0}{B_{ext}}[\frac{\pi}{2}X - \arctan(\frac{1}{1+\eta}\tan(\frac{\pi}{2}X))]\}^{-2}. \tag{15}$$

From Eq. (14) it follows that its solutions are not either symmetric or antisymmetric, so we look for its solutions in the form

$$\rho(\omega, k_x; X) = g_0(X) f_0(X) \{ \sum_{k=0}^{\infty} a_k(\omega, k_x) \left[ \cos(k\pi X) - \frac{1}{2}\delta_{k,0} \right] + \sum_{l=1}^{\infty} b_l(\omega, k_x) \sin(l\pi X) \}. \quad (16)$$

Then Eq. (14) is reduced to the system of linear homogeneous equations

$$W a_n - \{ \sum_{k_1=0}^{\infty} r_{n,k_1}^{a,1} a_{k_1} + \sum_{k_2=0}^{\infty} r_{n,k_2}^{b,1} b_{k_2} \} = 0, \quad W b_m - \{ \sum_{k_3=0}^{\infty} r_{m,k_3}^{a,2} a_{k_3} + \sum_{k_4=0}^{\infty} r_{m,k_4}^{b,2} b_{k_4} \} = 0, \quad (17)$$

where $n, k_1, k_3 = 0, 1, 2, 3, ...$ and $m, k_2, k_4 = 1, 2, 3, ...$; the matrix elements

$$r_{n,k}^{a,1} = \frac{\pi}{2} \int_{-1}^{1} dX \cos(n\pi X) \int_{-1}^{1} dX' R_g^{(0)}(|\tan(\frac{\pi}{2}X) - \tan(\frac{\pi}{2}X')|, |k_x d|)$$
$$\times f_0(X') \left[ \frac{1+\eta}{1+\eta(2+\eta)\cos^2(\pi X'/2)} - 1 \right] \left[ \cos(k\pi X') - \frac{1}{2}\delta_{k,0} \right], \quad (18)$$

$$r_{n,k}^{b,1} = \frac{\pi}{2} \int_{-1}^{1} dX \cos(n\pi X) \int_{-1}^{1} dX' R_g^{(0)}(|\tan(\frac{\pi}{2}X) - \tan(\frac{\pi}{2}X')|, |k_x d|)$$
$$\times f_0(X') \left[ \frac{1+\eta}{1+\eta(2+\eta)\cos^2(\pi X'/2)} - 1 \right] \sin(k\pi X'), \quad (19)$$

$$r_{m,k}^{a,2} = \frac{\pi}{2} \int_{-1}^{1} dX \sin(m\pi X) \int_{-1}^{1} dX' R_g^{(0)}(|\tan(\frac{\pi}{2}X) - \tan(\frac{\pi}{2}X')|, |k_x d|)$$
$$\times f_0(X') \left[ \frac{1+\eta}{1+\eta(2+\eta)\cos^2(\pi X'/2)} - 1 \right] \left[ \cos(k\pi X') - \frac{1}{2}\delta_{k,0} \right], \quad (20)$$

$$r_{m,k}^{b,2} = \frac{\pi}{2} \int_{-1}^{1} dX \sin(m\pi X) \int_{-1}^{1} dX' R_g^{(0)}(|\tan(\frac{\pi}{2}X) - \tan(\frac{\pi}{2}X')|, |k_x d|)$$
$$\times f_0(X') \left[ \frac{1+\eta}{1+\eta(2+\eta)\cos^2(\pi X'/2)} - 1 \right] \sin(k\pi X'). \quad (21)$$

Using Eqs. (17), in Fig. 6 for $\eta = 1$ we plot dimensionless phase velocities for four most fast MEMPs with negative chirality and four most fast MEMPs with positive chirality as function of $2M_0/B_{ext}$. Here the values of $W$ for $2M_0/B_{ext} \to 0$ coincide with the results obtained from Eqs. (9), (12).

To conclude, it has been shown that lateral inhomogeneous magnetic field allows the existence of the "magnetic gradient" or special magnetic-edge magnetoplasmons due to complex lateral structure of magnetic field distribution. We have obtained the eigen value problem that corresponds to the motion of charge density wave perpendicular to magnetic gradient. At low wave vector $k_x$ magnetoplasmons have acoustic type dispersion, $\omega \propto k_x$. For non-monotonic distribution of magnetic field "magnetic gradient" magnetoplasmon may move in both directions. To solve eigen value problem we have used two types of numerical approaches: first is the grid method that diagonalizes large Hermitian matrix and second is semi-analytical approach that expand each eigen mode on the set of orthogonal functions. The latter approach needs less computing time and provides better accuracy for fastest positive and negative modes. However, the grid method is more universal and gives reasonable accuracy for larger amounts of fastest modes.

**Acknowledgments**

We are grateful to Lloyd W. Engel for many useful discussions on a possibility of MEMP, due to inhomogeneity of magnetic field, relevant experiments and experimental conditions. The research leading to these results has received funding from the European Union Seventh Framework Programme (FP7/2007-2013) under grant agreement n° PCOFUND-GA-2009-246542 and from the Foundation for Science and Technology of Portugal. We also acknowledge support by Brazilian FAPEAM grants: Universal Amazonas (Edital 021/2011) and PVS.

## References

[1] Volkov V A and Mikhailov S A 1991 *Electrodynamics of Two-Dimensional Electron Systems in High Magnetic Fields*, in *Landau Level Spectroscopy, Modern Problems in Condensed Matter Sciences*, Ed. by G. Landwehr and E. I. Rashba (North-Holland, Amsterdam) vol 27.2 chapter 15 pp 855-907; Volkov V A and Mikhailov S A 1988 *Zh. Eksp. Teor. Fiz.* **94** 217 [1988 *Sov. Phys. JETP* **67** 1639]

[2] Kushwaha M S 2001 *Surface Science Reports* **41** pp 1-416

[3] Balev O G and Vasilopoulos P 1998 Phys. Rev. Lett. **81** 1481

[4] Khannanov M N, Fortunatov A A and Kukushkin I V 2009 *Pis'ma Zh. Eksp. Teor. Fiz.* **90** 740 [2009 *JETP Lett.* **90** 667

[5] Silva S and Balev O G 2010 *J. Appl. Phys.* **107** 104310

[6] Balev O G and Larkin I A 2013 *J. Phys.: Conf. Ser.* **456** 012023

[7] Larkin I A and Sukhorukov E V, Phys. Rev. **49**, 5498 (1994).

[8] Jackson J D 1999 *Classical Electrodynamics* 3d ed. (John Wiley&Sons, New York)

# Fluctuation-dissipation theorem in isolated quantum systems out of equilibrium

**Ehsan Khatami[1], Guido Pupillo[2], Mark Srednicki[3] and Marcos Rigol[4]**
[1]Department of Physics, University of California, Davis, California 95616, USA
[2]IPCMS (UMR 7504) and ISIS (UMR 7006), Université de Strasbourg and CNRS, Strasbourg, France
[3]Department of Physics, University of California, Santa Barbara, California 93106, USA
[4]Department of Physics, The Pennsylvania State University, University Park, Pennsylvania 16802, USA

E-mail: ekhatami@ucdavis.edu

**Abstract.** We study the validity of the fluctuation-dissipation theorem for an isolated quantum system of harmonically trapped dipolar molecules taken out of equilibrium by means of a quench, a sudden change in the Hamiltonian parameters. We find that the integrability of the system plays a crucial role in the validity of the fluctuation-dissipation theorem. Namely, the system thermalizes according to the eigenstate thermalization hypothesis and the theorem holds if the system is nonintegrable after the quench. However, it fails if the system is integrable, unless the initial state is an eigenstate of a nonintegrable Hamiltonian, in which case the system still thermalizes despite the eigenstate thermalization hypothesis failing to describe it.

## 1. Introduction

Non-equilibrium dynamics of isolated quantum many-body systems have attracted a great deal of attention in recent years, partly owing to the significant experimental advancements being made in the simulation of such systems by confining, and controlling the interactions of, ultracold atomic gazes. To generate non-equilibrium dynamics, these systems can be prepared in an initial state that is not an eigenstate of the Hamiltonian being simulated, a process called quenching. Among prominent experimental examples are the realization of "quantum Newton's cradle" using Bose gases confined in one-dimensional geometries [1], or their relaxation on a uniform lattice after being prepared in an initial state with alternating site densities [2].

Numerical studies have also played an important role in this regard by showing that, unlike generic nonintegrable systems, integrable systems do not thermalize after quench, and that observables after relaxation can be described by a generalized Gibbs ensemble, in which all constants of motion play a role similar to the temperature [3, 4]. In case of nonintegrable isolated systems, the thermalization can be understood via the eigenstate thermalization hypothesis (ETH) [5, 6, 4].

Recently, other questions related to the dynamics of these systems after quench have arisen, in particular, as related to the validity of the fluctuation-dissipation theorem (FDT). The FDT is an important theorem in statistical mechanics, which relates response functions to the dynamic correlations in the system. If a classical or quantum system at equilibrium experiences a small perturbation, FDT states that the perturbation dissipates in time in the same way as a random thermal or quantum fluctuation. The FDT is valid as long as the perturbation is in the linear response regime and that the system is in thermal equilibrium. However, the applicability of the theorem to systems out of equilibrium, especially to isolated quantum system taken far from equilibrium is much less clear.

Recent analytic studies have concluded that the FDT is not valid for integrable systems out of equilibrium as it does not lead to a unique definition for temperature [7, 8], and that for a subsystem of an infinite system, dynamics are described by the same ensemble that describes its static properties [9]. The validity of FDT for an entire isolated system, either integrable or generic nonintegrable, however, has been investigated only in a recent publication by us [10], where exact diagonalization has been used to study the properties of one-dimensional systems of hard-core bosons with dipolar interactions after quench. We have found that the FDT holds for the system out of equilibrium in two scenarios: (i) when the system after quench is nonintegrable, and (ii) when the initial state before the quench is an eigenstate of a nonintegrable Hamiltonian.

In this paper, we report new numerical results for the nonlocal observable of zero momentum occupation number and study the dynamics of correlation functions that allow us to determine the validity of the FDT at a higher temperature than in Ref. [10] to explore the role temperature may play. We start with the Hamiltonian for hard-core bosons with dipolar interaction trapped in a harmonic potential:

$$\hat{H} = -\sum_{j=1}^{L-1}\left(\hat{b}_j^\dagger \hat{b}_{j+1} + \text{H.c.}\right) + V\sum_{j<l}\frac{\hat{n}_j\hat{n}_l}{|j-l|^3} + g\sum_j x_j^2 \hat{n}_j, \tag{1}$$

where $\hat{b}_j^\dagger$ ($\hat{b}_j$) creates (annihilates) a hard-core boson ($\hat{b}_j^{\dagger 2} = \hat{b}_j^2 = 0$) at site $j$, and $\hat{n}_j = \hat{b}_j^\dagger \hat{b}_j$ is the number operator. The hopping amplitude is set to one as the unit of energy throughout this paper, $V$ is the strength of the dipolar interaction, $g$ is the strength of the confining potential, $x_j$ the distance of site $j$ from the center of the trap, and $L$ the number of lattice sites. We always choose the total number of bosons to be $L/3$. Also $\hbar = k_B = 1$, and we use open boundary conditions. To reduce the computational burden, we work in the subspace with even parity under reflection. The systems described by the above Hamiltonian can be realized in experiments using gas of bosonic ground state molecules such as, LiCs with dipole moment $\sim 5.6$ Debye, confined longitudinally by a one-dimensional optical lattice.

We perform three different quenches:

(i) $\{V_{\text{ini}} = 0, g_{\text{ini}} = \gamma\} \rightarrow \{V_{\text{fin}} = 0, g_{\text{fin}} = \gamma/10\}$ (integrable to integrable),

(ii) $\{V_{\text{ini}} = 8, g_{\text{ini}} = \gamma\} \rightarrow \{V_{\text{fin}} = 0, g_{\text{fin}} = \gamma\}$ (nonintegrable to integrable), and

(iii) $\{V_{\text{ini}} = 8, g_{\text{ini}} = \gamma\} \rightarrow \{V_{\text{fin}} = 2, g_{\text{fin}} = \gamma\}$ (nonintegrable to nonintegrable).

in which $\gamma$ is chosen such that $\gamma x_1^2 = \gamma x_L^2 = 4$. We fix the temperature to $T = 10$, in the sense that the initial state for different quenches, $|\phi_{\text{ini}}\rangle$, is selected such that the energy of the system $E_{\text{tot}} = \langle \phi_{\text{ini}}|\hat{H}_{\text{fin}}|\phi_{\text{ini}}\rangle$ after the quench (when the system is described by $\hat{H}_{\text{fin}}$) corresponds to the energy of the canonical ensemble, i.e., $E_{\text{tot}} = \text{Tr}\{e^{-\hat{H}_{\text{fin}}/T}\hat{H}_{\text{fin}}\}/\text{Tr}\{e^{-\hat{H}_{\text{fin}}/T}\}$ at $T = 10$.

Following the conventions used in Refs.[11, 10], we define three correlation functions. The first one is called the fluctuation correlation:

$$C_{\text{Fluc}}(t) \propto \sum_{\substack{\alpha\beta \\ \alpha\neq\beta}} |c_\alpha|^2 |c_\beta|^2 |O_{\alpha\beta}|^2 e^{i(E_\alpha - E_\beta)t}, \tag{2}$$

where $E_\alpha$ are eigenvalues of $\hat{H}_{\text{fin}}$, $c_\alpha$ are the overlap functions of the initial state and the eigenstates of the final Hamiltonian, and $O_{\alpha\beta}$ are the matrix elements of the observable $\hat{O}$. $C_{\text{Fluc}}(t)$ relates the expectation value of the observable at time $t + t'$ to its value at an earlier time $t'$ (when a fluctuation is assumed to have taken place), i.e., $O_{t'+t} = C_{\text{Fluc}}(t) O_{t'}$. The second correlation function we study is

$$C_{\text{Diss}}(t) \propto \sum_{\substack{\alpha\beta \\ \alpha\neq\beta}} \frac{e^{-E_\alpha/T} - e^{-E_\beta/T}}{E_\beta - E_\alpha} |O_{\alpha\beta}|^2 e^{i(E_\alpha - E_\beta)t}, \tag{3}$$

which is calculated using Kubo's linear response formula [12, 11] assuming that a perturbation has been applied to the system at time $t = 0$. $C_{\text{Diss}}(t)$ relates the expectation value of the observable at time $t$ to its

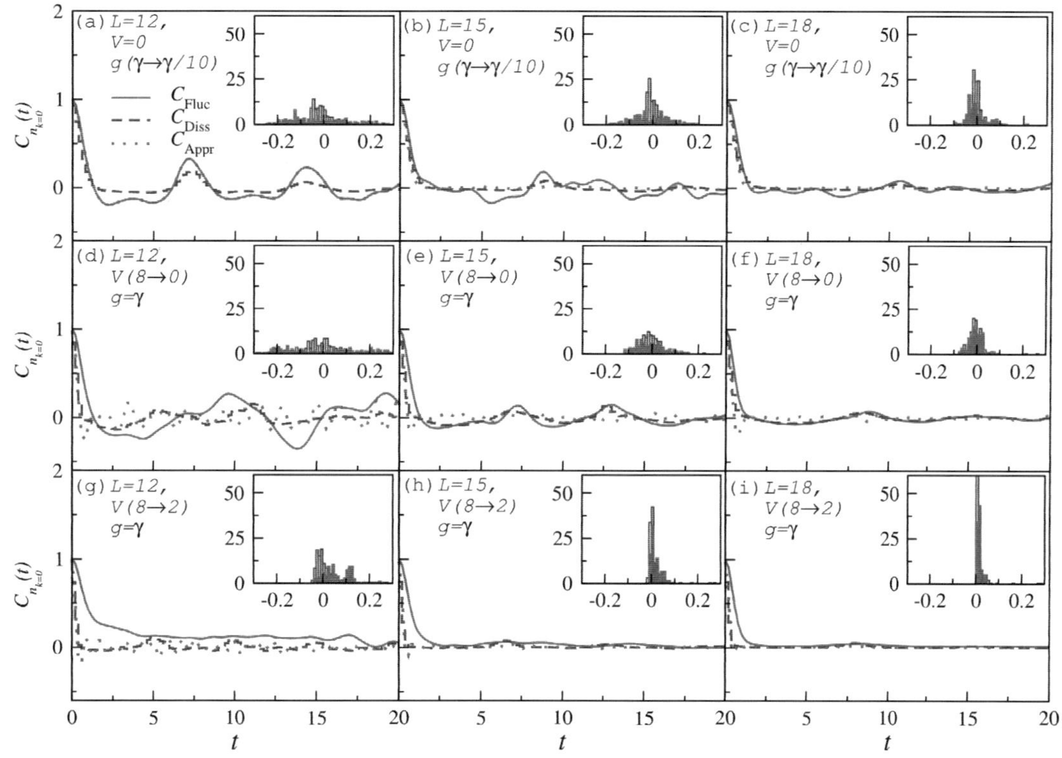


**Figure 1.** Correlation functions $C_{\rm Fluc}(t)$, $C_{\rm Diss}(t)$, and $C_{\rm Appr}(t)$ when the observable is the zero momentum occupation number, $n_{k=0}$ vs time $t$ at $T = 10$. Results are shown for the three quenches (i)-(iii) (from top to bottom, respectively), and three different sizes, $L = 12$, 15, and 18, (left to right). The insets show normalized histograms of $C_{\rm Fluc}(t)$ (filled red bars) and $C_{\rm Diss}(t)$ (empty blue bars) calculated for 200 data points between $t = 0$ and 40.

thermal value before the perturbation, i.e., $O_t = C_{\rm Diss}(t)O_{\rm Thermal}$. Finally, the third correlation function we calculate is

$$C_{\rm Appr}(t) \propto \int_{-\infty}^{+\infty} d\omega |f(E,\omega)|^2 e^{i\omega t}, \tag{4}$$

which can be shown to be valid if the observables follow the form suggested by the ETH [11], namely, $O_{\alpha\beta} = \Omega(E)\delta_{\alpha\beta} + e^{-S(E)/2} f(E,\omega)R_{\alpha\beta}$, where $E \equiv \frac{1}{2}(E_\alpha + E_\beta)$, $\omega \equiv E_\alpha - E_\beta$, $S(E)$ is the thermodynamic entropy at energy $E$, $e^{S(E)} = E\sum_\alpha \delta(E - E_\alpha)$, $\Omega(E)$ and $f(E,\omega)$ are smooth functions of their arguments, and $R_{\alpha\beta}$ is a random variable. The proportionality constant for all three correlation functions are such that $C(0) = 1$. Note that $C_{\rm Fluc}(t)$ depends explicitly on the choice of the initial state through the overlap functions, $c_\alpha$, whereas $C_{\rm Diss}(t)$ is a thermal quantity. If the system thermalizes after the quench and the ETH is valid, it can be shown that both of the above correlation functions reduce to $C_{\rm Appr}(t)$. In the following, we study these functions over time for the three quenches described above and analyze how similar or different they are. If the agreement of the fluctuation and dissipation correlation functions over time improves by increasing the system size, we conclude that the FDT is valid for the non-equilibrium system under investigation. If they also agree with the $C_{\rm Appr}(t)$, then the ETH form for the observables is a valid one.

We study three system sizes, $L = 12$, 15 and 18. Due to the large Hilbert space and the amount of computational expense arising from the need to compute and work with off-diagonal expectation values

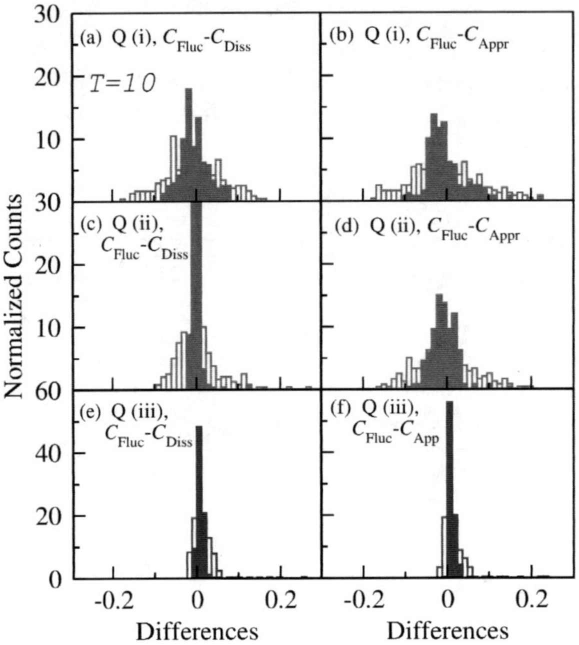


**Figure 2.** Normalized histograms of $C_{\rm Fluc}(t) - C_{\rm Diss}(t)$ (left panels) and $C_{\rm Fluc}(t) - C_{\rm Appr}(t)$ (right panels) for $n_{k=0}$, and for the three quenches (from top to bottom) and the two largest system sizes, calculated for 200 data points between $t = 0$ and 40. Empty bars are for $L = 15$, and filled bars are for $L = 18$.

of the observables, calculations for $L = 21$, which is the next size with one more particle, are currently unfeasible. Our computer program takes advantage of OpenMP parallelization for performing the sums in Eqs. (2) and (3).

In Fig. 2, we show these correlation functions for the nonlocal observable of zero momentum occupation number $n_{k=0}$. The system size increases in the panels from left to right and, from top to bottom, panels in different rows correspond to quenches (i) through (iii), respectively. In all cases, the correlation functions exhibit a fast drop from 1 followed by fluctuations over time. The time scale associated with the initial drop can be deduced from the width of the off-diagonal function $f(E_{\rm tot}, \omega)$ as a function of $\omega$. We also find that as the system size increases, the fluctuations in the correlation functions generally decreases. This is the case as the corresponding time scale is set by the average level spacing, which decreases exponentially by increasing the system size.

Those fluctuations are also suppressed in case of quench (iii) [Fig.1(g)-1(i)], due to the nonintegrable nature of the Hamiltonian. The magnitude of the fluctuations over time can be more carefully studied using histograms of the fluctuation and dissipation correlations, which are shown in the insets. As expected, we find that there is a qualitative different between the histograms in Fig. 1(c) and 1(f) and those in 1(i) for the largest system. For $L = 12$ finite-size effects prevent us from drawing such a distinction.

We also find that, at least for the two largest sizes, $C_{\rm Fluc}(t)$ better agrees with $C_{\rm Diss}(t)$ in case of quenches (ii) and (iii), in comparison to quench (i). This shows that, unlike for the integrable system after quench (i) from another integrable Hamiltonian, the FDT may be valid for the other two systems out of equilibrium. Namely, the nonintegrable system after quench (iii), and the integrable system after quench (ii), for which the initial state is an eigenstate of a nonintegrable Hamiltonian. $C_{\rm Appr}(t)$ on the other hand, generally agrees with the other two correlation functions only in the case of quench (iii), and the agreement improves by increasing the system size. These observations are similar to those made in Ref. [10], based on results for a local observable (the density at the center of the trap), and at a lower

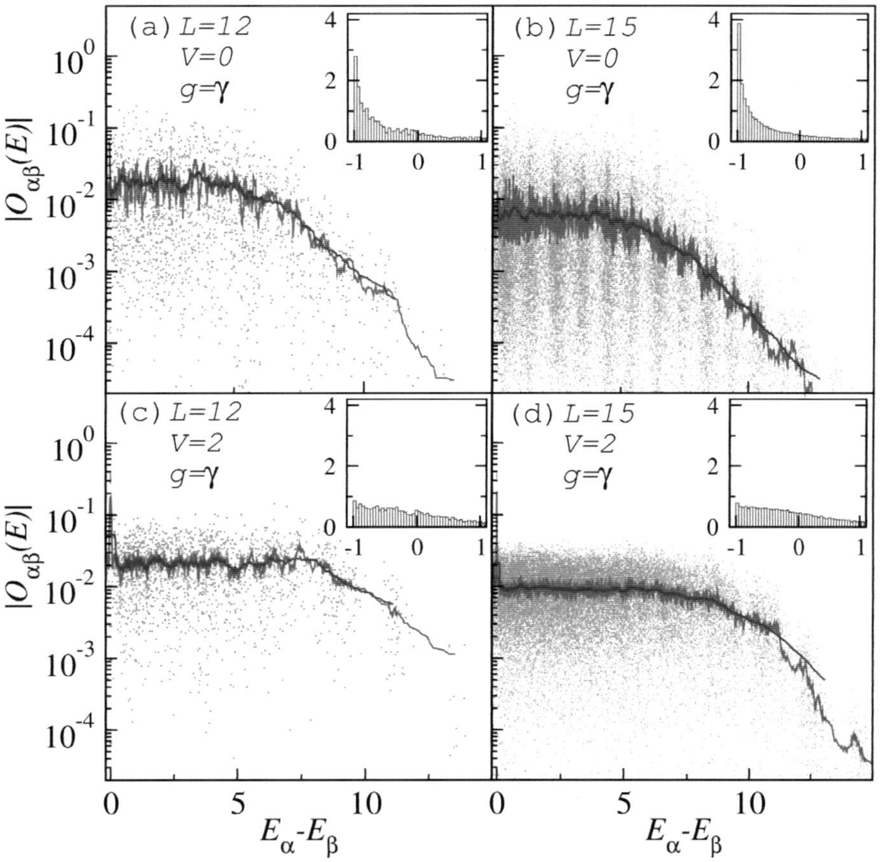


**Figure 3.** Absolute value of the off-diagonal matrix elements of $\hat{n}_{k=0}$ whose $E$ falls within 0.1 of $E_{\rm tot}$ vs $\omega$ for $L = 12$ (left panels) and $L = 15$ (right panels). (a) and (b) correspond to the final Hamiltonian in quench (ii), and (c) and (d) correspond to the final Hamiltonian in quench (iii). Lines are running averages with subset lengths of 20 and 100 for $L = 12$, and 50 and 1000 for $L = 15$. We call the one with the larger subset length $f_{\rm avg}(E,\omega)$. Insets show the histograms of the relative differences between the $n_{k=0}$ and $f_{\rm avg}$ The relative difference is defined as $(|O_{\alpha\beta}| - f_{\rm avg})/f_{\rm avg}$.

temperature of $T = 5$.

To better quantify these observations, in Fig. 2, we plot the normalized histograms of the differences between the fluctuation and dissipation, as well as the fluctuation and the approximate correlation functions for the two largest system sizes and for the three quenches. In case of quench (i), both the agreement between $C_{\rm Fluc}(t)$ and $C_{\rm Diss}(t)$, and between $C_{\rm Fluc}(t)$ and $C_{\rm Appr}(t)$ is poor as the histograms are relatively broad functions.

Quench (ii), however, shows a different trend [see Fig. 2(c) and 2(d)]. We find that the agreement between fluctuation and dissipation correlations improves as we increase $L$ from 15 to 18, and in the same time, the agreement between the fluctuation and the approximate functions is not as significant. This suggests that in the thermodynamic limit, the integrable system out of equilibrium thermalizes, and the FDT is applicable, provided that the state before the quench corresponds to a nonintegrable Hamiltonian. Moreover, the thermalization has occurred in this case without the ETH being valid. These are consistent with recent suggestions for alternative approaches to thermalization of integrable systems [13]. The failure of ETH can be seen in Figs. 3(a) and 3(b), where the off-diagonal matrix elements of $n_{k=0}$ in the energy eigenbasis do not exhibit a smooth function of the energy differences, $\omega$, as predicted by the ETH form. The off-diagonal values fluctuate wildly with significantly large number of them having very small values. This trend is clearer for the larger system size of $L = 15$ in Fig. 3(b). Histograms of the relative differences of the off-diagonal values and a running average, shown in the insets, show a sharp feature at -1, indicating these large fluctuations.

For the nonintegrable system in case of quench (iii), we find that all three correlation functions mostly agree with each other over time, inferred from the sharp features of the histograms in Fig. 2(e) and 2(f) and that the agreement improves by increasing the size. This suggests that both the ETH and the FDT are valid for nonintegrable systems out of equilibrium. In fact, in Figs. 3(c) and 3(d), we see that the fluctuations in the off-diagonal elements of the observable are dramatically reduced from those in Figs. 3(a) and 3(b), resulting in a smoother function of $\omega$, and offering more consistency with the ETH form. This is again reflected in the histograms, shown in the insets, which now have a more uniform functionality.

## Acknowledgments

We acknowledge support from the U.S. National Science Foundation, under Grants No. OCI-0904597 (E.K. and M.R.) and No. PHY07-57035 (M.S.), and from the European Commission, under ERC-St Grant No. 307688 COLDSIM, UdS, and EOARD (G.P.). E.K. also acknowledges partial support from U.S. National Science Foundation under award No. 1339172.

## References

[1] Kinoshita T, Wenger T and Weiss D S 2006 *Nature* **440** 900
[2] Trotzky S, Chen Y A, Flesch A, McCulloch I P, Schollwöck U, Eisert J and Bloch I 2012 *Nature Phys.* **8** 325
[3] Rigol M, Dunjko V, Yurovsky V and Olshanii M 2007 *Phys. Rev. Lett.* **98** 050405
[4] Rigol M, Dunjko V and Olshanii M 2008 *Nature* **452** 854
[5] Deutsch J M 1991 *Phys. Rev. A* **43** 2046–2049
[6] Srednicki M 1994 *Phys. Rev. E* **50** 888–901
[7] Mitra A and Giamarchi T 2011 *Phys. Rev. Lett.* **107**(15) 150602
[8] Foini L, Cugliandolo L F and Gambassi A 2011 *Phys. Rev. B* **84**(21) 212404
[9] Essler F H L, Evangelisti S and Fagotti M 2012 *Phys. Rev. Lett.* **109**(24) 247206
[10] Khatami E, Pupillo G, Srednicki M and Rigol M 2013 *Phys. Rev. Lett.* **111**(5) 050403
[11] Srednicki M 1999 *J. Phys. A* **32** 1163
[12] Kubo R 1957 *J. Phys. Soc. Japan* **12** 570
[13] Rigol M and Srednicki M 2012 *Phys. Rev. Lett.* **108**(11) 110601

# Novel method for identification of intrinsic vibration modes in piezoelectric crystals

**O A Ryabushkin, D V Myasnikov and A V Konyashkin**
Moscow Institute of Physics and Technology, Institutskiy per. 9, Dolgoprudnyy, 141700, Russia
Kotelnikov Institute of Radio-engineering and Electronics of RAS, Vvedensky Sq. 1, Fryazino, 141190, Russia

E-mail: roa228@mail.ru

**Abstract.** Novel method for identification of intrinsic vibration modes in piezoelectric crystals is introduced. This method is based on recently proposed theoretical model of piezoelectric resonance frequencies dependence on temperature. Calculation of crystal sample eigenmodes is based on Lagrange-Hamilton variation principle. Comparison of temperature dependencies of calculated resonance frequencies with measured ones reveals reliable method for identification of vast majority of the experimentally observed resonances. Influence of sample sizes variation on eigenmode frequencies is also considered.

## 1. Introduction

Resonant ultrasound spectroscopy (RUS) technique has been developed since 1960-th [1, 2]. At present one of the main task of RUS is determination of elastic and piezoelectric constants of various materials [3-5] as well as internal friction [6]. Elastic constants can be associated with free energy, Debye temperature, specific heat capacity and with other important parameters of the material. Decay of acoustic waves, which can be described by implementation of imaginary parts in elastic constants, gives information about unharmonicity and other relaxation mechanisms that occur in the material.

Obviously elastic constants can be determined by measuring velocities of sound waves in the given material. This principle is used in methods where propagation time of sound pulse in the sample is determined. Measurements of intrinsic vibration modes of the samples have been conducted since 1935. At that time there were no accurate methods for numerical calculations of eigenfrequencies of samples with complex geometry. However for isotropic noncrystalline cubic and spherical bodies solutions were found in 1880. In 1964 year the technique for determination of elastic properties of solids together with its dependence on temperature was demonstrated [1]. In that paper elastic and anelastic properties of sphere samples were determined by measuring its intrinsic vibration spectrum. Later samples in the form of rectangular parallelepiped were usually used in RUS, mostly due to the relative simplicity of manufacturing. More attention was paid to the low symmetry crystals.

In RUS the elastic properties of material are determined by measuring resonance vibration frequencies of sample. Substantial advantage of RUS technique is that whole set of elastic constants can be determined from single measurement, provided that experiment is performed correctly. Usually sample in the form of rectangular parallelepiped is fixed along its diagonal between two piezoelectric transducers. One of the transducers is used for excitation of sample vibrations at certain frequency, which can be tuned, and another one acts as a receiver. When excitation frequency coincides with any intrinsic vibration frequency of the sample the resonance peak can be observed in response. Intrinsic vibration frequencies of the sample depend on elastic constants, material density as well as sample form and sizes. Usually samples of millimeter dimensions are used and measured resonance frequencies lie in MHz range. Elastic constants can be determined from the resonance spectrum provided that all other necessary parameters are measured independently. It should be mentioned that mechanical contact of the transducers can considerably affect resonance frequencies. External influence is eliminated as much as possible in noncontact RUS [7]. For example if material possess piezoelectric properties the most convenient way to excite vibrations in it is to use external radiofrequency electric field, which can be applied avoiding mechanical contact between sample and electrodes.

From calculating point of view the problem of elastic constants determination is laborious enough. It includes both direct calculations of resonance frequencies for given set of elastic constants and inverse problem of identification of calculated eigenmodes with measured ones. That's why RUS methods became popular only after development of powerful computers. Milestone works concerning RUS appeared in early 1970-th [8, 9]. It was demonstrated that exact analytical solutions of eigenmode problem can be obtained only for samples with simple geometry such as plate, sphere, etc.

Following iteration procedure is used for calculation of elastic constants from experimentally measured resonance spectrum. Resonance frequencies of the sample are calculated using some initial set of elastic constant values. Then calculated spectrum is compared with measured one. By varying values of elastic constants one try to attain exact agreement between measured and calculated resonance spectra. Accuracy of calculated values of elastic constants depends on the amount of measured resonances. Its number should exceed 5-10 times that of independent material constants to be determined. Still in case of large quantity of detected resonances the convergence of described iteration procedure and as follows correspondence of calculated intrinsic mode with particular mode, measured experimentally, is under question. Besides, it is often happens so that in certain frequency range the number of experimentally observed modes turns out to be much less in comparison with calculated ones. It means that for accurate mode identification additional criteria should be applied. The most robust method is laser interferometry technique used for visualization of resonance vibration amplitude distribution at sample surface [10]. Despite this identification method is reliable it is also laborious enough because usually resonance amplitude values are of several nanometers order.

## 2. Mode identification

### *2.1. Calculation of intrinsic vibration modes*

For determination of resonance frequencies for bodies of different geometry most common approach is to use Lagrange-Hamilton variation principle. Eigenmodes of piezoelectric crystal can be found by varying spatial part of Lagrangian of the system that can be written as follows [3]

$$J_S = \iiint_{\Omega} \left( \rho w^2 u_i^2 - c_{ijkl} S_{ij} S_{kl} + \varepsilon_{ij} E_i E_j + 2 d_{ijk} S_{ij} E_k \right) d\Omega . \quad (1)$$

Here $F_{,j}=\partial F/\partial x_j$ denotes derivative; $c_{ijkl}$, $d_{ijk}$, $\varepsilon_{ij}$ are elastic, piezoelectric and dielectric tensors respectively; $\rho$ is density of the sample; $w$ is vibration frequency; $u_i(x_1,x_2,x_3)$ are the components of mechanical displacement of the sample points; $S_{ij} = 0.5\left(u_{i,j} + u_{j,i}\right)$ is strain tensor; $E_i = -\varphi_{,i}$ are components of electric field vector; $\Omega$ is the crystal volume.

In order to solve numerically problem (1) the Rayleigh-Ritz method is usually used. Following this method unknown functions of displacement $u_i$ and potential $\varphi$ are expanded using some set of orthonormal basis functions $\{\psi_p\}$ of the total number $N_b$. It means that equations for eigenmodes can be obtained by substitution of series expansions

$$u_i\left(x_1,x_2,x_3\right) = \sum_{p=1}^{N_b} C^{u}_{ip} \psi_p\left(x_1,x_2,x_3\right), \quad \varphi\left(x_1,x_2,x_3\right) = \sum_{q=1}^{N_b} C^{\varphi}_{q} \psi_q\left(x_1,x_2,x_3\right), \quad (2)$$

into Lagrangian (1). As a result we obtain function that depends on the expansion coefficients $C$

$$J_S\left(C\right) = w^2 C^{u}_{ip} C^{u}_{iq} \mathrm{W}_{pq} - C^{u}_{ip} C^{u}_{kq} \mathrm{Y}^{ik}_{pq} - 2 C^{u}_{ip} C^{\varphi}_{q} \Pi^{i}_{pq} + C^{\varphi}_{p} C^{\varphi}_{q} \mathrm{V}_{pq} . \quad (3)$$

Matrices W, Y, Π and V are

$$\mathrm{W}_{pq} = \iiint_{\Omega} \rho \psi_p \psi_q d\Omega, \quad \mathrm{Y}_{pq}^{ik} = \iiint_{\Omega} c_{ijkl} \frac{\partial \psi_p}{\partial x_j} \frac{\partial \psi_q}{\partial x_l} d\Omega,$$
$$\Pi_{pq}^{i} = \iiint_{\Omega} d_{ijk} \frac{\partial \psi_p}{\partial x_j} \frac{\partial \psi_q}{\partial x_k} d\Omega, \quad \mathrm{V}_{pq} = \iiint_{\Omega} \varepsilon_{ij} \frac{\partial \psi_p}{\partial x_i} \frac{\partial \psi_q}{\partial x_j} d\Omega. \tag{4}$$

After differentiation (3) with respect to expansion coefficients $C$ the minimum of the functional (1) corresponds to the condition of equal to zero derivatives $\partial J_S(C)/\partial C = 0$. In matrix form equations for eigenmodes can be written as

$$\begin{cases} \left(\mathrm{Y} - \Pi \mathrm{V}^{-1} \Pi^{\mathrm{T}}\right) C^{\mathrm{u}} = w^2 \mathrm{W} C^{\mathrm{u}} \\ C^{\varphi} = -\mathrm{V}^{-1} \Pi^{\mathrm{T}} C^{\mathrm{u}}. \end{cases} \tag{5}$$

Here 'T' and '-1' designate transposed and inverse matrix respectively and $C^{\mathrm{u}} = \left(C_1^{\mathrm{u}}, C_2^{\mathrm{u}}, C_3^{\mathrm{u}}\right)^{\mathrm{T}}$.

Solution of system (5) gives eigenfrequencies $w_n = 2\pi R f_n$ and pertinent spatial distributions of the displacement and potential the piezoelectric resonance thermal coefficients $K_i^{\mathrm{prt}}$ can be calculated provided that temperature dependence of the elastic constants $c_{ijkl}(\theta)$ is known. It was estimated that contribution to resonance frequency shift due to other parameters variation with temperature $\theta$ is not determinative.

When all parameters (mainly temperature) are uniform across the sample, then all the material constants in (4) can be taken outside the integral sign, and calculations are greatly simplified. Evaluation of the integrals in this case is reduced to evaluation of the following one:

$$\mathrm{B}_{pq}^{jl} = \iiint_{\Omega} \frac{\partial \psi_p}{\partial x_j} \frac{\partial \psi_q}{\partial x_l} d\Omega. \tag{6}$$

This integral can be further split into 1-dimensional integrals. More specifically, if we denote

$$g_{p_{xi} q_{xi}}^{00} = \int_{-L_i/2}^{L_i/2} \psi_p(x_i) \psi_q(x_i) dx_i; \quad g_{p_{xi} q_{xi}}^{01} = g_{p_{xi} q_{xi}}^{10} = \int_{-L_i/2}^{L_i/2} \psi_p(x_i) \frac{d\psi_q(x_i)}{dx_i} dx_i; \quad g_{p_{xi} q_{xi}}^{11} = \int_{-L_i/2}^{L_i/2} \frac{d\psi_p(x_i)}{dx_i} \frac{d\psi_q(x_i)}{dx_i} dx_i. \tag{7}$$

Then

$$B_{pq}^{jl} = \begin{pmatrix} g_{p_{x1} q_{x1}}^{11} g_{p_{x2} q_{x2}}^{00} g_{p_{x3} q_{x3}}^{00} & g_{p_{x1} q_{x1}}^{10} g_{p_{x2} q_{x2}}^{01} g_{p_{x3} q_{x3}}^{00} & g_{p_{x1} q_{x1}}^{10} g_{p_{x2} q_{x2}}^{00} g_{p_{x3} q_{x3}}^{01} \\ g_{p_{x1} q_{x1}}^{01} g_{p_{x2} q_{x2}}^{10} g_{p_{x3} q_{x3}}^{00} & g_{p_{x1} q_{x1}}^{00} g_{p_{x2} q_{x2}}^{11} g_{p_{x3} q_{x3}}^{00} & g_{p_{x1} q_{x1}}^{00} g_{p_{x2} q_{x2}}^{10} g_{p_{x3} q_{x3}}^{01} \\ g_{p_{x1} q_{x1}}^{01} g_{p_{x2} q_{x2}}^{00} g_{p_{x3} q_{x3}}^{10} & g_{p_{x1} q_{x1}}^{00} g_{p_{x2} q_{x2}}^{01} g_{p_{x3} q_{x3}}^{10} & g_{p_{x1} q_{x1}}^{00} g_{p_{x2} q_{x2}}^{00} g_{p_{x3} q_{x3}}^{11} \end{pmatrix}. \tag{8}$$

Weak temperature change of the sample can be treated as a perturbation of the initial problem, in which we rewrite first equation in linear system (5) as

$$\Gamma^{\mathrm{piezo}} C^{\mathrm{u}} = \rho w^2 C^{\mathrm{u}}, \tag{9}$$

with $\Gamma^{\mathrm{piezo}} \equiv \mathrm{Y} - \Pi \mathrm{V}^{-1} \Pi^{\mathrm{T}}$. If we denote a parameter that induces perturbation as $\mu$, then first-order correction to the $n$-th eigenmode of (9) gives

$$\frac{\partial w_n}{\partial \mu} = \frac{\left(C_n^{\mathrm{u}}\right)^{\mathrm{T}} \dfrac{\partial \Gamma^{\mathrm{piezo}}}{\partial \mu} C_n^{\mathrm{u}} - \dfrac{\partial \rho}{\partial \mu} w_n^2}{2\rho w_n}. \tag{10}$$

Relation (10) enables one to compare changes of the resonance frequency induced by the variation of elastic constants and thermal expansion (density change). If we denote average value of elastic constant as $\langle c \rangle$ and average sample size as $\langle L \rangle$ then $\left\langle \left(C_n^{\mathrm{u}}\right)^{\mathrm{T}} \Gamma^{\mathrm{piezo}} C_n^{\mathrm{u}} \right\rangle \sim \langle c \rangle / \langle L \rangle^2$, so

$$\left\langle \left(C_n^{\mathrm{u}}\right)^{\mathrm{T}} \frac{\partial \Gamma^{\mathrm{piezo}}}{\partial \mu} C_n^{\mathrm{u}} \right\rangle \approx \frac{1}{\langle L \rangle^2} \frac{d\langle c \rangle}{d\mu} - \frac{2\langle c \rangle}{\langle L \rangle^3} \frac{d\langle L \rangle}{d\mu}. \tag{11}$$

Journal of Physics: Conference Series **510** (2014) 012036 doi:10.1088/1742-6596/510/1/012036

Typically for crystals the following relation is valid

$$\frac{1}{\langle c\rangle}\frac{d\langle c\rangle}{d\theta} >> \frac{1}{\langle L\rangle}\frac{d\langle L\rangle}{d\theta}. \tag{12}$$

Now taking into account (12) the equation (10) is reduced to

$$\frac{\partial w_n}{\partial \mu} = \frac{C^{\mathrm{u}}_{ip,n}\left[\iiint\limits_{\Omega}\frac{\partial c_{ijkl}}{\partial \mu}\frac{\partial \psi_p}{\partial x_j}\frac{\partial \psi_q}{\partial x_l}d\Omega\right]C^{\mathrm{u}}_{jq,n}}{2\rho w_n}. \tag{13}$$

It was demonstrated that for common values of material piezoelectric and dielectric coefficients the contribution of piezoelectric effect to intrinsic vibration mode frequencies doesn't exceed 0.5% [11]. These frequency shifts occur mostly due to inevitable presence of external forces, conditioned by clamping of piezoelectric specimen between transducers in measurement procedure. Despite frequency shifts produced are small enough still they are several times higher than frequency measurement error typical for nonpiezoelectric crystals.

Dependence on temperature of each resonance of piezoelectric crystal can be characterized by its piezoelectric resonance thermal coefficient:

$$K^{\mathrm{prt}}_n = \frac{1}{2\pi}\frac{dw_n}{d\theta}. \tag{14}$$

In case crystal is uniformly heated its temperature derivatives of elastic coefficients can be taken out from integral sign in (13). Then piezoelectric resonance coefficients can be calculated as follows

$$K^{\mathrm{prt}}_n = \frac{\frac{\partial c_{ijkl}}{\partial \theta}C^{\mathrm{u}}_{ip,n}B^{jl}_{pq}C^{\mathrm{u}}_{kq,n}}{4\pi\rho w_n}. \tag{15}$$

Consequently full set of elastic constants of the crystal specimen can be deduced from a single frequency scan via multidimensional minimization of the error between the measured and calculated spectra. Spectrum is calculated from the known sample dimensions, its mass, and a set of surmised elastic constants. As it was stated the main problem is that usually number of calculated modes in narrow frequency range greatly exceeds number of resonances that are experimentally observed. Moreover measured resonance amplitudes considerably differ from one another and mathematical criteria for theoretical analysis of amplitude values has not been elaborated yet.

### *2.2. Piezoelectric resonance frequencies temperature dependence for mode identification*

Simple ways of the crystal intrinsic vibration modes identification can be realized by exploiting experimentally controlled variations of sample parameters or external factors that lead to change of resonance frequencies. We have recently introduced novel method for mode identification of piezoelectric crystals by analyzing piezoelectric resonance thermal coefficients $K^{\mathrm{prt}}_n$ of the excited resonances [12]. In first approximation elastic constants temperature derivatives can be taken from literature. For each $n$-th resonance the value of $K^{\mathrm{prt}}_n$ is measured in crystal uniform heating experiment. Then each experimentally observed piezoelectric resonance at $Rf_n$ frequency is attributed to the calculated one $Rf^{\mathrm{calc}}_n$ that has closest value of $K^{\mathrm{prt}}_n$, computed using equation (15).

Introduced algorithm allows closest fit of experimental vibration spectrum with calculated one because resonance frequency shift with temperature of the certain mode is directly connected with its dependence on each elastic constant. As follows frequency dependence on temperature is also specified by spatial configuration of displacement, i.e. by vibration mode structure.

Measure of difference between calculated and measured resonance spectra can be taken, for example, as follows

$$M = \sum_{n=1}^{N_e}\sigma_i\left(Rf^{\mathrm{calc}}_n - Rf_n\right)^2. \tag{16}$$

Here $N_e$ is total number of experimentally observed resonance modes and $\sigma_i$ is weighting factor.

In detail algorithm of mode identification and elastic constants determination includes steps that are listed below

- Processing of the piezoelectric resonance spectra measured for two crystal temperatures, i.e. determination of resonances (frequencies $Rf_n$ and corresponding amplitudes) at these temperatures and determination of piezoelectric resonance thermal coefficients $K_n^{\text{prt}}$ of each mode;
- Calculation of crystal intrinsic vibration modes frequencies $Rf_n^{\text{calc}}$ and corresponding $K_n^{\text{prt}}$ values using some set of elastic constants and its temperature derivatives.
- Selection of frequency interval $\Delta f$ (usually about 50 kHz), in which correspondence of certain measured resonance to the calculated one will be sought for.
- Selection from measured spectrum the resonance with highest amplitude and putting it in accordance with calculated mode in region $\Delta f$ that has closest calculated value of $K^{\text{prt}}$ coefficient to that of the given resonance. This step is performed consequently for other measured resonances in descending order of their amplitudes;
- Calculation of identification error using (16), in which weighting factors correspond to resonance amplitudes;
- Variation of elastic constants for finding best agreement between measured and calculated modes, i.e. minimum of (16).

*2.3. Resonance frequency dependence on sample sizes*

As it was estimated in (12) the shift of eigenmode frequencies due to variation of sample sizes, induced by crystal temperature change, is much less in comparison with contribution of elastic constants alteration with temperature. However any imperfection of sample form and uncertainty of sample size absolute values may result in substantial difference between measured and calculated resonance spectra even if elastic properties of the material are known for sure. Dependence of intrinsic vibration mode frequencies on variation of sample sizes is considered below.

For estimation we can use equation (10) with perturbation parameter $\mu$ representing any size $L$ of the crystal. However in this case derivative of $\Gamma^{\text{piezo}}$ in respect to size parameter, expressed using designations (4) as

$$\frac{\partial \Gamma^{\text{piezo}}}{\partial L} = \frac{\partial \mathrm{Y}}{\partial L} - \left( \frac{\partial \Pi}{\partial L} \mathrm{V}^{-1} \Pi^{\mathrm{T}} - \Pi \mathrm{V}^{-2} \frac{\partial \mathrm{V}}{\partial L} \Pi^{\mathrm{T}} + \Pi \mathrm{V}^{-1} \frac{\partial \Pi^{\mathrm{T}}}{\partial L} \right), \tag{17}$$

cannot be directly calculated because basis functions depend on crystal sizes itself. So $\mathrm{B}_{pq}^{jl}$ derivatives with respect to $L$ should be clarified.

Here we shall consider crystal sample in the form of rectangular parallelepiped with sizes $L_1 \times L_2 \times L_3$. It is common practice to use Legendre polynomials $P_{p_{xi}}(x_i)$ as basis functions [8, 9]

$$\psi_p(x_1, x_2, x_3) = \prod_{i=1}^{3} \left[ \left( \frac{2p_{xi} + 1}{L_i} \right)^{1/2} P_{p_{xi}} \left( \frac{2x_i}{L_i} \right) \right], \tag{18}$$

because for such sample shape they form orthogonal system:

$$\iiint_{\Omega} \psi_p \psi_q d\Omega = \delta_{p_{x1}q_{x1}} \delta_{p_{x2}q_{x2}} \delta_{p_{x3}q_{x3}}. \tag{19}$$

Here $p_{xi}$ is polynomial degree specified by $i = \overline{1,3}$ and $p = \overline{1, N_b}$ where $N_b$ is total number of basis functions.

It can be demonstrated that due to properties of Legendre polynomials the following expressions for derivatives of $\mathrm{B}_{pq}^{jl}$ can be obtained:

$$\frac{\partial B_{pq}^{jl}}{\partial L_1}=-\frac{1}{L_1}\begin{pmatrix} 2g^{11}_{p_{x1}q_{x1}}g^{00}_{p_{x2}q_{x2}}g^{00}_{p_{x3}q_{x3}} & g^{10}_{p_{x1}q_{x1}}g^{01}_{p_{x2}q_{x2}}g^{00}_{p_{x3}q_{x3}} & g^{10}_{p_{x1}q_{x1}}g^{00}_{p_{x2}q_{x2}}g^{01}_{p_{x3}q_{x3}} \\ g^{01}_{p_{x1}q_{x1}}g^{10}_{p_{x2}q_{x2}}g^{00}_{p_{x3}q_{x3}} & 0 & 0 \\ g^{01}_{p_{x1}q_{x1}}g^{00}_{p_{x2}q_{x2}}g^{10}_{p_{x3}q_{x3}} & 0 & 0 \end{pmatrix}, \tag{20}$$

$$\frac{\partial B_{pq}^{jl}}{\partial L_2}=-\frac{1}{L_2}\begin{pmatrix} 0 & g^{10}_{p_{x1}q_{x1}}g^{01}_{p_{x2}q_{x2}}g^{00}_{p_{x3}q_{x3}} & 0 \\ g^{01}_{p_{x1}q_{x1}}g^{10}_{p_{x2}q_{x2}}g^{00}_{p_{x3}q_{x3}} & 2g^{00}_{p_{x1}q_{x1}}g^{11}_{p_{x2}q_{x2}}g^{00}_{p_{x3}q_{x3}} & g^{00}_{p_{x1}q_{x1}}g^{10}_{p_{x2}q_{x2}}g^{01}_{p_{x3}q_{x3}} \\ 0 & g^{00}_{p_{x1}q_{x1}}g^{01}_{p_{x2}q_{x2}}g^{10}_{p_{x3}q_{x3}} & 0 \end{pmatrix}, \tag{21}$$

$$\frac{\partial B_{pq}^{jl}}{\partial L_3}=-\frac{1}{L_3}\begin{pmatrix} 0 & 0 & g^{10}_{p_{x1}q_{x1}}g^{00}_{p_{x2}q_{x2}}g^{01}_{p_zq_z} \\ 0 & 0 & g^{00}_{p_{x1}q_{x1}}g^{10}_{p_{x2}q_{x2}}g^{01}_{p_{x3}q_{x3}} \\ g^{01}_{p_{x1}q_{x1}}g^{00}_{p_{x2}q_{x2}}g^{10}_{p_{x3}q_{x3}} & g^{00}_{p_{x1}q_{x1}}g^{01}_{p_{x2}q_{x2}}g^{10}_{p_{x3}q_{x3}} & 2g^{00}_{p_{x1}q_{x1}}g^{00}_{p_{x2}q_{x2}}g^{11}_{p_{x3}q_{x3}} \end{pmatrix}. \tag{22}$$

Components $g$ can be calculated using relations (7).

It can be demonstrated that contribution of the first term in (17) is several orders higher than that of the parenthesized sum. So from variation theory the resonance frequency change due to sample size alteration will be given by the following relation

$$\frac{\partial w_n}{\partial L}=\frac{C^{\mathrm{u}}_{ip,n}\left[c_{ijkl}\dfrac{\partial B_{pq}^{jl}}{\partial L}\right]C^{\mathrm{u}}_{jq,n}}{2\rho w_n}. \tag{23}$$

Piezoelectric resonance dimensional coefficient $K^{\mathrm{prd}}$ that determines resonance frequency shift with the change of crystal size in $x_i$ direction is

$$K_{in}^{\mathrm{prd}}=\frac{1}{2\pi}\frac{\partial w_n}{\partial L_i}. \tag{24}$$

## 3. Experimental results

Simplified block-scheme of the experimental setup used for measurement of piezoelectric resonances in crystals is shown in figure 1. Low voltage from radiofrequency (RF) generator is applied to the crystal that is placed between two metallic electrodes forming capacitor. Load resistor R is connected in series with the capacitor. For each value of the electric filed $E^{\mathrm{rf}}$ frequency $f$ the crystal response is analyzed by measuring complex voltage $U_R$ on the load resistor via lock-in amplifier.

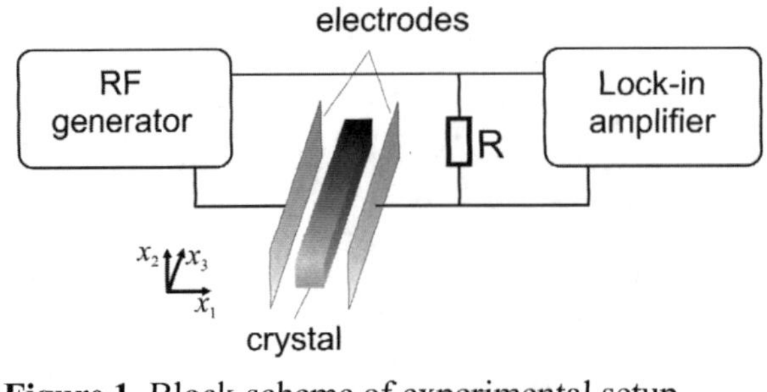


**Figure 1.** Block-scheme of experimental setup.

Resonances correspond to specific peaks that are observed at certain frequencies $Rf_n$ in measured $U_R$ signal, which is directly proportional to the electric current flowing through the crystal. For crystal temperature stabilization the capacitor with crystal in it is placed inside thermostat. Radiofrequency power that affects crystal is of tens microwatts order and doesn't change crystal temperature.

Experiment was performed with quartz crystal in the form of rectangular parallelepiped with dimensions 3×3×30 mm$^3$. Crystal is cut along crystallographic axes and its z-axis is directed along $x_3$.

Response spectrum measured for two orientation of quartz crystal at temperature 293 K is introduced in figure 2.

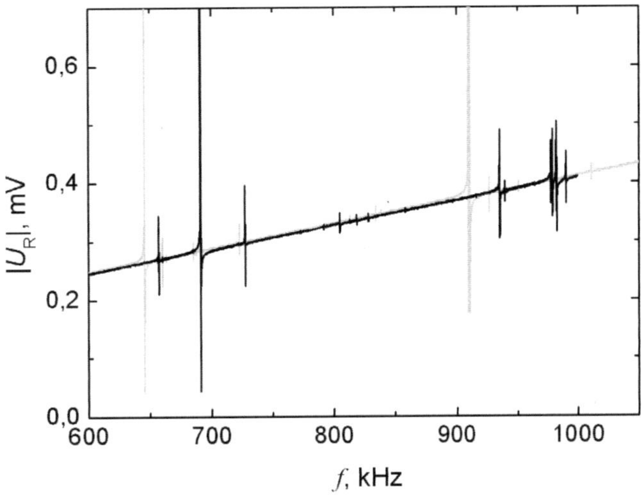


**Figure 2.** Frequency dependencies of the $U_R$ voltage amplitude measured for two orientations of quartz crystal in respect to electrodes.

Table 1 contains partial list of measured resonance parameters (frequencies and piezoelectric resonance thermal coefficients $K^{prt}$ together with calculated ones exploiting introduced identification method, based on analysis of piezoelectric resonance thermal coefficients.

**Table 1.** Parameters of piezoelectric resonances determined for quartz crystal

| Resonance number | Measured $Rf$, kHz | Calculated $Rf^{calc}$, kHz | Disagreement d$Rf$, kHz | Measured $K^{prt}$, Hz/K | Theoretical $K^{prt}$, Hz/K | $\Delta Rf$ induced by 3% increase of $L_3$, kHz |
|---|---|---|---|---|---|---|
| **1** | 645.377 | 642.1 | 3.28 | -30.0 | -32.5 | -21.1 |
| **2** | 655.485 | 578.1 | 77.39 | -50.4 | -50 | -29.1 |
| **3** | 660.002 | 700.7 | 40.7 | -30 | -31.1 | -23.1 |
| **4** | 663.0 | 630.3 | 32.7 | 12.5 | 18.4 | -29.4 |
| **5** | 804.758 | 805.3 | 0.54 | 20 | 23.6 | -26.8 |
| **6** | 909.592 | 897.1 | 12.49 | -45 | -49.7 | -33.3 |
| **7** | 911.181 | 909.2 | 1.98 | 10 | 7.9 | -36.2 |
| **8** | 926.8 | 905.9 | 20.9 | -27.5 | -26.5 | -30.2 |
| **9** | 935.333 | 948.3 | 12.96 | -13.3 | -12.5 | -30.1 |
| **10** | 939.548 | 871.3 | 68.24 | -23.3 | -23.4 | -36.2 |
| **11** | 950.605 | 949.8 | 0.81 | -27.5 | -27.6 | -0.2 |
| **12** | 982.016 | 977.2 | 4.82 | -56.7 | -56.6 | -5.9 |
| **13** | 982.338 | 988.5 | 6.16 | -36.7 | -36.4 | -0.05 |
| **14** | 1011.315 | 1018.3 | 6.98 | 0.5 | 6.6 | -1.26 |

It should be stressed that resonance frequencies are measured with high precision. However accuracy of sample geometry measurement is low enough and doesn't exceed 5%. Figure 3 represents calculated resonance frequency changes due to increase of crystal size $L_3$ by 3%. Corresponding values of frequency shift induced by $L_3$ change are listed in the last column of Table 1. It can be seen that in this case frequency shifts of certain resonances reach several tens of kilohertz. As an example, comparison between measured and calculated frequency values for four modes is introduced in Figure 3 inset. Error bars indicate calculated frequency intervals corresponding to 3% sample size $L_3$ uncertainty.

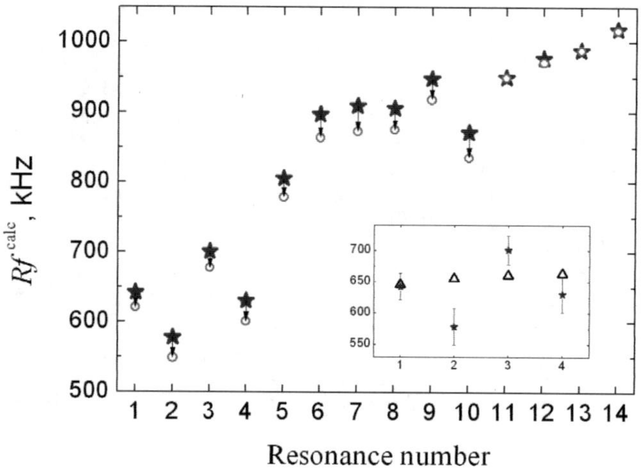


**Figure 3.** Calculated resonance frequency change induced by 3% increase of crystal sample size $L_3$. Stars and circles represent $Rf^{calc}$ initial and resulted values respectively. Inset shows measured (triangles) and calculated frequency values with error bars indicating interval produced by 3% sample size $L_3$ uncertainty.

This means that correct determination of sample geometry parameters is very important for RUS measurements. In our case main disagreement of measured and calculated resonance frequencies is explained by the high uncertainty of crystal sizes.

## 4. Conclusions

Two ways of identification of the sample intrinsic vibration modes are introduced. First one relies on association of measured and calculated piezoelectric resonance thermal coefficients $K_n^{prt}$ of vibration modes. These coefficients are easily measured. However for $K_n^{prt}$ numerical calculation the temperature dependence of elastic constants is required. Also sample geometry should be determined with high accuracy. It is demonstrated that even small uncertainty of sample sizes can result in substantial disagreement of absolute values of measured and calculated resonance frequency values. Calculations reveal that for certain vibration modes the frequency change of tenth kilohertz order is produced by 3% variation of sample size. In turn, frequency disagreement of measured and calculated modes, exploiting its frequency dependence on temperature, can be of the same order.

Second method of mode identification that supposes variation of sample sizes can be used without predetermination of elastic constants dependence on external parameters such as temperature, pressure, humidity etc. However, main disadvantage of the last method is irreversible change of sample sizes.

## Reference

[1] Fraser D B and LeCraw R C 1964 Novel method of measuring elastic and anelastic properties of solids *Rev. Sci. Instrum.* **35** 1113

[2] Holland R 1968 Resonant properties of piezoelectric ceramic rectangular parallelepipeds *J. Acoust. Soc. Am.* **43** 988

[3] Migliori A and Sarrao J 1997 *Resonant ultrasound spectroscopy (application to physics, materials measurements, and nondestructive evaluation)* (New York: John Wiley & Sons)
[4] Ogi H, Kawasaki Y, Hirao M and Ledbetter H 2002 Acoustic spectroscopy of lithium niobate: elastic and piezoelectric coefficients of solids *J. Appl. Phys.* **92** 2451
[5] Ogi H, Ohmori T, Nakamura N and Hirao M 2006 anelastic, and piezoelectric coefficients of α-quartz determined by resonance ultrasound spectroscopy *J. Appl. Phys.* **100** 053511
[6] Lee T, Lakes R S and Lai A 2000 Resonant ultrasound spectroscopy for measurement of mechanical damping: comparison with broadband viscoelastic spectroscopy *Rev. Sci. Instr.* **71** 2855
[7] Schwarz R B and Vuorinen J F 2000 Resonant ultrasound spectroscopy: applications, current status and limitations *J. of Alloys and Compounds* **310** 243
[8] Demarest H H 1971 Cube-resonance method to determine the elastic constants of solids *J. Acoust. Soc. Am.* **49** 768
[9] Ohno I 1976 Free vibration of a rectangular parallelepiped crystal and its application to determination of elastic constants of orthorhombic crystals *J. Phys. Earth* **24** 355
[10] Ogi H, Sato K, Asada T, Hirao M 2002 Complete mode identification for resonance ultrasound spectroscopy *J. Acoust. Soc. Am.* **112** 2553
[11] Ohno I 1990 Rectangular parallelepiped resonance method for piezoelectric crystals and elastic constants of alpha-quartz *Phys. Chem. Minerals* **17** 371
[12] Myasnikov D V, Konyashkin A V and Ryabushkin O A 2010 Identification of eigenmodes of volume piezoelectric resonators in resonant ultrasound spectroscopy *Tech. Phys. Lett.* **36** 632

Journal of Physics: Conference Series **510** (2014) 012037 doi:10.1088/1742-6596/510/1/012037

# Smoothed Particle Method for Real-Space Electronic Structure Calculations

**Soichiro Sugimoto and Yasunari Zempo**

Computer and Information Sciences, Hosei University,
3-7-2 Kajino, Koganei, Tokyo 184-8584, Japan

E-mail: 12t0007@cis.k.hosei.ac.jp

**Abstract.** We present a meshfree particle method for real-space electronic structure calculations based on symmetric smoothed particle hydrodynamics (SSPH). As a simple example, we applied SSPH to the Schrödinger equation for a one-dimensional harmonic oscillator. The results calculated using SSPH are in good agreement with analytical solutions. We compared results for different cases of kernel function, smoothing length, number of particles, and the order of the Taylor series expansion. The results using SSPH and the higher-order finite-difference method are also compared. Our results indicate that SSPH can be applied to real-space electronic structure calculations that require high accuracy.

## 1. Introduction

The real-space approach is widely used for electronic structure calculations based on density functional theory [1, 2, 3]. In particular, it is used for large systems that include many atoms, because a real-space mesh is suitable for large-scale parallel computing. The real-space approach also allows the capture of a clear physical image because it does not use an explicit basis such as a plane wave or Gaussian basis sets. For the simplest real-space implementation, the higher-order finite-difference method is employed as the discretization technique [1, 2]. A regular mesh is efficient because it produces a sparse and structured matrix. However, a regular mesh has equal spacing everywhere, which seems to be an inefficient distribution of computation points. There are also real-space implementations using finite-element and meshfree methods [1, 4, 5]. These methods have an advantage in their non-uniform distribution of computation points, which also reduces computational costs.

Smoothed particle hydrodynamics (SPH) [6, 7, 8, 9] is a typical meshfree particle method, in which the system is represented by a finite set of arbitrarily distributed particles without using any mesh. SPH is widely used to solve hydrodynamic equations in various areas such as astrophysics and magnetohydrodynamics [10]. In addition, the SPH-based method is applied to non-hydrodynamic partial differential equations such as the wave equation, diffusion equation, Poisson's equation, and Maxwell's equations [11, 12, 13, 14]. However, there are few studies addressing the use of SPH for electronic structure calculations.

The SPH-based method provides an efficient real-space technique because the computation points can be arbitrarily distributed. Since SPH is known for its low accuracy [12], the corrective smoothed particle method (CSPM) [15], the modified smoothed particle hydrodynamics (MSPH)

[16], and the symmetric smoothed particle hydrodynamics (SSPH) [17, 18] have all been proposed as alternatives. These methods improve the accuracy of SPH using the Taylor series expansion.

In this study, we employ the SSPH method because it is the simplest method in the higher-order series expansion. This study evaluates the accuracy of SSPH for electronic structure calculation applications. For simplicity, we applied SSPH to the Schrödinger equation for a one-dimensional harmonic oscillator, whose analytical solution is well known. In the evaluation, we focused on the relative error of the eigenvalue. The result is then compared to the higher-order finite-difference method.

The paper is organized as follows. The formulation of the SSPH method is presented in the next section. The numerical tests are shown in section.3. Finally, we summarize and discuss our conclusions in the last section.

## 2. Methods

First, we briefly introduce the SSPH method and then address the discretization of the Schrödinger equation.

### *2.1. Symmetric Smoothed Particle Hydrodynamics (SSPH)*

In SPH, the field function is approximated using integral representation, and the system is represented by a finite number of particles. The concept of the integral representation of a field function $\psi(x)$ starts from the following identity.

$$\psi(x') = \int_\Omega \psi(x)\delta(x - x')dx. \quad (1)$$

If the Dirac delta function is replaced by the function $W(x - x', h)$, which is called the kernel function, the integral representation is approximated as follows:

$$\psi(x') \approx \int_\Omega \psi(x)W(x - x', h)dx, \quad (2)$$

where $h$ is the smoothing length. A Gaussian function is a typical kernel function since the kernel function approximates the delta function.

$$W(x - x', h) = W(|x - x'|, h) = W(r, h) = \frac{1}{h\sqrt{\pi}}\exp\left(-\frac{r^2}{h^2}\right). \quad (3)$$

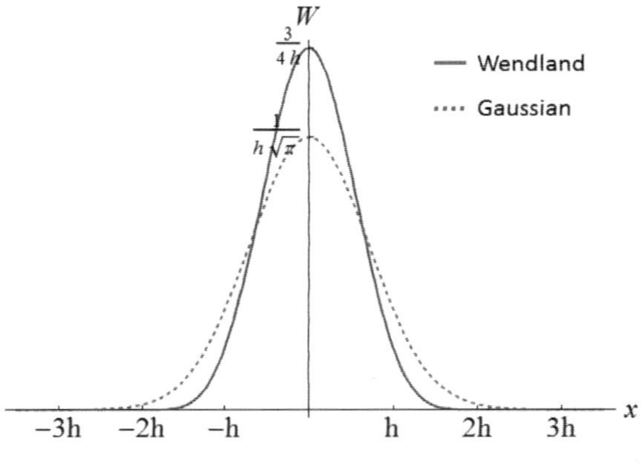


**Figure 1.** Wendland and Gaussian kernel functions in one-dimensional space, where $h$ is the smoothing length. The Wendland kernel is truncated at $2h$, whereas the Gaussian has a long tail.

In this study, we employ the Wendland kernel [19, 20].

$$W(r,h)=\begin{cases}\dfrac{3}{4h}\left(1-\dfrac{1}{2}\dfrac{r}{h}\right)^5\left(2\left(\dfrac{r}{h}\right)^2+\dfrac{5}{2}\dfrac{r}{h}+1\right) & (0\le\frac{r}{h}\le 2)\\ 0 & (\frac{r}{h}\ge 2)\end{cases}. \tag{4}$$

SSPH corrects the approximation of SPH using the Taylor series expansion. Using the Taylor series expansion of $\psi(x)$, the right-hand side of Eq.(2) and its moments can be written as follows:

$$\begin{aligned}\int_\Omega \psi(x)Wdx &= \psi(x_i)\int_\Omega Wdx+\left.\frac{\partial\psi}{\partial x}\right|_{x_i}\int_\Omega(x-x_i)Wdx+\frac{1}{2}\left.\frac{\partial^2\psi}{\partial x^2}\right|_{x_i}\int_\Omega(x-x_i)^2Wdx+\cdots\cdots,\\ \int_\Omega(x-x_i)\psi(x)Wdx &= \psi(x_i)\int_\Omega(x-x_i)Wdx+\left.\frac{\partial\psi}{\partial x}\right|_{x_i}\int_\Omega(x-x_i)^2Wdx+\frac{1}{2}\left.\frac{\partial^2\psi}{\partial x^2}\right|_{x_i}\int_\Omega(x-x_i)^3Wdx+\cdots\cdots,\\ &\vdots\\ &\vdots\\ \int_\Omega(x-x_i)^m\psi(x)Wdx &= \psi(x_i)\int_\Omega(x-x_i)^mWdx+\left.\frac{\partial\psi}{\partial x}\right|_{x_i}\int_\Omega(x-x_i)^{m+1}Wdx+\frac{1}{2}\left.\frac{\partial^2\psi}{\partial x^2}\right|_{x_i}\int_\Omega(x-x_i)^{m+2}Wdx+\cdots\cdots.\end{aligned} \tag{5}$$

Solving the following linear equation, $\psi(x_i)$ and its derivatives can be obtained.

$$\begin{bmatrix}K_{11} & K_{12} & \cdots & \cdots & K_{1(m+1)}\\ K_{21} & K_{22} & \cdots & \cdots & K_{2(m+1)}\\ K_{31} & K_{32} & \cdots & \cdots & K_{3(m+1)}\\ \vdots & \ddots & & & \vdots\\ \vdots & & & & \vdots\\ \vdots & & & & \vdots\\ \vdots & & & & \vdots\\ K_{(m+1)1} & K_{(m+1)2} & \cdots & \cdots & K_{(m+1)(m+1)}\end{bmatrix}\begin{bmatrix}\psi_i\\ \left.\frac{\partial\psi}{\partial x}\right|_{x_i}\\ \frac{1}{2}\left.\frac{\partial^2\psi}{\partial x^2}\right|_{x_i}\\ \vdots\\ \vdots\\ \frac{1}{m!}\left.\frac{\partial^m\psi}{\partial x^m}\right|_{x_i}\end{bmatrix}=\begin{bmatrix}\int_\Omega\psi(x)W(x-x_i,h)dx\\ \int_\Omega(x-x_i)\psi(x)W(x-x_i,h)dx\\ \int_\Omega(x-x_i)^2\psi(x)W(x-x_i,h)dx\\ \vdots\\ \vdots\\ \int_\Omega(x-x_i)^m\psi(x)W(x-x_i,h)dx\end{bmatrix}, \tag{6}$$

where

$$K_{IJ}=\int_\Omega(x-x_i)^{I-1}(x-x_i)^{J-1}W(x-x_i,h)dx \quad (I,J=1,2,\cdots,m+1). \tag{7}$$

In the SPH-based method, the system is represented by a finite number of particles. Therefore, the integral representation can be converted to a discretized form of summation over all particles. If the small region $dx$ at the location of the particle $j$ is replaced by the finite volume $\Delta V_j$,

$$\begin{aligned}\int_\Omega(x-x_i)^k\psi(x)W(x-x_i,h)dx &\approx \sum_{j=1}^N(x_j-x_i)^k\psi(x_j)W(x_j-x_i,h)\Delta V_j\\ &= \sum_{j=1}^N(x_j-x_i)^k\psi(x_j)W_{ij}\Delta V_j \quad (k=0,1,2,\cdots,m).\end{aligned} \tag{8}$$

The particle number density is defined as

$$n(x)\equiv\sum_{a=1}^N\delta(x_a-x). \tag{9}$$

It can be approximately expressed as

$$n(x_i) \approx \sum_{j=1}^{N} W(x_j - x_i, h). \tag{10}$$

The finite volume of the particle $i$ is equivalent to

$$\Delta V_i = \frac{1}{n(x_i)} = \frac{1}{n_i}. \tag{11}$$

Finally, the integral representation can be discretized as

$$\int_\Omega (x - x_i)^k \psi(x) W(x - x_i, h) dx \approx \sum_{j=1}^{N} (x_j - x_i)^k \psi(x_j) W_{ij} \frac{1}{n_j} \quad (k = 0, 1, 2, \cdots, m). \tag{12}$$

Thus, Eq.(6) is transformed to the discretized form.

### 2.2. *Discretization of the Schrödinger equation using SSPH*

The one-dimensional Schrödinger equation is

$$\left[-\frac{1}{2}\frac{\partial^2}{\partial x^2} + V(x)\right]\psi(x) = \varepsilon\psi(x) \ . \tag{13}$$

Considering Eqs.(6) and (12), we can easily obtain the discretized form of the wave function $\psi(x)$ and its second derivative. The Schrödinger equation at the location of the particle $i$

$$-\frac{1}{2}\frac{\partial^2 \psi}{\partial x^2}\bigg|_{x_i} + V(x)\psi(x_i) = \varepsilon\psi(x_i) \tag{14}$$

can be expressed using SSPH:

$$\begin{bmatrix} V(x_i) \\ 0 \\ -1 \\ 0 \\ \vdots \\ 0 \end{bmatrix}^{\mathrm{T}} \boldsymbol{K}^{-1} \begin{bmatrix} \sum_{j=1}^{N} \psi(x_j) W_{ij} \Delta V_j \\ \sum_{j=1}^{N} (x_j - x_i) \psi(x_j) W_{ij} \Delta V_j \\ \sum_{j=1}^{N} (x_j - x_i)^2 \psi(x_j) W_{ij} \Delta V_j \\ \vdots \\ \vdots \\ \sum_{j=1}^{N} (x_j - x_i)^m \psi(x_j) W_{ij} \Delta V_j \end{bmatrix} = \varepsilon \begin{bmatrix} 1 \\ 0 \\ 0 \\ 0 \\ \vdots \\ 0 \end{bmatrix}^{\mathrm{T}} \boldsymbol{K}^{-1} \begin{bmatrix} \sum_{j=1}^{N} \psi(x_j) W_{ij} \Delta V_j \\ \sum_{j=1}^{N} (x_j - x_i) \psi(x_j) W_{ij} \Delta V_j \\ \sum_{j=1}^{N} (x_j - x_i)^2 \psi(x_j) W_{ij} \Delta V_j \\ \vdots \\ \vdots \\ \sum_{j=1}^{N} (x_j - x_i)^m \psi(x_j) W_{ij} \Delta V_j \end{bmatrix}. \tag{15}$$

The matrix equation of the generalized eigenvalue problem can be obtained as

$$\boldsymbol{Hu} = \varepsilon \boldsymbol{Su}, \tag{16}$$

where

$$H_{ij} = \begin{bmatrix} V(x_i) \\ 0 \\ -1 \\ 0 \\ \vdots \\ 0 \end{bmatrix}^{\mathrm{T}} \boldsymbol{K}^{-1} \begin{bmatrix} W_{ij}\Delta V_j \\ (x_j - x_i)W_{ij}\Delta V_j \\ (x_j - x_i)^2 W_{ij}\Delta V_j \\ \vdots \\ \vdots \\ (x_j - x_i)^m W_{ij}\Delta V_j \end{bmatrix}, \quad S_{ij} = \begin{bmatrix} 1 \\ 0 \\ 0 \\ 0 \\ \vdots \\ 0 \end{bmatrix}^{\mathrm{T}} \boldsymbol{K}^{-1} \begin{bmatrix} W_{ij}\Delta V_j \\ (x_j - x_i)W_{ij}\Delta V_j \\ (x_j - x_i)^2 W_{ij}\Delta V_j \\ \vdots \\ \vdots \\ (x_j - x_i)^m W_{ij}\Delta V_j \end{bmatrix} \quad (17)$$

and $\boldsymbol{u} = \left[\psi(x_1), \psi(x_2), \cdots, \psi(x_N)\right]^{\mathrm{T}}$.

## 3. Numerical Tests

In this section, we discuss the accuracy of the eigenvalue. We first solve the Schrödinger equation for a one-dimensional harmonic oscillator

$$\left[-\frac{1}{2}\frac{\partial^2}{\partial x^2} + V(x)\right]\psi(x) = \varepsilon\psi(x), \quad V(x) = \frac{1}{2}\omega^2 x^2. \quad (18)$$

For simplicity, we use $\omega = 1$. The particles are uniformly distributed with constant spacing $\Delta x$ on the region $[-20, 20]$. The number of particles is $(40/\Delta x + 1)$.

Figure 2 shows the relative errors of the lowest eigenvalue, which is calculated using SSPH based on the $m$-th order Taylor series expansion. In this figure, $E_{\mathrm{SSPH}}$ is the eigenvalue of the SSPH method, and $E$ is the analytical solution. The results in Fig.2(a) are for the cases of $m = 2, 5, 8, 11$, and 14. The smoothing length $h$ is changed with the order $m$, $h = 1.5\Delta x, 3.2\Delta x, 4.5\Delta x, 6.0\Delta x$, and $7.5\Delta x$, respectively. The accuracy becomes higher as the order of the Taylor series expansion increases. Using the higher-order Taylor series

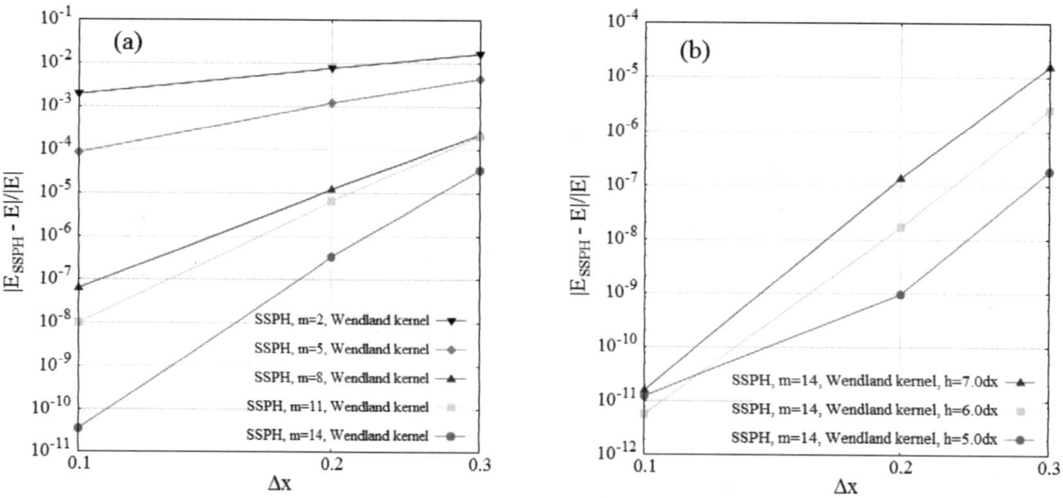


**Figure 2.** Convergence of the lowest eigenvalue for one-dimensional harmonic oscillator. (a) Results using SSPH based on the $m$-th order Taylor series expansion. ($m = 2, 5, 8, 11$, and 14.) (b) Results using SSPH based on the 14-th order Taylor series expansion. The smoothing length is $h = 5.0\Delta x, 6.0\Delta x$, and $7.0\Delta x$.

expansion with SSPH, we can obtain sufficient accuracy for electronic structure calculations. The computational cost to solve Eq.(17) increases with the order of the series expansion. As such order selection should be made according to the required accuracy.

A comparison of results using different smoothing lengths is shown in Fig.2(b). The Taylor series is expanded up to the 14-th order term, and the smoothing length is $h = 5.0\Delta x, 6.0\Delta x$, and $7.0\Delta x$. A shorter smoothing length gives higher accuracy at the same particle spacing $\Delta x$, since it gives a kernel function closer in shape to the delta function. The results indicate that a shorter smoothing length should be employed. However, a smoothing length that is too short makes the SSPH inconsistent and prevents us from solving the eigenvalue problem in Eq.(16). A smoothing length too long also gives incorrect results. We believe that the loss of locality causes these poor results. An appropriately short smoothing length should be employed to give correct results. The optimal selection of the smoothing length may be a goal in our future research.

Figure 3 shows a comparison of the SSPH results using Wendland and Gaussian kernels. The Taylor series of the SSPH is expanded up to the second-order term. The smoothing length is $h = 1.0\Delta x$. The results show that the Wendland kernel yields higher accuracy than the Gaussian kernel. The Wendland function gives almost the same accuracy as the three-point finite-difference method. As compared with the Gaussian kernel, the Wendland kernel is more

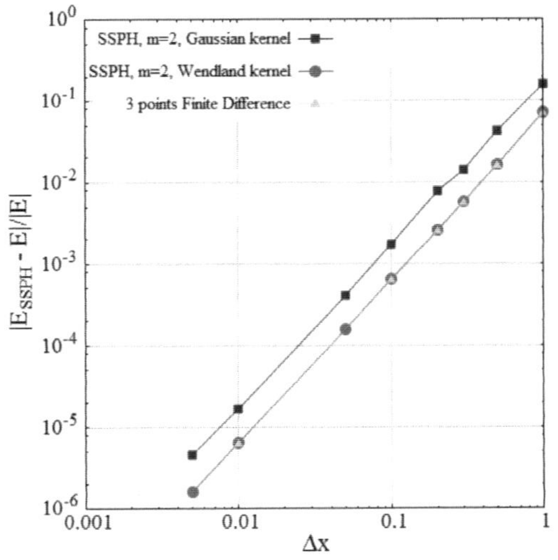


**Figure 3.** Convergence of the lowest eigenvalue for a one-dimensional harmonic oscillator. These are the results using SSPH and the three-point finite-difference method (triangle). The Taylor series of SSPH is expanded up to the second-order term. The Gaussian function (square) and the Wendland function (circle) are employed as the kernel functions.

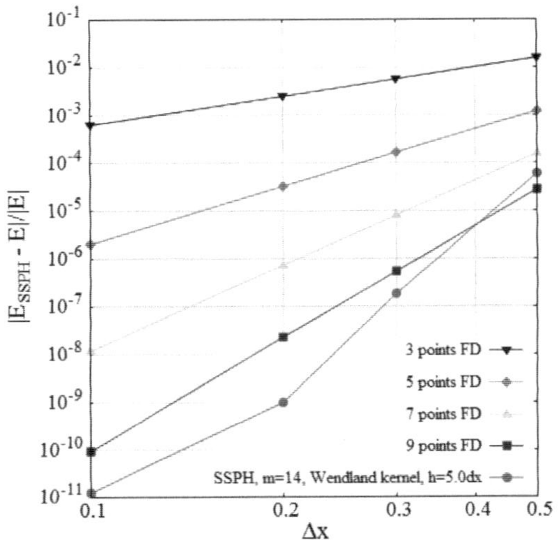


**Figure 4.** Convergence of the lowest eigenvalue for a one-dimensional harmonic oscillator. These are the results using SSPH (circle) and the higher-order finite-difference method, with three, five, seven, and nine point finite differences.

similar to the delta function. We consider this to be the reason that Wendland kernel gives higher accuracy.

In Fig.4, results using the higher-order finite-difference method and SSPH are compared because the higher-order finite-difference method is frequently employed for real-space electronic structure calculations. To achieve the same accuracy using SSPH, we need an SSPH Taylor series up to the 14-th order term and a smoothing length of $h = 5.0\Delta x$. The SSPH results give higher accuracy than the nine-point finite-difference method when $\Delta x = 0.3, 0.2$, and 0.1.

## 4. Conclusions and Discussion

We presented a method for real-space electronic structure calculations based on a meshfree particle method. The SSPH method has been applied to solve the Schrödinger equation of a one-dimensional harmonic oscillator. We investigated the effect of kernel function, smoothing length, and the order of the Taylor series expansion. The SSPH method based on the higher-order Taylor series expansion provides similar accuracy as the higher-order finite-difference method. SSPH may be applied in real-space electronic structure calculations that require high accuracy.

With the same particle spacing $\Delta x$, a shorter smoothing length gives higher accuracy, since it gives a kernel function closer in shape to the delta function. Our results indicate that we should employ an appropriately short smoothing length. The number of particles required in the support domain is determined so as to avoid any singularity of the SSPH system. We believe that a better smoothing length is determined from the number of particles and a nearest neighbor search algorithm. For our future study, we plan to study non-uniform particle distribution and methods for selecting an optimal smoothing length.

## Acknowledgments

The authors would like to acknowledge Professor S. S. Kano for his useful discussions and encouragement. This study was partially supported by a JSPS Grants-in-Aid for Scientific Research (C) Grant number 25390158, Sumitomo Chemical. Co. Ltd. and Simulatio Corporation.

## References

[1] Beck T L 2000 *Rev. Mod. Phys.* **72** 1041-80
[2] Chelikowski J, Troullier N, Wu K, and Saad Y 1994 *Phys. Rev.* B **50** 11355
[3] Zempo Y, Akino N, Ishida M, Ishitobi M, and Kurita Y 2008 *J. Phys. Cond. Matt.* **20** 064231
[4] Tsuchida E and Tsukada M 1995 *Phys. Rev.* B **52** 5573-78
[5] Jun S 2004 *Int. J. Numer. Meth. Engng.* **59** 1909-23
[6] Gingold R A and Monaghan J J 1977 *Mon. Not. Roy. Astr. Soc.* **181** 375-89
[7] Lucy L B 1977 *Astron. J.* **82** 1013-24
[8] Monaghan J J 1988 *Comput. Phys. Commun.* **48** 89-96
[9] Liu G R and Liu M B 2003 *Smoothed Particle Hydrodynamics: a meshfree particle method* (Singapole: World Scientific)
[10] Price D J 2012 *J. Comput. Phys.* **231** 759-94
[11] Laguna P 1995 *Astrophys. J.* **439** 814-21
[12] Stranex T and Wheaton S 2011 *Comput. Methods Appl. Mech. Engrg.* **200** 392-402
[13] Toscano E, Blasi G D and Tortorici A 2012 *Appl. Math. Comput.* **218** 8906-16
[14] Ala G, Francomano E, Tortorici A, Toscano E and Viola F 2006 *J. Comput. Appl. Math.* **191** 194-205
[15] Chen J K, Beraun J E, and Jih C J 1999 *Comput. Mech.* **24** 273-85
[16] Zhang G M and Batra R C 2004 *Comput. Mech.* **34** 137-46
[17] Batra R C and Zhang G M 2008 *Comput. Mech.* **41** 527-45
[18] Zhang G M and Batra R C 2009 *Comput. Mech.* **43** 321-40
[19] Wendland H 1995 *Adv. Comput. Math.* **4** 389-96
[20] Monaghan J J and Kajtar J B 2009 *Comput. Phys. Commun.* **180** 1811-20

# About the energy levels of GaAs

**Kulpash Iskakova**[1], **Rif Akhmaltdinov**[2], **Aisara Amanova**[3]
[1] Professor, *Physics and Mathematics Department, Kaz NPU by Abay*, Dostyk str 13, Almaty, 050010, RK
E-mail: ahmanovakulpash@yahoo.com
[2] ass. professor, *Chair of Automation and Computer Engineering, KOU*
[3] *student, KazNU by Al-Farabi, RK*

**Abstract.** Computer simulation of the properties of the crystal lattice gives to use up to $10^{10}\div10^{12}$ atoms. Using the described technique can be obtained from any of these atomic systems, as well as changes in the radii of coordination spheres at various distances. Simultaneously, you can get a number of nearby neighbors, the location of atoms on the coordination sphere. The initial structure is taken compound GaAs. The resulting model is given for different distances from the central atom. There are the distances from one to six. The result is an uneven growth of the number of atoms grouped along the radius of coverage. For the other elements of the possibility of uniform methods increase the number of atoms, such as Ga. We can develop an algorithm for constructing the structure of compound semiconductor-type $A^3B^5$, based on this principle. For example, to the type of $A^3B^5$ of semiconductor compounds it is used compound GaAs.

## 1. Introduction

To compound semiconductor type $A^3B^5$ is GaAs. Its crystal structure is a two face-centered cubic lattice, which are shifted relative to each other by ¼ the space diagonal. In the center of origin of the As atoms there are four nearest neighbors at the vertices of a tetrahedron. Each ion is surrounded by four nearest-neighbor ions, the antithesis of the mark. At a distance, there are four ion with $q_iq_j = -1$, eight ions antithesis sign $q_iq_j = +1$ at a distance. The system nods of the GaAs create the order of the atomic cores of the tetrahedral combinations. The source of difficulty in the calculation of band spectra is a priori is laid in these calculations the translational invariance, which involves the study does not end, real crystals, and the infinite, ideal. Manipulations with infinite objects are allowed in the initial stages of the abstraction of subject area (in our case, solid bodies), but they become simply mathematical trick when trying to obtain information about the physical properties of such complex objects as solid bodies. Solid body, as really existing system, and not a mathematical trick, requires a different approach than adopted for the simple monatomic isolated objects [1].

One of the advantages described technique — applicable to the systems with neighbor order, where Bloch's theorem is not applicable as the initial conditions, since the translational invariance in the real space and in the quasi-momentum space is not the necessary condition for the determination of the crystalline potential, the wave functions of the energy values in the space of wave vectors [1], [2].

Therefore the proposed technique is still the only one well-established technique that is applicable for the analysis of the band spectrum of finite imperfect crystals.

Journal of Physics: Conference Series **510** (2014) 012038 doi:10.1088/1742-6596/510/1/012038

## 2. Calculation of the dynamic three-dimensional structure $A^3B^5$

Consider the finite crystal with unbroken intervals (except for the crystal boundary violation). For simplicity, we take a large number of cores, when there is a nearly ideal periodicity of the crystal. In this case, the parameters obtained from the mathematical formalism for the analysis of the band spectrum are periodic functions, except for the boundary of the crystal. The crystal potential and wave functions have a period equal to the distance between the cores, depending on the chosen direction. The energies have multiple frequency modulus of the wave vector in the quasi-momentum space [1]-[3].

Now turn to the crystal with a very small number of cores. In such a crystal frequency is not as ideal as in the previous case. The simulation of such a crystal is insurmountable difficulties for traditional methods, but this is the case is taken by us as a basic demonstration version. The object in question is distant from a perfect crystal due to its limited size. And despite this, the technique presented by us showed the periodicity of the above parameters. Thus in our problems are not used the condition of periodicity and Bloch's theorem as the necessary initial conditions.

The case of uniform distribution of electrons in the space of interaction between them becomes energetically unfavorable. There is a possibility of an orderly configurations of the electrons in space. One more subtlety of crystal structure in calculating properties of the crystal related to the fact that the radii of the spheres accrue uniformly covering to a certain radius (for example, GaAs structure up to 498), then this is a violation of uniformity. Next violation of uniformity arises after a group of 918 atoms. Figures 1 and 2 are a group of atoms and their arrangement.

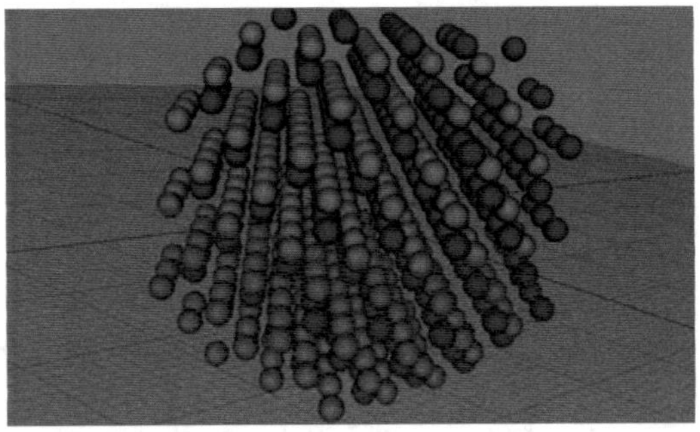

**Figure 1.** The grid with a radius of coverage 5, where there are 498 atomic cores.

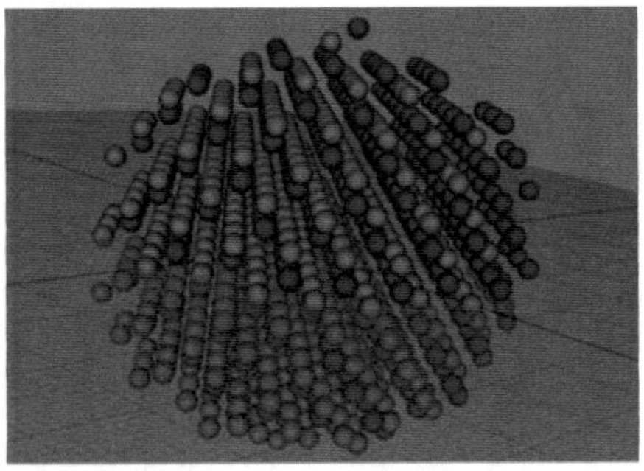

**Figure 2.** The As atoms and Ga for the radius of coverage 6. Red marked atoms As, in blue denote atoms Ga. Central atom is As.

But in fact, such a violation of uniformity means that it is impossible to carry out "and self-evident and obvious" extrapolation of the results and methods of calculation for the "normal covered by the scope" on large radii, which is doing traditionally.

In computer modeling of the crystal structure of GaAs, combined with three-and five-valent elements are observed the following regularities (see Table 1). When the radius of coverage is two in this range there are 38 cores. As the first atomic core in the center of origin, the four Ga atoms are located at a distance of 3 next 12 As atoms at a distance of 8. Nine Ga atoms at a distance there are 11. Six As atoms at a distance of 16, three Ga at a distance of 19 and three Ga at a distance of 27.

**Table 1.** The Combined lattice GaAs.

| $N$ | $x$ | $y$ | $z$ | $d^2$ | Ion | |
|---|---|---|---|---|---|---|
| 0 | 0 | 0 | 0 | 0 | As | |
| 1 | −1 | −1 | 1 | 3 | Ga | breaking |
| 2 | −1 | 1 | −1 | 3 | Ga | |
| 3 | 1 | −1 | −1 | 3 | Ga | |
| 4 | 1 | 1 | 1 | 3 | Ga | |
| 5 | 0 | 2 | −2 | 8 | As | breaking |
| 6 | 0 | 2 | 2 | 8 | As | |
| 7 | 0 | −2 | −2 | 8 | As | |
| 8 | 0 | −2 | 2 | 8 | As | |
| 9 | −2 | 2 | 0 | 8 | As | |
| 10 | −2 | −2 | 0 | 8 | As | |
| 11 | 2 | 2 | 0 | 8 | As | |
| 12 | 2 | −2 | 0 | 8 | As | |
| 13 | −2 | 0 | 2 | 8 | As | |
| 14 | −2 | 0 | −2 | 8 | As | |
| 15 | 2 | 0 | 2 | 8 | As | |
| 16 | 2 | 0 | −2 | 8 | As | |
| 17 | 1 | 3 | −1 | 11 | Ga | breaking |
| 18 | 1 | −1 | 3 | 11 | Ga | |
| 19 | −3 | 1 | 1 | 11 | Ga | |
| 20 | −1 | 3 | 1 | 11 | Ga | |
| 21 | 3 | 1 | −1 | 11 | Ga | |
| 22 | 1 | −3 | 1 | 11 | Ga | |
| 23 | −1 | 1 | 3 | 11 | Ga | |
| 24 | 3 | −1 | 1 | 11 | Ga | |
| 25 | 1 | 1 | −3 | 11 | Ga | |
| 26 | 0 | 4 | 0 | 16 | As | breaking |
| 27 | −4 | 0 | 0 | 16 | As | |
| 28 | 4 | 0 | 0 | 16 | As | |
| 29 | 0 | 0 | −4 | 16 | As | |
| 30 | 0 | 0 | 4 | 16 | As | |
| 31 | 0 | −4 | 0 | 16 | As | |
| 32 | 3 | 3 | 1 | 19 | Ga | breaking |
| 33 | 1 | 3 | 3 | 19 | Ga | |
| 34 | 3 | 1 | 3 | 19 | Ga | |
| 35 | 1 | 5 | 1 | 27 | Ga | breaking |
| 36 | 5 | 1 | 1 | 27 | Ga | |
| 37[1] | 1 | 1 | 5 | 27 | Ga | |

[1] Total number of cores with zero: 38

From this calculation, the differences in distance between the various groups are as (1):

$$\left.\begin{aligned} \Delta r_1 &= r_2 - r_1 = 3, \\ \Delta r_2 &= r_3 - r_2 = 5, \\ \Delta r_3 &= r_4 - r_3 = 3, \\ \Delta r_4 &= r_5 - r_4 = 5, \\ \Delta r_5 &= r_6 - r_5 = 3, \\ \Delta r_6 &= r_7 - r_6 = 8. \end{aligned}\right\} \qquad (1)$$

We observe quasiperiodic change in the difference. At the border radius of coverage this periodicity is broken. When increasing the radius of coverage this periodicity is restored to the previous radius, but is broken on the new range. Thus, the boundary conditions lead to a violation of the periodicity. Simulated lattice shown in the following figures. The first figure shows the four first neighbors As in the lattice. Neighbors' ions Ga (blue) form a tetrahedron with the center of As (red).

## 3. Modeling of the dynamic three-dimensional models of GaAs

The Figure 1 shows the grid with a radius of coverage 2. There is a periodic step structure. The structure of the central ion and its neighbors on the tetrahedron is stored and located in the center of this lattice. The Figure 2 shows the grid with a radius of coverage 6. There are 918 atomic cores in it. There is a periodic step structure. The structure of the previous lattice is preserved. And the structure of the central ion and its neighbors on the tetrahedron is stored and located in the center of this lattice. The Figure 3 shows the grid with a radius of coverage 1. There are 5 atomic cores in it. There is a periodic step structure. The structure of the central ion and its neighbors on the tetrahedron is stored and located in the center of this lattice.

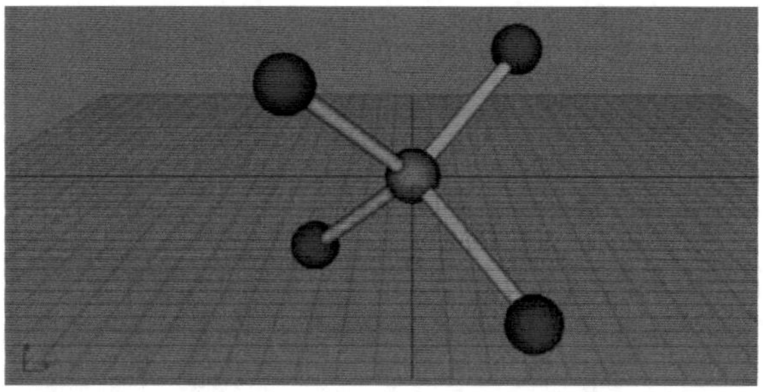

**Figure 3.** The grid with a radius of coverage where there are 5 atomic cores.

## 4. Discussion

It can be seen as an increase in the radius of coverage is restored periodicity in the lattice covered by the previous radius and there is a new violation of the periodicity on the boundary spanning lattice. The depth of violation increases with the radius of coverage. This is well observed in the tables you demonstrated for different cores.

On the basis of the direct lattice can construct the reciprocal lattice and define the properties of the Fermi surface. And the radius of coverage can be taken arbitrary (millions of skeletons).

The calculated structure allows to reach a real-time implementation and to take into account of defects and of real devices based on GaAs, because these are high-speed computing now by our proposed method. High-speed computing enables a dynamic computer simulation of the properties of real crystals of the GaAs.

Since the GaAs lattice is identical to the lattice of Si, sphalerite and diamond, we can apply the technology provided by and for these substances.

In calculating of coordination spheres cells of FCC- and BCC-lattices are presented as superlattice structure to fill own nodes by atoms of cells and additional nodes of the simple cubic lattices (SCL) — interstitial atoms [2]. The completing of coordination spheres with atoms is determined on the basis of a small SC-lattice, taking into account octahedral, tetrahedral internodes.

Links between the structural units play a dominant role. Coulomb attraction between electrons and nuclei combined into a solid unit. But again arise the question of the relationship of the spatial arrangement of the free electron gas with atomic nuclei in the crystal lattice. Contact in turn, arise from the nature of the electron distribution in space. In the existing models are impermeable ions crystal spherical. According to to the Pauli principle [1] and the theory of closed shells follows that atom is characterized by a stable electronic configuration. When crystal formation by being socialization electron leaving atomic orbital forms a defect in an electronic configuration resulting in an increase width of the unfilled valence [2-4]. In the crystal, when approaching to each other ions because of their impenetrability is broken quantum-mechanical effects. The main contribution to the total energy contributes strongly repulsive inter-ion Coulomb interaction. Much of the energy amounts electron-electron interaction in the calculation, which by existing methods there are big difficulties. The motion of the ions is inextricably linked with the movement of valence electrons [4, 5]. The total energy of the crystal depends on the location not only of ions and the electrons.

If atoms in the lattice arrange with the dense packing, P-nodes are centers of the inter-body lattices in the methods of presentment of the spatial distribution of atoms and internodes along the coordination spheres in the crystals. Tetrahedral and octahedral internodes in the ion-interionic terminology mean inter-body lattice space surrounded by the atoms in the crystal lattice nodes.

The Table 1 contains the number of atoms and the distance from the central atom of the diamond-like structure. The distance is given in units of the lattice constant a. Let us choose the maximum distance based on these conditions, forming a set of atoms contributions on which the calculation of the potential are taken into account.

In this work we consider the crystal potential for crystals with the diamond structure. The Coulomb potential is the sum of the capacity building of the nucleus and the electrons occupied levels:

$$U_c = U_n + U_0, \tag{2}$$

where $U_n = -2Z/r$. Here, $Z$ — number of the element, $r$ — distance from the atom to the specified point.

Potential $U_0$ is a solution of the Poisson equation:

$$\nabla^2 U_0 = -8\cdot\pi\rho_0(r), \tag{3}$$

where $\rho_0(r)$ — the density of the electrons of the occupied levels at the distance $r$ from the center of the atom. The density is calculated as follows:

$$\rho_0(r) = \sum_{n,l} r^2 R_{nl}^2(r). \tag{4}$$

Function $R_{nl}(r)$ — the radial part of the electron wave function for a hydrogen atom [4], [5]. Using the principle of superposition of the electronic densities and the additivity of the integral equation (3) can solve as follows:

$$U_0(r) = \sum_{i=0}^{I} u(r_i), \tag{5}$$

$$u(r_i) = \sum_{n,l} u_{nl}(r_i), \tag{6}$$

where $I$ — the set of the number of atoms minus one, $r_i$ — the distance from the center of the $i$-th atom to the specified point

$$\vec{r}_i = \vec{r} - \vec{a}_i. \tag{7}$$

Here $a_i$ — the radius-vector of the $i$-th atom with respect to the center of origin. Let us obtain the Poisson equation for an electron of $i$-th atom:

$$\nabla^2 u_{nl}(\vec{r}_i) = -8\pi r_i^2 R_{nl}^2(\vec{r}_i). \tag{8}$$

Solving (8), we obtain

$$u(\vec{r}_i) = -8\pi \sum_{n,l} \iint u_{nl}(\vec{r}_i)\, d\vec{r}_i^{\,2}. \tag{9}$$

Directly using the functions (5, 6, 7, 8, 9) we find the potential $U_0$ $(r)$. This approach allows us to calculate effectively and accurately the electronic potential. We rewrite the expression (2) accounted for the entire set of atoms:

$$U_c(\vec{r}) = \sum_{i} U_n(\vec{r}_i) + U_0(\vec{r}). \tag{10}$$

Thus, the Coulomb potential is calculated from all the surrounding atoms [1], [6].

This approach allows us to select any line capacity in any direction from the lattice points, and may be any length of line. Moreover, we can specify a curved line capacity.

The cell potential near the nucleus has a singularity, and the distance from the nucleus is close to zero in hyperbole. For a single atom inside the Wigner-Seitz sphere the main contribution comes from the nucleus of an atom [6], [7].

Consider the cell potential obtained at the line potential. In this case, clearly seen the potential frequency, which is created by all the atoms are now set. In passing through the core potential becomes singular. The potential between the atoms is identical and negligible in the magnitude. There is a little effect of the atoms located at the centers of the faces of the two face-centered cubic lattices [8].

The obtained simulation results show the crystal potential, the crystal can be modeled with all its features. In contrast to known methods of calculating of the crystal potential as described in the engineering calculations of the crystal potential is universal. It gives the results without making any correction factors, does not require the use of logarithmic scales, and other non-uniform, allows you to quickly calculate the crystalline potential for all elements of the given crystal structure. There are facilities for the further developments of such technology in the direction of accounting for various disturbances. It will allow a calculating the electronic potential even more precisely. However, even without taking into account the properties of an ideal crystal, this calculation provides sufficiently accurate results. To illustrate the power and versatility of the described technique for the calculating of the potential capabilities it is shown an interaction potential on line drawn along the main diagonals of the two face-centered cubic lattices in Figure 4.

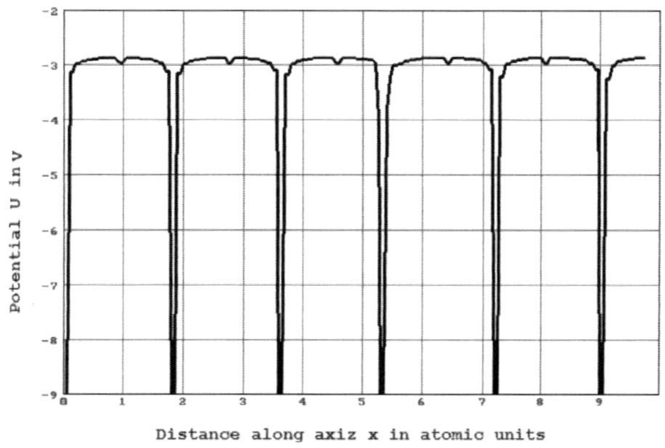


**Figure 4.** An interaction potential on line drawn along the main diagonals of the two face-centered cubic lattice.

In determining the energy spectra of quantum-mechanical problems for periodic structures, it often uses methods of the broadcast. In particular, semiconductors have a periodic structure. The crystal potential, having the grating period, usually modeled by the expression [1], [9]:

$$V(\vec{r}) = \left[\sum_{i=0}^{N} V_n(\vec{r}_i)\right] + U_0(\vec{r}) + V_{exh}(\vec{r}). \tag{11}$$

where $V_n$ $(r_i)$ — the potential created by the $i$-th nucleus at the point $\mathbf{r}$; $U_0$ $(\mathbf{r})$ — the electronic potential, formed by the electrons of the core at some point $\mathbf{r}$ and is determined from the Poisson equation

$$\nabla^2 U_0(\vec{r}) = -8\pi\rho_0(\vec{r}). \tag{12}$$

Here $\rho_0(\mathbf{r})$ — the electronic density of the occupied levels at a distance $r$ from the center of the atom; $V_{exh}(\mathbf{r})$ —the exchange potential at point $\mathbf{r}$; $\mathbf{r}_i = |\mathbf{r}-\mathbf{a}_i|$ — the distance from the $i$-th node to the point $\mathbf{r}$; $\mathbf{a}_i$ — the node location; $N$ — the number of neighbors taken into account relatively to the cores. The solution of Schrödinger equation based on the potential (11) is carried out by the Hartree-Fock-Slater [10].

The specificity of a solid compared to other bodies, is the decisive role of spatial-structural properties: a lattice type, a location and an interaction of the defects, a structure of core lattices and inter-ionic space. The period of the inter-ionic lattice coincides with the period of the core lattice. Spatially periodic structure of two weakly coupled and oppositely charged objects is the cause of the stability of rigid body — in this case the semiconductor GaAs — because between these objects appear spatially periodic Coulomb forces. In the geometric structure of each inter-ionic space surrounded on six sides by positive ion cores, and the ionic core in turn — intervals.

For a uniform distribution of electrons in the space of interaction between them becomes energetically unfavorable. There is the possibility of an orderly arrangement of electrons in space, the emergence of the so-called Wigner or "electron crystal".

The geometrical structure of the crystal consists not only of the atomic cores, but inter-ionic space intervals. The valence electrons in inter-ionic space intervals are also elements of the crystal, along with the atomic cores. The inter-ionic space is gap, i.e. the most probable location of the valence electrons. In addition, inter-ionic space intervals are also still a kind of "elements" of the crystal. But then the conclusion is that in the crystal lattice there are two — the well-known lattice consisting of the atomic cores, and another, equally important for the existence of the crystal lattice of the valence electrons are concentrated between the cores. The lattice periodicity of the atomic cores immediately implies the lattice periodicity of the inter-ionic space.

Even more natural in such a conclusion, if you do not forget that the atomic cores of the crystal lattice — not a point in space, but the figures, which occupy a significant part of the whole space of the crystal, and inter-ionic space also has a periodic structure, as well as skeletons. The period of this lattice coincides with the usual of the core of the lattice. Figure 5 shows a two-dimensional lattice with the lattice constant $a_0$. The coordinate origin of the system $Oxy$ is at the center of one of the cores, and the origin of the coordinate system $O'x'y'$ is located in the center of one of the inter-ionic cells. The coordinate system $O'x'y'$ is formed by the translation of the coordinate system $Oxy$ in $a_0n/2$, where $n = \{0, 1, 2, ...\}$. We assume inter-ionic intervals are the inter-ionic elements of the lattice, which is shifted with respect to a core of the lattice by half a lattice constant in all coordinate axes.

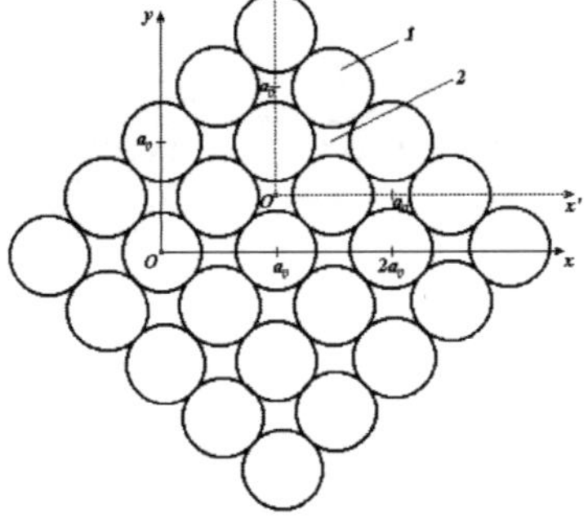


**Figure 5.** A two-dimensional lattice with the lattice constant $a_0$.

Cores and inter-ionic space two-dimensional lattice include 1) frame, 2) inter-ionic space.

The determining the location of the centers of the inter-ionic intervals can be easily receive from the definition of the basis vector for the nodes of the crystal (the core) of the lattice with any structure. But it should be noted that by increasing the radius of inter-ionic volumes (not just their centers) there is a change the number under the consideration inter-ionic space compared to the amount covered by the cores. Since you want to cover all inter-ionic space surrounding covered the skeletons, the radius of coverage inter-ionic space few more skeletons examined.

Crystal is in a stable condition thanks to the orderly arrangement of opposite charges. Most of the properties of crystals (eg, electrical conductivity, magnetic susceptibility), are caused not only property of the host lattice, but also the existence of inter-ionic lattice. This inter-ion lattice in addition to the core determines the physical properties of the crystal.

Specificity of a solid compared to other bodies is that in them play crucial role spatial structural properties: the lattice type, location and interaction of defects, the structure of the nucleus and inter-ion lattice. The period of the inter-ion lattice coincides with the period of the core lattice. Spatially periodic structure of two weakly coupled and oppositely charged objects is the cause of the stability of rigid body — in this case the GaAs — because between these objects appear spatially periodic Coulomb forces. In the geometric structure of each inter-ionic cell surrounded on six sides by positive ion cores, and the ionic core in turn — the inter-ionic gaps.

Therefore, in the calculation of the energy spectrum, it is necessary to consider the interaction of the electron and the probe with the cores and intercores. Oscillations are tightened to the centers of the cores and increase their amplitude with increasing levels of energy. This is explained the combined effects of strong impact of all cores and intercores on the test electron in the higher energy states. In the even states (2nd and 4th) are clearly visible the envelopes of the plane wave. In these states, the geometry of the crystal increases the total contribution of all the cores and the intercores in the amplitude of the envelope wave (the core and intercore lattice acts as a amplitude resonator). The energy levels obtained in the space of wave vectors are periodic functions:

$$E_j\left(\vec{kx}+\vec{g}\right)=E_j\left(\vec{kx}\right), \tag{13}$$

where $\vec{g}$ is a reciprocal lattice vector. The periodic dependence of the energy $E_j\left(\vec{x},\vec{kx}\right)$ from the one-dimensional wave vector $\vec{kx}$ with a period $2kx/a_0$ (where $a_0$ is the lattice period) throughout the entire crystal is obtained without regard the periodicity conditions.

Figures 6 and 7 shows the plots of the potential $U$ of the energy $E_j\left(\vec{x},\vec{kx}\right)$ of the coordinates of $x$ and the wave vector $\vec{kx}$ for a group of 498 atoms with an energy of 3.12 eV and 913 –2.93 eV.

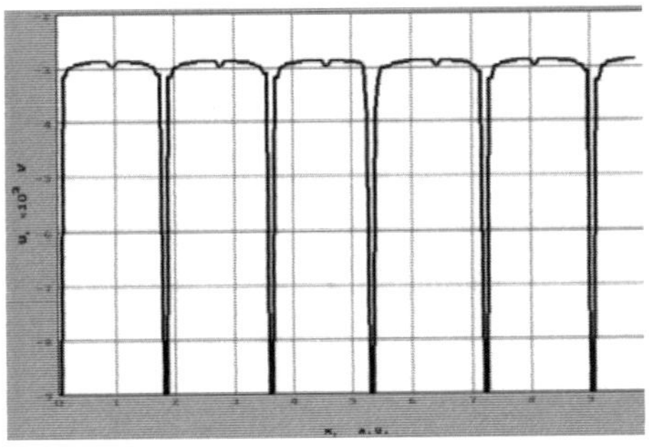

**Figure 6.** The grid with a radius of coverage 5, where there are 498 atomic cores with an energy of 3.12 eV.

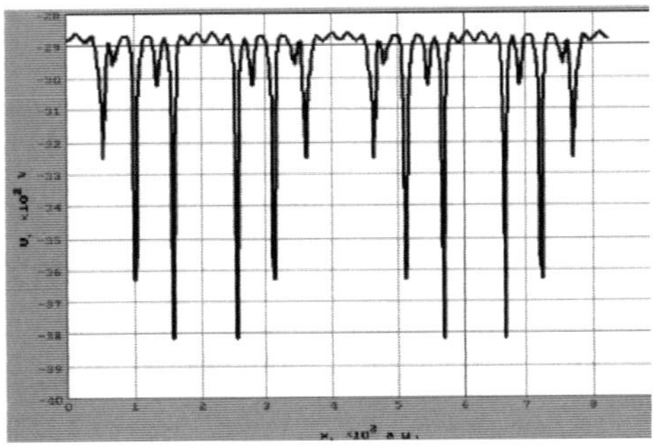

**Figure 7.** The As atoms and Ga for the radius of coverage 6, where there are 913 atomic cores with an energy of 2.93 eV.

## 5. Conclusion

We can develop an algorithm for constructing the structure of compound semiconductor-type $A^3B^5$, based on this principle. For example, to the type of $A^3B^5$ of semiconductor compounds it is used compound GaAs. It is crystal lattice of two face-centered cubic lattices shifted relative to each other at j spatial diagonal. Each Ga atom has four nearest neighbors located at the tops of a tetrahedron. Each ion is surrounded by four nearest neighboring ions of opposite sign. There are four ions with $q_i\, q_j= -1$ at a distance $\frac{1}{2}a_0 2^{\frac{1}{2}}$, eight ions of opposite sign of $q_i\, q_j = +1$ at the 2 distance $a_0 2^{\frac{1}{2}}$. The system of links of the GaAs structure establishes the order of the atomic body along tetrahedral combinations [2], [3]. For such compounds of the tetrahedral group of atoms is one-eighth of the basic lattice, so it is convenient to consider as a superstructure. The cell of the GaAs lattice consists of four tetrahedral sublattices.

One of the advantages of the described technique is its application to systems with short-range order, which does not use Bloch's theorem as the initial conditions. This was possible due to the fact that translational invariance follows as a consequence of the calculations, and not postulated initially. Thus, translational invariance in real space and in the quasi-momentum space is not a prerequisite for the determination of the crystal potential, wave functions and energies in wave vector space.

Journal of Physics: Conference Series **510** (2014) 012038 doi:10.1088/1742-6596/510/1/012038

This, in turn, allows you to define these parameters not only for infinite crystals, but also for the real limits on the size of the crystals. Therefore, it is possible to determine the properties of crystals with impaired structure of a single scheme, with no additional idealizations. Individual interest is a violation of the periodicity on the boundary of the crystal. In fact, it is inherent in all real crystals violation of the ideal (i.e. infinite) crystal structure.

The violation of a perfect crystal at the border no doubt affects the theoretical study of the band spectrum. Boundary effects appear in the form of decay phase portraits of the main parameters (the crystal potential, the wave functions and energies in wave vector space) in all directions of the crystal boundaries (in our case spherical).

In the study of solid-state we considered two sites within the crystal, virtually connected to each other. A theoretical study of solid-state is easily accomplished with the use of these facilities both qualitative and quantitative methods. But the most effective way of investigating solids is to use a direct computer simulation, and that we have done for simple structures.

Thus, the use of techniques discussed in the article allows the calculation of real properties of GaAs crystals with defects and other constraints.

## 6. References

[1] T. Loucks, Augmented plane wave method. New York, W. A. Bendjamin Inc., 1967, p. 12-62.

[2] Iskakova K.A., Ahmaltdinov R.F., Possibilities of the modifications of the method of calculation of wave function and zonal spectrum of crystal structure, *PCI-2010* «12th International Conference on the Physics and Chemistry of Ice». Sapporo, Japan.

[3] Iskakova K.A., Ahmaltdinov R.F., Modelling and calculation of FCC structures and principles of the algorithm structure Si, 88th annual meeting of the German Mineralogical Society, 19–22. September 2010 Munster, Germany.

[4] Bugaev V.N., Chepulski R. V. 1. Basic theory. The symmetry of interatomic lattice potentials in general crystal structures. – *Acta crystallogr.* 1995, v. 51, No **4**, p. 456–462.

[5] Sowa E.C., Gonis A. A multisite interaction expansion of the total energy in metals. *Instrum. and Meth. Phys. Res.* B. 1995, v. 102, No **1–4**, p. 1–2.

[6] Korhonen T., Puska M. I., Nieminen R. M. Vacancy formation energies for fcc and bcc transition metals. *Phys. Rev.* B. 1995, v. 51, No **15**, p. 9526–9532.

[7] Mirabella D.A., Aldao C.M. Effects of orbital nonorthogonality on band structure within the hight binding scheme. *Phys. Rev.* B. 1994, v. 50, No **16**, p. 12152–12155.

[8] Andrews Lawrence C., Bernstein Herbert I. Bravais lattice invariants. *Acta Crystallogr.* A. 1995, v. 51, No **3**, p. 413–416.

[9] Krimmel H.G., Ehmann I., Elsasser C. Calculation of atomic forces using the lineorized – augmented – plane wave method. *Phys. Rev.* B. 1994, v. 50, No 12, p. 8846–8848.

[10] Bodralexe E., Marksteiner P., Oh Y., Freeman A. J. Coupled channel equation for potential with a Coulomb singularity. *Comput. Phys. Commun.* 1994, N. **82**, p. 120–128.

Journal of Physics: Conference Series **510** (2014) 012039 doi:10.1088/1742-6596/510/1/012039

# *Ab initio* Molecular-Dynamics Study of Dissociation Mechanism of Highly Charged Molecules

**Satoshi Ohmura[1], Kiyonobu Nagaya[1], Fuyuki Shimojo[2] and Makoto Yao[1]**

[1] *Department of Physics, Kyoto University, Kyoto 606-8502, Japan*
[2] *Department of Physics, Kumamoto University, Kumamoto 860-8555, Japan*

E-mail: ohmura@scphys.kyoto-u.ac.jp

**Abstract.** Charge state dependence of dissociation mechanisms of the highly charged bromophenol $(C_6H_5OHBr)^{n+}$ ($n \leq 10$) is studied by *ab initio* molecular dynamics simulations based on density functional theory. When the charge state $n$ is 6 or 7, one or two hydrogen atoms dissociate in the first stage (at times shorter than 20 fs) while all hydrogen atoms dissociate when the charge state is over 7. After hydrogen dissociation, the carbon ring breaks at about 150 fs. There is also a difference on the mechanism of the ring breaking depending on charge states, in which the ring breaks with expanding ($n = 6, 7$) or shrink ($n \geq 8$).

## 1. Introduction

With the advent of free electron laser (FEL) such as Extreme Ultraviolet FEL and X-Ray FEL, many interesting phenomena in which atoms or molecules are highly excited due to the strong laser fields are observed. When a molecule is irradiated by hard x-ray, holes are proliferated through the Auger cascade and the molecule takes high charge. The highly charged molecular ions are quite unstable due to the Coulomb repulsive force and typically destroyed on a subpicosecond time scale (Coulomb explosion). The fragment ions may convey information about molecular shape because they reflect geometric structure of the parent molecule just before dissociation. Recently, dissociation of several highly-charged aromatic molecules has been investigated by position-sensitive time-of-flight measurements [1, 2]. Sugishima *et al.* investigated the momenta of ejected ions from highly charged benzene and fluorobenzene molecules by using position-sensitive time-of flight mass spectrometer [1]. They found that there exists clear angular correlation between the momenta of $H^+$-$H^+$ and $H^+$-$F^+$ fragment ions which reflects the structure of the parent molecules even for the dissociation channels for which the dissociations are not described by the simple Coulomb explosion model. In addition, Nagaya *et al.* have tried to investigate the conductivity properties of single molecules with using Coulomb explosion [2]. They determined the conductance of bromo aromatic molecules by ionizing the Br atom and detecting an opposite $O^+$ ion after the Coulomb explosion. Although many studies about Coulomb explosion have been reported, details of the Coulomb explosion are still unknown. In this paper, we investigate the dissociation mechanisms from highly charged bromophenol by *ab initio* molecular-dynamics simulations based on density functional theory (DFT) to clarify the atomic mechanisms of Coulomb explosion. Generally, relaxation of electron system is much faster than that of ionic (atomic) system. Therefore, we suppose that the dissociation of the charged molecules occurs on the ground state.

## 2. Methodology

### *2.1. Method of calculation or charged system*

To calculate charged molecular system in reciprocal space, we used cluster boundary conditions imposed with the method of Martyna and Tuckerman [3], which rigorously corrects for the error resulting from artificial repetitions of an aperiodic system by a screening function in the reciprocal space. In a fully periodic system, the potential described by the electron density $n(\mathbf{r})$ and the potential $\phi(\mathbf{r}-\mathbf{r}')$ is given by

$$\langle\phi\rangle = \frac{1}{2V}\sum_{\mathbf{g}} |n(\mathbf{g})|^2\tilde{\phi}(-\mathbf{g}), \tag{1}$$

where $n(\mathbf{g})$ and $\tilde{\phi}(-\mathbf{g})$ are the Fourier transform of the density and potential, respectively. For systems with fewer than three periodic dimensions,

$$\langle\phi\rangle^{(1)} = \frac{1}{2V}\sum_{\mathbf{g}} |\bar{n}(\mathbf{g})|^2\bar{\phi}(-\mathbf{g}). \tag{2}$$

where $\bar{\phi}(\mathbf{g})$ and $\bar{n}(\mathbf{g})$ denote a Fourier expansion coefficient of the potential and density, respectively in the non-periodic dimensions. In determining the form of $\bar{\phi}(\mathbf{g})$, we need to consider two functions $\phi^{\text{long}}(\mathbf{r})$ and $\phi^{\text{short}}(\mathbf{r})$, which are assumed to be the long and short range contributions to the total potential.

$$\phi(\mathbf{r}) = \phi^{\text{long}}(\mathbf{r}) + \phi^{\text{short}}(\mathbf{r}). \tag{3}$$

Assuming that the simulation cell is large enough so that $\bar{\phi}^{short}(\mathbf{g})\simeq\tilde{\phi}^{short}(\mathbf{g})$, the desired Fourier transform is expressed as

$$\begin{aligned}\bar{\phi}(\mathbf{g}) &= \bar{\phi}^{\text{long}}(\mathbf{g}) + \bar{\phi}^{\text{short}}(\mathbf{g})\\ &\simeq \bar{\phi}^{\text{long}}(\mathbf{g}) + \tilde{\phi}^{\text{short}}(\mathbf{g})\\ &= \bar{\phi}^{\text{long}}(\mathbf{g}) - \tilde{\phi}^{\text{long}}(\mathbf{g}) + \tilde{\phi}^{\text{short}}(\mathbf{g}) + \tilde{\phi}^{\text{long}}(\mathbf{g})\\ &= \hat{\phi}^{\text{screen}}(\mathbf{g}) + \tilde{\phi}(\mathbf{g})\end{aligned} \tag{4}$$

where $\tilde{\phi}(\mathbf{g}) = \tilde{\phi}^{\text{short}}(\mathbf{g}) + \tilde{\phi}^{\text{long}}(\mathbf{g})$ is the Fourier transform of the full potential $\phi(\mathbf{g}) = \phi^{\text{short}}(\mathbf{g}) + \phi^{\text{long}}(\mathbf{g})$ and

$$\hat{\phi}^{\text{screen}}(\mathbf{g}) = \bar{\phi}^{\text{long}}(\mathbf{g}) - \tilde{\phi}^{\text{long}}(\mathbf{g}). \tag{5}$$

Eq. (4) demonstrates that this formalism is equivalent to augmenting $\tilde{\phi}^{\text{screen}}(\mathbf{g})$ with a "screening potential" $\hat{\phi}(\mathbf{g})$ which cuts off the interactions from the periodic images of the simulation cell. This result leads to

$$\langle\phi\rangle = \frac{1}{2V}\sum_{\mathbf{g}} |\bar{n}(\mathbf{g})|^2[\tilde{\phi}(-\mathbf{g}) + \hat{\phi}^{\text{screen}}(-\mathbf{g})]. \tag{6}$$

Now, we consider the Coulomb potential, $\phi^{(Coul)}(\mathbf{r}) = 1/r$. Using an identity commonly employed in Ewald summation, the Coulomb potential can be written as

$$\begin{aligned}\frac{1}{r} &= \frac{\text{erf}(\alpha r)}{r} + \frac{\text{erfc}(\alpha r)}{r}\\ &= \phi^{(\text{long,Coul})}(r) + \phi^{(\text{short,Coul})}(r).\end{aligned} \tag{7}$$

where $\alpha$ is a convergence parameter which determines the region where the transition from short to long-range terms takes place.

The Coulomb potential for cluster systems in reciprocal space given by Eq. (4):

$$\bar{\phi}^{\text{Coul}}(\mathbf{g}) = \hat{\phi}^{(\text{screen,Coul})}(\mathbf{g}) + \tilde{\phi}^{(\text{Coul})}(\mathbf{g}). \tag{8}$$

The Hartree and local-pseudopotential terms in the DFT energy functional are of long-range nature. The Hartree term is given by

$$\begin{aligned} E_H[n] &= \frac{e^2}{2}\int_{D(V)} d\mathbf{r} \int_{D(V)} d\mathbf{r}' \frac{n(\mathbf{r})n(\mathbf{r}')}{|\mathbf{r}-\mathbf{r}'|} \\ &= \frac{e^2}{V}\sum_{\mathbf{g}}{}' |\bar{n}(\mathbf{g})|^2 \left[\frac{4\pi}{\mathbf{g}^2} + \hat{\phi}^{(\text{screen,Coul})}(\mathbf{g})\right] + \frac{e^2}{V}\hat{\phi}^{(\text{screen,Coul})}(0)\bar{n}(0) \end{aligned} \tag{9}$$

In a similar way, the local-pseudopotential term is obtained:

$$\begin{aligned} E_{local}[n] &= \sum_{I=1}^{N}\int_{D(V)} d\mathbf{r} V_{loc,I}(\mathbf{r}-\mathbf{R}_I)n(\mathbf{r}) \\ &= \frac{1}{V}\sum_{\mathbf{g}}{}'\sum_{I=1}^{N} \bar{n}(\mathbf{g})\exp(-i\mathbf{g}\cdot\mathbf{R}_I)[\tilde{V}_{loc,I}(\mathbf{g}) - eq_I\hat{\phi}^{(\text{screen,Coul})}(\mathbf{g})] \\ &\quad + \frac{1}{V}\sum_{I=1}^{N}\bar{n}(0)[\tilde{V}_{loc,I}^{(0)} - eq_I\hat{\phi}^{(\text{screen,Coul})}(0)] \end{aligned} \tag{10}$$

Here, the prime indicates that the $\mathbf{g} = 0$ term is eliminated and the term, $V_{local,\text{I}}^{(0)}$ is the nonsingular part of the local pseudopotential at $\mathbf{g} = 0$ and $q_{\text{I}}$ is the charge of atom I. The long-range term $\bar{\phi}^{(\text{long,Coul})}(\mathbf{g})$ is calculated with using a different supercell from that for the other terms (dual grid method) [4].

*2.2. Numerical details*

In our simulation, we used 4-bromophenol as a target system, in which a bromine atom and a hydroxyl group are located on the opposite sides of the carbon ring. Electronic states are obtained by using the projector-augmented-wave (PAW) method [5, 6], which is an all-electron electronic-structure-calculation method within the frozen-core approximation in the framework DFT. The generalized gradient approximation (GGA) [7] was used for the exchange-correlation function. The cutoff energies of plane waves were 30 and 250 Ry for the electronic pseudo-wave functions and pseudo-charge density, respectively. The energy functional was minimized using an iterative scheme [8, 9]. Projector functions were generated for the 2$s$ and 2$p$ orbitals of C and O, 1$s$ of H and 4$s$, 4$p$ and 4$d$ of Br. The equations of motion were integrated numerically using an explicit reversible integrator [10] with time step of 0.048 fs (hydrogen dissociation process) and 0.24 fs (carbon-ring breaking process). Initially, we began with carrying out an *ab initio* MD simulation of neutral molecule in the canonical ensemble using the Nosé-Hoover thermostat techniques [11, 12] for about 6 ps at 300 K. Then we picked several configurations from the MD simulation as initial configurations for MD simulations of the charged systems. The charge distribution within the molecule were determined by the self-consistent calculation in each MD step. Then the time evolution of the structure was obtained in the micro canonical ensemble.

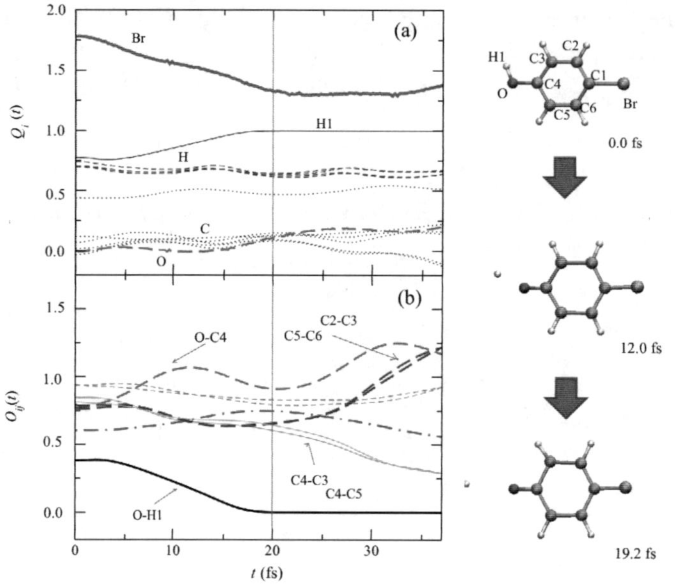


**Figure 1.** (Left panel) Time evolution of the gross charge (a) and bond overlap population (b) for $n$=6 (Right panel) Atomic configurations at t=0.0, 12.0 and 19.2 fs.

## 3. Result and discussion

When the charge state $n$ is more than 5, the molecule begins to break immediately after the molecule takes charges. From the simulation, it is found that the dissociation in this charge state region ($6 \leq n \leq 10$) occurs through a sequential process in a sense that the molecular dissociation occurs through the several-stage fragmentation. The first stage is hydrogen dissociation and the second stage is carbon-ring breaking. This dissociation process does not depend on initial configurations (we calculate each charge state with three different initial configurations); the molecular dissociation in all simulations occurs through hydrogen dissociation process and carbon-ring breaking process. However, we observed a sharp contrast in each process between the charge state is not more than 7 and over 7 . The typical examples of dissociation mechanisms of the +6 and +8 charged bromophenol molecules are shown as representatives in the following sections.

### *3.1. Hydrogen dissociaiton*

At times shorter than 20 fs after the molecule takes charges, Hydrogen dissociate from the molecule. In order to clarify the mechanism of this dissociation, we investigate the time evolution of the bonding and atomic charge nature by utilizing the population analysis [13, 14]. The gross charge $Q_i$ on the $i$th atom and the bond overlap population $O_{ij}$ for a pair of $i$th and $j$th atoms are given by

$$Q_i = Z_i - \sum_n \sum_{\mu \in i} \sum_\nu C^{\mu n} S_{\mu\nu} C_{\nu n}, \qquad (11)$$

$$O_{ij} = \sum_n \sum_{\mu \in i} \sum_{\nu \in j} \frac{1}{2} \Big[ C^{\mu n} S_{\mu\nu} C_{\nu n} + C^{\nu n} S_{\mu\nu} C_{\mu n} \Big], \qquad (12)$$

where $Z_i$ is the total number of valence electrons in the ground state of the free neutral atom and

$$S_{\mu\nu} = \langle \varphi_\mu | \varphi_\nu \rangle. \tag{13}$$

In this expression, $\varphi_\mu$ is the atomic wave function used as basis for the projection of the wave function of the interest $\chi_n$, and $C_{\mu n}$ is the coefficient of the projection, that is,

$$|\chi_n\rangle = \sum_\mu |\varphi_\mu\rangle C_{\mu n} \tag{14}$$

Figure 1 shows (a) the time evolution of the gross charge $Q_i(t)$ and (b) the bond overlap population $O_{ij}(t)$ with snapshots of atomic configurations when the charge state $n$ is 6. As displayed in the snapshot at $t = 12.0$ fs, a hydrogen atom in the hydroxyl group dissociates from the molecule. As shown in Fig. 1(b) $O_{\mathrm{O-H1}}$ (black solid line) becomes almost zero at $t = 20$ fs. During this dissociation, $Q_{\mathrm{H1}}$ (purple solid line) increases and becomes 1 at about $t = 20$ fs while $Q_{\mathrm{Br}}$ (magenta silid line) decreases with time till $t = 20$ fs. After this dissociation ($t > 20$ fs), $O_{\mathrm{C4-C3}}$ (green solid line) and $O_{\mathrm{C4-C5}}$ (cyan solid line) decrease while $O_{\mathrm{C2-C3}}$ (purple dashed line) and $O_{\mathrm{C5-C6}}$ (blue dashed line) increase with time. $O_{\mathrm{O-C4}}$ (red dashed line) also increases during this dissociation process.

When the charge state is 8, all hydrogen atoms dissociate from the molecule at the same time. This mechanism is shown in Fig. 2. $O_{ij}$ for all C-H bonds decrease with time and becomes almost zero at less than 20 fs. During this ejection, $Q_{\mathrm{Br}}$ (magenta silid line) decreases drastically with time. $Q_{\mathrm{O}}$ (red dashed line) also slightly decreases with time.

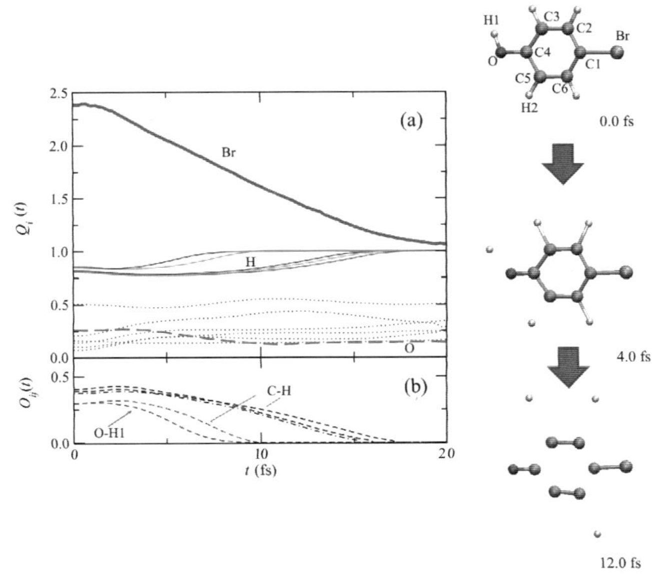


**Figure 2.** (Left panel) Time evolution of the gross charge (a) and bond overlap population for $n$=8 (b) (Right panel) Atomic configurations at t=0.0, 4.0 and 12.0 fs.

### *3.2. Carbon-ring breaking*

After the Hydrogen dissociation, the carbon ring breaks up into two fragments. In the case of $n$=6, C3-C4 and C1-C6 bonds break at about $t = 70$ fs and $t = 125$ fs, respectively, and the

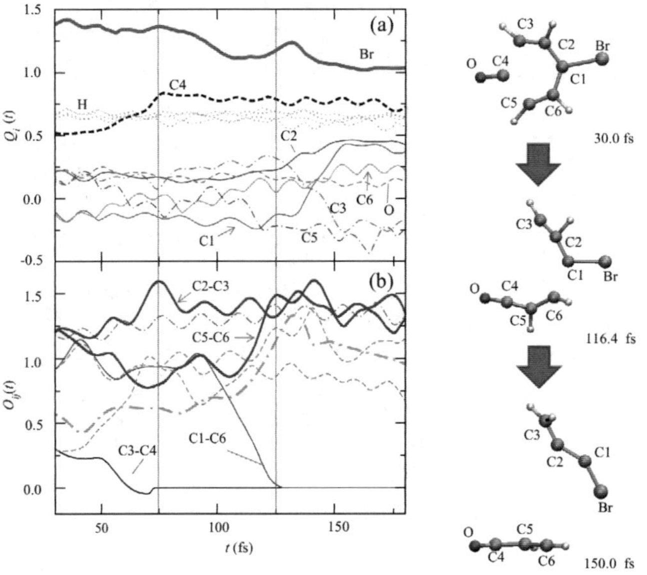


**Figure 3.** (Left panel) Time evolution of the gross charge (a) and bond overlap population (b) for $n$=6 (Right panel) Atomic configurations at t=40.0, 116.4 and 150.0 fs.

carbon ring splits into two fragments. These longer time behavior is shown in Fig. 3. The ionic charge of C4 (black dashed line) bonding to O becomes 0.75 after C3-C4 bond breaks. This value is the second largest after Br. This leads to a large Coulomb repulsive interaction between Br and C4. Therefore, Br atom is never located on the same fragment of O atom bonding to C4. After this fragmentation, the ionic charge of C2 (blue solid line) which does not have H atom and that of C6 (green solid line) which is the end of O-fragment have positive value. In the molecules with $n$=8, fragmentation occurs with a different mechanism from that for $n = 6$. This mechanism is shown in Fig. 4. During this fragmentation mechanism, some bond switching are observed. After C1-C6 bond breaks, new covalent bonds (C2-C6, C1-C3) are formed as displayed in the snapshot at $t = 70$ fs, and immediately after that, C1-C2 bond breaks. Through the bond switching process, the carbon ring becomes a chain-like structure as displayed in the snapshot at $t = 136.8$ fs. The Coulomb repulsive energy in the molecule is stronger than the covalent like interaction, the one of C-C or C-Br bonds is broken as seen in snapshot at 190 fs.

In order to clarify quantitatively the charge state dependence of the dissociation mechanism in the process of carbon-ring breaking, we calculate the time evolution of the areas of carbon ring $S_n$ at each initial charge state $n$, shown in Fig. 5. Here, we suppose that the area of carbon ring is equal to area of hexagon made of six carbon atoms. The residual charges after hydrogen ejection are noted in parentheses. The area of the carbon ring increases with time when $n = 6$ or 7. Since +5 charges are left on the molecule in these charge states, the ring expands due to the Coulomb repulsive force. On the other hand, $S_n$ decrease when $n$ is more than 7. Even though the residual charges of the molecule with $n = 10$ are the same of that with $n = 6$ or 7, the area decreases with time while $S_6$ and $S_7$ increase. This fact is attributed to the absence of C-H bonds in the molecule with $n \geq 8$. It is found from this figure that the carbon ring in lower charge state ($n < 8$) exhibits more Coulombic dissociation in a sense that the carbon ring breaks with simply expanding.

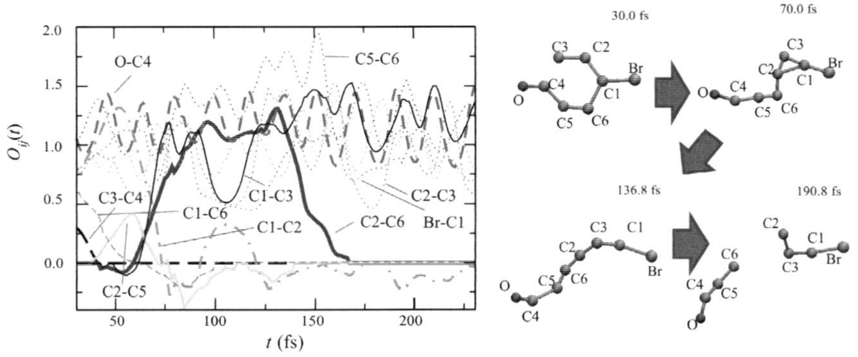


**Figure 4.** (Left panel) Time evolution of the bond overlap population for $n$=8 (Right panel) Atomic configurations at t=30.0, 70.0, 136.8 and 190.8 fs.

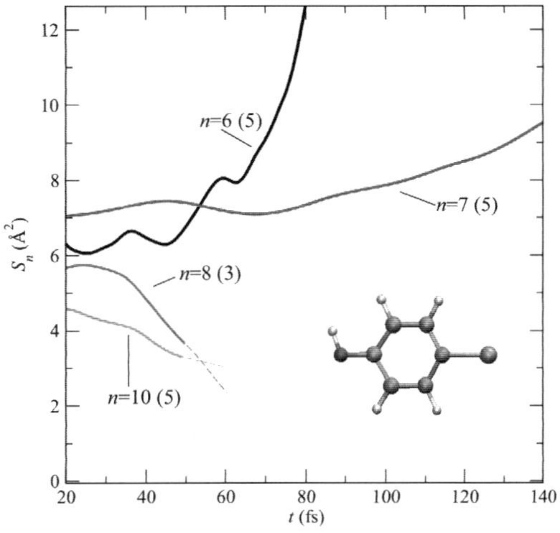


**Figure 5.** The time evolution of the area of aromatic ring $S_n$ after hydrogen ejection, where $n$ is the initial charge state.

## 4. Conclusion

We have investigated the dissociation mechanisms from highly charged bromophenol by *ab initio* MD simulations based on density functional theory with the cluster boundary condition method of Martyna and Tuckerman. It is found that dissociation mechanisms have clear charge state dependences. In the hydrogen dissociation at times shorter than 20 fs, hydrogen atom of OH group dissociates from the molecule with $n$=6 while all hydrogen atoms dissociate from the molecule $n$=8. In the process of carbon ring breaking, the charge state dependence is also observed. When the charge state is 6 or 7, the ring breaks with molecular expanding. On the other hand, molecular shrink during the ring breaking when the charge state is over 7.

## Acknowledgments

The authors acknowledge useful discussions with Professor Kazuhiro Matsuda at Kyoto University. This work was supported by KAKENHI [Grant-in-Aid for JSPS Fellows No. 24·5357, 23241033 and 23340106]. The computations were performed using the computer facilities at the

Supercomputer Center, Institute for Solid State Physics, University of Tokyo. The computations were also performed using the computer facilities at the Research Institute for Information Technology, Kyushu University for the use of its facilities.

## References

[1] Sugishima A, Nagaya K, Iwayama H, Yao M, Adachi J, Kimura Y, Yamazaki M and Yagishita A 2009 *J. Chem. Phys.* **131** 114309–6
[2] Nagaya K, Iwayama H, Sugishima A, Ohmasa Y and Yao M 2010 *Appl. Phys. Lett.* **96** 233101–3
[3] Martyna G J and Tuckerman M E 1999 *J. Chem. Phys.* **110** 2810–2821
[4] Yarne D A, Tuckerman M E and Martyna G J 2001 *J. Chem. Phys.* **115**
[5] Blchl P E 1994 *Phys Rev B* **50** 17953–17979
[6] Kresse G and Joubert D 1999 *Phys Rev B* **59** 1758–1775
[7] Perdew J P, Burke K and Ernzerhof M 1996 *Phys. Rev. Lett.* **77** 3865–3868
[8] Kresse G and Hafner J 1994 *Phys Rev B* **49** 14251–14269
[9] Shimojo F, Kalia R K, Nakano A and Vashishta P 2001 *Comput. Phys. Commun.* **140** 303–314
[10] Tuckerman M, Berne B J and Martyna G J 1992 *J. Chem. Phys.* **97** 1990–2001
[11] Nosé S 1984 *Mol Phys* **52** 255–268
[12] Hoover W G 1985 *Phys. Rev. A* **31** 1695–1697
[13] Mulliken R S 1955 *J. Chem. Phys.* **23** 1841–1846
[14] Shimojo F, Nakano A, Kalia R K and Vashishta P 2008 *Phys. Rev. E* **77** 066103

Journal of Physics: Conference Series **510** (2014) 012040 doi:10.1088/1742-6596/510/1/012040

# Validity of Molecular Dynamics for the Simulation of Soft Matter

**Sangrak Kim**

Department of Physics,
Kyonggi University,
154-42 Gwangyosanro, Youngtong-ku, Suwon 440-760, Korea

E-mail: srkim@kgu.ac.kr

**Abstract.** We examine the validity of molecular dynamics simulations for a simple one-dimensional system having a piecewise continuous linear repulsive potential wall with a constant slope $a$. The inevitable energy change $\Delta E$ that occurs in the collision with the potential wall is shown to be dependent on only two parameters $\alpha$ and $\mu \equiv a\tau/p_0$, where $\alpha$ is a fraction of the time step $\tau$ immediately after the contact with the potential wall and $p_0$ is the momentum just before the collision. The whole space of parameters $\alpha$ and $\mu$ can be divided into an infinite number of regions, where each region generates separately a positive or negative energy change $\Delta E$. The envelope of $|\Delta E|$ shows a power law behavior $|\Delta E| \propto \mu^\beta$ with the exponent $\beta \approx 1.08$. The average of $\langle|\Delta E|\rangle_\alpha$ over a uniform distribution of $\alpha$ shows a broken power law behavior $\langle|\Delta E|\rangle_\alpha \propto \mu^\beta$ with $\beta \approx 1.98$ at larger $\mu$ values and $\beta \approx 0.99$ at smaller $\mu$ values.

Hard core particles only feel repulsive impulses at the time when they collide. To simulate this system, we usually employ event-driven methods [1]-[2] to determine the time at which any two hard-core particles collide. At the time of collision, their longitudinal components of the velocities are exchanged, but their transverse components of the velocities do not change at all. Consequently, their total energy is completely conserved in the molecular dynamics simulations.

However, molecules usually have soft repulsive components in the potential function. This can be modeled by exponential or power functional forms [3]-[5]. For example, noble gases such as argons can be described by the Lennard-Jones potential

$$V_{LJ}(r) = 4\epsilon[(\frac{r}{\sigma})^{-12} - (\frac{r}{\sigma})^{-6}], \tag{1}$$

where $\epsilon$ and $\sigma$ characterize energy and length scales, respectively. In the simulation of a system with a soft potential [6]-[9], it is practically impossible to keep the energy of the system constant due to the inherent round-off errors involved in the digitization and truncation errors involved in the approximations of mathematical formula.

If the total energy of a system is different from the given initial total energy $E$, the system may sometimes show somewhat different thermodynamic and dynamic behaviors from what we originally aimed. Molecular dynamics corresponding to the NVE ensemble [10]-[11] generates phase trajectories moving on the constant energy surface $E =$ const. Therefore, it is important to know how the total energy $E$ depends on the control parameters of the system during the process of the molecular dynamics simulations.

To address this problem, we consider a very simple system as an example. The system is composed of a simple particle colliding with the soft potential wall in one-dimension. The soft potential wall is modeled as a piecewise linear potential:

$$V(q) = \begin{cases} 0, & \text{if } q \geq 0, \quad \text{(Region I)}, \\ aq, & \text{otherwise}, \quad \text{(Region II)}, \end{cases} \tag{2}$$

where $q$ is a coordinate of the particle and $a(< 0)$ is a constant that characterizes the slope of the potential. For simplicity, we take the unit mass of the particle. In region I where $q \geq 0$, the particle moves freely with the Hamiltonian $H(p,q) = \frac{p^2}{2}$. In region II where $q < 0$, it moves with the Hamiltonian $H(p,q) = \frac{p^2}{2} + aq$.

In this review, we summarize the previous results [12]-[13] about this problem and examine their implications. We present the analytical expression for energy change $\Delta E$ and its average effect over the initial conditions for NVE ensemble.

A velocity Verlet algorithm, one of the symplectic algorithms, is used to solve the Newtonian equation of motion. The equations for the algorithm for a particle moving from region I, with a coordinate $q_0$ immediately before the contact, to region II, with a coordinate $q_1$ immediately after the contact, are written as

$$\begin{cases} q_1 &= q_0 + \tau p_0, \\ p_1 &= p_0 - \frac{1}{2}a\tau. \end{cases} \tag{3}$$

For a particle moving in region II, we can write

$$\begin{cases} q_{n+1} &= q_n + \tau p_n - \frac{1}{2}a\tau^2, \\ p_{n+1} &= p_n - a\tau, \end{cases} \tag{4}$$

and thus, the trajectory is a parabola. The number $n_c$ denotes the number of steps immediately after the collision with the potential wall when the particle is leaving region II. Thus, $n_c - 1$ is the number of steps for the particle to stay in the potential wall during the collision. For $(n_c - 1)$-th step, we have

$$\begin{cases} q_{n_c-1} &= q_1 + (n_c - 2)\tau p_1 - \frac{1}{2}a\tau^2(n_c - 2)^2, \\ p_{n_c-1} &= p_1 - (n_c - 2)a\tau. \end{cases} \tag{5}$$

Furthermore, for a particle moving from region II, with a coordinate $q_{n_c-1}$ immediately before the exit of the potential wall, to region I with a coordinate $q_{n_c}$ immediately after the exit of the potential wall, the equations are written as

$$\begin{cases} q_{n_c} &= q_{n_c-1} + \tau p_{n_c-1} - \frac{1}{2}a\tau^2, \\ p_{n_c} &= p_{n_c-1} - \frac{1}{2}a\tau. \end{cases} \tag{6}$$

Finally, from Eqs. (3)-(6), we get

$$\begin{cases} q_{n_c} &= q_0 + \tau p_0 n_c - \frac{1}{2}a\tau^2 n_c(n_c - 1), \\ p_{n_c} &= p_0 - a\tau(n_c - 1). \end{cases} \tag{7}$$

From Eq. (7), $n_c$ can be calculated to be

$$n_c = ceil\left(\frac{1 + \frac{1}{2}\mu + \sqrt{(1 + \frac{1}{2}\mu)^2 + 2(\alpha - 1)\mu}}{\mu}\right), \tag{8}$$

where the function $ceil(x)$ denotes the minimum integer larger than $x$.

If we define $\Delta E$ as the relative energy change during the collision,

$$\Delta E \equiv \frac{{p_{n_c}}^2 - {p_0}^2}{{p_0}^2} = \mu(n_c - 1)(\mu(n_c - 1) - 2). \tag{9}$$

We can explicitly derive the analytical expression for an inevitable energy change $\Delta E$ due to the discrete process, which is only dependent on two parameters: (1) $\alpha \equiv 1 - \frac{q_0}{\tau p_0}$, which is a fraction of the time step $\tau$ immediately after the contact with the potential wall and (2) $\mu \equiv a\tau/p_0$, where $q_0$ and $p_0$ are the coordinate and momentum of the particle just before the collision with the wall.

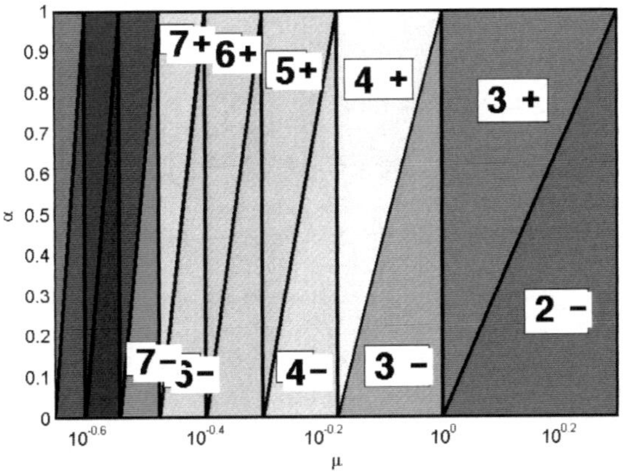


**Figure 1.** (Color on-line) Divisions of $\alpha$ and $\mu$ space up to $n_c = 10$. These divisions can continue up to $n_c \to \infty$. The cross points with abscissa are given by the formula $\mu = \frac{2}{n_c}, n_c = 2, \cdots, \infty$. All points on the vertical lines and diagonal lines, except $\alpha = 0$ show $\Delta E = 0$. Upper right-angled triangles show $\Delta E > 0$, denoted with the '+' symbol shown in the box. Lower right-angled triangles show $\Delta E < 0$, denoted with the '−' symbol shown in the box. The numbers shown in the box are the number $n_c$.

The whole space by the two parameters $\alpha$ and $\mu$ can be divided into an infinite number of regions where each region creates a positive or negative energy change $\Delta E$. Parameter $\alpha$ is in the range of $0 < \alpha \leq 1$, and $\mu$ is in the range of $0 < \mu \leq 2$ since $n_c$ is an integer greater than 1. From Eq. (8), for a fixed value of $\mu$, the number of collisions $n_c$ changes by 1 over the whole range of $\alpha$ from 0 to 1, since $n_c(\alpha = 1) - n_c(\alpha = 0) = 1$ for any value of $\mu$. If $\mu(n_c - 1) = 2$, then $\Delta E = 0$ for any value of $\alpha$. Thus, vertical lines $\mu = \frac{2}{n_c - 1}$ for $n_c = 2, \cdots, \infty$ divide the parameter space into an infinite number of subspaces, as shown in Fig. 1. Straight lines $\alpha = (n_c - 1)(\frac{1}{2}n_c\mu - 1)$ for $n_c = 2, \cdots, \infty$ divide the subspaces into two regions: upper triangles show a positive change in $\Delta E$, and lower triangles show a negative change in $\Delta E$. At points on the boundaries, $\Delta E = 0$. In Fig. 1, the regions are denoted with different colors and a pair of symbols including a number $n_c$ and a symbol, '+' or '-', up to $n_c = 10$.

Fig. 2 shows the results of $\Delta E$ as a function of $\mu$ for different values of $\alpha$. The upper bounds of individual values $|\Delta E|$ show a power law behavior $|\Delta E| \propto \mu^\beta$, with the exponent $\beta \approx 1.08$. This exponent governs the energy fluctuation from the round-off error at a discrete time step $\tau$. This implies that the round-off error in the energy, introduced by the discreteness, is nearly proportional to the time step $\tau$.

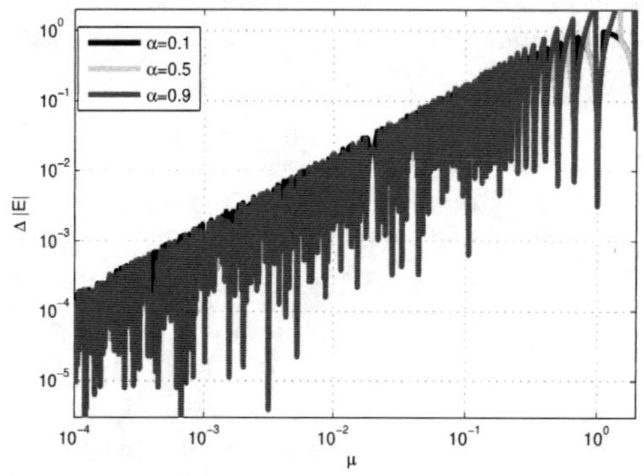


**Figure 2.** (Color on-line) Energy change $\Delta|E|$ *vs.* parameter $\mu$ along the horizontal lines in Fig. 1 for different values of parameter $\alpha = .1, .5$ and $.9$. The straight line corresponds to the envelope of upper bounds of $\log|\Delta E|$.

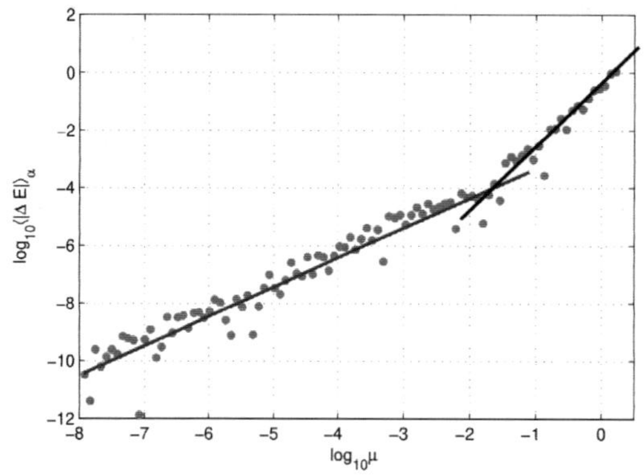


**Figure 3.** (Color on-line) Log-log plot of energy change $\langle|\Delta E|\rangle_\alpha$ *vs.* $\mu$. Averages are taken over a uniform distribution of parameter $\alpha$.

Let us now consider the analytical expression for the averages of energy change $\Delta E$ under appropriate distributions of the parameters $\alpha$ and $\mu$. Averages are taken along either the vertical lines (averages over $\alpha$) or the horizontal lines (averages over $\mu$ or $p_0$). $\langle|\Delta E|\rangle_\alpha$ is defined as the average of $|\Delta E|$ along a horizontal line with $\alpha$ = constant. In the simulations, we actually encounter many different realizations of $\alpha$; however, unfortunately, we do not know *a priori* the distribution of $\alpha$. Thus, we have to assume that the distribution of $\alpha$ is uniform on $(0, 1]$. The calculated results are shown in Fig. 3. To draw on the same side in Fig. 3, we take absolute values. As the value of $\mu$ decreases starting from $\mu = 2$ down to $\mu = 10^{-1.5} \approx 0.03$, the average $\langle\Delta|E|\rangle_\alpha$ decreases rather rapidly. Thus, it appears that they show two straight lines having exponents $\beta \approx 1.98$ at larger $\mu$ values and $\beta \approx 0.99$ at smaller $\mu$ values. The two straight lines intersect around $\mu \approx 0.03$. This implies that choosing a time step of $\tau \approx 0.03p_0/a$ can be thought of as an optimum for the simulations without generating large errors in the energy estimations.

In summary, the validity of molecular dynamics for soft matter has been examined analytically. The envelope of $\log|\Delta E|$ follow a power law with an exponent of 1.08. Furthermore, the average of $\langle|\Delta E|\rangle_\alpha$ shows a broken power law behavior $\langle|\Delta E|\rangle_\alpha \propto \mu^\beta$ having two exponents $\beta \approx 1.98$ at larger values of $\mu$ and $\beta \approx 0.99$ at smaller values of $\mu$. Thus, the error in the energy drift introduced by the discrete process is canceled significantly through the average process. Average over the initial conditions on $\alpha$ and $\mu$ can significantly increase the optimum time step up to several times. Although we have demonstrate the validity of molecular dynamics for a very simple system with a linear potential wall, the results can be generalized to a wide range of systems as well.

**Acknowledgment**
This research was supported by Basic Science Research Program through the National Research Foundation of Korea (NRF) funded by the Ministry of Education Science and Technology (2010-0011359).

## References

[1] Alder B and Wainwright T 1957 *J. Chem. Phys.* **27** 1208
[2] Haile J 1992 *Molecular Dynamics Simulation* (Wiley, New York)
[3] Frenkel D and Smit B 2002 *Understanding Molecular Simulation* (Academic, New York)
[4] Schlick T 2002 *Molecular Modeling and Simulation: An Interdisciplinary Guide* (Springer, New York)
[5] Griebel M, Knapek S and Zumbusch G 2007 *Numerical Simulation in Molecular Dynamics* (Springer, Berlin)
[6] Hairer E, Lubich C and Wanner G 2003 *Acta Numerica* **12** 399
[7] Cottrell D and Tupper P 2007 *BIT Numerical Mathematics* **47** 507
[8] McLachlan R and Perlmutter M 2004 *J. Phys. A: Math. Gen.* **37** L593
[9] Skeel R and Hardy D 2005 *J. Comput. Phys.* **206** 432
[10] Verlet L 1967 *Phys. Rev.* **159** 98
[11] Allen M and Tildesley D 1987 *Computer Simulation of Liquids* (Oxford University Press, Oxford)
[12] Kim S http:/arxiv.org.abs/1301.2768
[13] Kim S 2013 *Theo. Chem. Acc.* **132** 1332

# Ultrafast lasers and solids in highly excited states: results of hydrodynamics and molecular dynamics simulations

**Nail A. Inogamov [1], Vasily V. Zhakhovsky [2], Viktor A. Khokhlov [1], Sergey I. Ashitkov [2], Yusuf N. Emirov [3], Konstantin V. Khichshenko [2], Anatoly Ya. Faenov [2,4] Tatiana A. Pikuz [4], Masahiko Ishino [4], Masaki Kando [4], Nobory Hasegawa [4], Masaharu Nishikino [4], Pavel S. Komarov [2], Brian J. Demaske [6], Mikhail B. Agranat [2], Sergey I. Anisimov [1], Tetsuya Kawachi [4], Ivan I. Oleynik [7]**

[1] Landau Institute for Theoretical Physics, Russian Academy of Sciences, Russian Federation
[2] Joint Institute for High Temperatures of Russian Academy of Sciences, Russian Federation
[3] Advance Materials Research Institute, Florida International University, USA
[4] Quantum Beam Science Directorate, Japan Atomic Energy Agency, Japan
[5] University of Florida, USA
[6] Department of Physics, University of South Florida, USA

E-mail: nailinogamov@gmail.com (N. Inogamov)

**Abstract.** Action of ultrafast optical and X-ray lasers on metals is considered. It is known that under certain conditions surface structures appear as result of irradiation. Generation of nano-structures is usually associated with excitation of surface plasmons. But often structures do not have forms of ripples, and their spacial scales are order of magnitude less than optical wavelength. In the paper full description of surface nano-structures is given for the case of single shot laser action onto well polished boundaries. Plasmon effects are insignificant for this case and also for X-ray pulses. It is shown that structures are formed after laser illumination in a process of mechanical spallation of ultrathin surface layer of molten metal. Spallation is accompanied by a strong foaming of melt, breaking of foam, and freezing of foam remnants. Those remnants form chaotic nano-structures observed in experiments.

## 1. Introduction

Lasers with ultrashort pulse (durations $\tau_L$ from $\sim$ 10 fs to few ps) have a wide range of technological and bio-medical applications. There are applications where only ultrafast lasers solve problem. E.g., those lasers are irreplaceable in the studies of elastic-plastic [1, 2, 3] or polymorphic transformations near atomistic limit, and in studies of picosecond [4] chemical reactions (detonation under action of very high and very short mechanical load). Physics of energy transfer from photons to material heating and motion is not trivial. In metals we have to consider a two-temperature (2T) state [5] to understand origin of subsequent phenomena. Density functional theory (DFT) is necessary to approximate quantum electronic properties used in 2T thermodynamic and transport calculations. Hierarchy of scales from fs to ns have to be taken into account to follow formation of surface structures directly from a 2T stage.

In the paper we compare ultrafast actions of optical and X-ray lasers. Previously people have attributed structure formation solely to an interference between incident electromagnetic wave and surface plasmons [6, 7] (ripples, spacing between ripples $\sim 1$ um). Another insight starts from a paper [8]; it was shown that thermomechanical effects (nucleation and spallation) are responsible for structure formation after small number of shots $N$ of optical lasers $h\nu \sim 1$ eV onto well polished surfaces (the case of random structures, spatial scales $\sim d_T$ - thermal depth). Plasmons are absent in case of X-rays. Therefore thermomechanics is responsible for structure formation under action of ultrashort X-ray pulse at any number of shots. In future a hybrid approach including both a plasmon interference and a thermomechanics will be developed. It will allows to describe universally the cases with an arbitrary set of parameters $N$, $h\nu$, and incident fluence. In this paper we limit ourself to a thermomechanical ablation.

The mentioned spatiotemporal hierarchy is shown in Fig. 1. Electron thermal wave emits compression wave "compr" at a transonic stage. At a 2T stage a trajectory of the thermal wave corresponds to the bottom-right edge of the hatched region in Fig. 1(a). Compression wave runs away into the bulk, while a thermal layer remains near boundary. Gradually compression wave breaks and a shock wave appears. This event takes place at the right side relative to an area shown in Fig. 1. It isn't shown. Slightly above ablation threshold $F_{abl}$ the nucleation begins in a narrow layer near the point where a reflected acoustic wave intersects with compression wave running to the left, see Fig. 1 (a). Two lines going up from the intersection point of acoustic characteristics in Fig. 1 (a) present the left and right edges of a layer filled by two-phase mixture. Appearance of mixture strongly decreases dynamic and thermal contacts between the edges.

Significantly above an ablation threshold $F_{abl}$ a wide nucleation zone 2-3 appears, see Fig. 1 (b). Above evaporation threshold $F_{ev}$ the condensed layer 1-2 (spallation plate) disappears [9]. In the interval of fluences $F_{abl} < F < F_{ev}$ the layer 1-2 consists from hot liquid. The points a, m, and f in Fig. 1 (b) correspond to an instants
(a) When the last characteristics of compression wave leaves the thermal layer.
(m) Maximum penetration of melting front. In this point melting changes to solidification.
(f) Total freezing of molten layer between the curves 3 and 4 in Fig. 1 (b), see Section 3 below.
Molten metal in the layer 1-2 remains hot for a long time after freezing of the right edge.
Liquid in the point f freezes during $\sim$ns time interval. Near the point f the random surface structures form if breaking of foam in vicinity of the edge 3 in Fig. 1 (b) lasts longer than solidification of melt between the lines 3-4, see Section 3 below. We see that a process of formation of surface structures covers many orders of magnitudes - 7 orders if duration of pulse is 100 fs.

Short pulse laser ablation is better understood in case of optical lasers. In this case on metals the ablation threshold $F_{abl}$ is 2-3 times higher than a melting threshold $F_m$. Therefore if we are interested to study active motions[1] we have enough energy to melt a solid fast. In this case nucleation shown in Fig. 1 takes place in liquid. We can regard the bottom boundary of a 2T region and curve 4 in Fig. 1 (b) as a trajectory of a melting zone. Melting zone is filled by solid-liquid mixture. This zone is wide at a 2T stage, and becomes much more narrow at a 1T stage. Ablation by X-ray lasers has similarities and differences with optical lasers, we will discuss them below. Here let's mention few features. In metals the radiation energy of optical lasers $h\nu \sim 1$ eV is absorbed in intraband and interband transitions in the upper electronic shells. This energy almost entirely transfers to conduction electrons. Then 2T state appears since electrons are excited during shorter time interval $\tau_L$ than they will cool later $t_{eq}$. During the 2T stage the electrons gently transfer energy to ions: $h\nu \to T_e \to T_i$. This transfer proceeds through small steps of numerous e-i collisions. During the interval $t_{eq}$ the electrons are hot, and thermal energy spreads to a spatially significant depth $d_T$ many times larger than a skin depth.

[1] This needs $F > F_{abl}$, more exactly $F > F_{nucl}$, where $F_{nucl}$ is a nucleation threshold. The $F_{nucl}$ is 1-10% below $F_{abl}$.

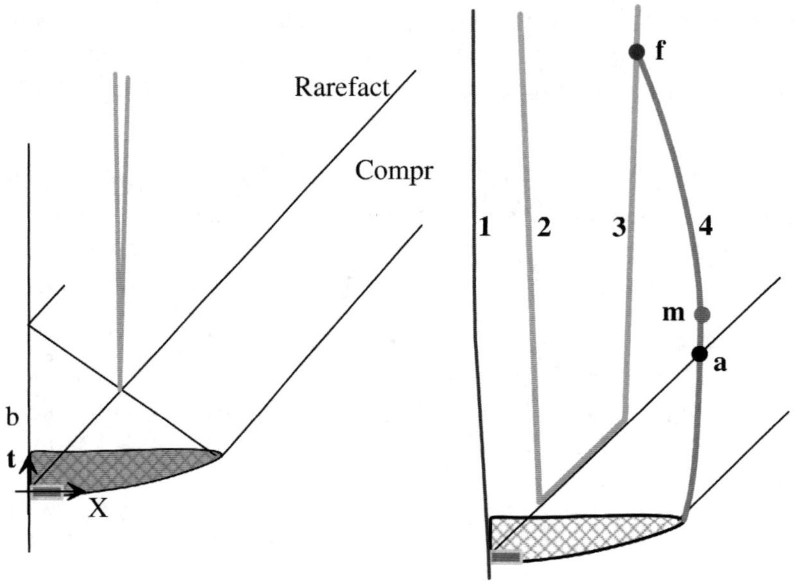


**Figure 1.** The $x, t$ diagram of sequence of processes initiated by ultrashort pulse. Time is reckoned from an instant when pulse intensity at a vacuum-target boundary "b" achieves maximum. Distance along a normal $x$ to the boundary is reckoned from initial (prior to pulse) position of a boundary. (a) The red rectangular is an absorption region $\tau_L \times \delta$, where $\delta$ is a skin depth for optical irradiation and an attenuation depth for X-rays. The rectangular where electrons were strongly excited is placed inside 2T region where system gradually returns to electron-ion equilibrium $T_e = T_i$. This is the hatched region with spatiotemporal dimensions $t_{eq} \times d_T$, where $t_{eq}$, $d_T$ are duration of temperature equilibration and thickness of thermal layer created during 2T stage, resp. Velocity of heat penetration is higher than speed of sound $c_s$ at a 2T stage. This velocity sharply drops down below $c_s$ during transition from 2T to 1T (one-temperature) stage. Compression nonlinear acoustic wave "compr" separates from a thermal layer during this transit. (b) The lines 1, 2-3, and 4 correspond to a vacuum boundary, edges of a mixture layer, and to melting-recrystallization front, resp.

This increases $F_{abl}$ since density of energy per unit volume necessary to overcome material strength weakly depends on value of $d_T$, see paper [10] where has been shown that $F_{abl} \propto d_{film}$ for thin films $d_{film} < d_T$.

In case of hard photons situation in metals is different. Very significant part of energy $h\nu$ directly transfers to ion energy [11]. Absorption of energy $h\nu$ by ion ionizes electron. This electron escapes from the ion thus changing charge of this ion. Ion with changed charge and its ion surrounding "immediately" (if we compare the transition period to an ion time scale which is an inverse Debye frequency $\sim 0.1$ ps) increase potential energy of their interaction ("immediately" means during the time of escape $\sim 1$ fs of an ionized electron from the ion). Fraction of this potential energy during the ion time scale transits to kinetic energy of ion motions. This rises internal energy and pressure of ion subsystem [11]. Such direct transfer $h\nu \to T_i$ decreases electron energy at a 2T stage. Thus decreasing importance of thermal layer created be electron heat conduction in metal at a "standard" 2T stage. Instead a thermal layer of directly heated ions appears. Its thickness is defined by attenuation depth $d_{att}$ of hard photons. The depth $d_{att}$ may be significantly thinner than a thermal depth $d_T$ linked with electron conduction. Expansion of heat from the layer $d_{att}$ is restricted since holes are sitting at deep electron shells (therefore tunneling probability is low) while mobility of ions typically is small.

Also additional heating of electron subsystem from ions from attenuation depth is small since temperature of ions is small in comparison with electron temperatures. Therefore spreading of ion energy from the depth $d_{att}$ through usual electron heat conduction is small. At comparable energies stored in electron subsystem and directly in ions the energy density per unit of volume may be significantly higher in the layer $d_{att}$. This will decrease ablation threshold on absorbed energy and ablation depth at a threshold in case of hard photons in comparison with optical lasers.

Let's mention that a thermal depth in dielectrics should be of the order of the attenuation depth $d_{att}$ since electron heat conduction is small. Let's also mention that it is difficult to study a range of intermediate absorbed energies in optical ablation of dielectrics because they are transparent and absorption begins only after optical breakdown when matter is too hot. While in the X-ray case the depth $d_{att}$ is fixed and small, matter is non-transparent and absorbing, and we can continuously create all states from below melting threshold $F_m$ and up to $F_{abl}$ and higher varying fluence $F$. Then regimes with foaming of molten dielectrics should exist. But it is difficult to freeze those foams because conduction cooling is slow.

## 2. Hydrodynamic expansion driven by electron pressure

2T states obeys 2T equation of state (EOS). This is a three dimensional EOS $T_e, T_i, \rho$. Pressure cold curves $p(\rho, T_e = 0, T_i = 0)$ for nickel together with a low temperature 1T isotherm $T = 0.529$ kK are shown in Fig. 2 (a). DFT cold curves satisfactory agree with each other and with the curves obtained from a wide-range EOS. DFT and EOS agree up to the point where pressure sharply increases to a value $p = 0$ on the EOS curve. This is a point of minimum of the (EOS, $T = 0$) curve in Fig. 2 (a). This point corresponds to the point in Fig. 3 (a) where continuation of spinodal, liquidus, and solidus touch the $T = 0$ axis. Fig. 2 (b) illustrates how excitation of electron subsystem increases total pressure. This increase is obvious since we see that pressure $p(\rho, T_i = 0, T_e)$ at equilibrium density $\rho = 8.9$ g/cc rises from zero at $T_e = 0$ to $\approx 75$ GPa at $T_e = 2.5$ eV. Similar DFT calculations of pressures of 2T crystals with cold ions (fixed in their lattice positions) was performed in works [12] (W), [13] (Ni), and [14] (Al,W).

Phase diagram of 1T EOS [15, 16] is shown in Fig. 3 (a). Two-phase region outlined by binodal consists from subregions of metastability and absolute instability separated by spinodal. Solidus and liquidus are prolonged through binodal in Fig. 3 (a). Solidus and liquidus are boundaries of solid-liquid mixture. It was found [16] that in most of metals liquidus intersects spinodal before the spinodal achieves the zero temperature limit. Aluminum is an example where liquidus and spinodal both independently reach the zero temperature limit. A zone of solid-liquid mixture can not exist if one of its boundary disappears. Therefore it is hypothesized that the spinodal achieves zero temperature at smaller degree of stretching (along a continuation of solidus in Fig. 3 a). Therefore a minimum of a cold curve taken from the 1T EOS in Fig. 2 (a) is achieved significantly earlier than the minimum of the pressure well obtained by DFT or taken from [17].

Our molecular dynamics (MD) operates with embedded atom model (EAM) interatomic potentials describing many-body interactions. Our EAM potentials for Al and Au were developed in paper [19], EAM potential for nickel has been obtained in [20]. Electron temperature dependent EAM have been constructed based on DFT results obtained on a dense grid of $(\rho, T_e)$ points. Electron pressure corresponding to this EAM is shown in Fig. 3 (b). 2T hydrodynamics code is based on a 2T EOS. In this 2T EOS the total pressure is presented as $p(\rho, T_e, T_i) = p_i(\rho, T_i) + p_e(\rho, T_e)$, where $p_e$ is found from equation $p_e(\rho, T_e, T_i = 0) = p(\rho, T_e, T_i = 0) - p(\rho, T_e = 0, T_i = 0)$ and it is supposed that $p_e(\rho, T_e, T_i)$ weakly depends on $T_i$, here $p$ is total pressure. In 2T EOS the ion pressure $p_i(\rho, T_i)$ is taken as pressure $p(\rho, T)$ from 1T EOS (without electronic Fermi type addition used in wide-range EOS) with substitution of $T_i$ in place of temperature $T$. In 2T EOS the electron pressure is

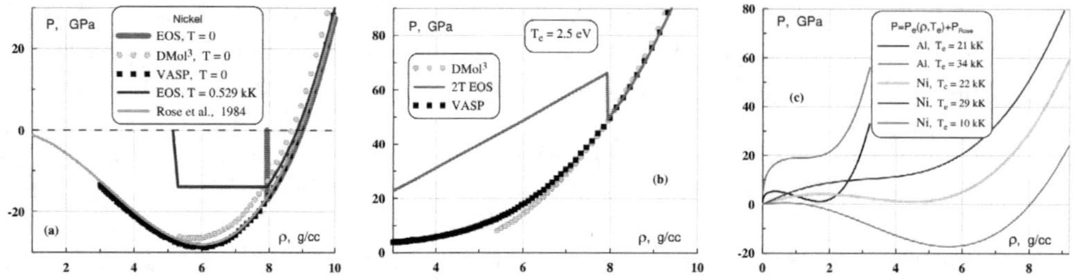


**Figure 2.** (a) Cold curves from density functional theory (DFT), wide-range 1T EOS [15, 16], and from analytical approximation [17]. DMol$^3$ (accelrys materials-studio) and VASP [18] packages are used for DFT calculations. (b) Dependence of total pressure of system on density at fixed ion $T_i = 0$ and electron $T_e = 2.5$ eV temperatures. (c) Isotherms of total pressure $p(\rho)$ for $T_i = 0$, $T_e =$fixed. They are calculated as a sum $p = p_e(\rho, T_e) + p_{Rose}(\rho)$, where $p_e$ is electron pressure and $p_{Rose}$ is taken from [17]. We see that there is an interval $0 < T_e < T_{min}$ where the profiles $p$ have minimums; $T_{min} = 32$ kK (2.7 eV) for Al and 26 kK (2.25 eV) for Ni. The minimum degenerates into an inflection point at $T_e \to T_{min} - 0$. There is also important subinterval $T_+ < T_e < T_{min}$ where $p$ in the minimum is positive; $T_+ = 20.5$ kK (1.8 eV) for Al and 21 kK (1.8 eV) for Ni. Isothermal ($T_e =$const) expansion flow has an equilibrium point on a right branch of dependence $p(\rho, T_e)$ if $0 < T_e < T_+$. Situation with a rarefaction jumps becomes possible in case $T_+ < T_e < T_{min}$, see text for explanations.

approximated as [13] ($\rho$ in g/cc, $T_e$ in K, approximation is based on DFT calculations)

$$p_e(\rho, T_e) = 1.1(\rho/8.91)^{1.1} KS(K+S)^{-1}, \qquad K = 1077\, T_e^2/2, \qquad S = 1.275 \cdot 10^5\, T_e^{1.3}. \qquad (1)$$

This approximation for $T_e = 2.5$ eV is presented as a curve "appr" in Fig. 3 (b).

There is an important difference between EAM and 2T EOS. The 2T EOS is based on 1T EOS. Approximation of metastable region in the 1T EOS shown in Figs. 2 and 3 (a) is influenced by measurements done at rather large spatiotemporal scales $\sim$mm and $\mu$s. Therefore the left slope of a "well" in a cold pressure profile "2T EOS" in Fig. 2 (a) is steep. In atomistic situation this slope should be more smooth, it should be more close to DFT data and to the curve "Rose et al." [17] as it is shown in Fig. 2 (c). The main question is: are there a well at a pressure profile when a whole profile is above the axis $p = 0$? Of course, there is a well in a case of a steep left slope at any electron temperature. Fig. 2 (c) demonstrates, first, that there are cases with a smooth well and, second, that there are cases with positive total pressure in the minimum of a well. Why this question is significant for us? We are searching for regime of *gaseous* rarefaction under action of dominating electron pressure. In this case the metal outflow is composed from atoms with excited electrons since matter remains in 2T state. Then a possibility to obtain macroscopic hydrodynamic *source of excited atoms* appears.

There are two scenarios. In the first of them the sharp rarefaction jump (RFJ) appears. Its existence is obvious from Figs. 3 (c) and 4. In the second scenarios there are or less productive[2] 2T evaporation from a boundary of 2T metal, or continuous flow without a rarefaction jump. Our 2T-HD code presents flows with a vacuum boundary if surface temperature $T_e$ is below the limit $T_+$ described in a caption to Fig. 2 (c). And in 2T-HD we always have a rarefaction jump above this limit since in a case of a well with steep left slope (see Fig. 2 a) we have a feature in the form of a well on the dependence $p(\rho)$ shown in Fig. 2 (b) at any temperature $T_e > T_+$; because the temperature $T_{min} = \infty$ for this case with steep slope. Corresponding

[2] In comp. with RFJ in the sense of amount of excited atoms.

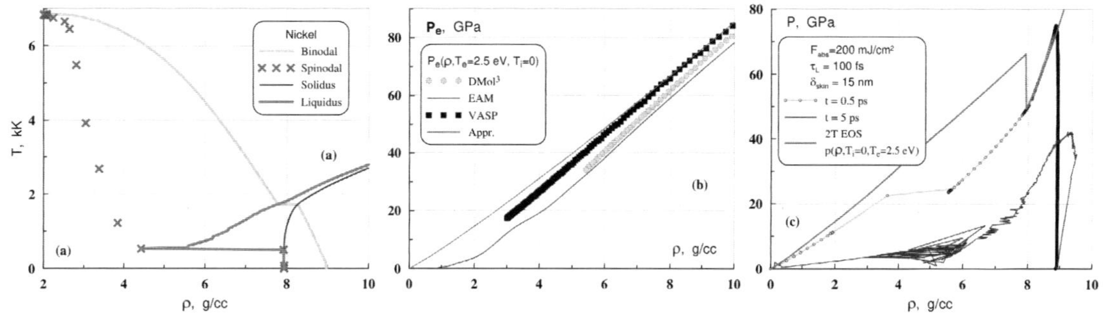


**Figure 3.** (a) The $\rho, T$ plane of a 1T EOS. The binodal bounds a two-phase region where equilibrium vapor contacts with condensed phase. A strip of a melting zone is continued inside a metastable region. Cold curves shown in Fig. 2 (a) go along the horizontal axis $T = 0$. (b) DFT data and their approximation with help of EAM and simple analytic approximation "appr." (1) taken from [13]. Electron pressure $p_e$ is defined from equation $p(\rho, T_e, T_i = 0) = p(\rho, T_e = 0, T_i = 0) + p_e(\rho, T_e, T_i = 0)$. (c) 2T EOS isothermal curve from Fig. 2 (b) and two instant profiles obtained by 2T hydrodynamics (2T-HD) code in case of nickel. Instant spatial distributions $\rho(x, t_{fix})$ and $p(x, t_{fix})$ are presented here as a parametric plot. In metal far from vacuum boundary we have $\rho \to \rho_0 = 8.9$ g/cc, $p \to 0$. Pressure increases when we approach to the boundary and after that decreases in the rarefaction flow. The first part of the rarefaction decrease of $p$ follows 2T EOS isothermal curve at the distributions corresponding to an early instants.

hypothetical flow is shown in Fig. 4. There are four main parts of this flow. They corresponds to profile $t = 0.5$ ps shown in Fig. 3 (c). The first part is an isochoric rise of pressure from initial value $p = 0$. It locates before the point where velocity changes sign in Fig. 4 (b) and where the decreases of pressure and density begin in Fig. 4 (c). In the second part the matter of nickel expands along the right slope of a well in Fig. 2 (b). The third part corresponds to a jump. It connects two points: one in the minimum of a well and other point is outside the jump. The fourth part corresponds to a freely expanding flow outside the jump. During the 2T stage the 2-3 monoatomic layers pass the jump for the conditions of pulse shown in Figs. 3(c) and 4.

Let us emphasize that as was said above we are interested here in a rarefaction flow going from a surface and on early 2T stage. We will not consider stretching of matter as result of deceleration of boundary, creation of tensile stress, nucleation, and spallation. Indeed, early rarefaction covers a small part of a heated layer $d_T$. At this stage $t \ll t_s = d_T/c_s$ and deceleration isn't significant. Spallation under influence of electron pressure has been considered in papers [21, 22]. see also papers [23, 24] where pressure $p_e$ was presented as a blast force.

We run MD simulations with EAM potential to check microscopic picture connected with expansion under action of electron pressure. Special EAM potential of nickel including electron temperature effects has been developed. This development is based on DFT calculations shown in Figs. 2 (a,b) and 3 (b). Resulting pressure dependencies $p(\rho, T_e = \text{fixed}, T_i = \text{fixed})$ (isotherms) are shown in Figs. 2 (c) and 3 (b). Figs. 2 (c) and 3 (b) present total pressure and electron contribution to total pressure, resp. Dependencies for Al are obtained in a Fermi model for conduction electrons, while dependencies for Ni are based on a set of DFT simulations. There are two approximations of pressure $p_e$ shown in Fig. 3 (b) both based on the DFT simulations. One is a simple approximation (1) taken from [13]. It is shown as the curve "appr." in Fig. 3

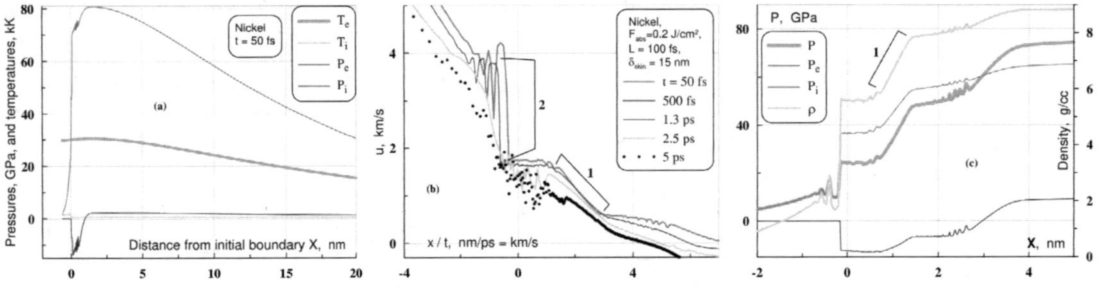

**Figure 4.** (a) Electron temperature and pressure dominate at an early 2T stage. Their distributions are shown at the end of laser pulse $t = 50$ fs; duration of pulse is $\tau_L = 100$ fs, see caption in Fig. 3 (c) where parameters of a laser pulse are given. Time is reckoned from a maximum in time of a laser pulse. Let us emphasize creation of significant tensile ion stress. This is a result of expansion stretching driven by huge electron pressure. (b) Velocity jump corresponding to RFJ is marked by digits 1 and 2. The head of a rarefaction wave runs with velocity $\approx 5-7$ km/s. At the end of a 2T stage this velocity drops to 4-5 km/s - speed of sound in Ni. (c) Pressure and density jumps, comp. with Figs. 4 (b), 3 (c) and 2 (b).

(b). Other is a curve "EAM" in Fig. 3 (b). It is calculated from the EAM potential. EAM potential is presented as a multi-parameters rational function minimizing deviations of pressure and energy surfaces from the dense set of the data points on a $\rho, T_e$ plane obtained thanks to DFT. Simplex approach based on $n$-dimensional geometrical figures (program ameba) is used for optimization of $n$ (few tens) parameters of the rational function.

The curves "appr." and "EAM" in Fig. 3 (b) are slightly different. Therefore their $\rho, p$-isotherms also differ. The $\rho, p$-isotherms corresponding to the approximation (1) for Ni and the curve "appr." in Fig. 3 (b) are shown in Fig. 2 (c). Isotherms for Ni are obtained by summing the approximation (1) and expression for cold pressure from the paper [17]. This family of isotherms has an interval $T_+ < T_e < T_{min}$ smaller than the same interval for Al. In degrees of softness of the pressure well on the isotherm there are 2T EOS (the most stiff well), Al in Fig. 2 (c), Ni in Fig. 2 (c), and Ni based on the EAM potential. The most soft well corresponds to the case of Ni described by EAM potential (a curve "EAM" in Fig. 3 b). It has the shortest interval $(T_+, T_{min})$, the minimum of a well goes far away in direction of small densities at the axis of $\rho$ before the minimum passes into halfspace $p > 0$.

The RFJ may appear in the interval $(T_+, T_{min})$. Our set of electron temperatures $T_e$ where MD simulations have been performed covers a range from 0 to 3 eV with a step 0.5 eV. We see that the value 2.5 eV is below the interval while the value 3 eV is above the interval. A shallow well means weak cohesion energy. A shallow well forms under stretching action of electron pressure. Metal with small cohesion is more volatile - evaporation proceeds more intensively at lower ion temperatures. Why the pressure well and the interval $(T_+, T_{min})$ are important are discussed in Fig. 5 (a). Initial state of Ni $rho = 8.9$ g/cc, $p = 0$ is shown as the point 1 in Fig. 5 (a). Isochoric heating rises pressure to a value corresponding to temperature $T_e$ - arrow 2. After that expansion begins. It decreases pressure along the arrow 3 in Fig. 5 (a). There are three cases in Fig. 5 (a) corresponding to different temperatures $T_e$. In the first case temperatures $T_e$ belongs to the interval $(0, T_+)$ - there is a minimum and it is below the axis $p = 0$. Then expansion stops in the point 4 when positive electron pressure and negative ion pressure balance each other.

The most interesting is the second case where temperature $T_e$ belongs to the interval $(T_+, T_{min})$. The isotherm for this case is shown in Fig. 5 (a) as a blue curve with a minimum

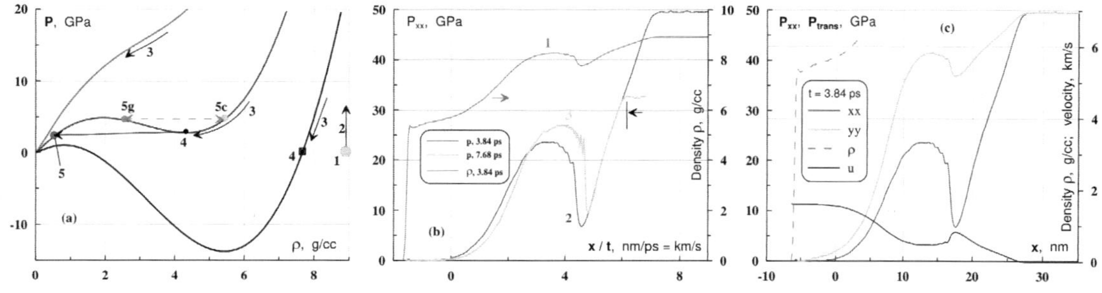


**Figure 5.** (a) Nickel $\rho, p$-isotherms taken from Fig. 2 (c) for other set of temperatures $T_e = 2.9, 2,$ and 1.1 eV. Dynamics of isothermal expansion depends on the particular isotherm along which matter expands. There are three qualitatively different ways of expansion depending on value $T_e$. These ways are shown here, see text for explanations. (b) Expansion of a fcc single nickel crystal 100 nm thick along direction 111. The left side of an expansion flow is shown. Point $x = 0$ is placed at the initial position of the left boundary of a film. Electron temperature $T_e \equiv 2$ eV distribution is homogeneous across the film. We suppose that values of $T_e$ is conserved in material particles during their motions because we consider an early stage. The left side expansion flow is self-similar $\rho(x,t) \to \rho(x/t)$ until arrival of the right side rarefaction wave. Approaching of the right wave is marked by the black arrow. (c) Unloading of transverse stress and liberation of transverse elastic energy. We see that transverse stress decays much slower than longitudinal stress during elastic uniaxial stretching. Therefore significant potential energy remains in transverse stresses. While liberation of longitudinal elastic energy accelerates matter at a first part of rarefaction. Drop of density on a density profile marks the vacuum-nickel boundary.

above the axis $p = 0$. A material particle near a vacuum boundary starts from the point 1, rises pressure along the arrow 2, begins to expand along the arrow 3, and arrives to the minimum 4. After that it has to expand further since pressure in the minimum 4 is higher than vacuum pressure. There are two ways to expand. In the first way the expansion proceeds along $4 \to (5_g, 5_c)$ - the condensed media decays into mixture of low and high density phases $5_g$ and $5_c$ due to acoustic instability, see Fig. 5 (a). In the other way from the minimum 4 is a rarefaction jump $4 \to 5$ decreasing pressure and density and increasing velocity in a style shown in Fig. 4 (b,c). It seems that the RFJ will exist in the case of a soft well shown in Fig. 5 (a) as it exists in a case of a well with a steep left slope. At last the high temperature state in Fig. 5 (a) freely expands as a gas along the path 1-2-3.

MD simulations performed with EAM with short or absent interval $(T_+, T_{min})$ are shown in Fig. 5, 6. We have began MD studies to understand a microscopic picture accompanying intriguing phenomenon of a RFJ - rarefaction shock. But instead the MD results shed light to another surprising phenomenon - the two-stage rarefaction when longitudinal stress release is followed by a spatially distant release of transverse stress. Longitudinal unloading is simply caused by expansion into vacuum, while the transverse release runs in direction from the vacuum-metal boundary into bulk. The transverse release propagates as a wave against an expansion flow. Thus lattice contribution manifests itself through this two-stage release. We must take into account that crystal survives a short period of time under conditions of heating from excited electrons. It will be interesting to consider an expansion of a 2T molten metal film in future MD simulations.

In Fig. 5 (b) expansion of a solid film under action of pressure $p_e$ overcoming atomic cohesion is shown. Profiles for two time instants are shown against self-similar axis $x/t$. We

Journal of Physics: Conference Series **510** (2014) 012041 doi:10.1088/1742-6596/510/1/012041

see that simple rarefaction profiles coincide. While the remarkable knoll is accelerated to the right, comp. longitudinal stress profiles 2 and 3. This knoll presents the right edge of a plastic wave propagating to the right side along the rarefaction wave. Propagation of plastic wave is supported by release of elastic energy in the space interval between the plastic front and the $p = 0$ layer near a vacuum boundary. Release of elastic energy is powerful enough to decrease expansion velocity, see Fig. 5 (c), and prevents drop of density - density even increases, see Fig. 5 (b). Fig. 5 (c) demonstrates two-step expansion. This is similar to the case described in papers [25, 26]. In the first step the matter accelerates under gradient of a longitudinal stress. During this acceleration the transverse elastic energy transforms into kinetic energy of expansion. At the first step longitudinal stress drops down much more than transverse stress. Transversely loaded state is unstable. In the second step the plastic transformations releases potential accumulated in transverse directions. This energy unloading send compression wave in both directions: downstream and upstream. The upstream wave as was said decelerates velocity of expansion and even rises density (!), see Fig. 5 (b). While the downstream part additionally accelerates nickel in direction to vacuum, see Fig. 5 (c) with a two-step velocity profile.

## 3. Freezing of nanostructures

Above the early 2T stage has been considered. Later equilibration of temperatures of electron and ion subsystem takes place, see Fig. 1. Typically duration of equilibration $t_{eq}$ is shorter than acoustic time scale $t_s = d_T/c_s$. Thus hydrodynamic motions on the scales $\sim d_T$ and $\sim t_s$ take place in 1T metal. Below ablation threshold $F_{abl}$ the back characteristics of compression wave running to the right and the back characteristics of compression wave running to the left intersects in the middle of a heated layer $d_T$, see Fig. 1 (a). The strip "Compr" in Fig. 1 (a) corresponds to the right compression wave. The strip "Rarefact" is filled with a reflected wave $p < 0$ running to the right. It forms during reflection of the left compression wave from a vacuum boundary. This reflection changes sign of stress. Shift of a boundary stops after reflection of the back characteristics of the left compression wave. Contact between the fluxes of the right running characteristics of the compression and tensile waves "Compr" and "Rarefact" is filled by a fan of characteristics outcoming from the small vicinity of the $x = 0, t = 0$ point. This fan isn't shown in Figs. 1. Spatiotemporal width of the fan gradually increases during its propagation. This process gradually decreases a slope of a pressure profile in the contact zone between the waves "Compr" and "Rarefact".

Below an ablation threshold the maximum stretching and tensile stress are achieved in the point of intersection of the back characteristics of the waves "Compr" and "Rarefact" (width of the fan is small at $t \sim t_s$). Therefore at $F = F_{abl}$ the nucleation starts in this point. Nucleation separates continuous metal in to two part. Nucleation zone evolve into a zone with vapor bubbles. They fill an interior between two blue straights in Fig. 1 (a). In papers [8, 27, 28] it was shown that nucleation threshold $F_{nucl}$ is few percents less than ablation threshold $F_{abl}$. Above ablation threshold the spallation plate separates and flies away from a main body. The spallation plate is a layer between the vacuum boundary and the left blue straight in Fig. 1 (a). In the interval $F_{nucl} < F < F_{abl}$ there are possibility that bubbles in the layer between the two blue straights in Fig. 1 (a) will remain frozen inside solid. This depend on a competition between the hydrodynamic process of shrinking of bubbles and thermal process of cooling and freezing [28]. The shrinking and bubbles collapses dominate at a final temporal part of cavitation when resistance to stretching overcomes inertia of matter accelerated at an initial expansion stage. This shrinking stage exists since we are below ablation threshold. For optical lasers the ablation threshold is few times higher than melting threshold. Therefore there are bubbles and liquid around the bubbles.

Results of large scale MD simulation including very important ingredient - a Monte-Carlo description of an electron thermal transport are presented in Figs. 6. Those calculations were made for aluminum. The sizes of matter in a computational domain were $500 \times 240 \times 24$ nm$^3$; 172 millions of atoms. Absorbed fluence is 80 mJ/cm2, the ratio to ablation threshold is 80/65 = 1.23.

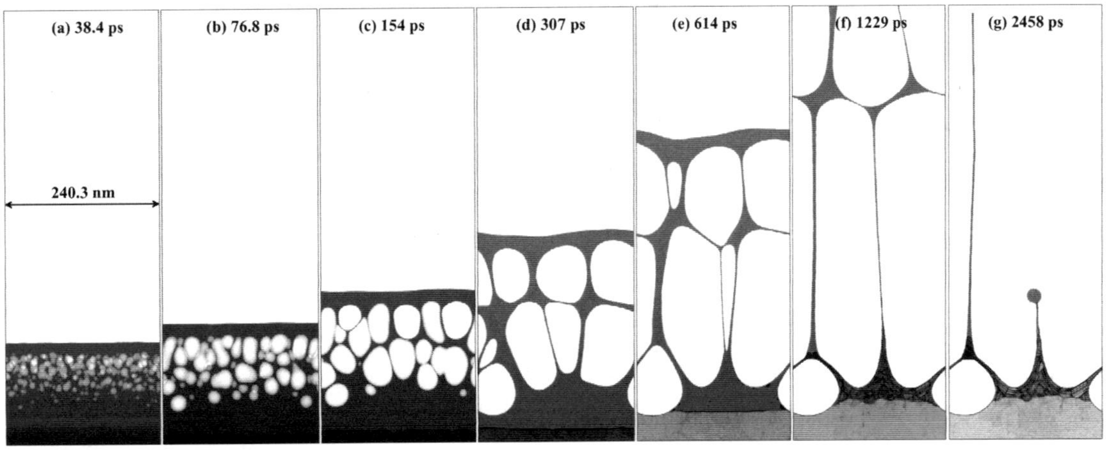


**Figure 6.** Nucleation, foaming, hydrodynamic decay of strongly stretched foam, appearance of droplets on the tip of jets, and freezing of remnants of broken liquid-vapor foam. Process covers two orders of magnitude in time. MD simulation spans scales from interatomic and up to micron size. Recrystallization front is clearly seen. It gradually approaches the foam region. Three first pictures present density maps. While the four next pictures show evolution of phase composition. The red and green colors correspond to liquid and solid aluminum, resp. Dark green color correspond to a region with ultradense concentration of dislocations. This region forms a "skin" including the frozen surface nanostructures.

Simulation shown in Figs. 6 is important for understanding of phenomena leading to appearance of chaotic structures. The key here is the processes of foam stretching and breaking in interplay with freezing of supercooled liquid. Hydrodynamic motions under action of inertia, tensile stress, and surface tension compete with an equally slow[3] solidification up to the end of motion - up to the final stop of motion in a thermally stressed state. A droplet at tip of jet demonstrates an amazing dynamics. Membranes surrounding bubbles in foam are ultimately stretched by inertial expansion up to 2-3 interatomic distances. Membranes break in their thinnest parts. This breaking forms a wall type feature made from a remnant of a membrane. The feature is topped with a droplet at its upper edge. The droplet is a formation elongated along the edge of membrane. It accumulates mass during contraction of broken membrane. It accumulates not only mass but also momentum of this mass. This momentum is created during contraction. Therefore velocity of a droplet differs from velocity of membrane itself. Complicated velocity field appears under action of surface tension and freezing/melting processes taking place in a contact between the liquid droplet and colder solidified membrane. The last picture in the series shown in Fig. 6 shows instant coexistence of liquid red droplet and frozen green jet type membrane. Experimental pictures in Figs. 9-11 below present plenty examples of such membranes, jets, and droplets.

Let us consider the remnant of a membrane attached to the bulk of a target. In Figs. 6 the bulk is located in the bottom of the pictures. As was said, after breaking this remnant begins to contract forming a long droplet along the edge of the broken membrane. In the last picture in Figs. 6 this droplet is elongated in direction perpendicular to the plane of picture.

[3] In a final stages they both slow their velocities down to velocities of the order of few tens of m/s.

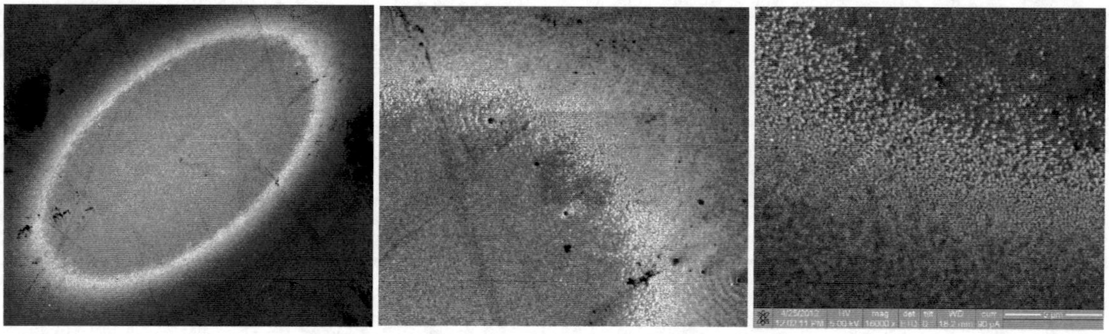

**Figure 7.** (a) General SEM view of typical crater after irradiation by optical near infrared Cr:forsterite laser. The spot has an elliptical shape since laser beam has been inclined, $p$-polarization, $45^o$ inclination, wavelength 1240 nm, duration 100 fs. Incident fluence was 480 mJ/cm$^2$, absorbed fluence was 220 mJ/cm$^2$. Ablation threshold for nickel for these conditions for incident fluence is 320 mJ/cm$^2$. The ratio of central to ablation incident fluences is $F_c/F_{abl} = 1.5$. (b) Ring type region of the lateral edge of the crater shown in previous figure. Fluence in this region was near ablation threshold. (c) Enlarged SEM view of the ring region. Length of the green marker going across the lateral ring is 7.04 um.

In a 3D picture this elongated droplet deforms as a result of inhomogeneity of velocity, stress, and temperature conditions along the droplet. Therefore 3D rounded droplets gradually form from the elongated one. They are clearly seen in the experimental Figures 10 and 11 below. Finally cooling due to pumping of heat from a hot surface layer into bulk totally solidifies liquid masses. Very slow creep continues during subsequent shrinkage of nanocrystallite solid under action of a field of residual stresses. Thus nanostructures appear on an irradiated target surface. Those structures are an interlacement of threads, threads with droplets, and broken and survived undersurface bubbles, see Figs. 9-11. They are mixed with a "skin" from a nanocrystallite solid. Broken and survived undersurface bubbles have narrow entrances going into the undersurface cavities visible in Fig. 10 (c).

Figures 7-9 correspond to ablation of a nickel film deposited onto glass substrate. Similar structures have been obtained in works [29, 30, 31, 32, 33]. The target from Ni has been irradiated by a $p$-polarized laser pulse from an air side. This is a Cr:forsterite laser. The first harmonics $\lambda = 1240$ nm has been used. Duration of pulse is $\tau_L = 100$ fs. Angle of inclination is $45^o$ degrees.

Important quantity is an ablation threshold. Let us give a data corresponding to a nickel target irradiated from an air side. Incident fluence at an ablation threshold is 0.32 J/cm$^2$. Of course this value depends on polarization and inclination. While a value of absorbed fluence at an ablation threshold after a single shot is independent on conditions of irradiation (polarization and inclination). Absorbed fluence at an ablation threshold for nickel is 0.14 J/cm$^2$. Absorption coefficient $A$ was specially measured. It depends on polarization, inclination, fluence, and duration. Near ablation threshold this coefficient is $A = 0.45$.

Experimental results (Figs. 9, 10) fully confirm our MD modeling of foaming and freezing. In experiments nickel film $\approx 1$ um thick has been deposited onto glass substrate. Film has been irradiated through air. Chains of single shot craters obtained on a film surface have been analyzed by optical and scanning electron microscopy (SEM). Fig. 7 (a) presents typical crater from such chain of craters.

Figs. 8 demonstrates few steps of FIB sample preparation for TEM (Transmission electron microscopy) - lamella technique. Figs. 9 presents pictures obtained at TEM. They clearly

**Figure 8.** (a) Two protective Pt layers was deposited across a crater ring for preparation of two TEM samples. The left "Sample A" contains two protective Pt layers: 1st - electron stimulated Pt deposition (the square area market by "e-Pt dep") to avoid Pt penetration into the Ni layer; and 2nd rectangular ion beam stimulated Pt deposition. The right region ("Sample B") has only ion beam Pt deposition with Pt penetration depth up to 50 nm into the nickel layer. (b) Milling of nickel from one side as the first step of future lamella preparation. (c) After two side milling of nickel the lamella has been welded to the nanomanipulator needle and lift out for transportation to TEM sample holder.

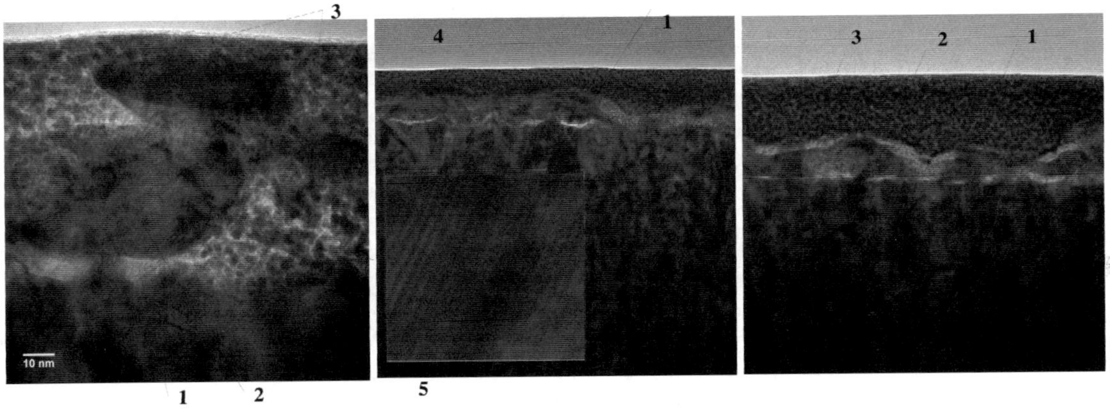


**Figure 9.** TEM images of irradiated surface of nickel in the region of lateral rings: 1- frozen undersurface bubbles, 2 - frozen jets, 3 - empty spaces filled with the Pt organic, 4 - a vacuum above the organic, 5 - nanocrystallite with visible atomic planes of the lattice. The disk type bubbles have a form of a crescent in a plane perpendicular to a target surface. Horizontal length of the bubble 1 marked in the left panel is 20 nm. The left panel size is $110 \times 110$ nm $^2$; 110 nm is $\approx 2700$ pixels.

demonstrate existence of frozen jets and undersurface bubbles in case of nickel. Previously these observations have been made for aluminum [28]. Now their presence is firmly established for transition metal.

SEM images of a crater and structures inside the craters are shown in Figs. 10 for the case of gold. We see that there is no tendency to form ripples even in the case of optical lasers, relatively high fluences, and limited number of shots, see Fig. 10 (c).

The X-ray lasers give us opportunity totally neglect plasmons contribution into formation of surface structures - no ripples at all. The reason is simple - photon frequency is many times higher than plasmon frequencies. Scales of structures enlarge with increasing of number of shots

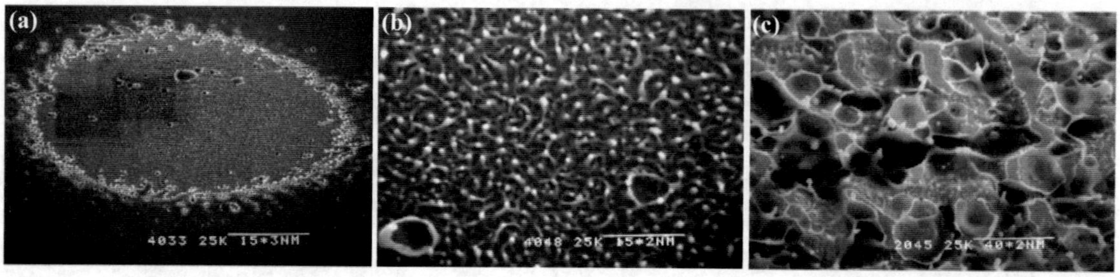


**Figure 10.** SEM images of surface structures at different magnifications. (a) Crater on a gold surface obtained after a single laser shot with pulse duration $\tau_L = 100$ fs. The scale is $15 \cdot 10^3$ nm = 15 um. Relative fluence in the center of elliptical spot is $F_c/F_{abl} = 1.5$. (b) Magnified view of a piece of the crater shown in the previous image. We see a homogeneous field of structures with sparse more large details. The scale is $15 \cdot 10^2$ nm = 1.5 um. (c) The small piece of a crater bottom formed by three identical shots into the same spot. Duration is $\tau_L = 100$ fs. Central relative fluence $F_c/F_{abl} = 1.1$ is smaller than in the previous pictures. The scale is $40 \cdot 10^2$ nm = 4 um.

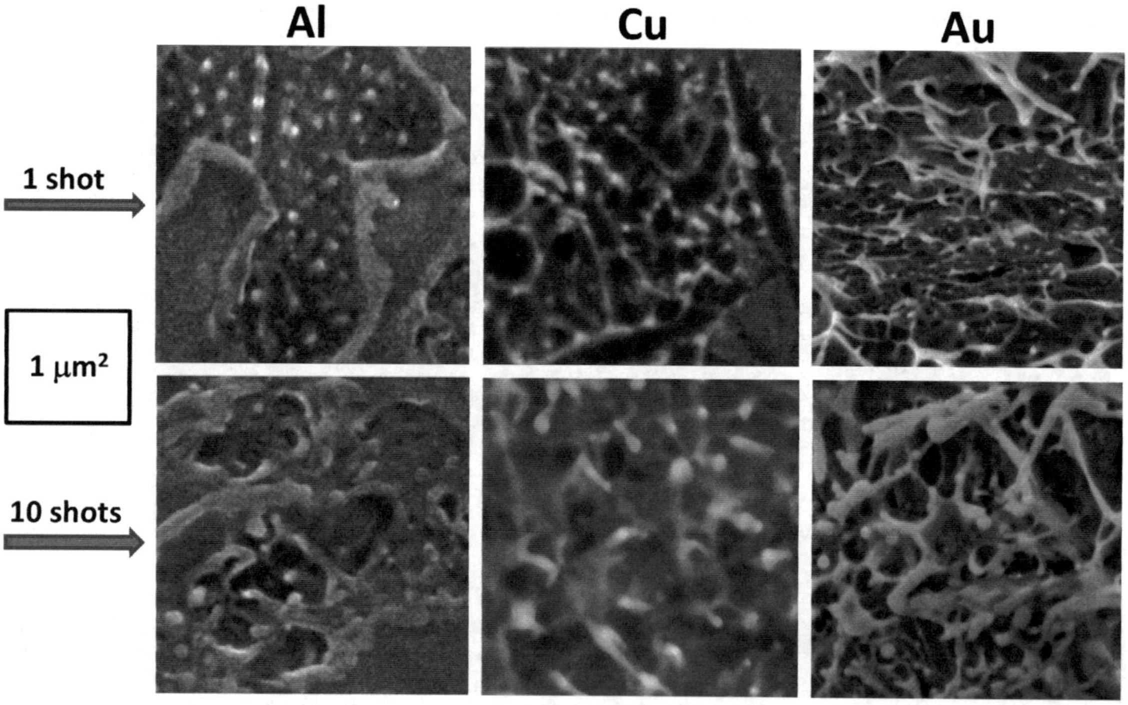


**Figure 11.** Surface structures formed by short pulse (duration 7 ps) soft X-ray ($h\nu = 90$ eV) laser irradiation. Small pieces from irradiated spots are shown. Area of those spots is $\approx 200$ um$^2$. Device is described in papers [34, 35]. Thresholds for ablation are: Al $14 \pm 5$ mJ/cm$^2$; Cu $16 \pm 5$ mJ/cm$^2$; Au $21 \pm 5$ mJ/cm$^2$.

$N$. See examples shown in Fig. 11. This is significant observation.

Ablation threshold is few times lower than in case of optical lasers. This is very important feature of short pulse X-ray ablation. We have to emphasize that we compare *absorbed* fluences (absorbed energy) in optical and X-ray cases necessary to overcome ablation threshold.

Explanation of this fact is linked to physics of 2T stage with excited electrons and to nucleation stage, see Section 1.

*Conclusion.* Above the early 2T stage in case of metals and optical lasers has been studied. New types of expansion dynamics have been described. Amount of ablated matter (few monoatomic layers) is small. But these atoms are ejected out in an unusual 2T state. This means that atoms remain excited during their separation from main body of a target. Theory and experiments on surface structure formation have been conducted and described. It is important to compare optical and X-ray irradiations. This allows to elucidate plasmons effects connected with interference of incident and surface electromagnetic waves. Surface structures exactly correspond to the picture developed above. The picture is based on phenomena of nucleation, foaming, and freezing of foam remnants appearing as result of mechanical breaking.

*Acknowledgments.* Authors (MBA, NAI, VVZh, VAKh, SIAsh) acknowledge support from RFBR grant 13-08-01095 A. This work was partly supported by a Grant-in-Aid for Scientific Research (B), No. 25289244, from the Ministry of Education, Culture, Sports, Science, and Technology (MEXT), Japan and by RFBR grants 12-02-00947 A (authors AYaF, TAP). This work was supported by ROSATOM contract H.4x.44.90.13.1111 (authors MBA, NAI, SIAsh).

## References

[1] Agranat M, Anisimov S, Ashitkov S, Zhakhovskii V, Inogamov N, Komarov P, Ovchinnikov A, Fortov V, Khokhlov V and Shepelev V 2010 *JETP Letters* **91** 471–477
[2] Ashitkov S, Agranat M, Kanel' G, Komarov P and Fortov V 2010 *JETP Letters* **92** 516–520
[3] Zhakhovskii V V and Inogamov N A 2010 *JETP Lett.* **92** 521–526
[4] Armstrong M R, Crowhurst J C, Bastea S and Zaug J M 2010 *J. Appl. Phys.* **108** 023511
[5] Anisimov S, Kapeliovich B and Perel'man T 1974 *Sov. Phys. JETP* **39**(2) 375–377
[6] Sipe J E, Young J F, Preston J S and van Driel H M 1983 *Phys. Rev. B* **27** 1141–1154 URL `http://link.aps.org/doi/10.1103/PhysRevB.27.1141`
[7] Akhmanov S A, Emel'yanov V I, Koroteev N I and Seminogov V N 1985 *Sov. Phys. Usp.* **28** 1084–1124 URL `http://iopscience.iop.org/0038-5670/28/12/R02`
[8] Zhakhovskii V V, Inogamov N A and Nishihara K 2008 *JETP Letters* **87** 423–427
[9] Inogamov N A, Zhakhovskii V V, Ashitkov S I, Petrov Y V, Agranat M B, Anisimov S I, Nishihara K and Fortov V E 2008 *JETP* **107** 1–19
[10] Demaske B, Zhakhovsky V, Inogamov N and Oleynik I 2010 *Phys. Rev.* B **82** 064113
[11] Cherednikov Y, Inogamov N A and Urbassek H M 2013 *Phys. Rev. B* **88** 134109
[12] Khakshouri S, Alfe D and Duffy D M 2008 *Phys. Rev.* B **78** 224304
[13] Inogamov N A, Petrov Y V, Zhakhovsky V V, Khokhlov V A, Demaske B J, Ashitkov S I, Khishchenko K V, Migdal K P, Agranat M B, Anisimov S I and Fortov V E 2012 *AIP Conf. Proc.* **1464** 593–608
[14] Sin'ko G V, Smirnov N A, Ovechkin A A, Levashov P R and Khishchenko K V 2013 *High Energy Density Physics* **9** 309–314
[15] Bushman A V, Kanel' G I, Ni A L and Fortov V E 1993 *Intense dynamic loading of condensed matter* (Taylor & Francis)
[16] *http://teos.ficp.ac.ru/rusbank/* URL `http://teos.ficp.ac.ru/rusbank/`
[17] Rose J H, Smith J R, Guinea F and Ferrante J 1984 *Phys. Rev.* B **29** 2963
[18] Kresse G and Furthmuller J 1996 *Computational Materials Science* **6** 15–50
[19] Zhakhovskii V, Inogamov N, Petrov Y, Ashitkov S and Nishihara K 2009 *Applied Surface Science* **255** 9592–9596
[20] Demaske B J, Zhakhovsky V V, White C T and Oleynik I I 2012 *AIP Conference Proceedings* **1426** 1211–1214
[21] Starikov S V, Stegailov V V, Norman G E, Fortov V E, Ishino M, Tanaka M, Hasegawa N, Nishikino M, Ohba T, Kaihori T, Ochi E, Imazono T, Kavachi T, Tamotsu S, Pikuz T A, Skobelev I Y and Faenov A Y 2011 *JETP Lett.* **93** 642–647
[22] Norman G E, Starikov S V, Stegailov V V, Saitov I M and Zhilyaev P A 2013 *Contrib. Plasma Phys.* **53** 129–139
[23] Gan Y and Chen J K 2010 *Mechanics of Materials* **42** 491–501
[24] Chen J, Chen W K and Rentzepis P M 2011 *J. Appl. Phys.* **109** 113522 URL `http://dx.doi.org/10.1063/1.3594732`
[25] Anisimov S, Zhakhovskii V, Inogamov N and et al 2003 *JETP Lett.* **77** 606–610

[26] Anisimov S I, Zhakhovskii V V, Inogamov N A, Nishihara K and Petrov Y V 2007 *Appl. Surf. Sci.* **253** 6390–6393

[27] Zhakhovskii V V, Inogamov N A and Nishihara K 2008 *J. Phys.: Conf. Ser.* **112** 042080

[28] Ashitkov S, Inogamov N, Zhakhovsky V, Emirov Y, Agranat M, Oleinik I, Anisimov S and Fortov V 2012 *JETP Lett.* **95** 176–181

[29] Vorobyev A Y and Guo C 2011 *J. Appl. Phys.* **110** 043102 [9 pages] URL `http://dx.doi.org/10.1063/1.3620898`

[30] Ionin A A, Kudryashov S I, Makarov S V, Seleznev L V, Sinitsyn D V, Ligachev A E, Golosov E V and Kolobov Y R 2013 *Laser Phys. Lett.* **10** 056004

[31] Hashida M, Ikuta Y, Miyasaka Y, Tokita S and Sakabe S 2013 *Appl. Phys. Lett.* **102** 174106

[32] Savolainen J M, Christensen M S and Balling P 2011 *Phys. Rev. B* **84** 193410

[33] Gurevich E L 2011 *Phys. Rev. E* **83** 031604 [5 pages] URL `http://link.aps.org/doi/10.1103/PhysRevE.83.031604`

[34] Ishino M, Faenov A Y, Tanaka M, Hasegawa N, Nishikino M, Tamotsu S, Pikuz T A, Inogamov N A, Zhakhovsky V V, Skobelev I Y, Fortov V E, Khohlov V A, Shepelev V V, Ohba T, Kaihori T, Ochi Y, Imazono T and Kawachi T 2011 *Journal of Applied Physics* **109** 013504

[35] Ishino M, Faenov A, Tanaka M, Hasegawa N, Nishikino M, Tamotsu S, Pikuz T, Inogamov N, Zhakhovsky V, Skobelev I, Fortov V, Khohlov V, Shepelev V, Ohba T, Kaihori T, Ochi Y, Imazono T and Kawachi T 2012 *AIP Conference Proceedings* **1465** 236–240

Journal of Physics: Conference Series **510** (2014) 012042 doi:10.1088/1742-6596/510/1/012042

# Implementation of the Kinetic Plasma Code with Locally Recursive non-Locally Asynchronous Algorithms

**A Yu Perepelkina[1], V D Levchenko[1,2] and I A Goryachev[1]**

[1] Keldysh Institute of Applied Mathematics, Moscow, Russia
[2] National Research Nuclear University "MEPhI", Moscow, Russia

E-mail: mogmi@narod.ru

**Abstract.**
Numerical simulation is presently considered impractical for several relevant plasma kinetics problems due to limitations of computer hardware even with the use of supercomputers. To overcome the existing limitations it is suggested to develop algorithms which would effectively utilize the computer memory subsystem hierarchy to optimize the dependency graph traversal rules. The ideas for general cases of numerical simulation and implementation of such algorithms to particle-in-cell code is discussed in the paper. This approach enables the simulation of previously unaccessible for modeling problems and the execution of series of numerical experiments in reasonable time. The latter is demonstrated on a multiscale problem of the development of filamentation instability in laser interaction with overdense plasma. One variant of the simulation with parameters typical for simulations on supercomputers is performed with the use of one cluster node. The series of such experiments revealed the dependency of energy loss on incoming laser pulse amplitude to be nonmonotonic and reach over 4%, an interesting result for research of fast ignition concept.

## 1. Introduction

The exponential growth in computational technologies during last decades open up new possibilities in solving problems relevant for contemporary physics. First of all it includes multiscale problems, the numerical solution of which was unachievable for a long time.

For this concept one of the most indicative fields is plasma kinetics simulation. Practical simulation problems require large amount of mesh cells and number of particles to satisfy the given accuracy criteria.

While the computational systems and simulation algorithms have evolved considerably since the invention of numerical schemes for plasma kinetics simulation, there remain problems which are still considered impossible to solve even with the use of massively parallel computations.

In this paper we present the CFHall code which is aimed at significantly multiscale problems in plasma kinetics simulations [1, 2].

For example, numerical modeling is seen as a most reliable tool to aid in the research of the causes of electron transport to anode in Hall thruster plasma. For this problem state-of-art codes perform 2D3V (two dimensional in space, three dimensional in velocity) simulation with the use of various approximations [3, 4]. These approximations, as well as a 2D3V geometry, may alter the physics of the studied processes, but the simulation without them is still thought to be impossible on modern equipment.

The other problem which was considered in the development of the presented code is the simulation of filametation instability in laser interaction with plasma. In this case the simulation domain should be large enough to cover the laser pulse envelope and the plasma layer, at the same time each filament appearing in the instability should contain several mesh steps. Numerical simulation has aided the research of the problem ever since it was carried out in 2D3V geometry [5]. But since the nonlinear plasma processes may be misrepresented when confined to two dimensions, fully 3D simulations are needed for more accurate results.

Such relevant problems of plasma physics require processing of large amounts of data. In the Hall thruster problem the fields are usually modeled by Poisson equation with implicit schemes. The presence of memory subsystem hierarchy leads to degradation of the absolute efficiency of calculations with the increase of the amounts of processed data due to the nonlocality of data access. It also results in inefficient parallelization.

Recent 3D3V numerical studies utilize the mesh with no more than $1.5 \cdot 10^6$ cells [6, 7]. The use of supercomputers may raise this number to $10^8$ [8]. Nevertheless with such grids either the mesh step approaches the size of the filament, or the simulation domain is not enough to simulate laser pulses with Gaussian envelope. The acceleration of calculations achieved by massively parallel computations is impeded by the necessity of a large number of synchronizations.

One of the methods to improve the situation is to use more efficient computational algorithms which could both take advantage of memory subsystem hierarchy for computations on systems with shared memory, and minimize the synchronization events in parallel codes. In this work we show how this can be achieved in a plasma kinetics code with the use of Locally Recursive non-Locally Asynchronous (LRnLA) [9] algorithms.

The algorithms are universal in the sense that one may use them for any numerical scheme with local stencil. The implementations with the use of algorithm family similar to the one described in this paper already exist for solution of Maxwell equations in optics simulation [10], solution of elasticity equations in seismic media [11], gas dynamics simulation with RKDG (Runge-Kutta Discontinuous Galerkin) method [12].

The principal ideas for the effective algorithms are locality and asynchrony. The greater the locality of the algorithm is, the closer the subsequently processed data is placed in computer memory. It also means that if some data is processed at some time, it will sooner be processed again. The better is the asynchrony of the algorithm, the larger portions of data may be processed independently before the message passing between their processors becomes necessary.

In this paper we present the basics of the LRnLA algorithms that are used in CFHall code, show the specifics of their application to plasma kinetics calculations, and demonstrate the efficiency achieved by such approach on the example of a problem of filamentation instability simulation.

## 2. Kinetic model

As a mathematical model of plasma the Vlasov-Maxwell system of equations is used. We use the dimensionless units. Reciprocal electron Langmuir frequency $\omega_p$ is used for time units, electron masses $m_e$ for mass, electron charges $e$ for charge, light speed $c$ for velocities. From these the units for distances $c/\omega_p$ and for fields $m_e\omega_p c/e$ follow. This unit are scalable by electron density, so the physical quantities the simulation value correspond to are also scalable in a definite range.

Electromagnetic field evolution is calculated in accordance with the FDTD (Finite Difference Time Domain) numerical scheme [13]. PIC (Particle-in-Cell) method is used for plasma simulation [14, 15] with second-order form-factor. Superparticle relativistic acceleration is determined by the Boris scheme.

The field evolution time step is limited by the Courant condition:

$$dt_{FLD} \leq \frac{dx}{\sqrt{3}c}. \tag{1}$$

Time step for particle motion should be small enough to resolve the Langmuir plasma frequency:

$$dt_{PIC} < \frac{1}{\omega_e}. \tag{2}$$

These two conditions may be of different scale. That is why it is possible to use 2 different time steps in a simulation.

We shall first demonstrate how the algorithms are constructed for the 3 dimensional FDTD scheme for Maxwell equations with the second order of approximation, which involves only the electromagnetic field evolution, without calculations for plasma particles. The two equations used explicitly in calculation are:

$$\frac{\partial \vec{B}}{\partial t} = -\nabla \times \vec{E}, \quad \frac{\partial \vec{E}}{\partial t} = \nabla \times \vec{B} - \vec{j}. \tag{3}$$

Electric and magnetic field components are shifted against each other both in time and space, according to a Yee cell [13]. This scheme is simple and can be considered independently, and the algorithms built on it will comprise the basis for the more complicated codes.

When proceeding to the algorithms for PIC method it is essential to note the width of its stencil. According to the method, the distribution function of plasma is approximated by a model function, and the integration of Vlasov equation then takes the form of Newtonian equations. For the construction of this model distribution we use the family of functions $\Lambda^{\ell}$ , parametrized by $\ell$.

$$\Lambda^{\ell}(x-x') = \int \Lambda^{\ell-1}(x-x'')\Lambda^{0}(x'-x'')dx'', \quad \Lambda^{0}(x) = \begin{cases} 1 & |x| < 0.5, \\ 0 & |x| > 0.5 \end{cases} \tag{4}$$

$$\Lambda^{\ell}(\vec{r}) = \Lambda^{\ell}(x)\Lambda^{\ell}(y)\Lambda^{\ell}(z). \tag{5}$$

With $\ell = 0$ $\Lambda^{\ell}$ is a rectangle function, with $\ell = 1$ it is a tent function, with $\ell = 2$ it is comprised of three stitched parabola pieces. To calculate the charge density in the point $\vec{r_i}$ knowing the positions of particles $\vec{r_j}(t)$ we use

$$\rho(\vec{r_i}) = \sum_{\alpha} e_{\alpha} \sum_{j} \Lambda^{\ell+1}(\vec{r_j}(t)-\vec{r_i}), \tag{6}$$

and the current densities are found from the continuity equation. With higher $\ell$ the order of approximation increases, and the stencil becomes wider. Since the FDTD scheme is of second order, it is reasonable to use the second order of approximation for PIC method. It also gives the computational advantage over the majority of PIC codes, since the required accuracy can be achieved with lower number of particles per cell. With the second order of approximation the stencil covers 3 cells in each direction. Since the particles can travel between cells during their motion, the actual size becomes equal to 4 if we restrict its movements by one cell radius.

## 3. Methods of calculation optimization

### *3.1. Data storage*

Firstly we shall show how we can take advantage of the idea of locality when choosing the data structures for the simulation.

The finite difference schemes common property is the fact that the cross-shaped stencil is usually involved. Let us note that when we use the 3D array in the calculation the data in the array is stored in a single line in computer memory. If we have an array of the size $Nx \times Ny \times Nz$, and attempt the calculation for the element with $(ix, iy, iz)$ indices in it (which means that it is the element with index $i = ix + Nx \cdot iy + Nx \cdot Ny \cdot iz$) if we use a scheme with a cross-shaped stencil we need its closest neighbors along the three axes. These neighbor cells on the $x$ axis are located 1 cell away from the point in question, but neighbor cells along $y$ and $z$ directions are located $Nx$ and $Nx \cdot Ny$ cells away correspondingly. It means that for large meshes each calculation is not local in itself, and such data structures lowers the effectiveness of calculations.

One way to avoid this is to store the data in Morton order Z-curve [16] (see fig. 1). While there may be other more efficient solutions for such problems, Z-curve is also easy to implement with the use of recursive templates in programming languages. In the current work it is utilized in a `cubeLR` structure. Since this works only when the dimensions of the array are equal to each other and may be represented as $2^n$, where $n$ is some integer number, when array dimensions differ, the simulation domain is subdivided into cubical blocks. The relations between blocks are determined by a structure of ordered lists `nLAnet`. The further discussion is related to cubical grid without the loss of generality.

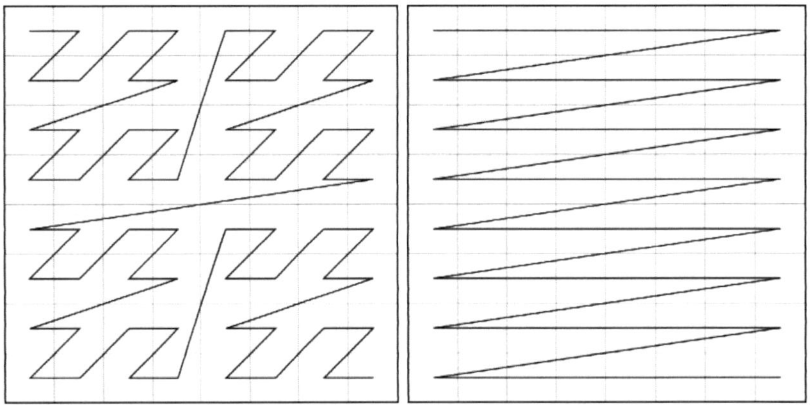

**Figure 1.** Mapping of a 2D data on Z-order curve and common array

### *3.2. LRnLA algorithms*

Z-curve may be seen as a recursive traversal rule in the 3D data space. Let us now consider a 4D operations-data space. In this space one can imagine a dependency graph of the calculations for the problem. In fig. 2 the dependency graph for 1D projection of the problem is illustrated. The arrows show the data exchanges, the operations are represented by points. It is convenient to place it on a 2D space. Vertical axis corresponds to operations, and can be thought of as a time axis, horizontal axis corresponds to data, and can be associated with coordinates. The first layers of points along the horizontal axis show data initialization operations, the following ones show the calculations on some time step.

LRnLA algorithms deal with optimizing dependency graph traversal rules. To understand how the algorithms of the LRnLA family are constructed one needs to be able to visualize the algorithms as shapes in the operation-data space.

Algorithm is a rule which determines the particular order in which the given calculations in the dependency graph are processed. Let us enclose several (or all) points of the dependency graph in a 4D polytope (Fig. 2). This will comprise some "calculation block". This polytope

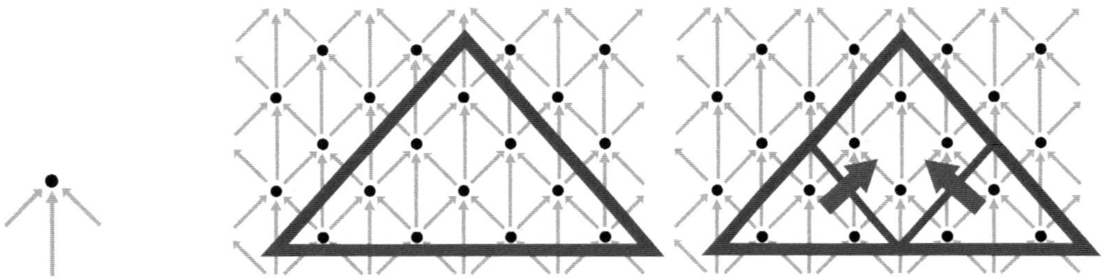

**Figure 2.** An example of a dependency graph for FDTD scheme, projected on two dimensional plane. The stencil of a numerical scheme (left). A polytope covers several operations (center). A polytope is subdivided into smaller ones so that the dependencies between calculation blocks is unilateral (right)

may be divided into smaller ones, each of which may (or may not) cover some points of the graph, and the data dependencies (represented by arrows) between blocks are unilateral. These dependencies determine to some extent the order in which these calculation blocks are carried out. The block which are not connected by arrows (asynchronous blocks) may be calculated either subsequently or in parallel. The resulting polytope may also be divided further, until they cover only 1 calculation. This way the shape of the polytope, along with the rule it is divided into smaller shapes with unilateral dependencies, visually illustrates and determines a calculation algorithm.

For further illustration let us explain how this concept works for conventional traversal rules. The conventional algorithm deals with the problem layer by layer, calculating all operation for one given time step before advancing to the next one. In terms of a polytope division illustration one can cover the whole dependency graph by a rectangular polytope, then subdivide it into layers. Each layer is represented by a rectangle with height equal to one time step in the fourth dimension and cover all simulation domain in other dimensions. This layers comprise the calculation blocks that are processed subsequently. This algorithm is not effective in terms of both locality and asynchrony. With the increase of mesh size the locality of such algorithm decreases significantly, and with the use of parallel calculations the data exchanges between processors is required on each time step.

One of the solutions provided by the LRnLA algorithms family is illustrated on fig. 2. The triangle (cone in greater dimensions) is subdivided recursively into triangle and diamond shapes. After the calculation in triangles is processed, the data is transfered to calculation in the diamond. The algorithm is easiest to implement if the shapes of the polytopes involved are similar, so the division becomes recursive.

### *3.3. ConeFold*

Since the polytope subdivision may produce new shapes where the rules of subdivision also differ, the algorithm types for multidimensional problems may become large and difficult to program. To cover the whole operation space while using only one shape and rule of its division the algorithm by the name of "ConeFold" (CF) is used in the presented plasma kinetics simulation code. The shape is illustrated in figs. 3. Its division is $(d+1)$-binary, where $d$ is the dimension of the simulation domain. We can associate a number, let it be called "rank", with the CF size. One CF with rank $R$ divides into $2^{d+1}$ CFs with rank $R-1$.

In the implementation the rank $R$ CF algorithm corresponds to some function call, which conducts $2^{d+1}$ function calls which correspond to CFs of rank $R-1$ in the order determined by the data dependencies. CF of rank zero performs calculation according to the numerical scheme

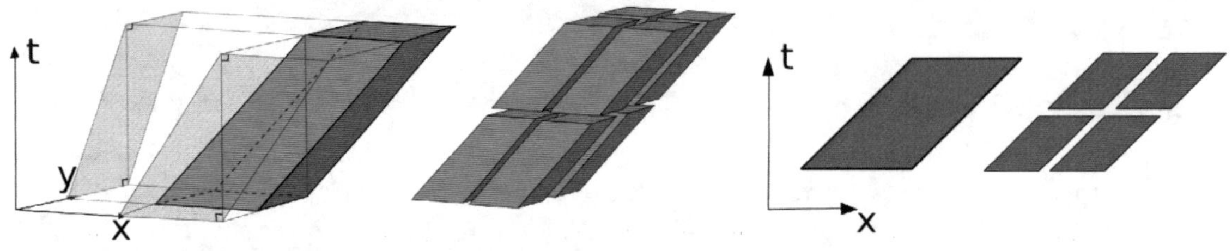


**Figure 3.** ConeFold algorithm shape and decomposition. Projection for 2D and 1D simulation

in one data cell. For FDTD scheme the cell is chosen as a Yee cell, so that it contains one complete set of electromagnetic field components: 3 electric field components and 3 magnetic field components. The minimal cell for the algorithm may also contain several Yee cells in case the stencil size of the scheme is larger.

The CF which has all the data of the given problem in its projection on the data space has rank `R = MaxRank`. Its height corresponds to $2^{MaxRank}$ time steps. As illustrated by the slant sides of the ConeFold shape, after one such algorithm is processed in the domain, the data cells involved are situated on the different time steps. To bring all the data to the single time evolution moment by covering the 4D rectangular domain $2^d$ CFs should be processed (fig. 4).

### *3.4. Algorithms for boundaries*

The first complication arises when we note that calculations may differ depending on the cell. The boundary cells are processed differently since boundary conditions applies. One should take note that different decomposition rules apply to shapes corresponding to the algorithms which intersect the boundary. Figure 5 shows how the decomposition works in 1D simulation (2D operation-data space). The first ConeFold of the maximal rank intersects the right boundary of the domain. Let us denote this algorithm with $X$ symbol. It divides into 2 CFs of the same type ($X$), one that is situated completely inside the domain (denoted by $D$ symbol), and one that is outside and thus should not be processed. This way the right boundary CF ($X$) calls 3 CFs. The $D$ CF consists of 4 CFs with $D$ symbol. On zero rank $X$ CF performs the calculation with the account of boundary conditions. The left boundary is considered likewise, and the boundary CF is denoted by $I$.

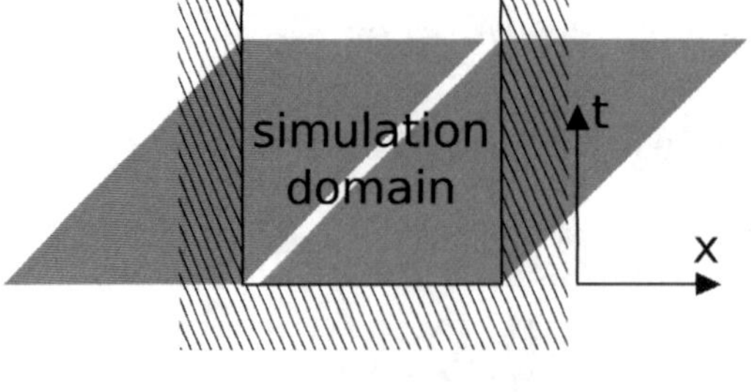


**Figure 4.** Coverage of the rectangular simulation domain by ConeFolds of the maximal rank, 1D simulation case

In a 2D case 9 cases should be considered corresponding to the internal calculation, calculation on each edge and each corner. The intersection of algorithm shapes with boundaries define the rules of their decomposition.

Journal of Physics: Conference Series **510** (2014) 012042 doi:10.1088/1742-6596/510/1/012042

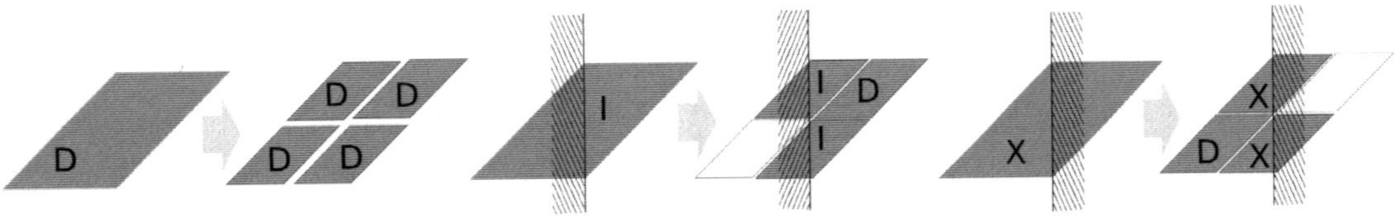


**Figure 5.** Decomposition of the algorithm inside the simulation domain and on the boundary

### *3.5. Vectorization*

It is not necessary to use the 4D LRnLA decomposition of the dependency graph for 3D simulation domains. One can use the conventional data storage and algorithms along one axis, and locally recursive decomposition along the others. In this case the calculation can be accelerated further with the use of SSE/AVX vectorization along this isolated axis in the following way.

The component values along z axis are stored in a vectorized array. Array elements are stored in vectors with `Nv` elements, where `Nv=4` if SSE with float data type or AVX with double data type is used or `Nv=8` for AVX with float. Operations on such vectors take the same amount of time as operations on scalars (double or float). Therefore it is possible make the calculation up to 8 times faster. The actual acceleration is slightly less due to costs of vector formation.

### *3.6. Other algorithms based on ConeFold*

Among the advantages of ConeFold algorithm is its versatility in a sense that it can be adapted to various numerical schemes, local stencil being the only limitation, and to various computer architectures.

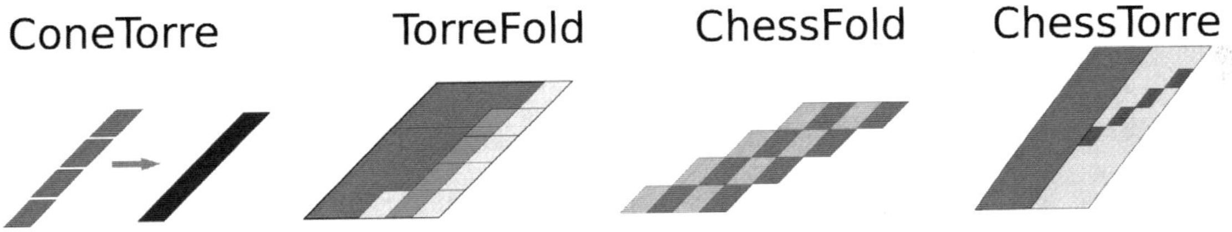


**Figure 6.** Algorithms based on ConeFold. In TorreFold segments of different color may be processed by different processor cores. In ChessFold the segments of same color in each diagonal row are asynchronous

The distinctive feature of shared memory systems is the bottleneck in memory subsystem hierarchy. To overcome the difficulty the algorithms locality may be further improved by using a tower-like polytope "ConeTorre" (see fig. 6), which consists of several CFs stacked onto each other in operations (time) dimension. For multicore processors it is convenient to use a polytope "TorreFold" that has a CF shape and subdivides into several ConeTorres, each of which can be processed by separate core.

For NUMA architectures it is more profitable to stack ConeFolds not in a single tower, but with an additional shift, so that all shifted CFs are asynchronous, and it constitutes the "ChessFold" algorithm.

The "ConeTorre" and "ChessFold" algorithms can be combined for the implementations on systems with several levels of parallelization, such as clusters, up to supercomputers.

### *3.7. Specifics of algorithms for PIC simulation*

As it was mentioned earlier, the stencil of PIC method is larger than the one in FDTD calculations, and this leads to the necessity of several modifications of the basic ConeFold algorithm. Also, as previously stated, for CFHall code we choose different time steps for field and plasma evolution. Here is how we modify the algorithm by taking these two points into consideration.

The subdivision of the ConeFold of maximal rank continues recursively to rank `R=PicRank+1` (instead of continuing to zero rank). `PicRank` here denotes a chosen number which controls the $dt_{PIC}/dt_{FLD}$ ratio. After the subsequent subdivision we get $2^{d+1}$ CFs. The $2^d$ CFs of `R = PicRank` from the top layer continue the subdivision until zero rank, and on the zero rank the calculation for field evolution according to FDTD scheme is performed on one data cell. The lower layer of CFs with `R=PicRank` are subdivided according to the same rule as described above for the CF with `R=PicRank+1`, until the rank reaches `R = FFrank`. This number is determined by the form-factor of PIC method and equals 2 in the current code. This CF performs the calculation for the PIC method.

The concept is illustrated on fig. 7 for `PicRank=3`. ConeFold with rank `R=FFrank` calculates the forces applied to particles from the field values in cells, performs the acceleration and movement of particles and calculates the contribution to charge current densities through cell faces from this movement. The current densities established after this movement are used in $2^{PICRank+1} - 2^{FFRank}$ calculations of field components. For `PicRank=3` the time step ratio becomes $dt_{PIC}/dt_{FLD} = 12$.

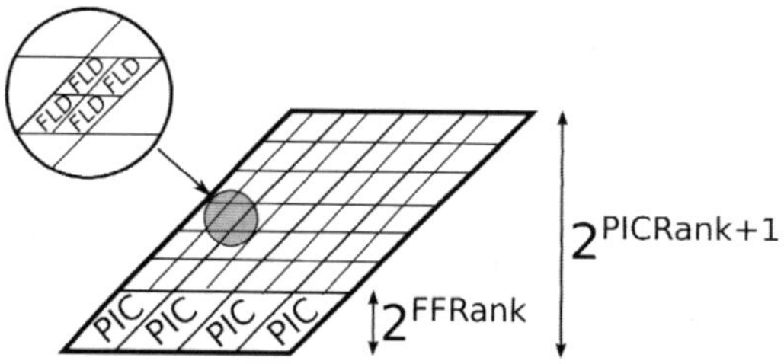


**Figure 7.** Subdivision of CF with rank $R = PicRank + 1$. Any algorithm with larger rank consists of some number of such shapes

The optimization methods applied in CFHall can be summarized as follows. The data is stored in a 2D locally-recursive (Z-order) array associated with $x$-$y$ plane. In each cell all values along $z$-axis are stored in a vectorized array. ConeFold LRnLA algorithm is applied for the two directions of locally recursive array. The subdivision includes special ConeFold type (fig. 7) to account for PIC calculations and field and particle time step difference. On higher ranks CFs combine into TorreFold algorithms.

## 4. Simulation examples

As an example where the described optimization is essential, in the scope of this paper the performance of CFHall code is demonstrated on the problem of laser interaction with overdense plasma.

The problem statement is illustrated on fig. 8. Plasma region occupies approximately half of the simulation domain. Laser pulse propagates in the positive direction of $x$-axis. Its transverse envelope is Gaussian and depends only on $y$ coordinate. Periodic boundary is applies on $z$ axis, other boundaries are reflective. The initial problem statement is 2D, but the simulation domain is 3D3V. The 3D structures may only arise due to nonlinear plasma effects, which are intended to be observed in the study.

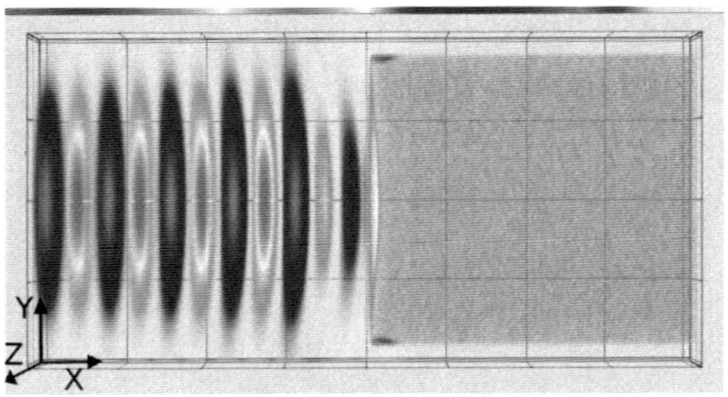


**Figure 8.** Illustration of the problem statement for the simulation

Such problem is relevant for the advance of fast ignition concept. The interaction of ultraintense laser pulse with plasma layer leads to magnetic field generation due to the nonlinear process of Weibel filamentation instability. The effect of the resulting stable magnetic field structures is crucial for the understanding of electron acceleration efficiency in the system.

To study the nonlinear process of filamentation instability with calculations on a grid, it is essential that one field oscillation covers several mesh steps. The filament size is usually considered to be of the order of skin depth length $c/\omega_p$ [17].

In the current modeling, mesh size is $2048 \times 1024 \times 128$ cells, spatial and time steps are $dx = dy = dz = 0.1c/\omega_p$, $dt = 0.025\omega_p^{-1}$. In each cell there are 4 electrons and one ion, about $0.6 \cdot 10^9$ particles in total. Ions displacement during the simulation time is negligible, however their movement is not restricted. Pulse frequency is three times lower than electron Lengmuir frequency, its longitudinal envelope covers approximately ten wavelengths. The amplitude was varied in $1/3 \div 9$ range in the chosen units. The simulation continues to $250 \div 400\omega_p^{-1}$.

The data for computation takes $\sim$ $50Gb$. The simulation has been performed on dual processor CPU Intel Xeon X5670 node. Asynchronous calculations are managed by POSIX threads. One time step in simulation takes $\sim$ 3 seconds, so the results for 1 variant are obtained after $\sim$ 16 hours.

During the simulation process the electrons are accelerated from the plasma layer boundary. 3D filamentary magnetic structures appear in areas through which accelerated electrons pass. The stable magnetic fields are formed on the plasma boundary.

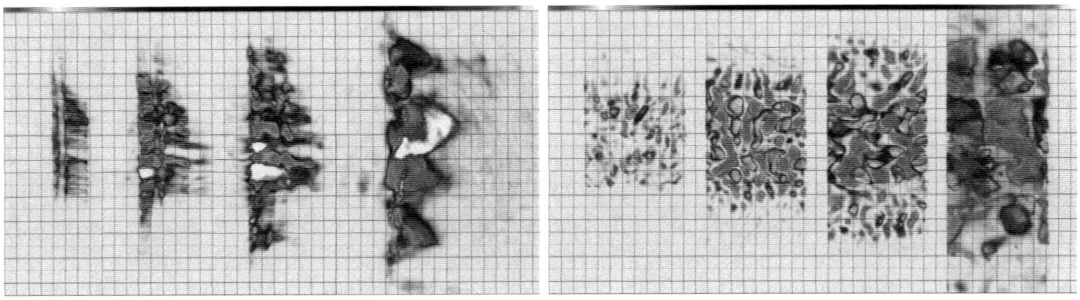

**Figure 9.** Simulation results. Evolution of filamentation in magnetic field components. $x$-$y$ projection on the left, $x$-$z$ projection on the right

As a separate study, trial simulations were carried out with 256 cells along $z$ axis and with 16 electrons in cell. The difference in the studied effects was negligible. Since the calculation time increases for these parameter values reaching up to two days, they weren't used for the final series.

The series of experiments was carried out for different amplitude of incoming laser pulse. The ratio of the energy transferred to magnetic field formation was studied in each variant. The nonmonotonic dependency of this ratio on pulse amplitude was discovered (fig. 10). Its maximum reaches a considerable value of $\sim 4.5\%$.

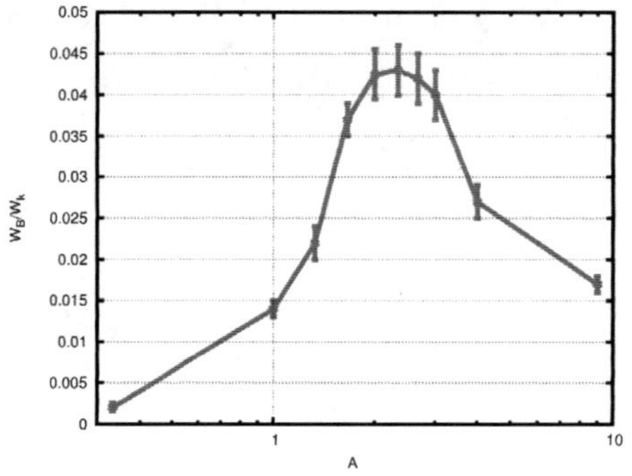


**Figure 10.** Simulation results. Dependency of the energy ratio transferred to magnetic field — energy in magnetic fields divided by particle kinetic energy

## 5. Conclusion

The LRnLA algorithm family "ConeFold" had been constructed and implemented in the plasma kinetics simulation code CFHall. The algorithms ensure the locality of subsequently processed data and minimize data synchronization, enabling more efficient use of computer hardware for numerical simulation. The calculation time in comparison with standard approaches is $\sim 10 \div 1000$ times lower. The precise value of the acceleration may vary depending on the specific implementation with which CFHall is compared, on the dimensions and characteristic sizes in problem statement and thus the amount of data being processed. The precise comparison requires separate work and is not included in the scope of the current paper.

The code performance had been demonstrated to achieve physically relevant results in the modeling of laser pulse interaction. From the simulation the non-monotonic dependency of the ratio of energy transferred to magnetic field growth was obtained. The ratio may become up to 4.5% for some values of laser intensity.

Omitting the further consideration of physical results obtained, in the scope of this paper we focus the attention on the benefit of the use of LRnLA algorithms for simulation. We have chosen enough mesh resolution for simulation to separate individual magnetic field filaments, used 3D3V geometry and thus we ensure that the increment in field growth would not be misrepresented.

On the other hand the calculation takes reasonable time, so the series of simulations with varying parameters, being at the same time relevant for front-line plasma physics questions may be carried out on the hardware available for individual researchers.

## Aknowledgements

The work is supported by RFBR grants 12-01-00708-a, 12-01-00490-a.

## References

[1] Perepelkina A Yu, Levchenko V D and Goryachev I A 2013 *J. Phys.: Conf. Series* **441** 012014
[2] Perepelkina A Yu, Levchenko V D and Goryachev I A 2013 *Matematicheskoe Modelirovanie* **25(11)** 98–110 (in Russian)
[3] Justin M 2007 *Advances in Fully-Kinetic PIC Simulations of a Near Vacuum Hall Thruster and Other Plasma Systems*, PhD thesis, MIT
[4] Cho S, Komurasaki K and Arakawa Y 2013 *Physics of Plasmas*, **20**, 063501
[5] Elkina N V and Levchenko V D 2003 *Appl. Phys.* **5** 16-23
[6] Okada T and Ogawa K 2007 *Phys. Plasmas* **14**, 072702
[7] Borthakur A and Das N 2013 *IJ BSS* **2(1)** 107-111
[8] Ruhl H 2002 *Plasma Sources Sci. Technol.* **11** A154
[9] Levchenko V D 2005 *J. of Inf. Tech. and Comp. Systems* (in Russian) **1** 68
[10] Zakirov A V and Levchenko V D 2009 *PIERS proceedings, Moscow, Russia* 580–584
[11] Ivanov A V, Kaplan S A, Karakin A V, Levchenko V D, Levchenko T V and Rok V E 2006 *Geoinformatika* **3** 59-61
[12] Korneev B A and Levchenko V D 2013 *A Locally Recursive non-Locally Asynchronous Algorythm to RKDG Methods for Gas Dynamics Equations* (preprint KIAM RAS, Moscow, Russia) 28
[13] Taflove A and Hagness SC 2000 *Computational Electrodynamics: The Finite-Difference Time-Domain Method* (2nd ed., Boston: Artech House)
[14] Sigov Yu S *Numerical Experiment - the Link Between the Past and the Future of Plasma Physics* (in Russian), (FIZMATLIT, Moscow, 2001)
[15] Birdsall C and Langdon A 1985 *Plasma Physics via Computer Simulation* (New York:Mc Graw Hill.)
[16] Morton G M 1966 *A Computer Oriented Geodetic Data Base; and a New Technique in File Sequencing* (Technical Report, Ottawa, Canada: IBM Ltd.)
[17] Treumann R A and Baumjohann W 2012 *Annales Geophysicae* **30** 427-431,

Journal of Physics: Conference Series **510** (2014) 012043 doi:10.1088/1742-6596/510/1/012043

# Model Kinetic Description For Many-Component Plasma

**V.V. Belyi**

**Abstract.** A consistent derivation of the model linearized collision operator for a multicomponent system is presented. In these results an ambiguity in the choice of coefficients is eliminated, in contrast to the BGK type models. A technique for reconstruction of the model collision integral form based on a known expression for the model linearized operator is proposed. It is shown that the model collision integral in the local (not complete) equilibrium approximation does not contain a complicated exponential, that is common for the BGK type integrals. Boltzmann's H-theorem is proved for our model.

IZMIRAN, Russian Academy of Sciences, Troitsk, Moscow, Russia

E-mail: sbelyi(at)izmiran.ru

## 1. Introduction

As it is well known the Balescu collision operator for plasma

$$I_a^{BALU}(\mathbf{p}) = \sum_b 2e_a^2 e_b^2 n_b \frac{\partial}{\partial p_i} \int \frac{k_i k_j \delta(\mathbf{kv} - \mathbf{kv}')}{k^4 \left|\varepsilon(\mathbf{kv}, \mathbf{k})\right|^2} \{\frac{\partial f_a}{\partial p_j} f_b - f_a \frac{\partial f_b}{\partial p'_j}\} d\mathbf{k} d\mathbf{p}' \tag{1}$$

satisfy general properties:
conservation laws:

$$\sum_a \int \varphi_a(\mathbf{p}) I_a(\mathbf{p}) d\mathbf{p} = 0, \text{ if } \varphi_a(\mathbf{p}) = 1, \ \mathbf{p}, \ \frac{\mathbf{p}^2}{2m_a} \tag{2}$$

and Boltzmann's H-theorem:

$$\frac{\partial}{\partial t} H(t) = -\frac{\partial}{\partial t} S(t) = \frac{\partial}{\partial t} \sum_a \int f_a(\mathbf{p},t) \ln f_a(\mathbf{p},t) d\mathbf{p} \leq 0, \tag{3}$$

where $S(t)$ - is the entropy of the system.

$$\frac{\partial}{\partial t} H^{BALU}(t) = -\sum_{ab} e_a^2 e_b^2 n_a n_b \int \frac{k_i k_j \delta(\mathbf{kv} - \mathbf{kv}')}{k^4 \left|\varepsilon(\mathbf{kv}, \mathbf{k})\right|^2} (\mathbf{k} \frac{\partial \ln f_a}{\partial \mathbf{p}} f_b - f_a \mathbf{k} \frac{\partial \ln f_b}{\partial \mathbf{p}'})^2 f_a f_b d\mathbf{k} d\mathbf{p}' d\mathbf{p} \leq 0. \tag{4}$$

The Balescu kinetic equations cannot be resolved exactly except for Maxwell molecules. In this case the eigenfuctions of the collision operator are Hermite polynomials. In all other cases, the modeling approach is used. The most widely used model kinetic equation, especially in the case of discrete simulation of gas and plasma dynamics, is the Bhatnagar, Gross and Krook (BGK) model equation [1]. We recall that in the BGK model the collision term for the one-component system is the deviation of the distribution function (d.f.) from the Maxwellian d.f. whose parameters are the moments of the d.f.:

$$I^{BGK}\{f\} = -\nu(f - f^0) \tag{5}$$

where

$$f^0 = \frac{n}{(2\pi mT)^{3/2}} \exp -\frac{m(\mathbf{v}-\mathbf{V})^2}{2T}; \tag{6}$$

and $n(\mathbf{r},t) = \int f d\mathbf{p}$; $\mathbf{V}(\mathbf{r},t) = \frac{1}{n}\int \mathbf{v} f d\mathbf{p}$; $\mathbf{T}(\mathbf{r},t) = \int \frac{m(\mathbf{v}-\mathbf{V})^2}{3n} f d\mathbf{p}$ are the local density, the mean velocity and the temperature in energy units, respectively. This model term vanishes at equilibrium and satisfies the conservation laws:

$$\int \varphi(\mathbf{p}) I^{BGK}\{f\} d\mathbf{p} = 0, \text{ if } \varphi(\mathbf{p}) = 1,\ \mathbf{p},\ \frac{\mathbf{p}^2}{2m} \tag{7}$$

and the Boltzmann's H-theorem:

$$\frac{\partial}{\partial t} H^{BGK}(t) = \nu \int (f - f^0) \log \frac{f}{f^0} d\mathbf{p} \leq 0, \tag{8}$$

The advantage of the BGK model is that the solution of the kinetic equation reduces to that of a system of algebraic equations [2]. A weak point is that the model implies that the Prandtl number (Pr) equals 1.

Holway and Cercignani [2, 3] independently introduced the so-called ellipsoidal statistical model in order to take into account real Prandtl numbers by substituting the local anisotropic Gaussian distribution for the local Maxwellian distribution:

$$f^0 = n\pi^{-3/2} (\det A)^{1/2} \exp - \sum_{i,j=1}^{3} \alpha_{ij} (\mathbf{v}_i - \mathbf{V}_i)(\mathbf{v}_j - \mathbf{V}_j) \tag{9}$$

$$A = \|\alpha_{ij}\| = \left\| (\mathrm{Pr})^{-1} (2T/m)\delta_{ij} - 2(1-\mathrm{Pr})\mathrm{p}_{ij}/n\,\mathrm{Pr} \right\|^{-1}$$

In the problems of linear transport and fluctuations one usually uses the linearized form of the BGK collision operator:

$$\delta \widehat{I} \left| h \right\rangle = -\nu \left( \left| h \right\rangle - \sum_{\alpha=1}^{5} \left| \Psi_\alpha \right\rangle \left\langle \Psi_\alpha | h \right\rangle \right), \tag{10}$$

where $|h\rangle$ is defined by $f = f^0 + \delta f = f^0\,(1+h)$, and $|\Psi_\alpha\rangle$ are the first five Hermite polynomials.

One of first works dedicated to derivation of the linearized collision integral is the paper by Gross and Jackson (GJ) [4]. Later, extension of the BGK kinetic model for the inclusion of higher order matrix elements was discussed and it was applied to investigate the generalized Enskog equation and the dynamic structure factor for gas and fluids [5-11]. The approximation consisted in taking into account exactly a finite part of the matrix operator, while the remaining part was only represented by the diagonal matrix elements. In the present paper we do not consider spatial inhomogeneities and assume the wave vector $\mathbf{k} = \mathbf{0}$. But we do take into account the non-diagonal components arising in the collision operator expansion with respect to the complete system of polynomials in the quadratic approximation. Making use of these non-diagonal elements allows us to obtain a new form for the model integral with the Spitzer corrections taken into account. In the case of a Boltzmann gas of hard spheres these corrections are insignificant.

A more grave situation arises in the case of many-component systems. According to the Gross and Krook (GK) model [12], the collision operator has the form of the deviation of the d.f. from a "mythical" exponent:

$$I_a^{GK}\{f_a\} = -\sum_b \nu_{ab}[f_a - \frac{n_a}{(2\pi m_a T_{ab})^{3/2}} \exp - \frac{m_a(\mathbf{v} - \mathbf{V}_{ab})^2}{2T_{ab}}], \tag{11}$$

were the parameters$\mathbf{V}_{ab}$ and $T_{ab}$ are related linearly to the d.f. moments:

$$\mathbf{V}_{ab} = \alpha_{aa}\mathbf{V}_a + \alpha_{ab}\mathbf{V}_b, \; T_{ab} = \beta_{aa}T_a + \beta_{ab}T_b, \tag{12}$$

Coefficients $\alpha_{aa}$, $\alpha_{ab}$, $\beta_{aa}$ and $\beta_{ab}$ are chosen in such manner that both conservation laws and balance equations for the momenta and energy for each component hold valid. Since the number of equations that should be satisfied by the parameters of the model (for a five-moment description of a two-component system there are four equations: 2 for the balance of moments and 2 for the balance of temperature) is less than the number of unknown parameters (in this approximation there are 5:$\nu_{ab}$; $\alpha_{aa}$; $\alpha_{ab}$; $\beta_{aa}$; $\beta_{ab}$), there is an arbitrariness in the choice of parameters. Therefore there exist various modifications (see, for example, [13]) of the collision model which correctly describe relaxation of the five moments. But, probably, the most dubious point of the GK model is the complicated exponential dependence on the d.f. Recently, a new type of G-K model for gas mixtures [10, 14] was proposed and Boltzmann's H- theorem was proven for this model.

## 2. Model Construction

### *2.1. One-component systems*

To correct the BGK model, following Sirovich [15] we introduce two projection operators $\widehat{H}$ and $\widehat{N}$ satisfying

$$\widehat{H}\widehat{N} = \widehat{N}\widehat{H} = 0; \; \widehat{H} + \widehat{N} = \widehat{Id}. \tag{13}$$

Here $\widehat{Id}$ is the identity operator, $\widehat{H}$ is the operator of projection onto the 'hydrodynamical subspace' spanned by kets corresponding to the polynomials of the lowest order in the moment variable. In the BGK model these kets are the first five polynomials which correspond to the collisional invariants: density, momentum and kinetic energy. However, one may include in this subspace polynomials of higher order. Their number and order depend on the physical processes that one wishes to treat "exactly". Thus, one may take into account non-invariant values like those of the pressure tensor and heat flux. The projection operator $\widehat{N}$ maps the state vector onto the remaining "non-hydrodynamical subspace". Since we are interested in a model operator describing the first 13 moments correctly we take the operator $\widehat{H}$ in the following form:

$$\widehat{H} = \sum_{i=1}^{13} |\Psi_i\rangle \langle \Psi_i| \,. \tag{14}$$

The linearized collision operator is the following:

$$\delta\widehat{I} = \widehat{H}\delta\widehat{I}\widehat{H} + \widehat{H}\delta\widehat{I}\widehat{N} + \widehat{N}\delta\widehat{I}\widehat{H} + \widehat{N}\delta\widehat{I}\widehat{N}. \tag{15}$$

Since the first five Hermite polynomials are the eigenfunctions of the collision operator for identical particles corresponding to the zero eigenvalue, it follows that for $1 \leq i \leq 5$

$$\widehat{H}\delta\widehat{I}\widehat{H} = \widehat{H}\delta\widehat{I}\widehat{N} = \widehat{N}\delta\widehat{I}\widehat{H} = 0. \tag{16}$$

The higher Hermite polynomials are eigenfunctions of the collision operator only for a Maxwell molecule. In this case, non-diagonal matrix elements $\widehat{H}\delta\widehat{I}\widehat{N}$ and $\widehat{N}\delta\widehat{I}\widehat{H}$ vanish:

$$\widehat{H}\delta\widehat{I}\widehat{N} = \widehat{N}\delta\widehat{I}\widehat{H} = 0. \tag{17}$$

For any other interaction potentials the Hermite polynomials are not the eigenfunctions of the collision operator, the equality(17) does not hold and the collision operator matrix elements contain non-diagonal elements. Our first approximation is that we accept (17) as a valid formula for the Boltzmann gas of hard spheres and Coulomb plasma. However, the first approximation is not sufficient for describing real gas and plasma. In the second approximation we take into account only the non-diagonal terms closest to the diagonal. As we will show below, the corrections for a Boltzmann gas of hard spheres turns out to be small, but for Coulomb systems they are not small and play a major role in the Spitzer corrections to transport coefficients. We can continue this process and take into account in the third approximation the next, non-diagonal, terms more distant from the diagonal elements. We performed these calculations and found that the third approximation yields very small corrections (compared to the Spitzer approximation), that can be neglected.

Since for one-component systems the operator is Hermitian and isotropic, Wigner-Ekkart theorem implies

$$\left\langle Y_l^m h_1(u^2) \right| \delta\widehat{I} \left| Y_{l'}^{m'} h_2(u^2) \right\rangle = \delta_{ll'} \delta_{mm'} \left\langle\left\langle Y_l^m h_1 \right| \delta\widehat{I} \left| Y_l^m h_2 \right\rangle\right\rangle, \tag{18}$$

where $h_1(u^2)$,$h_2(u^2)$ are arbitrary functions , $Y_l^m$–are spherical harmonics and $\langle\langle \,|\,|\, \rangle\rangle$ means the reduced matrix element independent of m. From (18) the selection rule follows: the contribution to the non-diagonal matrix elements $\widehat{H}\delta\widehat{I}\widehat{N}$ and $\widehat{N}\delta\widehat{I}\widehat{H}$ is given only by polynomials with identical pairs of orbital numbers $l$ and $m$. For example, for the polynomial $|\Psi_6\rangle = \frac{\sqrt{3}}{2}(u_x u_x - \frac{1}{3}u^2)$ defining the $xx$ component of the pressure tensor, the non-zero contribution to the non-diagonal matrix elements is given by non-hydrodynamical polynomials of higher order in $u^2$ but with the same values of $l$ and $m$ $(l = 2; m = 2)$. For example, $\left|\Psi_6^{(2)}\right\rangle =$

$\sqrt{\frac{3}{14}\frac{1}{2}}(u^2-7)(u_x u_x - \frac{1}{3}u^2)$'The main modeling procedure consists of approximating the non-hydrodinamical contribution. If the operator $\widehat{H}$ involves the first 13 Hermite polynomials, then the neglect of the term $\widehat{N}\delta\widehat{I}\widehat{N}$ does not affect calculations for such transport coefficients as viscosity and heat conductivity. Nevertheless the approximation

$$\widehat{N}\delta\widehat{I}\widehat{N} = -\nu\widehat{N} \tag{19}$$

allows one to describe at least qualitatively the 'tails' of neglected 'non-hydrodynamical' terms ($\nu$ corresponds to the longest non-hydrodynamical relaxation time). An account of these 'tails' may be important at the kinetic level of fluctuation description. Using this approximation one may rewrite, in a first approximation, that corresponds to the Maxwell's molecule, the model operator as follows:

$$\delta\widehat{I} = -\nu\widehat{Id} + \widehat{H}(\delta\widehat{I} + \nu)\widehat{H}. \tag{20}$$

For a 13 moment basis for $\widehat{H}$ in the first approximation

$$\delta\widehat{I}\,|h\rangle = -\nu\,|h\rangle + \nu\sum_{i=1}^{5}|\Psi_i\rangle\,\langle\Psi_i|h\rangle + \sum_{i=6}^{13}|\Psi_i\rangle\,(\langle\Psi_i|\,\delta\widehat{I}\,|\Psi_i\rangle + \nu)\,\langle\Psi_i|h\rangle\,. \tag{21}$$

In the second approximation, in (21) appear the nearest non-diagonal entries:

$$\sum_{i=6}^{13}|\Psi_i\rangle\,\langle\Psi_i|\,\delta\widehat{I}\left|\Psi_i^{(2)}\right\rangle\left\langle\Psi_i^{(2)}|h\right\rangle, \tag{22}$$

where the non-hydrodynamical polynomials, which we take into account, are

$$\left|\Psi_i^{(2)}\right\rangle = \frac{1}{\sqrt{14}}(u^2-7)\,|\Psi_i\rangle\,,\ 6 \le i \le 10 \tag{23}$$

$$\left|\Psi_{r+10}^{(2)}\right\rangle = \frac{1}{\sqrt{280}}(u^4 - 14u^2 + 35)\,|\Psi_r\rangle\,,\ 1 \le r \le 3. \tag{24}$$

To close the terms (22) we use the following equation for the non-hydrodynamical moments in the Fourier presentation

$$(-i\omega - \left\langle\Psi_i^{(2)}\right|\delta\widehat{I}\left|\Psi_i^{(2)}\right\rangle)\left\langle\Psi_i^{(2)}|h\right\rangle_\omega = \left\langle\Psi_i^{(2)}\right|\delta\widehat{I}\,|\Psi_i\rangle\,\langle\Psi_i|h\rangle_\omega \tag{25}$$

Thus in the second approximation, linearized model collision operator has the form

$$\delta\widehat{I}\,|h\rangle_\omega = -\nu\,|h\rangle_\omega + \nu\sum_{i=1}^{5}|\Psi_i\rangle\,\langle\Psi_i|h\rangle_\omega - \sum_{i=6}^{13}|\Psi_i\rangle\,({}_i\Lambda_i^{(2)}(\omega) - \nu)\,\langle\Psi_i|h\rangle_\omega\,, \tag{26}$$

where

$${}_i\Lambda_i^{(2)}(\omega) = -\,\langle\Psi_i|\,\delta\widehat{I}\,|\Psi_i\rangle - \frac{\langle\Psi_i|\,\delta\widehat{I}\left|\Psi_i^{(2)}\right\rangle^2}{-i\omega - \left\langle\Psi_i^{(2)}\right|\delta\widehat{I}\left|\Psi_i^{(2)}\right\rangle} \tag{27}$$

contains the square of the non-diagonal entries and the projection of the kinetic equation resolvent to the non-hydrodynamical subspace. Here we take into account non-stationarity of non-hydrodynamical moments. Thus, although the original collision integral is Markovian, the part projected onto the supspace of 13 moments becomes, in the second approximation, a frequency dependent operator. A similar situation occurs in quantum-mechanical perturbation theory. Note that in the Markov approximation the second order corrections in (27)are negative for any interaction potentials.

Calculate now matrix elements of the operator for a concrete interaction potential, namely for Coulomb plasma and a Boltzmann hard sphere gas. For the Coulomb interaction take the linearized collision operator in the Balescu-Lenard form. In this case the matrix elements are of the form

$$\langle\Psi_i|\,\delta\widehat{I}\,|\Psi_j\rangle = -\int d\mathbf{p}d\mathbf{p}'d\mathbf{k}\frac{e^4k_rk_s\delta(\mathbf{kv}-\mathbf{kv}')}{k^4\left|\varepsilon(\mathbf{kv},\mathbf{k})\right|^2}f^0(\mathbf{p})f^0(\mathbf{p}') \tag{28}$$

$$\times(\frac{\partial\Psi_i}{\partial\mathbf{p}_r}-\frac{\partial\Psi_i}{\partial\mathbf{p}'_r})(\frac{\partial\Psi_j}{\partial\mathbf{p}_s}-\frac{\partial\Psi_j}{\partial\mathbf{p}'_s}).$$

For i=j the entries are

$$\langle\Psi_i|\;\delta\widehat{I}\,|\Psi_i\rangle \leq 0 \tag{29}$$

for any polynomials. Equality to zero corresponds to the five-time-degenerate zero eigenvalue. It is easy to see that the matrix elements of the operator $\widehat{H}\delta\widehat{I}\widehat{H}$ have following values:

$$\langle\Psi_i|\;\delta\widehat{I}\,|\Psi_j\rangle = -\;\delta_{ij}[\;\Lambda_1\sum_{k=6}^{10}\delta_{ik}+\;\Lambda_2\sum_{k=11}^{13}\delta_{ik}], \tag{30}$$

where $\Lambda_1,\Lambda_2$ are the relaxation frequencies of the pressure tensor and the heat flux vector which in terms of plasma parameters are as follows:

$$\Lambda_1 = \frac{8}{5}\frac{ne^4\sqrt{\pi}}{m^{1/2}T^{3/2}}Ln;\;\; \Lambda_2 = \frac{2}{3}\;\Lambda_1. \tag{31}$$

The matrix elements of the operators $\widehat{N}\delta\widehat{I}\widehat{N}$ and $\widehat{H}\delta\widehat{I}\widehat{N}$ defining 'tails' and the second approximation for the Coulomb plasma are equal to [16]

$$\langle\Psi_{14}|\;\delta\widehat{I}\,|\Psi_{14}\rangle = -\frac{2}{3}\;\Lambda_1;\;\Psi_{14} = \frac{1}{\sqrt{120}}(u^4-10u^2+15)$$

$$\langle\Psi_{15}|\;\delta\widehat{I}\,|\Psi_{15}\rangle = ... = \langle\Psi_{21}|\;\delta\widehat{I}\,|\Psi_{21}\rangle = -\frac{3}{2}\;\Lambda_1;\;\Psi_{15} = u_xu_yu_z$$

$$\langle\Psi_{22}|\;\delta\widehat{I}\,|\Psi_{22}\rangle = ... = \langle\Psi_{35}|\;\delta\widehat{I}\,|\Psi_{35}\rangle = -\frac{191}{16}\;\Lambda_1;$$

$$\Psi_{22} = \frac{1}{\sqrt{105}}\frac{1}{8}(35u_x^4-30u^2u_x^2+3u^4);$$

$$\left\langle\Psi_i^{(2)}\middle|\,\delta\widehat{I}\,\middle|\Psi_i^{(2)}\right\rangle = -\frac{201}{168}\Lambda_1;\;\left\langle\Psi_i\middle|\,\delta\widehat{I}\,\middle|\Psi_i^{(2)}\right\rangle = \frac{3}{2\sqrt{14}}\Lambda_1;\;6\leq i\leq 10$$

$$\left\langle\Psi_{10+r}^{(2)}\middle|\,\delta\widehat{I}\,\middle|\Psi_{10+r'}^{(2)}\right\rangle = -\frac{15}{14}\delta_{rr'}\Lambda_1;\;\left\langle\Psi_{10+r}\middle|\,\delta\widehat{I}\,\middle|\Psi_{10+r'}^{(2)}\right\rangle = \frac{1}{7}\delta_{rr'}\Lambda_1;\quad 1\leq r\leq 3. \tag{32}$$

This estimate implies that the higher tensor character of the polynomial leads to higher values of the matrix elements and for polynomials with the same tensor character the matrix elements are greater for polynomials of higher order in $u^2$. The smallest value of the diagonal matrix elements for the non-hydrodynamical polynomial is achieved for the $\Psi_{14}$ polynomial and equals $\Lambda_2$ (the heat flux relaxation frequency). The same holds for Maxwell's molecule and the Boltzmann gas hard sphere at least. Thus in the first approximation the collision operator is of the form:

$$\delta\widehat{I}\,|h\rangle = -\frac{2}{3}\Lambda_1\,|h\rangle + \frac{2}{3}\Lambda_1\sum_{i=1}^{5}|\Psi_i\rangle\,\langle\Psi_i|h\rangle - \frac{1}{3}\Lambda_1\sum_{i=6}^{10}|\Psi_i\rangle\,\langle\Psi_i|h\rangle\,. \tag{33}$$

We see that the polynomials corresponding to the heat flux disappear. In the second approximation

$$\delta\widehat{I}\,|h\rangle_\omega = -\nu(|h\rangle_\omega - \sum_{i=1}^{13}|\Psi_i\rangle\,\langle\Psi_i|h\rangle_\omega)$$
$$-\sum_{i=6}^{10}\Lambda_1^{(2)}(\omega)\,|\Psi_i\rangle\,\langle\Psi_i|h\rangle_\omega - \sum_{i=11}^{13}\Lambda_2^{(2)}(\omega)\,|\Psi_i\rangle\,\langle\Psi_i|h\rangle_\omega\,, \tag{34}$$

where

$$\Lambda_1^{(2)}\,(\omega) = \Lambda_1(1 - \frac{\Lambda_1 9/56}{-i\,\omega + \Lambda_1 205/168});\ \Lambda_2^{(2)}\,(\omega) = \Lambda_2(1 - \frac{\Lambda_1 3/14}{-i\,\omega + \Lambda_1 15/14}). \tag{35}$$

In the Markov approximation ($\omega = 0$) this leads to Spitzer values [17] of transport coefficients.

$$\Lambda_1^{(2)} = \Lambda_1(1 - \frac{27}{205});\ \ \Lambda_2^{(2)} = \Lambda_2(1 - \frac{1}{5}). \tag{36}$$

This correction are rather significant. In the third approximation the relaxation frequencies vary by no more than one per cent. Thus our model leads to Spitzer kinetic coefficients.

The second-order corrections for Boltzmann gas are one order less than for Coulomb plasma and one may stop at the first approximation. Thus in the case of a Boltzmann gas of hard sphere the collision operator may be represented in the form [18]

$$I\{f\} = -\nu\{f - f^0(1 - P_{ij}\frac{\delta v_i \delta v_j}{4PT}m)\}, \tag{37}$$

where

$$P_{ij} = m\int \mathbf{dp}f(\delta v_i\delta v_j - \delta_{ij}\frac{\delta v^2}{3}) \tag{38}$$

is the pressure tensor.

In the equilibrium state, $P_{ij} = 0$ and $I\{f^0\} = 0$. In the equation for the heat flux only the first term in (37) contributes. The relaxation of the pressure tensor is determined by both the first and the last terms in this equation.

Now we will prove the $H$ theorem for our model (37):

$$\frac{\partial}{\partial t}H(t) = \nu\int(f-\Phi)\log\frac{f}{\Phi}\mathbf{dp} + \nu\int(f-\Phi)\log\Phi\mathbf{dp},\quad \text{where}\ \ \Phi = f^0(1 - P_{ij}\frac{\delta v_i\delta v_j}{4PT}m)$$
$$= \nu\int(f-\Phi)\log\frac{f}{\Phi}\mathbf{dp} + \nu\int[f - f^0(1-P_{ij}\frac{\delta v_i\delta v_j}{4PT}m)][\log f^0 + \log(1-P_{ij}\frac{\delta v_i\delta v_j}{4PT}m)]\mathbf{dp}, \tag{39}$$

The first term in the second square bracket in (39) vanishes, while the second term can be expended as: $\log(1+x) = x - x^2/2 + x^3/3...$; for $-1 < x \leq 1$. Hear only the first term contributes after the integration.

Therefore:

$$\frac{\partial}{\partial t}H(t) = \nu\int(f-\Phi)\log\frac{f}{\Phi}d\mathbf{p}+\nu\int[f-f^0(1-P_{ij}\frac{\delta v_i\delta v_j-\delta_{ij}\frac{\delta v^2}{3}}{4PT}m)](-P_{ij}\frac{\delta v_i\delta v_j-\delta_{ij}\frac{\delta v^2}{3}}{4PT}m)d\mathbf{p}. \tag{40}$$

Taking into account that

$$\int f^0(\delta v_i\delta v_j-\delta_{ij}\frac{\delta v^2}{3})(\delta v_k\delta v_l-\delta_{kl}\frac{\delta v^2}{3})d\mathbf{p} = \frac{T^2}{m^2}(\delta_{ik}\delta_{jl}+\delta_{il}\delta_{jk}-\frac{2}{3}\delta_{ij}\delta_{kl}), \tag{41}$$

we obtain the $H$ theorem:

$$\frac{\partial}{\partial t}H(t) = \nu\int(f-\Phi)\log\frac{f}{\Phi}d\mathbf{p}-\frac{\nu}{8P^2}P_{ij}P_{ij} \le \nu\int(f-\Phi)\log\frac{f}{\Phi}d\mathbf{p} \le \mathbf{0} \tag{42}$$

Thus our collision integral in the form (37) possesses all necessary properties and is free from the drawbacks of the one-component model of the BGK model mentioned above. The linearized form of (37) is congruent with the linearized ellipsoidal statistical model [2-3].

Earlier, another model correctly describing the viscosity and thermal conductivity relaxation was proposed ad hoc [19]:

$$I\{f\} = -\nu\{f-f^0[1-\mathrm{Pr}\frac{\mathbf{J}\delta\mathbf{v}}{5P}(\frac{\delta v^2}{mT}-5)]\}, \tag{43}$$

where $\mathbf{J}$ is the heat flux. But this model does not give a correct description of non-hydrodynamic 'tails'.

### *2.2. Many-component systems*

Using the technique described above for a one-component system, one may get the following expression for the linearized model collision operator of a many-component system in the five-moment approximation [18]:

$$\delta\widehat{I}_{a\mathbf{p}}\delta f_a(\mathbf{p}) = -\nu_a\delta f_a(\mathbf{p})+\sum_{j=1}^{5}\nu_a f_a^0(\mathbf{p})\Psi_j^a(\mathbf{p})\int\Psi_j^a(\mathbf{p}')\delta f_a(\mathbf{p}')d\mathbf{p}'$$

$$+\sum_b\sum_{i,j=1}^{5} f_a^0(\mathbf{p})\Psi_i^a(\mathbf{p})\left\langle\Psi_i^a\right|\delta\widehat{I}\left|\Psi_j^b\right\rangle\int\Psi_j^b(\mathbf{p}')\delta f_b(\mathbf{p}')d\mathbf{p}', \tag{44}$$

where $f_a^0(\mathbf{p})$ -is the local equilibrium distribution function (with different temperatures and mean velocities), $\nu_a$ -is the inverse time of the heat flux relaxation of component $a$, and $\left\langle\Psi_i^a\right|\delta\widehat{I}\left|\Psi_j^b\right\rangle$ represents the matrix elements of the linearized collision integral (of the Balescu-Lenard integral, for example).

In order to recover the form of the model collision integral from its linearized form it suffices to use conservation of the number of elastically interacting particles. This property, as well as total momentum and energy conservation, is valid for both mean and fluctuating quantities. Consequently, the expression for Langevin's source intensity in the kinetic equation for the distribution function fluctuation [20]

$$(\frac{\partial}{\partial t}+\mathbf{v}\frac{\partial}{\partial\mathbf{r}}+\delta\widehat{I}_{a\mathbf{p}})\delta f_a(\mathbf{x},t)+\delta\mathbf{F}\frac{\partial}{\partial\mathbf{p}}f_a(\mathbf{x},t) = y_a(\mathbf{x},t) \tag{45}$$

should satisfy the conditions

$$\sum_b \int \Psi^b(\mathbf{p}_2)(y_a y_b)_{\omega,\mathbf{k},\mathbf{p}_1,\mathbf{p}_2} d\mathbf{p}_2 = 0 \text{ for } \Psi^b(\mathbf{p}_2) = 1,\ \mathbf{p}_2,\ p_2^2/2m_b\ . \tag{46}$$

The spectral function of the Langevin source in a non-equilibrium state is given [21, 22] by following form:

$$(y_a y_b)_{\omega,\mathbf{k},\mathbf{p}_1,\mathbf{p}_2} = -(\delta\widehat{I}_{a\mathbf{p}_1} + \delta\widehat{I}_{b\mathbf{p}_2})\delta_{ab}\delta(\mathbf{p}_1 - \mathbf{p}_2) f_a(\mathbf{p}_1) + \delta_{ab}\delta(\mathbf{p}_1 - \mathbf{p}_2) I_a(\mathbf{p}_1) + I_{ab}(\mathbf{p}_1, \mathbf{p}_2)\ , \tag{47}$$

where$I_{ab}(\mathbf{p}_1, \mathbf{p}_2)-$ is the so-called "not integrated"' collision operator:$\sum_b \int I_{ab}(\mathbf{p}_1, \mathbf{p}_2) d\mathbf{p}_2 = I_a(\mathbf{p}_1)$.
In the case of plasma the "not integrated" collision operator has the form:

$$I_{ab}(\mathbf{p}_1, \mathbf{p}_2) = 2e_a^2 e_b^2 n_b\left(\frac{\partial}{\partial p_i} - \frac{\partial}{\partial p_i'}\right) \int \frac{k_i k_j \delta(\mathbf{kv} - \mathbf{kv}')}{k^4 \left|\varepsilon(\mathbf{kv}, \mathbf{k})\right|^2} \left(\frac{\partial}{\partial p_j} - \frac{\partial}{\partial p_j'}\right) f_a f_b d\mathbf{k}. \tag{48}$$

Summing (47) over $b$ and integrating over $\mathbf{p}_2$, and taking into account (46), we get

$$I_a(\mathbf{p}_1) = -\frac{1}{2}(\delta\widehat{I}_{a\mathbf{p}_1} + \delta\widehat{I}_{b\mathbf{p}_2})\delta_{ab}\delta(\mathbf{p}_1 - \mathbf{p}_2) f_a(\mathbf{p}_1) d\mathbf{p}_2. \tag{49}$$

Since(46) and(47) are of a general character, the relation (49) is valid both for "exact" and model collision integrals. Substituting (44) into (49), we obtain a quite simple and at the same time sufficiently rigorous form of the model collision integral for many-component plasma:

In the local equilibrium state the collision integral assumes the fairly simple form:

$$I_a(\mathbf{p}_1) = -\sum_b \nu_{ab} f_a^0(\mathbf{p}) \left[\delta\mathbf{v}_a m_a \frac{\mathbf{V}_a - \mathbf{V}_b}{T_a} + \left(\frac{m_a}{T_a}\delta\mathbf{v}_a^2 - 3\right)(T_a - T_b)\frac{m_a}{m_a + m_b}\right], \tag{50}$$

where $\nu_{ab}$ is the momentum relaxation frequency for plasma:

$$\nu_{ab} = \frac{4}{3}\sqrt{2\pi} e_a^2 e_b^2 n_b L \sqrt{\frac{m_a}{m_b}} \frac{(m_a + m_b)^{1/2}}{(m_a T_b + m_b T_a)^{3/2}},\ \text{and } \delta\mathbf{v}_a = \mathbf{v}_a - \mathbf{V}_a. \tag{51}$$

In the second term in (50) we took into account the isothermal case and in the third term we neglected corrections of the square mean velocities. The first term in (50) describe relaxation to the local equilibrium state, the second term describes relaxation of the momenta and the last term the temperature relaxation. In this case it is easy to verify the Boltzmann H-theorem for this form of the collision integral.

$$\frac{\partial}{\partial t} H(t) = -\sum_a \int \frac{\delta\mathbf{p}^2}{2m_a T_a} I_a(\mathbf{p}_1) d\mathbf{p} = -\frac{3}{4}\sum_{ab} \nu_{ab} \frac{m_a n_a}{m_a + m_b} \frac{(T_a - T_b)^2}{T_a T_b} \leq 0, \tag{52}$$

Thus, the complicated exponential dependency typical for the GK model appears to be unfounded and does not hold for states remote from the full equilibrium.

$$I_a(\mathbf{p}_1) = -\nu_a[f_a(\mathbf{p}) - f_a^0(\mathbf{p})(1 - P_{ij}\frac{\delta v_i \delta v_j}{4PT}m)]$$
$$-\sum_b \nu_{ab} f_a^0(\mathbf{p}) \left[\delta \mathbf{v}_a m_a \frac{\mathbf{V}_a - \mathbf{V}_b}{T_a} + \left(\frac{m_a}{T_a}\delta \mathbf{v}_a^2 - 3\right)(T_a - T_b)\frac{m_a}{m_a + m_b}\right] \quad (53)$$

Thus, the time evolution of the two-component gas up to the hydrodynamic stage can be described as follows: first, the lighter component achieves equilibrium, then the balanced state of the heavy component, and finally a balance across the gas mixture is achieved. At all these stages Boltzmann's H-theorem holds.

## 3. Conclusion

Using the well-known projection technique, a new form of the collision operator for a Boltzmann gas of hard spheres and for Coulomb plasma has been developed. The proposed collision operator takes into account relaxation of the first 13 hydrodynamic moments properly and accounts for the contribution of non-diagonal components in the expansion of the linearized collision operator in the complete system of Hermite polynomials. The non-diagonal components accounted for in this basis in the quadratic approximation contribute to the diagonal components. It is shown that for a system of charged particles with a Coulomb interaction potential, these contributions are essential and lead to Spitzer corrections to the transport coefficients. In the case of a Boltzmann gas of hard spheres these corrections are insignificant. In the case of a many-component system, the nonlinear model collision integral is constructed on the basis of the linearized one. Unlike previous cases, it does not exhibit any complicated exponential dependence and avoids coefficient ambiguity in the many-component collision integral.

## 4. References

[1] Bhatnagar P L, Gross E P and Krook M 1954 *Phys. Rev.* **94** 511
[2] Holway L H 1966 *Phys. Fluids* **9** 1658
[3] Cercignani C and Tironi C 1967 *Rarefied Gas Dynamics vol 1* ed C L Brundin (New York: Academic); Cercignani C 1969 *Mathematical Methods in Kinetic Theory* (New York: Plenum)
[4] Gross E P and Jackson E A 1959 *Phys. Fluids* **2**, 432
[5] Furtado P M, Mazenko G F and Yip S 1976 *Phys. Rev.* A **13** 1641
[6] Groome L, Dufty J W and Lindenfeld M J 1979 *Phys. Rev.* A **19** 304
[7] Kamgar-Parsi B, Cohen E G D and I. M. de Schepper 1987 *Phys. Rev.* A **35** 4781
[8] Dufty J W 1990 *Lectures on Thermodynamics and Statistical Mechanics* ed M. Lopez de Haro and C. Varea (New Jersey: World Scientific)
[9] Santos A, Montanero J M, Dufty J W and Brey J J 1998 *Phys. Rev.* E **57** 1644
[10] Groppi M, Monica S, and Spiga G 2011 *Euro-Phys.Lett.* **96**, 63002
[11] Brull S, Pavan V, and Schneider J 2012 *Eur. J. Mech.*-B/Fluids **33** 74
[12] Gross E P and Krook M 1956 *Phys. Rev.* **103** 593
[13] Green J M 1973 *Phys. Fluids.* **16** 2023
[14] Akasaki I, Sota S, Sakai H, Tanaka T, Koike M and Amano H 1996 *Electron. Lett.* **32** 1105
[15] Sirovich L 1966 *Phys. Rev.* **9** 2323
[16] Belyi V V and Paiva-Veretennicoff I 1985 *Theor.Math. Phys.* **62**, 196
[17] Spitzer Jr.L 1956 *Physics of Fully ionized Gases* (New York: Interscience)
[18] Belyi V V 2009 *J. Stat. Mech.* P06001
[19] Shahov E M 1968 *Izv. AN USSR* (Soviet) ser.MJG **1** 156
[20] Lifshitz E M and Pitaevskii L P 1979 Physical Kinetics **X** (Moscow: Nauka)
[21] Kogan S M and Shulman A I 1969 *Sov. Phys. JETPH* ***56*** *862*
[22] Gantsevich S A, Gurevich V L and Katilius R 1979 *Riv. Nouovo Cimento* **2** 1

Journal of Physics: Conference Series **510** (2014) 012044 doi:10.1088/1742-6596/510/1/012044

# Developments of large eddy simulation for compressible space plasma turbulence

**A.A. Chernyshov**[1], **K.V. Karelsky**[1] **and A.S. Petrosyan**[1,2]

[1] Theoretical section, Space Research Institute of Russian Academy of Sciences, 84/32 Profsoyuznaya, 117997, Moscow, Russia

[2] Moscow Institute of Physics and Technology (State University), 9 Institutskii per., 141700, Dolgoprudny, Moscow Region, Russia

E-mail: apetrosy@iki.rssi.ru

**Abstract.** In this paper, brief review of studies of compressible space plasma turbulence is given. Magnetic fields and compressibility are present that are different from properties of hydrodynamic incompressible turbulence of neutral fluid. This considerably complicates studies of characteristics of compressible magnetohydrodynamic turbulence with direct numerical simulation. Large eddy simulation method developed as alternative of direct modeling is discussed in the review. The method is based on filtrating procedure of governing equations of magnetohydrodynamics and subgrid-scale modeling of universal small-scale turbulence. Descriptions of both the method and different subgrid-scale models for compressible magnetohydrodynamic turbulent flows are present. It is shown that large eddy simulation has good future trends for research of compressible magnetohydrodynamic turbulence.

## 1. Introduction

Numerical simulation of turbulent magnetohydrodynamic (MHD) flows is an effective tool for the study of the flows of the charged fluid of astrophysical, helio- and geophysical plasma (for instance, solar corona expansion, solar wind, flows in the solar convection zone, turbulence in interstellar matter), that is beyond the reach of direct experimental study. Besides, MHD turbulence are widely encountered in various applied areas, for example, it is possible to mention actual studies of the noise decrease, using plasma actuator for control of large vortex structures in jets, investigations of decrease of turbulent flow resistance in aerospace industry, MHD flow in a channel for steel-casting process and in pipe for cooling of nuclear fusion reactors.

Complete information about turbulent fluid flow could be obtained by means of direct numerical simulation (DNS) that is in the numerical solution of full nonstationary system of magnetohydrodynamic equations. This approach allows resolving all scales of charged fluid flows and does not require special closures for magnetohydrodynamic equations. However, direct numerical computation of MHD-turbulence faces fundamental difficulties related to the large hydrodynamic and magnetic Reynolds numbers typical for studied processes. In that case the number of degrees of freedom of turbulent flow is large and minimal number of mesh points must be so large that application of direct numerical simulation for study of turbulent flows with real Reynolds numbers is limited by available computational resources.

Large eddy simulation (LES) approach describes approximate turbulence dynamics, where the large-scale part of turbulent flow is computed directly, while the small-scale one is modeled.

Possibility of using filtration operation in LES for decomposition of turbulent flow characteristics into large-scale and small-scale parts is due to sufficient isotropy, homogeneity and universality of small scales of turbulent flow. Under these conditions small-scale flow is eliminated from the initial system of flow equations with the use of filtration procedure and then their influence is taken into account by means of subrgrid-scale (SGS) models (otherwise called by subfilter scale (SFS)), parameterizing all the magnetohydrodynamic values of small-scale turbulent flow through the filtered ones. Thus, large-scale motion is computed by means of solving of the filtered non-stationary equations of magnetohydrodynamics. LES is a method for simulation of flows with large Reynolds numbers, since LES assumes that the energy is transferred within the inertial subrange that results in reduction of number of degrees of freedom and significant increase of computation efficiency as compared with DNS.

Recently, LES approach for study of compressible forced and decaying MHD turbulence has been developed for the cases of polytropic gas and heat-conducting plasma [1, 2, 3, 4, 5, 6, 3, 7, 8, 9]. This paper is a review latest achievements in the development of LES method of compressible MHD turbulence. Here we restrict ourselves to the case of MHD turbulence of polytropic fluid. In this review, we briefly describe the basic concepts and principles of LES, the subgrid-scale modeling, dynamic procedure in MHD and the important results and conclusions.

## 2. Governing equations and Subgrid-scale models

In this section, we formulate the theory of large eddy simulation for compressible magnetohydrodynamics and obtain filtered MHD equations. Subgrid scale modeling in LES is discussed and various closure models for compressible MHD case are proposed.

To obtain the MHD equations governing the motion of the filtered (that is resolved) eddies, the large scales from the small are separated. LES approach is based on the definition of a filtering operation: a resolved (or large-scale) variable, denoted by an overbar in the present paper, is defined as

$$\bar{f}(x_i) = \int_{\Theta} f(\acute{x}_i) G(x_i, \acute{x}_i; \bar{\triangle}) d\acute{x}_i, \tag{1}$$

where $G$ is the filter function satisfying the normalization property, $\Theta$ is the domain, $\bar{\triangle}$ is the filter-width associated with the wavelength of the smallest scale retained by the filtering procedure and $x_j = (x, y, z)$ are axes of Cartesian coordinate system.

It is convenient to use the Favre filtration (it is also called mass-weighted filtration) to avoid additional SGS terms in compressible flow. Therefore, Favre filtering will be used further on. Mass-weighted filtering is determined as follows: $\tilde{f} = \overline{\rho f}/\bar{\rho}$, where the tilde denotes the mass-weighted filtration.

Thus, applying the Favre-filtering operation, we can rewrite the MHD equations for compressible fluid flow as:

$$\frac{\partial \bar{\rho}}{\partial t} + \frac{\partial \bar{\rho}\tilde{u}_j}{\partial x_j} = 0; \tag{2}$$

$$\frac{\partial \bar{\rho}\tilde{u}_i}{\partial t} + \frac{\partial}{\partial x_j}\left(\bar{\rho}\tilde{u}_i\tilde{u}_j + \frac{\bar{\rho}^{\gamma}}{\gamma M_s^2}\delta_{ij} - \frac{1}{Re}\,\tilde{\sigma_{ij}} + \frac{\bar{B}^2}{2M_a^2}\delta_{ij} - \frac{1}{M_a^2}\,\bar{B}_j\bar{B}_i\right) = -\frac{\partial \tau^u_{ji}}{\partial x_j}; \tag{3}$$

$$\frac{\partial \bar{B}_i}{\partial t} + \frac{\partial}{\partial x_j}\left(\tilde{u}_j\bar{B}_i - \tilde{u}_i\bar{B}_j\right) - \frac{1}{Re_m}\frac{\partial^2 \bar{B}_i}{\partial x_j^2} = -\frac{\partial \tau^b_{ji}}{\partial x_j}; \tag{4}$$

$$\frac{\partial \bar{B}_j}{\partial x_j} = 0, \tag{5}$$

Here $\rho$ is the density; $u_j$ is the velocity in the direction $x_j$; $B_j$ is the magnetic field in the direction $x_j$; $\sigma_{ij} = 2\mu S_{ij} - \frac{2}{3}\mu S_{kk}\delta_{ij}$ is the viscous stress tensor; $S_{ij} = 1/2\left(\partial u_i/\partial x_j + \partial u_j/\partial x_i\right)$

Journal of Physics: Conference Series **510** (2014) 012044 doi:10.1088/1742-6596/510/1/012044

is the strain rate tensor; $\mu$ is the coefficient of molecular viscosity; $\eta$ is the coefficient of magnetic diffusivity; $\delta_{ij}$ is the Kronecker delta.

The filtered magnetohydrodynamic equations (2) - (5) are written in the dimensionless form using the standard procedure [10] where $Re = \rho_0 u_0 L_0/\mu_0$ is the Reynolds number, $Re_m = u_0 L_0/\eta_0$ is the magnetic Reynolds number. $M_s = u_0/c_s$ is the Mach number, where $c_s$ is the velocity of sound defined by the relation $c_s = \sqrt{\gamma p_0/\rho_0}$, and $M_a = u_0/u_a$ is the magnetic Mach number, where $u_a = B_0/(\sqrt{4\pi\rho_0})$ is the Alfvén velocity.

To close the equations (2) - (4) it is assumed that the relation between density and pressure is polytropic and has the following form: $p = \rho^\gamma$, where $\gamma$ is a polytropic index and it is supposed that $\gamma = 5/3$ .

The effect of the SGSs appears on the right-hand side of the governing equations (3) - (4) through the SGS stresses:

$$\tau_{ij}^u = \bar{\rho}\left(\widetilde{u_i u_j} - \tilde{u}_i \tilde{u}_j\right) - \frac{1}{M_a^2}\left(\overline{B_i B_j} - \bar{B}_i \bar{B}_j\right); \tag{6}$$

$$\tau_{ij}^b = \left(\overline{u_i B_j} - \tilde{u}_i \bar{B}_j\right) - \left(\overline{B_i u_j} - \bar{B}_i \tilde{u}_j\right). \tag{7}$$

Consequently, the filtered system of magnetohydrodynamic equations contains the unknown turbulent tensors: $\tau_{ij}^u$ and $\tau_{ij}^b$. To determine their components special turbulent closures (models, parameterizations) based on large-scale values describing turbulent magnetohydrodynamic flow must be used. The main idea of any SGS closures used in LES is to reproduce the effects of the subgrid scale dynamics on the large-scale energy distribution, at that as a matter of fact Richardson turbulent cascade is simulated. In order to close the system of MHD equations we should find such parameterizations for $\tau_{ij}^u$ and $\tau_{ij}^b$ that would relate these tensors to the known large-scale values of the flow parameters.

In this paper we consider five SGS models and three of them (the Smagorinsky model for MHD case, Kolmogorov model for MHD-turbulence, model based on cross-helicity) use eddy-viscosity assumption to try to simulate the diffusive transport and dissipation of kinetic and magnetic energy. One of them (scale-similar model) is based on the assumption that the most active SGSs are those closer to the cutoff, and that the scales with which they interact are those immediately above the cutoff wave number. Also, the last SGS closures considered in present work is mixed model for MHD turbulence.

The first model is the Smagorinsky model. In [4], we represent its extension to the MHD-case for compressible turbulence, given by

$$\tau_{ij}^u - \frac{1}{3}\tau_{kk}^u \delta_{ij} = -2C_1 \bar{\rho} \bar{\triangle}^2 |\tilde{S^u}| \left(\tilde{S_{ij}} - \frac{1}{3}\tilde{S_{kk}}\delta_{ij}\right), \tag{8}$$

$$\tau_{ij}^b = -2D_1 \bar{\triangle}^2 |\bar{j}| \bar{J_{ij}}, \tag{9}$$

$$\tau_{kk}^u = 2Y_1 \bar{\rho} \bar{\triangle}^2 |\tilde{S^u}|^2 \tag{10}$$

where $\bar{j}$ is the resolved electric current density, $\tilde{S_{ij}} = \left(\partial\tilde{u_i}/\partial x_j + \partial\tilde{u_j}/\partial x_i\right)/2$ is a large-scale strain rate tensor, $\bar{J_{ij}} = \left(\partial\bar{B_i}/\partial x_j - \partial\bar{B_j}/\partial x_i\right)/2$ is a large-scale magnetic rotation tensor and $|\tilde{S^u}| = \left(2\tilde{S_{ij}^u}\tilde{S_{ij}^u}\right)^{1/2}$.

The generalized Kolmogorov scaling subgrid model for MHD compressible turbulence is

$$\tau_{ij}^u - \frac{1}{3}\tau_{kk}^u \delta_{ij} = -2C_2 \bar{\rho} \bar{\triangle}^{4/3} \left(\tilde{S_{ij}} - \frac{1}{3}\tilde{S_{kk}}\delta_{ij}\right), \tag{11}$$

$$\tau_{ij}^b = -2D_2 \bar{\triangle}^{4/3} \bar{J_{ij}}, \tag{12}$$

$$\tau_{kk}^{u} = 2Y_2\bar{\rho}\bar{\triangle}^{4/3}|\tilde{S^u}| \tag{13}$$

The third model examined in this paper uses the local cross-helicity change instead of the energy dissipation functions. Therefore, the model based on cross-helicity is defined as:

$$\tau_{ij}^{u} - \frac{1}{3}\tau_{kk}^{u}\delta_{ij} = -2C_3\bar{\rho}\bar{\triangle}^2|\tilde{f}|\left(\tilde{S_{ij}} - \frac{1}{3}\tilde{S_{kk}}\delta_{ij}\right), \tag{14}$$

$$\tau_{ij}^{b} = -2D_3\bar{\triangle}^2 sgn(\bar{j}\ \tilde{\omega})|\bar{j}\ \tilde{\omega}|^{1/2}\bar{J}_{ij}, \tag{15}$$

$$\tau_{kk}^{u} = 2Y_3\bar{\rho}\bar{\triangle}^2|\tilde{f}||\tilde{S^u}| \tag{16}$$

Here, $\bar{S_{ij}^b} = \left(\partial\bar{B}_i/\partial x_j + \partial\bar{B}_j/\partial x_i\right)/2$, $\tilde{S_{ij}} = \left(\partial\tilde{u}_i/\partial x_j + \partial\tilde{u}_j/\partial x_i\right)/2$, $\tilde{f} = |\tilde{S_{ij}}\ \bar{S_{ij}^b}|^{1/2}$, $\tilde{\omega} = \nabla\times\tilde{u}$ is the vorticity and function $sgn(.)$ determines the sign of the argument.

The scale-similarity model can be calculated in a LES by means of the filtered variables in contrast to other models. In a priori tests, the scale-similarity model reproduces the correlation between this model and actual turbulent stress tensor very well even when the flow is highly anisotropic [11]. This indicate that the similarity model predicts important structures of the turbulent stress at the right locations. One of the main advantages of the scale-similarity model is that it does not require a determination of special model constants in contrast to the eddy-viscosity parameterizations. The scale-similarity model are based on the assumption that the most active subgrid scales are those closer to the cutoff wave number, and that the scales with which they interact most are those right above the cutoff [12]. The "largest subgrid scales" can be obtained by filtering the SGS parameter, for example, velocity $\acute{u}_j = u_j - \tilde{u}_j$ to yield $\tilde{\acute{u}}_j = \tilde{u}_j - \tilde{\tilde{u}}_j$ The scale-similarity model for MHD compressible case is written in the following manner:

$$\tau_{ij}^{u} = \bar{\rho}\left(\widetilde{\tilde{u}_i\tilde{u}_j} - \tilde{\tilde{u}}_i\tilde{\tilde{u}}_j\right) - \frac{1}{M_a^2}\left(\overline{\bar{B}_i\bar{B}_j} - \bar{\bar{B}}_i\bar{\bar{B}}_j\right) \tag{17}$$

$$\tau_{ij}^{b} = \left(\overline{\tilde{u}_i\bar{B}_j} - \tilde{\tilde{u}}_i\bar{\bar{B}}_j\right) - \left(\overline{\bar{B}_i\tilde{u}_j} - \bar{\bar{B}}_i\tilde{\tilde{u}}_j\right) \tag{18}$$

The mixed model for MHD compressible turbulence is a combination of two models: the scale similarity and the Smagorinsky models for MHD case:

$$\tau_{ij}^{u} - \frac{1}{3}\tau_{kk}^{u}\delta_{ij} = -2C_5\bar{\rho}\bar{\triangle}^2|\tilde{S^u}|\left(\tilde{S_{ij}} - \frac{\delta_{ij}}{3}\tilde{S_{kk}}\right) + \bar{\rho}\left(\widetilde{\tilde{u}_i\tilde{u}_j} - \tilde{\tilde{u}}_i\tilde{\tilde{u}}_j\right) - \frac{1}{M_a^2}\left(\overline{\bar{B}_i\bar{B}_j} - \bar{\bar{B}}_i\bar{\bar{B}}_j\right) \tag{19}$$

$$\tau_{ij}^{b} = -2D_5\bar{\triangle}^2|\bar{j}|\bar{J}_{ij} + \left(\overline{\tilde{u}_i\bar{B}_j} - \tilde{\tilde{u}}_i\bar{\bar{B}}_j\right) - \left(\overline{\bar{B}_i\tilde{u}_j} - \bar{\bar{B}}_i\tilde{\tilde{u}}_j\right) \tag{20}$$

$$\tau_{kk}^{u} = 2Y_5\bar{\rho}\bar{\triangle}^2|\tilde{S^u}|^2 \tag{21}$$

The parameters $C_k$, $Y_k$ and $D_k$ ($k = 1, 2, 3$ or $5$) are model constants, their values being self-consistently computed during run time with the help of the dynamic procedure that explains in next section.

## 3. Dynamic procedure in MHD

In order to solve the problem concerning the choice of the constants, Germano suggested a dynamic procedure [13, 14] that determines the value of the constant at every time step. Then, this technique for determination of model constants has been applied for incompressible [15] and compressible MHD turbulence [1].

In this method an additional test filter with width more than the width of the basic one is used. Let $T_{ij}^u$ and $T_{ij}^b$ obtained with use of test filter be the test subgrid tensors for $\tau_{ij}^u$ and $\tau_{ij}^b$, respectively.

$$T_{ij}^{u} = \tilde{\tau_{ij}^{u}} + L_{ij}^{u}, \tag{22}$$

$$T_{ij}^{b} = \bar{\tau}_{ij}^{b} + L_{ij}^{b}, \quad (23)$$

where Leonard tensors:

$$L_{ij}^{u} = \left(\frac{\widehat{\overline{\rho u_i}\,\overline{\rho u_j}}}{\bar{\rho}}\right) - \frac{\widehat{\overline{\rho u_i}}\,\widehat{\overline{\rho u_j}}}{\hat{\bar{\rho}}} - \frac{1}{M_a^2}\left(\widehat{\bar{B}_i\bar{B}_j} - \hat{\bar{B}}_i\,\hat{\bar{B}}_j\right), \quad (24)$$

$$L_{ij}^{b} = \left(\left(\frac{\widehat{\overline{\rho u_i}\,\bar{B}_j}}{\bar{\rho}}\right) - \frac{\widehat{\overline{\rho u_i}}\,\hat{\bar{B}}_j}{\hat{\bar{\rho}}}\right) - \left(\left(\frac{\widehat{\bar{B}_i\,\overline{\rho u_j}}}{\bar{\rho}}\right) - \frac{\hat{\bar{B}}_i\,\widehat{\overline{\rho u_j}}}{\hat{\bar{\rho}}}\right). \quad (25)$$

Here, the sign $\widehat{\ }$ designates a test filtration of variable. The Leonard tensors $L_{ij}^{u}$ and $L_{ij}^{b}$ can be found from the large-scale values of velocity and magnetic field. The relations (24) and (25) connect initial subgrid and test tensors.

Let us define the turbulent viscosity and magnetic diffusion coefficients as follows: $\nu_t = C_s\alpha_{ij}^{u}$ (for $\tau_{ij}^{u}$) ), $\acute{\nu}_t = C_I\alpha^{u}$ (for $\tau_{kk}^{u}$), $\eta_t = D\varphi_{ij}^{B}$ (for $\tau_{ij}^{b}$).

The least-squares method for determination of the constants to minimize errors of functionals is applied [16]:

$$Z^{u} = \langle\left(T_{ij}^{u} - \hat{\tau}_{ij}^{u} - L_{ij}^{u}\right)^2\rangle, \quad (26)$$

$$Z^{b} = \langle\left(T_{ij}^{b} - \hat{\tau}_{ij}^{b} - L_{ij}^{b}\right)^2\rangle. \quad (27)$$

where the angle brackets denote spatial averaging.

Consequently, the constants are determined dynamically at every time step according to the following expressions:

$$C_s = \frac{\langle L_{ij}^{u}\ M_{ij}^{u}\rangle}{\langle M_{ij}^{u}\ M_{ij}^{u}\rangle}, \quad (28)$$

$$C_I = \frac{\langle L_{kk}^{u}\rangle}{\langle\hat{\alpha^{u}}|\hat{\bar{S}}^{u}| - \widehat{\alpha^{u}|\bar{S}^{u}|}\rangle}, \quad (29)$$

$$D = \frac{\langle L_{ij}^{b}\ m_{ij}^{B}\rangle}{\langle m_{ij}^{B}\ m_{lk}^{B}\rangle}, \quad (30)$$

where

$$M_{ij}^{u} = \hat{\alpha}_{ij}^{u}\left(\hat{\tilde{S}}_{ij}^{u} - \frac{\delta_{ij}}{3}\hat{\tilde{S}}_{kk}\right) - \widehat{\left[\alpha_{ij}^{u}\left(\tilde{S}_{ij}^{u} - \frac{\delta_{ij}}{3}\tilde{S}_{kk}\right)\right]}, \quad (31)$$

$$m_{ij}^{B} = \hat{\varphi}_B\hat{\bar{J}}_{ij} - \widehat{\varphi_B\bar{J}_{ij}}. \quad (32)$$

Here $\widehat{[\ .\ ]}$ indicates that test filtering procedure applies to the expression in square brackets. It should be noted, that the Favre-filtered task variables in the initial equations also have to be Favre-filtered in case of use of test filter. Here, denotation $\hat{\tilde{u}}$ signifies $\hat{\tilde{u}} = \widehat{\overline{\rho u}}/\hat{\bar{\rho}}$. The characteristic width of the test filter as a rule is usually taken twice more, i.e. $\hat{\Delta} = 2\bar{\Delta}$.

It is important to note, that the values of constants can be negative that correspond to energy backscatter, i.e. to the local gain of energy at the resolved scales.

Journal of Physics: Conference Series **510** (2014) 012044 doi:10.1088/1742-6596/510/1/012044

## 4. Aspects of numerics

In our works, numerical code of the fourth order of accuracy for MHD equations written in the conservative form is used. The skew-symmetric form is applied for the nonlinear terms, because the skew-symmetric form of the equation is a form obtained by averaging divergence form and convective form of the nonlinear term. In spite of analytical equivalence of all the three of these forms, their numerical computations give different results. To reduce error of discretization when finite difference scheme is used and to improve computational accuracy the skew-symmetric discretization is applied for nonlinear terms. For numerical solution of system of MHD equations, in this work we use the numerical code based on non-spectral finite-difference schemes. To separate the turbulent flow into large and small eddy components, discrete Gaussian filter of the fourth order of accuracy is applied.

For the performance evaluation of various subgrid-scale closures in the LES approach, direct numerical simulation results obtained by modeling of MHD turbulence over a various range of magnetic Reynolds number, hydrodynamic Reynolds number and Mach number are presented in detail in [4]. DNS calculations are performed on a sufficient grid size so that the accuracy of the obtained results provide real dynamics of three-dimensional compressible MHD turbulence and allow comparison with LES calculations. Initial data for all numerical calculations were chosen so that to provide a regime of developed turbulence and express the role of magnetic effects in MHD flow (see paper [4] for detailed information).

The results obtained by large eddy simulation are compared with DNS computations and performance of large eddy simulation is examined by difference between LES- and filtered DNS-results. The initial conditions for LES are obtained by filtering the initial conditions of DNS. The initial velocities and magnetic fields are chosen to resemble a MHD mixing layer by adapting the hydrodynamic case for study of decaying turbulence. The shear for velocity and magnetic field in the initial conditions is given $z-$ coordinate depending [4]. The Courant-Friedrichs-Levy stability condition (CFL-condition) is imposed on the time step. The uniform mesh with $64^3$ grid cells is used for LES and $256^3$ for DNS. For time integration the third order low-storage Runge-Kutta method is used. Periodic boundary conditions for all the three dimensions are applied, the simulation domain is a cube with dimensions of $2\pi \times 2\pi \times 2\pi$. The model constants in subgrid-scale closures are determined by means of the dynamic procedure and test filter with filter width twice as large as initial one is used.

For the case of forced MHD turbulence, the initial isotropic turbulent spectrum close to $k^{-2}$ with random amplitudes and phases in all three directions was chosen for kinetic and magnetic energies in Fourier space. The choice of such spectrum as initial conditions is due to velocity perturbations with an initial power spectrum in Fourier space similar to that of developed turbulence [17]. This $k^{-2}$ spectrum corresponds to spectrum of Burgers turbulence. Initial conditions for the velocity and the magnetic field have been obtained in the physical space using inverse Fourier transform. In this case, the simulation domain is a cube $\pi \times \pi \times \pi$.

## 5. LES of decaying compressible MHD turbulence

We perform three-dimensional numerical simulation of decaying compressible MHD turbulence. For the performance evaluation of various subgrid-scale closures in the LES approach, direct numerical simulation results obtained by modeling of MHD turbulence over a various range of magnetic Reynolds number, hydrodynamic Reynolds number and Mach number are presented in this survey. DNS calculations are performed on a sufficient grid size so that the accuracy of the obtained results provide real dynamics of three-dimensional compressible MHD turbulence and allow comparison with LES calculations.

Time dynamics of the kinetic and magnetic energy respectively are displayed in Fig. 1 and Fig. 2. Fig3 shows time evolution of the cross-helicity. The computation results are presented for the following initial parameters of compressible MHD turbulence: $\lambda = 0,84$ is Taylor microscale,

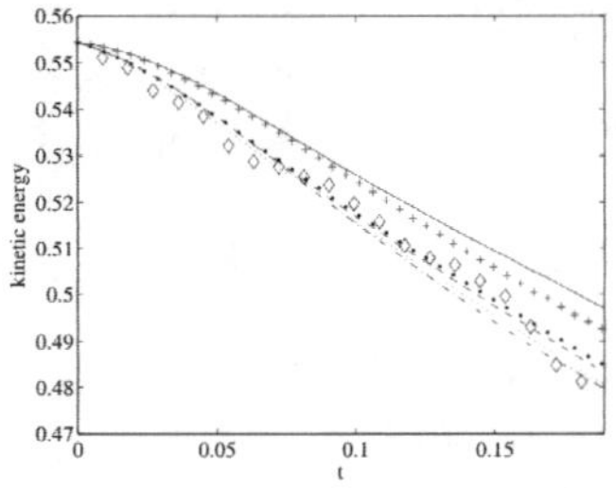


**Figure 1.** Time evolution of the kinetic energy.

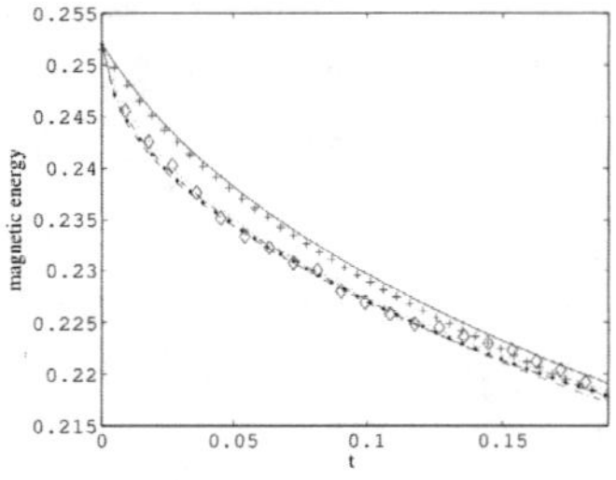


**Figure 2.** Time evolution of the magnetic energy.

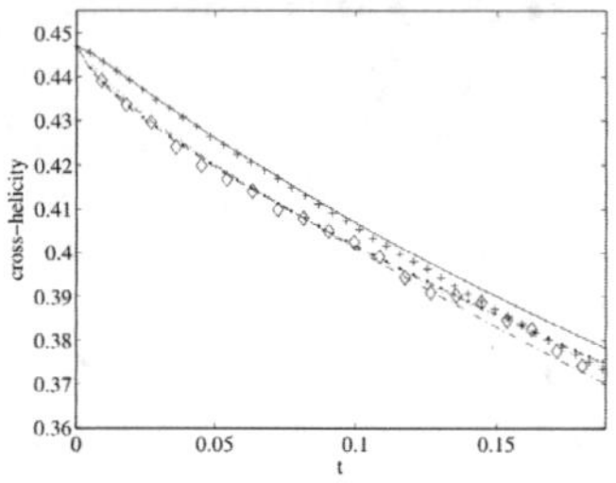


**Figure 3.** Time dynamics of the cross-helicity.

$Re = 390$ is hydrodynamic Reynolds number, $Re_m = 10$ is magnetic Reynolds number, $M_s = 0.6$ is Mach number, $\kappa = 0.8$ is Kolmogorov scale, $\mu = 0.01$ is coefficient of molecular viscosity and $\gamma = 5/3$ is polytropic index. The solid line corresponds to the case when tensors $\tau^u_{ij}$ and $\tau^b_{ij}$ are omitted, which means that LES, in fact, is DNS on the coarse LES-grid starting from filtered initial conditions. The solid line is included here for a more detailed analysis and understanding of an influence SGS models on the modeling of MHD flow for comparison DNS results with LES results made by various SGS models. The dashed line denotes the Smagorinsky model for MHD-turbulence, the dotted line is the Kolmogorov model, the black point line is the model based on cross-helicity. The marker + is the scale-similarity model and the dashed-dot line represents the mixed model for MHD-case. The DNS results are represented by the diamond line. As expected, the results of model without subgrid parametrization exhibit the largest deviation from DNS results. It follows from Fig. 1 and Fig. 2, that the similarity model gives the worst predictions among SGS closures. The other subgrid models significantly improve the calculation accuracy.

The differences between SGS models for magnetic energy are shown to decrease with reducing magnetic Reynolds number and all models discussed above demonstrate good agreement with DNS results at small magnetic Reynolds number. The effect of subgrid-scale closures increases with magnetic Reynolds number for modeling of compressible MHD turbulence, but rate of dissipation of the magnetic energy decreases with increasing magnetic Reynolds number. The best results are shown for the Smagorinsky, the Kolmogorov and the cross-helicity models for evolution of the magnetic energy. The same behaviour of plots was observed for the cross-helicity: the influence of subgrid-scale parameterizations increases with magnetic Reynolds number.

For kinetic energy larger divergence of LES results is observed with decrease in magnetic Reynolds number using various SGS closures. The scale-similarity model shows the worst results, however, the other SGS closures increase calculation accuracy. For time dynamics of both magnetic turbulent intensities and kinetic turbulent intensities, the impact of SGS parameterizations on results of modeling of MHD flow increase with magnetic Reynolds number. The role of anisotropy in computations and the divergences between LES and DNS results for anisotropy increase with decreasing magnetic Reynolds number.

Sonic Mach number exerts essential influence on results of modeling [4]. The divergence between DNS and LES results for kinetic energy increases with Mach number. The Smagorinsky model and the cross-helicity model yield the best accordance with DNS under various Mach number. The deviations in results for magnetic energy, on the contrary, decrease with increasing Mach number. It is necessary to notice, that magnetic energy reaches a stationary level more rapidly with reducing Mach number. The Smagorinsky model shows the best results for cross-helicity both for high and for low Mach numbers. The skewness of velocity components, calculated with the use of LES, coincides better with DNS results with increasing Mach number. The choice of SGS parametrization virtually did not have an effect on skewness of magnetic field

Journal of Physics: Conference Series **510** (2014) 012044 doi:10.1088/1742-6596/510/1/012044

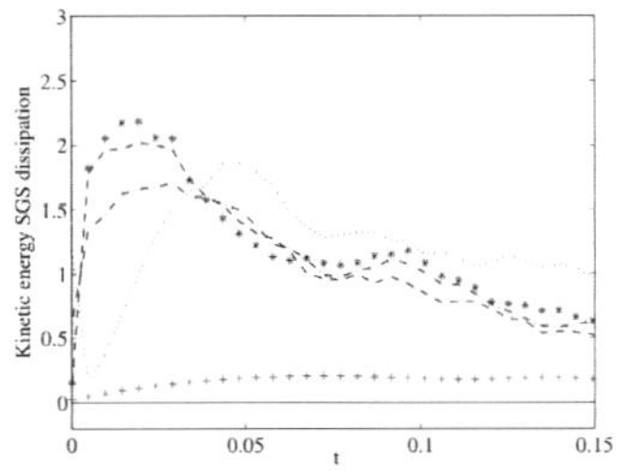


**Figure 4.** Time evolution of the kinetic energy subgrid dissipation.

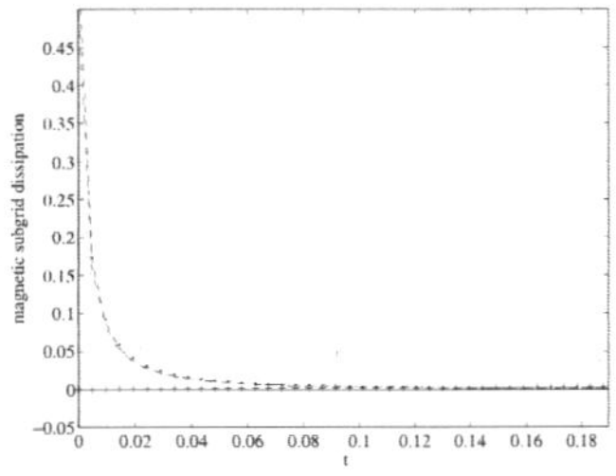


**Figure 5.** Time evolution of the magnetic energy subgrid dissipation.

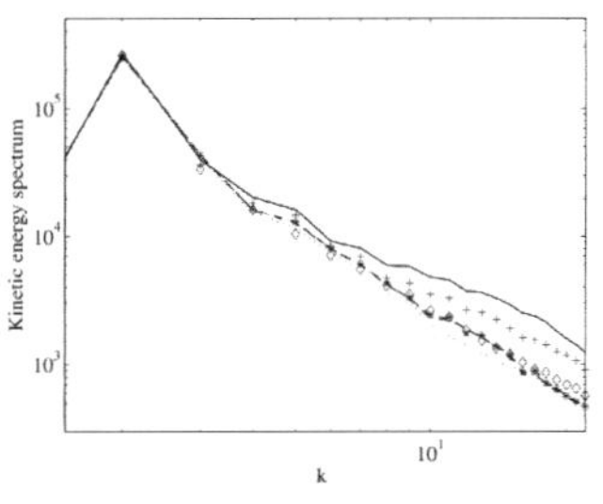


**Figure 6.** Kinetic energy spectrum.

components. The turbulent intensities obtained by LES technique agree better with DNS results with increasing Mach number.

Time evolution of the kinetic energy subgrid scale dissipation is plotted in Fig. 4. Here notations are the same as in the Fig. 10-Fig. 3 but the black point line is the model based on cross-helicity. This term, defined here as $\chi_u = -\tau^u_{ij}\tilde{S}_{ij}$, is written so that to be a sink of filtered kinetic energy when $\chi_u > 0$ and a source when $\chi_u < 0$ (i.e., backscatter). The value $\chi_u$ defines the amount of energy which transfers from large scale to subgrid scale and $\chi_u$ depends from SGS models that is used to find the subgrid tensor $\tau^u_{ij}$. It is clear that SGS dissipation for scale-similarity model till the time $t = 0.13$ is the minimum, therefore this SGS model does not supply with the dissipation of kinetic energy properly and in Fig. 1 scale-similarity model demonstrates the worst results. The most value of the subgrid dissipation of kinetic energy is available for Kolmogorov and mixed models. The magnetic energy subgrid scale dissipation is determined as $\chi_b = -\tau^b_{ij}\bar{J}_{ij}$, and represents the large-scale energy drained by the subgrid scales, forward scatter corresponds to $\chi_b > 0$ and backscatter to $\chi_b < 0$ (see Fig. 5). Magnetic subgrid energy is much less for scale-similarity closure, than for other SGS models [4].

One more test for the LES provides the spectral distribution of the kinetic and the magnetic energies that shows redistribution of energy depending on wavenumber, i.e. at different scales. Besides that, spectrum allows to estimate influence of subgrid closure models on properties of compressible turbulent flow. The Fig. 6 and Fig. 7 present curves for kinetic and magnetic energy spectra respectively ($k$ is the wavenumber). It is clear that at large scales (small values of wavenumber) plots of curves practically coincide, and the dependence from subgrid parameterizations virtually disappears. The largest scales are mostly independent of the applied subgrid-scale modeling approach. The distinction become apparent mainly at small scales (corresponding to large wave numbers). In general, the study of the energy spectra confirm the observations of the subgrid model performances made earlier. The spectrum of the filtered DNS is in good agreement with the Smagorinsky, Kolmogorov, cross-helicity and mixed models, with a slight preference for cross-helicity model and mixed model. Moreover, we observe that the simulations with the scale-similarity closure and no SGS model are not able to generate the requisite amount of the dissipation, in other words, these cases exhibit a energy pile-up due to the lack of dissipation and the contributions of high wavenumber are too high. These outcomes confirm the results obtained from the time evolution of the kinetic and magnetic energy [5, 1].

Interesting and important results were obtained in [7] where validation of LES method for study of flatness (or kurtosis) and skewness of decaying compressible magnetohydrodynamic turbulence was investigated. The skewness and the kurtosis of a velocity component are, respectively, defined as:

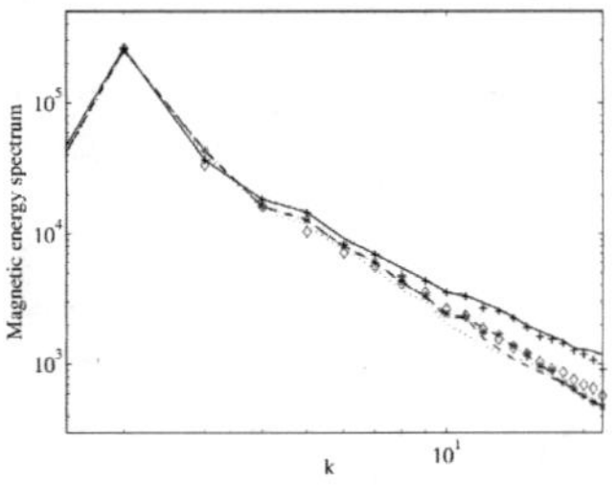


**Figure 7.** Magnetic energy spectrum.

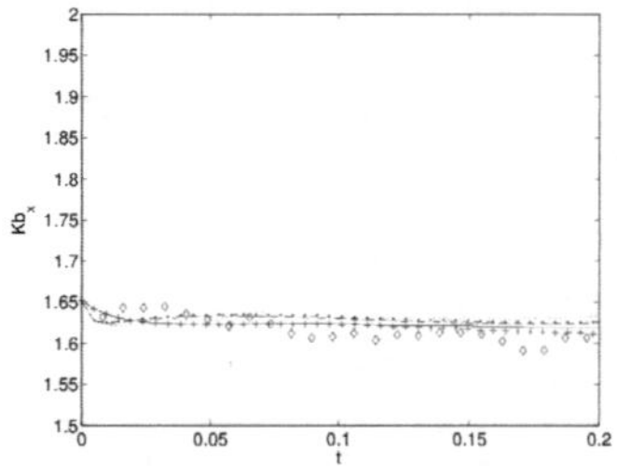


**Figure 8.** Time dynamics of flatness of the magnetic field

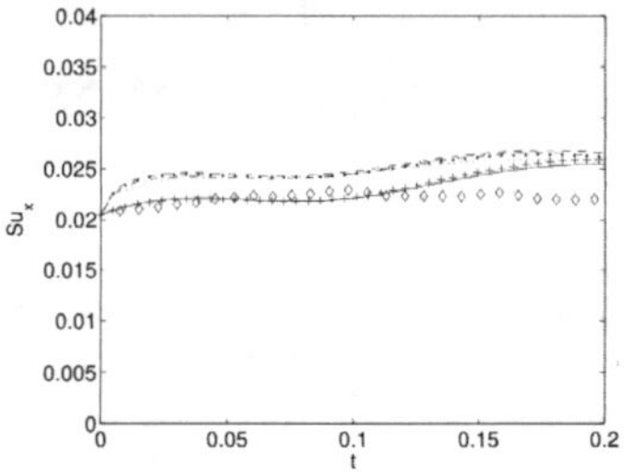


**Figure 9.** Time dynamics of skewness of the velocity

$$Su_j = \frac{< u_j^3 >}{(< u_j^2 >)^{3/2}} \qquad Ku_j = \frac{< u_j^4 >}{(< u_j^2 >)^2} \tag{33}$$

The sample of comparison of various subgrid-scale models of LES and DNS for flatness of the magnetic field $Kb_x$ and for skewness of the velocity $Su_x$ is represented in Fig. 8 and Fig. 9, respectively. The flatness and the skewness define properties of probability density for the turbulent process. It is shown that the subgrid-scale parametrization choice for the flatness and the skewness of the velocity is more essential than for the same characteristics of the magnetic field. In consequence of the investigation carried out, it is shown that LES method provides adequate results and LES technique may be used for studying of PDF characteristics (for instance, structural functions of various moments of velocity and magnetic field) of compressible MHD fluid flow at various similarity numbers [7].

## 6. LES of forced compressible MHD turbulence

Magnetohydrodynamic turbulence, similar to hydrodynamic turbulence, possesses scale-similarity properties in the inertial range. A fully developed turbulent state is expected when the integral and dissipation scales are separated by several orders of magnitude.

Validity of the LES method for studies of physical processes in forced compressible MHD turbulence until recently remained unidentified. The problem is that in the case of compressible MHD turbulence for numerical solutions of filtered equations in LES (as well as for the solutions of governing equations in the DNS approach) the finite-difference and finite-volume schemes in coordinate (or physical) space are the most proper. Moreover, the solutions of the basic equations by finite-difference methods allow to investigate inhomogeneous and non-stationary turbulent flows directly. The traditional way of driving force implementation for the compressible MHD flows relies on studies of incompressible turbulent fluid and is based on the spectral representation of the external force and a subsequent re-calculation of this force in physical space. It should be noted that in compressible MHD there are four types of waves: Alfven, fast magnetosonic, slow magnetosonic and entropy. In this case the interactions between the aforementioned types of waves can lead to the richer picture of the turbulence spectra in the inertial range. The traditional forms of external forces (based on local spectral representations of a turbulence source) can strongly oversimplify the flow pattern in compressible MHD turbulence and so simplify the description of such turbulence under various similarity parameters. In this case the validation of the LES method requires proving the possibility of reproducing of the scale-similarity Kolmogorov and Iroshnikov-Kraichnan spectra in natural conditions. The work [9] is devoted to study of this important aspect of the LES method.

For the study of forced compressible MHD turbulence in the inertial range we apply the

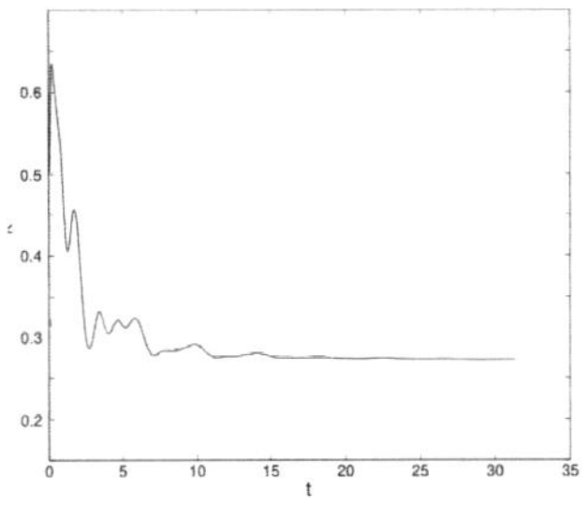

**Figure 10.** Time evolution of the kinetic energy with linear forcing.

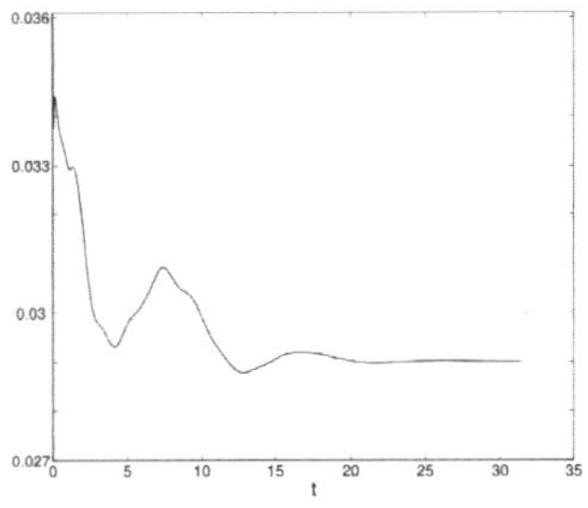

**Figure 11.** Time evolution of the magnetic energy with linear forcing.

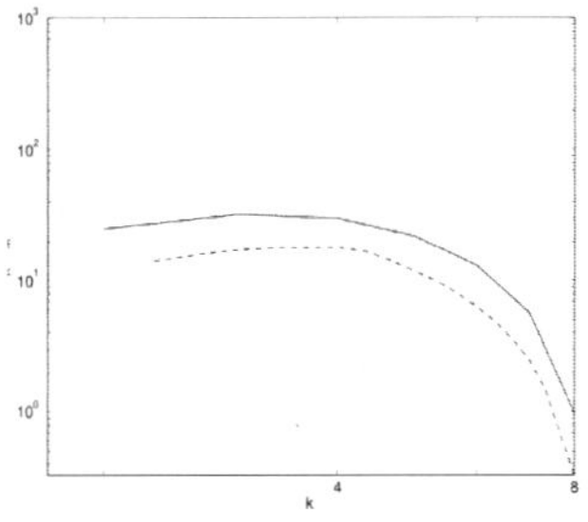

**Figure 12.** Normalized spectra of kinetic energy (solid line) and magnetic energy (dashed line).

linear driving force [9, 8]. The idea of linear forcing is to add a force proportional to the fluctuating velocity that resembles a turbulence which forced with a mean velocity gradient, that is, a shear. This force appears as a term in the equation for the fluctuating velocity that corresponds to the production term in the equation of turbulent kinetic energy. In this work, we generalize this approach to the case of compressible MHD flow of electrically conductive fluid. Such approach is called "linear forcing". Since compressible MHD turbulence is considered, the system of MHD equations includes the magnetic induction equation, and in this case the driving force is proportional to the magnetic field in the induction equation. In contrast to a spectral representation of the external force, the linear driving force acts on all scales of the physical space, and thereby it can reproduce more precisely a picture of mode interactions in compressible turbulent MHD flow. Actually, linear forcing corresponds to a turbulence with the driving force being caused by a mean velocity gradient, that is, by velocity shear. In the magnetic induction equation, the linear external force can be interpreted as being responsible for the production of magnetic energy due to the interaction between the magnetic field and the mean flow-velocity shear. In the works [9, 8], it is shown that linear forcing is a proper method for modeling forced MHD turbulence in physical space. Detailed information concerning linear forcing for compressible MHD turbulence one can find in [9]. The expressions of the external force, which allows obtaining a statistically stationary regime of turbulence, are derived. The formulas used for the formulation of large-eddy simulation approach are obtained. The potential possibilities of the LES method to reproduce physics of the flow under investigation in a stationary regime both for polytropic and for heat-conducting fluids are studied in detailed [9].

In Fig.10 and Fig.11 temporal evolution of kinetic energy and magnetic energy is presented for case of polytropic plasma. It is visible from the numerical results that statistically stationary turbulence arises after an initial time interval when large fluctuations are observed and the values of $E_K$ and $E_M$ practically do not vary with time. This means the balance between dissipation and energy injected in the system is occurred. It is interesting to note that $E_K$ reaches a stationary regime slightly faster than $E_M$. The study of inertial range properties is one of the main tasks in studies of scale-similarity spectra of MHD turbulence. Inertial range properties are defined as time averages over periods of stationary turbulence conditions [18]. Kolmogorov-like spectrum $-5/3$ is observed when MHD turbulence is examined and magnetic energy is much less than kinetic energy, that is, nonlinear interactions are much more considerable than magnetic ones and the fluid is practically neutral hydrodynamic. In Fig.12, spectra of the kinetic and the magnetic energies are shown (solid line is the kinetic energy spectrum and dashed line is the magnetic energy spectrum). The spectra are normalized by factor $k^{5/3}$. Dot line represents Kolmogorov scaling. The spectra of the kinetic and magnetic energy are obtained after time averaging of values in statistically stationary regime and there is clearly marked inertial turbulent range with the Kolmogorov-like spectrum $k^{-5/3}$ for both kinetic energy and magnetic energy as follows from Fig.12. The residual energy spectrum [19] that is determined as $E_R^K = |E_M^K - E_K^K|$ was studied in [9]. This spectrum is of interest because it gives an insight into the spectral interplay of kinetic and magnetic energies and exhibits self-similar scaling. It is demonstrated the normalized smoothed spectrum of residual energy and $E_R^K \sim k^{-7/3}$ in inertial range of turbulence which was theoretically obtained and was numerically confirmed for incompressible MHD turbulence [19]. It is apparent that there are no differences observed in statistical properties in inertial range due to weak compressibility (such result has been obtained recently for compressible hydrodynamic turbulence of neutral gas [20]). Accordingly, the proposed linear forces in LES provide correct results in studies of MHD turbulence.

It is shown in [9] that for the case when the kinetic energy of the flow is initially much larger than the magnetic energy, a Kolmogorov-like spectrum is obtained. When the magnetic energy is initially larger than the kinetic energy, the Iroshnikov-Kraichnan spectrum occurs. Thus, we have demonstrated the efficiency of the LES method for studies of scale-invariant properties of compressible MHD turbulence. The linear representation of the driving force developed for the LES method may also be of use in the DNS approach.

The obtained results can be used to study turbulent flows in various problems of space and astrophysical plasmas, in thermonuclear plasmas, plasma aerodynamics and problems in engineering applications. Thus, LES method has good future trends for research of compressible magnetohydrodynamic turbulence.

## Acknowledgments

The work was supported by the program P-22 of Russian Academy of Science Presidium "Basic problems in solar system studies", Russian Foundation for Basic Research (13-05-90436, 14-02-31848) and Grant of President of Russian Federation for supporting young Russian scientists (MK-267.2014.5).

## References

[1] Chernyshov A A, Karelsky K V and Petrosyan A S 2006 *Phys. Plasmas* **13** 032304
[2] Chernyshov A A, Karelsky K V and Petrosyan A S 2006 *Phys. Plasmas* **13** 104501
[3] Chernyshov A A, Karelsky K V and Petrosyan A S 2008 *Physics of Fluids* **20** 085106
[4] Chernyshov A A, Karelsky K V and Petrosyan A S 2007 *Physics of Fluids* **19** 055106
[5] Chernyshov A A, Karelsky K V and Petrosyan A S 2008 *Flow, Turbulence and Combustion* **80** 21–35
[6] Chernyshov A A, Karelsky K V and Petrosyan A S 2008 *Astrophysical Journal* **686** 1137–1144
[7] Chernyshov A A, Karelsky K V and Petrosyan A S 2009 *Theor. Comput. Fluid Dyn.* **23** 451–470
[8] Chernyshov A A, Karelsky K V and Petrosyan A S 2012 *Flow, Turbulence and Combustion* **89** 563–587
[9] Chernyshov A A, Karelsky K V and Petrosyan A S 2010 *Phys. Plasmas* **17** 102307
[10] Biskamp D 2003 *Magnetohydrodynamic turbulence* (United Kingdom: Cambridge University Press)
[11] Ferziger J 1996 *Simulation and Modeling of Turbulent Flows* ed Gatski T, Hussami Y and Lumley J (Oxford University Press, New York (NY)) pp 109–154
[12] Bardina J, Ferziger J H and Reynolds W C 1980 *AIAA 13th Fluid and Plasma Dynamics Conference, Snowmass, Colo* p 10
[13] Germano M, Piomelli U, Moin P and Cabot W 1991 *Phys. Fluids A* **3** 1760–1765
[14] Germano M 1992 *J. Fluid. Mech.* **238** 325–336
[15] Müller W C and Carati D 2002 *Phys. Plasmas* **9** 824–834
[16] Lilly D 1992 *Phys. Fluids A* **4** 633–635
[17] Low M M M, Klessen R S, Burkert A and Smith M D 1998 *Phys. Rev. Lett.* **80** 2754–2764
[18] Biskamp D and Müller W 2000 *Physics of Plasmas* **7** 4889–4900
[19] Müller W C and Grappin R 2004 *Plasma Phys. Control. Fusion* **45** B91B96
[20] Benzi R, Biferale L, Fisher R T, Kadanoff L P, Lamb D Q and Toschi F 2008 *Physical Review Letters* **100** 234503–+

Journal of Physics: Conference Series **510** (2014) 012045 doi:10.1088/1742-6596/510/1/012045

# Exploring phase space turbulence in magnetic fusion plasmas

**T.-H. Watanabe[1], Y. Idomura[2], S. Maeyama[3], M. Nakata[3], H. Sugama[1], M. Nunami[1], and A. Ishizawa[1]**

[1]National Institute for Fusion Science, Toki, Gifu 509-5292, Japan
[2]Japan Atomic Energy Agency, Kashiwa, Chiba 277-8587, Japan
[3]Japan Atomic Energy Agency, Rokkaho, Aomori 039-3212, Japan

E-mail: watanabe.tomohiko@nifs.ac.jp

**Abstract.** Plasma turbulence accompanied with fluctuations of the distribution function and the electromagnetic elds develops on the phase space composed of the con guration space and the velocity space. Detailed structures of the distribution function in magnetic fusion plasmas are investigated by means of gyrokinetic simulations performed on massively parallel supercomputers. The gyrokinetic simulations of drift wave turbulence have demonstrated entropy transfer in the phase space, zonal flow enhancement by helical elds and the resultant transport reduction. The state-of-the-art high performance computing is utilized for a multi-scale turbulence simulation covering ion- and electron-scales and for a global-scale simulation of turbulent transport in a sub-ITER sized plasma.

## 1. Introduction

Plasma turbulent transport is one of the most important issues in the magnetic fusion research, and the most challenging target in numerical simulations of fusion plasmas. While fusion plasma confinement has reached the break-even condition or beyond, comprehension of particle and heat transport, of which levels are *anomalously* higher than estimates of collisional transport, is still demanded for a more reliable prediction of future experiments [1]. Direct numerical simulations of plasma turbulence are expected to provide physics-based understandings on the transport properties.

Magnetic fusion plasma with a high temperature of several keV is almost collisionless and beyond applicability of the conventional fluid approximation. In order to overcome the limitation, and to reveal the turbulent transport mechanisms in fusion plasmas, gyrokinetic simulations have been developed [2], where time-evolutions of distribution functions are solved on the phase space. The new approach extends a concept of turbulence onto the velocity space. Indeed, drift wave turbulence simulations manifest cascades of fluctuations of the distribution function $f$ not only in the configuration (real) space but also in the velocity space [3, 4, 5]. Deformation of the distribution function in the kinetic plasma turbulence leads to development of fine-scale structures of $f$ through collisionless particle motions preserving the phase-space volume, where production of the entropy variable defined by a quadratic integral of a perturbed part of $f$ balances with the turbulent transport [6]. Then, the fine structures of $f$ are mixed in the phase space, which is a unique feature of collisionless or weakly collisional plasma turbulence. A similarity to a mixing process in the neutral fluid turbulence is discussed in this article.

Journal of Physics: Conference Series **510** (2014) 012045 doi:10.1088/1742-6596/510/1/012045

In recent studies of plasma turbulent transport, zonal flows, which are radially sheared plasma flows with poloidal and toroidal symmetry and are driven by the electric ($\boldsymbol{E} \times \boldsymbol{B}$) drift motions of ions and electrons, have attracted many researcher's attention [7]. The zonal flows self-generated by turbulence play a crucial role in regulating turbulence and in reducing transport. A collisionless response function of zonal flow to a source term gives the kernel for the zonal flow evolution, and consists of two parts; A short time scale response is related to a geodesic acoustic mode (GAM) oscillation [8], while a long time scale response provides a residual zonal flow component [9]. The GAM oscillation is accompanied with reversal of the electrostatic potential, and suffers from the collisionless (Landau) damping. The Landau damping of plasma waves, including the GAM oscillation, is associated with the aforementioned collisionless mixing process of the distribution function in the velocity space. The residual part of the response function, which influences an efficiency of zonal flow generation, is related to an unmixed (or coherent) structure of the distribution function. Therefore, exploration of phase-space structures of $f$ is expected to provide a key idea for achieving a better plasma confinement.

The gyrokinetic simulation of plasma turbulent transport has been widely utilized as a powerful tool for evaluating anomalous transport fluxes in fusion plasmas. Successful comparisons with current experiments demonstrate a potential for predicting future experiments. Although fully predictive simulations for future devices demand a leap in the physical and numerical modelings as well as in high performance computing (HPC), recent developments in the both fields, computational plasma physics and the HPC, are enhancing the capability toward the burning plasma simulation.

In the present article, we shortly review our kinetic simulations of phase space turbulence in magnetic fusion plasmas, and introduce the state-of-the-art peta-scale gyrokinetic simulations performed on the K computer [10]. The paper is organized as follows. The gyrokinetic equations are summarized in the next section. In section 3, generation of fine structures of the distribution function in the velocity space is discussed. Section 4 deals with simulations of zonal flows and turbulence controlled by confinement fields. Recent gyrokinetic simulation results obtained by the use of the K computer will be shown in section 5. A summary is given in the last section.

## 2. Gyrokinetic equations

The anomalous transport in magnetic fusion plasmas is attributed to turbulence induced by drift waves of which frequencies are much lower than the ion cyclotron (gyro-) frequency. Parallel and perpendicular wavelengths of drift waves are characterized by the device size and the thermal gyroradius, respectively. As the Vlasov or Boltzmann equation for a one-body distribution function involves multiple spatio-temporal scales, such as the Debye length, the plasma frequency, the gyrofrequency ($\Omega$), etc, a reduced set of formulae is quite useful to describe the drift wave turbulence with low frequencies ($\omega \ll \Omega$).

Nonlinear gyrokinetic equations are originally derived by taking a gyrophase average of the Vlasov equation and by applying a recursive formulation [11] under the gyrokinetic ordering of

$$\frac{\omega}{\Omega} \sim \frac{k_\parallel}{k_\perp} \sim \frac{v_D}{v_t} \sim \frac{\rho}{L} \sim \frac{\delta n}{n_0} \sim \epsilon \tag{1}$$

with a smallness parameter $\epsilon$, where $k_\parallel$, $k_\perp$, $v_D$, $v_t$, $\rho$, $L$, $\delta n$, and $n_0$ are the parallel and perpendicular wavenumbers, a guiding-center drift velocity, a thermal speed, a thermal gyroradius ($= v_t/\Omega$), a typical scale-length of equilibrium, a perturbed density, and an averaged density, respectively. In an electrostatic limit, it is written as

$$\begin{aligned} \frac{\partial}{\partial t} + v_\parallel \boldsymbol{b} \cdot \nabla + \boldsymbol{v}_d \cdot \nabla - \frac{\mu}{m_s} (\boldsymbol{b} \cdot \nabla B) \frac{\partial}{\partial v_\parallel} \quad \delta f_s + \frac{c}{B_0} \boldsymbol{b} \cdot \nabla \Phi \times \nabla \delta f_s \\ = \boldsymbol{v}_* - \boldsymbol{v}_d - v_\parallel \boldsymbol{b} \quad \cdot \frac{e_s \nabla \Phi}{T_s} F_{Ms} + C \end{aligned} \tag{2}$$

where $F_{Ms}$ and $\delta f_s$ ($= f_s - F_{Ms}$) mean the equilibrium Maxwellian and perturbed parts of a gyrocenter distribution function $f_s$ for $s$ species. Here, the independent variables of the velocity space coordinates are a velocity component parallel to the confinement field ($v_\parallel$) and a magnetic moment ($\mu$). The magnetic and diamagnetic drift velocities are denoted by $\boldsymbol{v}_d$ and $\boldsymbol{v}_*$, respectively. The electrostatic potential acting on the gyrocenter is represented by $\Phi$. Also, $c$, $m_s$, $e_s$, $T_s$, $B_0$, and $C$ mean the speed of light, mass and charge of a particle, temperature, the confinement field strength, and a collision term, respectively. A unit vector parallel to the confinement field is given by $\boldsymbol{b}$. The $\delta f$ formulation is useful for a local simulation of turbulence and transport.

A Hamiltonian formulation of the gyrokinetic equation in a collisionless limit is also obtained by means the Lie transform to the gyrocenter coordinates [12],

$$\frac{Df_s}{Dt} = \frac{\partial f_s}{\partial t} + \{f_s, H_s\} = 0 \tag{3}$$

where $H_s$ means a perturbed Hamiltonian independent of the gyrophase. The Poisson brackets are denoted by $\{,\}$. Equation (3) is rewritten in a conservation form as the Hamiltonian flow is incompressible, and preserves the phase volume. It is a great advantage in a global-scale turbulence simulation which solves evolution of the total distribution function (that is, so-called full-$f$ approach).

In the local limit, the $\delta f$ gyrokinetic equation derived form the recursive method coincides with that obtained from the Hamiltonian formulation [13]. In the both cases, the electromagnetic fluctuations are calculated from the Poisson's and Ampère's laws of which source terms include velocity-space integrals of the distribution function.

## 3. Phase space turbulence and transport

To emphasize a basic property of the phase space turbulence in a magnetized plasma, that is, turbulent mixing of the distribution function in the velocity space, we briefly review results of a simple two-dimensional gyrokinetic simulation of ion temperature gradient (ITG) turbulence [3, 4]. Model setting in the real space with a uniform magnetic field is schematically shown in Fig. 1. The velocity space dimension is further reduced by assuming the Maxwellian distribution for the perpendicular velocity space, and Eq. (2) is integrated over the $\mu$ space. We solve the reduced gyrokinetic equation for ions while assuming the adiabatic electron response ($\delta n_e/n_0 = e\phi/T_e$) in the electrostatic limit. In the simulations presented in this section, zonal flow components are neglected so that interactions of turbulent eddies and generation of fine structures of the distribution function are clearly captured in a fully developed turbulence.

The ITG modes are destabilized when the ion temperature gradient exceeds a critical value, and drive the ion heat transport perpendicular to the magnetic field (that is, in the $x$ direction in Fig. 1). A snapshot of the electrostatic potential fluctuation at $t = 600L_n/v_{ti}$ is plotted in Fig. 2, where the instability growth has been saturated and the statistically steady turbulence is sustained. Here, $L_n$ means the density gradient scale length, and $v_{ti}$ is the ion thermal speed. The positive and negative contours of $\phi$ are, respectively, represented by thick and thin solid curves. The largest eddy mainly drives the ion heat transport, while smaller scale fluctuations are generated in the turbulence. While the collision frequency in the present simulation is fixed to zero, a statistically steady transport is observed in the ITG turbulence. This fact is associated with the mixing of distribution function in the phase space.

In the collisionless (or in a weakly collisional) plasma turbulence, fluctuations develop not only in the real but also in the velocity space. Each panel in Fig. 3 shows velocity-space structures of the ion distribution function for the linearly most unstable mode, where complex-valued amplitudes of $\delta f_{\boldsymbol{k}}$ are normalized by the electrostatic potential $\phi_{\boldsymbol{k}}$, and the real and imaginary parts are plotted by green and blue lines, respectively. Right after saturation of the

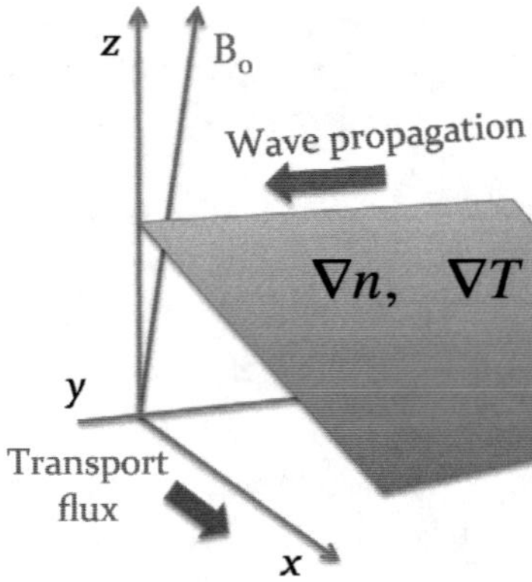


**Figure 1.** Model setting for ion temperature gradient turbulence simulation with a uniform magnetic field and density and temperature gradients.

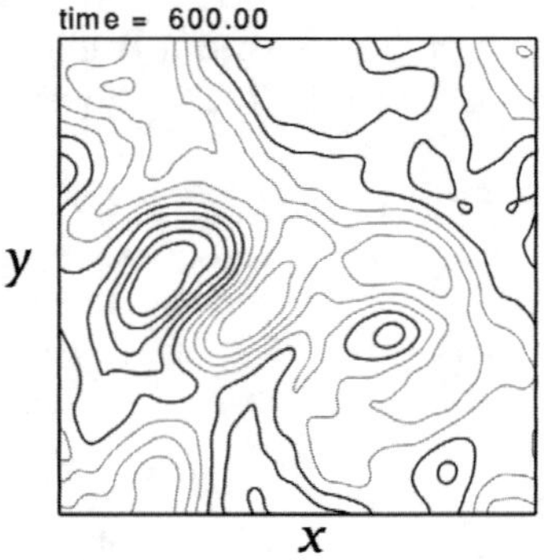


**Figure 2.** Contour plot of electrostatic potential found in ion temperature gradient turbulence simulation, where thick and thin lines represent positive and negative values, respectively.

linear instability growth, at $t = 200L_n/v_{ti}$, one finds a smooth profile of $\delta f_{\boldsymbol{k}}$ which reflects the eigenfunction of the ITG mode. As the time advances, fine-scale fluctuations develop in the velocity space. While 8,192 grid points are used for the $v_\parallel$ coordinates, the finest scale length in the velocity space reaches the grid size at $t \sim 600L_n/v_{ti}$. Then, the numerical simulation is stopped. The obtained result manifests spontaneous and continuous generation of fine structures of $\delta f$ which is attributed to the advection term, $v_\parallel \boldsymbol{b} \cdot \nabla \delta f$, coupled with the turbulent stretching of $\delta f$ in the real space (namely, effective increase of the parallel wavenumber) [4].

Generation of the fine structures of $\delta f$ is related to the turbulent transport through an entropy balance. From the gyrokinetic equation for the perturbed distribution function, Eq. (2), with a help of the quasi-neutrality condition and the adiabatic electron response, one derives a balance equation,

$$\frac{d}{dt}(\delta S_n + \delta_{n,1}W) = J_{n-1/2} - J_{n+1/2} + \delta_{n,2}\frac{L_n}{L_T}Q_i - 2\nu n \delta S_n \ , \tag{4}$$

where

$$\delta S_n = \sum_{\boldsymbol{k}} \delta S_{\boldsymbol{k},n} = \sum_{\boldsymbol{k}} \frac{1}{2} n! |\hat{f}_{\boldsymbol{k},n}|^2 \tag{5}$$

$$J_{n-1/2} = \Theta k_y n! \mathrm{Im} \ \ \hat{f}_{\boldsymbol{k},n-1}\hat{f}^*_{\boldsymbol{k},n} \quad , \tag{6}$$

and $\delta_{n,m} = 1$ for $n = m$ but $\delta_{n,m} = 0$ for $n \neq m$. The electrostatic potential energy and the ion heat flux are denoted by $W$ and $Q_i$, respectively. Also, $L_T$ is the temperature gradient scale length, and $\Theta = \theta L_n/\rho_i$ with the inclination angle ($\theta \ll 1$) of the magnetic field. Here, $\hat{f}_{\boldsymbol{k},n}$ ($n = 0, 1, 2, ...$) are defined by coefficients of the Fourier and Hermite-polynomial expansion of the perturbed distribution function,

$$\delta f_{\boldsymbol{k}}(v_\parallel) = \sum_{n=0}^{\infty} \hat{f}_{\boldsymbol{k},n} H_n(v_\parallel) F_M(v_\parallel) \ . \tag{7}$$

Journal of Physics: Conference Series **510** (2014) 012045 doi:10.1088/1742-6596/510/1/012045

The model collision term is given by the Lenard-Bernstein operator,

$$C\left(\delta f\right) = \nu \frac{\partial}{\partial v_{\parallel}} \quad \frac{\partial}{\partial v_{\parallel}} + \frac{v_{\parallel}}{v_t^2} \quad \delta f(v_{\parallel}) \tag{8}$$

with the collision frequency $\nu$.

Summation of Eq. (4) for $n$ provides a total balance equation of an entropy variable, $\delta S$,

$$\frac{d}{dt}\left(\delta S + W\right) = \frac{L_n}{L_T} Q_i + D \ , \tag{9}$$

where

$$\delta S = \sum_{n}^{\infty} \delta S_n \quad \text{and} \quad D = -\sum_{n}^{\infty} 2\nu n \delta S_n \ . \tag{10}$$

When the ITG instability growth is saturated and the turbulent flows are statistically steady, $dW/dt \approx 0$ and $Q_i$ is approximately constant. In the collisionless turbulence shown above, therefore, the growth of fine structures of $\delta f$, leading to increase of $\delta S$, balances with the transport, that is,

$$\frac{d}{dt}\delta S \approx \frac{L_n}{L_T} Q_i \ . \tag{11}$$

This is a *quasi-steady state* of collisionless turbulence. In case with finite collisionality ($\nu \neq 0$), fine structures of $\delta f$ developed in the velocity space are dissipated by ion-ion collisions, and the increase of $\delta S$ is saturated at a finite level. Then, a *statistically steady state* of weakly collisional turbulence is realized where

$$\frac{L_n}{L_T} Q_i + D \approx 0. \tag{12}$$

A spectral theory of velocity-space fluctuations has been developed for the statistically steady state of the weakly collisional turbulence [4]. In the steady turbulence with finite collisionality, the left hand side of Eq. (4) vanishes. For $n \geq 3$, thus, difference of the entropy transfer functions, $J_{n-1/2} - J_{n+1/2}$, balances with the collisional dissipation. However, if $\nu L_n / v_{ti} \ll 1$, the dissipation term plays a role only for large values of $n$, namely, it is localized in the high-$n$ regime. Thus, an intermediate region, which is free from the entropy production by the transport or the collisional dissipation, appears where the transfer function $J_{n-1/2}$ is nearly constant and independent of $n$. Figure 5 depicts the entropy production, transfer, and dissipation processes in the $n$-space. The obtained results suggest a similarity to the passive scalar transport in a neutral fluid turbulence with a large Prandtl number [4], and exhibit an unique picture of the plasma turbulence which develops in the phase space.

## 4. Zonal flows and turbulence control

The velocity-space mixing of the distribution function also plays a significant role in dynamics of the zonal flow which is considered to regulate the ITG turbulence in toroidal fusion plasmas. Evolution of a zonal flow is given by a turbulent (nonlinear) source term as well as short- and long-time response kernels which are derived from the linear gyrokinetic equation [9]. The short-time response describes the GAM oscillation of which collisionless Landau damping creates fine striation structures of $\delta f$ due to passing particles. A coherent structure of $\delta f$ associated with the neoclassical polarization shielding is produced due to trapped particles in the long-time response of zonal flows [14] which determines the residual level after the GAM damping [9]. It is expected that the higher residual level of zonal flows contributes to the more effective generation, and leads to the lower turbulent transport. Thus, it is preferable to find a magnetic configuration in which the zonal flow generation is enhanced and induces a better confinement. One of the

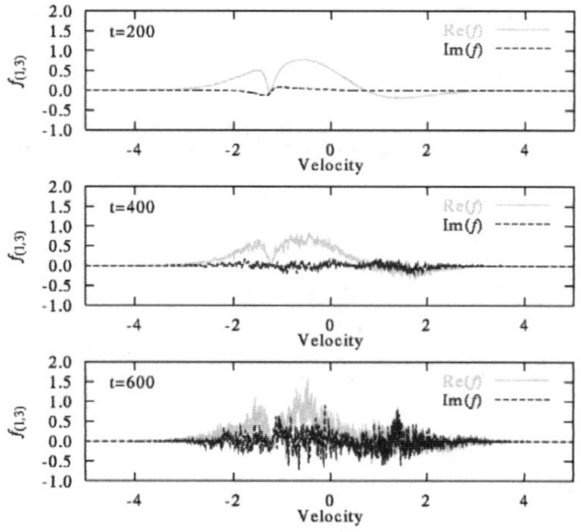


**Figure 3.** Snapshots of velocity distribution function at different times where green and blue lines, respectively, show real and imaginary parts of a Fourier amplitude (normalized by the potential) for the most unstable mode.

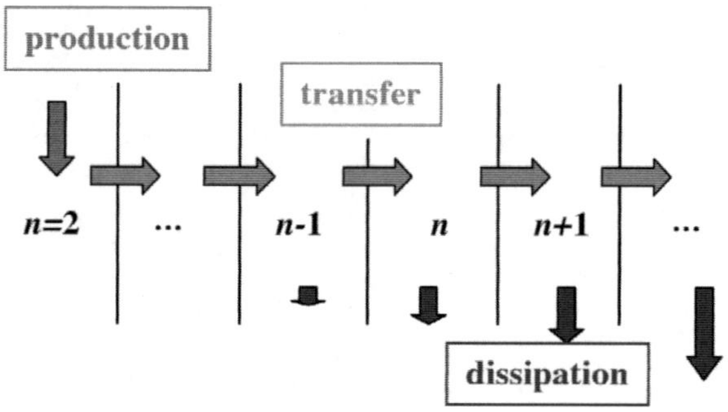


**Figure 4.** Schematic view of entropy production, transfer, and dissipation in macro-, intermediate, and micro-scales of the velocity space.

examples is shaping of a poloidal cross section of a tokamak [15]. A non-axisymmetric toroidal confinement provides us more degrees of freedom in optimization for the zonal flow response. Here, we discuss the zonal flow response in a helical confinement system, that is, the Large Helical Device (LHD) [16].

In the non-axisymmetric systems, helical-ripple-trapped particles making a radial drift motion provide additional shielding of zonal flow potential [17, 18]. A gyrokinetic simulation of the zonal flow response in a helical system clearly captures a velocity-space structure of $\delta f$ associated with the zonal flow potential shielding. A snapshot of the ion distribution function for a zonal flow component is shown in Figure 5 where a color contour shows the real part of $\delta f$. Vertical striation patterns are generated during the Landau damping of GAM due to particle motions passing through the toroidal and helical ripples of the confinement field strength. In a trapped region marked by a white dashed oval, one finds a negative part of $\delta f$ (colored by blue) which is caused by the radial drift motion of trapped particles and is responsible to shielding of the zonal flow potential. Thus, the higher zonal flow response is found in helical configurations with the slower radial drift velocity.

The above theoretical and numerical studies of zonal flows provide an important implication to the helical plasma confinement. If the field configuration could be optimized so as to lower the radial drift motion responsible for the neoclassical ripple transport, it simultaneously improves the anomalous transport though the zonal flow enhancement [19]. Gyrokinetic simulations of the ITG turbulence in helical systems, by means of the GKV code [14], confirm that the self-generated zonal flows enhanced in an optimized configuration with the slower radial drift lead to reduction of the ion heat flux [20], which is qualitatively consistent with experiments. More recently, the gyrokinetic simulations have been extended for direct comparisons with the LHD experiments [21, 22], showing a quantitative agreement on the ion heat transport flux [23]. The above simulations are carried out by means of a local flux tube model, and the obtained results

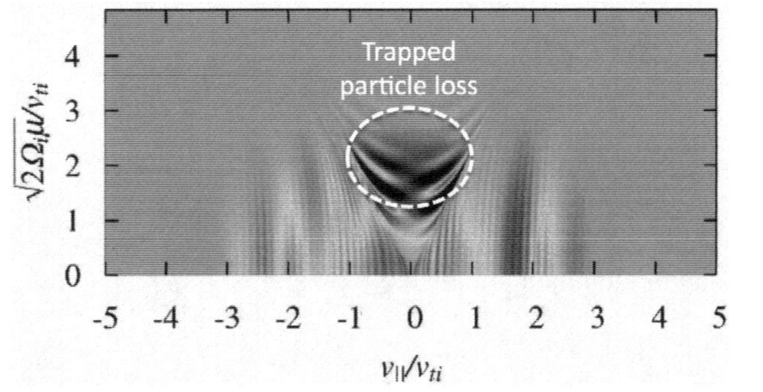


**Figure 5.** Color contour of velocity distribution function of a zonal flow component, where a negative part colored by blue (marked by a dashed circle) means loss of helical ripple trapped particles in a non-axisymmetric confinement system.

are utilized for development of an anomalous transport model that takes into account the effect of turbulence regulation by zonal flows [24]. It is a practical outcome of exploring the phase space turbulence which involves not only large scale structures but also small scale fluctuations of the velocity distribution function.

## 5. Peta-scale simulation of fusion plasma turbulence

Phase space turbulence simulations shown in the previous sections demand large computational costs. For example, over $5 \times 10^{10}$ grid points [that is, $128 \times 128 \times 768 \times 128 \times 48$ in the $(x, y, z, v_\|, \mu)$ coordinates] are employed for numerically resolving the ITG turbulence fluctuations on the five-dimensional phase space of a non-axisymmetric system [20], and more than 100 hours of computational time were spent using 256 nodes of the Earth Simulator [25] [which was a huge vector-parallel machine with the peak performance of $4 \times 10^{13}$ (40 tera) flops]. Thanks to the recent development of supercomputer technology, gyrokinetic simulations with the similar resolution are able to be more frequently executed on current systems with the peak speed of several hundreds tera flops. For more comprehensive understandings of plasma turbulence, nevertheless, peta-flops scale simulations are highly demanded. In the followings we discuss the two examples of peta-scale turbulent transport simulations performed on the K computer.

### *5.1. Ion- and electron-scale turbulences*

In burning fusion plasmas, electron heat transport is a critical issue, because alpha particles produced by D-T reactions mainly interact with electrons. But, it is still an open issue in which scale of turbulence drives the anomalous electron heat flux. Typical wavenumber and frequency of the electron-scale turbulence are, respectively, characterized by the electron gyroradius and transit time which are smaller than those in the ion-scale turbulence by a factor of $v_{ti}/v_{te}$ (that is, $\sqrt{m_e/m_i}$ for $T_e = T_i$). Thus, the anomalous electron heat transport is intrinsically a multi-scale problem where one needs to consider the ion- and electron-scale turbulences simultaneously. Smallness of the mass ratio, $m_e/m_i$, requires the peta scale computing, since the computational cost is proportional to $(m_i/m_e)^{3/2}$, e.g. $(m_i/m_e)^{3/2} \sim 8 \times 10^4$ for a hydrogen plasma.

A multi-scale plasma turbulence simulation using the gyrokinetic simulation code, GKV [14, 26], has been carried out on the K computer, where the electron and ion temperature gradient modes are simultaneously destabilized by the temperature gradients. A color contour of the electrostatic potential fluctuation is shown in Figure 6 where a small panel shown on the right is a magnified plot at the center of the simulation box. In the main panel, one finds growth of the ITG mode with the wavelength of $15 \sim 20\rho_i$. As shown in the magnified plot, fine scale turbulent fluctuations are embedded in the ITG mode perturbations. Because of the faster

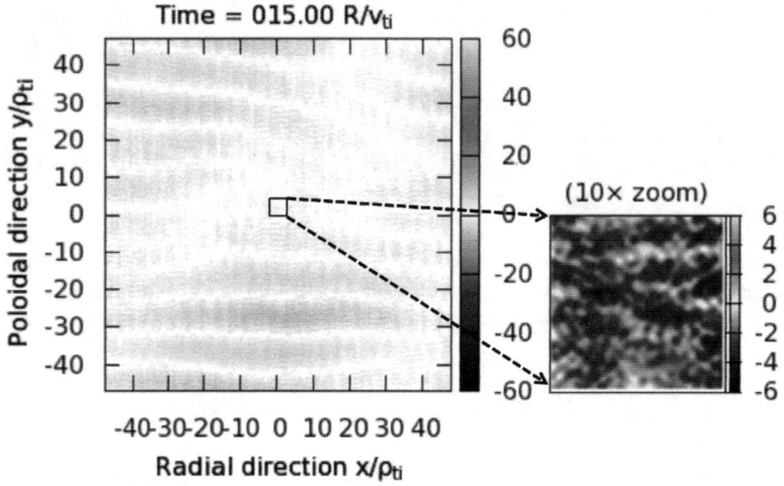


**Figure 6.** Color contour of electrostatic potential of multi-scale turbulence where small-scale fluctuations associated with electron motions are embedded in ion-scale drift waves. A magnified plot is shown on the right.

electron motions than ions, the electron temperature gradient (ETG) mode primarily grows in the early phase of the simulation, driving the electron heat transport. Then, the ITG modes are excited with longer spatio-temporal scales, and their growths are saturated by the self-generated zonal flows. Once the ion-scale turbulence dominates, the electron scale turbulence is regulated, and the electron heat is mainly carried by the ion-scale fluctuations. More detailed results of the multi-scale turbulent transport simulation will be published elsewhere.

The highly optimized GKV code enables us to perform the multi-scale simulation of the ion and electron scale transport with the real mass ratio ($m_i/m_e = 1836$). Several optimization techniques have been employed, such as a process mapping suitable to the topology of inter-node connections, overlapping of communications with computations, and so on [27]. Finally, the parallelization efficiency of 99.99994% is achieved on $\sim$ 600 k processor cores of the K computer, which implies further potential of gyrokinetic simulations to elucidate the anomalous transport in fusion plasmas with multi-scale and multi-physics turbulence.

*5.2. Global-scale simulation of plasma turbulent transport*

The gyrokinetic turbulence simulations discussed above deal with local transport phenomena where the simulation domain perpendicular to the confinement field is scaled with the thermal ion gyroradius ($\rho_i$), and is assumed to be much smaller than the equilibrium scales. This assumption may be validated for large-sized devices, such as ITER where the ratio of the minor radius of a torus ($a$) to $\rho_i$ is about $a/\rho_i \gtrsim 600$. However, it is necessary to take account of a global radial domain, if formation or relaxation of the equilibrium profiles are focused on, or if the scale separation of fluctuations and background quantities is non-trivial. The GT5D code [28], solving the full-$f$ gyrokinetic equation, like Eq. (3), with finite collisionality and energy source and sink terms, has been developed and applied to the ion scale turbulent transport problems in tokamaks [29]. Recent results from the GT5D simulations have demonstrated a plasma size scaling of the ion heat transport, momentum transport and toroidal rotation, intermittent transport events, and a role of source and sink models [30, 31, 32, 33].

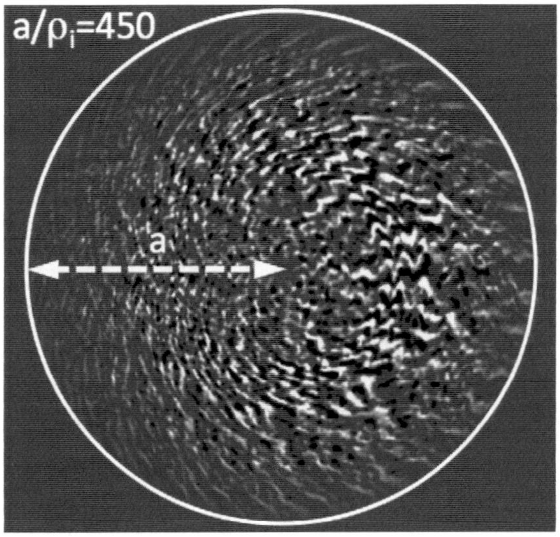


**Figure 7.** Color contour of electrostatic potential of ion temperature gradient turbulence in a sub-ITER scale plasma with $a/\rho_i = 450$.

As expected, the GT5D simulations demand huge computational costs, and have been performed by utilizing 100 tera flops systems or beyond [34]. So far, the ITG turbulent transport for a plasma size of $a/\rho_i \leq 450$, which is relevant to the present day tokamaks, has been simulated by the GT5D code [33]. However, it is still not enough for predicting performance of ITER of which plasma volume is more than ten times larger than that of the current experiments. Thus, it demands the use of a peta-scale machine, like the K computer. The GT5D code is highly optimized also for the K computer by applying a hybrid parallel model and communication-overlap techniques [35], where an excellent strong scaling is confirmed even with a sustained computational performance of $\sim$ 10 %. The achieved parallelization efficiency is 99.99989 % for $\sim$ 600 k cores of the K computer, which enables the ion-scale turbulence simulations for a sub-ITER scale plasma of $a/\rho_i = 600$.

A snapshot of the electrostatic potential found in the ITG turbulence simulation for the case of $a/\rho_i = 450$ is shown in Figure 7, where one clearly finds the turbulent fluctuations scaled by the ion thermal gyroradius. It means that the fusion plasma confinement itself is a multi-scale problem of the macro-scale profile relaxation and the micro-scale turbulent transport. It is also observed that non-local transport events influence a global confinement of ion thermal energy. Detailed results on the statistical property of turbulent transport and the size scaling, including the case of $a/\rho_i = 600$, will be reported elsewhere.

## 6. Summary

Phase space turbulence and transport in magnetic fusion plasmas has been investigated by means of gyrokinetic simulations where time-evolutions of a one-body distribution function coupled with fluctuations of electromagnetic fields are solved numerically. In this article the recent progress is briefly reviewed where we focus on mixing processes in the phase space and on multi-scale plasma turbulence.

Generation of fine velocity-space structures of the perturbed distribution function $\delta f$ is characterized by increase of a quadratic integral (the entropy variable) of which transfer from the macro to micro velocity scales balances with the cross-filed transport driven by the drift wave turbulence. Velocity-space structures of $\delta f$ are also influential to zonal flows which regulate the ion temperature gradient turbulence. The zonal flow response is controlled by the helical confinement field which determines the radial drift motion of trapped particles. This is confirmed by the gyrokinetic simulations resolving the velocity-space structure of $\delta f$ for a zonal flow component. Gyrokinetic turbulence simulations have verified the transport reduction due to zonal flow enhancement controlled by the helical field, and contribute to development of a reduced transport model.

The recent progress of high performance computing enables us to utilize peta-scale systems, such as the K computer. A multi-scale gyrokinetic simulation clarifies a whole spectrum of turbulent fluctuations in the case where the ion- and electron-scale turbulences are simultaneously excited. Also, a global gyrokinetic simulation has been extended to resolve the turbulent transport problem in a sub-ITER scale plasma where non-local transport events are identified.

Production and control of a burning plasma is a critical issue in the ITER experiment, where prediction of the plasma confinement performance is an important subject in plasma physics. But, there are still some open issues associated with the turbulent transport, such as the isotopic effect, the impurity transport, and so on. Understandings of the unresolved issues are believed to be brought by future developments of gyrokinetic turbulence simulations with a help of the state-of-the-art of the high performance computing.

**Acknowledgments**
This work is partly supported by the MEXT grant for HPCI Strategic Program Field No.4, by grants-in-aid of MEXT (# 2760660, 21560861, and 24561030), by the International Fusion Energy Research Center projects, and by the National Institute for Fusion Science collaborative research programs.

## References
[1] ITER Physics Expert Group 1999 *Nucl. Fusion* **39** 2175
[2] Garbet X, Idomura Y, Villard L and Watanabe T H 2010 *Nucl. Fusion* **50** 043002
[3] Watanabe T H, Sugama H 2002 *Phys. Plasmas* **9** 3659
[4] Watanabe T H, Sugama H 2004 *Phys. Plasmas* **11** 1476
[5] Tatsuno T et al 2009 *Phys. Rev. Lett.* **103** 015003
[6] Sugama H et al 1996 *Phys. Plasmas* **3** 2379
[7] Diamond P H et al 2005 *Plasma Phys. Control. Fusion* **47** R35
[8] Winsor N et al 1968 *Phys. Fluids* **11** 2448
[9] Rosenbluth M N and Hinton F L 1998 *Phys. Rev. Lett.* **80** 724
[10] Miyazaki H et al 2012 *FUJITSU Sci. Tech. J.* **48** 255
[11] Frieman E A and Chen L 1982 *Phys. Fluids* **25** 502
[12] Brizard A J and Hahm T S 2007 *Rev. Mod. Phys.* **79** 421
[13] Sugama H and Horton W 1998 *Phys. Plasmas* **5** 2560
[14] Watanabe T H and Sugama H 2006 *Nucl. Fusion* **46** 24
[15] Xiao Y and Catto P J 2006 *Phys. Plasmas* **13** 082307
[16] Motojima O et al 2003 *Nucl. Fusion* **43** 1674
[17] Sugama H and Watanabe T H 2005 *Phys. Rev. Lett.* **94** 115001
[18] Sugama H and Watanabe T H 2006 *Phys. Plasmas* **13** 012501
[19] Ferrando-Margalet S, Sugama H and Watanabe T H 2007 *Phys. Plasmas* **14** 122505
[20] Watanabe T H, Sugama H and Ferrando-Margalet S 2008 *Phys. Rev. Lett.* **100** 195002
[21] Nunami M, Watanabe T H, and Sugama H 2010 *Plasma Fusion Res.* **5** 016
[22] Nunami M, Watanabe T H, Sugama H, and Tanaka K 2011 *Plasma Fusion Res.* **6** 1403001
[23] Nunami M, Watanabe T H, Sugama H and Tanaka K 2012 *Phys. Plasmas* **19** 042504
[24] Nunami M, Watanabe T H and Sugama H 2013 *Phys. Plasmas* **20** 092307
[25] Habata S, Yokokawa M and Kitawaki S 2003 *NEC Res. Develop.* **44** 21
[26] Maeyama S et al 2013 *Comput. Phys. Commun.* **184** 2462

[27] Maeyama S et al 2013 *Plasma Fusion Res.* (in press)
[28] Idomura Y et al 2008 *Comput. Phys. Commun.* **179** 391
[29] Idomura Y et al 2009 *Nucl. Fusion* **49** 065029
[30] Camenen Y et al 2012 *Nucl. Fusion* **51** 073039
[31] Jolliet S and Idomura Y 2012 *Nucl. Fusion* **52** 023026
[32] Idomura Y 2012 *Comput. Sci. Disc.* **5** 014018
[33] Nakata M et al 2013 *Nucl. Fusion* (in press)
[34] Idomura Y and Jolliet S 2011 *Proceedings of SC11* (Seattle, USA, 2011).
[35] Idomura Y et al 2013 *Int. J. High Perform. Comput. Appl.* 1094342013490973 (in press)

Journal of Physics: Conference Series **510** (2014) 012046 doi:10.1088/1742-6596/510/1/012046

# Nonlinear oscillations in the Knudsen plasma diodes

**A. Ya. Ender, V. I. Kuznetsov**
Ioffe Physical-Technical Institute
194021 Saint Petersburg, Russia

E-mail: victor.kuznetsov@mail.ioffe.ru

**Abstract.** Under analytical and numerical study are the kinetic phenomena inherent for nonlinear oscillations in 1D diodes in which charged particles are supplied from an emitter and move with no collision in the interelectrode spacing. An original numerical method ($E, K$ code) is used to build up the ion and electron velocity distribution functions as well to study the structures forming in the course of the nonlinear oscillations both in the vacuum diode with monoenergetic electron beam and in the Knudsen diode with surface ionization. The oscillations are shown to occur owing to the Bursian-Pierce instability.

Time-dependent processes occurred in planar plasma diodes are considered. We suppose that the electrons and ions with a prescribed velocity distribution functions (DF) leave the emitter surface and travel collisionlessly between electrodes (the Knudsen regime) at a self-consistent electric field. Two type of devices are considered:

1. The vacuum diode with no ions and DF of the emitted electrons being near the monoenergetic one (the Bursian diode),
2. The Knudsen diode with surface ionization, where the emitted electrons and ions have the half-Maxwellian DF (the KDSI).

The devices exhibit the large amplitude oscillations of the electron current across the diode when a current of the entered electrons exceeds a certain threshold value. The paper presents a theory of such oscillations.

## 1. Numerical code

The highly nonlinear oscillations arisen in the diode are accompanied by an intense energy exchange between charged particles (electrons and ions) and the time-dependent electric field. Such process results in a strongly non-equilibrium DF and even its disruption. In order to describe correctly such processes the kinetic and Poisson's equations should be solved. Such a problem can be solved analytically only for very particular cases. Therefore, we solve this problem numerically.

The kinetic part of the problem is the most difficult one. The best known and used in nowadays plasma simulations are codes based on plasma model as a system of macroparticles (see, e.g., XPDP1 code [1]). These codes, however, have a disadvantage of being rather noisy, which is mainly due to smaller number of particles within each cell. They are presented as rather high-speed ones. However, as it is shown in Reference [2], only a strong decrease in size of space and time steps of computational grid and, consequently, an appreciable increase in the number of macroparticles makes it possible to obtain the reliable results with XPDP1 code, in particular, in situations involving the particle reflection.

We have developed a new numerical code, which is called the $E, K$ code in the literature [3]–[6]. The code involves the fact that, in the collisionless case, the relationship takes place:

$$f(\tau, \zeta(\tau), u(\tau)) = f_0^+(\tau_0, u_0). \tag{1}$$

Here $f$ and $f_0^+$ are the DF at the point $(\zeta, \tau)$ and at the emitter, respectively, $\zeta(\tau) = \zeta(\tau; \tau_0, u_0)$, $u(\tau) = u(\tau; \tau_0, u_0)$, and $\dot{\zeta}(\tau_0) = u_0$. In calculating the DF at the time $\tau = t$, the electric field at all the points of the interelectrode gap is known at all preceding times, i.e., over entire time interval $(0, t)$. The values of the field distribution $\varepsilon(t, \zeta)$ at the nodes of the spatio-temporal mesh are stored in computer memory. Calculating the DF at a mesh node is reduced to computing a number of the trajectories of test particles.

The main feature of the method is that the trajectory of each particle is calculated backward in time until it reaches the electrode surface. As a result, from a given arrival velocity $u^1$, the velocity $u_0^1$ and time $\tau_0^1$ at which a particle is injected from the emitter are determined, as well as the value of the DF at the velocity $u^1$. Next, the code gives the trajectory for a new arrival velocity $u^2 = u^1 + \Delta u$, and so on. The velocity step $\Delta u$ is chosen so that the difference between the values of the DF at the neighbouring trajectories does not exceed a given value:

$$|f_0^+(\tau_0^1, u_0^1) - f_0^+(\tau_0^2, u_0^2)| < \varepsilon_f. \tag{2}$$

As a result, the step $\Delta u$ is chosen automatically from the gradient of the DF. This choice of $\Delta u$ provides the required accuracy in the calculation of the DF and its moments and, in particular, ensures a high accuracy in the determination of the points in which the DF has discontinuities.

In calculating the trajectories, the electric field strength within each cell is approximated by a linear function of the coordinate and time, and the position of a particle and its velocity are calculated as the power series:

$$\begin{gathered}\varepsilon(\tau, \zeta) = A_0 + B_0(\tau - \tau_1) + [A_l + B_l(\tau - \tau_1)] \cdot \zeta, \\ \zeta = \sum_{s=0}^{n} a_s(\tau - \tau_1)^s, \quad v = \sum_{s=1}^{n} s a_s(\tau - \tau_1)^{s-1}.\end{gathered} \tag{3}$$

For the coefficients $a_s$, the simple recurrence relations were obtained:

$$\begin{gathered}a_0 = Z, \quad a_1 = U, \\ a_2 = -(A_0 + A_1 \cdot a_0)/2, \\ a_3 = -(B_0 + B_1 \cdot a_0 + A_1 \cdot a_1)/6, \\ \cdots\cdots\cdots\cdots\cdots\cdots \\ a_s = -(B_1 \ldots a_{s-3} + A_1 \cdot a_{s-2})/[s(s-1)], \quad s \geq 4.\end{gathered} \tag{4}$$

Here, $Z = \zeta(\tau_1)$ and $U = \dot{\zeta}(\tau_1)$. This method for calculating the trajectories ensures that the electric field is continuous when the boundaries of the cells are crossed – a circumstance that is especially important for the trajectories of reflected particles – and simultaneously provides high accuracy in the calculation of the parameters of the trajectories.

In Reference [7], $E, K$ code was checked for several electric fields, which allow to obtain analytical formulas for trajectories of particles, as well as for self-consistent time-dependent processes, which are described by analytical theory [8]. All cases under consideration demonstrated high accuracy of the code. Comparison of E,K code and XPDP1 code shown, that the former exceed considerably XPDP1 code in accuracy [7]. Besides, in order to obtain reliable results using XPDP1 code one has to make an average over a large number of time steps. However, its value is a free unknown parameter.

Journal of Physics: Conference Series **510** (2014) 012046 doi:10.1088/1742-6596/510/1/012046

## 2. The Bursian diode

About 30 years ago, the vacuum diode with directed electron beam attracted an attention in connection with the creation of high power microwave generators, such as vircator, reditron, and so on. Under certain conditions, a potential barrier, reflected a portion of electrons, is formed, named the virtual cathode (VC), and highly nonlinear oscillations arise in the diode.

### *2.1. Steady solutions*

In plasma diodes with an electron beam it is convenient to use the kinetic energy of the electrons at the emitter $W_b$ and the beam Debye length $\lambda_D$ as energy and length units, respectively [9]:

$$\lambda_D = \left[\frac{2\epsilon_0 W_b}{e^2 n_0}\right]^{1/2} = \left(\frac{8\epsilon_0^2}{e^2 m}\right)^{1/4} \frac{W_b^{3/4}}{J_b^{1/2}} \approx 0.3238 \cdot 10^{-2} \frac{V_b^{3/4}}{J_b^{1/2}} [cm], \quad W_b = mv_0^2/2. \tag{5}$$

Here, the current density of the beam $J_b$ and the acceleration voltage $V_b = W_b/e$ are expressed in amperes per square centimeter and volts, respectively.

The steady solutions of the Bursian diode are determined only by two dimensionless values: an inter-electrode distance $\delta = d/\lambda_D$ and the external voltage $V_C = e\Phi_C/(2W_b)$.

The PD with a single potential minimum is typical for the steady solutions of the diode. When the minimum potential $|\eta_m|$ turns into the initial electron energy, a portion of electrons is reflected and returns to the emitter. This is the solution equipped with a virtual cathode. It names a solution of electron reflection. If $V_C$ is fixed, the steady solutions lie on the dependence $\varepsilon_0(\delta)$ on the $\{\varepsilon_0, \delta\}$ plane (Figure 1). The branches 1 and 2 correspond to solutions without electron reflection, and 3 – with reflection [9].

A stability of steady solutions without reflection with respect to the small perturbations has been studied by using a dispersion equation. The solutions lay on branch 1 are stable, whereas those on branch 2 are unstable. However, there is no dispersion equation for solutions with reflection. We have investigated the stability properties of such solutions with $E, K$ code [6]. In calculations, the time scale is selected to be the time needed an electron to overcome $\lambda_D$. The code is the exact one giving an opportunity to determine the eigenvalues (growth rate $\Gamma$ and frequency $\Omega$) of the main mode from simulation of a small perturbation evolution.

We have calculated such process for a set of gap values, and obtained the dependencies of eigenvalues on $\delta$. One can see from Figure 2, that a threshold $\delta_l$ exists below which all solutions on branch 3 are stable to small electron perturbations. For $\delta > \delta_l$, the steady solutions are unstable to the small perturbations. As $\delta$ increases, the alternation of regions of stable and unstable steady solutions occurs. Thus, the steady-states with a partial reflection of electrons from the VC can be realized in the Bursian diode.

### *2.2. Nonlinear oscillations*

In order to see the process termination when a small perturbation increases, it is necessary to study its nonlinear stage. We studied a regime with the VC at zero external voltage.

In numerical simulations, the time step was $\Delta\tau = 0.05$; the coordinate step was $\Delta\zeta = \delta/N$ with $N = 200$; and the electron DF at the emitter was chosen to be

$$f_0^+(u_0) = (2\Delta)^{-1}\Theta\left[\Delta^2 - (1-u_0)^2\right] \tag{6}$$

with a small velocity spread $\Delta$.

At each time moment $\tau^p$, firstly the DF of electrons is calculated with $E, K$ code. Then having calculated the DF and electron density, we solve the Poisson's equation with the following boundary conditions for the potential:

$$\eta(0) = 0, \quad \eta(\delta) = V_C. \tag{7}$$

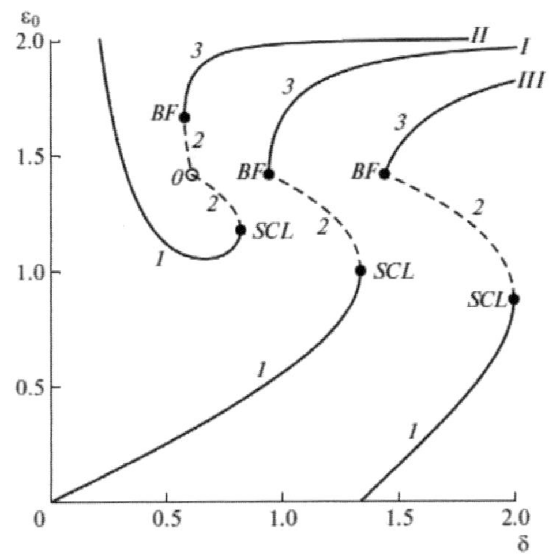


**Figure 1.** The dependence $\varepsilon_0$ on $\delta$. Curve I – $V_C = 0$, II – -0.4, 3 – +0.2.

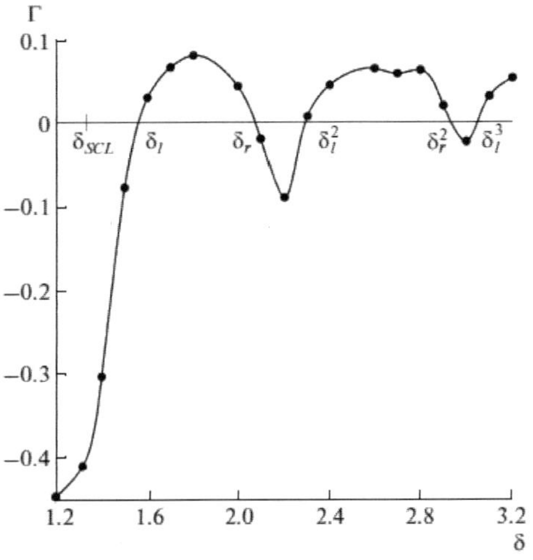


**Figure 2.** Dispersion curves; $V_C = 0$, $\Delta = 0.02$.

The problem reduces to solving the set of linear algebraic equations

$$\begin{gathered} \varepsilon_k^p = \varepsilon_{k-1}^p - (\Delta n)_k^p\,, \\ \eta_k^p = \eta_{k-1}^p - (1/2)\left(\varepsilon_k^p + \varepsilon_{k-1}^p\right)\left(\zeta_k^p + \zeta_{k-1}^p\right), \\ \eta_0^p = 0, \quad \eta_N^p = V_C. \end{gathered} \tag{8}$$

Here, $1 \leq k \leq N$ and $(\Delta n)_k^p$ is the electron charge in the layer $(\zeta_{k-1}^p, \zeta_k^p)$, which is calculated with allowance for the electron fluxes $\nu$ through the cell boundaries,

$$(\Delta n)_k^p = (\Delta n)_k^{p-1} - (1/2)\left(\tau^p - \tau^{p-1}\right)\left(\nu_{k-1}^{p-1} + \nu_{k-1}^p + \nu_k^{p-1} + \nu_k^p\right). \tag{9}$$

At each time step $\tau^p$, the self-consistent solution is achieved as follows. For the time $\tau^p$, the electric field distribution is known only at all preceding time moments $\tau \leq \tau^{p-1}$, but not yet at $\tau = \tau^p$. The field is extrapolated to the time interval $(\tau^{p-1}, \tau^p)$, and the DF and electron density at the time $\tau^p$ are then calculated. The density being known, the refined field distribution at the time $\tau^p$ is calculated and the density is recalculated for the new field. As a rule, it is sufficient to carry out only one to two such iterations.

The nonlinear process ends, as a rule, with the periodic oscillations. The dependencies of the potential $\eta_m$ at the top of the VC on its position $\zeta_m$, when growth of a small perturbation finished, are closed curves (Figure 3,a-d) [10]. As the electrode gap increases, the shape of the curve changes from elliptic (near the threshold) to nearly triangular. Figure 3 shows such curves for the 1st instability region $(\delta_l, \delta_r)$.

The $E, K$ code gives an opportunity to plot the electron velocity DF and density $n_e(\zeta)$ at any time with high accuracy. Figure 4,a shows the DF projection on the velocity-coordinate plane at a fixed time. The DF has a very complicated structure. Integration of the DF over velocities yields the electron distribution within the gap. Figure 4,b shows that the electron density varies by several orders of magnitude. It has a sharp peak at the point corresponding to the left boundary of the region from which the electrons are reflected.

For each instant of time, we can obtain such fine parameters of the process as a position $\zeta_M$ and potential $\eta_M$ of the point corresponding to the maximum of the density distribution and to plot the attractor in the $\{\eta_M, \zeta_M\}$ plane (Figure 3,e-h). The attractor is a closed curve of an oval shape. In analogy with the theory of finite-dimensional dynamic systems, we call such a curve the limiting cycle. In Figure 3, the heavy dot within the cycle corresponds to a steady solution. At the final stage of the process, the dependence $\eta_M$ on $\zeta_M$ winds onto the limiting cycle from inside (Figure 3,g).

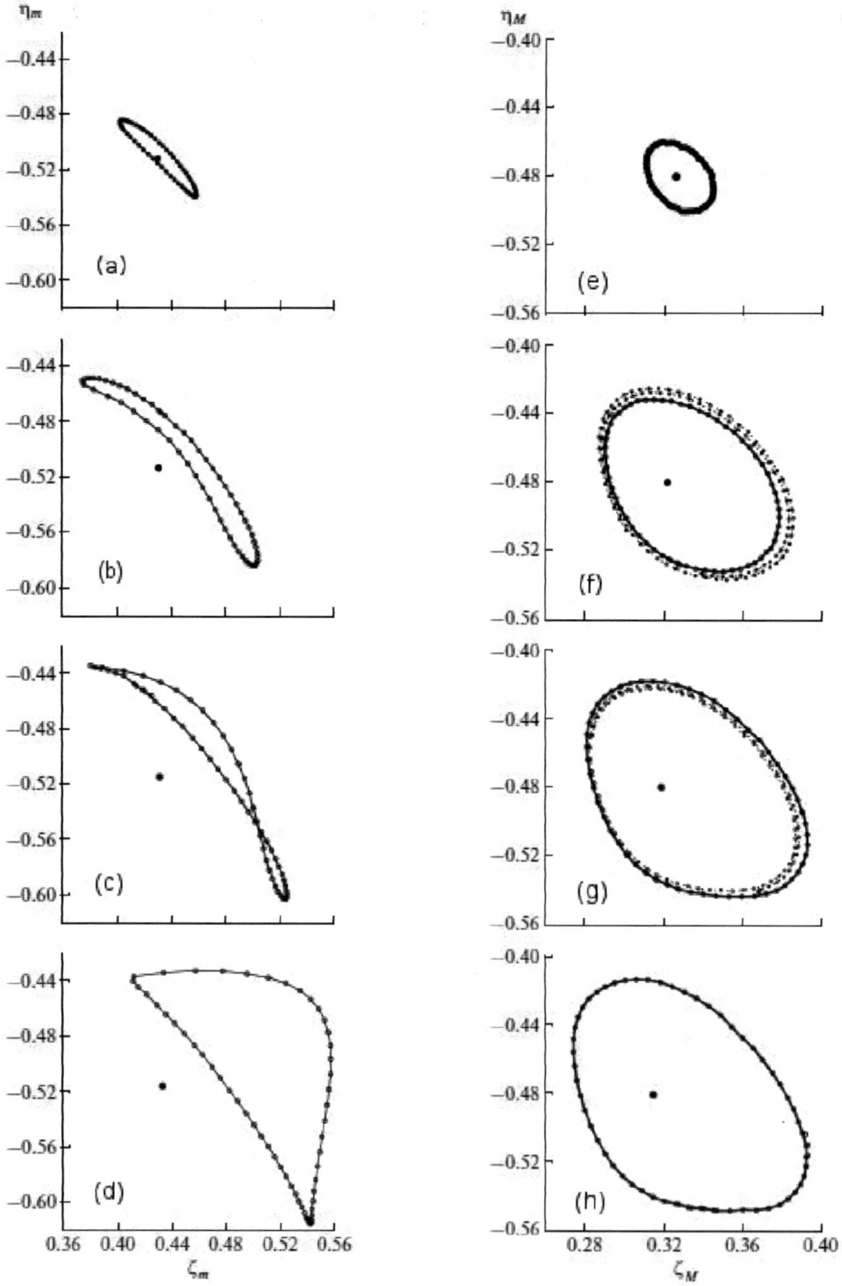


**Figure 3.** Dependences $\eta_m(\zeta_m)$ (a-d) and limiting cycles $\eta_M(\zeta_M)$ (e-h); $\delta = 1.6$ (a,e), 1.7 (b,f), 1.8 (c,g), 2.0 (d,h). The fat dots show steady solutions.

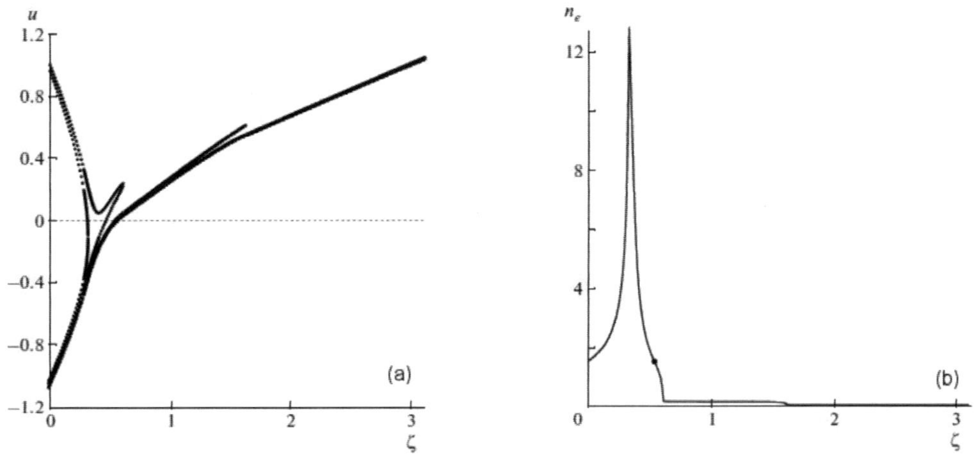

**Figure 4.** Projection of the electron DF onto the plane $\{u, \zeta\}$ at the time $\tau = 134.6$ (a) and the corresponding electron density distribution (b); $\delta = 3.1$. The heavy dot in a density shows the position of the top of a VC.

Further, we should check:
i) whether the relevant solutions are stable or not, and
ii) whether they are unique for given external parameters.

In order to solve these problems we selected the initial conditions responsible for the points located beyond the cycles. In these processes, the dependence $\eta_M(\zeta_M)$ wound like a helix onto the limiting cycle from outside (Figure 3,f). This cycle coincides with the limiting cycle obtained in simulating the development of a small perturbation from the steady solution. Thus, in the 1st instability region, the steady limiting cycles correspond to the regime of nonlinear periodic oscillations and each such cycle is unique.

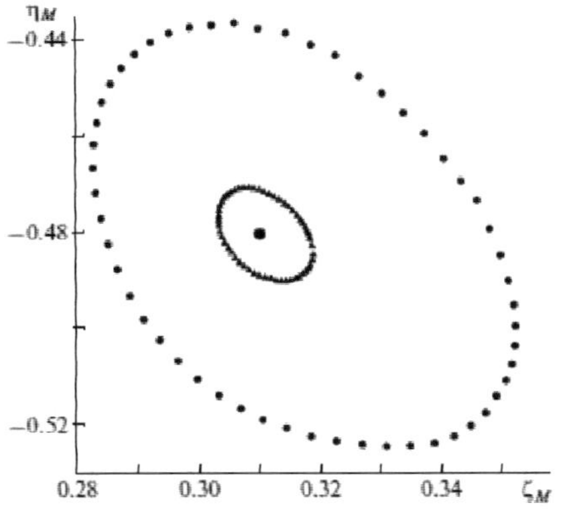

**Figure 5.** Two limiting cycles for $\delta = 2.4$.

We calculated a set of the transient processes started from initial conditions located beyond the cycles, for the electrode distances beyond the instability threshold $\delta_l$. We found that, in these cases, the processes ended also with nonlinear periodic oscillations of the same type as those in the 1st instability region.

In the 2nd instability region $(\delta_l^2, \delta_r^2)$, we revealed a new class of oscillations of which amplitudes are substantially smaller than that of the oscillations described above. In Figure

5, the new cycle is shown as the minor oval. However, for these gap values, large-amplitude oscillations exist also.

We found that, at certain times, the convective current density increases in a jump like manner (Figure 6,a). The jump in the current develops on a time scale much shorter than the electron time-of-flight through the electrode gap. Such behaviour of the current density stems from the character of the evolution of the electron distribution in the electrode gap during the oscillations. The reason is that, at the times when the height of the VC has a minimum, a fairly large group of electrons overcomes the potential barrier. Further, these electrons form a sharp front in their spatial distribution (Figure 6,b). The front moves toward the collector, and when it reaches the collector, the convective current density increases sharply.

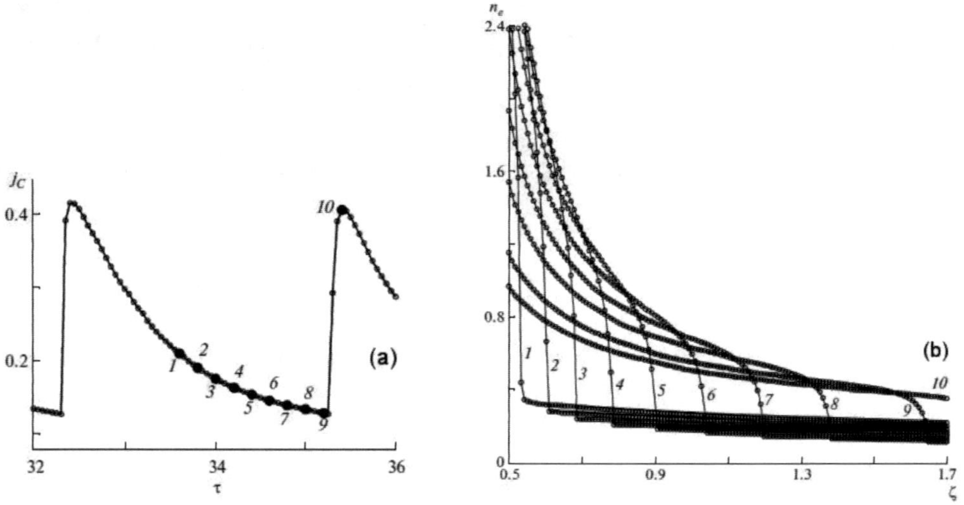


**Figure 6.** The convective current (a) and electron (b) density evolution during an oscillation period.

### *2.3. Long-lived electrons*

Earlier in numerical simulations an unusual phenomenon, being characterized for oscillations, was revealed: existence of so-called long-lived electrons [11]. These are the electrons that reach the VC, which serves as a potential barrier for them, and oscillate with it during several periods. We have found a reason for the onset of long-lived electrons, and analysed their properties [10]. In order to study the features of the long-lived electrons we have considered a model of the electric field, which describes properly trajectory of the VC oscillation:

$$\eta(\zeta,\tau) = -\alpha^2/2\left[(\zeta_m^0)^2 - (\zeta-\zeta_m^0)^2\right] - \alpha^2\zeta_m^0\kappa\cos(\Omega\tau)\zeta. \tag{10}$$

Here, $\alpha^2 = -2\eta_m^0/(\zeta_m^0)^2$, $\eta_m^0$ and $\zeta_m^0$ are the height and the position of the VC top in the steady-state case ($\kappa = 0$), and the parameters $\Omega$ and $\kappa$ characterize the frequency and amplitude of the VC oscillations. The quantity $\alpha$ is determined by the curvature of the PD in the vicinity of the VC top. For an electron that is injected from the emitter with the velocity $v_0$ at the time $\tau_0$, we obtain the following expressions:

$$\zeta(t) = A_+\exp[\alpha(t-\tau_0)] + A_-\exp[-\alpha(t-\tau_0)] + \zeta_m^0\left[1 + B\kappa\cos(\Omega\tau)\right], \tag{11}$$
$$\dot{\zeta}(t)/\alpha = A_+\exp[alpha(t-\tau_0)] - A_-\exp[-\alpha(t-\tau_0)] - \zeta_m^0 B\varphi\kappa\sin(\Omega\tau).$$

Here, $B = (1+\varphi)^{-1}$, $\varphi = \Omega/\alpha$, $A_\pm = \left\{\zeta_m^0 \mp v_0/\alpha + \zeta_m^0 \sqrt{B}\kappa \cos\left[\Omega\tau_0 \mp \arctan(\varphi)\right]\right\}/2$. These formulas show that, at $\alpha(t-\tau_0) \gg 1$, the electrons can either overcome the potential barrier and reach the opposite electrode ($A_+ > 0$) or they can be reflected from the barrier and return to the emitter ($A_+ < 0$) that depends on the relationship between $v_0$ and $\tau_0$. It is also seen that a group of electrons exists that can oscillate in the vicinity of the VC for a fairly long time until they fall into the 1st or the 2nd group. These are just long-lived electrons. The emitter values of velocity $v_0$ and time $\tau_0$ of such electrons lie in a very small vicinity of the universal curve

$$v_0 = \alpha\zeta_m^0 \left\{1 + \sqrt{B}\kappa \cos\left[\Omega\tau_0 + \arctan(\varphi)\right]\right\}. \tag{12}$$

This relationship is deduced from the condition $A_+ = 0$, and does not depend on parameters of the arrival point $(\zeta, \tau)$.

Trajectories of long-lived electrons are determined by last term in formulas (11). Hence, long-lived electrons are those beam electrons, the kinetic energy of which is close to zero when they reach the vicinity of the potential minimum. As a result, such electrons are reflected from the potential barrier, but they move slower than the VC ($B < 1$), which thus overtakes them, so they again catch up with the barrier and are again reflected from it, and so on. It is the reason why long-lived electrons exist. More over, any electron "lives" only a finite time. Indeed, the formulas (11) contain an exponentially increasing term. Therefore electron trajectory is unstable to small variations in the initial conditions. On sufficiently long time scales $\tau$, this term will become greater than the remaining terms and the electron will come to one of the electrodes, no matter how close the point $(v_0, \tau_0)$ is to curve (12) or how small the coefficient $A_+$ is.

## 3. Nonlinear oscillations in the KDSI

The Knudsen diode with surface ionization (KDSI) is a collisionless, one-dimensional, single-emitter plasma diode. The emitter serves as a source of electrons by Richardson emission and ions by surface ionization of neutral alkali atoms. The emitted electron and ion DFs are the half-Maxwellian ones. The Knudsen thermionic energy converter [12] and Q machine [13] are examples of the KDSI.

### *3.1. Steady solutions*

The self-consistent steady solutions of the KDSI are determined by three dimensionless values: emitter neutralization ratio $\gamma = n_i^{0,+}/n_e^{0,+}$, electrode distance $\delta = d/\lambda_D$, and external voltage $V_C = e\Phi_C/kT_E$. Typical PD exhibits a quasineutral plasma region with practically constant potential, embedded between two space-charge sheaths adjacent to the electrodes. We have analysed stability properties of such solutions in overneutralizated regime ($\gamma > 1$), and built up a boundary of the region of stable solutions (Figure 7) [14]. Fat points show data of experiment carried out to determine the boundary beyond which the oscillations initiate [15]. One can see the excellent agreement between our theory and this experiment.

### *3.2. Nonlinear oscillations*

A number of experiments demonstrated the large amplitude oscillations of electron current in the KDSI (see, e.g., [12, 13]). Their period is comparable with the ion transit time of the electrode gap $d$. We have shown that the oscillations are characterized by the periodic sequences of two types of stages: the slow (related to the slow motion of the ions) and the fast (compared with the time for an electron to overcome the electrode distance) ones [4]. In calculating the slow stage, we assume that the electrons overcome the gap $d$ before the ions cross the distance of the Debye length $\lambda_D$. Then, we consider that to the moment when ions moved over distance $\lambda_D$, the electrons and electric field in the electrode gap have already a time to redistribute and conform

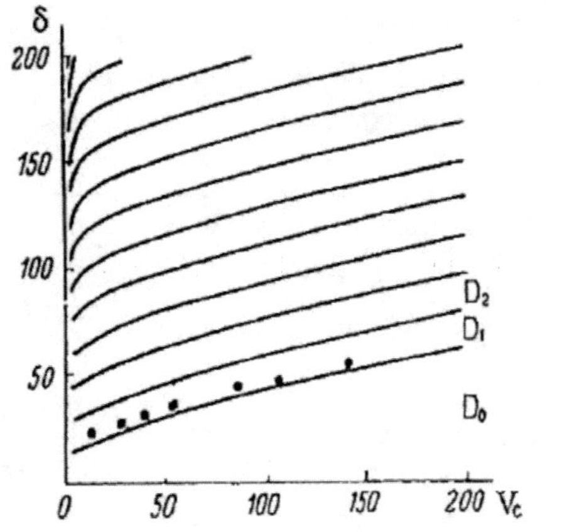


**Figure 7.** Boundaries of regions, contained stable and unstable solutions; $\gamma = 5$.

with a given ion distribution. Therefore the time scale is selected to be the time needed an ion to overcome $\lambda_D$. The ion DF is calculated by $E, K$ code.

At each time-step $\tau^p$, the electron and electric field distributions for known ion background $n_i(\tau^p, \zeta)$ are found from the stationary problem. In this case, an explicit expression for electron density can be obtained for PD of any type. It depends on a potential value at the given point, and also at the points of PD local minimum, being the points of electron reflection: $n_e = n_e(\eta, \eta^1, \ldots, \eta^l)$. Inserting the electron density in the Poisson equation, we reduce the problem to solution of nonlinear ordinary differential equation of a 2nd-order:

$$d^2\eta/d\zeta^2 = n_e(\eta, \eta^1, \ldots, \eta^l) - n_i(\tau^p; \zeta) \qquad (13)$$

with boundary conditions:

$$\eta(0) = 0, \quad \eta(\delta) = V_C. \qquad (14)$$

As a rule, this problem has several solutions. The PDs that can be realized as well as their stability properties, are analysed by the $\eta, \varepsilon$–diagram technique [14].

During the slow stage, the ions are redistributed in the electrode gap. At certain instants of this stage, the conditions of the Pierce type instability onset appear [16]: a potential jump forms in the emitter sheath, which accelerates strongly the electrons, and electron beam with a small spread in velocities propagates through the plasma. At such instants KDSI plasma differs from the Pierce diode one only due to the ion background, through which the electron beam goes. The ion background is nonuniform in the KDSI.

During the fast stage, the ion distribution may be considered as the steady one. And the time scale is selected as the time needed an electron to overcome $\lambda_D$. Self-consistent simulations of this stage are similar to those in the vacuum diode. Figure 8,a is an example of calculation of the electron transient process which finishes in a state of the VC (curve 4). The initial PD is a distribution of the monotonic growth. Relevant state belongs to the boundary of a stable region (curve 1). Figure 8,b shows the maximum deviation $\Delta\Phi$ of a potential from the initial one in time [17].

The instability gives an interesting physical phenomenon – a cut-off of current. It occurs in the case when the electron transient process finishes in a state for which the PD with the VC realizes: a region with negative potential forms at a distance from the emitter (Figure 9,a). As a result a strong limitation of electron current, passing through the diode, occurs. Because a time of VC formation is about the mean time-of-flight of the electrons over the diode gap, one

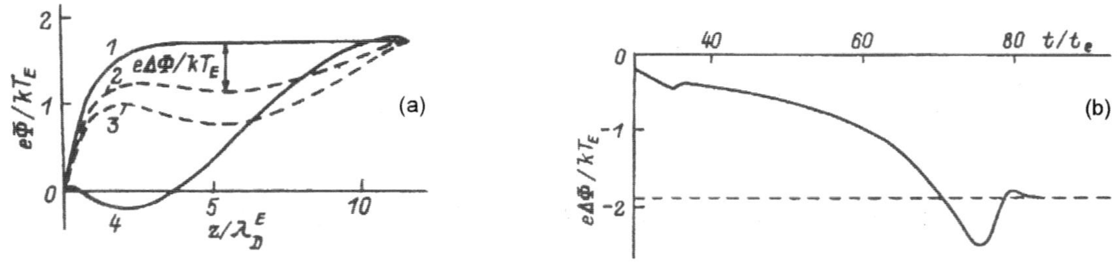


**Figure 8.** Evolution of PD during the fast stage in the KDSI with $\gamma = 5$, $\delta = 11.4, V_C = 5$.

can speak that the current changes instantaneously in diode. This structure is stable. During the slow stage it moves to the collector with a characteristic ion velocity $\sim (2kT_E/m_i)^{1/2}$ and drives an electron current, passing through the diode. In Figure 9,b, these structures correspond to the parts of the solid curves with the current density less than $j_E$. The vertical dashed lines denote the fast electron transitions. Oscillation period is about a time-of-flight of the thermal ions over the diode gap, and modulation depth $\beta = [j_{max}(t) - j_{min}(t)] / j_{max}(t) \sim 0.5$. These facts are corroborated by experiments [12, 13].

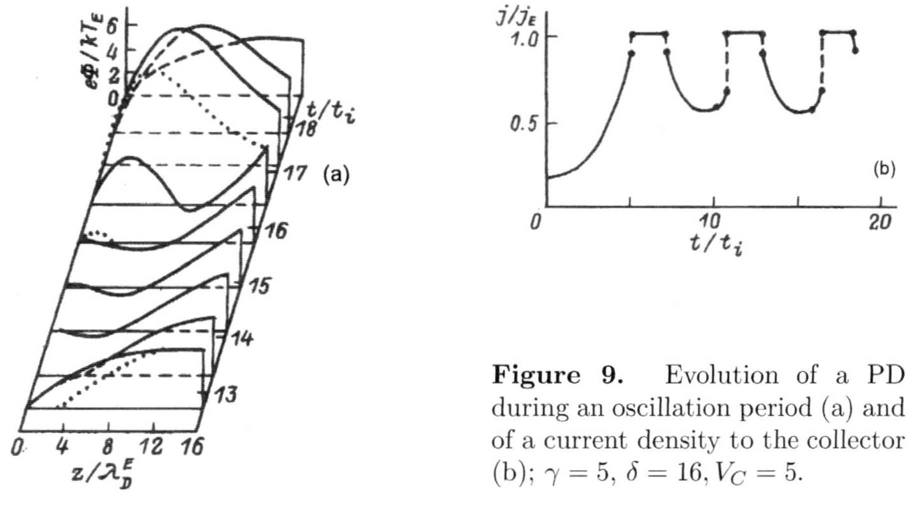


**Figure 9.** Evolution of a PD during an oscillation period (a) and of a current density to the collector (b); $\gamma = 5$, $\delta = 16, V_C = 5$.

From the DP evolution during an oscillation period (Figure 9,a) one can see that, within the time interval ($12.7 < \tau < 16.4$) to the right of a potential minimum, a double layer exists, within which the potential changes by an amount of about an external voltage $\Phi_C$, and an essential separation of charges occurs at the distance of several $\lambda_D$. The electron current, of which magnitude depends on a potential barrier height for electrons, passes through the layer.

Development of electron instability gives rise to a number of interesting physical phenomena. For example, during the fast stage, a potential well is formed (Figure 9,a, $\tau > 16.4$) and accumulation of potential energy occurs. Then, during a slow stage, the latter converts into directed kinetic energy of the ions and thus results in ion acceleration up to relatively high energy. More over, the ions as beams had a very small velocity spread (Figure 10,a), and the ion kinetic energy localization occurs (Figure 10,b).

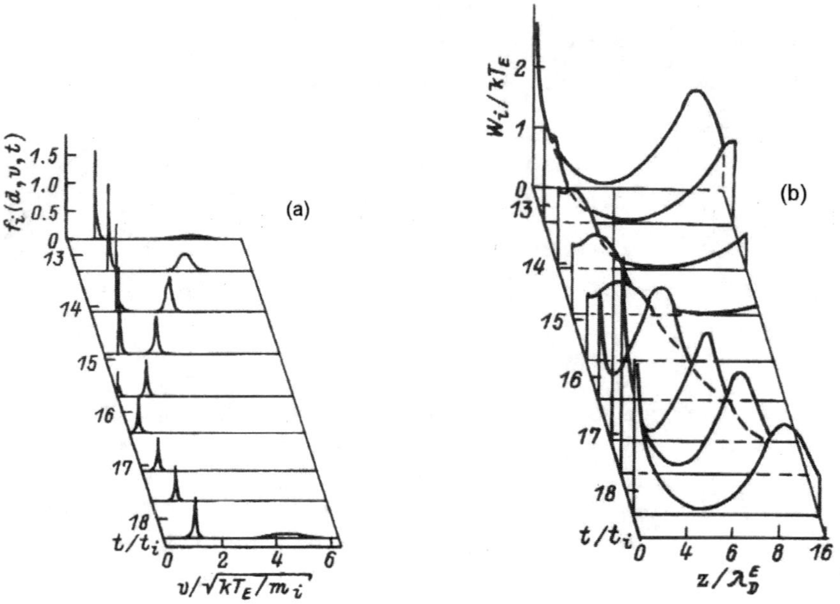


**Figure 10.** Evolution of an ion DF at the collector (a) and of ion kinetic energy during an oscillation period; $\gamma = 5$, $\delta = 16, V_C = 5$.

### *3.3. Collisionless trap of electrons in the potential well*

The PD with a hump is a potential well for electrons (Figure 10,a). Self-consistent collisionless electron trapping in the well, may result due to the energy exchange between electrons and the time-dependent electric field arising during the well formation. Really, when the well depth increases, an electron, moving in the well, loses its energy, to the electric field. With increasing gap length, the energy losses growth, and a threshold gap $d^*$ exists, in exceeding which these losses become larger than the initial energy of electron at the emitter. As a result, the electron reflects from a potential barrier in the collector region, and turns out to be trapped in the potential well [18].

The trapped electrons do not spread over the well as a time-independent structure, but they are a dynamic structure: forming a localized spatial bunch with sharp fronts, which bounces between the electrodes, being broaden, when passing the center region of the gap (Figure 11,a, curve 2), and shrank in the vicinity of electrodes (curves 1 and 3).

During the process, a charge of the bunch electrons decreases due to some electrons get away the electrodes. Such a time-dependent behaviour of the trapped electrons leads to oscillations of the PD (Figure 11,b), so that the well depth oscillations also (solid curve), being less than this one if the trapped electrons were absent (dashed curve). The electron trapping plays the crucial role in self-consistent calculations of nonlinear KDSI dynamics.

## 4. Conclusion

Thus, as in the diode with the beam-like velocity distribution function of emitted electrons, as in the Knudsen diode with surface ionization the large amplitude oscillations of the electron current can exist. They arise when a current of the incoming electrons exceeds certain threshold value which is of the order of the plasma beam Debye length. As a result, the Bursian-Pierce instability develops resulting in a current cut off. During the oscillation process a set of nonlinear structures form. For example, dynamical VC and DL exist, the long-lived and trapped electrons

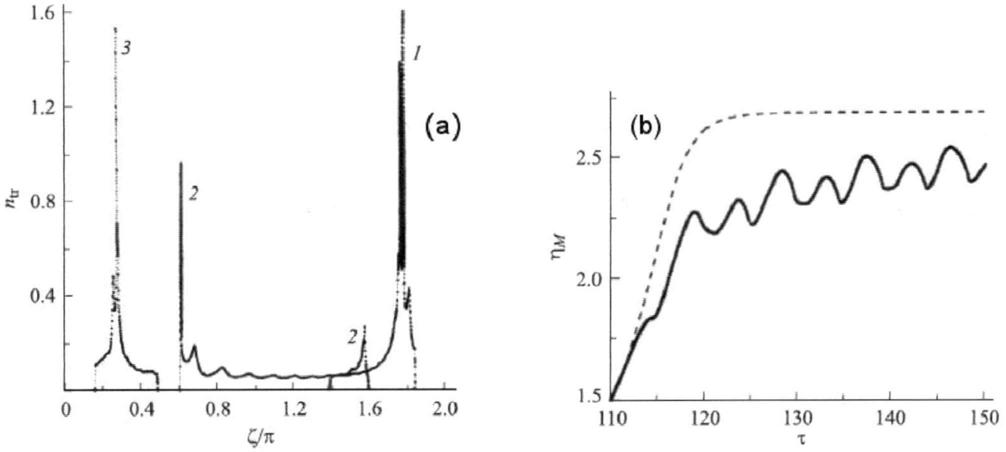


**Figure 11.** Spatial distribution of trapped electrons at three time moments (a) and evolution of a well depth (b); $\delta = 1.97\pi$.

appear, bunching charge particles occurs, and so on.

We proposed to create fast electron switches on the basis of theory of nonlinear oscillations in the Bursian diode [8]. Results of this theory can be useful for microwave generators which convert oscillatory energy of the self-consistent electric field into microwave radiation. Used the theory of nonlinear oscillations in the KDSI we proposed new type of the themionic energy converter which converts heat energy directly to alternate current power [19].

**Acknowledgements**
This work was supported in part by the Russian Foundation for Basic Research (projects no. 12-08-00490 and no. 13-08-00149).

## References

[1] http:// ptsg.eecs.berkeley.edu/ April 2008
[2] Kolinsky H 1998 Das Stablitatsverhalten von Plasmadioden (*PhD thesis*) Bayreuth University (in German)
[3] Kuznetsov V I, Ender A Ya 1979 *Sov. Phys. Tech. Phys.* **24** 1199
[4] Kuznetsov V I, Ender A Ya 1983 *Sov. Phys. Tech. Phys.* **28** 1431
[5] Kuznetsov V I, Ender A Ya, Schamel H and Akimov P. V 2004 *Phys. Plasmas* **11** 3224
[6] Kuznetsov V I, Ender A Ya 2010 *Plasma Phys. Rep.* **36** 226
[7] Kuznetsov V I 2006 Theory of nonlinear oscillations and transient processes in plasma diodes (*Doctor's thesis*) Ioffe Institute (in Russian)
[8] Akimov P V, Schamel H, Ender A Ya, and Kuznetsov V I 2003 *J. Appl. Phys.* **93** 1246
[9] Kuznetsov V I, Solov'ev A V and Ender A Ya 1994 *Tech. Phys.* **39** 1207
[10] Kuznetsov V I, Ender A Ya 2010 *Plasma Phys. Rep.* **36** 236
[11] Selemir V D, Alekhin B V, Vatrunin V E, Dubinov A E et al. 1994 Plasma Phys. Rep. **20** 621
[12] Babanin V I, Kolyshkin I N, Kuznetsov V I, Mustafaev A S, Sitnov V I and Ender A Ya 1982 *Sov. Phys. Tech. Phys.* **27** 793
[13] Iisuka S, Michelsen P, Rasmussen J J et al. 1985 *J. Phys. Soc. Jap.* **54** 2516
[14] Kuznetsov V I, Ender A Ya 1977 *Sov. Phys. Tech. Phys.* **22** 1295
[15] Kucherov R Ya et al 1989 *Sov. J. Plasma Phys.* **15** 766
[16] Pierce J R 1944 *J. Appl. Phys.* **15** 721
[17] Kuznetsov V I, Ender A Ya 1981 *Sov. Phys. Tech. Phys.* **26** 1326
[18] Kuznetsov V I, Ender A Ya and Kuhn S 2006 *Tech. Phys.* **51** 1257
[19] Babanin V I, Ender A Ya, Kolyshkin I N, Kuznetsov V I, Sitnov V I, Paramonov D V 1997 Proc. 32nd IECEC (Honolulu) vol 1 (NY AIChE) p 427

Journal of Physics: Conference Series **510** (2014) 012047 doi:10.1088/1742-6596/510/1/012047

# MMALE numerical simulation for multi-material large deformation fluid flows

**Qinghong Zeng**

Institute of Applied Physics and Computational Mathematics, Beijing 100094, China

zeng_qinghong@iapcm.ac.cn

**Abstract**. Multi-material arbitrary Lagrangian Eulerian (MMALE) method, coupled with moment of fluid (MOF) interface reconstruction, was studied and applied to multi-material large deformation flow problems. In MMALE method, the mesh lines are allowed to cross the material interfaces and mixed cells are introduced. In mixed cells, the MOF method is used to reconstruct the interfaces. The numerical examples show the high accuracy and resolution of MOF-MMALE, it is validated that MOF-MMALE method is an effective way to simulate flow problems of multi-material large deformation.

## 1. Introduction

Multi-material large deformation fluid flows exist frequently in astrophysics, weapon physics and inertial confinement fusion (ICF) fields. There are two important characteristics in these kinds of flow phenomena. One is the violent movement of the fluid, which leads to large deformation of fluid field, the other is the existence of multi-material and the distortion of the interface between different materials. Up to now, numerical simulation is one of the major ways to study these problems. In numerical simulation of multi-material large deformation fluid flows, the most important thing is tracking the material interfaces accurately while dealing with the large deformation of fluid flows simultaneously.

The numerical methods for multi-material large deformation fluid flows are Lagrangian method, Eulerian method or arbitrary Lagrangian-Eulerian (ALE) method. In these methods, such-and-such difficulties are exist. Lagrangian method can track the material interfaces sharply, but it can't deal with the large deformation of computational mesh. Eulerian method can hold the large deformation of fluid flows, but it will bring large numerical diffusion of physical variables. If the deformation of fluid is small, ALE method is the best selection, but if the deformation is large, ALE method is helpless.

To alter this predicament of numerical simulation of multi-material large deformation fluid flows, many research teams (LLNL, LANL, AWE, CEA, IAPCM, etc.) start to develop newly numerical frame to overcome this difficulty and multi-material ALE (MMALE) method[1-3] is presented. In MMALE, mixed cells are introduced and the material interfaces are permitted to cross computational cells. Here mixed cells are cells including two or more kinds of materials. The introduction of mixed cells make the computational code robust while keep the computational results as accurate as possible. To handle the material interfaces in mixed cells, moment of fluid (MOF) interface reconstruction method[4-6] is used in MMALE simulations. MMALE method is employed widely in the computation of multi-material and large deformation fluid flows.

The MMALE method coupled with MOF interface reconstruction (MOF-MMALE) was studied, and applied to multi-material large deformation flow problems. This paper is organized as follows. In section

2, we present the theoretical system of MMALE numerical simulation. The theoretical system is consisted of single velocity multi-material model, Lagrangian computing, rezoning, remapping, closure model and MOF interface reconstruction. Next, some numerical examples include more than two materials to emphasize key properties of MMALE method in Section 3. Finally, conclusions and perspectives about future works are given in Section 4.

## 2. MMALE method

Because of the introduction of mixed cells in MMALE, extra steps are needed to accomplish this framework, such as capturing the interface, closing the system of equations, and computing the equation of state in mixed cell.

The flowchart of MMALE method is shown in figure 1. First, the Lagrangian computing is performed, in this step the velocity, mesh, density variables are updated. Then the closure model is called to update the volume fraction. Next, the mesh is rezoned to obtain new mesh with good quality. Subsequently, MOF interface reconstruction is employed to determine the position of material interfaces. Finally, the physical variables are remapped to transfer the old variables from old mesh to new mesh. Hereto, one step of MMALE method is finished.

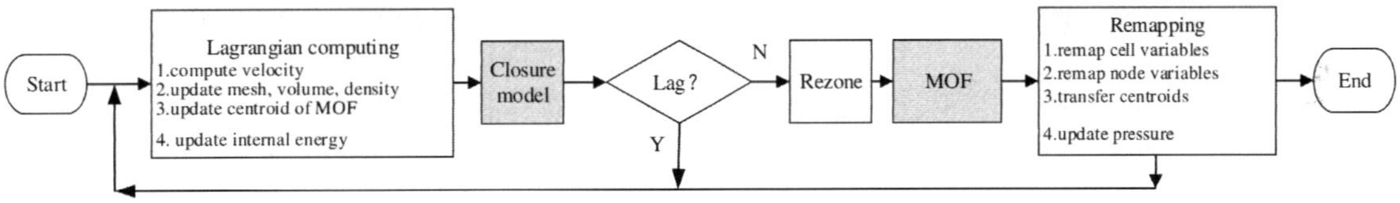


**Figure 1.** The flowchart of MMALE method.

### 2.1. *Single Velocity multi-material model*

In mixed cell, single velocity multi-material model[7] is adopted. In this model, different materials in one mixed cell have same velocity but different thermodynamic state, as shown in figure 2.

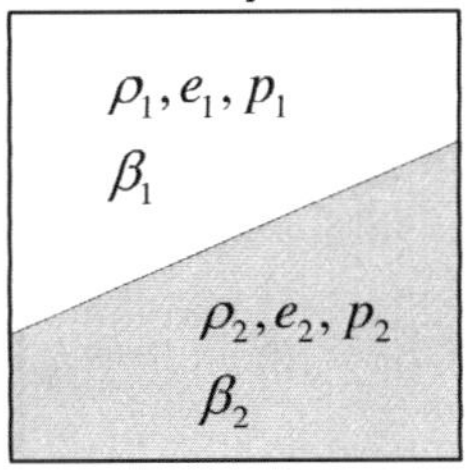


**Figure 2.** Mixed cell.

The system of equations of single velocity multi-material model can be written as

$$\begin{aligned}
\frac{\mathrm{d}\boldsymbol{u}}{\mathrm{d}t} &= -\frac{1}{\rho}\nabla p \\
\frac{\mathrm{d}\rho_i}{\mathrm{d}t} &= -\rho_i \nabla \cdot \boldsymbol{u}_i \\
\frac{\mathrm{d}\beta_i}{\mathrm{d}t} &= \beta_i \left(\nabla \cdot \boldsymbol{u}_i - \nabla \cdot \boldsymbol{u}\right) \\
\frac{\mathrm{d}e_i}{\mathrm{d}t} &= -\frac{p_i}{\rho_i}\nabla \cdot \boldsymbol{u}_i \\
\frac{\mathrm{d}\boldsymbol{x}}{\mathrm{d}t} &= \boldsymbol{u}
\end{aligned} \tag{1}$$

Here $\rho_i, \boldsymbol{u}_i, \beta_i, e_i$ and $p_i$ are density, velocity, volume fraction, specific internal energy and pressure of the $i$ th material, respectively. $\boldsymbol{u}$ is the velocity, $\rho$ is the average density and $p$ is the average pressure. the equation of state of the material is given by

$$p_i = p_i(\rho_i, e_i) \tag{2}$$

The system of equation (1) for a multi-material medium is not closed and one needs to use some closing relations in mixed cells. The closure model will be discussed in subsection 2.4.

### 2.2. *Lagrangian computing*

Lagrangian computing is performed in staggered mesh. In staggered mesh, the coordinate and velocity are defined at cell nodes and state variables, such as density, pressure, are defined at cell centers. For space discretization, compatible hydrodynamic algorithms are used and extended to multi-material case in our MMALE method. Compatible hydrodynamic algorithms[8] are classical algorithms to solve the system of hydrodynamic equations in staggered mesh, and applied in many large scientific code. Predictor-corrector algorithms are employed for time discretization, which can reach two order accuracy in time.

### 2.3. *Rezoning and remapping*

Rezoning is an important step in MMALE to get new mesh with good quality. In current implementation of our MMALE method, only simple rezoning strategy is used, for example, rezoning to initial mesh or Winslow rezoning algorithm. Small adjustment is applied according to actual problems. More refined rezoning algorithms will be integrated into our code to improve the robustness and capacity of our code in future works.

The remapping must be split into two steps, one for cell variables and the other for node variables. For cell variables remapping, Sutherland-Hodgman polygon intersection algorithm is used. For node variables remapping, the subcell remapping algorithm[9] is used.

### 2.4. *Closure model*

In Lagrangian computing step, not only the mesh, velocity, density, but also the volume fraction are needed to update. After remapping step, the average pressure in mixed cell is also resolved. Extra assumptions are added to compute the material divergence and average pressure in mixed cell.

Currently, equal compressibility is provided. Equal compressibility model is based on assumption of that the compressibility of each material in mixed cell is equal, and the different material divergence is equal each other.

$$\nabla \cdot \boldsymbol{u}_i = \nabla \cdot \boldsymbol{u} \tag{3}$$

We can deduce easily that

$$\frac{\mathrm{d}\beta_i}{\mathrm{d}t} = 0 \tag{4}$$

Which shows that the volume fraction is not changed in equal compressibility model. The average pressure is given by

$$p = \sum_i \beta_i p_i \tag{5}$$

### 2.5. *MOF interface reconstruction*

MOF interface reconstruction method was suggested by Dyadechko in 2005[4], many feathers of which are similar to VOF. In figure 3, MOF interface reconstruction in a mixed cell including two materials is shown. The curve segment is true interface between materials and the straight is

Journal of Physics: Conference Series **510** (2014) 012047 doi:10.1088/1742-6596/510/1/012047

reconstructed interface. The region Ω with shadow is reference material, Φ is angle between the interface and the x-axis, $0 \le \Phi < 2\pi$. $(n_x, n_y) = (\cos\Phi, \sin\Phi)$ is the direction of interface, which points from the interface to the inside of reference material. $(x_c, y_c)$ is centroid of the reference material.

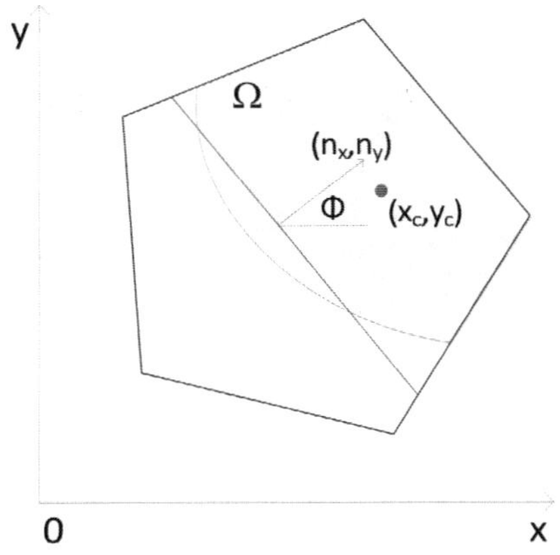


**Figure 3.** MOF interface reconstruction.

MOF achieves second-order accuracy by tracking and using the volume fractions and first moments or centroid of each material region in a computational cell. In plane coordinates, centroid is defined as

$$x_c = A_\Omega^{-1} \int_\Omega x\mathrm{d}x\mathrm{d}y, \quad y_c = A_\Omega^{-1} \int_\Omega y\mathrm{d}x\mathrm{d}y \tag{6}$$

Here $A_\Omega = \int_\Omega \mathrm{d}x\mathrm{d}y$ is the area of region Ω.

The location of the linear interface in each mixed cell is chosen to preserve the volumes and provide the best possible approximation to the material centroids. The object function is minimized using a optimizing process to obtain the direction of interface

$$E(\phi) = \left[x_c^{\mathrm{comp}}(\phi) - x_c\right]^2 + \left[y_c^{\mathrm{comp}}(\phi) - y_c\right]^2 \tag{7}$$

Here $\left(x_c^{\mathrm{comp}}, y_c^{\mathrm{comp}}\right)$ is computational centroid.

## 3. Numerical examples

All the numerical results presented in this section are performed in plane geometry. The material are characterized by a perfect gas equation of state which writes $p = \rho e(\gamma - 1)$ where $\gamma$ stands for the specific heat ratio of the gas.

### *3.1. Two-material shock tube problem*

Consider a two-material shock tube problem. The computational domain is [0,1] with an interface initially located at x=0.5. The initial state for left material are $(\rho_l, p_l, \boldsymbol{u}_l, \gamma_l) = (1, 1, 0, 1.4)$ and $(\rho_r, p_r, \boldsymbol{u}_r, \gamma_r) = (0.125, 0.1, 0, 5/3)$ for the right material. This is a true multi-material problem because of the different specific heat ratio.

We set all the physical boundaries as solid walls, and paved 200, 400 cells on the computational domain. The problem is simulated by MMALE method and Lagrangian method. In rezoning step of MMALE simulation, the new mesh is rezoned back to initial mesh which is actually equivalent with Eulerian method. We compared the MMALE numerical results with exact solution and Lagrangian results in figure 4. We have found from these figures that not only MMALE results but also Lagrangian results are both shown good agreement with exact solutions. Some inaccuracy appears in specific internal energy figure near the contact discontinuity for Lagrangian results, but doesn't for MMALE results.

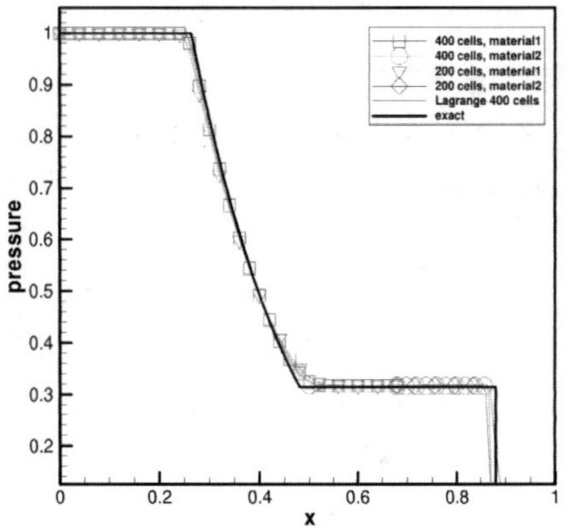


(a) Pressure

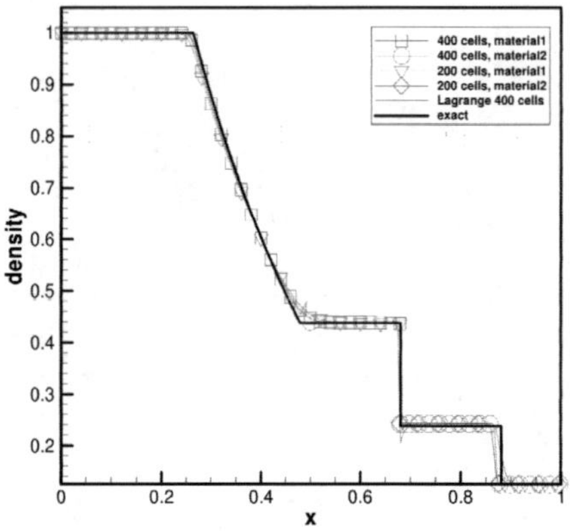


(b) Density

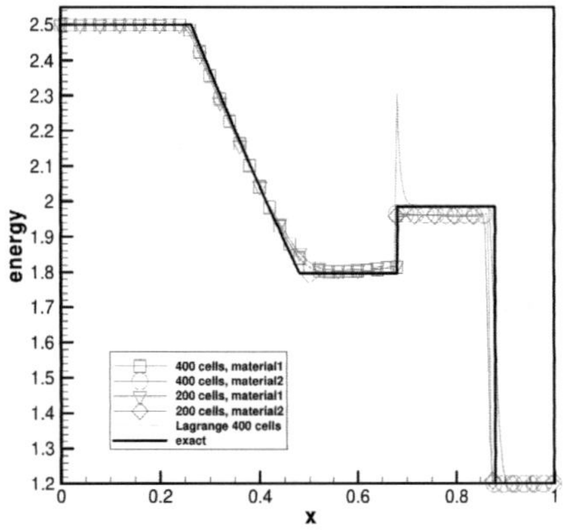


(c) Energy

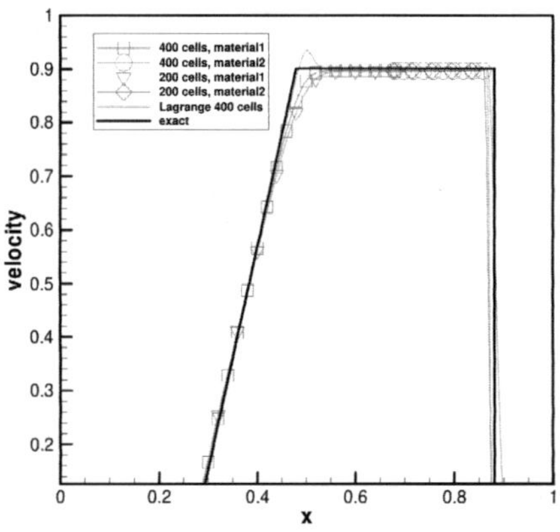


(d) Velocity

**Figure 4.** Two-material shock tube problem. Comparison of MMALE results with exact solutions and Lagrangian results.

### 3.2. *Rayleigh-Taylor instability*

This test case deals with the well-known Rayleigh-Taylor instability. The computational domain is the rectangular box [0,1/3]x[0,1] which is paved with 34x100 cells. The initial set up consists of two immiscible fluids which are separated by a perturbed interface, whose equation writes $y_i(x) = 0.5 + a_0 \cos(6\pi x)$, $a_0 = 0.01$. The heavy fluid is located above the light one. The densities of the two fluids are $\rho_h = 2$ and $\rho_l = 1$. The same polytropic index $\gamma_h = \gamma_l = 1.4$ is shared by the two fluids. A downward gravity field is applied, $\boldsymbol{g} = (g_x, g_y)^t = (0, -0.1)^t$.

This configuration is not stable, the heavy fluid will sink and light fluid will rise. We set all the physical boundaries as solid walls and set time step dt=0.001. Mixed cells are existed at initial time. The MMALE numerical results for ten moments from time 1 to 10 is shown in figure 5, the time evolution of the instability can be seen clearly from these ten figures. It can also be seen from these figures that MOF-MMALE can identify small or tiny structure, for example, the "hook" structure near the edge of the heavy fluid "mushroom" is reconstructed successfully by MOF-MMLAE. This capacity of high resolution is attributed to MOF interface reconstruction.

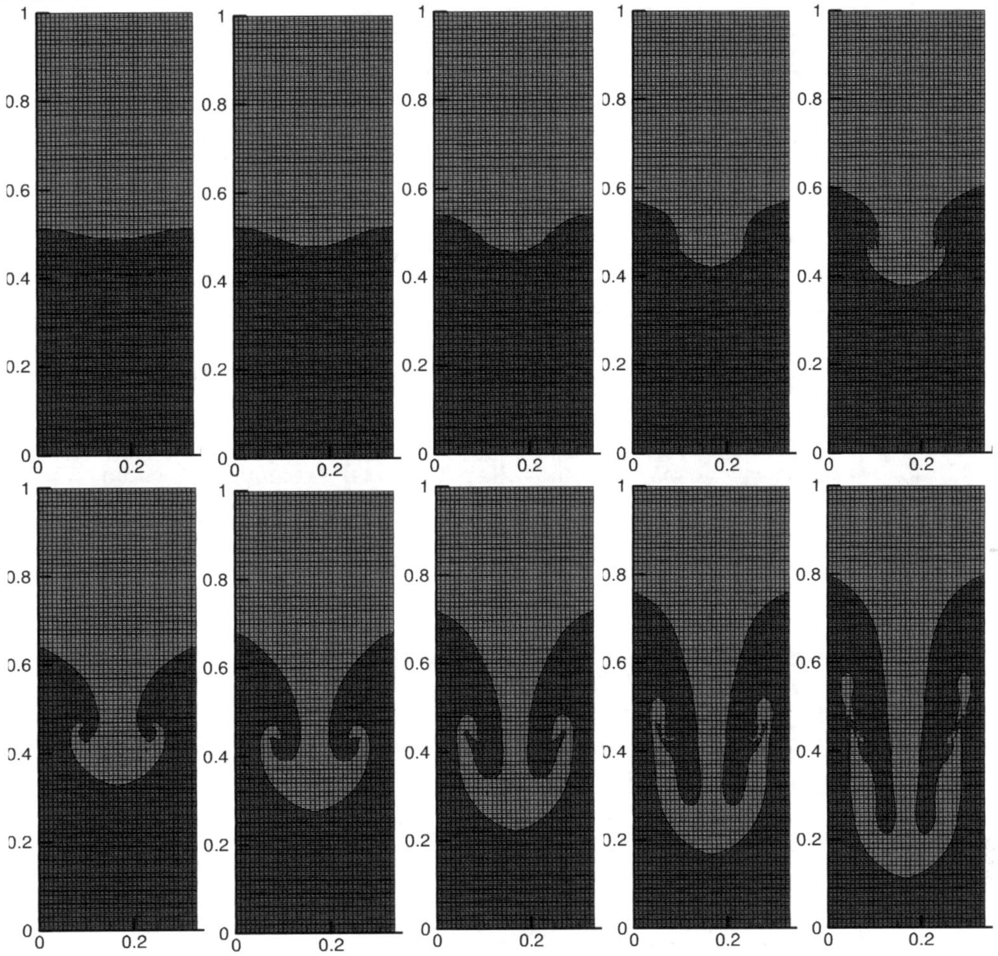

**Figure 5.** MMALE numerical results of Rayleigh-Taylor instability problem for ten moments from time 1 to 10.

### 3.3. *Interaction of shock wave with Helium bubble*

This test case corresponds to the interaction of shock wave with a cylindrical Helium bubble surrounded by air at rest. Consider a rectangular domain of dimensions [0,0.65]x[-0.089,0.089] initially full of air of data $(\rho_a, p_a, \boldsymbol{u}_a, \gamma_a) = (1, 10^5, 0, 1.4)$ displayed in figure 6. The cylindrical bubble is represented by a disk characterized by its center $(x, y) = (0.32, 0)$ and its radius $r = 0.025$. The initial data of Helium bubble are $(\rho_h, p_h, \boldsymbol{u}_h, \gamma_h) = (0.182, 10^5, 0, 1.648)$. We set all boundary conditions as solid wall except at the right end where we impose a piston boundary condition defined by the inward velocity $u^* = -124.824$. The moving of the piston will produce a shock wave, and the shock wave will interact with the Helium bubble at time $t_i = 668.153 \times 10^{-6}$. MOF-MMALE method is used to simulate this problem.

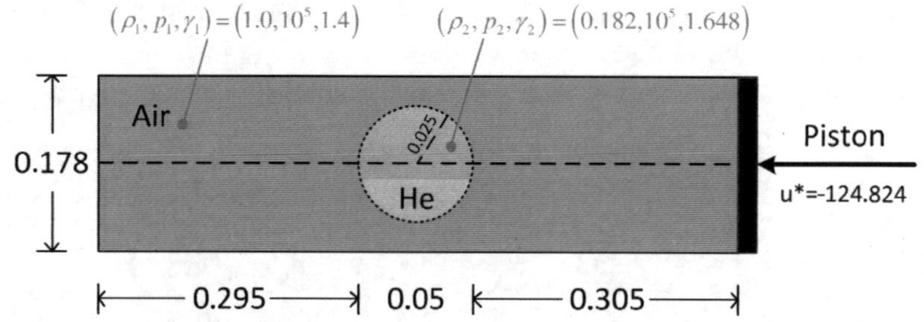


**Figure 6.** Interaction of shock wave with Helium bubble.

To compute this problem, we paved the computational domain with 260x72 cells and fixed the time step dt=$10^{-7}$. In this case, the initialization of the volume fraction is performed by computing the intersection between the circle, which corresponds to the bubble boundary, and the initial mesh. The last time of computation is set to 1342 x$10^{-6}$. We have performed computations with MMALE method and displayed in figure 7 the time evolution of the bubble at three different moments of 800 x$10^{-6}$, 1100 x$10^{-6}$ and 1342 x$10^{-6}$. At each step, we distribute the cells equally in the computational domain compressed by the right end piston.

The comparison of an experimental image[10] and the numerical results of interface at the final time has been displayed in figure 8. It shows a good agreement that validates our MOF-MMALE method from a qualitative point of view.

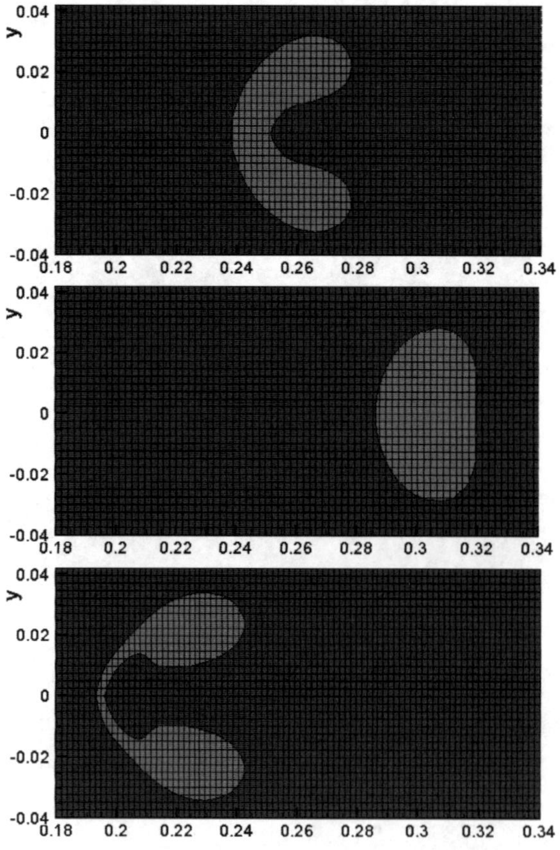


(a) t = 800 x$10^{-6}$

(b) t = 1100 x$10^{-6}$

(c) t = 1342 x$10^{-6}$

**Figure 7.** Interface of Helium bubble at three different moments.

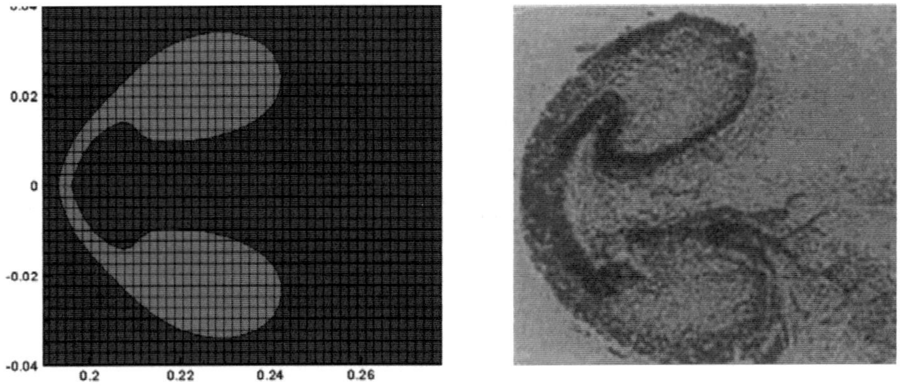


**Figure 8.** Comparison of MOF-MMALE interface (left) with experiment (right) at final time 1342 x$10^{-6}$.

## 4. Conclusions and future works

We have described MOF-MMALE method devoted to the numerical simulation of multi-material large deformation fluid flows. The basic ideal of MOF-MMALE is to introduce mixed cells based on traditional ALE method. The introduction of mixed cells can increase the freedom of mesh rezoning and reduce the deformation of computational mesh, which make the MMALE method deal with the large deformation of fluid flows easily. On the other hand, MOF interface reconstruction is employed to determine the material interface in mixed cell.

Though some models used in current MOF-MMALE method are simple, the numerical results have shown the outstanding capacity, accuracy and efficiency in simulation of multi-material large deformation fluid flows. MOF-MMALE method is an effective way to simulate flow problems of multi-material large deformation.

The future works incude improving the thermodynamic closure model, increasing mesh scale and performing large-scale parallel computing.

**Acknowledgements**

This work was supported by the National Natural Science Foundation of China under Grant No. 11001026. The author thanks Dr. Wenjun Sun for providing me some algorithms in the implementation of the MOF-MMALE code.

**References**

[1] Galera S, Maire P H and Breil J 2010 *J. Comput. Phys.* **229** 5755
[2] Breil J, Galera S and Maire P H 2011 *Computer and Fluids* **46** 161
[3] Anbarlooei H R and Mazaheri K 2009 *Comput. Methods Appl. Mech. Engrg.* **198** 3782
[4] Dyadechko V and Shashkov M J 2005 Moment-of-fluid interface reconstruction *Los Alamos National Laboratory Report* LA-UR-05-7571
[5] Ahn H T and Shashkov M J 2007 *J. Comput. Phys.* **226** 2096
[6] Dyadechko V and Shashkov M J 2008 *J. Comput. Phys.* **227** 5361
[7] Goncharov E A, Kolobyanin V Y, Sadchikov V V and Yanilkin Y V 2006 Methods for computation of thermodynamic states of mixed cells in Lagrangian gas dynamics *New Models and Hydrocodes for Shock Wave Processes in Condensed Matter*, Dijon, France
[8] Caramana E J, Burton D E, Shashkov M J and Whalen P P 1998 *J. Comput. Phys.* **146** 227
[9] Loubere R and Shashkov M J 2005 *J. Comput. Phys.* **209** 105
[10] Haas J F and Sturtevant B 1987 *Journal of Fluid Mechanics* **181** 41

Journal of Physics: Conference Series **510** (2014) 012048 doi:10.1088/1742-6596/510/1/012048

# Nonlinear generation of surface waves against the wind in a limited fetch growth model

**A Pushkarev**[1,2,3]

[1] LPI, 53 Leninskij Prospekt, 119991, Moscow, Russia
[2] Novosibirsk State University, Novosibirsk, Russia
[3] Waves and Solitons LLC, Phoenix USA

E-mail: dr.push@gmail.com

**Abstract**. We describe the model for excitation of ocean surface gravity waves by the wind blowing away from the shore. Numerical simulation shows that wave turbulence behaviour can be split in time into two different physical regimes: classical and non-classical one. The classical regime corresponds to self-similar solution developing for the characteristic times defined by ratio of the channel width to characteristic advection velocity of the wave field. The non-classical regime is happening later in time when the energy starts to reflect from the opposite boundary condition and corresponds to energy advection against the wind. It exhibits itself in waves generating predominantly parallel to the shore line, tending to slant toward the shore, as approaching to the beginning of the fetch, at 15 degrees.

## 1. Model description

We study kinetic equation for waves (Hasselmann equation [1]) for energy spectral density $\varepsilon(\vec{r},\vec{k},t)$, describing nonlinear waves propagation in geometric optics approximation:

$$\frac{\partial \varepsilon}{\partial t}+\frac{\partial \omega_k}{\partial \vec{k}}\frac{\partial \varepsilon}{\partial \vec{r}}=S_{nl}+S_{wind}+S_{diss} \quad (1)$$

for the simplified, but real physical situation, which includes the most important component of the general statement - non-stationarity in time, advection, exact nonlinearity $S_{nl}$, wind forcing term $S_{wind}$ and wave dissipation term $S_{diss}$ due to wave-breaking.

Such simplified formulation is called the limited fetch wind growth situation, where Hasselmann equation is reduced to

$$\frac{g}{2\omega}\cos\theta\frac{\partial \varepsilon}{\partial x}=S_{nl}+S_{wind}+S_{diss} \quad (2)$$

where $\boldsymbol{x}$ is the spatial coordinate, orthogonal to the shore, and $\theta$ is the angle between individual wavenumber $\boldsymbol{k}$ and the axis $\boldsymbol{x}$. This situation is presented schematically on the figure 1:

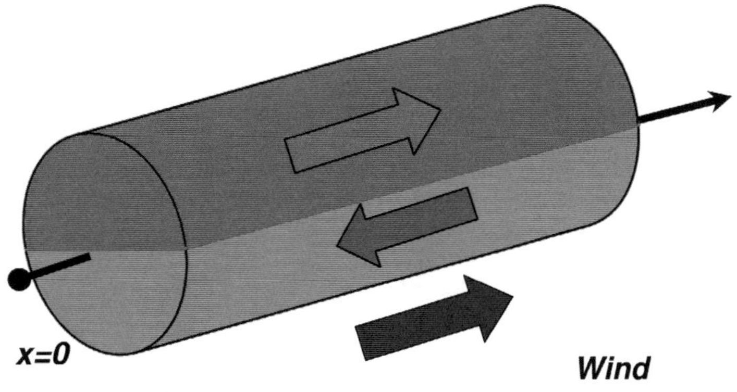


**Figure 1.** Schematic description of the energy fluxes along the fetch in real and Fourier spaces.

Here black axis presents the real space axis $\boldsymbol{x}$ orthogonal to the shore line, the red and green cylinder show Fourier space of the system. Red part of the cylinder corresponds to the positive advection velocity in front of time derivative of Hasselmann equation (2) (red arrow is directed away from the shore), while green part of the cylinder corresponds to negative advection velocity (green arrow directed toward the shore), in correspondence with the $\cos\theta$ in equation (2). Such schematic picture suggests that limited fetch growth consists of essentially three different processes happening in real and Fourier spaces: (i) energy advection away the shore in real space (red tube); (ii) energy advection toward the shore in real space (green tube); (iii) nonlinear interaction of the waves between red and green parts of the tube in Fourier space for any given point of the fetch and time.

It is important to use proper boundary conditions for the solution of Caughy problem. We used zero value of the wave field for the red portion of the tube on the left boundary a green portion of the tube on the right boundary. For green portion of the tube on the left and red portion of the tube on the right we used the free energy flux boundary conditions. Physically, such choice of the boundary conditions corresponds to simulation of the waves excitation in the strait of final width by uniform wind blowing orthogonally to the shore.

For the solution of equation (2) we need to know the source terms in its right-hand side. The procedure of exact calculation of $S_{nl}$ term is well-known and given by Webb-Resio-Tracy algorithm [2]. The parameterization of the wind input term $S_{in}$ is given by (see [3]) :

$$S_{wind} = 0.159\frac{\rho_{air}}{\rho_{water}}\omega\left(\frac{\omega}{\omega_0}\right)^{4/3} f(\theta), \quad f(\theta) = \begin{cases}\cos\theta \text{ for} -\pi/2 \le \theta \le \pi/2 \\ 0 \text{ otherwise}\end{cases} \tag{3}$$

$$\omega_0 = \frac{g}{u_{10}}, \qquad \frac{\rho_{air}}{\rho_{water}} = 1.3\cdot 10^{-3}$$

and is based on the fact that equation (2) has self-similar solution [3]

$$\varepsilon = x^{p+q} F(\omega x^q) \tag{4}$$

with parameters

$$q = \frac{3}{10}, \quad p = 1 \tag{5}$$

along with the fact of fitting the experimental data by specific regression line, see [4], [5]. The contribution of the dissipation term $S_{diss}$ was calculated similar to [5], where white-capping dissipation term was introduced implicitly through $f^{-5}$ ($f = \frac{\omega}{2\pi}$) energy spectral tail stretching in frequency range from $f_d = 1.1$ to $f_{\max} = 2.0$.

## 2. Numerical experiments

We solved Cauchy problem for the Eq.(2) with the following set parameters

- Fetch size 40 km
- Number of nodes 40
- 72 logarithmic frequencies
- 36 points angle resolution
- Wind speed $U_{10} = 5m/\sec$

which translates into:

- Characteristic angular frequency $\omega_{cr} = g/U_{10} = 2rad/\sec$
- Characteristic linear frequency $f_0 = 1.1Hz$
- Dimensionless fetch $\chi = \frac{L}{U_{10}^2} g = 1.6 \cdot 10^4$

Initial conditions were chosen in the form of low-amplitude noise in the red part of the Fourier space and zero in the green part of the phase space cylinder (see figure 1).

Figure 2 represents total energy of the fetch as a function of time. One can see that energy growth evolution can be split into two parts -- first of relatively fast growth for time less than 4 hr, and the second part of relatively slow growth.

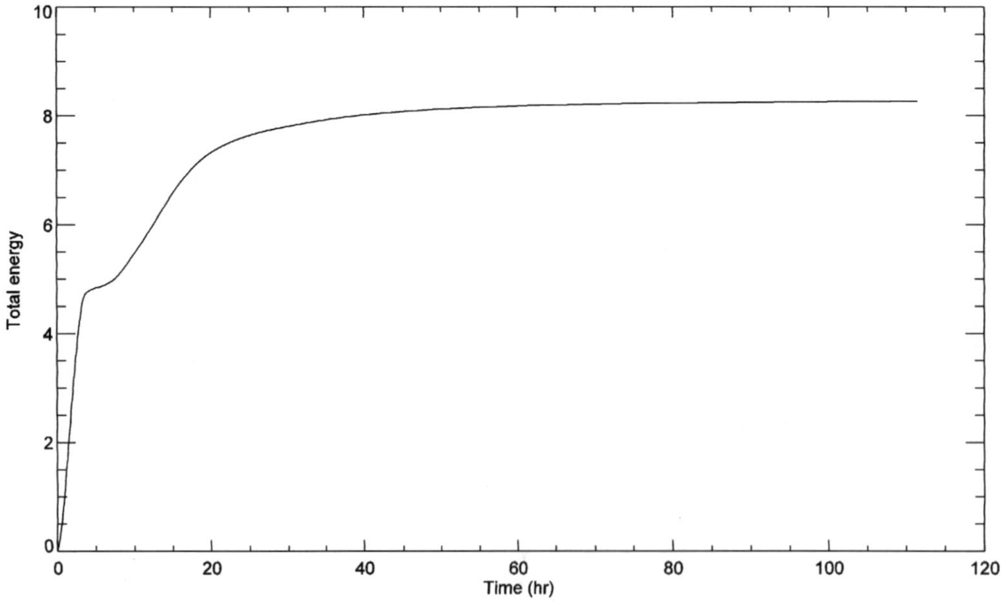


**Figure 2.** Total energy of the fetch as a function of time.

Comparison of the phase space distributions of the spectrum for *t=2hr* ( see figure 5 a,b ), and *t=24hr* (see Fig.6 a,b) shows qualitative difference in their shape, too.

For earlier times *t=2hr* the shape of the spectrum has the form of the single hump, growing with the distance from the shore along the fetch. For later time *t=24hr* the shape of the spectrum is more complex: besides classical single-hump energy spectrum, growing away from the shore, one can see side satellites in the form of spikes, corresponding to waves propagating along the shore.

Furthermore, the direction of those waves propagating along the shore is getting slanted toward the shore with the distance diminishing toward the shore. Closer to the shore, the waves are propagating at 15 degrees toward the shore line. This observation is quite remarkable: it means that long enough excitation of the waves in time by the wind blowing away from the shoreline finally excites the waves coming against the wind toward the shore.

Figure 3 represents distribution of energy along the fetch for different times. One can see that energy behavior for time less than *4 hr* is described by threshold-like behaviour propagating along the fetch, in correspondence with [6].

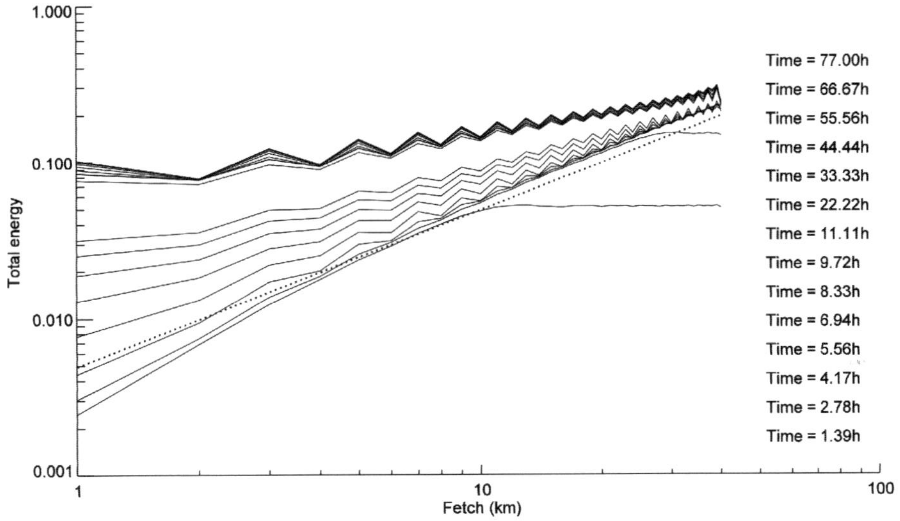


**Figure 3.** Energy distribution along the fetch. Dotted line -- linear function.

At time around *4 hr*, the energy growth is approximated by linear function of the fetch distance. This dependence was predicted in [7] through self-similarity analysis: it was shown that for fetch-limited self-similar solution the dimensionless energy $\widetilde{E} = \frac{E \cdot g^2}{U^4} = a\chi$. For time $t = 4\ hr$, we get from numerical simulation results $a = 2.4 \cdot 10^{-7}$. Table 1 represents experimental data (see **[7]** ) on measuring the constant ***a*** for different experiments. One can see good correspondence of our numerical simulation and experimental predictions in table 1:

| Experiment | a |
|---|---|
| **Nakata Bay (Mitsuyasu et al., 1971)** | **2.89 × 10−7** |
| **JONSWAP (Hasselmann et al., 1973)** | **1.6 × 10−7** |
| **Lake St. Clair (Donelan et al., 1992)** | **1.7 × 10−7** |
| **Bothnian Sea (Kahma, 1981)** | **2.6 × 10−7** |

**Table 1**. Values of the constant **a** for different experiments

Figure 4 shows mean frequency distribution along the fetch for different moment of time less than *4 hr*. One can see that the portion of the "threshold-like" function is closely described by self-similar solution (4)-(5) :

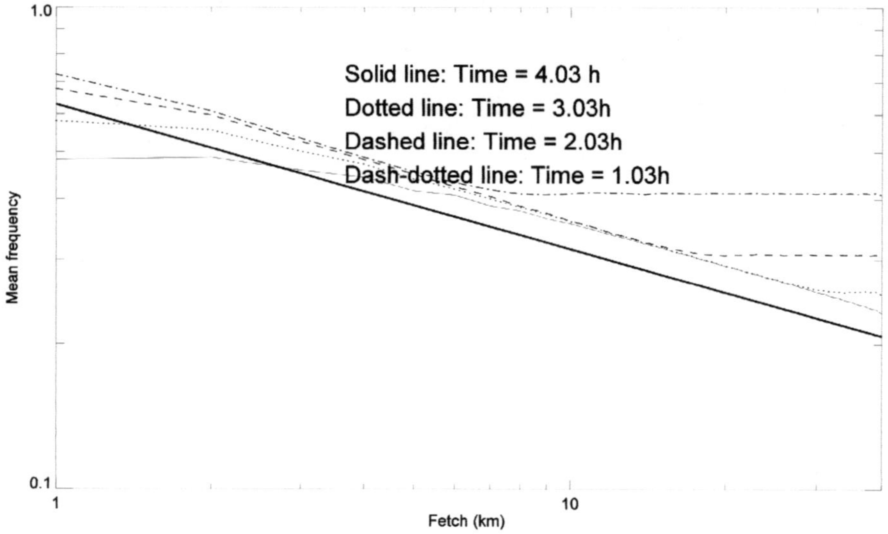


Figure 4. Logarithm of mean frequency distribution as a function of the logarithm of fetch for four different moments of time: 1 hr, 2 hr, 3 hr and 4 hr. Solid line - self-similar prediction q=0.3 from equations (4)-(5)

## 3. Conclusion

All the above facts clearly show that limited fetch growth in the channels of final width can be split in time into two different physical regimes.

The first regime corresponds to self-similar solution developing for the characteristic times defined by ratio of the channel width to characteristic advection velocity of the wave field and could be called classical. This regime corresponds mainly to energy advection in the red part of the cylinder on figure 1.

The second regime is happening later in time when the energy starts to propagate from the opposite boundary condition and corresponds to energy advection against the wind in the green cylinder on figure 1. It exhibits itself in waves generating predominantly parallel to the shore line, tending to slant toward the shore, as approaching to the beginning of the fetch, at 15 degrees.

One of the most important results of our experiments is the possibility of generation of wind waves against the wind due to nonlinear effects.

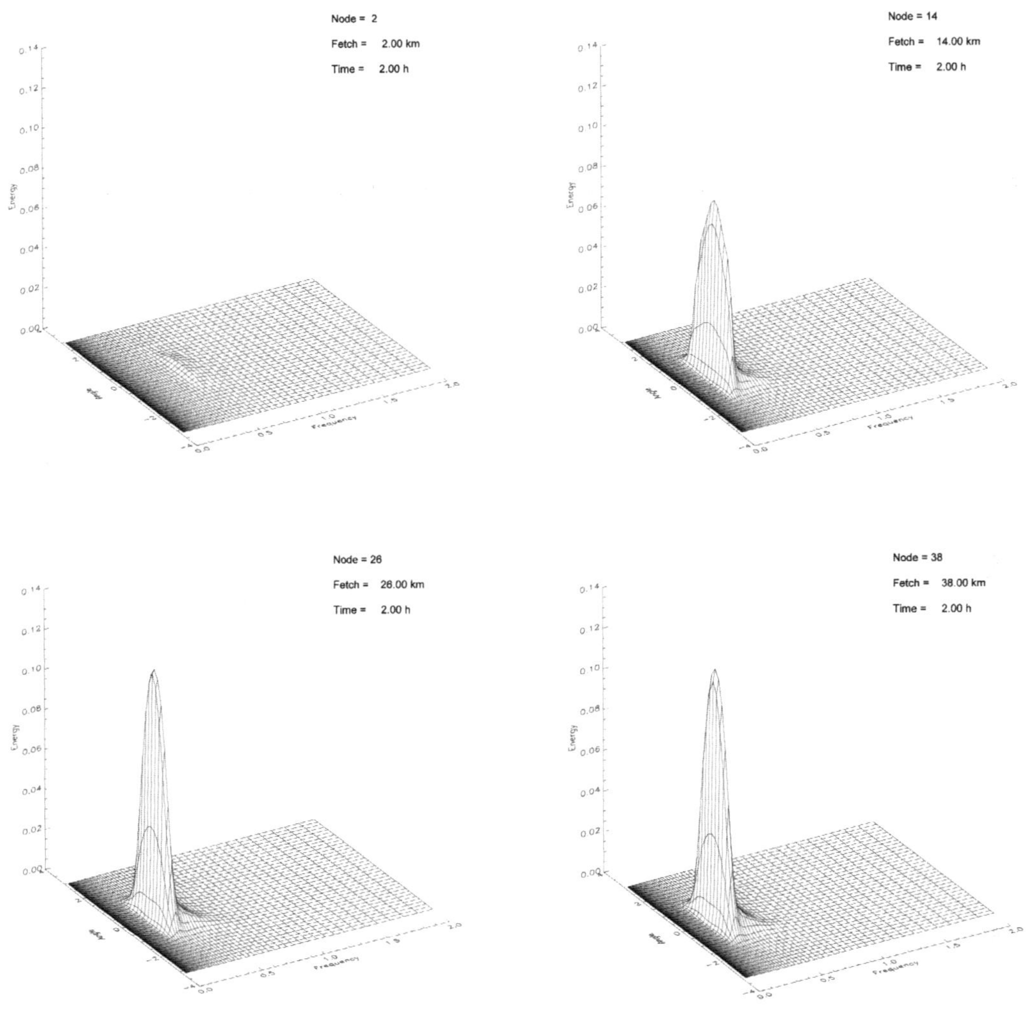


**Figure 5a**. Spectral energy distribution as a function of frequency f and angle $\theta$ at the fetch distances 2 km, 14 km, 26 km, 38 km for time 2hr.

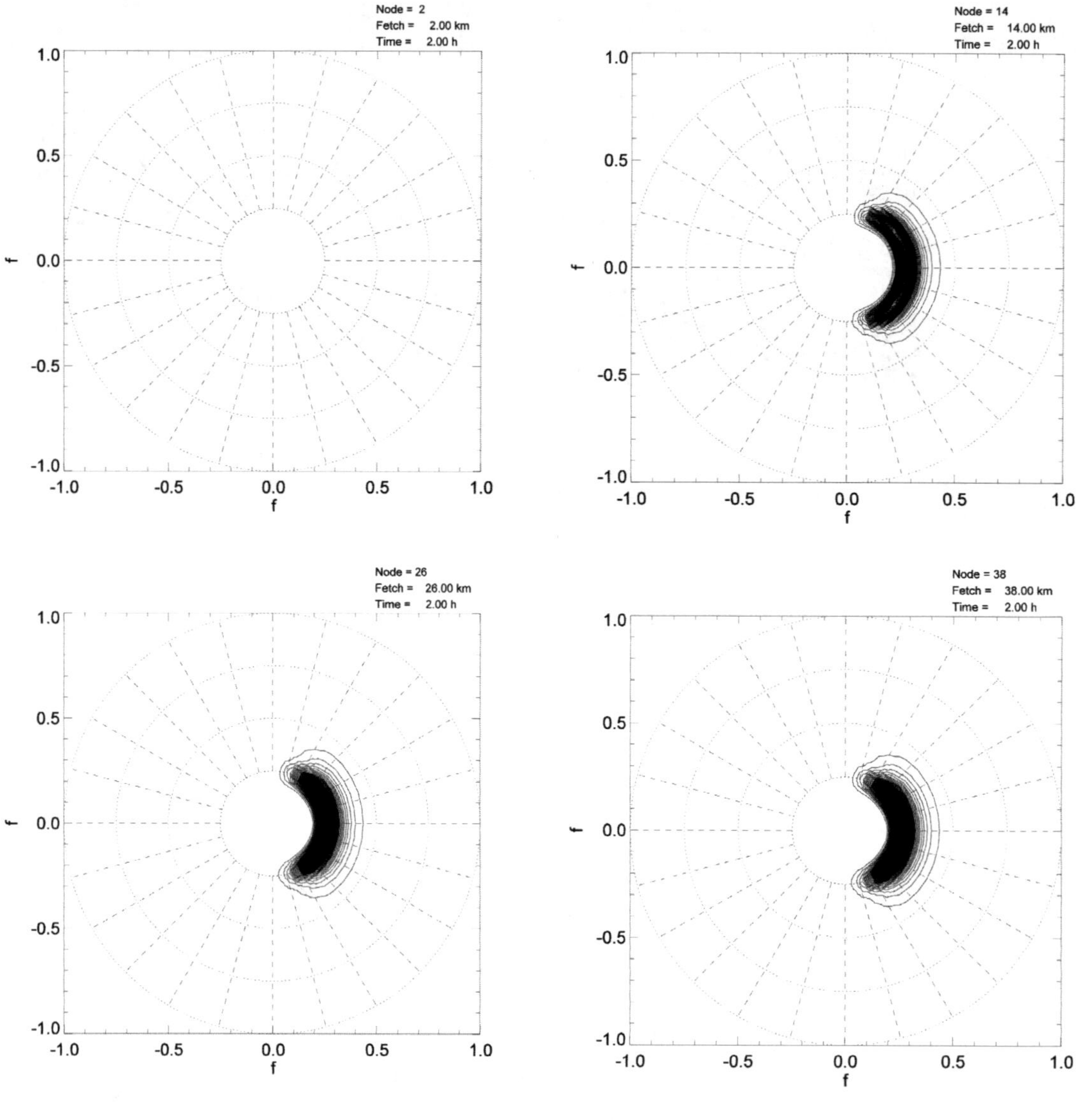


**Figure 5b**. Spectral energy distribution as a function of frequency f and angle $\theta$ in polar coordinates at the fetch distances 2 km, 14 km, 26 km, 38 km for time 2hr.

Journal of Physics: Conference Series **510** (2014) 012048 doi:10.1088/1742-6596/510/1/012048

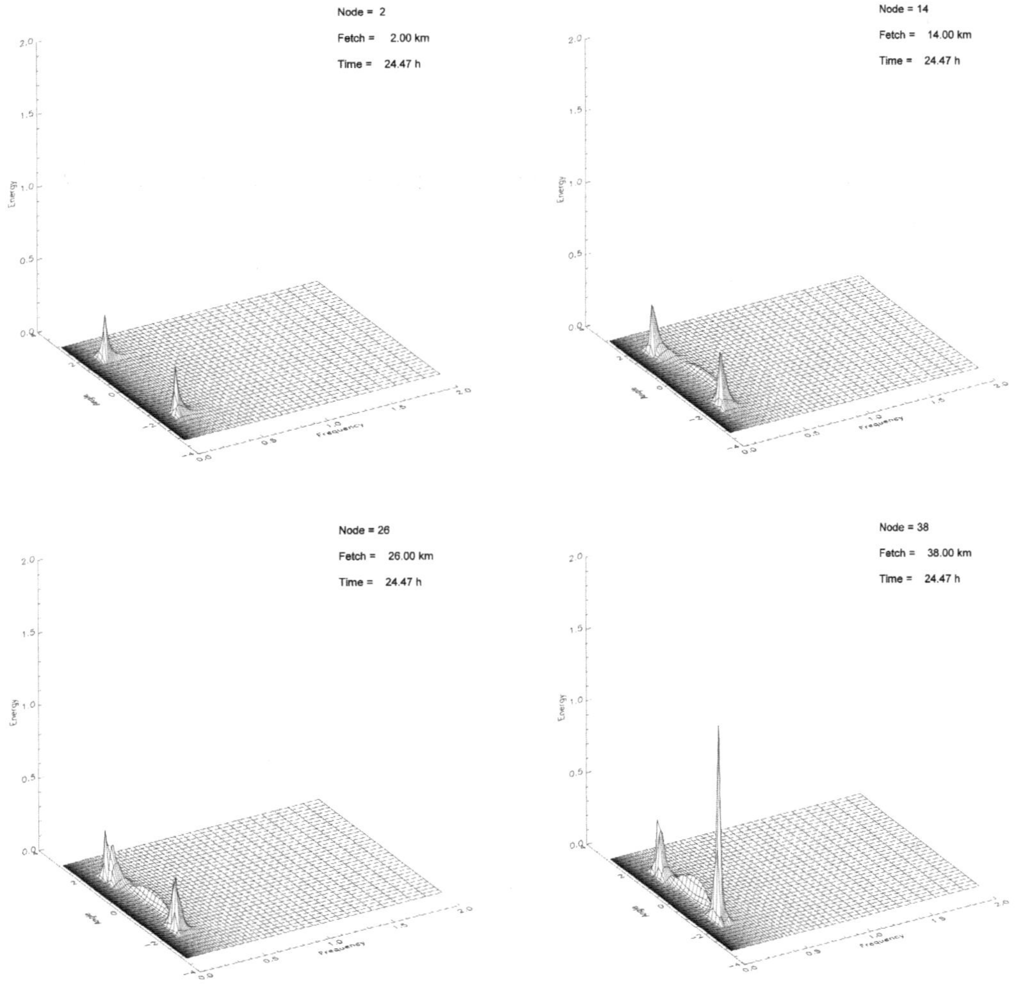


**Figure 6a.** Spectral energy distribution as a function of frequency f and angle $\theta$ at the fetch distances 2 km, 14 km, 26 km, 38 km for time 24.47 hr.

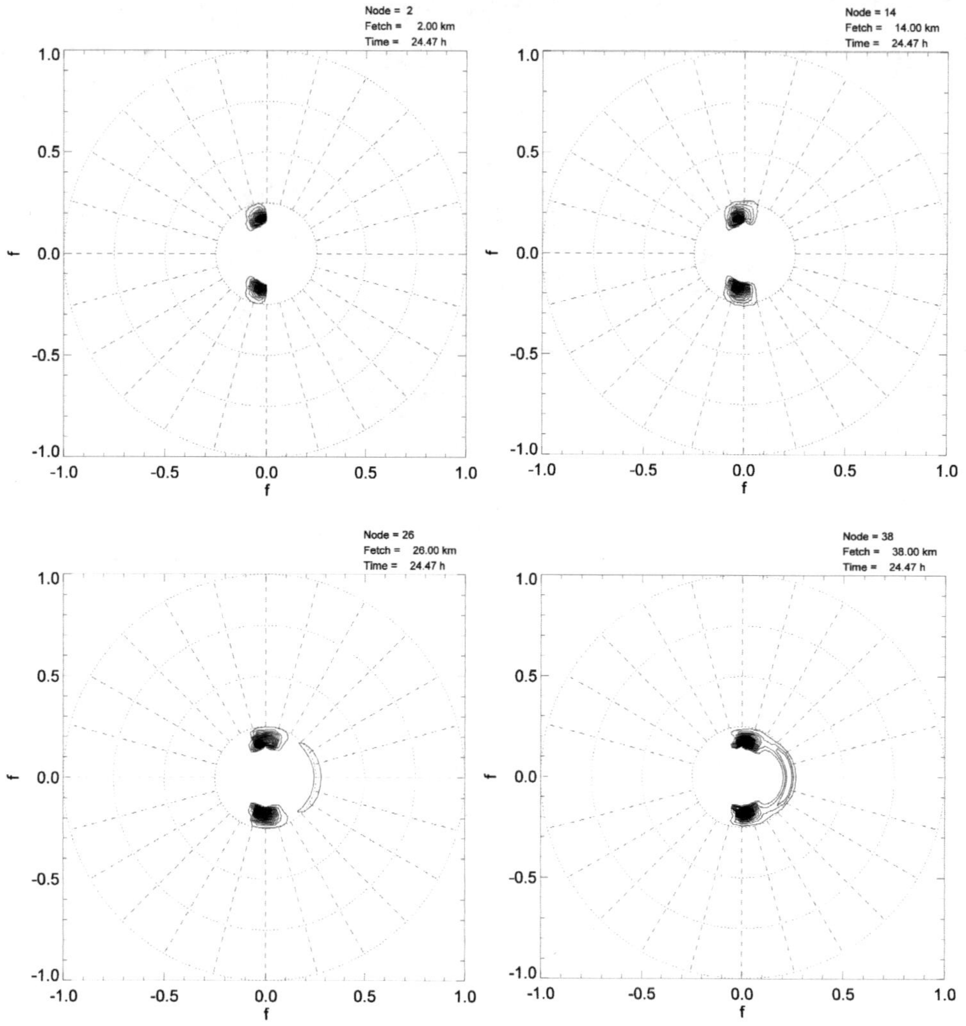


**Figure 6b.** Spectral energy distribution as a function of frequency f and angle $\theta$ in polar coordinates at the fetch distances 2 km, 14 km, 26 km, 38 km for time 24.47 hr.

## References

[1] Hasselmann K 1962 On the nonlinear energy transfer in a gravity wave spectrum. Part I *J. Fluid Mech.* **12** 481-500

[2] Tracy B and Resio D 1982 Theory and calculation of the nonlinear energy transfer between sea wavesin deep water *WES Report* **11** U S Army Corps of Engineer

[3] Zakharov V Resio D Pushkarev D 2012 New wind input term consistent with experimental, theoretical and numerical considerations *arXiv:1212.1069 [physics.ao-ph]*

[4] Resio D. and Long C 2004 Equilibrium-range constant in wind-generated spectra *Journal of Geophysical Research* **109** CO1018

[5] Long C and Resio D 2007 Wind wave spectral observations in Currituck Sound, North Carolina *Journal of Geophysical Research* **112** CO5001

[6] Zakharov V and Pushkarev A April 21-25, 2013 Nonlinear generation of surface waves against the wind in a limited fetch growth model *20th meeting WISE*, College Park, Maryland, USA

[7] Zakharov V 2005 Nonlinear Processes in Geophysics **12** 1011–1020

[8] Badulin S, Babanin A, Resio D and Zakharov V 2007 Weakly turbulent laws of wind-wave growth *J. Fluid Mech.* 591 339–378

Journal of Physics: Conference Series **510** (2014) 012049 doi:10.1088/1742-6596/510/1/012049

# Optimization of solver for gas flow modeling

**D Savichkin[1], O Dodulad, Yu Kloss**

National Research Centre "Kurchatov Institute"
1 Akademika Kurchatova pl, Moscow 123182, Russian Federation

E-mail: denis.savichkin@gmail.com

**Abstract.** The main purpose of the work is optimization of the solver for rarefied gas flow modeling based on the Boltzmann equation. Optimization method is based on SIMD extensions for x86 processors. Computational code is profiled and manually optimized with SSE instructions. Heat flow, shock waves and Knudsen pump are modeled with optimized solver. Dependencies of computational time from mesh sizes and CPU capabilities are provided.

## 1. Introduction

In present work we solve the Boltzmann equation by finite-difference method by using the conservative projection method for calculation the collision integral [1]. The method states that the laws of mass conservation, momentum and energy are strictly observed, and the collision integral of the Maxwellian distribution equals to zero. Real molecular potentials can be used in simulation. The method can be applied to modeling a single component gas, a gas mixture and a gas with internal degrees of freedom.

A multidimensional modeling uses much memory and computing resources. Accuracy of computational method depends on mesh of covering. To model the rarified gas we use Boltzmann equation [2]. Problem-Solving Environment [3] designed for simulation of rarefied gas flows was developed on the basis of conservative projection method and supports multi-processor systems based on MPI [4]. The Problem-Solving Environment (PSE) allows one to perform calculations on personal computer, including the use of graphics processors [5].

Due to limited resources some bottlenecks are located using performance analyzers. The code is optimized with SSE and AVX instructions. A version for Intel MIC architecture is now under development. The work presents results of simulation of some slightly disturbed flows as well as supersonic flows. Examples of the slightly disturbed flows show modeling of Knudsen pump [6] and its' multistage modification. The accuracy of the obtained results is proved by parametric computations and is only limited by computational resources.

## 2. Kinetic equation

The Boltzmann equation is written as [2].

$$\frac{\partial f}{\partial t} + \boldsymbol{\xi}\frac{\partial f}{\partial \boldsymbol{x}} = I(\boldsymbol{\xi}) \tag{1}$$

Equation (1) is solved on a grid $\boldsymbol{x}_i$ and on a uniform grid consisting of $N_0$ equidistant velocity nodes $\boldsymbol{\xi}_\gamma$ with mesh size of $h$ introduced in domain $\Omega$ of volume $V$. The velocity grid is constructed as follows. A cube is defined in the velocity space and a uniform grid is constructed in it. Then, a sphere is inscribed in the cube, and all the outer nodes are eliminated. As a result, the number of grid nodes is roughly halved.

We use of the free-stream density $n_0$ and $T_0$ temperature values as the reference values [1]. The scales of velocity $\nu_0$, length $\lambda$, and time $\tau_0$ are defined as

$$\nu_0 = \sqrt{kT_0/m}, \lambda = 1/\sqrt{2}\pi n_0 \sigma_{eff}^2, \tau_0 = \frac{\lambda}{v_0} \tag{2}$$

Here, $v_0 = v_T/\sqrt{2}$, where $\nu_T$ – is the mean thermal velocity of molecules according to [7], $\lambda$ – is the free path of molecules, $\tau_0 = \tau_T/\sqrt{2}$, where $\tau_T$ – is the mean free time. Knudsen number is $Kn = \lambda/L$, where $L$ – is the characteristic size of the flow. Proceeding to dimensionless variables and a dimensionless distribution function $\boldsymbol{\xi}^* = \boldsymbol{\xi}/v_0$, $t^* = t/\tau_0$, $b^* = b/\sigma_{eff}$, $\boldsymbol{x}^* = \boldsymbol{x}/\lambda$, $f^* = f/(n_0 v_0^{-3})$, and retaining the previous notation, we obtain the following equation in dimensionless variables: [6]

$$\frac{\partial f}{\partial t} + \boldsymbol{\xi}\frac{\partial f}{\partial \boldsymbol{x}} = \frac{1}{\sqrt{2}\pi} I(\boldsymbol{\xi}) \tag{3}$$

The hard sphere model of a molecular potential is considered in this paper. The effective diameter $\sigma_{eff}$ equals to $\sigma_\infty$ and $b_m$ equals to $\sigma_\infty$ for this model.

Equation (1) is splitted into advection equation and collision operator. Splitting is written as $S_\tau = A_{\tau/2} C_\tau A_{\tau/2}$ where $S_\tau$ is operator for iterations $t_0$ through $t_0 + \tau$. $A_{\tau/2}$ is free molecular flow

$$\frac{\partial f}{\partial t} + \boldsymbol{\xi}\frac{\partial f}{\partial \boldsymbol{x}} = 0 \tag{4}$$

with time step $\tau/2$ and $C_\tau$ is operator of collision relaxation

$$\frac{\partial f}{\partial t} = I(f, f) \tag{5}$$

with time step $\tau$. The boundary conditions are used to solve advection equations. The same cycle is repeated at the next time step, but a new grid $S_\nu$ of integration nodes is generated for computing the collision operator. This grid is obtained by periodic shift of the original one through a random 8-dimensional vector. The advection equation is solved using first- and second-order (TVD scheme [7]) accurate conservative difference scheme. The method for solving Eq. (5) ensures that solution is nonnegative.

The collision integral is written as

$$I(f, f) = \int_{-\infty}^{\infty}\int_0^{2\pi}\int_0^{b_m} \phi\,(f' f'_* - f f_*)\, g b db d\varepsilon d\xi \tag{6}$$

Where $g = |\xi - \xi_*|$,$b_m$ is the target distance, $\varepsilon$ is the collision scattering angle. After introducing $\phi(\xi_\gamma) = \delta(\xi - \xi_\gamma) + \delta(\xi_* - \xi_\gamma) - \delta(\xi' - \xi_\gamma) - \delta(\xi'_* - \xi_\gamma)$ at the point $\xi_\gamma$ we obtain

$$I(\xi_\gamma) = \int_{-\infty}^{\infty}\int_{-\infty}^{\infty}\int_0^{2\pi}\int_0^{b_m} \phi\,(f' f'_* - f f_*)\, g b db d\varepsilon d\xi d\xi_* \tag{7}$$

To evaluate Eq. (7), we define the domain $\Omega \times \Omega \times [0, 2\pi] \times [0, b_m]$ with mesh $\xi_{\alpha_\nu}$, $\xi_{\beta_\nu}$, $\varepsilon_\nu$, $b_\nu$ and use Korobov's uniform grid [8]. Note that, points after collision $\xi'_{\alpha_\nu}$, $\xi'_{\beta_\nu}$ do not concide with velocity nodes. For velocity $\xi'_{\alpha_\nu}$ two nearest points $\xi_{\lambda_\nu}$, $\xi_{\mu_\nu}$ are determined. Two points $\xi_{\lambda_\nu + s_\nu}$

and $\xi_{\mu_\nu - s_\nu}$ are located symmetrically from sum $g_\nu = \xi_{\alpha_\nu} + \xi_{\beta_\nu}$ and determined for velocity $\xi'_{\beta_\nu}$. Since that delta-functions in Eq. (7) turn to $\delta(\xi'_{\alpha_\nu} - \xi_\gamma) = (1 - r_\nu)\delta(\xi_{\lambda_\nu} - \xi_\gamma) + r_\nu \delta(\xi_{\mu_\nu} - \xi_\gamma)$. Where the coefficient $r_\nu$ is obtained from the conservation laws: $(\xi'_{\alpha_\nu})^2 + (\xi'_{\beta_\nu})^2 = (1 - r_\nu)(\xi^2_{\lambda_\nu} + \xi^2_{\mu_\nu}) + r_\nu(\xi^2_{\lambda_\nu + s_\nu} + \xi^2_{\mu_\nu - s_\nu})$.

The distribution functions $f'_{\alpha_\nu} f'_{\beta_\nu}$ after collision can be found by interpolation

$$(f_{\lambda_\nu} f_{\mu_\nu})^{1-r_\nu} (f_{\lambda_\nu + s_\nu} f_{\mu_\nu + p_\nu})^{r_\nu} \tag{8}$$

This interpolation gives us zero for the collision operator if $f_\beta$ are the values of Maxwellian function at the velocity nodes. Using of the logarithmic interpolation Eq. (8) is important to modeling slightly disturbed flows. Modeling supersonic flows with existing computational resources becomes extremely time consuming. Eq. (9) is a transformation to replace the resource-intensive operation of multiplication by less expensive sum.

$$\exp\left[(1 - r_\nu)\ln(f_{\lambda_\nu} f_{\mu_\nu}) + r_\nu \ln(f_{\lambda_\nu + s_\nu} f_{\mu_\nu + p_\nu})\right] \tag{9}$$

After a predetermined number of steps PSE saves the set of macro-parameters which are calculated according to the formulas for density $n$, velocity $\boldsymbol{u}$, temperature $T$ and pressure $P$

$$n = \int f d\boldsymbol{\xi},\ \boldsymbol{u} = \frac{1}{n}\int \boldsymbol{\xi} f d\boldsymbol{\xi},\ T = \frac{1}{3n}\int \xi^2 f d\boldsymbol{\xi} - u^2, P = nT \tag{10}$$

## 3. Optimization methods

### *3.1. Compilers' comparison and operation time for GPU-based solver*

Structure of PSE is shown on Figure 1. There are some components of multiform input, GPU-adapted calculation core and solver with unstructured meshes and various outputs.

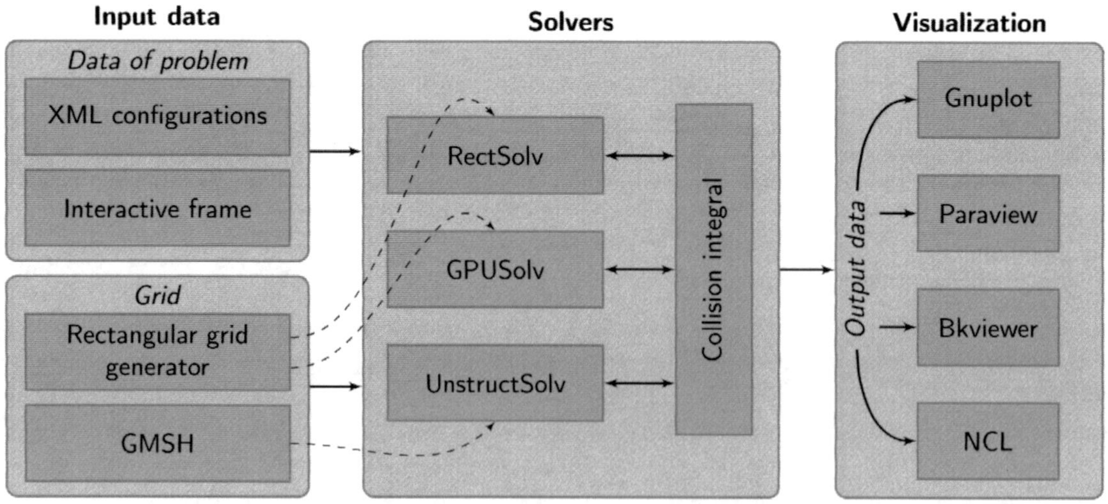


**Figure 1.** Structure of the problem solving environment.

Heat flow problem [9] modeling was compiled with GNU g++ 4.4.3 with -O3 and Intel icpc 13.0.1 with a similar option. Korobov's grid equaled to $10^5$. The number of spatial cells varied from 30 to 1000. Operation time of the code is shown in Table 1. The comparison result shows that there are no significant differences between two compilers. However, there are many optimization options in modern compilers some parts of code have to upgrade manually.

**Table 1.** Compilers' comparison.

| Compiller \ Nodes | 30 | 50 | 100 | 200 | 500 | 1000 |
|---|---|---|---|---|---|---|
| g++ | 130$s$ | 173$s$ | 272$s$ | 463$s$ | 1066$s$ | 2124$s$ |
| icpc | 122$s$ | 163$s$ | 272$s$ | 479$s$ | 1109$s$ | 2193$s$ |

For GPU-based code there were pre-generated collision meshes and one of them was randomly chosen on each time step. Comparison of optimization based on graphics processors is listed in Table 2.

**Table 2.** Operation time for GPU-based solver.

| mesh in physical space | 64x64 | 96x96 | 128x128 | 160x160 |
|---|---|---|---|---|
| acceleration of transfer | 210 | 347 | 395 | 422 |
| acceleration of relaxation | 32.5 | 35.6 | 39.5 | 33 |
| total acceleration (times) | 105 | 139 | 150 | 135 |

Result was produced with a single core CPU (2.66GHz) and 240 cores GPU (GTX285). Speedup for calculations transfer part of solver grown for mesh sizes 64x64, 96x96 and 128x128. However, mesh size didn't govern on speedup for the relaxation part of solver. For small meshes transfer part influenced on general operation time. Then operation time of transfer part became negligible. And due to weak paralyzation potential for collision integral algorithm there was speedup decreasing for 160x160 mesh.

### *3.2. SIMD optimization*

Based on profiling shown on Figure 2 the function ci::iter must be taken into account. That function calculates Equation (8). Function calc_int_node generates Korobov's grid and collision nodes. Function layer calculates transfer with boundary conditions by Eq. (4). Function main saves macro-parameters by Eq. (10). A pow is a function from standard C++ library which calculates power in Eq. (8).

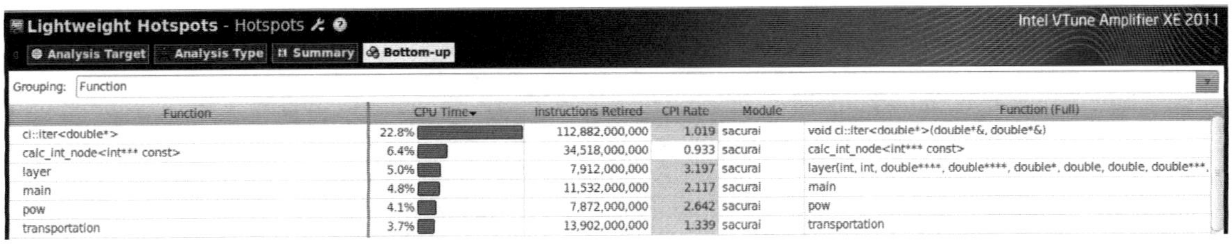


**Figure 2.** Intel VTune profiling output.

There are two distribution functions are needed for algorithm. The data storage of each distribution function is 6-dimensional array in the physical and velocity space. A linear array matches to a 3-dimensional array for each space. Solver has modules for transfer and relaxation part of solver. The code solvers Eq. (5) is given bellow

```
...
//loop by the physical space
for(size_t it = 0; it < Nr; ++it)
        ...
        //loop by the velocity space
        for (typename Nodes::const_iterator p = nodes.begin();
        p != nodes.end(); ++p)
                ...
                //get data
                double g3 = f(i1m + it);
                double g4 = f(i2m + it);
                ...
                //main calculation
                double d = (- g3*g4 + g1*g2) * n.c;
...
```

According to profiling in Fig. 2 the hardest part of calculation is Eq. (9), which is optimized manually. The collision integral algorithm have weak thread parallelism. It's explain speedup limit of GPU-optimized solver for Eq. (5). SIMD extensions provides us to use data with parallelism. In listing the loaded data are not contiguous in the application memory space. For the use SSE instructions the data access pattern have to be a unit-stride. The data storage structure must be inverted. The order of loop by the velocity and physical space is changed. This process allows to operate with the prepared vector of data and shown in listing bellow

```
...
//loop by the velocity space
for (typename Nodes::const_iterator p = nodes.begin();
p != nodes.end(); ++p)
    ...
        //loop by the physical space
        for(size_t it = 0; it < Nr; it += 2 )
        ...
                //load data in registers
                __m128d fast1l = _mm_load_pd(&f(i1l));
                __m128d fast2l = _mm_load_pd(&f(i2l));
                ...
                //main calculation
                __m128d ex = sse::exp(_mm_add_pd(_mm_mul_pd(
                        sse::log(f1lf2l), mrf
                ), _mm_mul_pd(sse::log(f1mf2m), rf)));
                ...
                //store data from registers
                _mm_store_pd(&f(i1l), fast1l);
                _mm_store_pd(&f(i2l), fast2l);
...
```

In each memory access, the core loads not one byte, but a whole cache line, which is 16 bytes wide. With SSE instructions processor core load two different cache lines for double precision numbers. The dimension of the data vector depends on the device processor architecture. Using

SSE halves the length of the loop in a physical space. This optimization let us to avoid the constant movement of the data from memory to the cache and back. Thus more CPU time is used optimally. The theoretical speedup for Eq. (9) is 2 (SSE), 4 (AVX), 8 (MIC).

The dependence of the operation time from the number of cores is shown in Figures 3 and 4 Calculation is done until the hundredth time step. The standard Linux tool (time) is used to determine parameters.

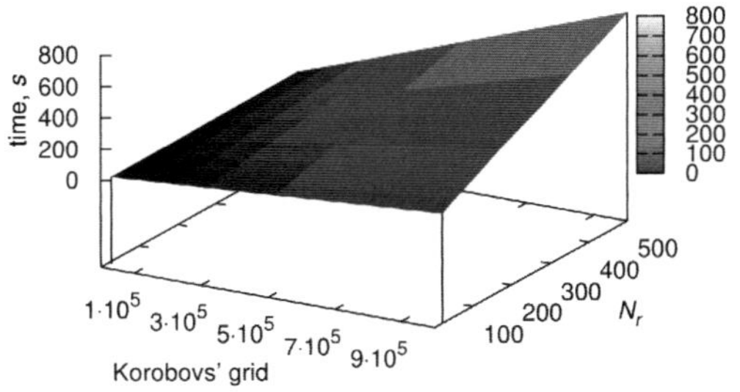


**Figure 3.** Operation time for different meshes of standard algorithm based code.

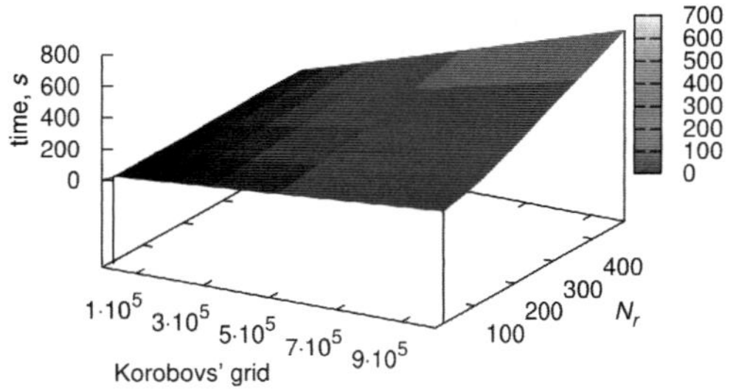


**Figure 4.** Operation time for different meshes of tuned algorithm based code.

Figure 3 is a picture of the old unmodified algorithm and Figure 4 is a tuned solver core. The average gain in time is about 10-20%. The linear growth can be seen in the number of spatial cells for both cases. A similar dependence is observed on the power of Korobov's grid [8].

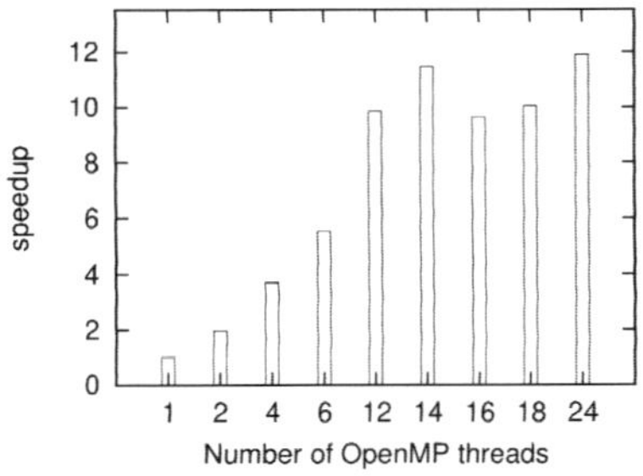


**Figure 5.** Acceleration of solver from number of processor threads.

Figure 5 shows dependencies of acceleration from number of threads. Data for this plot is obtained on one node of cluster (two Xeon X5680 processors) as a solution to a shock wave problem.

## 4. Physical problems

The problem of calculating the parameters of the shock wave was examined in detail in [1]. Density jump from $n_1 = 1$ to $n_2 = 3$ happened in the space of a lot bigger than the common sizes. Parameters of gas before and after the shock associated by Rankine-Hugoniot conditions. There are same conditions the border of the Maxwell function. The simulation result is shown in Figure 6. Korobov's grid equaled to $10^6$, the number of spatial cells is $N_x = 480$. The transfer equation was approximated by the first order scheme. The calculation was performed by 16 cores (2 cluster nodes).

Next problem was modeling parameters of the gas in thermos with the wall being heated. The temperature of the left wall increased linearly according to $T = T_0 + \alpha t$, where $\alpha = 0.1, T_0 = 1$. Korobov's grid equaled to $10^5$. The number of spatial cells was $10^3$. The transfer equation was approximated by the first order scheme. The Knudsen number was $Kn = 0.0015$. The result for propagating density at different time steps are shown on Figure 7. The $x$-axis was normalized by 1 and started from point 0.06. Similar results with the work of A. Sacurai et al. [10] was gained.

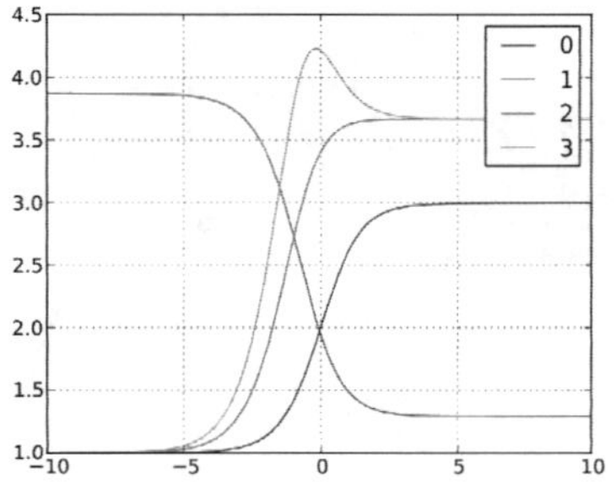


**Figure 6.** Density(0), velocity(1), temperature(2) and $T_{xx}(3)$ depend on space coordinates for steady shock wave.

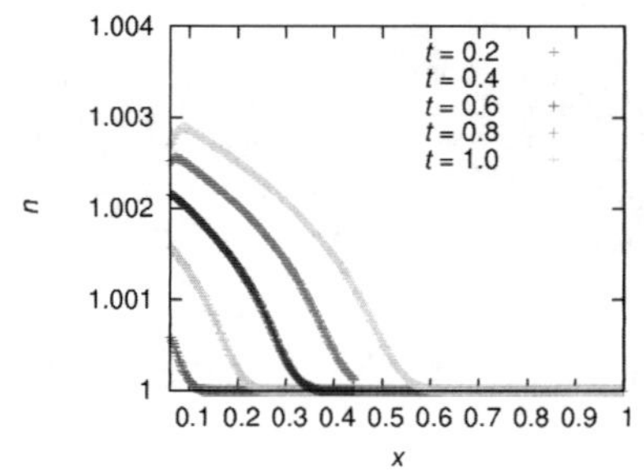


**Figure 7.** Propagating density wave from x at different times for heat flow problem.

The calculation of the parameters of the gas at the Knudsen pump [6] is shown on Figure 10. There was a complex geometry of Knudsen pump consisting of two tanks connected by a set of eighteen narrow tubes and tubes with a tenfold larger diameter. The central (narrow and wide) tubes had temperature gradient with a linear doubling. The device had a monatomic gas inside with the initial density equal 1 and temperature with a leaner doubling. Capacity was divided into four parts mirror boundary conditions due to the symmetry. On the remaining walls of the diffuse boundary conditions with the accommodation coefficient equaled to 1. Korobov's grid equaled to $5 * 10^4$ the number of spatial cells of the order was $1.5 * 10^4$. The transfer equation was approximated by the first order scheme. The calculation was performed by 120 cores (10 cluster nodes).

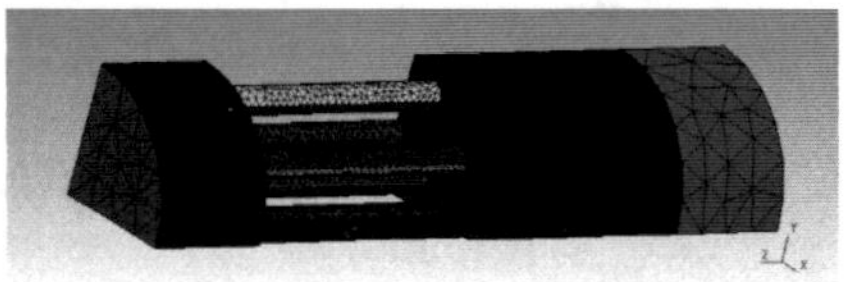

**Figure 8.** Geometry of Knudsen pump with unstructured mesh.

The task of calculating the parameters of the gas in the cavern was modeled. The bottom wall of the container had a 0.001 of the speed of thermal molecular velocities. Mesh of speed was 20 grids by radius, the total number of nodes was 4224, spatial grid was 128X128. Power Korobov's grid equaled to 50000.

Simulation time of all the tasks can be predicted by comparing the characteristics of computational grids in Figure 4. The results of measurements of the code speed are shown in Table 3.

There are satisfactory result for heat flow problem. Small cash size and old processors architecture may have a significant effect on the performance gain. For Knudsen pump modeling

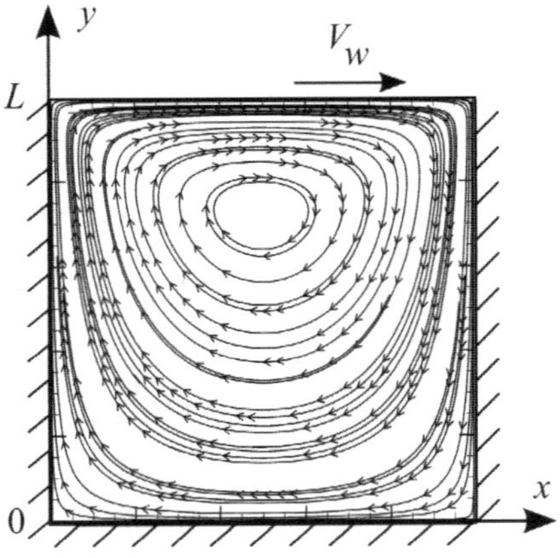


**Figure 9.** Conditions of gas modeling into cavern and streamlines in equilibrium.

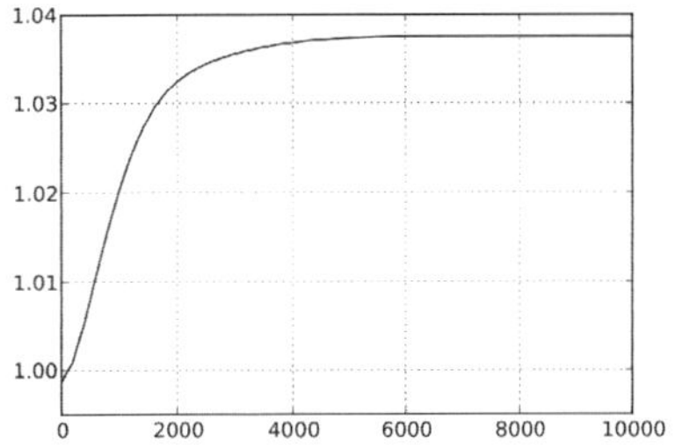


**Figure 10.** The ratio of pressure in the two boxes of Knudsen pump from time steps.

**Table 3.** Operation time of solver.

| Problem | simple | tuned | acceleration | node |
|---|---|---|---|---|
| Heat flow | $2h\ 7min$ | $1h\ 29min$ | 30% | Core-i5 |
| Shock wave | $10h\ 44min$ | $9h\ 13min$ | 14% | 2× Xeon 5365 |
| Knudsen pump | $53min$ | $59min$ | – | 20× Xeon 5680 |

with $5*10^4$ Korobov's grid means that main part of calculation is transfer and data transposition function may result in a significant performance loss. It leads to grow the operation time for such problems.

## 5. Conclusions

More efficient code was obtained and tested on different problems. Simulations of slightly disturbed flows and supersonic flows were provided. Examples of the slightly disturbed flows were shown by modeling of Knudsen pump and its multistage modification.

The total operation time achieved on the architecture of Intel Xeon was 3min 18s which is 1.44 times faster than in previous code version (4min 46s). Speedup 1.73 for the architecture of the Intel Phi with wide vector processing units gained from 29min 37s up to 17min 8s. As a result, approximately 30 - 40 % performance increase of the total time of the code was received for supersonic flows.

## Acknowledgments

Results were obtained with using resources Supercomputer Simulation Laboratory SUSU (http://supercomputer.susu.ac.ru) and resources MCS NRC Kurchatov institute (http://computing.kiae.ru). Special thanks are extended to D. Babokin of Intel for providing Intel Phi coprocessor.

## References

[1] Tcheremissine F G 2006 *Comp. Math. and Math. Phys.* **46** 329–343

[2] Kogan M N 1967 *Rarefied Gas Dynamics* (Moscow: Nauka) p 440
[3] Anikin Y A, Dodulad O I, Kloss Y Y, Martynov D V, Shuvalov P V and Tcheremissine F G 2012 *Vacuum* **86** 1770–1777
[4] Chapman B, Jost G and van der Pas R 2008 *Using OpenMP* (MIT press) ISBN 978-0-262-53302-7
[5] Kloss Y, Shuvalov P and Tcheremissine F 2010 *Procedia Computer Science* **1** 1077–1085
[6] Takata S and Umetsu H 2011 *AIP Conference Proceedings*
[7] LeVeque R J 2007 *Finite Volume Methods for Hyperbolic Problems* (Cambridge University Press) ISBN 978-0-521-00924-9
[8] Korobov N M 1989 *Trigonometric Sums and Applications* (Moscow: Nauka)
[9] Ohwada T 1996 *Phys. Fluids* **8**
[10] Sakaguchi G, Tsukamoto M and Sakurai A 2012 *28th International Symposium on Rarefied Gas Dynamics* **1501** 131–136

Journal of Physics: Conference Series **510** (2014) 012050 doi:10.1088/1742-6596/510/1/012050

# Freak waves at the surface of deep water

**A.I. Dyachenko[1,2], D.I. Kachulin[1] and V.E. Zakharov[1,2,3,4]**

[1]Novosibirsk State University, Novosibirsk, Russia
[2]Landau Institute for Theoretical Physics, Moscow, Russia
[3]Physical Institute of RAS, Moscow, Russia
[4]Department of Mathematics, University of Arizona, Tucson, USA

E-mail: d.kachulin@gmail.com

**Abstract.** In the paper [1] authors applied canonical transformation to water wave equation not only to remove cubic nonlinear terms but to simplify drastically fourth order terms in Hamiltonian. After the transformation well-known but cumbersome Zakharov equation is drastically simplified and can be written in X-space in compact way. This new equation is very suitable for analytic study as well as for numerical simulation. At the same time one of the important issues concerning this system is the question of its integrability. The first part of the work is devoted to numerical and analytical study of the integrability of the equation obtained in [1]. In the second part we present generalization of the improved Zakharov equation for the "almost" 2-D water waves at the surface of deep water. When considering waves slightly inhomogeneous in transverse direction, one can think in the spirit of Kadomtsev-Petviashvili equation for Korteveg-de-Vries equation taking into account weak transverse diffraction. Equation can be written instead of classical variables $\eta(x,y,t)$ and $\psi(x,y,t)$ in terms of canonical normal variable $b(x,y,t)$. This equation is very suitable for robust numerical simulation. Due to specific structure of nonlinearity in the Hamiltonian the equation can be effectively solved on the computer. It was applied for simulation of sea surface waving including freak waves appearing.

## 1. Compact Zakharov equation

It is well known that the Hamiltonian describing the propagation of gravitational waves in deep water can be represented by an infinite series expansion. Hamiltonian variables here are the elevation of the free surface and the velocity potential at the surface. Analytical and numerical analysis are often limited to only a few terms of the expansion up to the fourth order as usual:

$$H = \frac{1}{2}\int g\eta^2 + \psi\hat{k}\psi dx - \frac{1}{2}\int \{(\hat{k}\psi)^2 - (\psi_x)^2\}\eta dx + \frac{1}{2}\int \{\psi_{xx}\eta^2\hat{k}\psi + \psi\hat{k}(\eta\hat{k}(\eta\hat{k}\psi))\}dx \quad (1)$$

In the paper [1] authors applied canonical transformation to water wave equations not only to remove cubic nonlinear terms but to simplify drastically fourth order terms in Hamiltonian. This transformation explicitly uses the fact of vanishing exact four waves interaction for water gravity waves for 2D potential fluid. After the transformation well-known but cumbersome Zakharov equation is drastically simplified and can be written in $X$-space in compact way.

$$i\frac{\partial b}{\partial t} = \hat{\omega}_k b + \frac{i}{4}\hat{P}^+\left[b^*\frac{\partial}{\partial x}(b'^2) - \frac{\partial}{\partial x}(b^{*\prime}\frac{\partial}{\partial x}b^2)\right] - \frac{1}{2}\hat{P}^+\left[b\cdot\hat{k}(|b'|^2) - \frac{\partial}{\partial x}(b'\hat{k}(|b|^2))\right], \quad (2)$$

Journal of Physics: Conference Series **510** (2014) 012050 doi:10.1088/1742-6596/510/1/012050

Here $\hat{P^+}^2 = \hat{P}^+ = \frac{1}{2}(1 - i\hat{H})$ - projection operator to the upper half-plane. Hamiltonian in X-space is the following:

$$\mathcal{H} = \int b^* \hat{\omega}_k b dx + \frac{1}{2} \int \left|\frac{\partial b}{\partial x}\right|^2 \left[\frac{i}{2}\left(b\frac{\partial b^*}{\partial x} - b^*\frac{\partial b}{\partial x}\right) - \hat{k}|b|^2\right] dx. \quad (3)$$

This equation is very suitable for analytic study as well as for numerical simulation. At the same time one of the important issues concerning this system is the question of its integrability. The next subsection is devoted to numerical and analytical study of the integrability of the equation obtained in [1].

### 1.1. *Analysis of integrability of the compact Zakharov equation*

At first we developed the spectral algorithm to simulate the equation. Numerical integration of the equation were carried out on the base of Runge-Kutta method 4th order accurate in time. Then we try to find breather solutions of the equation (2). Breather is the localized solution of the form:

$$b(x,t) = B(x - Vt)e^{i(k_0 x - \omega_0 t)}, \quad (4)$$

Here $k_0$ is the wavenumber of the carrier wave, $V$ is the group velocity. In the Fourier space breather can be written as follow:

$$b_k(t) = e^{-i(\Omega + Vk)t}\phi_k, \quad (5)$$

where $\Omega$ is close to $\frac{\omega_{k0}}{2}$. For $\phi_k$ the following equation is valid:

$$(\Omega + Vk - \omega_k)\phi_k = \int \tilde{T}_{kk_1}^{k_2k_3}\phi_{k_1}^*\phi_{k_2}\phi_{k_3}\delta_{k+k_1-k_2-k_3}dk_1dk_2dk_3. \quad (6)$$

Where $\tilde{T}_{kk_1}^{k_2k_3}$ four-wave interactions coefficient. Localized in space breather-type solutions with different group velocities and amplitudes were found by iterative Petviashvili method ($n$ - is the number of iteration):

$$(\Omega + Vk - \omega_k)\phi_k^{n+1} = M^n \int \tilde{T}_{kk_1}^{k_2k_3}\phi_{k_1}^{*\ n}\phi_{k_2}^n\phi_{k_3}^n\delta_{k+k_1-k_2-k_3}dk_1dk_2dk_3, \quad (7)$$

Petviashvili coefficient $M^n$ is the following:

$$M^n = \left[\frac{< \phi_k^n(\Omega + Vk - \omega_k)\phi_k^n >}{< \phi_k^n \int \tilde{T}_{kk_1}^{k_2k_3}\phi_{k_1}^{*n}\phi_{k_2}^n\phi_{k_3}^n\delta_{k+k_1-k_2-k_3}dk_1dk_2dk_3 >}\right]^{\frac{3}{2}}. \quad (8)$$

In the limit of NLSE breather solution (4) is nothing but well-known NLSE soliton. Numerical simulations of collisions of such breathers were studied. If the Eq.(2) is integrable collisions of breathers would be pure elastic. However, numerical simulation shows the collisions are not elastic (Figs. 1 and 2). Figure 2 shows the zoomed Fig. 1. To confirm that, we study analytically amplitudes of six-wave interactions for this equation (see for details [2]). It was found that six-wave amplitude is not canceled for this equation. So, the kernel of six-wave element of scattering matrix is nonzero $\tilde{T}_{q_1q_2q_3}^{p_1p_2p_3} \neq 0$ on the resonant manifold:

$$p_1 + p_2 + p_3 = q_1 + q_2 + q_3$$

$$\omega_{p_1} + \omega_{p_2} + \omega_{p_3} = \omega_{q_1} + \omega_{q_1} + \omega_{q_3} \quad (9)$$

Thus, 1-D Zakharov equation is not integrable. However the question about integrability of fully nonlinear system is still unclear. Exact equation for deep gravitational waves has his own six wave term which would change total six wave coefficient.

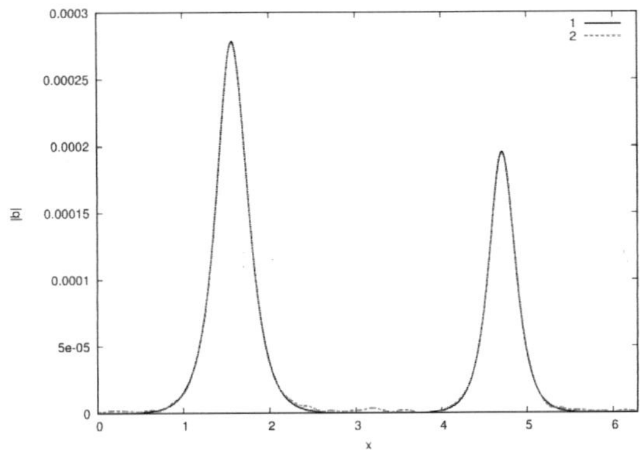


**Figure 1.** Modulus of $b(x,t)$ for two point of time. Curve 1 corresponds the initial statement ($t = 0$), curve 2 corresponds to the statement after 100 breather collisions ($t$ 88000).

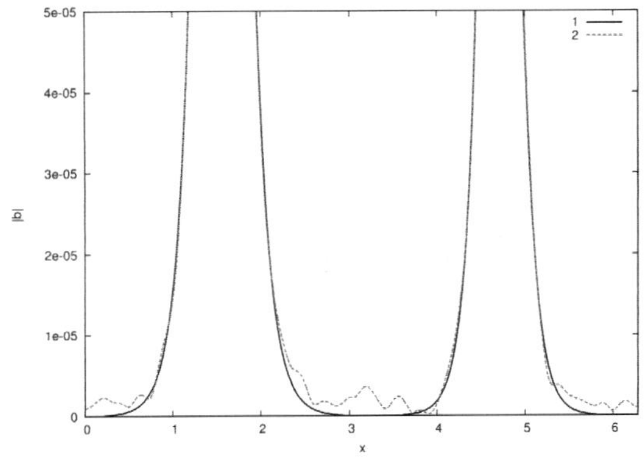


**Figure 2.** Modulus of $b(x,t)$ for two point of time. Curve 1 corresponds the initial statement ($t = 0$), curve 2 corresponds to the statement after 100 breather collisions ($t$ 88000).

## 2. Zakharov equation for the "almost" 2-D water waves at the surface of deep water

The equation (2) can be generalized for almost 2D waves. Waves that slightly inhomogeneous in the transverse direction can be considered in the spirit of the Kadomtsev-Petviashvili equation for the Korteweg-de-Vries equation: the frequency $\omega_k$ can be treated as 2D $\omega_{k_x,k_y}$, while the four-wave interactions coefficient $\tilde{T}^{k_2k_3}_{kk_1}$ not dependent on $y$ and $b(x,y,t)$ is a function of $x$ and $y$:

$$\mathcal{H} = \int b^*\hat{\omega}_{k_x,k_y} b dx dy + \frac{1}{4}\int |b'_x|^2\left[\frac{i}{2}(bb'^*_x - b^*b'_x) - \hat{K}_x|b|^2\right] dx dy. \qquad (10)$$

Corresponding equation of motion is the following:

$$i\frac{\partial b}{\partial t} = \hat{\omega}_{k_x,k_y} b + \frac{i}{4}\hat{P}^+\left[b^*\frac{\partial}{\partial x}(b'^2_x) - \frac{\partial}{\partial x}(b^{*\prime}_x\frac{\partial}{\partial x}b^2)\right] - \frac{1}{2}\hat{P}^+\left[b\cdot\hat{k}(|b'_x|^2) - \frac{\partial}{\partial x}(b'_x\hat{k}(|b|^2))\right]. \quad (11)$$

For Fourier harmonics Hamiltonian can be written as following:

$$H=\int\omega_{k_x,ky}|b_{k_x,k_y}|^2dk_xdk_y + \frac{1}{2}\int T^{k_{x3}k_{x4}}_{k_{x1}k_{x2}} b^*_{k_{x1},y_1} b^*_{k_{x2},y_2} b_{k_{x3},y_3} b_{k_{x4},y_4}\delta_{k_{x1}+k_{x2}-k_{x3}-k_{x4}}\times$$
$$\times\delta_{k_{y1}+k_{y2}-k_{y3}-k_{y4}}dk_{x1}dk_{x2}dk_{x3}dk_{x4}dk_{y1}dk_{y2}dk_{y3}dk_{y4} \qquad (12)$$

Here

$$T^{k_xk_{x1}}_{k_{x2}k_{x3}} = \frac{\theta(k_x)\theta(k_{x1})\theta(k_{x2})\theta(k_{x3})}{16\pi^2}\left[(k_xk_{x1}(k_x+k_{x1}) + k_{x2}k_{x3}(k_{x2}+k_{x3})) - (k_xk_{x2}|k_x-k_{x2}| + \right.$$
$$\left. + k_xk_{x3}|k_x-k_{x3}| + k_{x1}k_{x2}|k_{x1}-k_{x2}| + k_{x1}k_{x3}|k_{x1}-k_{x3}|)\right]. \qquad (13)$$

$\theta$ - functions in (13) is the manifestation of waves moving in the same direction. Corresponding evolution equation is the following:

$$i\frac{\partial b_{k_x,k_y}}{\partial t} = \omega_{k_x,k_y}b_{k_x,k_y} + \int T^{k_{x3}k_{x4}}_{k_{x1}k_{x2}} b^*_{k_{x1},y_1} b^*_{k_{x2},y_2} b_{k_{x3},y_3} b_{k_{x4},y_4}\delta_{k_{x1}+k_{x2}-k_{x3}-k_{x4}}\times$$

$$\times \delta_{k_{y1}+k_{y2}-k_{y3}-k_{y4}} dk_{x1} dk_{x2} dk_{x3} dk_{x4} dk_{y1} dk_{y2} dk_{y3} dk_{y4}. \tag{14}$$

Let try to find solutions:

$$b(x,y,t) = f(y,t) e^{i(k_{x0}x - \omega_0 t)} \tag{15}$$

where $\omega_0 = \omega_{k_{x0},0}$.Expand $\omega_{k_x,k_y}$ around $\omega_{k_{x0},0}$:

$$\omega_{k_{x0},k_y} = \omega_0 + \frac{\omega_0}{4k_{x0}^2} k_y^2 + O(k_y^3) \tag{16}$$

Substituting (15) and using (16) into Eq.(11) yields defocusing NLS equation:

$$i\frac{\partial f(y,t)}{\partial t} = -\frac{\omega_0}{4k_{x0}^2} \frac{\partial^2 f(y,t)}{\partial y^2} + k_{x0}^3 |f(y,t)|^2 f(y,t) \tag{17}$$

The solutions of equation (17) are well-known (kink, grey soliton). Also, the equation (11) is very suitable for robust numerical simulation. Due to specific structure of nonlinearity in the Hamiltonian (10) viz. four-wave interactions coefficient $\tilde{T}_{kk_1}^{k_2k_3}$ not dependent on $y$, the equation (11) can be effectively solved on the computer by using 1D Fast Fourier Transform in the periodic domain $x \in [0, 2\pi], y \in [0, 2\pi]$.

### *2.1. Modulational instability of monochromatic wave*

Monochromatic wave

$$b(x,y,t) = B_0 e^{i(k_{x0}x + k_{y0}y - \omega_0 t)} \tag{18}$$

is the simplest solution of (11). Indeed, plugging (18) in to the equation (11) one can get the following relation

$$\omega_0 = \omega_{k_{x0},k_{y0}} + k_{x0}^3 |B_0|^2. \tag{19}$$

We consider perturbation to the solution

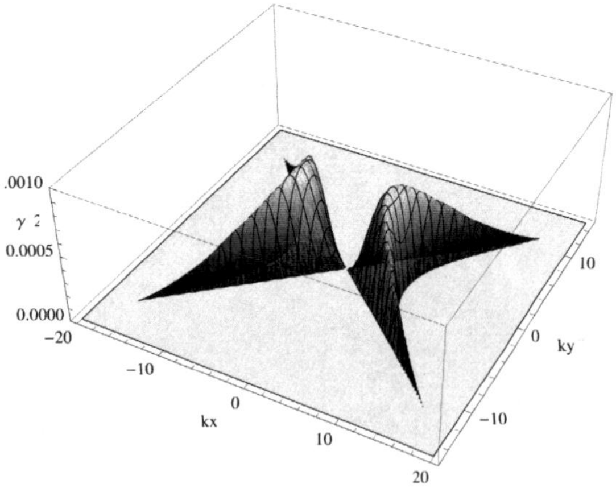


**Figure 3.** The growth rate $\gamma^2_{k_x,k_y}$ for case $k_{x0} = 25, k_{y0} = 0, |B_0| = 0.0015 (\mu \sim 0.12)$ .

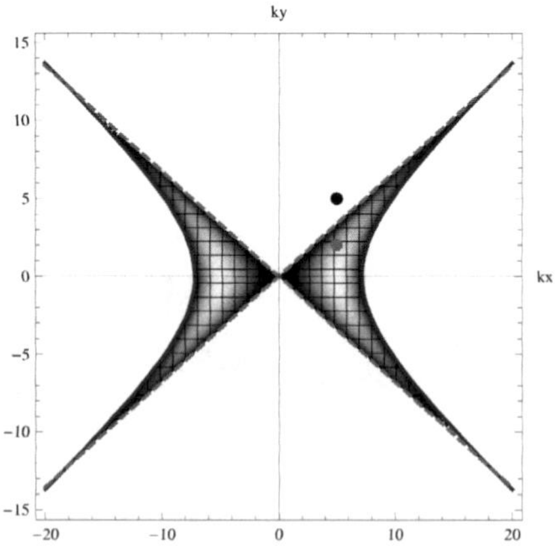


**Figure 4.** The growth rate $\gamma^2_{k_x,k_y}$ for case $k_{x0} = 25, k_{y0} = 0, |B_0| = 0.0015 (\mu \sim 0.12)$ .

$$b(x,y,t) = B_0 e^{i(k_{x0}x + k_{y0}y - \omega_0 t)}, \tag{20}$$

or, in $K$-space:

$$b(k_x,k_y,t) = 2\pi B_0 \delta_{k_x - k_{x0}} \delta_{k_y - k_{y0}} e^{-i\omega_0 t}, \tag{21}$$

Perturbed solution has the following form:

$$b \Rightarrow (b_{k_{x0},k_{y0}} + \delta b_{k_{x0}+k_x,k_{y0}+k_y} e^{-i\Omega_+ t + \gamma_{k_x,k_y} t} + \delta b_{k_{x0}-k_x,k_{y0}-k_y} e^{-i\Omega_- t + \gamma_{k_x,k_y} t}) e^{-i\omega_0 t}. \quad (22)$$

with the following condition:

$$\Omega_+ = -\Omega_- \quad (23)$$

If we introduce steepness of the carrier wave $\mu^2 = 2|B_0|^2 \frac{k_{x0}^3}{\omega_{k_0}}$ then for growth rate is

$$\gamma^2_{k_x,k_y} = \frac{1}{8}\frac{\omega^2_{k_0}}{k_0^4} k_x^2 \left[ \left(1 - 2\left(\frac{k_y}{k_x}\right)^2\right) - 6\mu^2 \right] \left[ \mu^2 \left(k_0 - \frac{|k_x|}{2}\right)^2 - \frac{k_x^2}{8}\left(1 - 2\left(\frac{k_y}{k_x}\right)^2\right) \right]. \quad (24)$$

There is the growth rate function (24) on figures 3, 4 for the case $k_{x0} = 25, k_{y0} = 0, |B_0| = 0.0015 (\mu \sim 0.12)$. The black dot ($k_x = 5, k_y = 5$) on the figure 4 corresponds to "stable" region and red ($k_x = 5, k_y = 2$) dot corresponds to "modulational unstable" region.

*2.2. Numerical simulation of the freak wave appearing*

We have performed the numerical simulation of the freak wave formation in the framework of equation (11). Initial condition was chosen as slightly perturbed monochromatic wave (18) with the steepness of the carrier wave $\mu \sim 0.12$ and $k_{x0} = 25$ as shown in figure 5. After several

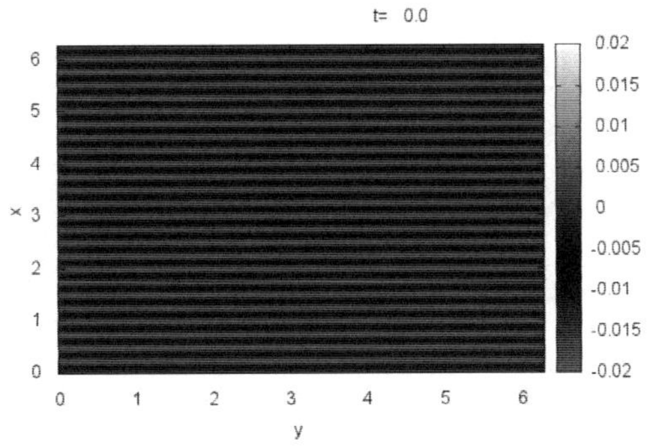


**Figure 5.** Initial free surface $\eta(x, y, t = 0)$

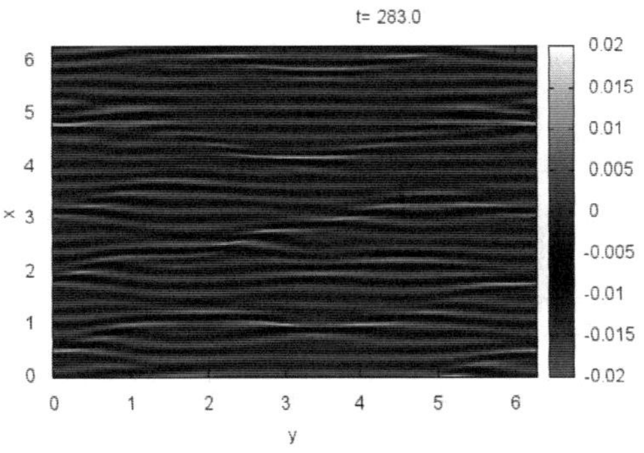


**Figure 6.** Extreme wave formation at the surface $\eta(x, y, t = 565)$

hundreds of carrier wave periods we observed the freak wave formation with the steepness of $\mu \sim 0.5$ (shown in figure 6).

## Acknowledgments

The work partly is supported by the Government of the Russian Federation (the grant for supporting of research conducted at higher schools and headed by leading scientists, no. 11.G34.31.0035, led by V. E. Zakharov, Novosibirsk State University).

## References

[1] Dyachenko A.I. and Zakharov V.E. 2012 *Europ. J. Mech.* B **32** 17
[2] Dyachenko A.I., Kachulin D.I. and Zakharov V.E. 2013 *JETP. Lett.* **98(1) 43**

# AUTHOR INDEX

# AUTHOR INDEX